Graphics in Engineering Design

Graphics in Engineering Design

Third Edition

Alexander Levens
University of California, Berkeley

William Chalk
University of Washington

John Wiley & Sons
New York   Chichester   Brisbane   Toronto

Library of Congress Cataloging in Publication Data:
Levens, Alexander Sander, 1904–
 Graphics in engineering design.

 Second ed. published in 1968 under title: Graphics,
analysis and conceptual design.
 Includes indexes.
 1. Engineering graphics. 2. Design, Industrial.
I. Chalk, William, joint author. II. Title.
T353.L628 1980 604'.2 79-17291
ISBN 0-471-01478-8
Printed in the United States of America

10 9 8 7 6 5 4 3

This book was set in Electra by Progressive Typographers, Inc.,
and printed and bound by Kingsport Press.
The designer was Ben Kann. Vivian Kahane was the copy editor.
Lily Kaufman supervised production.

To Ethel

Preface

The core of engineering is, in a broad sense, the design of machines, processes, structures, and circuits and the combination of these components into plants and systems. Professional engineers must be capable of predicting the costs and performances of the components, plants, and systems to meet specific requirements. Design teams composed of technicians, technologists, designers, drafters, engineers, and nontechnical personnel create the components, plants, and systems through the application of scientific, technical, and economic principles, tempered by judgment based on experience. In the creation of components, plants, and systems to solve technical problems, the design engineers use the design process which, stated simply, is: the definition of the problem; the generation of alternative solutions; the selection of the "best" solution; and the implementation of the solution.

Graphics plays significant roles in the design process in almost every step, from the designer's conceptual sketches to the drafter's final technical drawings.

Designing is a conceptual process that is done mainly in the mind, and the making of sketches is a recording process, a reliable memory system, that designers use for self-communication—talking to themselves—to help them "think through" the various aspects of their project. Graphics is an integral part of the conceptual phase because, frequently, the making of a simple sketch to express a design concept suggests further conceptual ideas. Often the conceptual sketches are redrawn as scaled layout drawings as the ideas are developed further; the selected concepts ultimately appear in the formal technical drawings (and specifications) that become the prime movers for starting production to create a physical reality of the design. Thus graphics is the common mode of communication among the members of the team that was organized to produce a design that satisfies the specified requirements of the customer (that is, a government agency; a manufacturer; a corporation; an individual; or group of individuals).

Technical education must contribute significantly to the development of young, well-qualified persons who can and will face challenging engineering situations with imagination and confidence.

Students who prepare for the stimulating and exciting profession of engineering must have an adequate education in mathematics, physics, chemistry, graphics, and the engineering sciences, and a good background in the humanities and social sciences. With respect to graphics in design, this means good facility in freehand sketching; thorough knowledge of the fundamental principles of orthogonal projection and experience in applying these principles to the solution of space problems that arise in engineering and science; knowledge and use of the graphical methods of computation; and knowledge of the development of the design process and the ability to cope with the "many solutions" type of problems that are so characteristic of professional engineering.

Students who prepare for careers in the additional technological areas represented on the design team must have adequate education in the same areas as engineering students, but more emphasis usually must be placed on the operation and applications of instruments and hardware than on the engineering sciences.

This textbook provides well-qualified students with an up-to-date treatment of engineering graphics and a sound introduction to design that will help them to become "graphically literate" so they can confidently apply graphics to the synthesis, analysis, and solutions of problems that arise in design, development, and research.

Graphics-Design courses offered in the first or second year of engineering or technology programs must fulfill a vital role; they must provide a good foundation for their application in other courses (that is, mechanics, kinematics, machine design, and hardware design as needed in research projects) as well as later, when graduates enter industry.

We believe that the selection and arrangement of the topics also fulfill the needs of students in their preparation for the applications in other courses and, later, on the job.

Our approach accomplishes the following goals.

1. It points out society's real need for individuals (and teams) who can solve societal-technical problems in areas such as communication, transportation, health care, energy, housing, processing, and agriculture.
2. It introduces an effective problem-solving approach—the design process—as used by engineers, designers, and technologists.
3. It clearly demonstrates, through discussions, numerous examples, and exercises, that engineering graphics is, in the broadest sense, an essential part of the design process, especially during the conceptual, analysis, and communication stages.
4. It provides complete and readily understood explanations of the two fundamental principles of orthogonal projection; it demonstrates the numerous applications of these principles to solve three-dimensional problems (tra-

ditionally known as descriptive geometry) that arise in design projects and later in the preparation of technical orthographic drawings for the purpose of implementing the design.
5. It shows that effective communication in written, spoken, and graphical modes is the key to having a design solution understood and ultimately accepted by concerned persons, whether the presentation is in a classroom setting or in a conference room in industry.

The text is organized into three main parts—introduction, application, and implementation.

PART I INTRODUCTION TO DESIGN AND FUNDAMENTALS OF ENGINEERING GRAPHICS

Chapter 1 provides students with an insight into engineering and engineering technology and into the significant achievements—sophisticated communication systems, transportation systems, health systems, housing systems, and energy systems—largely accomplished by design intent.

The continuing need of our society for qualified engineers and technologists to assist in the solutions of our "survival problems"—pollution, transportation, housing, and energy—provides enduring opportunities for young persons who have the interest and background to prepare for the dynamic professions of engineering and technology.

It is pointed out that young persons who have an aptitude for mathematics and the sciences and are motivated toward solving our societal problems, most of which include technology, should seriously consider these professions.

The design process and the role of graphics in that process are introduced. The material in this chapter

sets the stage for the other chapters. Freehand sketching is introduced in Chapter 2 as the first form that a designer uses to record creative ideas. Then, in Chapter 3, a brief introduction to the use of drawing instruments and scales precedes a section on layout drawings. Layout drawings are freehand sketches, redrawn to scale (with instruments), usually for analysis purposes. The use of layout drawings in motion and clearance studies (linkages and connecting-rod motion, for instance) and in the transmission of design information to drafters who prepare the finished working drawings is clearly illustrated.

Details of the design process and its phases and steps follow in Chapter 4. Design examples of first-year students and of professionals are included to illustrate a variety in the approaches to solutions of design projects.

The discussion of the design process appears early in the text; this enables instructors who plan to include a design project in a course to start the project near the beginning of the term and continue its progress while further topics in graphics (such as applications of the fundamental principles of orthogonal projection, graphical solutions and computations, vectors and vector diagrams) are studied and used. Toward the end of the project, topics such as fasteners, dimensioning, and working drawings could also be undertaken. In the final stages of the design project, technical report writing, design documentation, and patenting can be explored.

PART 2 APPLIED GRAPHICS FOR DESIGN, ANALYSIS, AND COMMUNICATION

The applications of the two fundamental principles of orthogonal projection and of the four basic problems (true length of a line segment, point view of a line, edge view of a plane surface, and true shape of a plane surface) to the solution of three-dimensional problems that arise in design work are clearly and amply illustrated in Chapter 6.

Chapter 7, *Developments and Intersections in Design*, and Chapter 8, *Vector Quantities and Vector Diagrams* logically follow the topics in Chapter 6; they demonstrate further applications of the four basic problems. The vector diagrams are especially useful in force analysis in design. Three chapters, Chapter 9 (*Pictorials*), Chapter 10 (*Charts and Graphs*), and Chapter 12 (*Computer Graphics*) are devoted primarily to communication topics. The new computer graphics material includes a variety of applications in areas such as aircraft design, biomechanics research, surface display mapping, mathematical surfaces, computer-aided design of nomograms, and a solution to a typical geometry problem (the perpendicular distance from a point to a plane). The effective use of computer graphics in aircraft design by the McDonnell Aircraft Co., St. Louis, Missouri, gives students an excellent example of the use of interactive computer graphics in modern design work.

Chapter 11, *Graphical Solutions and Computations*, includes graphical calculus, empirical equations, and an introduction to nomography; it complements undergraduate mathematical, scientific, and technical courses and provides tools for later use in industry. Students will find this chapter stimulating and useful. Experience with the material on graphical calculus will greatly enhance students' understanding of integration and differentiation. It will be seen that, in several cases, it is not convenient to express the relation between two variables algebraically, and that a graphical so-

lution is best suited to the problem. This is often true in dealing with experimental data.

After presenting the material on graphical, numerical, and graphonumerical methods of integration and differentiation, it is appropriate to introduce *computer solutions*.

PART 3 GRAPHICS FOR DESIGN IMPLEMENTATION

The three ways to communicate technical information—graphically, in writing, and orally—are discussed here. Chapter 13 includes information on sections, fasteners (both English and metric), and conventional practices, with many examples that illustrate clearly their meanings and applications. The material on selection of manufactured components from vendors' catalogs (see Chapter 13) informs students that it is often better to purchase some components than to design and build them. Chapter 14 deals with dimensions and specifications. It discusses general dimensioning, tolerancing, limits and fits (both English and metric units), and *dimensioning to ensure a single interpretation of a design*; several examples of metric design drawings prepared by industry are also included. The text emphasizes use of the metric system.

One reviewer of the original manuscript stated that: "Dimensioning has been the most difficult topic to get across to students. This chapter (14) follows a traditional approach to the subject. . . . There is, however, one innovation that impresses me—dimensioning to ensure a single interpretation of a design. This, essentially, is what dimensioning is all about."

Technical report writing, design documentation, and patent information are discussed in Chapter 15.

We hope that the material on technical report writing will help to overcome the often-heard criticism that "engineers can't write well."

The appendix is quite comprehensive. It is more like a handbook; this is necessary when one is assigning design projects to students.

The chapters in the text can be arranged in various sequences to meet the needs of most courses. For example, courses such as (or similar to) the following ones could include the listed chapters.

1. *Introduction to Design*. Chapters 1, 2, 3, and 4, and selected topics from Chapters 5, 6, 8, 9, 11, 13, 14, and 15.
2. *Graphics and Design*. Chapters 1, 2, 3, 4, 5, 6, 7, 8, 9, 11, and selected topics from Chapters 13, 14, and 15.
3. *Engineering Graphics with an Emphasis on "descriptive Geometry."* Chapters 1, 2, 3, 5, 6, 7, and 8, and selected topics from Chapters 9, 10, 11, and 12.
4. *Engineering Graphics with a Technical Drawing Emphasis*. Chapters 1, 2, 3, 5, 9, 10, 12, 13, 14, and selected material from Chapter 15.

Other combinations of chapters can provide good material for elective courses such as nomography, graphical analysis, and graphical mathematics. Such electives could be offered at the upper division level.

To augment the use of the text and to facilitate the conduct of graphics-design courses, three workbooks are available, each with a different emphasis. Basic material, however, is included in all three.

We cannot start too early to give students the experience of confronting and solving the "open-ended" type of problem. The first year in engineering education is not too soon! We recognize that students' backgrounds are quite limited at this stage of their careers; therefore, the proposed projects are relatively simple. Nevertheless, in principle, they are of the same character as some of the more involved projects that arise in engineering practice. The students' experience in coping with open-ended projects will be stimulating, challenging, and rewarding. They will find a good spirit of competition among their classmates, especially with regard to who has developed the "best" design solution.

As students progress in their engineering education and begin to study mechanics, strength of materials, design (in the broad sense), and research, they should continue to employ, whenever appropriate, graphical methods to solve problems that arise in these areas.

Graphics in Engineering Design, as presented here, reflects today's thinking and our continual effort to develop a meaningful and worthwhile treatment that we believe is consistent with the needs of a scientific and engineering era. Our experiences with students continue to be most gratifying. They are stimulated to learn and apply fundamental principles; to "think through" a problem instead of depending on rote learning; to learn and use graphical methods of computation; and to appreciate, through their own experience with design projects, that the qualified engineer must have the necessary education and experience to cope with real engineering situations that arise and continue to arise in a constantly growing, dynamic, technological era.

A. Levens
W. Chalk

Acknowledgments

We deeply appreciate the excellent cooperation we received from the many organizations that supplied photographs, graphs and charts, examples, and commercial metric design drawings. In each case courtesy lines have been included to identify the source.

We especially thank the following colleagues: Professors F.H. Moffitt, E.F. Popov, and J.M. Raphael of the Civil Engineering Department (University of California) for material on photogrammetry, metric tables for structural shapes, and the Ross Dam data, respectively; C.W. Radcliffe, professor of mechanical engineering and director of the Biomechanics Laboratory at Berkeley, L.W. Lamoreux, senior development engineer, and Frank Todd, research associate, of the Biomechanics Laboratory, for their supply of drawings and computer outputs relating to human locomotion, Professor P.J. Pagni for the flame propagation data used in empirical equations and J.F. Schon, Lecturer, for valuable suggestions concerning dimensioning, both in the Mechanical Engineering Department at Berkeley, and Professor M.E. Childs, chairman of the Mechanical Engineering Department, and Professor A.L. Babb, chairman of the Nuclear Engineering Department, both at the University of Washington, Seattle, for their continual support and encouragement.

We are grateful to Jack Ryan, president of the Jack Ryan Group, for material on his philosophy of design, and to Andrew Dubois, chief mechanical engineer for Airco Temescal in Berkeley, California, for the design case study example.

We thank these reviewers — Professors Duane Ball, University of Colorado, Boulder, Robert Britton, University of Missouri at Rolla, Fred T. Fink, Michigan State University, East Lansing, Percy H. Hill, Tufts University, Medford, Massachusetts, and Robert Mickadeit, Allen Hancock College, Santa Maria, California—for helping us to produce a better book.

A.L.
W.C.

Contents

Appendices

PART 1

Introduction to Design and Fundamentals of Engineering Graphics

Chapter 1 Introduction

Our natural desire to survive and deliberately improve the physical means of our well-being gave impetus to the development of technology. In early days this technology, which was very simple, provided crude shelters made from available local materials or from carvings in suitable earth formations. The need for food stimulated creativity in fashioning simple implements for fishing and hunting.

Engineering and Technology

Early efforts to improve the physical means of our well-being signaled the beginning of the "Art of Engineering." Since those early days, human efforts have become much more efficient through the accumulation of technical knowledge and experience over many centuries. More recently, the explosion of technological knowledge has enabled us to walk on the moon; to design, build, and operate space laboratories; to design, construct, and use sophisticated *communication systems*, such as telephones, radios, televisions, satellites, and computers, *transportation systems*, such as highways, bridges, motor vehicles, automated rapid transit trains, ships, airplanes, and pipelines, *health sys-*

tems, such as hospitals, health care monitoring equipment, kidney machines, air and water quality controls, and noise controls, *housing systems*, such as houses, office buildings, structures, appliances, furniture, and utilities, and *energy systems*, such as fossil fuels and hydroelectric, nuclear, and solar energies; and on and on.

Our sophisticated technical knowledge and experience have allowed us to achieve these physical accomplishments more by "design intent" than by accident.

"Design intent," to continually improve our physical well-being through a planned, technical, goal-directed action, is engineering.

Defined somewhat differently, "Engineering is the profession in which a knowledge of the mathematical and natural sciences gained by study, experience, and practice is applied with judgment to develop ways to utilize, economically, the materials and forces of nature for the benefit of mankind."*

Along with our achievements in technology, we have suffered several "side effects," e.g., water, air,

---

* Engineering Council for Professional Development definition.

---

and noise pollution (in trying to meet the demands of a growing, affluent population); transportation problems (congested freeways, inadequate parking facilities, etc.); environmental encroachments; power shortages; and fuel shortages.

The technology to solve our "survival" problems is available now! We can succeed in solving these problems once we establish realistic priorities and then fund (tax ourselves directly or indirectly) our decisions to fulfill our needs.

Engineering and engineering technology are essential to our society. They are alive and exciting professions that will continue to identify our technical problems and apply our constantly expanding engineering knowledge and experience to their solutions.

Your Role

You who have an aptitude for mathematics and the sciences and are motivated toward solving our societal problems, most of which involve technology, should seriously consider the dynamic professions of engineering or engineering technology as a career.

Engineering education must stress the importance of the basic sciences—mathematics, physics, and chemistry—and the engineering sciences. The technologists of the present and future eras must be well grounded in these areas. Of course, there is also a great need for practical experience and the development of good judgment. Humanities and social sciences must be stressed more than in the past.

Modern curricula in engineering and engineering technology should clearly provide students with the opportunity to achieve both technical and social goals.

Engineering Design

At the core of the possible solutions to societal-technological problems of our times is *engineering design*, which broadly includes circuits, machines, structures, and processes, and their combinations, in the design of plants and systems.

It is the responsibility of the qualified engineer to use these components effectively and to predict the performance and costs of plants and systems to meet specified requirements.

Engineering design deals with the application of science, tempered by judgment based on experience, to solve practical problems. Successful solutions of design problems result in physical and economical reality and not merely in a sketch, idea, or written report. In most cases there is no unique solution to a design problem. There are usually several adequate solutions, some of which are better than others. "The formulation of most design problems is incomplete and the designer must himself define the problem. The data available for a design process are almost always incomplete, and often contain much irrelevant matter. The data may be redundant and even contradictory. The designer must use judgment in the selection of data and in deciding when the available information is sufficient for accomplishing the task in hand."*

Journal of Engineering Education, "Report on Engineering Design," April 1961, pp. 647–648.

Designing and Sketching

Designing is a conceptual process that is done largely in the mind; making a sketch is a recording process, a reliable memory system, that engineers use to "talk to themselves," to help themselves "think through" the various aspects of their project. Typical conceptual sketches are shown in Figure 1.1.

The creative art of conceiving a physical means of achieving a technological objective is the essential and most crucial step in an engineering project. The conceptual process (synthesis) precedes analysis, which is the very important second step required to refine the conception.

Figure 1.1 Conceptual design sketches help designers communicate with themselves and provide an excellent format for recording ideas.

Engineering Graphics

Engineering graphics is an integral part of the conceptual phase since, more often than not, making a simple sketch to express a design conception suggests further conceptual ideas.

Engineers who have developed the ability to communicate graphically, to *visualize* geometrical and physical configurations, and to "think graphically" (spatial visualization) have a decided advantage in creating a physical means of achieving an objective. Figure 1.2 shows a technical drawing, which is one form that engineers utilize to communicate graphically. The "thinking-through" process is an important exercise of the mental powers of judgment, conception, and reflection for the purpose of reaching a result—a design that is the "best compromise" solution to a given project. Of course, the study of analytical courses is essential because it is difficult to predict performance and costs of a design without analysis while it is in the paper stage.

Figure 1.2 A technical drawing usually contains information to describe the shape, size, and special features of a component, device, mechanism, etc., in sufficient detail to enable a skilled technician to create the object.

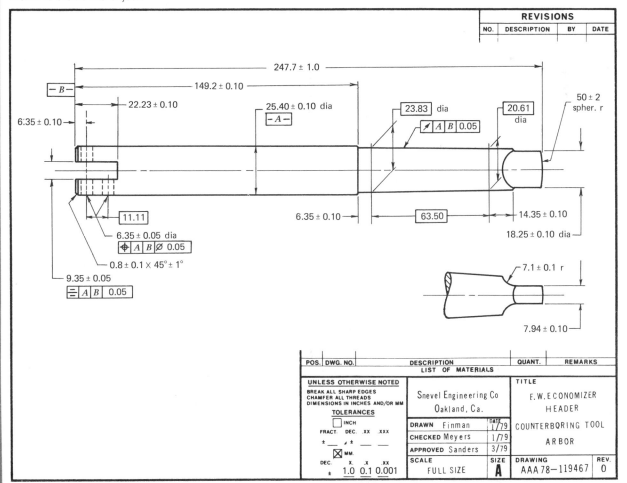

POS.	DWG. NO.		DESCRIPTION		QUANT.	REMARKS
			LIST OF MATERIALS			

UNLESS OTHERWISE NOTED
BREAK ALL SHARP EDGES
CHAMFER ALL THREADS
DIMENSIONS IN INCHES AND/OR MM

TOLERANCES

☐ INCH
FRACT.   DEC. .XX   .XXX
±   ⌀ ±

☒ MM.
DEC.   X.   .X   .XX
±   1.0 0.1 0.001

Snevel Engineering Co	TITLE
Oakland, Ca.	F.W. ECONOMIZER HEADER

DRAWN Finman	DATE 1/79	COUNTERBORING TOOL
CHECKED Meyers	1/79	ARBOR
APPROVED Sanders	3/79	

SCALE FULL SIZE	SIZE A	DRAWING AAA 78−119467	REV. 0

Physical Reality

Transforming an idea into reality is a great engineering and technology achievement. The result may be a new structure, device, system, material, or process that enhances comfort, health, and knowledge. For example, Figure 1.2 is a detailed drawing of an idea for a tool arbor. Figure 1.3 shows the physical reality (pictorially) of the idea.

All about us we see the work of the engineering team. We have already mentioned several—the telephone, radio, television, computer, structures, motor vehicles, etc.—*all of which would not have been achieved without engineering design*. Figure 1.4 shows some of the products of engineering.

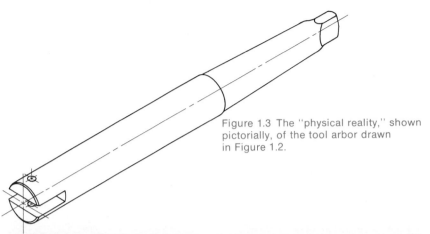

Figure 1.3 The "physical reality," shown pictorially, of the tool arbor drawn in Figure 1.2.

Figure 1.4 Typical products evolved through the engineering design process, fabrication, and construction. Courtesy Bethlehem Steel Corp.

Engineers and Scientists

Usually engineers do not work alone. For example, engineers and scientists are dependent on each other. Scientists investigate the basic laws of nature and try to define the principles by which they are governed; engineers apply these findings to the creation of something that is both useful and beneficial to us. The famous engineer, the late Dr. Theodor von Kármán, stated it very well when he said, "The scientists explore what is, and the engineers create what has never been."

The new creation may result from a novel combination of existing components or from new materials and processes that are spinoffs from advanced technology. For example, part of our future electrical energy needs may come from the efficient use of the sun's heat. We may have to know much more about the direct conversion of the sun's heat to electricity, on a large enough basis for an adequate supply, to compete *economically* with alternative energy sources and methods. We may need new inputs from the scientists in order to advance our engineering design capability. And, similarly, the engineers must provide the tools that are required to make scientific discovery possible. A good example of this is the space program, which required sophisticated launching equipment, instrumentation, computer systems for space-travel guidance, and monitoring equipment to record the temperatures and pulse and respiration rates of the astronauts while in travel and in their performance of experiments.

The Technological Team

The solution of complex engineering problems usually involves the skills and knowledge of artisans, technicians, scientists, and engineers, organized as a team to accomplish the tasks of the various technological aspects of the problems. In recent years the highly qualified engineering technician has become a very important member of this team. The tasks of routine design and the preparation of design and production drawings are usually assigned to such technicians. Others, such as political scientists, sociologists, and economists, may be used as part of the team or as consultants before the best "compromise" solution is selected.

DESIGN PROCESS

Engineers approach the solution of technological problems in a systematic way that we can describe as a "design process." This process, simply described, would include: (1) establishing criteria for accepting solutions, (2) generating possible solutions, and (3) selecting the best solution. All three would involve some form of graphics.

Our text describes in more detail this design process, its phases and steps, and the important role that engineering graphics plays in the process. Our intent is not to make the process seem more complex than it is, but to separate the numerous overlapping activities so we can discuss them.

Phase I   Recognize and Define the Problem. Phase I begins with the recognition that a problem exists. Further efforts are aimed at searching for information and ultimately defining the objectives, requirements, and constraints for the problem. A progress report normally follows.

Phase II   Generate and Evaluate Alternative Solutions. Phase II entails the exhaustive generation and consideration of possible alternative solutions to the problem, leading to a preliminary decision. A design review of the decision refines or sometimes rejects the decision.

Phase III   Communicate the Solution. Phase III includes delineation and communication of the end product in all of its details in a report that includes a summary of Phases I and II, design drawings, and specifications for manufacturing. Models and prototypes are utilized to develop the product. The models and prototypes are often used to augment oral presentations.

Phase IV   Develop, Manufacture, and Market the Final Product. Phase IV comprises the transformation of the descriptions of the end product generated in Phase III into a commercial, physical reality. Sales, marketing, and consumer feedback are also included in Phase IV.

Most schools provide opportunities for students to complete Phase III, but few can afford Phase IV. We focus here on the first three phases and refer to Phase IV whenever the discussion leads to consideration of topics such as consumer needs, manufacturing practices, sales, etc.

Design Process Steps

The *steps* describe the design process in more detail than the *phases*. The sequence of the steps as listed below *is not necessarily followed in that order and, in most cases, they will overlap.* The experience of the designer usually determines the extent to which variations in the sequence will occur. In fact, many experimental design engineers develop their own patterns. *For the beginning student in engineering, the order shown should be quite helpful;* ultimately, students develop their own.

Phase I   Recognize and Define the Problem

1. *Initiate Planning*. Once you have received the problem, allocate the available time for the phases with approximately 25, 25, and 50% for Phases I, II, and III, respectively. In industry Phase IV would also be scheduled. Keep a record of your activities and time spent, develop preliminary plans, and start a file for idea sketches, notes, correspondence, etc

2. *Gather Information*. Use available resources such as a technical library, faculty, engineers, and shop technicians.

3. *Define the Problem*. List the objectives, functions, constraints, and special requirements.

4. *Prepare a Brief Progress Report*. Outline your definition of the problem and your immediate plans for solving it.

Phase II   Generate and Evaluate Alternative Solutions

5. *Collect Ideas*. Many ideas are generated during Phase I.

6. *Prepare Additional Conceptual Sketches*. Prepare sketches of new ideas and combinations of

new and old ideas. Brainstorming helps.

7. *Organize*. Arrange ideas and concepts into potential alternative solutions to the problem.

8. *Select*. Select the most promising ones. These will be the most probable candidates.

9. *Develop and Analyze*. Develop and analyze the selected alternatives. Prepare layout drawings. Use mathematical models and computers if available. Conduct experiments.

10. *Decision*. Decide on the most promising solution that you have available.

11. *Design Review*. Conduct a design review; i.e., present your ideas to knowledgeable individuals. In school, present them to your classmates and instructor, in industry, to your immediate supervisor and, if requested, to management.

Phase III Communicate the Solution

12. *Refine*. Refine your solution based on comments and questions during the *design review* or, if you do not have a review, consult someone experienced for an objective evaluation of your project. (Sometimes designers must start over while mumbling the old cliché, "Well, I guess it's back to the drawing board.")

13. *Prepare Drawings and Specifications*. Prepare appropriate drawings and specifications that satisfy the needs of the user(s). In school sketches, assemblies, and some details with specifications are often satisfactory. In industry the three types are usually required, but in greater quantity and in more detail. Often the specifications (materials, quantities, hardware, etc.) are separate documents.

14. *Model or Prototype*. Build a model and/or prototype if facilities are available.

15. *Final Report*. Prepare a final, typewritten report, enclosed in a folder, which could include the following.

Introduction. How the problem originated.

Definition. Objectives, functions, constraints, costs, etc.

Alternatives Considered. Sketches or drawings and a description of alternative solutions that you considered but discarded and why.

Selected Alternatives. Drawings and text that completely describe the alternative you selected.

Conclusions. Suggestions for further action.

Appendix. Appropriate calculations, reference materials, brochures, sketches, etc.

Prepare a *final oral report*. Use visual aids such as transparencies, slides, flip-charts, and simple cardboard-glue models.

We mentioned earlier that schools often require only the first three *phases*, so we have directed most of our comments toward that end. Thus, the following discussion of the steps in Phase IV, which relate more to industry, is somewhat brief. Furthermore, industrial practices include such an extensive variety of approaches, depending on the product, the market, the quantity to be manufactured, the materials, the existing technology, etc., that we do not have space to describe them all adequately.

Phase IV Develop, Manufacture, and Market the Final Product

16. *Review Specifications and Drawings*. Review the drawings and specifications for completeness. Review the availability of personnel, materials, and hardware.

17. *Plan for Manufacturing*. Time, materials, financing, budgeting, scheduling, and personnel planning are the keys to successful product development and manufacturing, whether the product is a complicated spacecraft system or a relatively simple household appliance. Product liability considerations such as keeping good records, following codes, and standards should be incorporated.

18. *Build, Test, and Revise*. Constructing working prototypes and testing them uncovers facets of designs that often cannot be predicted on paper.

19. *Packaging*. Sometimes shipping loads and environmental conditions during transit are more severe than the actual end use.

20. *Marketing, Sales, and Distribution*. The ultimate goal of most businesses is to market their products at a profit.

21. *Consumer Feedback and Redesign*. Appropriate redesign should be incorporated to satisfy the ultimate user for satisfaction and safety.

Figure 1.5 shows the phases and steps of the design process and the relationships of engineering graphics to the design process

```
       DESIGN PROCESS: PHASES AND STEPS

   ┌─────────────────────────────────────────────────────┐
   │   Phase I    RECOGNIZE AND DEFINE THE PROBLEM        │
   │                                                     │
   │         1. Start Planning (Chart)                   │
   │      *┌─> 2. Gather Information (Sketches)           │
   │       └─ 3. Define the Problem                      │
   │         4. Prepare A Progress Report                │
   └─────────────────────────────────────────────────────┘
                          │
                          ▽
   ┌─────────────────────────────────────────────────────┐
   │   Phase II    GENERATE SOLUTIONS                     │
   │                                                     │
   │         5. Collect Ideas (Sketches)                 │
   │     ┌─> 6. Generate New Ideas (Sketches)            │
   │    ┌┼─> 7. Organize Ideas                           │
   │    │└─ 8. Select Promising Ideas                    │
   │    │┌─> 9. Develop and Analyze Promising Ideas      │
   │    ││      (Layout Drawings, Applied Graphics)      │
   │    │└─ 10. Decide on Best Idea                      │
   │    └── 11. Conduct A Design Review (Visual Aids)    │
   └─────────────────────────────────────────────────────┘
                          │
                          ▽
   ┌─────────────────────────────────────────────────────┐
   │   Phase III COMMUNICATE THE SOLUTION                 │
   │                                                     │
   │     ┌─>12. Refine and Revise Final Solution         │
   │     └─ 13. Prepare Drawings and Specifications      │
   │            (Technical Drawings, Computer Graphics)  │
   │        └─ 14. Make A Model or Prototype             │
   │        15. Prepare Final Reports--Written, Graphics,│
   │            and Oral (Figures, Technical Drawings)   │
   └─────────────────────────────────────────────────────┘
                          │
                          ▽
   ┌─────────────────────────────────────────────────────┐
   │   Phase IV DEVELOP, MANUFACTURE, AND MARKET          │
   │            THE FINAL PRODUCT                         │
   │                                                     │
   │     ┌─>16. Check Drawings and Specifications        │
   │     │      (Technical Drawings, Computer Graphics)  │
   │     │  17. Plan for Manufacturing                   │
   │     └─ 18. Build Prototypes                         │
   │        19. Design Packaging (Sketches, Layout Drawings,│
   │            Technical Drawings)                      │
   │        20. Conduct Sales Research                   │
   │        21. Obtain Customer Feedback                 │
   └─────────────────────────────────────────────────────┘

    *The arrows indicate iterative paths between steps.

    ( ) The parentheses indicate graphics most likely
        used.
```

Figure 1.5 Flowchart showing the design process phases and steps.

ROLE OF ENGINEERING GRAPHICS IN THE DESIGN PROCESS

Engineering graphics is an essential part of creative design, and creative design is the core of engineering. For example, in Phases I and II, the truly creative phases of the design process, the engineer exercises mental power to conceive a physical means of achieving a technological objective. The making of *freehand sketches* (orthographic, isometric, oblique, or perspective) to express conceptual designs is a very important recording process; it is a reliable memory system that enables designers to "communicate with themselves" and their co-workers. The recording of the initial conceptual design sketches stimulates the designer's thinking toward other ideas of a conceptual nature. Freehand sketching is one of the most effective means of communication among the members of a design team—the engineers, designers, technicians, drafters, and production personnel. Techniques for making freehand sketches are discussed in Chapter 2.

Also in Phase II, there are factors such as function, reliability, safety, costs, and eye appeal that affect the design of the end product. Here, again, engineering graphics plays a role; layout drawings, graphs, and charts may be needed for studies related to interrelationships of components to functions and to costs. More will be said about graphs and charts in Chapter 10. Layout drawings are discussed in Chapter 3. Eye-appeal evaluation will often require the preparation of several perspective drawings to show the three-dimensional appearance of the more promising design choices. Chapter 9 includes this type of drawing.

Presentation of the results obtained in Phase II generally is necessary in industry in a design review; it enables management to reach a decision as to the merits of the proposed approach to the design project. Such presentations may consist of a written report that is quite complete or, most likely, of a shorter report that highlights the results and is accompanied by an oral discussion that provides a good opportunity to "sell" management on the recommended solution. The presentation must be well prepared to be successful. Invaluable suggestions for the preparation of technical reports and oral presentations are discussed in Chapters 4 and 15.

And, of course, in Phase III engineering graphics plays a large role in connection with the preparation of working drawings of each component (detail drawings) and of the relationship among the components (assembly drawings). A formal report and an oral presentation of the results usually follow Phase III, particularly in student-classroom environments. These reports are also discussed in Chapters 4 and 15.

The drawings prepared in Phase III set the production machinery in motion. Purchasing agents utilize the drawings and specifications to procure materials. Shop superintendents schedule their machines and personnel to perform the proc-

esses indicated on the drawings. Assembly crews use drawings to check relationships and completeness.

Production plans are frequently presented in charts and in critical path schedules. Progress can be noted on the charts and the schedules.

Several chapters stressing design, such as sketching, drawings, and reporting, have been noted with reference to the role of graphics in the design process but, in fact, each chapter discusses topics in engineering graphics that are utilized in the *design process*. Several chapters emphasize analysis, including Chapters 8 ("Vector Quantities and Vector Diagrams"),

Chapter 11 ("Graphical Solutions and Computations"), and Chapter 12 ("Computer Graphics").

APPLIED ENGINEERING GRAPHICS

Typical Technical Problems

Angle Between Plane Surfaces

In connection with the preparation of the working drawings there are usually a number of technical problems that need to be solved graphically. For example, in the design of an aluminum structure for an aircraft, there may be two members, a rib and a strut, that are to be held in position by a connection angle, as shown in Figure 1.6.

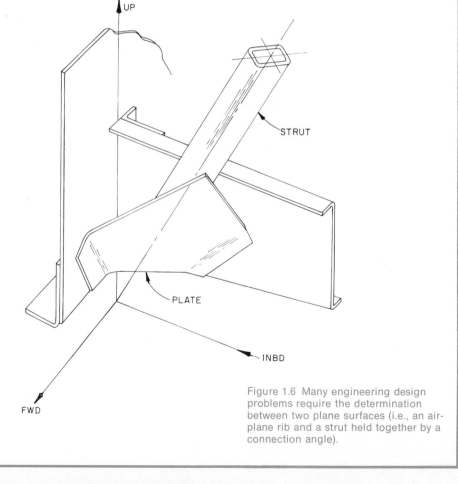

Figure 1.6 Many engineering design problems require the determination between two plane surfaces (i.e., an airplane rib and a strut held together by a connection angle).

In order to bend a flat aluminum plate to the proper angle, it is necessary to determine the angle between the plane surfaces of the rib and the strut. Similar problems arise in the determination of the angle between intersecting surfaces in a number of applications, such as the one shown in Figure 1.7. We observe the same basic problem in the edge-reinforcing angle members of the "hopper" shown in Figure 1.8 and again in the supporting members of the boat windshield shown in Figure 1.9. All of these problems reduce to one basic problem: *finding the angle between two plane surfaces*.

Figure 1.7 An application of the angle between plane surfaces in the design of the corner support angle irons for the glass plates of the control tower.

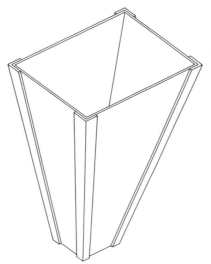

Figure 1.8 Edge-reinforcing angle members of a hopper. Still another example of the need to determine the angle between two plane surfaces.

Figure 1.9 Boat pilots windshield supports. An additional example of the "angle between two plane surfaces problem."

True Lengths

In addition, technical problems arise that require the determination of the true lengths of members (rods, cables, shafts, etc.). For example, in Figure 1.10, the true length of the reinforcing steel bar *AB*, which is one of several bars that may be required in the design of the concrete *wing wall*, must be determined in order to provide information for ordering the correct lengths of the bars. Again, in the design of the abutment wall, some bars continue into the wing wall (i.e., bar *M*). The angle of bend of the bar before bending must also be determined.

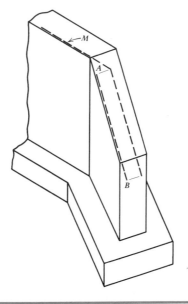

Figure 1.10 Reinforced concrete wing wall. Bar *AB* requires true length determination. Bar *M* requires determination of "angle of bend."

True Shapes

In a number of design projects, problems arise that require the determination of the true shapes of plane surfaces. For example, in Figure 1.11 it is necessary to show the true shape of the sloping surface in order to disclose the geometry of the surface and the location of the holes.

For another example, let us refer back to the hopper shown in Figure 1.8. Let us assume that the sides of the hopper are to be made from steel plate. It will be necessary to determine the true shapes of the sides to provide patterns that would be used to cut the plates properly.

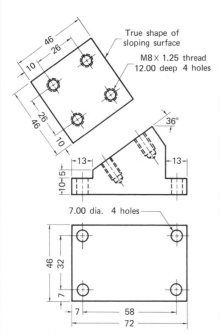

Figure 1.11 True shape of sloping surface needed to show the geometry of the surface and hole locations.

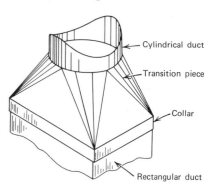

Figure 1.12 Typical products that require patterns (developments of surfaces) for fabrication.

Funnels and pails

(a)

Developments

Also, in many designs, parts of products are made from sheet metal or other materials that can be readily shaped or formed. For example, various types of funnels, transition pieces, etc., as shown in Figure 1.12, require patterns (developments, or flat-sheet layouts) for cutting the material to the proper shape so that it can be formed to meet the three-dimensional requirements in the design working drawings.

Cylindrical duct
Transition piece
Collar
Rectangular duct

(b)

(c)

Intersections

In other design projects problems arise that deal with the intersection of surfaces. For example, in gasworks design, the piping system, shown in Figure 1.13, involves the intersection of cylindrical surfaces.

Figure 1.13 Intersection of cylindrical pipes—gasworks park.

The intersection of a cylindrical surface and a spherical surface occurs in the tank design shown in Figure 1.14, and the intersection of an airplane wing surface with the

Figure 1.14 Intersection of cylinder and sphere in tank design. (Courtesy Chicago Bridge and Iron Co.)

fuselage surface, as shown in Figure 1.15, occurs in virtually all aircraft design. Of course, there are many problems in which the intersection of plane surfaces must be determined. An example is shown in Figure 1.16.

Figure 1.15 Intersection of aeroplane wing and fuselage. (Courtesy The Cessna Aircraft Co.)

Figure 1.16 Intersections of plane surfaces. Outdoor art display.

Force Polygons

In Bridge design, for example, once a decision has been made as to the type of structure (i.e., a steel through-truss) (Figure 1.17) that will best fulfill the requirements, calculations of the forces acting in the members of the bridge must be made in order to select the steel shapes (I beams, angle irons, bars, etc.) that will safely carry the loads. Although the calculations can be made with the aid of computers, desk calculators, and pocket calculators, a graphical solution is not only simple but also provides a convenient method for assuring reliable results. It is a fairly common practice, for example, to use graphical methods (vector, or Maxwell, diagrams) to determine the forces acting in the members of a truss. Many engineering firms often use graphical solutions as a check on the analytically obtained results.

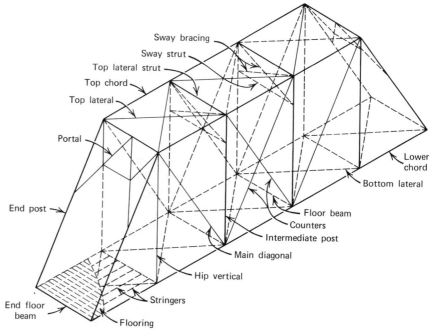

Figure 1.17 Through bridge truss. (*Design of Steel Structures,* B. Bresler, T. Y. Lin, and J. B. Scalzi, 2nd Ed., John Wiley & Sons, Inc., 1968.)

In determining the forces that act on the members of three-dimensional frames, such as those shown in Figure 1.18, the graphical method, which is discussed in Chapter 8, is an excellent, simple, reliable, and sufficiently accurate one. In many cases this method is superior to nongraphical solutions.

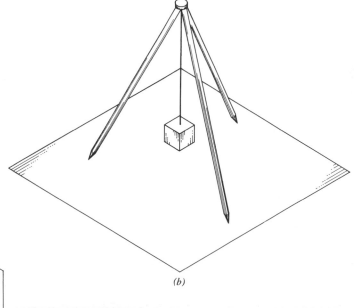

(b)

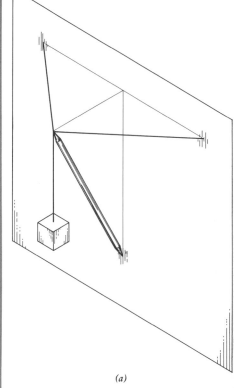

(a)

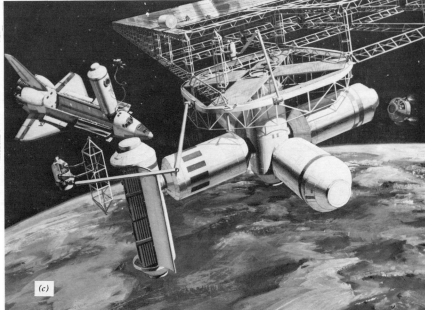

(c)

Figure 1.18 (a,b) Three-dimensional space frames. (c) Large space structures. (Courtesy Rockwell International's Space Division.)

A thorough knowledge of the fundamental principles of orthogonal projection and their application to the solution of the preceding problems and to others that are discussed in Chapters 5 to 8 will greatly enhance the ability to "think graphically" and will strengthen the powers of visualization; this combination we can identify as "imagineering."

Working Drawings

Working drawings disclose the geometry and the dimensions of all the components. In Chapters 13 and 14 we will devote some time to dimensioning and to conventional drawing practices to show sections, threads, fasteners, etc.

Engineering students should recognize that it is important to learn and use this material in preparing the working drawings (and specifications) to implement Phase III of the design process. After graduation *and good design experience on the job*, you will most likely *supervise* the work of qualified technicians and drafters who will be concerned with most of the details that are so important to achieving success in an engineering design project.

ADDITIONAL APPLICATIONS OF ENGINEERING GRAPHICS

Topographic Maps

In the planning of building projects (e.g., housing developments and high-rise structures), it is necessary to have a topographic map of the site in order to study grading and drainage problems and the locations of the proposed units. Similar problems occur in highway design. An example is shown in Figure 1.19.

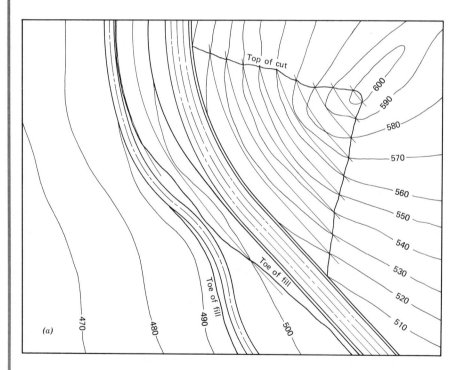

Figure 1.19 Topographic maps are used in highway design. (*a*) Location of the "toe of the fill" and the "top of the cut." (*b*) Typical photograph of the "toe of the fill" and the "top of the cut."

Pictorial Representation

Previously we mentioned isometric, oblique, and perspective *freehand sketches* in connection with the recording of conceptual designs (Phase I). Pictorial representations (Chapter 9)—isometric, dimetric, trimetric, oblique, and perspective sketches—are also used to simplify and clarify the material in parts manuals (Figure 1.20), installation manuals (Figure 1.21), and technical reports. In technical reports it is quite usual to find charts and diagrams that greatly enhance the value of the reports.

SEARS 26" MEN'S 10-SPEED LIGHTWEIGHT BICYCLE — MODEL 505.474531
SEARS 26" LADIES 10-SPEED LIGHTWEIGHT BICYCLE — MODEL 505.474541

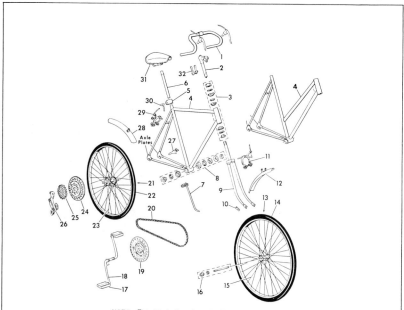

NOTE: This illustration shown for Parts Identification only.

KEY NO.	Model 505.474531 PART NO.	Model 505.474541 PART NO.	PART NAME	KEY NO.	Model 505.474531 PART NO.	Model 505.474541 PART NO.	PART NAME
1	28607-C	28607-C	Handlebar	18	14745	14745	Crank
2	26023	26023	Stem	19	28165	28165	Front Chainwheel Assembly
3	16107	16107	Head Fitting Set	20	10151-57	10151-57	*Chain (114 Links - 56½")
4	24159	24160-F	Frame	21	25282	25282	Rear Wheel
5	1C5017	1C5017	Saddle Post Clamp	22	24171	24171	*Rear Tire
6	8514-BW	8514-BW	Saddle Post	23	25216	25216	*Rear Spokes -36 (10-31/32")
7	10680	10680	Kickstand	24	28245	28245	Spoke Protector
8	5506-B	5506-B	Hanger Fitting Set	25	28253	28253	Rear Freewheel Assembly
9	28250	28250	Fork	26	28244	28244	Rear Derailleur
10	1C5281	1C5281	Wheel Retainer	27	28243	28243	Front Derailleur
11	25150	25150	Front Caliper Brake Assembly	28	12784	12784	Rear Fender
				29	25151	1C5280	Rear Caliper Brake Assembly
12	12784	12784	Front Fender	30	12158	12158	Saddle Post Clamp (Quick Release)
13	14437	14437	Front Wheel	31	1C6460	1C6460	Saddle
14	24171	24171	*Front Tire	32	1C5102	1C5102	Shift Lever Assembly
15	25215	25215	*Front Spokes-36(11-1/32")	—	24463	24463	Cable Organizer
16	2496-B8	2496-B8	Front Axle Set	—	1C6491	1C6491	Reflector Kit
17	1C5245	1C5245	Pedals	—	1C5318	1C5318	Brown Handlebar Tape W/Plugs

Always state Tire size when ordering Tire, Wheel, Fork, or Fender.

— Not Illustrated
* Standard Hardware Items - May be purchased locally

Standard Accessories and Hardware Items may be purchased from any Sears Retail or Catalog Sales Store. Specify color on all enameled parts. NOTE: If unavailable locally, order nuts, bolts, and washers and note where used.

Figure 1.20 Pictorial used to identify parts only.

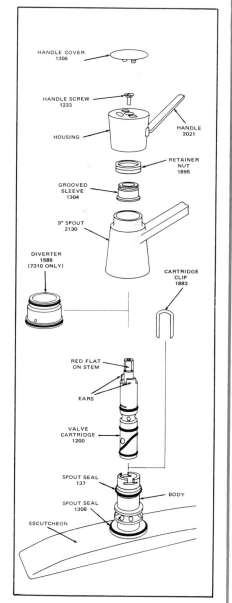

Figure 1.21 Pictorials in an installation example.

Data that are presented in a way that makes understanding and analysis easier is a great plus for the person who needs to use the data. For example, a chart that shows the increase in population as a function of time might be of some interest to the layperson. However a company that is concerned with the future marketability of its products could be more interested in knowing *the rate* at which the population was increasing. How should the data be plotted to reveal such information? Chapter 10 discusses such problems and tells how to present the data to the best advantage.

Graphical Calculus

In the design of large or complex systems, it is often necessary to carry on some research and development (R & D) work to achieve answers to a number of essential subproblems before good progress can be made in the overall design of the project. Modeling the problem physically, mathematically, or analogically and collecting, presenting, and analyzing the experimental data are essential to an identification and understanding of the variables and their interrelationships. For example, in a project "to design an artificial leg with which amputees would be satisfied—good function, comfortable, reliable, and fairly priced," a number of problems arose that re-

quired investigations before the actual design work could even be started. No usable information was available at the time the project was started as to the magnitude of motion at hip, knee, and ankle joints. Such information was needed in the design of an artificial leg. What could be done to obtain the magnitudes of motion in normals (persons with both legs functioning)? A study of this problem by the team of orthopedic surgeons and engineers resulted in the development of a method* to solve the problem. Once the data were collected and analyzed (not an easy job), the ranges of the magnitudes of motion were determined and made available as input information for the design of prototypes. Other problems dealt with the determination of displacement as a function of time during a walking cycle; from such data it was possible to determine velocity as a function of time and also of acceleration by employing a simple graphical method of differentiation (see Figures 1.22 and 1.23). The graphical methods for integration and differentiation are discussed in Chapter 11.

---

* Paper by A. S. Levens, V. T. Inman, and J. Blosser, "Transverse Rotation of the Segments of the Lower Extremity in Locomotion," in the *Journal of Bone and Joint Surgery*, November 1948. Numerous similar studies have been made since then.

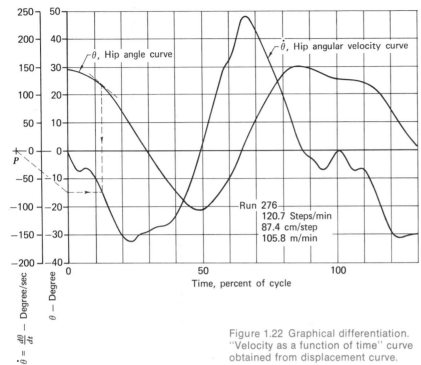

Figure 1.22 Graphical differentiation. "Velocity as a function of time" curve obtained from displacement curve.

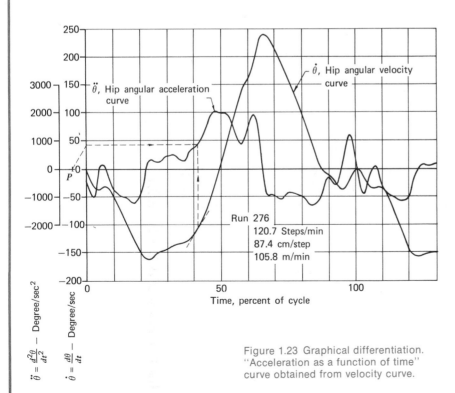

Figure 1.23 Graphical differentiation. "Acceleration as a function of time" curve obtained from velocity curve.

Empirical Equations

In the fields of research, development, and design, laboratory experiments are often conducted to observe the physical behavior between related variables. Graphical representation of test data obtained from experiments and the determination of the empirical equation that best represents the data help the designer to gain an understanding of the relationships between the variables.

Plotting data and the methods used to determine empirical equations are discussed in Chapter 11.

Nomography

A very useful tool in engineering design (and in other fields) is nomography, the graphical representation and solution of mathematical expressions. The theory involved in the design of nomograms is simple and easily understood, and is explained in Chapter 11.

Nomograms provide a convenient, time-saving means for repetitive solutions of mathematical formulas and for the study and analysis of the interrelationships among the variables. A nomogram is an efficient, low-cost computer that can be applied to a variety of problems that arise in engineering design, the health sciences, the physical and biological sciences, statistics, manufacturing, business, etc.

Nomograms are utilized in numerous technical and nontechnical applications. Technical examples are shown in Figures 1.24 to 1.26.

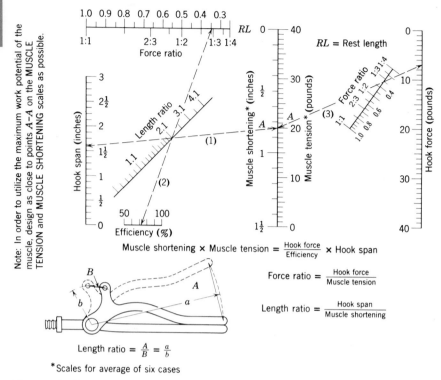

Note: In order to utilize the maximum work potential of the muscle, design as close to points A–A on the MUSCLE TENSION and MUSCLE SHORTENING scales as possible.

$$\text{Muscle shortening} \times \text{Muscle tension} = \frac{\text{Hook force}}{\text{Efficiency}} \times \text{Hook span}$$

$$\text{Force ratio} = \frac{\text{Hook force}}{\text{Muscle tension}}$$

$$\text{Length ratio} = \frac{\text{Hook span}}{\text{Muscle shortening}}$$

$$\text{Length ratio} = \frac{A}{B} = \frac{a}{b}$$

*Scales for average of six cases

Figure 1.24 Design chart for artificial hook-type hand. (Below-elbow amputees.)

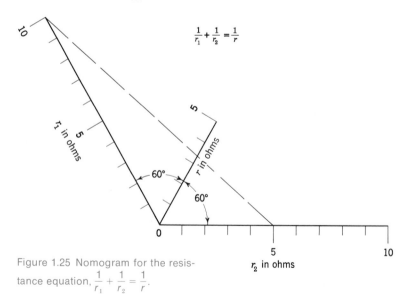

$$\frac{1}{r_1} + \frac{1}{r_2} = \frac{1}{r}$$

Figure 1.25 Nomogram for the resistance equation, $\dfrac{1}{r_1} + \dfrac{1}{r_2} = \dfrac{1}{r}$.

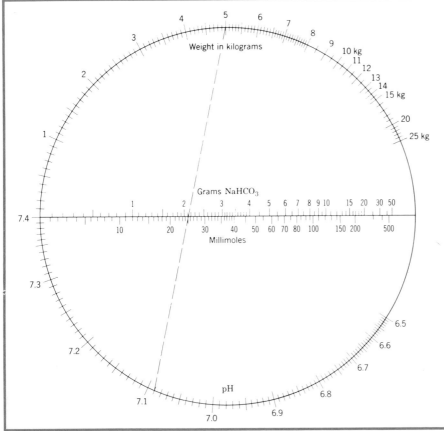

Figure 1.26 Nomogram for approximation of alkali requirements of patients in acidosis. (Courtesy E. C. Varnum, Barber-Colman Co.)

Trade-Offs

In engineering design, once engineers have determined the mathematical relation among several design parameters, they could design a nomogram for the mathematical expression and then study the interrelationships among the parameters to determine the best combinations (trade-offs) that satisfy the specific requirements.

Computer Graphics

We are witnessing a tremendous increase in the use of computers in conjunction with peripheral equipment such as digitizers, X-Y plotters, and cathode-ray tubes (CRT) with light pens to *input graphical data* and to *output graphical information* (charts, diagrams, contour maps, orthographic and pictorial drawings, etc.).

The increased use of computer graphics has been greatly enhanced by the availability of reasonably priced minicomputers and peripheral equipment (hardware) and the development of ingenious programs (software) that have significantly enlarged the areas of application.

Engineering design, architectural design, urban planning, and manufacturing have greatly bene-fited from computer graphics. Many problem areas in those fields have yielded readily to the employment of computer graphic solutions.

A unique application of computer graphics is shown in Figure 1.27, which reveals a perspective

Figure 1.27 Computer graphics output example showing a portion of the San Francisco skyline, without the pyramid structure. (Courtesy Dynamic Graphics, Berkeley, California)

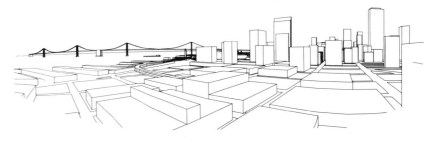

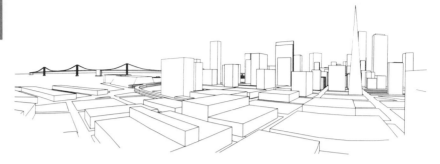

Figure 1.28 Computer graphics output example showing the inclusion of the pyramid structure. (Courtesy Dynamic Graphics, Berkeley, California)

view of a portion of San Francisco before a proposed pyramid-type structure was constructed. It was of interest to concerned citizens to see what the area would look like in a perspective that included the proposed structure. It was a simple matter to do this by a slight modification of the program. The result is shown in Figure 1.28. Removing or adding structures was handled quite readily. The ingenious software package solved the hidden-edge problem effectively and economically. The input information consisted of a plan view of the structures and height data of each structure. Details such as doors, windows, etc., were not needed in this study.

The planning stage of housing complexes or industrial structures, studies of building locations, and preparation of perspectives from various points of view can be made quickly and economically. Changes can be made in little time. A few examples are shown in Figures 1.29 to 1.33. Additional material on computer graphics is presented in Chapter 12.

Figure 1.29 Computer-drawn plan view of development indicating locations and angles of subsequent perspective views. (Courtesy Dynamic Graphics, Berkeley, California)

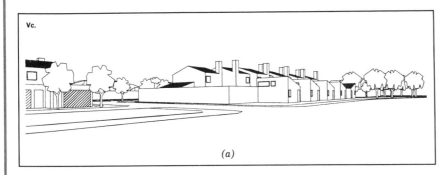

(a)

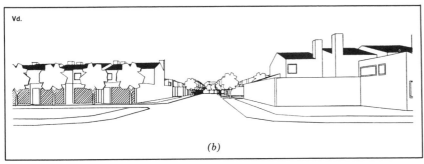

(b)

Figure 1.30 (a) Perspective view as driver approaches the main entry street into the development. (b) Perspective view from main entry street. (Courtesy Dynamic Graphics, Berkeley, California)

Figure 1.31 Aerial perspective view of housing development. (Courtesy Dynamic Graphics, Berkeley, California)

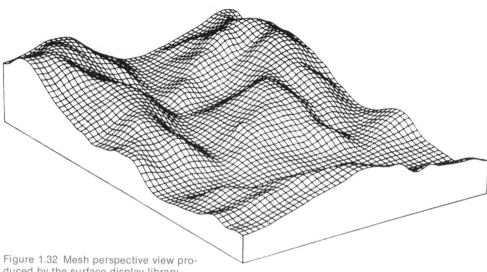

Figure 1.32 Mesh perspective view pro-
duced by the surface display library.
(Courtesy Dynamic Graphics, Berkeley,
California)

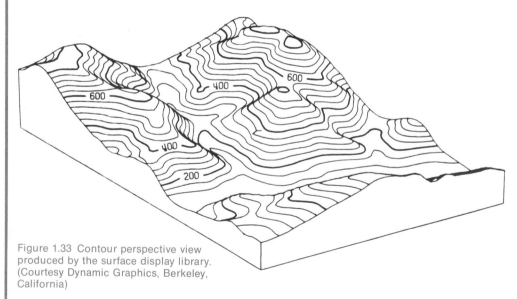

Figure 1.33 Contour perspective view
produced by the surface display library.
(Courtesy Dynamic Graphics, Berkeley,
California)

SUMMARY

This chapter describes the important role of engineering design and engineering graphics in our continuing effort to improve the physical means of our well-being through a planned, technical, goal-directed action—engineering.

Engineering and technology will continue to identify the technical problems of our society and to apply the ever expanding technical knowledge and experience to their solutions through engineers and technicians.

Engineers approach the possible solutions to such problems in a systematic way that we have described as the "design process." The role of engineering graphics in that process has been described, and it clearly indicates that engineering and engineering technology students need: (1) to develop reasonable proficiency in freehand sketching; (2) a thorough understanding of the fundamental principles of orthogonal projection and their application to the solutions of three-dimensional problems that arise in design work; (3) to communicate graphically through properly prepared sketches and orthographic working drawings to express design concepts and information needed for manufacture and fabrication of prototypes, structures, etc; (4) to know about charts and graphs and their use in the preparation of various types of reports and manuals; and (5) to understand and use graphical mathematics and other graphic methods of solving problems wherever appropriate in engineering.

In addition, you should be capable of expressing yourselves effectively in written and oral forms. The experience gained in writing design proposals and project reports and in making oral presentations will help you to achieve reasonable proficiency in writing and speaking.

The material in the following chapters will help you gain an insight to engineering design. In addition, it will prepare you to use engineering graphics, in all of its aspects, in coping with relatively elementary design projects that are in keeping with the limited technical knowledge of first- and second-year engineering and technology students.

However, you already have had some experience in design. For example, most new students can read drawings to some degree from working on models of boats, cars, and airplanes or from other hobbies such as weaving, sewing, or leathercraft. All students have some skill in sketching and writing, and a few have had some experience in technical drawing. Many have solved problems that have more than one solution, such as modifying a car, preparing for a debate, writing a term paper, or deciding on which school to attend. While reading these comments, you may be asking yourselves, "Why then should I continue to study graphics and design?" We believe that this text will enhance the skills you may have, and it will provide a basis for learning new ones. You will experience the thrill of working on real "open-ended" design projects, and you will enhance your ability to solve engineering design problems.

Chapter 2 focuses on sketching and introduces fundamentals of orthogonal projection, which are the bases for many of the drawings you will use in analysis, design, and communication. Chapter 3 ("Layout Drawings") logically follows sketching because designers usually redraw their conceptual sketches to scale to check relationships and to prepare information for drafters. Chapter 4 discusses the design process and provides guidelines to help you get involved in using it; we feel that you will understand better what designers and engineers do by using the design process. Chapter 5 continues with the fundamentals of orthogonal projection and provides an excellent foundation for the topics in Part 2 ("Applied Graphics for Design, Analysis, and Communication").

Exercises

Orientation to Technology

1. Start an idea file (scrapbook) relating to your interests in technology. (We know an individual who keeps, among other files, an "Aw Shucks" file. Most of the entries are from newspaper accounts of accidents: roof collapsing, brake failing, load tipping, etc.) An idea file is useful because it develops and expands your interests and provides a source of projects not only for this course, but for future design activities. Inventions may grow out of your collection of ideas.

2. (a) Visit a consulting engineers' office (civil, electrical, mechanical, etc.) and get information concerning their activities (design work, supervision of projects, environmental reports, etc.). Submit a report to your instructor.

(b) Visit a construction company. Learn about the projects that are in progress. Try to arrange for a trip to a construction site. Ask the engineer in charge to explain the activities that you do not understand. Submit a report to your instructor.

(c) Visit a manufacturing company. Learn about their current activities. Do not be bashful; ask questions, such as "What is the product used for?" and "How many are produced?" Get information about cost data and other related areas. Submit a report to your instructor.

3. Plan and prepare a display for your school or classroom that includes items relating to one or all of the systems mentioned early in this chapter, such as communication, transportation, health, housing, energy, etc. Engineers' Week (coincident with George Washington's birthday) would be an excellent time to show your display to the public.

Design Process

1. The following exercise will help you to obtain a preliminary understanding of the design process. (Groups of three to five students are recommended.)

(a) Select a problem statement from the list below or develop a statement of your own.

(1) Devise a better means to get people efficiently from town to an airport or from an airport into town.

(2) Design a better means of removing snow.

(3) Design a better way to extricate people from wrecked vehicles.

(4) Design a kitchen-size device for crushing cans.

(5) Design a better way to get people and health care services together (or people and *other* services).

(6) Devise means to make use of alternate forms of energy.

(b) Follow the steps in the design process that would be appropriate for a discussion group. Select topics from the following choices.

(1) Discuss and list where you would search for *information*.

(2) *Define* the problem as completely as you can. You may need to make some assumptions. Include cost, function, constraints, etc.

(3) Prepare a brief written *progress report* for your instructor. Include your tentative definition of the problem.

(4) Generate a number of *alternative* ways to solve the problem you defined in part b2.

(5) *Select* the alternative you believe will best solve the problem as you defined it.

(6) Prepare and give a design review to your classmates. Tell them the problem and how you defined it. Ask for suggestions and comments.

(7) Discuss and list what you believe needs to be done to finish the project, such as drawings, written portions, testing, etc.

(8) Describe how you believe the object would be manufactured.

Graphics

1. Peruse a technical report in an area of your interest and list the types of graphics utilized.

2. Visit an engineering firm or office to inquire about the types of graphics they use. Use the design process as a framework for organizing your questions and report.

Applied Graphics

1. Search for examples in periodicals, reports, textbooks, etc., that illustrate the graphical concepts discussed in this chapter, such as *true length, true size, angle between planes, intersections, nomography, computer graphics, graphical calculus,* etc.

Chapter 2 Sketching for Technical Design

INTRODUCTION

Sketching is the usual method that a designer uses for initially recording and communicating ideas. Words alone are inadequate to record three-dimensional concepts, and verbal communication is often misunderstood.

We do believe that developing sketching skills early in your studies is important for two reasons: (1) sketching three-dimensional objects on a two-dimensional surface will prepare you for learning the fundamental principles of orthogonal projection; and (2) achieving even beginner's skills in sketching will help you get started in the design process.

Incidentally, another practical benefit to be derived from learning and doing technical sketching is to help you understand more clearly figures in science, math, and engineering texts and to help you in the preparation of figures and charts for homework assignments.

TWO TYPES OF SKETCHES

Sketches are an extension of the creative designer's mind, just as tools are an extension of your muscles. Two types of sketches used by designers to extend their thoughts are (1) incomplete, rapidly drawn sketches used for *communicating directly* with themselves or with someone else, and (2) more complete, carefully drawn sketches prepared for *future use*.

DIRECT COMMUNICATION

"Let me show you what I mean," are words that often precede a direct-communication sketch while the designer simultaneously reaches for a pen or pencil. Then any available surface (including a tablecloth, but not at home, of course) becomes the designer's sketch pad. Characteristically the sketch grows rapidly and is incomplete, but it conveys an idea because the designer verbally fills in the blanks during the growth of the sketch. An example of the direct-communication sketch is shown in Figure 2.1. In communication with themselves, designers follow a similar pattern and often fill many pages with partially completed sketches. Figure 2.2 is a page of sketches from a student designer's notebook; observe how incomplete they are and how they suggest that the student was recording ideas on the use of levers.

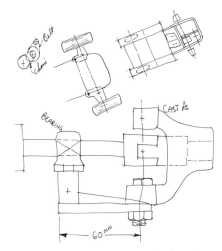

Figure 2.1 A direct-communication, freehand sketch is incomplete because the designer explains the sketch as it develops.

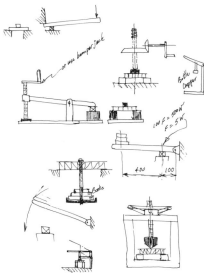

Figure 2.2 Freehand sketches in a student's file.

FUTURE USE

Sketching for future use often begins with an overall, light-lined outline of a design concept in which the designer fills in the details, darkens selected lines, locates key dimensions, and adds notes. Figure 2.3 shows a more developed concept of one of the lever assemblies from Figure 2.2. Some notes are written in longhand; however, most designers prefer lettering.

Freehand lettering and freehand sketching are skills that are easy to learn and valuable to have. We assume that you have done some sketching and lettering before, but have had a minimum of formal instruction; therefore, we present the following topics to help you further develop your skills.

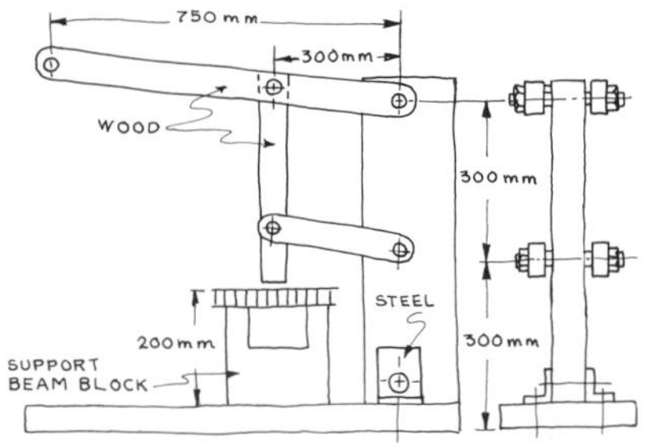

Figure 2.3 A more developed freehand sketch in the student's file.

A. *Freehand Lettering and Freehand Sketching*
1. *Freehand Lettering.* You can do acceptable and effective lettering if you observe a few practical suggestions.
   (*a*) *Use 2H or softer lead for lettering.* Make lines approximately $\frac{1}{2}$ mm thick.
   (*b*) *Print Large.* Letters and numbers should be at least 5 mm high because various departments and individuals in an industrial organization need to obtain information easily and rapidly from sketches and technical drawings. This is especially true for shop personnel who may have trouble reading prints that are blotched with grease and perforated with small holes by hot metal chips. Also, many drawings are photographed and reduced in size; therefore, a letter size that can be read easily is essential.
   (*c*) *Use all vertical or all inclined letters.* A slight slant of the letters forward or backward is probably the easiest to do and looks the best. Examples of good lettering appear in Figure 2.4, and further examples are in Appendix A.

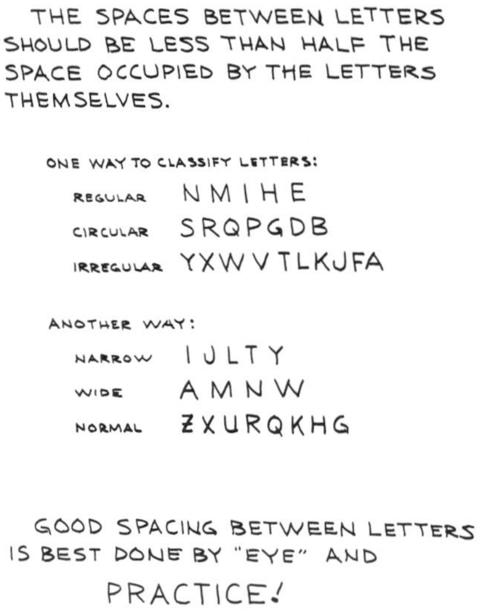

THE SPACES BETWEEN LETTERS SHOULD BE LESS THAN HALF THE SPACE OCCUPIED BY THE LETTERS THEMSELVES.

Figure 2.4 Freehand lettering with a slight slant.

ONE WAY TO CLASSIFY LETTERS:

REGULAR   N M I H E

CIRCULAR   S R Q P G D B

IRREGULAR Y X W V T L K J F A

ANOTHER WAY:

NARROW I J L T Y

WIDE   A M N W

NORMAL Z X U R Q K H G

GOOD SPACING BETWEEN LETTERS IS BEST DONE BY "EYE" AND

PRACTICE!

   (*d*) *Space the letters in a word.* Make the areas of the spaces between letters approximately equal.
2. *Freehand Sketching*
   (*a*) *Materials.* The simple tools employed in making freehand sketches are pencils, paper, and erasers. In many cases any pencil and any piece of paper will do.

   When an accurate record is necessary (e.g., modification of members of an existing design), prepare dimensional sketches using a sketch pad of square-grid paper.

   If you find it necessary to prepare freehand sketches on plain sheets, use a transparent tracing paper with a square-grid *underlay* as an aid in preparing fairly accurate sketches. Also, there is available a paper known as *fade-out-blue* that can be used when copies of sketches are required. This paper is a transparent sheet with coordinate lines that will not show when copies are made.

   Use the HB or F grade of lead. In general, it is best to use the softer lead for black lines and the harder lead for light lines. Use a conical point for drawing light lines such as dimension lines, extension lines, and centerlines. Visible, hidden, and cutting-plane lines should be drawn with the softer lead and a rounded point. Relative weights of lines are shown in Figure 2.5.

(b) *Basic Techniques*. Most engineers, designers, drafters, and production illustrators develop their own techniques in making free-hand sketches. You will find, however, that the following suggestions are quite useful in developing skills for preparing technical sketches, because virtually all "hardware" components of a plant or system consist of simple geometric shapes that usually contain *straight lines*, *circles*, *arcs*, and *ellipses*.

(1) *Sketching Vertical Lines* (see Figure 2.6). The hand, forearm, and elbow should rest on the drawing surface. For trial lines and construction lines, the pencil is held lightly. The pencil point should extend 35 to 50 mm beyond the fingers. Line segments about 60 mm long can be drawn with a finger motion. The forearm should be shifted downward to a position that will permit the extension of the line when necessary. Longer segments can be drawn by using the edge of the drawing board or the edge of a sketch pad as a guide, as shown in Figure 2.7.

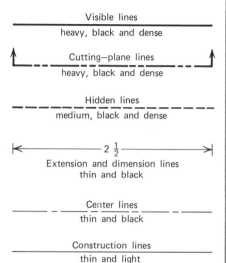

Figure 2.5 Relative weights of lines for sketching.

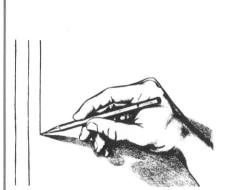

Figure 2.6 When sketching vertical lines, the hand, forearm, and elbow should rest on the drawing surface.

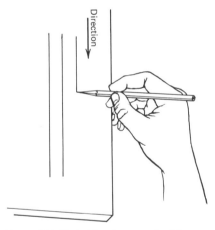

Figure 2.7 Sketching long vertical lines may be done easily by using the edge of a pad of paper, a drawing board, or a desk as a guide.

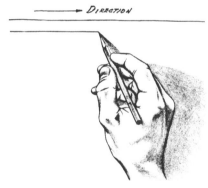

Figure 2.8 Horizontal lines are sketched sliding the forearm from left to right.

(2) *Sketching Horizontal Lines*. Lightly drawn line segments, approximately 60 to 75 mm long, are produced by using a sliding action of the forearm from left to right (see Figure 2.8). A set of horizontal lines equally spaced, for example, can be produced by first drawing two vertical lines that are placed as shown in Figure 2.9. The spaces between the horizontals are then marked off as A-1, 1-2, B-1', 1'-2', etc. Starting at point A and sighting point B, light segments are drawn, and so on for the remaining horizontals. Finally, the light lines are corrected if necessary and are strengthened by using the softer lead to obtain dense black lines.

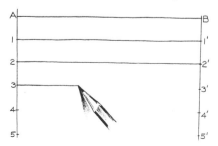

Figure 2.9 Sketching horizontal lines equally spaced.

(3) *Sketching Inclinded Lines*. The hand, forearm, and elbow rest on the drawing surface. Figure 2.10 shows hand and pencil positions for sketching lines upward and to the right. Figure 2.11 illustrates the position for sketching lines downward and to the right. Since most people find it easy to sketch horizontals, inclined lines can be drawn as horizontals by simply rotating the drawing sheet to accommodate this position.

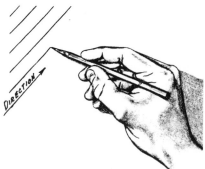

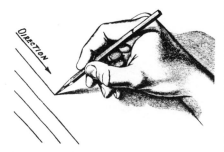

Figure 2.10 Sketching inclined lines upwardly.

Figure 2.11 Sketching inclined lines downwardly.

(4) *Sketching Circles*. First draw a light sketch of a square whose sides are equal to the diameter of the intended circle. Now sketch the diagonals and the horizontal and vertical midlines of the square. Points are then located on the diagonals at a distance from the center that is closely equal to the radius of the circle. A fine-line circle is sketched through these points and the end points of the midlines. Finally, the circle is made black and dense by using the softer lead. (see Figure 2.12).

  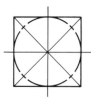 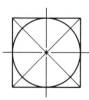

Figure 2.12 Steps in sketching circles.

(5) *Alternative Methods.* Other methods for sketching circles are shown in Figures 2.13 and 2.14. In Figure 2.13 a paper strip is used to locate points for the circle. As the strip is rotated about point A, light dots are placed at points such as B, which is at a distance from A equal to the radius of the circle. A fine-line circle is sketched through the dots and then made black and dense using a softer lead. In Figure 2.14 a rotation method is shown. The middle joint of the little finger is used as the pivot point while the paper is rotated as shown. Sketching circular arcs can be handled in a similar manner.

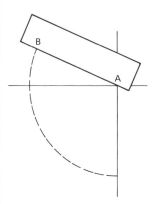

Figure 2.13 Sketching a circle using a marked radius.

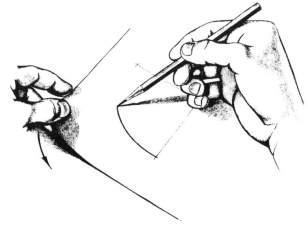

Figure 2.14 Sketching a circle by rotating the paper.

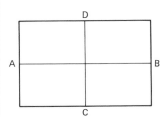

Figure 2.15 First step in sketching an ellipse: an outline.

(6) *Sketching Ellipses.* One method used to sketch an ellipse is first to sketch a rectangle whose long and short sides are equal, respectively, to the desired major and minor axes of the ellipse. Sketch in the axes AB and CD, as shown in Figure 2.15. Next, sketch relatively flat curves through C and D and sharper curves at A and B, as shown in Figure 2.16. Then add smooth connecting curves to complete the ellipse (see Figure 2.17). Finally, strengthen the ellipse and correct if necessary by using the softer lead to form a black, dense curve, as shown in Figure 2.18.

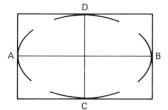

Figure 2.16 Second step in sketching an ellipse: add curves.

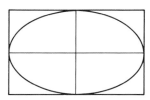

Figure 2.17 Third step in sketching an ellipse: complete the curves.

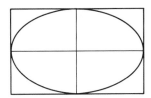

Figure 2.18 Final step in sketching an ellipse: darken the ellipse.

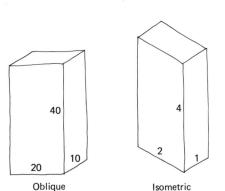

Figure 2.19 Trammel method of
sketching an ellipse.

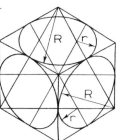

Figure 2.20 Use of arcs in drawing
an ellipse.

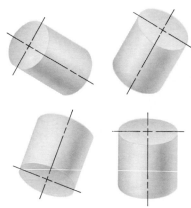

Figure 2.21 Relation between axis of
cylinder and major axis of an ellipse.

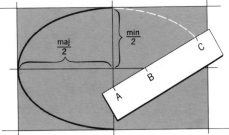

Oblique   Isometric

Figure 2.22 Proportion in sketching:
good representation.

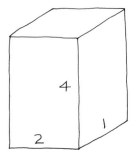

Figure 2.23 Poor representation of
sketching in proportion.

Another method used quite frequently is to locate points on the ellipse using the "trammel method" and then sketch a curve through these points using a fine-line technique. This is then followed by slight corrections if necessary, and a strengthening of the final curve. The trammel method is illustrated in Figure 2.19. Lay off on a strip of paper a distance such that AC equals half of the major axis and distance BC equals half of the minor axis. As A and B take different positions on the minor and major axes, respectively, C locates points on the ellipse.

Circles usually appear as ellipses in pictorial sketches, as shown in Figure 2.20 on three faces of a cube. The approximate parts of the ellipses are clearly indicated.

Further applications of the uses of ellipses appear in Figure 2.21, which shows four cylinders in different positions. *Note that the axis of the cylinder is perpendicular to the major axis of the ellipse in each case.*

(7) *Proportion.* You should sketch objects in true proportion but not necessarily to a specified scale, although in some cases a particular scale may be required. We cannot overemphasize the importance of true proportion in making freehand sketches. What do we mean by proportion? Consider a building that is 40 m high, 20 m wide, and 10 m deep. The ratio of these dimensions, 40:20:10 or 4:2:1, shows no detail but clearly preserves the true proportions of the building (see Figure 2.22). Simple as ratios are, you may be careless and prepare a sketch that may have the correct ratios indicated but, nevertheless, may leave the viewer with a false impression (see Figure 2.23).

(8) *Perspective Sketching.* Proportion in a perspective sketch is another matter. A perspective sketch closely resembles the observer's actual view of a given object. It is as though you were standing close to a window, and sighted a rectangular-shaped building through the window and then sketched a fairly accurate outline of the building on the glass, just as you

saw it. Maintaining fairly accurate proportions of the various measurements of objects are important in perspective sketching. For example, the vertical edges of a rectangular-shaped structure *will not appear* to be of equal length, and horizontal lines will not appear parallel. To compare their *apparent* lengths, a pencil can be used as a measuring instrument. Hold the pencil at arm's length and at right angles to the line of sight so that a portion of the pencil covers the apparent length of the vertical edge of the structure nearest you (see Figure 2.24). Record that length on your sketch pad and then repeat this technique to obtain the apparent lengths of the other vertical edges. In this manner you have a basis for comparing the relative apparent lengths. Other measurements can be obtained in a similar way. The theory of perspective and its applications are discussed in Chapter 9.

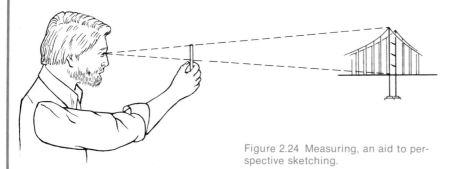

Figure 2.24 Measuring, an aid to perspective sketching.

(c) *Applied Techniques*

(1) *Orthographic Sketches*. Orthographic sketches are produced as though you were looking through a window at an object and had sketched the object on the glass, just as you saw it, but your sketch would appear as though you were a long distance from the window and were observing the object through a telescope. Equal vertical heights would appear equal in length, and all horizontal lines would appear parallel. Then, if you could view the object from different locations and sketch it in a similar manner (i.e., on a window glass), you could collect a set of views of the object that would describe it fairly well. In practice orthographic views are usually arranged in positions in which you would most likely want to see the actual object in order to study or measure it. As an illustration, consider a rectangular toolbox 600 mm wide, 200 mm deep, and 250 mm high, such as the one shown in Figure 2.25. You would look at the top of the box while you measured the width and depth (Figure 2.26a).

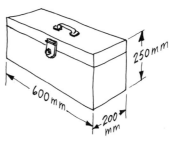

Figure 2.25 Toolbox in perspective sketch.

Figure 2.26 Orthographic views of the toolbox.

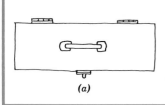

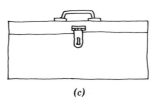

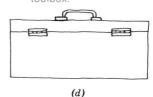

(a)   (b)   (c)   (d)

To measure the height, you would probably look at the end (Figure 2.26*b*) or the front (Figure 2.26*c*). For the back, see Figure 2.26*d*. Note how the relatively rectangular views in Figure 2.26 are different from the drawing in Figure 2.25, which shows the toolbox as it would actually appear.

Figure 2.27 shows the three views of the toolbox as you would probably arrange them in an orthographic drawing; the light construction lines were left in the drawing to illustrate the development of the views. In general, develop each view by sketching a light outline first, fill in details and, finally, darken the appropriate lines. Note the use of a section view (i.e., a view as though the object were cut in half) to *see* it more clearly.

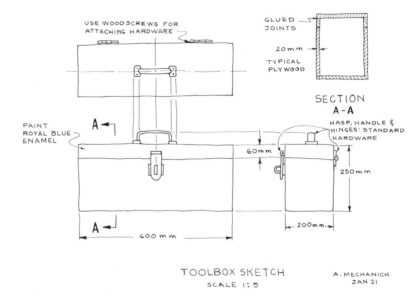

Figure 2.27 Completed sketch of toolbox. Note the use of sectioning to convey information. Refer to Chapter 13 for additional topics on sectioning.

Chapter 5 contains further information on orthographic drawings, such as laying out reference planes (similar to the windows mentioned earlier), using the intersections of reference planes, and determining a minimum number of views to describe an object. Section views are discussed in detail in Chapter 13.

(2) *Talking Sketches for Direct Communication*. Three kinds of talking sketches are used frequently. The first, an *oblique sketch* (see Figure 2.28) is probably used the most because of the manner in which it is developed. Start an oblique, talking sketch as though you were going to draw only an orthographic view, such as the front view of the receiver shown in Figure 2.28*a*. Next, depict depth by drawing oblique lines at an angle to the horizontal, as shown in Figure 2.28*b* and, finally, finish visible outlines of the rear of the object with lines the same length as those in the front view. The second kind, *orthographic drawing*, has already been discussed in the preceding section. If the designer and the listener-observer have

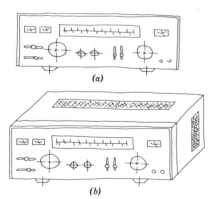

Figure 2.28 A talking sketch for direct communication: oblique freehand.

the time, a set of orthographic views may be developed such as those shown in Figure 2.29, which are more complete than those contained in Figures 2.1 and 2.2. The third kind is a *combination of a variety of forms* such as oblique, isometric (see Chapter 9), and perspective, since the talking sketch is frequently impromptu and attention to exact form is not necessary. Figure 2.30 illustrates this third kind.

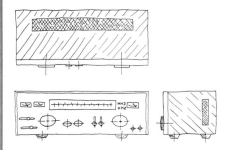

Figure 2.29 A talking sketch for direct communication: orthographic freehand.

Figure 2.30 A talking sketch for direct communication: a mixture of techniques, freehand.

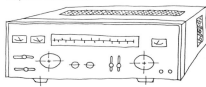

(3) *Board Sketches*. Direct communication with more than a few persons often requires the use of a chalkboard or a large pad of paper on an easel. As a student in a design-graphics course, you may be required to present a brief progress report to your class using one or both of these surfaces. Later, as a practicing designer or engineer, you may be asked to present your ideas to colleagues, sales personnel, management, or customers using a chalkboard, a special surface for felt pens, or an easel. The tools you need for chalkboard or easel sketching are essentially the same as those used for pencil sketching, except that the drawings and lettering are, of course, larger. To get started in board sketching, try the following: make letters at least a height equal to 1/250 of the distance to the viewer farthest from the board. For example, for viewers 10 m from the board, print letters at least 40 mm high. Before the presentation, establish light outlines of objects or figures you want to draw; during the presentation, fill in the details and heavy-in the lines you want to emphasize. Practice your sketches at least once before presenting them; if you are to use chalk, practice until you can draw lines without "screeching!"
(4) *Overlay Sketches*. Overlay sketches, which are primarily timesavers, are made simply by placing a sheet of tracing paper over an already completed sketch or drawing and then tracing a part of it in order to visualize a change or to visualize particular relationships. They are time-savers because with the use of overlays the whole view need not be drawn. The following two examples illustrate the utility of overlay sketches.

A designer completed the sketch of a hi-fi speaker cabinet (Figure 2.31*a*) and wanted to see how a minor change would look if the "housing" were modified slightly. Instead of redoing the whole sketch, the designer just placed a piece of tracing paper over the original and sketched the modifications, as illustrated in Figure 2.31*b*. While sketching the modification, the designer could have been talking to himself or to someone else. Architects make extensive use of this approach when dealing with clients who want to see how slight changes in the drawings of a house would affect various living spaces.

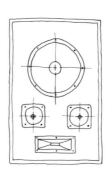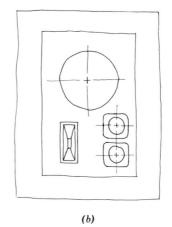

(a)   *(b)*

Figure 2.31 An overlay sketch is used to visualize changes without redrawing the complete original sketch.

Another use of an overlay sketch appears in Figure 2.32*a*, which shows a sketch of the four-bar linkage arrangement of a front-end loader. Relative movement of link *AB* with respect to the scoop *CD* can be analyzed using an overlay of the scoop (Figure 2.32*b*).

B. *Lettering and Sketching Using Aids.* The materials and techniques described and recommended for freehand lettering and freehand sketching may also be used with lettering guides and templates. Lettering guides and templates are employed frequently in the preparation of a variety of technical communications, such as engineering drawings, homework assignments, and technical reports.

1. *Engineering Drawings.* Lettering guides and templates are helpful in preparing sketches with a more finished look than freehand sketches, such as orthographic drawings, pictorials, and graphs that do not have to be drawn to scale. Of course, the guides and templates may be used for scaled drawings, too, and the variety of templates is impressive, because almost any technical field that uses drawings seems to have its own templates to save time in drawing repetitive symbols and shapes.

2. *Homework Assignments.* Most students prefer a finished look for their homework figures, especially in engineering science courses. They might also use guides and templates for mathematics, physics, and computer science courses. For example, a special template is used for drawing flow charts for computer programs. Other tem-

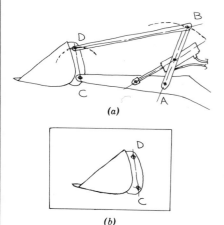

(a)

(b)

Figure 2.32 An overlay sketch is used to visualize relative movements of components.

plates are available for electronic circuitry, logic diagrams, architectural drawings, etc.

3. *Technical Reports*. Sketches, charts, and graphs augment the written text in technical reports and frequently do not need to be drawn accurately to scale, but they do require a finished look. Straight edges, triangles, circle templates, ellipse templates, and lettering guides are used, and ink is usually preferred to pencil. Commercially made letters, lines, and symbols with an adhesive backing may be pressed on drawings to attain an even more finished appearance.

CONCLUSION

We believe that the ability to draw freehand sketches and to use drawing aids will help you in design activities in two important ways: to provide you with a means to record ideas quickly, which will free you for further creative activity, and to give you a form to communicate your ideas efficiently to others. Furthermore, we believe that you can achieve an effective level of skill in sketching in a relatively short time.

Sketching three-dimensional concepts on a two-dimensional surface will help to prepare you for learning the principles of orthogonal projection, because you will begin to appreciate the interrelationships of the different views of an object. Drawing these same sketches to scale will further your understanding. This type of scaled drawing is the subject of the next chapter.

Exercises

Exercises for Chapter 2 fall into two categories, freehand lettering, and sketching, and lettering and sketching with the aid of guides. We have included topics in both categories that we believe will be of interest to you but, more important, will be of value to you in learning engineering graphical techniques in becoming a designer and in expanding your engineering technical vocabulary. Generally, the topics in the exercises will follow in order the development of the topics in the chapter.

Freehand Lettering and Sketching

LETTERING

Certainly content and accuracy should be the prime objectives for information on drawings, homework, or reports whether they are done freehand, with guides, or by machines. Nevertheless, lettering that is easy to read and well done seems to have as much impact, at least initially, on a reader as the content and accuracy. We believe first impressions are important and surely encourage substance and accuracy; therefore, these first few exercises focus on developing a style of lettering.

1. Use a 2H lead pencil with a relatively dull point (tends to encourage larger letters) and do the following.
(a) Obtain a lined pad of paper or prepare parallel guidelines about 5 mm apart.
(b) Refer to Appendix A for samples of both vertical and slant lettering and select one style.
(c) Copy phrases from Appendix C while attempting to space letters so that the area of the "spaces" between them are approximately equal.
2. Prepare samples of acceptable vertical or slant lettering of assigned passages in this text or a text of your instructor's choice. For example, prepare a page that states the two fundamental principles of orthogonal projection (Chapter 5).
3. Prepare samples of acceptable vertical or slant lettering of a paragraph of your own choice. (Include some technical terms that are new to you.)
4. In acceptable vertical or slant lettering technique prepare a statement of what you hope to achieve in your graphics and design courses.
5. In acceptable vertical or slant lettering copy specified representative statements and information from a technical drawing that is furnished by your instructor, found in a library, found in a local manufacturing facility, or found in a technical environment appropriate to your locale.

SKETCHING: STRAIGHT LINES

1. Refer to the weight of lines in Figure 2.5 and prepare a page of straight lines of your choice; for example:
(a) A chessboard pattern (Use closely spaced parallel lines for the "black" squares and start with a "black" square in the lower left corner.)
(b) An artistic pattern of various sizes of rectangles and squares.
2. Prepare a page of freehand estimates for angles usually found in most problem courses (e.g., 10°, 15°, 20°, 30°, 45°, 60°, and 90°, plus combinations of these). (Superimpose the correct angles drawn with the aid of protractors or drawing machines.)

3. Sketch an actual small cube in a number of positions; for example,

(a) Rotating the cube horizontally.
(b) Rotating the cube vertically.
(c) Combining (a) and (b).

SKETCHING: CURVED LINES

1. Refer to the discussions of circles and ellipses in this chapter and fill a page with at least five circles of varying diameters and five ellipses of varying "fatness."

2. The basis for sketching the golden section is shown in Figure E-2.1. Select a basic square of approximately 50 mm and sketch the golden section shown. Then combine several golden sections of varying size in a pleasing pattern.

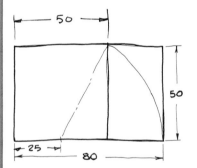

Figure E-2.1

3. Select one of the following equations, calculate the dependent variable y for various values in the independent variable x, and plot the results on cartesian coordinates. For example, $y = x^2$.

x	-2	-1	$-\frac{1}{2}$	0	$+\frac{1}{2}$	$+1$	$+2$
y	4	1	$\frac{1}{4}$	0	$\frac{1}{4}$	1	4

(a) $\dfrac{x^2}{10} + \dfrac{y^2}{9} = 1 \ (-4 \le x \le 4)$

(b) $xy = 9 \ (1/9 \le x \le 9)$

(c) $y = e^{-x^2} \ (-2 \le x \le 2)$

(d) $y = \cos 2\pi x \ (0 \le x \le 2)$

(e) $y = e^{-x} \ (-1 \le x \le 2)$

(f) $y = e^{-2x} \cos 2\pi x \ (0 \le x \le 2)$

(g) $y = 1 - e^{-x} \ (0 \le x \le 2)$

(h) Your instructor's choice.

4. Sketch a view showing the curved portions of common engineering products as assigned by your instructor, material from references provided by your instructor, or from actual hardware).

(a) Spur gear.
(b) Belt pulley.
(c) V-belt sheave.
(d) Hoist drum (for wire rope).
(e) Roller bearing.
(f) Ball bearing.
(g) Sleeve bearing.
(h) Compression spring.
(i) Roll pin.
(j) Retaining ring.
(k) Electric motor.
(l) Hi-fi speaker.
(m) High-frequency directional antenna.
(n) Parabolic reflector.
(o) Selected portion of an aircraft.
(p) Selected portion of a spacecraft.
(q) Further representative products related to engineering and technical disciplines.
(r) Products relating to medicine.
(s) Further products relating to a variety of industries, food, transportation, housing, health care, etc.

5. Visit a museum and sketch representative curves from the objects displayed (the profiles of a vase, etc.).

SKETCHING: PROPORTION

1. Select a familiar object with a predominance of straight lines and do the following steps for a one-view sketch of the object.

(a) Estimate "by eye" the relative height and width and sketch an "outline box" on a sheet of paper of your choice. (Grid paper for technical use is helpful.)

(b) Then lightly sketch within the "box" the important lines of the object.

(c) Finally, "heavy-in" the lines you want to emphasize.

2. Select several cylindrical objects and make isometric sketches as follows.

(a) Estimate the height and diameter and sketch an outline box.

(b) Then lightly sketch the ellipse at the top and bottom.

(c) Finally, heavy-in the lines you wish to emphasize. (See Figure E-2.2 for steps a, b, and c.)

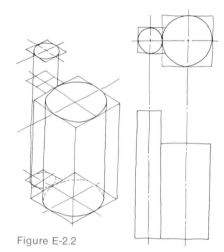

Figure E-2.2

3. Select combinations of rectangular, cylindrical, and other shapes as assigned and prepare isometric sketches following the outline box approach. Suggestions:

(a) Items related to engineering disciplines (A.E., BioE, CerE, ChE, CE, EE, ME, MetE, NucE, OceanE, PetroleumE, TextileE, etc.).

(b) Items selected by your instructor to expand your technical vocabulary.

(c) Your choice.

SKETCHING: PERSPECTIVE

1. Large rectangular objects or multiples of identical objects are probably the easiest to sketch in perspective for beginners. Examples include buildings, trucks, desks, boxes, telephone poles, railroad ties, and picket fences. Select objects or have them assigned and do the following.

(*a*) Estimate by eye the relative sizes of lines nearest and farthest from you and sketch them in appropriate locations on your paper utilizing an outline box.

(*b*) Sketch appropriate details.

(*c*) Heavy-in the lines of your choice for emphasis.

SKETCHING: ORTHOGONAL

1. Sketch two or three orthographic views of objects selected from the list below. Utilize an "outline box," lightly sketch lines of details within the box, and then heavy-in the appropriate details.

(*a*) Objects provided by your instructor.

(*b*) Objects suggested by the glossary, to expand your technical vocabulary.

(*c*) Selected objects from larger systems, such as:

  (*1*) The steering wheel of an automobile.

  (*2*) The landing gear of a small airplane.

  (*3*) A portion of an earth-moving machine.

  (*4*) A component in an electronic piece of equipment.

  (*5*) A component of a food-processing device (e.g., blender).

(*d*) Objects in your other courses (laboratory equipment, demonstrators, tools, etc.).

(*e*) Objects from hobbies.

(*f*) Conceptual ideas of your own for technical or nontechnical devices. For example, design one or more pieces of a chess set using more flat surfaces and angles than

are usually on the traditional humanlike pieces.

2. Prepare orthographic views of assigned objects and then prepare perspective, isometric, or oblique sketches of the same object.

SKETCHING: TALKING SKETCHES

1. Investigate an assigned topic and then prepare orthographic, isometric, oblique, or perspective sketches of one or more concepts relating to the topic.

(*a*) For your instructor.

(*b*) For a classmate.

(*c*) For a class (at the board).

Topic sources could vary as noted.

(*a*) Your text.

(*b*) Other textbooks.

(*c*) Common mechanisms from everyday living such as a ball-point pen, retractor-ejector, automatic door, automobile component, refrigerator cycle, solar heater cycle, typewriter, or movie camera.

(*d*) Laboratories.

(*e*) Shops.

(*f*) Hardware stores.

2. Prepare a sketch of an object of your choice or an assigned one and be able to demonstrate the use of an *overlay* to explain to someone how the object or a portion of it could be modified.

3. Prepare a sketch of an assembly of components, some of which move with respect to others, and use an overlay to show relative movements. Suggested examples: 10-speed bicycle brake (side or center pull), linkages under the hood of an automobile (carburetors, hood, etc.), piston and crank mechanisms, assemblies shown in technical magazines or texts, etc.

Lettering and Sketching with the Aid of Guides

Most of the exercises for freehand lettering and sketching may be assigned to be done with the aid of guides depending, of course, on the guides available. Further exercises with

guides could be related to course work topics such as mathematics, physics, statics, computer flowcharting, etc.

Try the format for homework, shown in Figure E-2.3, which contains the essential information for most technical problems: a heading, the given information in words, numbers, and graphical representation, what is to be determined, the method of solution, the answer underlined, and an arrow to the answer from an appropriate symbol in the right hand margin.

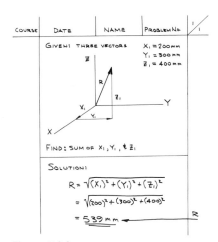

Figure E-2.3

DESIGN SKETCHES

The following design exercises provide an opportunity for generating ideas for beginning designers. Use any appropriate combinations of formats for freehand sketches, such as orthographic views combined with an isometric drawing, etc., of your choice or as assigned by your instructor. Include your name, the date, a title, an indication of the size of the object, and any pertinent notes.

FURNITURE/STRUCTURE

1. Design a chair to suit individual tastes and dimensions.
2. Design a water bed.
3. Design a school desk.
4. Design a desk or work space for a specific activity.
(*a*) Hobbies.
(*b*) Games.
(*c*) Manufacturing.
5. Design a garage, carport, etc.

HOBBIES/SPORTS

1. Design a hi-fi speaker cabinet.
2. Design a backpack frame.
3. Design a tennis racket or other rackets, paddles, etc.
4. Design a small boat.
5. Design a golf cart.
6. Design a small furnace for firing ceramics.

7. Your choice or as negotiated with your instructor.

SCHOOL

1. Design a new school logo.
2. Design a display board or cabinet.
3. Plan an area for a new facility.
(*a*) Pool.
(*b*) Gymnasium.
(*c*) Auditorium.
(*d*) Parking lot.
(*e*) Personnel tower at entrance to campus.

CITY

1. Design equipment for a playground.
2. Plan a playground.
3. Plan an interchange between a freeway and a city street.

Chapter 3 Layout Drawings

INTRODUCTION

After freehand sketches of design ideas are made, as described in Chapter 2, the design sketches are redrawn to scale in layout drawings. This chapter on layout drawings precedes the chapter on the design process because we believe that the ability to make freehand sketches and to prepare layout drawings are necessary skills for the effective analysis of ideas.

The drawing techniques for laying out design ideas are the same as those used for laying out orthographic drawings (Chapter 5) and for laying out applied graphics problems, which occur throughout the remainder of the text. Thus, we encourage you to concentrate on these techniques and skills now, before you continue on to Chapters 4 and 5.

LAYOUT DRAWINGS IN DESIGN

Designers, engineers, and students usually enjoy this phase of engineering graphics because of the thinking-through nature of the design layout drawing and because of the relatively informal format used. Moreover, we believe that design layouts will capture your interest, particularly the first category of the two that we include in this chapter. In *assemblies, movements, and clearances*, freehand sketches of ideas are redrawn to scale to determine how the parts fit together or move in relation to each other. The second category is *design development and communication*, in which design ideas are drawn to scale, analyzed for performance, modified as necessary, and ultimately used as a document for communicating the design information to drafters, who usually redraw the components in the more formal technical drawings. However, before discussing these two categories of design layouts, we will concentrate on drawing techniques.

INSTRUMENTS FOR DRAWING

The instruments required to prepare layout drawings include the pencils and tools already mentioned in Chapter 2, plus a few additional items. We will suggest a minimum of tools required to do layouts for design and encourage you to acquire additional items as you need them; for example, dividers are needed in the applied graphics problems described in later chapters. Suggested minimum tools and items are:

1. *Drawing Surface and Paper.* Almost any flat surface on which to attach backing paper plus drawing paper will do. Some students prefer the portability and straight edges of a desk-size drawing board.

2. *Equipment for Drawing Parallel Lines.* Horizontal lines on either plain paper or paper with a fade-out grid are commonly drawn with the aid of a T square (or drafting machine). Vertical lines may be drawn on fade-out grid paper with the aid of any straight edge such as a triangle but, if plain paper is used, a T square (or an equivalent horizontal guide) in conjunction with a triangle is usually used. Inclined parallel lines are easily drawn with the help of two triangles, one slid along an edge of the other. Perpendicular lines other than horizontal-vertical ones may be drawn with two triangles, one rotated 90° with respect to the other (see Figure 3.1).

3. *Drawing Triangles.* Two drawing triangles, 45° and 30–60°, are recommended. The 45, 30–60

combinations will provide you with a means to draw angles in 15° increments (see Figure 3.1). The sizes of the triangles may vary from 150 to 250 mm along the largest leg, depending on the size of the drawing paper. Most classroom drawing paper sizes will be either 216×280 mm or 280×432 mm.

4. *Protractor*. A protractor for laying out or measuring angles is recommended; however, the tangent method of "rise over run" for calculating angles is usually satisfactory, particularly with ubiquitous hand calculators. This technique, however, requires the use of a scale or a grid to measure the rise and the run.

5. *Scales*. Three different types of scales would be useful; nevertheless, only one needs to be purchased as a starter and, because we emphasize metric units, we recommend a metric scale. The other two scales, engineer's and architect's, should be in your possession eventually, because this country will be in a transition from English to metric units for a number of years.

6. *Compass*. A compass capable of scribing a radius of approximately 200 mm and having a threaded-type adjustment for stability is required.

7. *Tape*. Any tape that will not tear the paper when it is removed is acceptable; masking tape or drafting tape is most often used. Staples may be used if a suitable soft backing is available.

Figure 3.1 shows the suggested instruments and items with which to get started; those that you may need soon or may wish to have (shown in Figure 3.2) are:

1. Dividers.
2. Erasing shield.
3. Pencil sharpener.

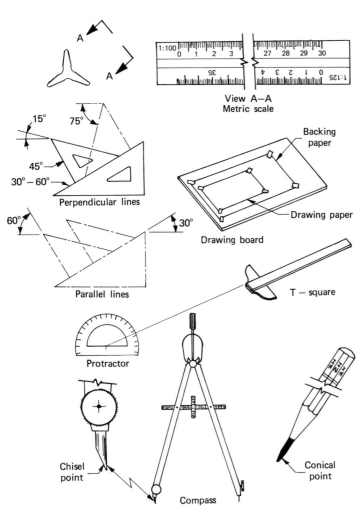

Figure 3.1 Instruments and items for drawing preliminary layout drawings.

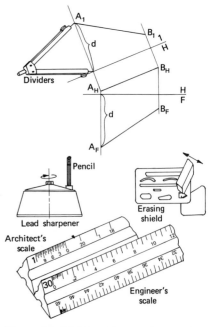

Figure 3.2 Additional instruments and items for doing applied graphics.

4. Brush (not shown)

5. Architect's and Engineer's scales.

We assume that the items on this list are somewhat familiar to most of you; nonetheless, we will comment briefly on their use in the order in which they would be used in starting a layout drawing.

SUGGESTIONS FOR GETTING STARTED

1. *Drawing Surface, Paper, and Tape.* For reference purposes label the four corners of both the backing sheet and a drawing sheet number 1 on the lower left corner and the other corners 2, 3, and 4 in a counterclockwise direction. Tape the backing sheet to the drawing board near the left edge (near the T of the T square) and align it vertically and horizontally with the board. Do corners 1 and 3 first; then pull and tape corners 2 and 4 sufficiently to have a smooth sheet. Repeat this procedure for taping the drawing sheet to the backing sheet, but use the T square to align the paper first.

2. *Pencils and Compass.* Use a pencil lead that has a cone-shaped point that is dulled slightly at the point. Compass lead works best if a chisel shape is used, because the chisel edge is dulled slightly and rounded at the corners.

3. *Lines.* A simple way to draw a line of constant width is to roll the pencil in your fingers as you follow a straight edge, tilting the pencil slightly in the direction in which the pencil is moving. The slower the speed, the wider the line, and the movement should generally be in a direction away from your body. Lines for objects should be wider than centerlines which, in turn, should be wider than dimension lines, but they all should be of the same blackness.

4. *Scales and Layouts.* We will assume now that you have taped a 216 × 280-mm sheet of paper on a drawing surface with the 280-mm edge horizontal and that a metric scale is available. We will now "talk-through" two layouts and then discuss scales in general.

Full-size layout, or full scale, is a natural layout to do first. The free-hand sketch of a limit switch shown in Figure 3.3 will be used for our initial layout. The metric scale 1:100, which may be used for a large number of scales in multiples of 10, is selected; the one we need

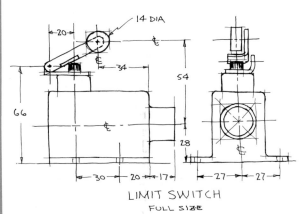

Figure 3.3 A freehand sketch of a limit switch.

is the one-to-one scale of 1:1.00. Note that shifting the decimal point also provides us with 1:10.0 and 1:100 reductions, or 1: 0.100 (10×) magnification. Try to center the two views of the limit switch on the paper and to leave room between them for notes. First locate the centerlines of the switch body, the ends of the arm, and the roller in both the front and side views, as shown in Figure 3.4*a*. Then draw light outlines of the three components of the limit switch. Finally, heavy-in the lines of the components, as illustrated in Figure 3.4*b*. Note that some of the light lines of the outline remain; this is a common practice for layout drawing. Also observe that some notes are in longhand.

Half-size layout, or half scale, is our next example. We will reduce a full-size checkerboard pattern to half size. Tape a new sheet of 216 × 280-mm plain drawing paper to your working area and arrange the 280-mm edge in a horizontal position. The full-size checkerboard is 400 mm square with 64 equal-size squares, 32 white and 32 black, alternatively. To draw the checkerboard to half size, use the 1:20 scale and mentally place a decimal point to get 1:2.0. Locate 400 mm on the half scale, which should just fit the narrow dimension of the drawing paper. Center a 400-mm square on your paper and lay out the 64 (50-mm) squares. Then starting at the 50-mm square in the lower left corner, draw a number of parallel lines of your choice in it to make that square a "black" one. Similarly, make every other square a "black" one.

The primary advantage of using

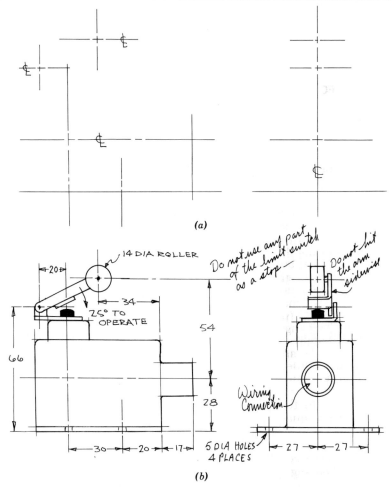

Figure 3.4 Stages of development of a layout drawing for the limit switch in Figure 3.3.

a *scale* should be apparent: to draw objects, whether large or very small, to a convenient size. An alternative way to scale objects would be to transpose the actual dimensions of an object by dividing by a scale factor, such as 0.1, 2, 3, 5, 8, 10, 50, or 100, but the time spent would be prohibitive and the chances for errors would increase. *A scale does all the transposing automatically and thus saves time and decreases the probability of errors*.

Several metric scales are tabu-lated in Figure 3.5 to illustrate their versatility, and Figure 3.6 contains several metric scales with the same linear dimensions noted for comparisons. Also, scales may represent other quantities, such as force, velocity, and acceleration, as discussed in Chapter 8.

A mental exercise to help understand scales falls into a category that engineers occasionally use called "back-of-the-envelope" calculations; some call it "order-of-magnitude" estimating. For example, ask yourself what scale,

Designation on scale	Scale designation on drawing		Actual reduction or magnification
1:100	Full size or	1:1.00	0
		1:1000	1/1000
		1:100	1/100
		1:0.10	10 times
		1:0.01	100 times
1:20		1:20	1/20
	Half size or	1:2.0	1/2
		1:0.20	5 times
1:40		1:40	1/40
		1:4.0	1/4
		1:0.40	$2\frac{1}{2}$

Figure 3.5 Metric scales for drawings may be used for reduction or magnification.

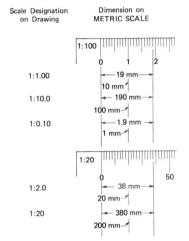

Figure 3.6 A comparison of several metric scales.

roughly, could be used for drawing one view of the objects listed in Figure 3.7 on a large portion of a 216 × 280-mm sheet of paper. To verify your estimates of the objects in Figure 3.7, try drawing one view of a similar object from your own experience or locale.

The first exercises at the end of the chapter permit further practice in using instruments and scales and in making preliminary layout drawings.

DESIGN LAYOUTS

Design layouts are usually orthographic drawings in which the designer uses as many drawing techniques as necessary to develop ideas, resolve problems, and communicate information to others. The drawings are made to scale; however, designers use freehand work and handwritten notes liberally where accuracy and formality are not required. The more frequent uses of these layout drawings are now discussed.

1. *Assemblies, Movements, and Clearances.* Assemblies of component parts, housings to contain the parts, and frames to support the total assemblies, all drawn to scale, provide the designer with information that is difficult to obtain from any other procedure except actually building a full-scale model. Of course, full-scale models, or prototypes, are preferred if the means to produce them are available. An illustration of an assembly of a number of components is

Object and estimated size	Estimated scale for drawing object on 216 × 280-mm paper
Gear for wristwatch (5 mm)	10 times
Electronic component from hand calculator (4 to 7 mm)	5 to 10 times
Hand calculator (150 mm long)	Full size 1:1.0
Hifi receiver (150 × 400 mm)	Half size 1:2.0
Skate board (150 × 600 mm)	1:4 or 1:5
Sports car (3 × 12 m)	1:60 or 1:75
Truck (15 × 40 m)	1:200
Large warehouse (15 × 60 m) (front view only)	1:400
Skyscraper (200 m tall)	1:1000
Small town, map (15,000 sq meters)	1:100,000

Figure 3.7 Preliminary scale estimates for drawing objects on standard paper.

shown in Figure 3.8. Note the mixture of formal and informal notes and drawing techniques.

Analyzing relative movements between components and determining clearances for moving parts are probably the most interesting and useful applications of layout drawings. The jet blast deflector shown in Figure 3.9 is an assembly of components designed to move the deflector into an upright position on the deck of an aircraft carrier to protect personnel from the hot gases generated by the aircraft engines. The hydraulic cylinder, *AB*, extends in length; this rotates *BC* about point *C*, which in turn rotates *CD* about point *C* and causes linkage *DE* to move upward, which finally causes the deflectors *EF* to rise. Figure 3.10*a* shows the deflector and the components in several positions. To show the movements more effectively, designers employ two additional complementary techniques: overlays and cardboard "articulated" models. Figure 3.10*b* shows one of the overlays, used in conjunction with Figure 3.10*a*. Figure 3.11 shows a drawing of a cardboard articulated model of the jet blast deflectors, mounted on a cardboard base so that the moving components will have fixed points where needed such as *A*, *C*, and *F*.

Clearances are readily analyzed using design layouts, as illustrated in Figure 3.12, which shows the movement of a connecting rod *BC* in an air compressor moving in relation to the proposed internal housing geometry. An overlay was used to simulate movement of the connecting rod *BC*; *B* followed the circular path of the crank shaft *AB*,

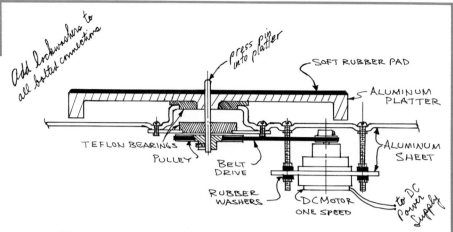

Figure 3.8 A layout drawing for checking an assembly of components.

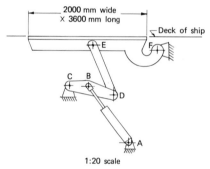

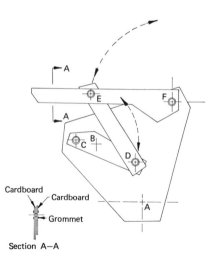

Figure 3.9 An assembly of components to perform a function.

Figure 3.11 A cardboard, articulated model of the jet blast deflector.

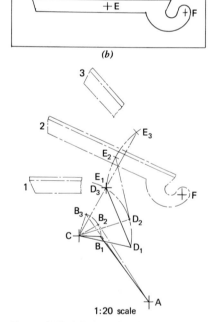

Figure 3.10 A layout drawing to analyze the relative movements of the jet blast deflector system in Figure 3.9.

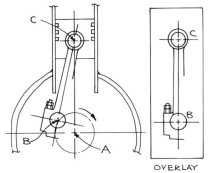

Figure 3.12 A layout drawing to analyze clearances by utilizing an overlay.

and C was confined to follow the linear path of the piston centerline.

2. *Design Development and Communication.* A design layout complete with critical dimensions and notes regarding the function of the design, geometrical sizes, location of features such as holes, special instructions for finishing, some analyses of applied loads and reactions, and any further appropriate information provides designers with records for their own uses, or the layouts may be passed to drafters for formal detailing into technical drawings. The variety of layout drawings is limited only by the needs of the designer or the recipient of the drawings; a design layout could be a component, an assembly of components, an electronic circuit, a plan of a playground, the route of a highway, the path of a system of piping for power hydraulics, the cross section of a furnace for ceramics, the movements of a landing gear of an aircraft, the cross section of a proposed mining or drilling operation, etc.

A design layout for a drafter prepared from a designer's freehand sketch is shown together with the sketch in Figure 3.13*a* and 3.13*b*. A design layout for the designer's own files would probably be less complete.

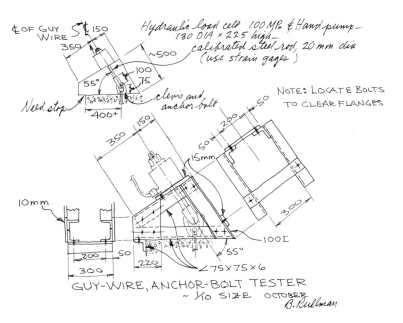

Figure 3.13 (*a*) The freehand sketch of a design idea.

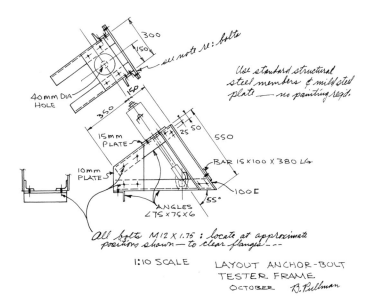

Figure 3.13 (*b*) The sketch redrawn in a design layout prepared for a drafter.

CONCLUSION

The drawing materials and techniques discussed in this chapter provide the basis on which you can proceed to further engineering graphics in Chapter 5 and beyond.

The materials and techniques applied in design layouts will help you to analyze and develop design ideas while following the design process, which is discussed in the next chapter.

Exercises

Drawing Instruments

1. Refer back to "Suggestions for Getting Started"; tape a backing sheet and a drawing sheet on a flat surface and do:

(a) The *full-scale* layout of the limit switch.

(b) The *half-scale* layout of the checkerboard.

2. Refer to Appendix A and try the "Geometrical Constructions," either your choice or as assigned.

3. Refer to Figure 3.7. Select an object in your own locale similar to one in the figure; draw one view of it to *scale* (estimate the dimensions if they are not available) on a 216 × 280-mm sheet of paper. Some examples are:

(a) A small electric component 10 times its size.

(b) A sports car 1/50 its size.

(c) A large building 1/1000 its size.

(d) Your choice or as assigned.

4. Lay out to scale one of the traverses described below and determine the closing distance *CA*. North is always toward the top of the drawing sheet unless otherwise indicated; however, to avoid confusion, always draw an appropriate symbol on your paper to indicate north. (*Suggestion*. Make a rough, freehand sketch of the problem first to obtain an approximate "size" of the solution so the problem can be started and finished without going off the paper.)

(a) *AB* is 30 m at N 60° E. (*Note*. Compass directions are always measured with respect to either the north or the south, whichever one is indicated.) *BC* is 50 m at S 50° W. (*Suggestion*. Use the 1:50 scale with each major division equal to 10 m, which is equivalent to having a 500 to 1 scale, or, 1:500.) Note that *CA* (C to A) lies in the direction N( )E; *AC* would be S( )W.

(b) *AB* = 300 mm, North, *BC* = 175 mm N 80° W. (Try a 1:2 scale.)

(c) *AB* = 35 km S 20° E, *BC* = 50 km N 45° W.

5. Make a freehand sketch of an existing object or a new design and make a layout drawing of the object or design. (An alternative exercise would be to exchange freehand sketches with another student and prepare a layout drawing from the information given to you.) Several suggestions for objects to sketch or new design ideas follow.

(a) Objects.

(1) A portion of an automobile.

(2) A piece of laboratory apparatus.

(3) A hi-fi speaker cabinet.

(4) An attachment for a bicycle, an automobile, a boat, etc.

(5) A skateboard.

(b) New design.

(1) A short stepladder.

(2) A means for elevating (safely) one end of an automobile for home use while servicing it.

(3) A cabinet for a hi-fi set plus records, tapes, etc.

(4) A modification to an existing product.

(5) A skateboard for two people.

6. Vendors' catalogs of mechanical, electrical, and electronic products usually include one figure to represent a "line" of their products. The figure will be dimensioned with letters for the dimensions, which are listed in tables. For example, one catalog is published for the Boston Gear Works, in Quincy, Massachusetts. It includes a section on "pillow blocks" (a frame that contains a standard antifriction bearing) that has figures like those just mentioned. Have your instructor assign one or more sizes from the Boston Gear Works catalog or from catalogs of other companies. (*Note*. These catalogs are excellent sources of technical information for expanding vocabulary and filling idea banks for future design activities.)

Design Layouts

1. Refer to Figure E-3.1, which is a freehand sketch of a limit switch and an actuating shoe. The shoe is to be attached to the sliding door so that when the door is closed, the shoe will activate the limit switch (i.e., cause arm *AB* to rotate), causing a light to go on (or off) at a remote location. Prepare a layout drawing to determine if the shoe causes the arm to rotate at least 18° but not more than 32°. An overlay will be helpful.

2. A four-bar linkage is a mechanism found in a variety of products such as automobiles, production machinery, farm machinery, aircraft, and homes. Frequently springs are incorporated in the linkage assembly to assist in operating the mechanism. Figure E-3.2 is a four-bar linkage for converting rotary motion to reciprocating motion (*Note*. One of the "bars" is represented as a solid base, commonly shown schematically as a line with "whiskers."). Prepare a layout drawing of the mechanism as shown and determine

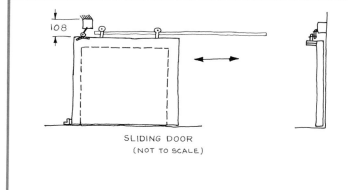

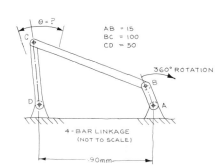

Figure E-3.2

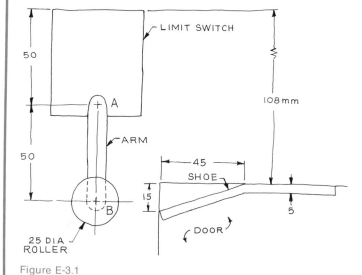

Figure E-3.1

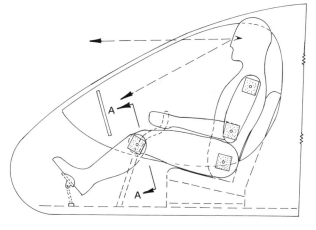

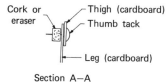

Section A—A

Figure E-3.3

the degree of travel of bar *CD* for a 360° rotation of bar *AB*.

3. Locate, sketch, and measure a four-bar linkage. Then prepare a layout drawing of it to show its total movement. (*Suggestions.* A four-bar linkage plus a spring is commonly utilized in conjunction with a hood of an automobile, hydraulically powered backhoes have a four-bar linkage arrangement, and powered tailgates on trucks use a four-bar linkage.)

4. Make a cardboard, scaled, articulated manikin similar to the one in Figure E-3.3, but utilizing your own dimensions. Then prepare a conceptual layout drawing of a chair and a desk to suit you. That is, show a profile of the chair and the desk plus several key dimensions.

5. Refer to Figure E-3.4, which shows an assembly of a cylinder housing, a piston, a connecting rod, and a crankshaft for an air compressor. Use the information given to prepare a *layout drawing* of the assembly rod to determine the actual minimum clearance between the cylinder housing and the connecting rod. Use an overlay of the connecting rod.

6. Select a parking area near a building on your campus and lay out the area to scale on drawing paper. Make an overlay of an average size automobile and its turning radius and attempt to redesign the parking spaces for more capacity.

7. *Sheave and Bracket.* Refer to Figure E-3.5, which shows a freehand sketch for a sheave and a bracket for use with a tow rope on a ski slope. From the information given in the sketch, prepare a layout drawing of the sheave and bracket and include enough information so that a drafter could prepare the formal detail drawings needed to have

the items manufactured and assembled. One layout drawing prepared by a student from the sketch in Figure E-3.5 is shown in Figure E-3.6.

8. *Bottom Loading Furnace.* Figure E-3.7 is a freehand sketch of a concept of a mechanism for loading ceramic samples into a furnace by entering through the bottom. Prepare a layout drawing of the mechanism and platform. Include enough information so a drafter could prepare the formal detail drawings needed to have the parts made. Refer to Figure 3.13b for a sample layout drawing and the information that a layout drawing should contain.

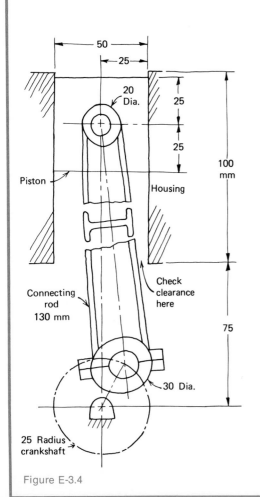

Figure E-3.4

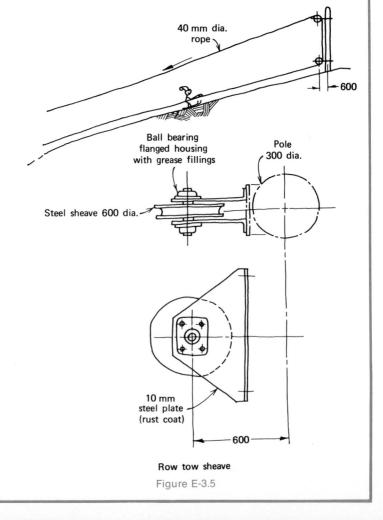

Row tow sheave

Figure E-3.5

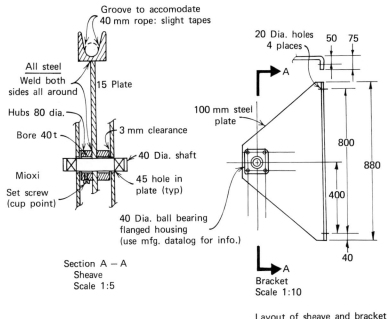

Groove to accomodate
40 mm rope: slight tapes

All steel
Weld both
sides all around   15 Plate

Hubs 80 dia.
Bore 40t   3 mm clearance

Mioxi   40 Dia. shaft

Set screw   45 hole in
(cup point)   plate (typ)

40 Dia. ball bearing
flanged housing
(use mfg. datalog for info.)

Section A – A
Sheave
Scale 1:5

All dimensions in
millimeters

Figure E-3.6

20 Dia. holes
4 places   50   75

100 mm steel
plate

800

880

400

40

Bracket
Scale 1:10

Layout of sheave and bracket
for rope tow

Jan. 12   Chris T. Snow

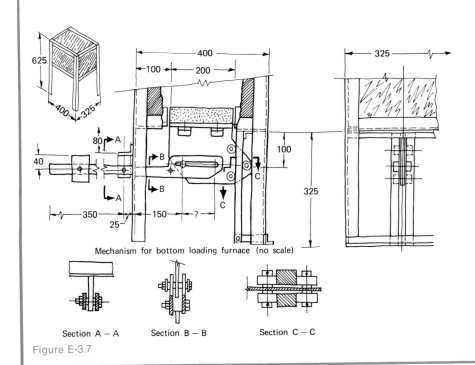

625

400   325

400

100   200

80

40

100

325

350   150   ?

25

Mechanism for bottom loading furnace (no scale)

325

Section A – A   Section B – B   Section C – C

Figure E-3.7

Chapter 4 Design Process: Phases and Steps

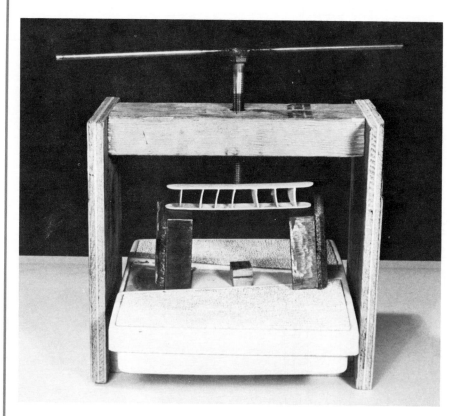

bies, at home, and on the job. We therefore believe that beginning students can build on their earlier experiences while learning new skills in engineering graphics and design. We have observed that beginning students often ask, "What is design?", even though they have done design in the past, but were not cognizant of the steps they were following. In this chapter we describe briefly how students like yourselves have already done design, and we then discuss the steps taken in the more comprehensive design process. After the discussion on the design process we include comments and design summaries from three individuals who developed a design process of their own (as you will do) and applied their approach to solving technical design problems. The first individual is an engineering student, and the other two persons are practicing engineers.

ROUTINE DESIGNING

Early experiences in designing, whether for hobbies, home projects, or job situations, follow what we describe as routine designing, which is not as inclusive as the formal design process, but is an ap-

INTRODUCTION

In Chapter 1 we introduced the design process, its phases, and its steps as the systematic way in which engineers and designers approach design problems, and we emphasized the functional role that engineering graphics plays in design. We also noted that essentially all students have had some experience in solving design problems and in using graphics in hob-

propriate approach to use in some design situations. For example, in designing a simple product such as a wooden ladder for use around the home, you probably would have done the following.

1. Recalled and sketched ladders that you had already seen.
2. Searched for new ideas in likely places, such as hardware stores and catalogs.
3. Determined materials and sizes of members on the basis of old and new information, experience, judgment, and intuition, and sized the rungs on the basis of human factors rather than strength of the material alone.
4. Priced the materials for the ladder and purchased them.
5. Built the ladder, revised it during its construction, and then tested it.

In our course we have assigned this same ladder problem to beginning students as an introduction to design and as a demonstration of how their experiences and judgment, plus intuition, can be used to create a satisfactory solution. The students usually produced a design on paper and then had a stress analysis done with the help of their instructor. The ladders consistently have had a safety factor of about two, and the rungs were more than adequate because they were sized to fit the human hand as compared to being sized to withstand a force. The most severe use of the ladders was thought to be an individual carrying a toolbox and standing in the middle of the ladder, which spanned the distance between a garage and a house.

Thus, routine designing in general is a response to a need or a problem that requires the synthesis of ideas into a solution that is based primarily on the designer's experiences, judgment, and intuition. Some analyses of materials and costs are done, the drawings of the solutions are made, and the actual product is built.

FLEXIBILITY OF THE DESIGN PROCESS

You may now be wondering when you should use routine designing and when you should use the design process. We believe that once you have followed the steps in the design process to solve a design problem, you will discover that the process can be adapted to routine as well as complex, open-ended problems, and that the flexibility of the process is a direct function of the experiences and skills of the individual designer.

We also believe that the design process will provide you with the following:

1. An introduction to a "professional" approach to problem solving (i.e., an approach that considers a problem from various points of view and bases decisions on as much information as possible in the time available).
2. A systematic checklist of items to consider to be certain that "all bases are covered."
3. A plan for systematically keeping good records, some of which may be useful for patenting and product safety considerations.
4. An appreciation of the importance of planning, costs, and allocating time.
5. A collection of suggestions for incorporating creative approaches in generating possible solutions.
6. An appreciation of the use of engineering graphics to develop, analyze, and communicate technical information.

DESIGN PROCESS

The design process may be divided into four separate phases, each phase entailing a number of steps. Actually, the *phases* overlap in time and the *steps* occur in varying sequences; nevertheless, they are separated and labeled for discussion purposes and for providing a checklist for beginners.

We concentrate on the first three of the four phases for practical and economic reasons. The first three phases can be accomplished effectively in a classroom situation within a reasonable period of time and at a reasonable expense. The third phase normally ends in a design on paper, complemented by simple models or prototypes.

The fourth phase is briefly discussed in order to remind the reader that the usual objective in technical problem solving is to achieve the physical realization of the solution and deliver the product to the customer. However, the end products of some consulting engineering organizations are actually proposed solutions to design problems. These proposed solutions are on paper and occasionally are accompanied by scaled models, such as a model of a proposed water-treatment plant from a civil engineering consulting firm. Therefore, their products essentially are produced by the end of Phase III.

Now let us turn our attention to Phase I and the steps that it entails.

Phase I Recognize and Define the Problem

Phase I is essentially a recognizing, questioning, defining, and planning stage where you ask as many preliminary questions relating to the problem as possible and search for answers to those questions. Questioning is systematic; you ask *who, when, where, why, how,* and *how many.* Formulate plans to make the best use of the time available for Phase I, which is approximately one-quarter of the total project time in classroom situations. The major goal in Phase I is to define the problem accurately by setting objectives, recognizing constraints, specifying functions, establishing cost limits, and identifying any further requirements. Phase I usually ends in a progress report.

Step 1 Start Planning

Initial planning includes several substeps that all occur essentially at the same time. Soon after you are assigned a design problem, start a log to record what you do and the time you spend to do it. Entries should include the date, a newspaper-style summary (who, where, what, . . .) of your design activities, and the time spent on each activity. Each student should keep a log whether working alone or as a member of a design team.

Concurrently with the log, keep a file of notes, sketches, drawings, copies of information from journals, texts, etc., and any other related materials.

Successful designers keep good records, so beginners are encouraged to start this practice early in their problem-solving careers. Not only will good records become

useful on future assignments, but they will provide your instructor (or supervisor, or patent attorney) with a basis for evaluating your performance. Most designers admit that immediately after being given a problem they start thinking and generating ideas for its solution; therefore, keep some means of recording ideas nearby at all times, and place these ideas in your file. Additional ideas will come to you while you search for information, and the originator of the problem will often offer some initial suggestions. Also, when you are investigating what has been done before, ideas for modifying existing designs to meet new requirements will emerge. Ideas will come when they are least expected. Consequently, carry a note pad everywhere.

Recognize the Problem

The term "recognize" refers to the several ways in which you might acquire a problem to solve. As a beginning design student, you will probably be assigned a problem by

your instructor or be given a choice of a problem from: (1) those collected by your instructor, (2) a textbook, or (3) your own ideas. For example, a team of beginning students, whom we will refer to throughout this discussion of the design process, were assigned a problem, to design an inexpensive device that would apply a force to a small wooden beam until it was broken and, at the same time, show the breaking force to an audience of sixth-, seventh-, and eighth-grade science students. The demonstration was to show qualitatively how the arrangement of the wooden pieces affected the load-carrying capacity of the beamlike structure. The same kind of wood was used in each case. An organization that had contacted the instructor was planning a number of science-engineering demonstrations for these youngsters and had offered one design concept (Figure 4.1) for the beam-breaking demonstration. The team thought it was too complicated both for the students to observe and for a teacher to build

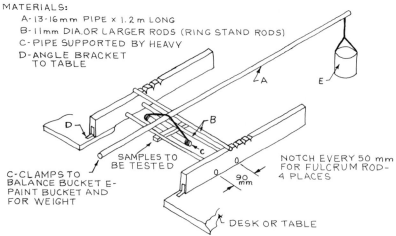

Figure 4.1 Recognition of a problem for a student design team: to design a better beam-breaking demonstrator than the one shown.

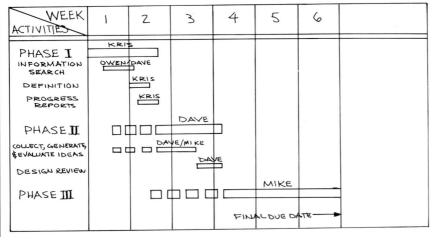

PRELIMINARY PLANNING CHART   GROUP 4

Figure 4.2 A preliminary planning chart establishes a tentative schedule of the team's time and effort toward achieving a design goal.

and operate and, therefore, they wanted to develop a better design.

Prepare a Preliminary Planning Chart
Before making this preliminary chart you should skim through the first three phases of the design process to get an overall picture of what needs to be done. Essentially Phases I, II, and III will require about 25, 25, and 50%, respectively, of the time you have available; one way to prepare a chart is to plan both forward from the starting date and backward from your final due date and allocate the time you have available based on those approximate percentages. The preliminary planning chart is a graphic display of what needs to be done, when it is to be done, and by whom. This provides all members of a design team or anyone else with a timely reminder of the needs of the project, as illustrated in Figure 4.2. Observe that the student design team had 6 weeks in which to complete their project; the names of the individuals responsi-

ble for each phase were noted on the time bars.

Step 2 Gather Information
Search sources that will most likely produce information; during this searching prepare questions that will become the basis for a preliminary definition. The student design team started with the obvious source, the organization that had contacted their instructor, and generated the following questions.

1. What constraints, such as safety, would restrict the design of the device?
2. Who would built it?
3. How much *could* it cost?
4. What materials would be available?
5. When was the device needed?
6. What should the device do?

They continued their search by contacting junior high school science teachers, a wood shop, and the library; someone even consulted their instructor. The science

teachers indicated that they would use wooden tongue depressors, since a large supply was available to them at no cost, so the team decided to find out just how strong depressors are as a single beam and glued together in various configurations. Figure 4.3 is a summary of their investigation, in which they tested beams to failure; failure was defined as the inability to support a further increase in load. The team was allowed to cut depressors into smaller pieces before gluing in order to have more variety of configurations; however, four depressors was the maximum number permitted per beam, whether whole or cut into parts.

The team's preliminary list of questions relating to objectives, constraints, function, costs, personnel involved, materials, etc., not only provided them with suggestions of possible sources of information to be investigated, but also helped them to arrive at a final definition, which is the next step.

BEAMS (MADE OF WOOD TONGUE DEPRESSORS)	FORCE REQUIRED TO CAUSE BEAM TO FAIL (NEWTONS)	ADHESIVE
∏ (BOX)	800	CONTACT CEMENT (CC)
∏ (BOX)	630	WHITE GLUE (WG)
|||| (4 VERTICALS)	520	CC
I (EYE)	400	WG
≡ (4 HORIZ.)	360	
⧄ (RIBS)	200	CC
⊞ (RIBS)	90	CC
⊞ (RIBS)	55	CC
— (1 HORIZ.)	55	CC

Figure 4.3 Gathering information included testing example beams to failure.

Step 3 Define the Problem

Before establishing the final definition of your problem, you will probably become involved in iteration, or repeating steps already taken, which occurs throughout the life of a design project. Iteration is mentioned here to focus on the fact that designers are continually going back to do something better or to do something not considered earlier. Each time you iterate, the definition (or whatever phase you are in) will probably improve to some degree.

However, one always present frustration that student designers and practicing designer-engineers alike must face is that, at a given time, they must make decisions based on the information that they have been able to collect by that time and go on to the next step or phase. The definition must be completed at the end of Phase I.

The student team used their preliminary list of questions plus the following questions to complete their definition.

1. How much time should a typical demonstration take?
2. How visible should the demonstration be (i.e., how many students should observe it)?

Their final definition at this time was as follows. The device must be:

1. Portable (i.e., easily carried by a teacher).
2. Able to be built by school maintenance personnel.
3. Made of common materials.
4. Safe and easy to use by teachers and students (sixth to eighth graders).
5. Able to demonstrate principles in a class period (about 50 min).
6. Visible to at least 20 students.
7. Built for less than $20 (at current prices).
8. Painted and pleasing to look at.

GROUP 4 MACHINE DESIGN, INC.

January 18

Classroom Demonstrations
Co-Ordinator of Beam Machine Project

Dear Sir:

This progress report is to inform you of the results of our analysis of the beam project. Our group has researched and discussed the functions, objectives, and constraints of the project. A summary of our conclusions follows.

The overall function of the project is, as we understand it, to create an interest in engineering at the junior high school level. Since engineering is an application of math and science, the project will focus on an experiment that demonstrates an application of a physics principle. Moreover, the project will show the relative strengths of tongue depressor beams of various designs.

The specific objective of our analysis is to design a machine to test the relative strengths of the tongue depressor beams. Specifically, we will attempt to simplify the machine design shown on page 8A attached.

We have identified several design criteria for the beam testing machine.

1. The machine must be capable of exerting up to 900 Newtons of force on test beams. This limit was established from the results of experiments we performed, and is confirmed by the results of other experiments. (Because of high forces required to break the sample beams, we have decided to incorporate a mechanical advantage in the machine design.)

2. The machine must demonstrate a physics concept that can be understood by sixth, seventh, and eighth grade science students. Options discussed to achieve a mechanical advantage include lever, winch, pulley, spring and gears. These options may be used in combination with one another.

3. Materials must be free since there is no budget allotment. Materials found in the school woodshop are viable possibilities (e.g., 2x4's, 2x6's, plywood, dowels, glue, clamps, etc.). Since the schools are in lower income areas, it is assumed that no materials will be brought from home.

4. The machine must be relatively lightweight, and easy to handle and store. A maximum of four feet was established for length.

5. The machine and physical concepts demonstrated must be easily visible by students.

Having made these decisions, they were anxious to contact their client, the organization, for confirmation. They did so in the next step.

Step 4 Prepare a Brief Progress Report

At this stage in a project it is very important to communicate with

Page 2

We have also established criteria for the tongue depressor beams.

6. Maximum of four tongue depressors may be used per beam.

7. Beam should be tested with the flat side on top.

With Phase I now complete, we are beginning work on Phase II. During this phase, which will last one and one-half weeks, we will develop several alternative machine designs. Using the criteria from Phase I, we will select two or three alternatives to analyze in depth. The conclusion of Phase II will be to discuss designs with our colleagues and select a final solution.

Phase III will include finalizing design details and preparing oral, written and graphical analysis. We will also build a prototype for use in demonstrations. We expect to present our final design during the second week of February.

We hope our preliminary definition of the problem will fulfill the needs of the project. If you have any comments, feel free to contact us.

Sincerely,

Kristen Chadwick
Progress Report Chairman
Group 4 Machine Design, Inc.

Figure 4.4 Phase I of the design process usually ends with a progress report to the client.

your client (preferably in writing) for several reasons.

1. Before spending more time on the problem, it is advisable to be certain that you and your client agree on what the problem is and that your general conception of possible solutions satisfies the client's expectations.

2. Decisions made orally by both parties should be in writing to prevent unnecessary misunderstandings later.

3. Clients usually want to know "what they are getting for their money," so tell them what is being done and your immediate plans.

This written response to the cli-

ent is called a *progress report*, and it should contain at least the following information.

1. Your current final definition, which should be accompanied by sketches or drawings if they are needed.

2. Brief statements summarizing your current and expected activities.

3. Written summaries of oral decisions, such as due dates and expected expenses.

4. A request for confirmation as soon as possible.

The term "client" could be defined as being one of several categories of individuals or groups. In a classroom the client is usually the instructor, who wears several hats —client, coach, technical consultant, teacher, and friend. However, if the problem solution is for a private individual or a group outside the classroom, then the client will most likely be someone else, such as a private citizen, a company, a representative from a public institution (e.g., a hospital), a representative from local, county, or state government, or a representative from an organization like the one that contacted the team's instructor. The team's progress report to the organization is shown in Figure 4.4.

Your instructor, who functions in a classroom somewhat like a manager does in industry, may

FROM: Group 4 Machine Design, Inc.

Figure 4.5 A brief progress report to the team's instructor told what the individuals had done and what their future responsibilities were.

Group 4 Chairpersons (revised)

Phase I   Progress Report   Kris Chadwick

Phase II   Design Review   Dave Strohm

Phase III   a) Oral Presentation   Mike McRae
             b) Written Portion   Owen King
             c) Graphics   Dave Evans

Current Status

Phase I is completed. We have defined the function, objectives, and constraints of the project, and established a time schedule. The progress report has been given to the client. See copy enclosed.

Phase II is now in progress. Five alternative designs have been drawn and we are in the process of discussing their applicability to the criteria established in Phase I. We hope to narrow the alternatives to 3 options and discuss them with the class on January 29, Thursday. After choosing one design, we may build a prototype and continue on with Phase III.

need additional information about the individual responsibilities and accomplishments of each student member. Therefore, the design team presented their instructor with the report shown in Figure 4.5.

The two reports usually end Phase I for a classroom project.

Phase II   Generate and Evaluate Alternative Solutions

Phase II includes collecting the problem-solving ideas generated during Phase I, generating new ideas, sifting through all of them to select the ones that appear the most feasible, deciding on the best candidate for fulfilling the definition, and conducting a design review that highlights the best candidate. Phase II usually consumes approximately one-quarter of the project time for classroom problems.

Step 5   Collect Ideas

Of the ideas sketched and collected during Phase I, some were probably revised as the project developed, others were combined, and a few were discarded. Sometimes the client changes the direction of the project as a result of the progress report, or new requirements demand new considerations. Nevertheless, once the definition of the problem has been established and confirmed, tentative solutions to satisfy the definition are needed. At this time your efforts should be devoted to developing ideas that build on the ideas you recorded in conceptual sketches made during Phase I and to generating new ideas. Sometimes you may find

that your mind seems to go blank and you cannot produce anything new. You are not alone. Other student designers and engineers also find that their idea banks seem to be drained; they, too, need help in generating new ideas. Thus, we present several techniques for stimulating the flow of new ideas.

Step 6   Prepare Additional Conceptual Sketches

Routine designing, discussed at the beginning of this chapter, produces ideas based on the designers' experiences and imagination; the variety of the ideas is directly related to the extent of experience and the willingness of designers to use their imagination. The following techniques have been developed to stimulate imaginations and increase the flow of new ideas.

Innovation Techniques

We assume that everyone has some innate talent for being imaginative and creative, and that these talents

can be developed as effectively as other thinking processes. Our discussion includes two categories.

1. *Developing Talent for Perceiving Ideas*. Enhances perception for seeing from an imaginative point of view.

2. *Developing Talent for Generating Ideas*. Helps to bridge the "gap" between steps 6 and 7.

CATEGORY 1   DEVELOPING TALENT FOR PERCEIVING IDEAS.   The two freehand sketches in Figure 4.6 illustrate an *observing-sketching* technique for "seeing more." A student was asked to recall from memory the details of an entrance to a Student Union Building that the student had passed through hundreds of times and to record the details in an impromptu sketch (Figure 4.6*a*). The sketch in Figure 4.6*b* was drawn at a later time by the same student, from memory, after being directed to observe the entrance again, "as though you were going to sketch it later." The lesson here is that if you study an object as though you were sketching it, you will see and retain more than if you just casually look at it.

Attribute Listing. The waxed-paper cup in Figure 4.7 was selected to illustrate the concept that if you systematically describe objects, you will obtain more information than if you casually observe them. This familiar, relatively uncomplicated object was chosen because of its commonness. A tool or a relatively simple manufactured product could be more appropriate in a design course; however, the cup satisfactorily illustrates our discussion of attributes and also provides a good example of what we call *unset*.

The *attributes* of an object can be classified as:

1. Geometry and size.
2. Materials.
3. Manufacturing processes.
4. Function.
5. Special features such as aesthetics, surface treatment, etc.

(a)

(b)

Figure 4.6 One innovation technique for developing talent for perceiving ideas is to observe objects as though you were going to sketch them later.

Attributes		
Geometry & size	Material	Manufacturing process
Truncated cone	Paper	Print design on paper
70 mm large diameter	Wax	Wax paper
45 mm small diameter	Pigment	Cut side and bottom
100 mm high		Roll and seal (glue)
Thickness of cup approximately 0.1 mm		

Function and special features

To hold cool liquids while drinking
Lip rolled to prevent paper cuts on lips
Bottom approximately 5 mm above bottom of cone to isolate liquid from surface and ease of manufacturing
Color looks cool and pattern suggests waves.

Figure 4.7 Attribute listing.

The list of attributes of the cup in Figure 4.7 was produced by first-year students. As an exercise, try listing the attributes of a tool such as a screwdriver.

Functional Set/Unset. Some individuals restrict themselves to using items for the purpose for which they were originally intended, such as using the waxed-paper cup only for containing cool liquids for drinking. In other words, they have *set* their minds and confined their thinking. The class members who listed the attributes of the paper cup were also asked to generate as many ideas as they could for other uses, or *unset* uses, of the paper cup. The list in Figure 4.8 summarizes most of their suggestions. How many *unset* uses can you list for the screwdriver?

Puzzles with Lessons. Puzzles that include a lesson can illustrate the arbitrary constraints that some people place on their thinking. The puzzle in Figure 4.9 is repeated three times so that you can practice on the first two and put your final solution on the last one. The stipulation is that you must place a pencil down on any point on the paper and draw four straight lines without lifting the pencil from the page and, while drawing the lines, go through each of the nine dots at least once. Try the puzzle before reading on.

A word description of how one solution would appear when completed is as follows. Start at dot 1, go right through 2 and 3 and con-

Flower pot
Store small items
Sand mold
Two with string attached between the bottoms, for communications
Remove the bottom and use as a megaphone
Start a fire
Shell game
Dice shaker
Water scoop for bailing
Paint holder
Vase
Use four with sticks to make a wind speed indicator
Cookie cutter

Figure 4.8 An example of unset (for waxed-paper cup).

tinue right until lined up with 6 and 8; go through 6 and 8 and beyond until lined up vertically with 7-4-1; go vertically upward through 7-4 and hesitate at 1; then go directly through 5 and 9 to complete the puzzle.

The lesson to be learned here is that if one arbitrarily establishes a boundary consisting of the eight outside points, 1-2-3-6-9-8-7-4-1, the puzzle cannot be solved as stated. In other words, do not set arbitrary constraints on thinking before attacking a problem.

Imagination processes, just as other thinking processes, need exericse in order to remain flexible and to grow. Several innovation techniques, such as sketching, attribute listing, functional set, and puzzles, already discussed exercising imagination. A further type of activity designed to exercise your imagination is to view simple geometric shapes and list as many items as possible that the shape could represent. As an exericse, try listing items that the shapes in Figure 4.10 could represent.

There are additional puzzles and imagination exercises in the problems at the end of the chapter.

The foregoing discussions encourage the reader to see ideas and

1	2	3						
0	0	0	0	0	0	0	0	0
4	5	6						
0	0	0	0	0	0	0	0	0
7	8	9						
0	0	0	0	0	0	0	0	0

Figure 4.9 Try this puzzle with a lesson before reading further.

situations from different viewpoints, to avoid placing arbitrary boundaries on thinking, and to exercise imaginative thinking. The next section continues in a similar vein, but is more specific regarding techniques for generating ideas.

CATEGORY 2 DEVELOPING TALENT FOR GENERATING IDEAS.

Creative Process. The creative process described here is one that most people have experienced, but have not analyzed. The steps in the creative process are discussed in the next few paragraphs in the firm belief that if you know how the creative process works, you can increase your chances of having it work for you.

Individuals who claim to have

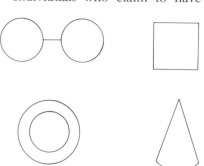

Figure 4.10 Exercises for imagination development.

experienced the creative process come from a wide range of backgrounds, such as technology, art, literature, music, medicine, law, and business. Descriptions of how the creative phenomenon worked for them differ considerably. Some have claimed that their ideas came to them at night, and they would awaken to record their thoughts. Other persons have had new thoughts emerge while they were performing diverse activities such as walking, reading, bathing, shaving, etc.; all of these persons were engaged in relaxing activities, but most acknowledged that they needed to be almost totally involved in their respective problem or project before the phenomenon worked for them. Even though their individual descriptions differed, the basic ingredients seemed to be the same; thus we will summarize the basic ingredients in four steps for the *creative process*.

For the first step, *accumulation*, it seems that total involvement means acquiring as much information as possible regarding the problem to be solved. Then, in the second step, time should be allowed for the information to be processed in the brain (i.e., to *incubate*). Following an incubation period, relaxing activities often allow ideas to come "out of the blue." This third step is called *illumination*. Finally, the ideas must be analyzed and verified; this is the fourth step, *verification*.

The solid lines in Figure 4.11 illustrate the three areas of the brain being considered: *conscious thinking*, *memory storage*, and *subconscious thinking*. The solid lines also depict the communication links that most humans have with their world: hearing, sight, smell, taste, and touch. Imagine now that the dashed line 1-2 represents *accumulation* of information to be

stored. At the same time, assuming that a considerable amount of energy to resolve a problem is being expended, the subconscious mind is apparently notified that an attempt to resolve a conflict is presented. Line 1-3 represents this link.

Presume now that a considerable amount of energy to collect information, facts, and ideas, to resolve the problem in the time available, has been expended; the next step is to divert conscious attention to other activities or to just plain "goof off." (Some supervisors may need to be convinced of the necessity of this step.) Dashed line 4-5 represents *incubation*, the subconscious mind's sorting, shifting, and combining of ideas. During this relaxed period new combinations of ideas to resolve the problem are generated.

Line 6-7 depicts *illumination*, the "out-of-the-blue" ideas, the "ah-ha" experience. An ability to sketch rapidly would certainly be helpful during this step.

The last step is *verification*, comparing the ideas with the definition of the problem. Ideas may have to be verified through experiments, prototypes, models, layout drawings, or discussions with knowledgeable persons.

Knowing that the creative process exists and believing that everyone has the potential for making it work should help you to be more effective at generating ideas. To summarize, the necessary ingredients seem to be to (1) conscientiously expend energy on *accumulating* information, (2) allow time for *incubation*, and (3) schedule mind-relaxing activities by doing something different or by just plain relaxing.

Obviously, your subconscious mind cannot be controlled; there will be times when the process has

been faithfully followed, but *illumination* does not occur. Nevertheless, there are activities available to you to get to your idea banks.

One activity to encourage ideas, and popularized by Alex Osborn in *Applied Imagination*, is brainstorming. If you find that your design team needs to generate ideas for finding solutions to a problem you have defined, try the following brainstorming techniques.

1. No one should criticize anyone else, no matter how ridiculous a suggestion may sound. The far-out idea may trigger a new thought in another member's mind. A positive atmosphere encourages creativity, while a negative atmosphere discourages creativity.

2. One member of the group should jot down or sketch all the ideas, preferably on a surface that is visible to everyone. In this way earlier suggestions may be combined with later ones to form new solutions.

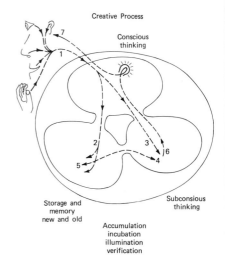

Figure 4.11 Creative process steps.

3. A comfortable, friendly atmosphere is conducive to idea generation, and a session should not last much longer than an hour at any one time.

Brainstorming sessions can be followed by a critique period to cull out obviously nonfeasible ideas or to ask individual participants to expand on their ideas.

Another technique for systematically jogging your memory is to use lists of triggering words or lists of design questions, such as the ones that follow.

1. *Verb Lists.* Verbs are action words, and some type of action is usually required in the function of a product. For example, assume that a requirement exists to design a means to shut or to close a device. The list in Figure 4.12 could help you to generate ideas to close the device.

2. *Design Checklists.* Design checklists function somewhat like grocery lists or checklists for taking a trip. The typical grocery or planning list is formulated before becoming immersed in the sea of grocery shelves or before entering in the hectic process of packing, so that an important item will not be forgotten. Design checklists are reminders to cover all facets of a design; they also help to generate ideas such as the trigger verbs. Figure 4.13 is an illustration of a design checklist for product design.

Solving the Physics. This approach helps to generate ideas by temporarily shelving the definition of a problem. Phrased differently, it is easier to generate a partial solution to a problem first and then modify it to meet the definition than it is to generate an idea that, at the outset, meets all the requirements, constraints, and functions. Thus, solving the physics entails two steps; the first is to ignore temporarily the more restrictive objectives and requirements from the problem definition while generating ideas that will work and that partially solve the problem, and the second is to revise and modify the ideas to meet the specific requirements and objectives. Checklists are very helpful during this second step.

This extensive discussion of techniques for generating ideas for alternative solutions was included in Step 6 because students admit that it is the most difficult step to do thoroughly. For example, most students seem to want to develop the first idea that surfaces in order to get the job done; they are reluctant to push themselves further to generate additional ideas for alternative solutions. Other beginning students admit that their idea banks are quickly depleted and they need help. All students hope that in the future, as they attain

slam	zip	flood	encase
close	slide	pivot	coat
shut	button	rivet	paint
push	sew	weld	jacket
pull	cover	choke	press
clamp	pack	crush	roll
squeeze	jolt	bend	bury
buckle	nail	cork	entomb
draw	screw	stop	can
seal	enclose	staple	stuff
wrap	envelop	cap	bundle
tie	crank	cement	

Figure 4.12 A verb list for shutting or closing a device.

1. Review possible methods of solutions.

electrical	pneumatic
mechanical	electronic
hydraulic	chemical
magnetic	ceramic
nuclear	

2. Review possible facets of the design to be certain you have covered everything.

shape	thermal
size	manufacturing process
stiffness	cost
strength	safety
failure modes	weight
reliability	noise
fatigue	style
life	controls
corrosion	maintenance
paint	number to be made
surface treatment	accessability
wear	vibration
friction	locking devices
lubrication	

3. Once you have a design that you believe will work, look for improvements by asking questions such as the following.

Can the design be simplified?
Can the size change?
Can sequences be varied?
Can substitutions be made?
Does the object look right (styling)?

Figure 4.13 A design checklist helps you to be thorough and may suggest further ideas.

A1 = C-clamp B1 = fish scale C1 = Wood
A2 = Lever B2 = Known weights C2 = Metal
A3 = Threaded screw B3 = Numbers of books
A4 = Handcrank B4 = Bathroom scale
B5 = Calibrated spring

Figure 4.14 Subalternatives for the
beam-breaking device.

more technical experience in school and on the job, the more likely it will be that they will have more design ideas from which to draw.

Step 7 Organize the Ideas into Alternative Solutions

The next task is to sift through and cull the far-out ideas that often surface during brainstorming sessions and that usually fail to fulfill the criteria that were established in the definition of the problem. Brainstorming sessions are usually free-wheeling, which tends to encourage the participants to let their imaginations run wild. For example, the student design team mentioned in Phase I produced the following ideas for breaking the beam structure: use a karate chop, a cleaver, or an axe; run over the beam with a bicycle; slowly use a crowbar; close a door on it; use an automobile jack; employ a home bottle capper; crush the beam in a vise; or use a modified cider press. All of these methods could be used to break the beam, but only a few would allow the observer to see some qualitative differences in how strong one beam was in comparison to another. The slower methods would be more visible to the audience and would allow a readout means to be employed.

All the ideas that survive the first cut should be organized into reasonable alternatives and subalternatives. The subalternatives for the beam-breaking device were organized in three subcategories: (1) methods to break the beam, (2) readout means, and (3) construc-

tion materials. A number of overall alternatives were formulated from the subalternatives. Figure 4.14 shows the arrangement of the subalternatives for the beam-breaking device. The most promising combinations must be identified next.

Step 8 Select the Most Promising Alternative Solutions

The most promising alternative solutions are the ones that seem to fulfill best the objectives and meet the constraints set forth in the definition of the problem from Phase I.

For example, the student team developing the beam-breaking device selected the following combinations of subalternatives for further study. Figure 4.15 shows a sketch of a handcrank and a fish scale (A4-B1-C2). Figure 4.16 is a sketch of the threaded screw and a

bathroom scale (A3-B4-C1). And Figure 4.17 shows a sketch of a lever with weights (books) to supply the load (A2-B3-C1).

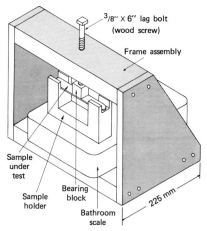

Strength of Materials Test Machine

Figure 4.16 Alternative B: threaded screw and bathroom scale.

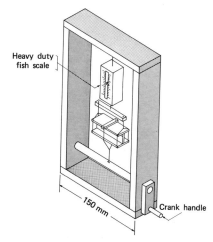

Figure 4.15 Alternative A: handcrank and fish scale.

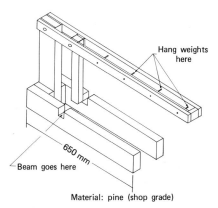

Beam Tester Preliminary Design

Figure 4.17 Alternative C: lever and known weights.

Step 9 Develop and Analyze Selected Alternatives

Developing and analyzing selected alternatives illustrate the use of layout drawings, which were discussed in Chapter 3, to develop ideas from sketches. Figures 4.15, 4.16, and 4.17 are freehand sketches that the design team used to show approximate proportions and to indicate which materials were tentatively selected. Figure 4.18 shows one of the alternatives redrawn to scale in a layout drawing that was used to check the assembly of the components.

A further analysis done by the design team with the help of their instructor was to estimate what the expected loads might be on the structure and then compare the loads with the strengths of the materials. The team learned that their design was structurally sound and that analyzing the loads was a topic they would study in a course on *statics*, and that investigating the strength of the materials was part of a course on *strength of materials*. Their instructor wanted to show them an overall approach to synthesis and analysis that included geometrical models, mathematical models, and laboratory experiments, so he drew a representative flow diagram, which is reproduced in Figure 4.19. The discussion was concluded when the design team was told that during the analytical calculations of Step 9 and before decisions are made in Step 10 designers frequently must repeat some previous steps because of information obtained during Step 9 or the lack of information needed in Step 10. Some designers call this "fine-tun-

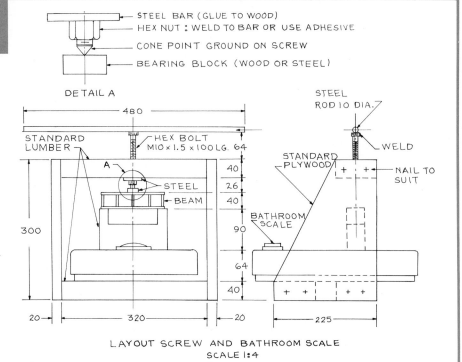

Figure 4.18 A layout drawing of an alternative for the beam-breaking device.

ing" their designs, but this retracing of earlier steps is just another example of iteration.

Step 10 Make a Decision

In Step 3 we noted that designers and problem solvers are usually faced with the dilemma that decisions must be made at a particular time based on the information available at that time, even though it is felt that given more time better decisions could be made. One technique to help expedite the chore of having to make a decision is to use a decision table. Figure 4.20 shows such a table; it was used by the student design team. Each criterion from the definition of the problem (specification, constraint, objective, etc.) in the left column is given a weighting factor based on a scale of 1 to 10. Next, each alterna-

tive solution is compared with the others against each criterion. For example, "How well does solution A compare with B and C to satisfy criterion 1?" Then an appropriate score (again 1 to 10) is awarded to each solution. Finally, the products of the weighting factors times the score for each alternative are summed. The highest total is the best solution based on the analysis and judgments at that time.

The decision table in Figure 4.20 shows that solution B scored the highest, but C was a close second and should be investigated further.

Step 11 Conduct A Design Review

A design review is an in-house function that allows you to discuss your design decisions with classmates or colleagues (but not with your client) and to receive some

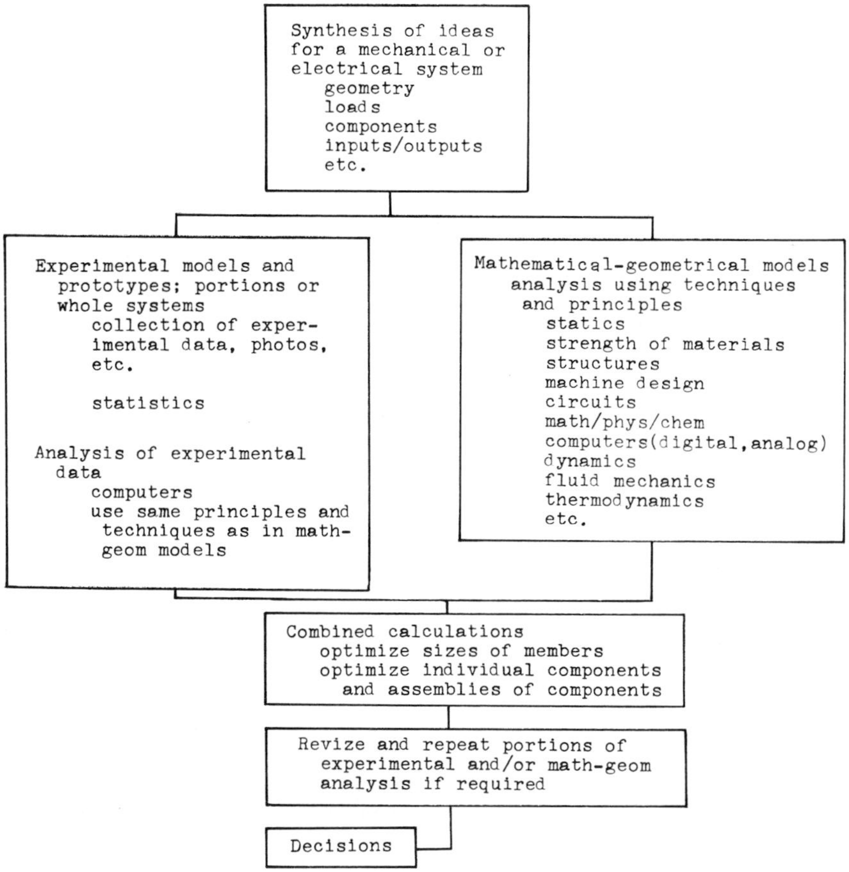

```
                    Synthesis of ideas
                    for a mechanical or
                    electrical system
                        geometry
                        loads
                        components
                        inputs/outputs
                        etc.
```

```
Experimental models and          Mathematical-geometrical models
  prototypes; portions or          analysis using techniques
  whole systems                     and principles
      collection of exper-             statics
      imental data, photos,            strength of materials
      etc.                             structures
                                       machine design
      statistics                       circuits
                                       math/phys/chem
Analysis of experimental               computers(digital,analog)
  data                                 dynamics
      computers                        fluid mechanics
      use same principles and          thermodynamics
      techniques as in math-           etc.
      geom models
```

```
                    Combined calculations
                        optimize sizes of members
                        optimize individual components
                        and assemblies of components
```

```
                    Revize and repeat portions of
                    experimental and/or math-geom
                    analysis if required
```

```
            Decisions
```

Figure 4.19 A flow diagram of a general
approach—synthesis and analysis
leading to a decision.

Criteria[a]	Weighting factor	Handcrank and fish scale A		Threaded screw and bathroom scale B		Lever and books C	
Cost	10	9	90	8	80	8	80
Ease to make	8	7	56	10	80	8	64
Visibility	8	9	72	8	64	10	80
Safety	6	6	36	10	60	8	48
Storage (size and weight)	5	8	40	10	50	8	40
Availability of parts	7	8	56	9	63	9	63
TOTALS			350		397		375

[a] From definition of the problem

Figure 4.20 A decision table used by the
student design team to help them select
an alternative.

immediate feedback. The design review presentation should provide the audience with sufficient information to appreciate why the problem originated and what the client wanted. Your current definition of the problem, the objectives, the function, and the constraints, should be stated. Next, demonstrate that you considered the problem from a number of approaches by discussing some of the alternatives you discarded and why you discarded them. Finally, describe in detail the alternative you selected and how it best meets your definition of the problem.

One successful format you can use for a design review includes the following elements.

1. Request that the review be informal and encourage questions at any time.

2. Ask your classmates to play the role of devil's advocate; that is, have them constructively ask questions such as, "Why did you do that?", "Did you think of the following . . . ?", and "How much does it weigh, cost, etc.?"

3. Appoint a member of your team or ask a classmate to tape-record all the questions and comments offered during the design review so that your group can respond to them before the final report is due at the completion of Phase III.

In some instances it may be appropriate to contact your client after the design review with an interim report to keep the client informed of your progress and to provide an opportunity for further feedback. If a report is made to your client at the end of Phase II, submit a written report and follow any conference with a letter summarizing important decisions made orally to avoid any future misunderstanding.

At the end of Phase II you will probably find yourself on the upward slope of what we call the *morale versus time curve*, shown in Figure 4.21. At the beginning of a project, your enthusiasm is high and you can hardly wait to do design. You soon realize that there is more to the problem than you originally thought, and you may bog down in getting information. New design ideas may be slow in emerging. Arguments erupt in group meetings, and your ideas may get thrown out. The status of the project seems hopeless, and your self-confidence hits a low point. However, if you persevere and keep working on problems as they arise, you will resolve them, gain back your confidence, and forge on to complete the project successfully. Students invariably finish their projects with a higher degree of morale than when they started, because they have successfully conceived and developed a solu-

tion *and* gained new confidence in themselves for solving problems that have more than one possible answer.

The student design team had a design review of their beam-breaking device, made a few revisions, and then communicated their final design.

Phase III Communicate the Solution

Classroom projects usually end with Phase III. Phases I and II will have used approximately half the time alloted; Phase III requires the remainder of the time to prepare written, graphical, and oral communications. Simple cardboard-wood-string-wire models should be built when appropriate, but prototypes (working, full-scale models) are preferred if facilities and funds are available.

Step 12 Refine the Solution/Final Plans

Now is the time to review "what you have done" and "what you are doing," and to prepare a list of "what you need to do." A typical list for a student project could include:

1. Refine your design by responding to comments and questions from the design review.

2. Complete the analyses for sizes and materials.

3. Conduct experiments or tests for final verification of design decisions.

4. Decide how many and what type of technical drawings will be needed.

5. Estimate the types and numbers of illustrations, graphs, tables, sketches, and layout drawings required for the final report.

6. Design and write the final report.

7. Plan and build a simple model

that represents your whole design or an appropriate part of it.

8. Prepare and practice on oral presentation.

9. Give the oral presentation.

Final Planning Chart

The activity list may be incorporated in a final planning chart, as illustrated in Figure 4.22. Names of the individuals on the design team who were responsible for the activities are included above the appropriate activity bar. The person responsible for coordinating the group's efforts for Phase III should obtain or make estimates of the time required to finish each activity and attempt to balance the remaining hours against the estimated need. *Remember that drawings and written portions require about twice the time originally estimated!*

Step 13 Prepare Drawings and Specifications

Technical drawings that evolve from the layout drawings of Step 9 communicate the necessary information to the proper recipients (the machine shop, the structural shop, the foundry, purchasing, your client, etc.) in order to achieve the physical realization of the design and to insure its proper function. The drawings are usually done with instruments, to scale, and incorporate the accepted drawing standards of the industry. Specific information about form, symbols, standards, content, and conventions may be found in Chapters 13 and 14. An example of a technical drawing is shown in Figure 4.23, which is a detailed-assembly drawing for the threaded screw and bathroom scale alternative.

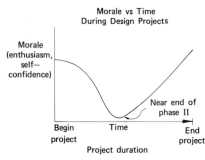

Morale vs Time
During Design Projects

Morale
(enthusiasm,
self—
confidence)

Near end of
phase II

Begin
project

Time

End
project

Project duration

Figure 4.21 A typical cycle of morale versus time during a project.

FINAL PLANNING CHART

MACHINE DESIGN INCORPORATED GROUP 4

ACTIVITY \ WEEK DAY	4 M	4 W	4 F	5 M	5 W	5 F	6 M	6 W	6 F
DESIGN REVIEW RESPONSE	KRIS								
EXPERIMENTS/ TESTING & PROTOTYPE			MIKE/DAVE				MIKE/DAVE		
WRITTEN INTRO DEFINITION	KRIS KRIS			OWEN					
ALTERNATIVES DISCARDED				MIKE/KRIS					
ALTERNATIVE SELECTED WRITTEN GRAPHICS	(OWEN)					DAVE			
MODEL/ PROTOTYPE AND MATERIALS					OWEN/MIKE/DAVE				
PLAN AND PRACTICE ORAL							MIKE/KRIS		
GIVE ORAL									MIKE

Figure 4.22 Final planning chart for completing Phase III.

Figure 4.23 Detailed-assembly technical drawing for the threaded screw and bathroom scale alternative.

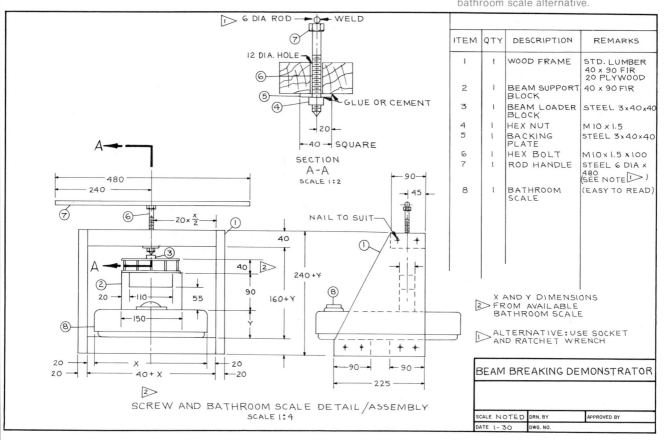

SCREW AND BATHROOM SCALE DETAIL/ASSEMBLY
SCALE 1:4

ITEM	QTY	DESCRIPTION	REMARKS
1	1	WOOD FRAME	STD. LUMBER 40 x 90 FIR 20 PLYWOOD
2	1	BEAM SUPPORT BLOCK	40 x 90 FIR
3	1	BEAM LOADER BLOCK	STEEL 3 x 40 x 40
4	1	HEX NUT	M10 x 1.5
5	1	BACKING PLATE	STEEL 3 x 40 x 40
6	1	HEX BOLT	M10 x 1.5 x 100
7	1	ROD HANDLE	STEEL 6 DIA x 480 (SEE NOTE 1)
8	1	BATHROOM SCALE	(EASY TO READ)

2 ▷ X AND Y DIMENSIONS FROM AVAILABLE BATHROOM SCALE

1 ▷ ALTERNATIVE: USE SOCKET AND RATCHET WRENCH

BEAM BREAKING DEMONSTRATOR

SCALE NOTED	DRN. BY	APPROVED BY
DATE 1-30	DWG. NO.	

Step 14 Construct Simple Models or Build a Prototype

Simple models fulfill two important functions. They can help you to visualize overall relationships and generate new ideas. Building a three-dimensional model from a two-dimensional drawing often involves the synergistic use of hands as well as the brain and eyes, and doing this activity frequently produces new ideas. A simple model that may be built for an oral presentation can represent the whole design or just a significant portion of the whole. The materials can be anything available, such as paper, wood, tape, glue, cardboard tubes from rolls of paper, soda straws, etc. Note that constructing simple "cardboard and glue" models can be effective during any phase of the design process, particularly when new ideas are needed or the design involves the possibility of interferences of parts.

Prototypes also fulfill several functions. They enhance oral presentations and help to visualize relationships, and they demonstrate whether or not a design will work as expected. When students build prototypes, they soon discover what tolerancing means. Furthermore, they begin to appreciate the amount and nature of information required to complete a useful technical drawing. Another important function of building a prototype and checking whether or not it works is that it is a worthwhile design-debugging process.

Figure 4.24 shows a photograph of a rough, working prototype of the threaded screw and bathroom scale alternative. Rough prototypes

Figure 4.24 A prototype of the threaded screw and bathroom scale alternative.

or models have been constructed by students in dorm rooms, homes, shops, and classrooms. Most students seem to be experts at scrounging materials and tactfully obtaining assistance from shop personnel.

Students have been known to comment "I sure learned a lot!" after building a prototype. When questioned further, they admit that their first impressions, based on what they had read and heard regarding manufacturing processes, are somewhat different in the "real" world. "There's a world of difference between the clean pages in a book and the feel, sounds, and smells in a shop," one student said. Therefore, we recommend that students build prototypes of design ideas to gain this additional insight into manufacturing processes if the means are available and if it is appropriate to do so.

Furthermore, we highly recommend that students consult experienced, skilled, shop personnel before and during a prototype construction phase. They can assist in selecting appropriate machine tools, electrical and electronic equipment, and measuring devices and in selecting materials. Shop personnel have been known to bail out students the day before a prototype is due, but they resent being asked for help in a crisis when they know that the student had time to contact them earlier.

Step 15 Prepare Final Reports

Writing a report can be approached in a manner similar to that used in solving design problems. For example, once the purpose of the report (recognition of the goal) is established, search for and collect information that seems appropriate. Next, attempt to establish definitely (definition) the boundaries for the report, such as who are the recipients, what should they be told, and what organization of topics seems to be most appropriate. Then sift through all the information collected to select those that best fit the definition.

Different arrangements and orders of the topics are proposed next (generation of alternatives), followed by a decision as to which of these organizations to use (decision). For example, some reports have the conclusions stated early instead of at the end.

Once the organization is established, an outline should be made; related topics can be grouped to follow your outline logically. Some portions of the report will have been written previously, and new portions will have been added; now the parts can be forged into a whole (communication). For your first overall draft, write as well as possible and let the words and sentences flow; try not to interrupt your writing to resolve questions that you can take care of later, such as looking up words, using correct grammar, selecting the right word, etc. Finally, the report should be revised for smoothness, continuity, and accuracy.

The student design team used the following organization for their final report. (Note that a letter of transmittal to their client was attached to the outside of their report. It was typed in a business letter form, saying that the enclosed report contained a solution to the problem presented to them, the date they received it, that they were pleased to have had the opportunity to work for the client, and that they anticipated further good relations. In industry the letter could also include a fee statement.)

FINAL REPORT

Title Page
Abstract
Table of Contents
Body of the Report
   Introduction
   Final Definition
   Alternatives Considered, but
      Discarded
   The Alternative Selected
      (Completely Described in
        Writing and in Drawings)
   Technical Analyses
   Cost Information
   Instructions for Building and
      Operating the Demonstrator
   Recommendations
References
Appendix

Considerations for enhancing and checking their report were suggested by their instructor.

1. Is there sufficient information to satisfy your client?
2. Typed or printed?
3. Organization and readability (headings and subheadings)?
4. Evaluation of audience?
5. Spelling and grammar?
6. Technical vocabulary?
7. Report in a folder?
8. Are the technical drawings complete? (*Note.* The student design team needed technical drawings completed sufficiently so that a maintenance person in a junior

high school could build the device.)
  (*a*) Is the geometry of the device completely described?
  (*b*) Are all the important lines located, such as the centerline of the device, centerline of holes, key surfaces, etc.?
  (*c*) Are appropriate tolerances indicated?
  (*d*) Are all the materials specified?
  (*e*) Are there any special notes or instructions required?
  (*f*) Is the usual information on the drawings, (i.e., name, date, title, etc.?

(For further information on technical drawings, refer to Chapters 13 and 14; for further information on report writing, refer to Chapter 15).

Selected portions of the beam-breaking team's final report are now included to illustrate some of the items noted under Final Re-

port. Figure 4.25 contains the abstract; Figure 4.26*a* and 4.26*b* contain the definition of the problem and conclusions and recommendations respectively.

ABSTRACT

Our group was formed to design a simpler working model of a strength-of-materials testing machine for demonstration to junior high school students than had heretofore been designed. This report covers all aspects of our group's selected design, from background and definition of problem through alternatives considered and selected, and finally plans for construction and operation of the device. There is also a technical analysis of various parts of the design, included so that basic physical principles could be more fully understood.

Figure 4.25 The *abstract* from the design team's written report.

DEFINITION OF THE PROBLEM

The final definition of the problem is as follows.

There exists a need for a device that demonstrates to junior high school students the strength of various arrangements of wooden tongue depressors formed into beamlike structures. The device should be:
1. Easily carried by a teacher.
2. Made of common materials.
3. Easily constructed by average school maintenance personnel.
4. Able to operate within a class period.
5. Easy to see by a group of 20 students and understandable to sixth to eighth graders.
6. Less than $20 to build.
7. Safe to use by either students or teachers.
8. Readily stored.

(*a*)

Figure 4.26 (a) *definition of the problem.*

CONCLUSIONS AND RECOMMENDATIONS

We believe our design fulfills the definition of the problem because sixth, seventh, and eighth graders can certainly understand how it works, and it is inexpensive. This machine can be built by almost anyone in a couple of hours and, so far, our prototype seems to work just fine.

We recommend that the prototype and instruction manual be used by a teacher several different times and then be revised, if necessary, into a more finished form.

(*b*)

Figure 4.26 (b) the *conclusions and recommendations* from the design team's report.

Occasionally students include their decision table but, for the most part, the decision table is an in-house activity and should not be in the report to the client. The students' instructor should receive a copy of the decision table (if one is used), and the students should place a copy in their files.

The written and graphical description of the threaded screw and bathroom scale demonstrator was presented in enough detail to describe completely the design and its operation and to outline the steps for constructing it. Most projects do not include such a detailed description; however, one of the beam-breaking team's goals was to produce a teacher's manual for individuals who had only limited knowledge of reading drawings and shop practices.

The third mode of communicating, now that the written and graphic forms are completed, is the oral presentation.

Oral Presentation

One of the most cogent comments we have heard about giving an oral presentation is, "Tell 'em what you're going to say; say it; and then tell 'em what you've said." The following are hints and suggestions to help accomplish those three steps effectively.

Before you "Tell 'em what you're going to say," you need to know who you will be addressing. Your client will most likely be present, and your instructor may invite colleagues and others from various fields, such as education instructors and shop personnel, to your oral presentation. Also, your instructor may invite technical representatives from local industries for two main reasons; first, practicing designers and engineers can give you feedback on how your design ideas compare to those in industry; and second, the representatives can be sources of design projects and design information for the future. Thus, you may have a wide variety of backgrounds in your audience to direct your comments to, and you should plan accordingly.

The time available for your presentation determines how much detail you can cover: twelve minutes for presenting and two to five minutes for questions are typical for classroom presentations. Longer presentations may be appropriate if a variety of guests are present or if there are only a few presentations. *In any case, practice your presentation at least once before giving it!*

An effective mode of presentation is to use visual aids as an outline; see Figure 4.27, which is a copy of an overhead transparency used by a member of the student team. The transparency became the outline, and all the student had to do was point at each line as he indicated what he was going to "tell 'em." Then, to "tell 'em," he displayed separate transparencies of each item listed on the outline transparency. The definition of the problem was in tabular form, and each alternative that was discarded was presented as a sketch. The selected alternative was also presented as a sketch along with a brief list of the reasons for selecting it. A price list and parts list were on the next transparency; the last transparency contained concluding comments.

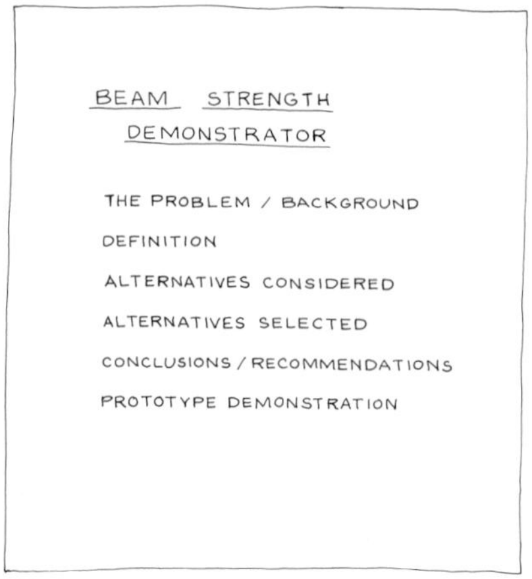

Finally, the student requested that the lights be turned on so he could demonstrate his prototype. Then he summarized briefly what he had just presented; that is, "he told 'em what he had said." His concluding statement was a request for questions.

During the presentation, the guests, the client, and the other students were asked to provide some written feedback in addition to asking questions at the end of the presentation. On a form provided by the instructor, each member of the audience recorded comments, suggestions, and some questions on the design plus an overall score on the presentation alone. Suggestions for judging the effectiveness of the presentation were: effective introduction, use of visuals, discussion of alternatives, completeness, technical vocabulary, and finishing on time. The written feedback and the overall score, unsigned, were given to the design team the next day during an instructor-team follow-up conference.

In industry the oral report at the end of Phase III could be a request to management to go ahead, to build, develop, and manufacture a prototype; however, in the usual classroom situation, the report finishes the project.

BEAM STRENGTH
DEMONSTRATOR

THE PROBLEM / BACKGROUND

DEFINITION

ALTERNATIVES CONSIDERED

ALTERNATIVES SELECTED

CONCLUSIONS / RECOMMENDATIONS

PROTOTYPE DEMONSTRATION

Figure 4.27 Transparency No. 1 used in the design team's oral presentation.

The next phase continues the design process through to marketing of the final product.

Phase IV Develop, Manufacture, and Market the Final Product

The remaining steps in the design process are listed here for completeness; however, only brief discussions of the steps are included because of the extensive variety of approaches used in industry. Nevertheless, Step 17 is emphasized due to its relevance in helping to generate ideas in any phase of the design process.

Step 16 Review Specifications and Drawings

Often individuals who have had experience in preparing drawings and in manufacturing processes objectively and carefully check drawings and specifications before they are released to purchasing and manufacturing departments.

Step 17 Plan for Manufacturing

Most devices are fabricated from pieces of common engineering materials that have been formed into standard shapes and are held together by fasteners, adhesives, or welds. Most manufacturing companies incorporate one or more of the basic processes in producing their product.

Most of the items contained in the following table may be found either in the glossary of technical terms in the Appendix or in Chapters 13 and 14. The list, of course, is not complete, but it does contain the more common items. Esoteric and state-of-the-art materials, shapes, and processes such as integrated circuits, laser beam cutting, and chemical milling are recommended for your future study.

Once the items in the preceding table become part of your vocabulary, you will discover that you see more when looking at manufactured products, such as the attributes of the product, and you incorporate the shapes and materials in your idea-generating activities.

Finances, Scheduling, and Personnel
Plans for manufacturing also include allocation of funds and personnel assignments in reasonable schedules. These are usually unique for any one project.

Step 18 Build, Test and Revise

Constructing working prototypes and testing them uncovers facets of designs that often cannot be predicted on the drawing board. For example, airplane manufacturers build two important prototypes; one is utilized to locate cables, instruments, and equipment in a full-scale mock-up (sometimes made of plywood in place of metals, etc.); the second is a full-scale prototype fabricated, outfitted, and flight-tested as a precursor to assembly-line planning.

Step 19 Packaging the Product

Sometimes shipping loads and environmental conditions during transit are more severe than the actual end use. Electronic apparatus, for example, is usually well packed in styrofoam or spongy, rubberlike materials for shipping purposes.

Step 20 Marketing, Selling, and Distributing

The automobile industry is one example of these three activities. This industry spends millions on marketing and advertising; they organize sales outlets; and they ship by truck, rail, and freighters.

Step 21 Consumer Feedback and Redesign

Appropriate redesign should be incorporated to satisfy the ultimate user for utility and safety. Observe how energy conservation, antipollution laws, and consumer protection laws have affected the designs of houses, automobiles, toys, clothes, etc. This is consumer feedback on a national scale.

Step 21 concludes our discussion of the phases and steps in the design process. An example of an engineering student's solution to a design problem is presented next.

John Kuhta's Approach to Solving a Design Problem

John Kuhta, an engineering student at the University of California, tackled a "real" problem that dealt with the lack of reliable safeguards in the operation of top-burner gas stoves. It was pointed out that each year hundreds of persons are asphyxiated by gas that escapes when the pilot goes out. In many instances, older persons for-

Common Engineering Materials, Shapes, and Processes

Materials	Shapes	Processes	Materials	Shapes	Processes
Metals	Bars	Casting	Adhesives	Construction	Bending
Plastics	Sheets	Forging	Fasteners	lumber	Welding
Woods	Plate	Cutting	Rubber	Plywood	Threading
Ceramics	Angles	Punching		Brick	Plating
Organic	Channels	Turning		Molded	Stamping
liquids	Beams	Drilling		plastic	
Synthetic	Pipe	Facing			
liquids	Tubing	Extruding			

get to turn off a burner on which a pan containing a liquid (water, soup, etc.) has been placed. As a consequence, the liquid evaporates and the pan is scorched. There is also the possibility of a fire when one leaves a burner unattended.

The problem was to create a safe system that would eliminate the cited hazards and yet make it possible to incorporate the design in a standard, top-burner gas stove. Kuhta was quite successful. In fact, his design won first prize in a contest sponsored by the Engineering Design Graphics Division of the American Society for Engineering Education.

Highlights of his report are presented here to acquaint you with his *approach* to the solution of this problem. In his first letter to his instructor, after considering the problem, he proposed the design of a safety valve that would completely shut off the gas flow if the pilot goes out for any reason. He further proposed the design of a weight-sensitive burner that would automatically control the operation of the valve to avoid the hazard of leaving the burner on and unattended. Significant portions of his report included the following data.

Introduction
In an effort to find a feasible, low-cost solution to valve operation, several designs were investigated. Those in the heat-sensitive area were the bimetal and relay systems; the self-generating, low-current thermocouple; the fluid mercury switch; the infrared sensor relay; and the thermistor and solenoid relay system. Also considered was

the mechanical vane and relay system, which is activated by the rising hot vapors of the burned gas. And in the field of light operation, the use of photocells and light-sensitive diodes was considered. Pressure switches, photocells, completing electric circuits, and mechanical systems were investigated for use as an automatic burner control.

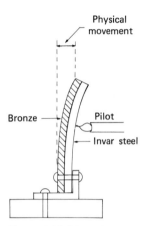

Figure 4.28 Bimetal element for automatic shutoff. Hot position shown. Cold is straight up.

Preliminary Research
Kuhta undertook a very careful study of the state of the art. He was quite curious about the current methods used in automatic shutoff devices, so his search was quite extensive. He investigated 15 significant methods that were in current use. A few of those are shown in Figures 4.28 to 4.31.

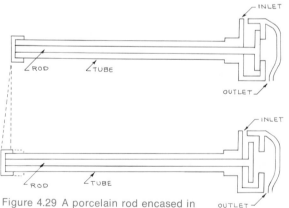

Figure 4.29 A porcelain rod encased in a metallic tube functions as an automatic shutoff device. The upper view is the cold position. The lower view is the hot condition; that is, when the pilot light is burning, the tube elongates and allows gas to flow.

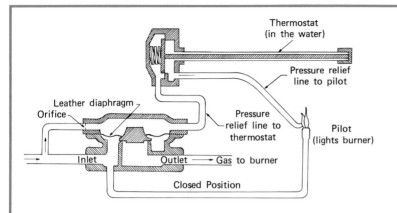

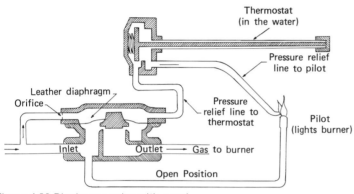

Figure 4.30 Diaphragm valve with a rod and tube thermostat as the controlling device in a large water heater. In the lower view the tube has cooled and allowed gas to flow past the diaphragm and on to the burner.

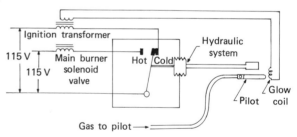

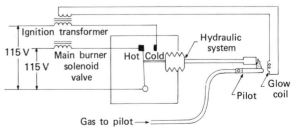

Figure 4.31 Glow-coil ignition controlling device. In the cold position (upper view) the ignition transformer circuit is closed. The glow coil ignites the pilot, which heats the hydraulic system, which closes the main burner circuit (lower view).

Synthesis

After completing his research, Kuhta proceeded to create several alternative concepts among which, hopefully, there might be a few promising ones. Some of the alternatives are shown in Figures 4.32 to 4.37. Note the questions that he put to himself and his remarks about costs (Figures 4.33 to 4.37). Further concepts that dealt with the burner control are shown in Figures 4.38 and 4.39.

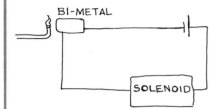

Figure 4.32 Sketch of bimetal and relay concept to shut off gas flow when pilot goes off.

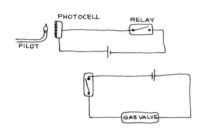

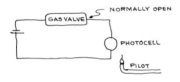

MAYBE I CAN REPLACE THESE TWO CIRCUITS WITH ONE ?
CAN PHOTOCELL ITSELF TRIP VALVE - ELIMINATE RELAY ?

CADMIUM SULFIDE PHOTOCELL ⌐ $2.00

SILICON PHOTOCELL - VARIABLE VOLTAGE ⌐ $3.00

    BUT SILICONE BETTER THAN CAD-SUL BECAUSE IT IS FASTER REACTING
        (MUST BE PLACED AWAY FROM DIRECT HEAT), BECAUSE HEAT CAN RUIN THE PHOTOCELL

Figure 4.33 Sketch of photocell and relay system to shut off gas flow if pilot fails.

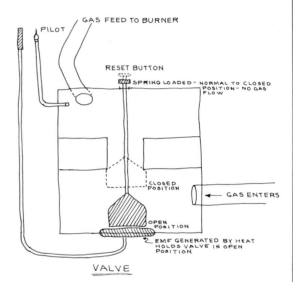

THERMOCOUPLE COSTS ⌐ $8.00)* COST TO HIGH FOR UNIT WHERE
SAFETY VALVE COST ⌐ $15 -$20) ELECTRICITY IS CONVENIENT

Figure 4.34 Sketch of thermocouple system to shut off gas flow.

POS. PROB: IS THE INFRARED SENSITIVE TO THE BLUE LIGHT FLAME OF BURNING GAS ?
    IS IT SENSITIVE TO SUCH SMALL QUANITIES OF LIGHT ?

FROM CONVERSATION WITH ELECTRICAL ENGINEER:
    INFRARED SENSOR IS IMPRACTICAL, TOO COSTLY, NOT WIDELY USED.

Figure 4.35 Sketch of infrared sensor concept to shut off gas flow.

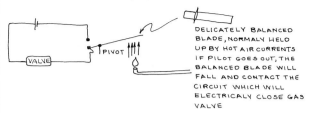

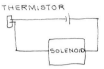

Figure 4.36 Sketch of blade type mechanical vane operated by hot air currents.

DELICATELY BALANCED BLADE, NORMALY HELD UP BY HOT AIR CURRENTS. IF PILOT GOES OUT, THE BALANCED BLADE WILL FALL AND CONTACT THE CIRCUIT WHICH WILL ELECTRICALLY CLOSE GAS VALVE

THERMISTOR

YSI LOW COST   1% PRECISION THERMISTOR
COST ~ $5.00

MOST OTHERS:
PRECISION 1/2 %
WITH HEAT SPAN – 80°C TO +150°C
EPOXY INCAPSULATED
COST OVER $7.00

TEFLON   1% PRECISION THERMISTOR
COST ~ $10.00

Figure 4.37 Sketch of thermistor concept to shut off gas flow if pilot light fails.

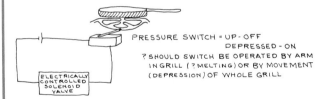

PRESSURE SWITCH = UP–OFF
DEPRESSED – ON
? SHOULD SWITCH BE OPERATED BY ARM IN GRILL (? MELTING) OR BY MOVEMENT (DEPRESSION) OF WHOLE GRILL

PHOTOCELL (FLUSH WITH BOTTOM OF PANS)

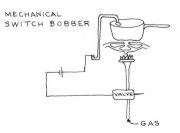

MECHANICAL SWITCH BOBBER

Figure 4.38 Sketches of automatic burner controls. For example, as the fluid in the pan boils away, the pan get lighter (upper view), glows (middle), or empties (lower).

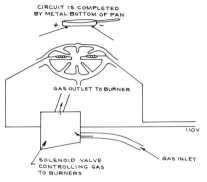

CIRCUIT IS COMPLETED BY METAL BOTTOM OF PAN

GAS OUTLET TO BURNER

SOLENOID VALVE CONTROLLING GAS TO BURNERS

GAS INLET

Figure 4.39 Sketch of burner control. If the pan melts the circuit is opened.

Evaluation, Optimization, and Decision

Having carefully analyzed each of the few promising concepts, obtained cost data, determined ease of installation, etc., Kuhta reached the conclusion that the "best" design "is a photocell and solenoid valve combination to be used as an automatic shut-off for the pilot, and a mechanical weight trip relay operating a solenoid valve in the gas line to the top burner." He supported his choice by pointing out that:

The bi-metal, the thermocouple, and Rod and Tube system popularly used now cost (*Author's note:* The costs quoted do not reflect today's prices.) approximately $15.00 to $20.00. The combination photocell and solenoid considered here would cost approximately 60 percent less. The cost of the automatic burner control is approximately $5.00 to $6.00. Since practically all stoves today have electrical wiring to operate clocks, timers, and lights, the need for electricity to power the solenoid valve is easily resolved.

Brief descriptions of the operation of the photocell and solenoid, the Robertshaw FM Automatic Pilot, and the automatic burner control were presented in the following form.

Photocell and Solenoid Operation

The photocell acts as a light sensitive relay and is used here to operate a solenoid. When the pilot is burning, the light produced energizes the photocell which acts as a closed relay which keeps the electrically controlled solenoid valve open. When the pilot goes out for any reason, the photocell acts

as an open relay and no current flows through the normally closed solenoid, and all gas flow to the burners and the pilot is shut off. So long as the pilot is burning, the solenoid is held open (see Figure 4.40).

Robertshaw FM Automatic Pilot Operation

When the main control valve is turned ON the gas flows to the actuating pilot (2) and to the inlet (3) of the safety valve. The actuating pilot (2) then ignites from the constant burning pilot (1). The actuating pilot (2) heats the thermal element of the hydraulic system (capillary tube) (4), vaporizing the heat sensitive fluid in the element. The vapor pressure transmitted through the capillary tube (4) causes the diastat (5) to expand. The movement of the diastat (5) is multiplied by

a system of levers (6) exerting force against the spring-loaded control valve (7). This action opens the spring-loaded safety valve (7), permitting the gas to flow through the adjustable orifice (8) to the main burner (9) (see Figure 4.41).

Automatic Burner Control Operation

The automatic burner control consists of a weight sensitive relay and a "normally closed" solenoid. When a pan is placed on the burner gate, it depresses, by means of an extension rod, the weight-sensitive relay which activates the solenoid thus permitting gas to flow to the burners. It should be noted that the weight-sensitive relay can be used in two ways. It can be set so that the weight of just any pan will ignite the burners; or in case a pan of only one weight were used, a variable

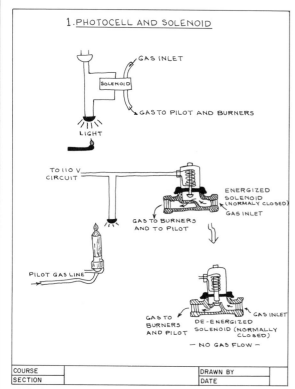

Figure 4.40 Sketch of photocell and solenoid concept.

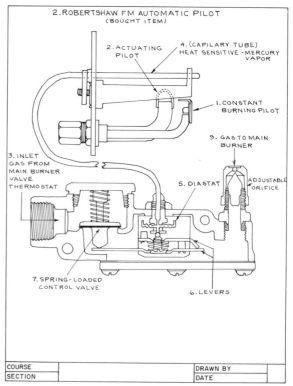

Figure 4.41 Sketch of Robertshaw FM Automatic Pilot.

activation force relay could be used so that a pan of any other weight would not activate the burner. This could be an important safety feature in avoiding the boiling or burning out of the contents of the pan or its scorching or warping. The design of this type of relay system can easily be incorporated with existing devices, such as the thermostatic burner which is activated when the sensor is depressed, then avoiding the need for a second device in the grate of the burner.

Finally, Figure 4.42 shows a sketch of the automatic shutoff and automatic burner control system placed in a typical gas burner.

Authors' Comments

1. The time constraint on the project precluded the building and testing of a prototype. Nevertheless, Kuhta was quite successful in producing a creative design that was novel, useful, and economical. Moreover, he had a most satisfying experience in coping with a "real" design problem (i.e., he gained confidence in his ability to apply an approach to solving technical problems).

2. A very important additional plus was the writing experience Kuhta gained in the preparation of the final report. He realized that it is quite essential to learn how to write well! More will be said about writing and preparing the project report in Chapter 15.

The next section contains comments from an engineering graduate who is a successful, practicing engineer.

Comments by Jack Ryan

Jack Ryan, president of the Jack Ryan Group, expressed the following thoughts, which were excerpted from his presentation to an engineering class. The comments reflect what he had observed while applying his own approach to solving technical problems.

Product Superiority

Product superiority in a highly competitive business situation tends to contribute more to sales growth and profit improvements than just about any other technique available to management.

The reasons for this are really quite simple. From a return-on-investment standpoint, good and bad products cost just about the same to introduce, e.g., each situation requires preliminary design, product engineering, tooling, promotional allocations, inventory build-up, etc. Secondly, opportunities to more effectively manage a business independent of product and people decisions tend to be rather limited, as the primary acquisition costs of labor, materials, and money are almost identical for all companies engaged in the process of doing business. If we additionally accept the concepts that value is determined by the ultimate product purchaser and/or user, and not by the accountant who applies fixed percentages to accumulated costs, the real opportunity for maximum gross margin attainment rests with the superiority of the product in the marketplace.

The achievement of product superiority in the marketplace is a result of *proper product definition,* the *creative response to that marketing mandate,* and the *correct execution of the idea in relation to the needs of the business plan,* especially to time and budgetary constraints.

Preliminary Design

Critical to program success is the preliminary design phase, for it's here where the real commitments to future

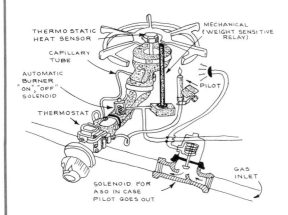

THERMO STATIC HEAT SENSOR

MECHANICAL (WEIGHT SENSITIVE RELAY)

CAPILLARY TUBE

AUTOMATIC BURNER "ON" "OFF" SOLENOID

PILOT

THERMOSTAT

GAS INLET

SOLENOID FOR ASO IN CASE PILOT GOES OUT

PHOTOCELL A.S.O. AND AUTOMATIC BURNER CONTROL PLACED IN A TYPICAL GAS BURNER SET UP

Figure 4.42 Sketch of photocell and automatic burner control placed in a typical gas burner stove.

expenditures and income are determined. The product is shaped, features are enumerated, major material considerations are evaluated, preliminary costing is completed, etc. From here, the major expenditures for engineering, tooling, trade introduction, inventory and promotion are made. The opportunities to make product changes after preliminary design, whether they be identified prior to consumer usage or not, are difficult to implement and most often will have an adverse effect on the planned contribution for that particular item. In addition to these direct costs, the so-called opportunity cost must also be considered, for creative energies employed in failure quite obviously precludes the use of that talent for profitable new product activities.

Apart from its impact on down-the-line costs, preliminary design takes less in calendar time to finish than does each of the activities that it sets into motion. Most typically, preliminary design, including proper engineering turnover takes little more than 120 days, while the entire cycle of idea origination to first production should on the average take approximately 18 months to complete. In short, preliminary design while taking little in dollars and time to support should be more correctly identified as the major investment point in the entire new product flow.

Elegant Simplicity

Elegant simplicity results when an invention is reduced to its simplest form. The simplest form is a design that has been streamlined, optimized and redesigned so that all components are utilized, that no addition can make an improvement and any reduction will degrade the operation of the invention.

The 80%-20% Rule

Eighty percent of the progress on a product is accomplished with 20% of the budget. Then 80% of the budget is spent perfecting the last 20% of the project. This well known scientific principle has been proven to apply to many aspects of business, scientific phenomena and product development.

The 20% of the budget that produces 80% of the results appears to be 16 times as effective as the 80% of the budget that produces 20% of the results.

We try to encourage making simple looks-like and works-like models of as many alternative product possibilities as possible using these high efficiency (16 to 1) dollars.

For example, a looks-like model of a new professional blower dryer might be developed to 80% completion for $1000 to make sure that the direction is correct. If it proves not to be, another 80% complete model might be built for an additional $1000, but when the 100% finished model is to be ready for production, the last 20% will probably cost an additional $4000.

The 80%-20% rule often results in what appears to be a startling escalation of expenditures at the end of a product, especially if the work is expedited to rush some projects into production against the next season's deadline. And then old charges for services on the product continue to trickle in after the project has been done.

Design Process/Models

Remember that the design process is one of successive approximation. This implies change and experimentation, so build your models with changes, experimentation, adjustment and rearrangement in mind. Make them easy to "play" with, without fear of "harming" them. That's what they're for.

An important advantage of a quick, crude model assembled with the im-

portant elements in *separate* pieces is that it *encourages* rearrangement and experimentation because the consequence of any change is "It couldn't hurt." A beautifully perfect model actually inhibits design. The best models are made quickly and economically by cannibalizing existing products and patching them together with sleazy mock-ups of the other parts.

Some personalities are offended by sawing a perfectly good product in two when they're not exactly sure of the final design, but it's much cheaper to waste samples than time. Try it, you'll get used to it and enjoy it. It's actually much more economical and produces a better design more quickly because the design is being done in 3-D instead of 2-D, which is always better, and the model, no matter how crude, is being built by the designer's own hands so there is a two-way communication—designer to model, and from model back to designer as it develops. (Take photos because these models are not durable or storable!)"

The foregoing observations were philosophical and are not a detailed description of a design solution. The next section is a detailed description of a design solution by a practicing engineer.

Summary of a Design Solution by A. O. Dubois

A. O. (Andy) Dubois, Chief Engineer, Airco Temescal Division, Berkeley, California, contributed the following design description as an illustration of how he approached the development of a design within challenging criteria.

Design of a Pneumatically-Controlled, Shutter Blade for Use With an Optical Coating System Employing an Electron Beam

Background

The lenses and mirrors of modern cameras and optical instruments are coated with thin layers of various metallic and dielectric materials. These thin layers are applied for a variety of reasons—depending upon the application. For example, it may improve the reflectivity of a mirror, or reduce the reflectivity of a lense, or reflect all except a narrow band of wavelengths for an optical filter. These thin coatings (about 1×10^{-7} meters thick) are applied in a vacuum chamber, because a vacuum environment is required if the coatings are to have the desired optical and mechanical characteristics.

The coating is applied by evaporating the material and allowing the vapor to condense on the surface of a mirror, lense, etc. When this evaporation is done in a vacuum, the atoms of the evaporating material travel in a straight line from the molten pool to the first surface upon which they can condense (as shown in Figures 4.43 and 4.44).

The material (usually in solid form) one wishes to apply may be heated to evaporation temperature by any of several types of heaters—resistance, induction, or electron beam. As the material is heated to the correct temperature, the optics must be protected from exposure to the vapor source because the source may split as the material becomes liquid and high vapor pressure impurities are evaporated out of the pool. A metal-blade shutter is interposed between the vapor source and the optics, to protect these optics until the pool is at the equilibrium condition to evaporate smoothly at the desired rate. This shutter is then moved aside rapidly to allow the coating to proceed. After the proper thickness is applied, the shutter is rapidly closed again to terminate the coating.

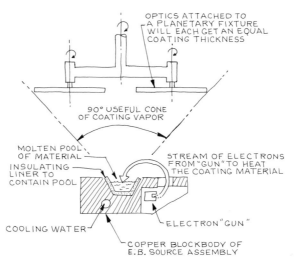

Figure 4.43 Electron-gun heating material to be vaporized; vapor travels in straight line to the optics on the planetary fixture.

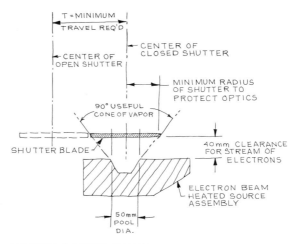

MINIMUM SHUTTER RADIUS

$$R_{MIN} = \frac{50}{2} + 40 \, TAN \, 45°$$

$$= 65 \, mm \quad \underline{USE \, 75 \, mm}$$

MINIMUM TRAVEL

$$T_{MIN} = 2R$$

$$= 150 \, mm \quad \underline{USE \, 170 \, mm}$$

Figure 4.44 A shutter blade may be interposed between the vapor source and the optics.

Customer Feedback

Customers have complained that earlier shutter designs were not convenient to service. When a shutter blade gets heavily coated with condensate, it must be removed and cleaned. Then when replacing the shutter blade the blade position should be self locating and it should be held by a single fastener. Preferably, one should not need to remove any loose pieces such as pins or keys since these can be lost or could be inadvertently left out during reassembly. Also, the shutter blade assembly should be portable so that the user has flexibility in changing his setup. Customer feedback and other design considerations led to the following design criteria for the current shutter blade design which was to be utilized in conjunction with an electron beam heater source.

*Design Criteria—Shutter
Blade Assembly*

1. The assembly should mount inside the vacuum chamber, be easily moved and preferably should be attached to the electron-beam source itself.
2. The lines transmitting the power to the shutter actuator should be flexible, so that if the power source is relocated no major changes would be required in gears, bearings, etc.
3. The assembly must not cause even the smallest leak of any kind into the vacuum chamber, and there should be no trapped pockets of air in its construction.
4. Construction should utilize as little non-metallic material as possible. Organic materials outgas and contaminate the high-vacuum environment. Ceramic and glass are brittle.

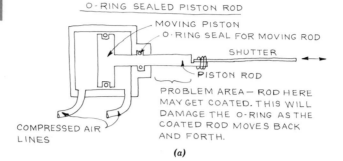

POSSIBLE CYLINDER SEAL CONCEPTS

O-RING SEALED PISTON ROD

MOVING PISTON
O-RING SEAL FOR MOVING ROD
SHUTTER
PISTON ROD
PROBLEM AREA — ROD HERE MAY GET COATED. THIS WILL DAMAGE THE O-RING AS THE COATED ROD MOVES BACK AND FORTH.
COMPRESSED AIR LINES

(a)

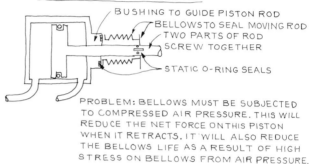

BELLOWS SEALED PISTON ROD

BUSHING TO GUIDE PISTON ROD
BELLOWS TO SEAL MOVING ROD
TWO PARTS OF ROD SCREW TOGETHER
STATIC O-RING SEALS
PROBLEM: BELLOWS MUST BE SUBJECTED TO COMPRESSED AIR PRESSURE. THIS WILL REDUCE THE NET FORCE ON THIS PISTON WHEN IT RETRACTS. IT WILL ALSO REDUCE THE BELLOWS LIFE AS A RESULT OF HIGH STRESS ON BELLOWS FROM AIR PRESSURE.

(b)

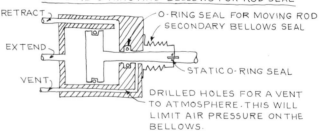

COMBINE O-RING AND BELLOWS FOR ROD SEAL

RETRACT
O-RING SEAL FOR MOVING ROD
SECONDARY BELLOWS SEAL
EXTEND
STATIC O-RING SEAL
VENT
DRILLED HOLES FOR A VENT TO ATMOSPHERE. THIS WILL LIMIT AIR PRESSURE ON THE BELLOWS.

THIS ROD SEAL DESIGN LOOKS GOOD; HOWEVER IT WOULD REQUIRE A VERY EXPENSIVE BELLOWS IF THE TRAVEL IS LONG. THEREFORE, USE SOME KIND OF MOTION AMPLIFICATION ON PISTON ROD TRAVEL.

(c)

Figure 4.45 Alternatives to satisfy the no-leakage criterion.

5. The assembly should be compact; and, as the shutter opens and closes it should occupy a minimum of space, because space is very valuable in the vacuum chamber.
6. The shutter should not obstruct the technicians view of the evaporant.

7. The shutter actuator should be designed to withstand high temperatures since the optics are usually heated in place before the coating is evaporated onto them. Also, the pool of evaporating material can be as high as 2,000° C.

8. The shutter blade should be easily removed for cleaning.

Alternatives to satisfy the no leakage criterion are sketched in Figure 4.45a, 4.45b, and 4.45c.

Alternatives for possible motion-amplification (an earlier design criterion) are sketched in Figure 4.46a, 4.46b, and 4.46c. The yolk from Figure 4.46c is shown in detail drawing in Figure 4.47.

Evaluate Various Power Sources for Shutter Actuation

Actuator	Advantages	Disadvantages
Electric	a. Leads very flexible b. Any stroke possible	a. Heat sensitive b. Needs to be canned, so as not contaminate the the vacuum c. Open/close position limits complicated
Electric solenoid	a. Leads very flexible b. Simple to service	a. Heat sensitive b. Needs to be canned c. Low force at one end of travel d. Short stroke
Hydraulic cylinders	a. Very compact b. Simple to service c. Lots of force available d. Any stroke available e. Simple control	a. Needs a very reliable shaft seal b. Messy if a leak develops c. No source of hydraulic pressure is already available, therefore expensive d. Needs shaft seal
Air cylinder	a. Compact b. Simple parts c. Lots of force available d. Any stroke available e. Simple control f. Clean	a. Needs shaft seal

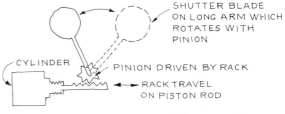

RACK AND PINION

O.K., BUT RACK AND PINION ARE SENSITIVE TO JAMMING.

(a)

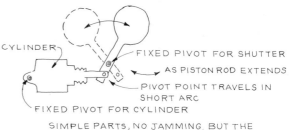

PIVOT AND LEVER

SIMPLE PARTS, NO JAMMING, BUT THE COMPRESS AIR TUBES MUST FLEX EACH CYCLE WHEN CYLINDER PIVOTS.

(b)

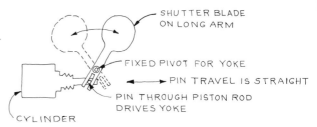

YOKE AND PIN DRIVE

THIS IS BEST - NO JAMMING, SIMPLE PARTS, AND CYLINDER DOES NOT MOVE.

(c)

Figure 4.46 Alternatives for possible motion amplification.

A final sketch of the shutter assembly is in Figure 4.48. A photograph of the final shutter assembly design is in Figure 4.49.

Finally, Figures 4.50 and 4.51 show photographs of the shutter assembly installed in the vacuum chamber.

CONCLUSIONS

This chapter completes our plans to implement one of the objectives of our text: getting you involved in the design process. Sketching in Chapter 2 encouraged you to communicate graphically with others and to record design ideas quickly. Layout drawings in Chapter 3 illustrated a transitional step in the graphical development and analysis of design concepts. And, in Chapter 4, we provided guide lines for taking necessary steps in the design process, knowing that once you have followed our order of steps, you will soon develop your own. Furthermore, we included examples of design activities from beginning students and from successful designer-engineers to broaden your perspective on the design process.

To provide you with opportunities to become further acquainted with the phases and steps, we have included a number of exercises that fall into two general categories: the first focuses on the phases and steps in the design process, the second on suggestions for design projects.

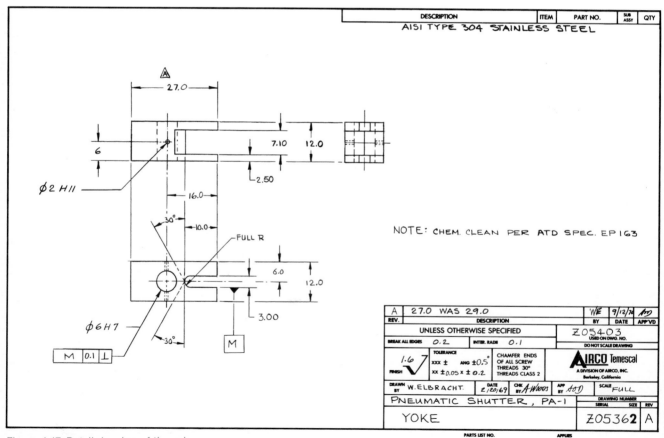

Figure 4.47 Detail drawing of the yoke from Figure 4.46c.

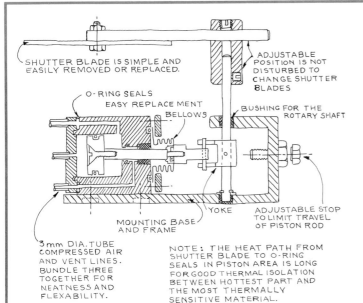

SHUTTER BLADE IS SIMPLE AND EASILY REMOVED OR REPLACED.

O-RING SEALS EASY REPLACEMENT

BELLOWS

ADJUSTABLE POSITION IS NOT DISTURBED TO CHANGE SHUTTER BLADES

BUSHING FOR THE ROTARY SHAFT

YOKE

ADJUSTABLE STOP TO LIMIT TRAVEL OF PISTON ROD

MOUNTING BASE AND FRAME

3 mm DIA. TUBE COMPRESSED AIR AND VENT LINES. BUNDLE THREE TOGETHER FOR NEATNESS AND FLEXABILITY.

NOTE: THE HEAT PATH FROM SHUTTER BLADE TO O-RING SEALS IN PISTON AREA IS LONG FOR GOOD THERMAL ISOLATION BETWEEN HOTTEST PART AND THE MOST THERMALLY SENSITIVE MATERIAL.

Figure 4.48 Final sketch of the shutter assembly.

Figure 4.49 Final shutter assembly design.

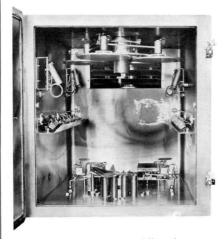

Figure 4.50 Shutter assemblies shown installed in vacuum chamber (lower left and lower right).

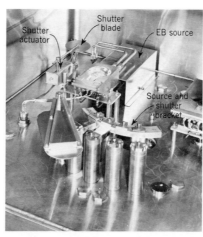

Shutter actuator

Shutter blade

EB source

Source and shutter bracket

Figure 4.51 Shutter assembly in vacuum chamber.

Exercises

Phase I Recognize and Define the Problem

1. For practice in recognizing and defining problems, select one or more of the possible design problem categories listed in "d" and do parts a, b, and c.

(*a*) In one or more sentences propose a modification to an existing product to either improve it or adapt it to another use.

(*b*) Prepare sketches of the existing product.

(*c*) List any objectives, constraints, requirements, etc., that you believe to be appropriate for the modification.

(*d*) Categories:

(*1*) *Personal.* Hobbies, home, sports, equipment, vehicles, energy.

(*2*) *Instructor's.* Research, teaching aid, consulting, other.

(*3*) *School.* Laboratories, shops, sports, classroom equipment, building access, medical, science, cultural, drama.

(*4*) *City and County.* Scouting, Junior Achievement, YWCA, YMCA, playgrounds, hospitals, transportation, industry, garbage, fire department, police department, stores, conservation, advertising, pedestrian traffic.

(*5*) *State.* Parks, highways, agriculture, fairs, conservation, energy.

(*6*) *National and International.* Peace Corps problems, military, space, natural resources, energy, communications.

2. For practice in *planning activities and time,* select one or more of the following projects and prepare a *planning chart* similar to Figure 4.2. You must decide (or be assigned) on the time available, the due date, if you are alone or working in a group, and the general activities required to accomplish the project.

(*a*) Build a garage or a carport.
(*b*) Assemble a toy or a commercial product such as:
  (*1*) A bicycle.
  (*2*) A model (boat, airplane, vehicle, etc.).
  (*3*) A prefabricated cabin.
  (*4*) A small appliance.
  (*5*) An automobile system (power train, brakes, lights, etc.).

3. For practice in obtaining information or becoming acquainted with sources of information that may be new to you, do one or more of the following.

(*a*) Investigate one of the listed sources of information and give a 2 to 3 minute report to your classmates telling what you found, where, how, etc.
  (*1*) City or state code or standard.
  (*2*) Federal code.
  (*3*) Technical journal.
  (*4*) Government report (ERDA, NASA, DOD, HEW, etc.).
  (*5*) Engineering index or science and technology index.
  (*6*) Vendor's catalogs.
  (*7*) Patent publications.
  (*8*) Skilled technicians-shop personnel.
  (*9*) Practicing engineer.
(*b*) Propose an approach for obtaining costs and technical information over the telephone from a source listed in the *Yellow Pages*. Present your ideas to your classmates and instructor for their comments. Then do it.
(*c*) Propose an approach for obtaining

information by letter from a manufacturing company, an agency, etc., and present your ideas to your classmates and instructor for their comments.

4. For practice in preparing a *progress report,* do one of the exercises in Exercise 1d, and prepare a simulated progress report to an imaginary client. Refer to Figure 4.4 for an example.

Phase II Generate and Evaluate Alternative Solutions

The most frequent response in a survey of beginning engineering students on what they believed to be the most difficult *step* in the *design process* was "coming up with alternatives." Practice helps to overcome this hurdle; however, before tackling specific design assignments, we recommend the following exercises on innovation techniques.

DEVELOPING TALENT FOR PERCEIVING IDEAS AND GENERATING IDEAS

1. Ask your instructor to select an object or a location, which you look at casually almost daily, and have you sketch what you can from memory. Then return to the object or area and observe it, knowing you will be asked to resketch it later. Compare your two sketches.

2. List the attributes of an assigned object using the guidelines given in this chapter. Then list "unset" uses of the object.

3. Do the same as in Exericse 2, but include a sketch of the object.

4. Exercises to stimulate creativity or flex your imagination can be arranged by your instructor. Several suggestions follow.

(*a*) Select one of the open-ended questions listed and brainstorm it as a class.

  (*1*) How could news be better transmitted to the public?
  (*2*) What can an individual do to conserve energy?
  (*3*) What design ideas can be learned from animals, plants, and insects in nature?
  (*4*) What else can be done to decrease the number of car thefts, home burglaries, traffic accidents, etc.?
(*b*) Refer to articles such as Eugene Raudstepp, "Games That Stimulate Creativity," *Machine Design,* July 21, 1977, and try some of the exercises listed. Additional articles may be located by using the American Science and Technology Index or the Engineering Index.

5. A *system analysis approach* is similar to attribute listing. For example, select a system such as the telephone headpiece and represent it as a square with an *input* (voice) and an *output* (electrical signal). Then break the system into as many subcomponents as you can, such as sound waves striking a diaphragm (*input,* sound waves, and *output,* vibrations), etc. The process of breaking down complex systems into smaller parts will help you to understand the system better and give you ideas for designing. Try some of these systems.

(*a*) Telephone.
(*b*) Automobile.
(*c*) Furnace.
(*d*) Radio.
(*e*) Television.
(*f*) Production process (i.e., a raw material goes in the plant and products come out—what happens in between?).

6. *Trigger Verbs.* Generate a list of verbs to accomplish the following.

(*a*) Open or take apart an object.
(*b*) Join or assemble at least two parts.
(*c*) Move an object.
(*d*) Rotate an object.

7. Select an existing design and ask questions that might lead to improvements in the design. Submit your recommendations to your instructor. Some typical questions are:

(a) Can the *size* be changed (larger, smaller, etc.)?

(b) Can the *order* be changed (arrangement, sequence, etc.)?

(c) Can the form be changed (curved, straight, harder, softer, etc.)?

(d) Can the *state or condition* be changed (open, closed, hotter, colder, etc.)?

(e) Can the *motion* be changed (speeded, slowed, rotated, etc.)?

(f) Can *substitutions* be made (plastics, metals, ceramics, organics, etc.)?

8. Sketch as many ideas as you can to utilize the energy stored in a common household-type mousetrap to:

(a) Power a vehicle the farthest horizontal distance. [This activity could be a contest. Your classmates or your instructor can establish any rules you believe to be appropriate, (cost, materials, etc.).]

(b) Power a climbing device to climb a surface such as a screen, a ladder, a woven material, etc.

9. Sketch as many ideas as you can for packaging an egg so that the egg in the package can be dropped from an established height without being broken. The landing surface, target area, etc., should be specified by your classmates or instructor.

10. Propose as many ideas as you can for building model beams or frames out of wood or paper to be tested to failure. Then test them for maximum load sustained per weight of the beam or frame.

(a) Propose alternative methods of testing the beams or frames.

ANALYSIS AND DECISIONS

1. For experience in mathematical modeling, testing, and becoming acquainted with characteristics of standard structural shapes, do any of the following.

(a) Measure the deflection of a simply loaded *cantilevered beam* of a structural shape of your choice such as flat bars, rods, tubes, angles, channels, I beams, etc., and record deflection at given loads. The mathematical model for describing deflections of cantilevered beams (for relatively small deflections) is:

$$y = \frac{Wl^3}{3EI}$$

where

y = deflection at the point of application of the load

W = load

l = length of the beam

E = elastic modulous of the material

I = moment of inertia (second moment of area); this is a resistance-to-bending term based on the cross-sectional geometry of the beam

You will need help to locate values of E and I; they are usually found in technical handbooks.

(b) Do the same as part a for other loading configurations specified by your instructor.

2. Review John Kuhta's solution to a design problem and comment on how he approached the problem, the analyses he did, and the decisions he made.

3. Review John Ryan's comments and compare his philosophy to the design process.

4. Review Andy Dubois' design problem and solution and comment on the definition he used and how he used it.

5. As an introduction to the use of a decision table, refer to Figure 4.20 and then do the following.

(a) List some requirements (definition) for a means of transportation based on your own locale.

(b) Give each requirement a weighting factor relative to all the other requirements.

(c) Propose at least three different alternatives for meeting the requirements of the definition.

(d) Rate each alternative against the others as to how well it meets each requirement.

(e) Finally, calculate the appropriate products and sums to see which alternative scores highest.

6. Select a decision situation or be assigned one and prepare a decision table.

DESIGN REVIEW

A *design review* should occur during a project so you can benefit from the expertise of your audience. Your instructor can schedule design reviews as part of the exercises for Phase II; for example, you could report on your tentative selection of a design for the mousetrap-energy exercise, the egg package, or the model beam.

Phase III Communicate the Solution

1. The following suggestions for exercises relate to communicating solutions to design problems.

(a) Prepare a short written report on a design solution after reviewing Step 15. Include subheadings and appropriate drawings as specified by your instructor.

(b) Give an oral presentation on a design solution and use as many different visual aids as practical. Note that most individuals can read letters and numbers that are at least 1/250 of the distance from the individual to the words (e.g., 10 m/250 would indicate letters 40 mm high as a minimum). Use models of your ideas, also.

2. Prepare an oral presentation on an

existing design from any one of the following sources. Include as many new (to you) technical words as you can. Refer to Step 15 for comments and suggestions on oral presentations.

(a) Technical journals.
(b) Technical periodicals.
(c) Textbooks.
(d) Visit an installation relating to your general field of interest.
  (1) Bridge.
  (2) Dam.
  (3) Radio or TV station.
  (4) Automobile facility.
  (5) Manufacturing plant.
  (6) Airplane.
  (7) Chemical plant.
  (8) Nuclear facility.
  (9) Other.

Phase IV Develop, Manufacture, and Market the Final Product

1. Refer to Step 17 and investigate assigned topics in the three categories listed, *materials, shapes,* and *processes.* The information you obtain could be in a number of forms, such as:

(a) Sketches and notes.
(b) Copies of articles.
(c) Photographs.
(d) Summaries of interviews.

2. Investigate in depth a product or process in your chosen field of interest. Learn what you can about the first three phases. What need was satisfied? Who designed the product or process? Are any documents available relating to decisions made during the development? What can you learn about the marketing, sales, or distribution procedure? (*Engineering Case Studies,* published by Stanford University, Palo Alto, CA, are excellent sources for this exercise.)

Project Suggestions

Short Projects

1. The number of bicycles on the streets has increased dramatically in recent years. Therefore, to plan for better handling and merging of bicycle traffic with automobile traffic, municipalities need some means of counting bicycles passing through certain areas. Design a means to accomplish this.

2. Design an attachment for an average-size wheelchair that would provide one of the following.

(a) A surface for working on a hobby.
(b) A special table for eating.
(c) A compartment for carrying small packages or small purchases.

Attempt to contact an individual who needs one of these attachments or some other requirement.

3. Design a special carrying case for components relating to any of the following.

(a) A hobby.
(b) A job.
(c) A recreation.
(d) A vacation.
(e) An education.

4. Design a toy or game for any one of the age groups listed, assuming the individual who will use the toy or game is confined to bed.

(a) Ages 3 to 5.
(b) Ages 6 to 9.
(c) Ages 10 to 11.
(d) Ages 12 to 15.
(e) Ages 16 and older.

5. Design an educational toy for children aged 5 to 12 for any of the following subjects.

(a) Mathematics.
(b) Physics.
(c) Chemistry.
(d) Biology.
(e) Geology.

Short or Long Projects

1. Design a device to help handicapped individuals open canned goods with one hand.

2. Contact an individual who is responsible for a research project, a laboratory, a shop facility, or a studio and ask if there are devices needed to augment their existing equipment. Designing a bracket to support some equipment is an excellent way to obtain design experience and, at the same time, learn something about an overall experiment, process, or system.

3. Electronic games that connect to TV sets are difficult to play for individuals with little or no strength in their hands, but their arms are relatively strong. Propose a modification to the levers or dials for a typical game to accommodate these individuals.

4. Design a teaching aid for a course in your chosen major. Confer with an instructor for suggestions and consultation.

5. Almost every year a variety of organizations sponsor student design contests. Ask your instructor for information. Several representative groups in the past have included ASME, SAE, James F. Lincoln Arc Welding Foundation, Student Competitions On Relevant Engineering (SCORE), and the Engineering Design Graphics Division of ASEE.

6. Contact your local municipal government and ask if there are any problems you might attack in recreation areas, solid waste collection, firefighting, crime, etc.

Long Projects

1. Design a means to lift a person in a wheelchair a vertical distance of at least 2 m either vertically or on an incline. Attempt to locate someone or an organization in your locale who would like to have a lift design.

(a) *Option 1.* The device should be designed so that an average homeowner could build and install such a device with the help of a neighbor who has a small shop in the basement or garage.

(b) *Option 2.* The device should be designed so that skilled professionals could build and install it.

2. Design a walkway for pedestrians to span a distance of 10 m. Be certain that you meet the standards of your local building codes. (Ask for help on strengths and other properties of common engineering materials.)

3. Design a collapsible, portable, bulletin board display device that can be carried and transported easily by one individual. Individuals who represent schools and who travel frequently often express the need for display material at fairs, conferences, etc., to advertise their school. Try to locate someone in your school who may need a portable display.

4. Design an apparatus to measure the performance of a relatively uncomplicated solar energy collector. For example, for an air heating system, you would want to measure air temperature in and out and rate of flow in air. You may want to be able to vary the size of the collector. An alternative would be to consider water as the medium or to substitute a wind energy collector and measure electrical power for various configurations and sizes of windmills.

5. Design a hand-powered can crusher so the average homeowner can compact recyclable cans. (A related project would be to design an apparatus for measuring the forces required to crush an average recyclable can.)

6. Design an electronic clock to be used at neighborhood swimming or track meets to show the time elapsed and the winner's time.

7. Design an electronic scoreboard for neighborhood basketball games.

8. Design a pogo stick for a particular age group. Assume 10,000 units will be made. Ask for assistance on spring characteristics.

Chapter 5 Fundamental Principles of Orthogonal Projection

INTRODUCTION

In the first chapter we learned that engineering graphics plays a most important role in the phases of the design process, not only in the preparation of freehand sketches, graphs, and charts, but also in the preparation of working drawings, which are essential to the construction of models and prototypes and the manufacture and fabrication of the end product.

In connection with the preparation of working drawings, it was shown that a number of technical problems (true lengths of rods, cables, shafts, and structural members, and angles between plane surfaces, intersections of surfaces, and forces in space frames, etc.) must be solved. Several practical examples were cited in Chapter 1.

The graphical solutions of the cited problems and similar ones that arise in engineering design depend on the engineer's ability to analyze the problems and to apply the fundamental principles of orthogonal projection to determine the required information.

ANALYSIS

Analysis of the problems is the "thinking-through" process. It is the mental visualization of the problem and its solution. For example, suppose it is necessary to determine the clearance between two high-tension wires, between two structural members, between two pipes, etc. Basically, all these problems belong to one family, skew lines (lines that are neither parallel nor intersecting.) *Essentially, our problem is to find the perpendicular distance between two skew lines*. Analysis of the problem should lead us to the conclusion that the distance will be apparent when one of the two skew lines appears as a point. The view of the two lines on a plane that is perpendicular to one of them will show the perpendicular distance between the two skew lines (see Figure 5.1).

Now let us consider another problem, the determination of the "angle of bend" of the connection angle between the structural members shown in Figure 5.2. *The problem is actually, reduced to the "determination of the angle between two intersecting planes."* The solution of this problem is quite simple. Again, "thinking-through the problem" (or analysis) reveals that the angle between the two planes will be seen in a view that shows the

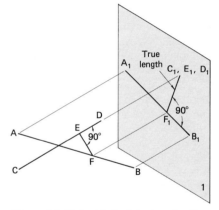

Figure 5.1 Pictorial sketch showing the distance (E_1F_1) between two skew lines. Plane 1 is perpendicular to line *CD*.

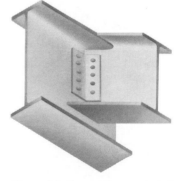

Figure 5.2 Connection angle between two I beams.

line of intersection of the two planes as a point. In this view, the two planes will appear as straight lines. The angle between the lines (edge views of the planes) is the required angle. See Figure 5.3 which, in general, typifies the solution pictorially. *Observe that, although the last problem appears to be quite different from the clearance problems first cited, the analyses of the problems reveal that their solutions are basically the same.*

Thorough mastery of the fundamental principles of orthogonal projection, development of analytic (thinking) power, and reasonable proficiency and accuracy in recording the graphical solutions to the cited problems, as well as similar problems that arise in engi-

neering design, are essential to your progress toward an engineering career.

What do we mean by orthogonal projection? Before we answer this question, we first consider the elements that are common to the systems of projection: (1) central or perspective projection; and (2) parallel projection.

The common elements are a plane of projection (picture plane), a point of sight, and a given object.

Central or Perspective Projection

Let us observe Figure 5.4, which shows an object, a plane of projection ("picture plane"), and a point of sight, or station point, S.

Suppose that a line (projector), r, is drawn from the station point, S, to some point, P, of the object. The intersection, P', of line r with the "picture plane" is the "perspective projection" of point P. If this process of projection was repeated for the other corners of the object, the lines joining the projections of these points would form the "projection of the object," in this case *a central or perspective projection.*

The three-dimensional effect, shown in perspective, results in a "pictorial" view. It lends itself to an easy understanding of the object.

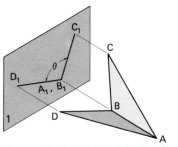

Figure 5.3 Pictorial sketch showing the angle, θ, between planes *ABC* and *ABD*. Plane 1 is perpendicular to edge *AB*, the line of intersection of the two planes.

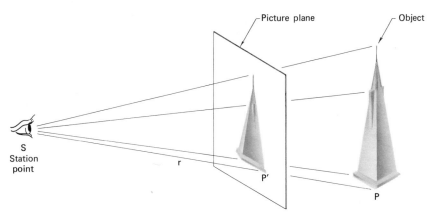

Figure 5.4 Central or perspective projection.

On the other hand, it contains distortions of linear and angular magnitudes, the true values of which may be necessary for computation, design, and production.

Parallel Projection—General Case

When the station point, S, is an infinite distance from the object, the projectors will be parallel to each other, as shown in Figure 5.5. Lines joining the points of intersection of these parallels with the plane of projection will form a view having a three-dimensional effect. If the object is oriented so that certain surfaces are parallel to the plane of projection, their true shapes will be revealed. Although this may be advantageous, it should be recognized that certain distortions still exist (i.e., elements not parallel to the plane of projection will not be shown in their true magnitudes).

Parallel Projection—Special Case, Orthogonal

When the parallel projectors are oriented at right angles to the "picture plane," the resulting projection is known as an orthogonal projection. This is the system that is most frequently employed in technical fields, because it gives accurate and complete information. It is true that in a number of relatively simple cases complete information can be given in a pictorial drawing; nevertheless, it has been found that in most instances, orthographic solutions are best adapted to engineering practice.

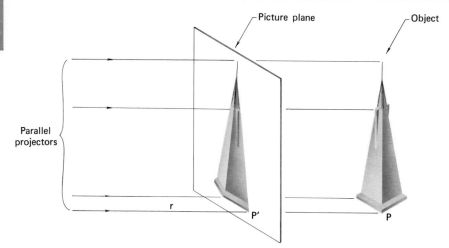

Figure 5.5 Parallel projection —general case.

Orthographic Views of an Object

EXAMPLE 1

Let us consider the object shown in Figure 5.6. Suppose we introduce a reference plane, F, known as the *frontal plane*, parallel to surface A of the object (Figure 5.7). Perpendiculars from the various corners of the object will intersect plane F in points that are properly connected to form the *front view* of the

Figure 5.6 Pictorial of the selected object.

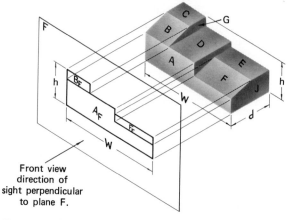

Figure 5.7 Pictorial of the selected object (Figure 5.6) and its orthographic view on plane F, which is parallel to surface A.

object. *It is the view of the entire object, obtained by viewing the object in a direction of sight, that is perpendicular to the F plane.* Orienting the plane parallel to surface A enables us to obtain a front view that includes the true shape of the surface. Note carefully that the height, h, and the width, w, are available in this view. Depth, d, however, cannot be seen in the front view. To disclose the depth, d, we will introduce the *horizontal reference* plane, H, which is perpendicular to the F reference plane, as shown in Figure 5.8. Perpendiculars from the salient corners of the object to the H plane will intersect that plane in points that are properly connected to form the *top view* of the object. *This view is obtained by viewing the object in a direction of sight that is perpendicular to the* H *plane.* Again, note carefully that the width, w, and the *depth*, d, are available in this view. In addition, observe that the top view includes the true shapes of surfaces C, D, and E. Why is this true?

Let us now revolve the H plane about the line of intersection of the H and F planes, so that the two views, H and F, lie in one surface, as shown in Figure 5.9. Since the sizes of the reference planes are arbitrary, we can omit their boundary lines and simply show the H and F views of the object, as drawn in Figure 5.10. If we were given these two views, would we have sufficient information to describe the shape of the object? Offhand, we would be inclined to say yes; however, it is possible to have other objects that would have the same top and front views. For example, one object could look like the pictorial shown in Figure 5.11; another like the one shown in Figure 5.12.

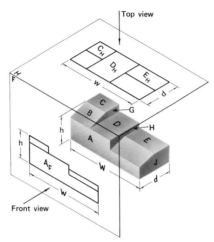

Figure 5.8 Pictorial showing the views of the object on planes H and F, which are perpendicular to each other.

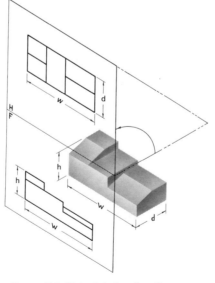

Figure 5.9 Pictorial showing the revolved position of plane H.

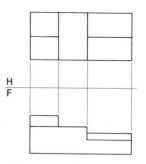

Figure 5.10 The orthographic views of the object on reference planes H and F.

Figure 5.11 One possible interpretation of the views in Figure 5.10.

Figure 5.12 Another possible interpretation of the views in Figure 5.10.

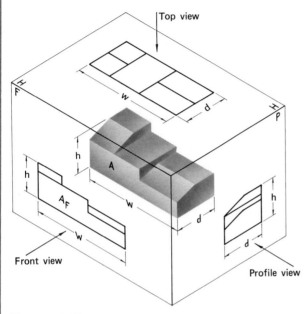

Figure 5.13 Pictorial showing the views of the object on planes *H*, *F*, and *P*.

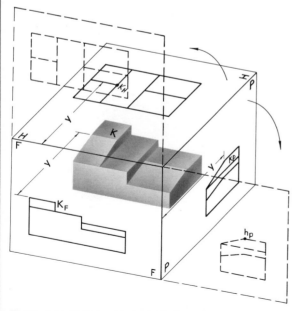

Figure 5.14 Flat-sheet representation of the three views shown pictorially.

To make certain that the intent of the design-engineer is disclosed *without ambiguity*, we add another view, in this case a *profile view*. In Figure 5.13 a profile reference plane, *P*, has been added on which the *profile view* of the object is shown. It should be observed that the *P* plane is perpendicular to both the *H* and *F* reference planes. The flat-sheet representation of the three views is shown pictorially in Figure 5.14 and orthographically in Figure 5.15.

Basic Relationships

Now let us *carefully study* the relationship of the *H* plane to the *F* plane, the relationship of the *P* plane to the *F* plane, and of any point of the object, such as *K*, to the three reference planes. *Please note the following observations.*

1. Planes *H* and *F* are perpendicular to each other. (These are *adjacent* planes.)

2. Planes *P* and *F* are perpendicular to each other. (These are adjacent planes.)

3. Planes *H* and *P* are each perpendicular to plane *F*; therefore:

(a) *The distance, y, that point* K *is behind the* F *plane will be seen twice, once in the top view,* H, *and again in the profile view,* P.

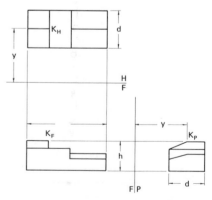

Figure 5.15 Flat-sheet representation of the views shown orthographically.

(b) In the flat-sheet representation the top and front views of point K lie on a line that is perpendicular to the line of intersection of the H and F planes.

(c) In the flat-sheet representation the front and profile views of point K lie on a line that is perpendicular to the line of intersection of the F and P planes.

The full significance of these observations will be realized as we consider other problems later. At this point, however, let us return to Figure 5.15 and consider the meaning of lines marked $\frac{H}{F}$ and $F|P$.

First, with respect to the horizontal line marked $\frac{H}{F}$, notice that:

1. The top view includes both the top view of the object *and the top or edge view of the F plane*. The latter is represented by the horizontal line marked $\frac{H}{F}$. The distance, y, shows how far point K is *behind* the F plane.

2. *When we observe the front view, the same horizontal line*, H, *represents the front or edge view of the* H *plane*.

Second, with respect to the vertical line marked $F|P$, observe that:

1. The front view includes both the front view of the object *and the front, or edge view of the P plane*.
2. When we look at the profile plane, P, the same vertical line represents the profile view of the F plane; therefore, the *distance, y,* again shows how far point K is behind the F plane.

Alternative Position of the Profile View

Let us study the pictorial shown in Figure 5.16. Observe that the P plane was first rotated into a position coincident with the H plane, and then coincident with the F plane. The orthographic representation of the views (without the boundaries of the reference planes) is shown in Figure 5.17.

Again, we note the following relationships.

1. Planes H and F are mutually perpendicular. (These are adjacent planes.)
2. Planes H and P are mutually perpendicular. (These are adjacent planes.)
3. Planes F and P are perpendicular to plane H.
4. Therefore, the distance, Z, that point K is below the H plane is seen twice, once in the front view and again in the profile view.

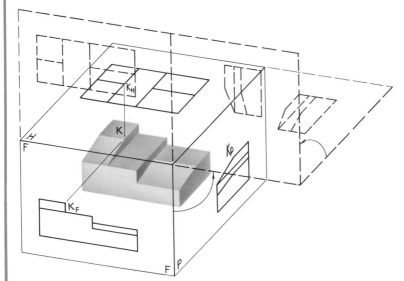

Figure 5.16 Alternative position of the profile views shown pictorially.

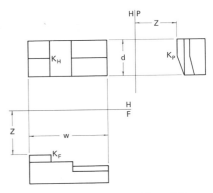

Figure 5.17 Alternative position of the profile view shown orthographically.

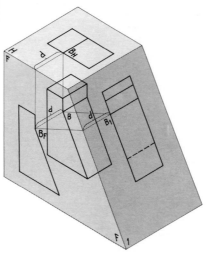

Figure 5.18 Pictorial of object and its
orthographic views on planes *H*, *F*, and
1, which is parallel to the inclined
surface.

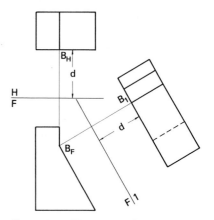

Figure 5.19 Flat-sheet representation of
the orthographic views of the object.

EXAMPLE 2

Now, let us consider the object shown pictorially in Figure 5.18, and its orthographic representation in Figure 5.19.

Again, it should be observed that:

1. Planes *H* and *F* are perpendicular to each other. (These are adjacent planes.)

2. Planes *F* and 1 are perpendicular to each other. (These are adjacent planes.)

3. *The* H *and* 1 *planes are each perpendicular to the* F *plane.*

4. *Therefore, the distance,* d, *that any point such as* B *is behind the* F *plane will be seen twice: once in the top view,* H, *and again in the supplementary view,* 1.

5. In the orthographic representation (Figure 5.19) the top and front views of point *B* lie on a line that is perpendicular to the horizontal line, which represents the intersection of the *H* and *F* planes.

6. In the orthographic representation the front and supplementary views of point *B* lie on the perpendicular to the inclined line, which represents the intersection of the

front (*F*) and supplementary (1) planes.

7. Since the *H* plane and the supplementary plane, 1, are perpendicular to the *F* plane, the distance, *d*, that any point such as *B* is behind the *F* plane will be seen twice, once in the top view, and again in the supplementary view on plane 1.

The Two Fundamental Principles of Orthogonal Projection

The two fundamental principles of orthogonal projection may now be stated as follows.

1. *In a two-dimensional representation the orthographic views of a point on two mutually perpendicular planes lie on a line that is per-*

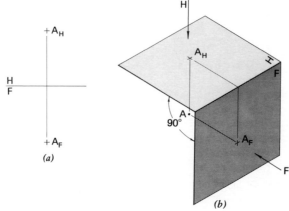

Figure 5.20 In the two-dimensional representation, (*a*) the orthographic views of point *A*, shown pictorially in Figure 5.20 (*b*), lie on a perpendicular to the line of intersection of planes *H* and *F*.

pendicular to the line of intersection of the two planes (see Figures 5.20 and 5.21).

2. *When two planes are perpendicular to a third plane, the distance that a point is from the third plane will be seen twice, once in each of the views on the other two planes* (see Figures 5.22 and 5.23).

The relationships among the views in Figures 5.20 to 5.23 are consistent with the two fundamental principles. To strengthen your understanding and use of the two principles further, consider the problems in the following examples.

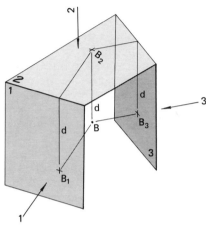

(a) Pictorial

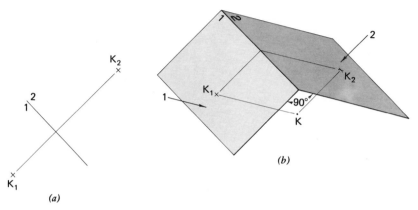

(b)

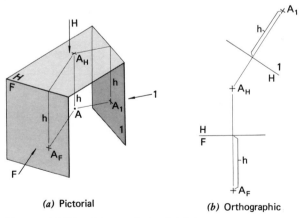

(a)

Figure 5.21 In the two-dimensional representation, (*a*) the orthographic views of point *K*, shown pictorially in Figure 5.21 (*b*), lie on a perpendicular to the line of intersection of planes 1 and 2.

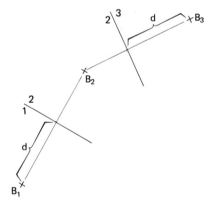

(b) Orthographic

Figure 5.23 The distance that point *B* is from plane 2 is seen twice, once in the view on plane 1 and again in the view on plane 3.

(a) Pictorial

(b) Orthographic

Figure 5.22 The distance that point *A* is below plane *H* is seen twice, once in the view on plane *F* and again in the view on plane 1.

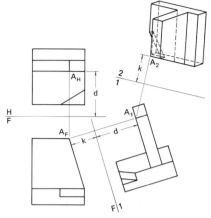

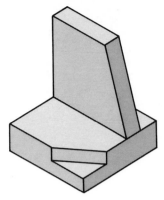

Figure 5.24 Application of the two fundamental principles of orthogonal projection.

Figure 5.25 Pictorial of object in Figure 5.24.

EXAMPLE 1

Suppose we wish to obtain the views of the object shown in Figure 5.24 on supplementary planes 1 and 2. We can readily understand what the object looks like from the pictorial drawing shown in Figure 5.25. Let us select point A, one of the salient corners of the object. The H and F views of A are easily identified in the orthographic representation in Figure 5.24.

Now the view of point A on plane 1 lies on the perpendicular drawn through A_F to the line of intersection of the two mutually perpendicular planes F and 1. *This step is an application of the first fundamental principle of orthogonal projection.*

We observe that planes H and F are mutually perpendicular, and that planes F and 1 are also mutually perpendicular; therefore, planes H and 1 are perpendicular to plane F. Hence, the distance, d, that point A is behind plane F is seen twice, once in the top view, H, and again in the view on supplementary plane 1. Since distance, d, is available in the top view, H, we can lay off this distance, as shown in Figure 5.24, to locate the view A_1. *This is an application of the second fundamental principle of orthogonal projection.*

In order to locate the view A_2 we again make use of the two fundamental principles. Since planes 1 and 2 and planes F and 1 are mutually perpendicular, it is quite clear that planes F and 2 are each perpendicular to plane 1; therefore, the distance that point A is from plane 1 is seen twice, once in the F view, distance k, and again in the

view on plane 2, thus locating the position of view A_2. Of course, we know from the first fundamental principle that views A_1 and A_2 lie on the perpendicular drawn through A_1 to the line of intersection of the two planes 1 and 2.

Similarly, we can obtain the views of additional points to complete the views of the object on planes 1 and 2, as shown in Figure 5.24.

Note that in locating the view A_1, we were concerned with the views on planes H and F, and that

in locating the view A_2, we were concerned with the views on planes F and 1. In each case we were concerned with only *three views*. Each set of three views consisted of two views, which were on planes that were perpendicular to the plane that showed the other view.

Stated simply, the relationship among the views on planes H, F, and 1 is basically the same as the relationship among the views on planes F, 1, and 2.

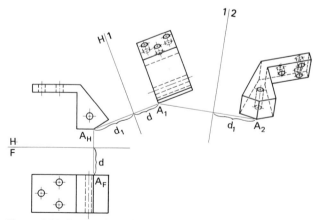

Figure 5.26 Application of the two fundamental principles of orthogonal projection to supplementary views on planes 1 and 2.

EXAMPLE 2

Let us obtain the views of the object shown in Figure 5.26 on supplementary planes 1 and 2. We assume that the H and F views are given.

Now let us analyze the problem of locating the view of the object on plane 1.

We observe that:

1. Planes H and F are mutually perpendicular.
2. Planes H and 1 are mutually perpendicular.
3. Therefore, planes F and 1 are perpendicular to plane H.

Employing the *first fundamental principle* of orthogonal projection, we know that A_F and A_H lie on the perpendicular to the line marked $\frac{H}{F}$ and, similarly, A_H and A_1 lie on the perpendicular to the line marked $H|1$. Employing the *second fundamental principle*, we know that the distance, "d", that point A is below the H plane will be seen again in the view on plane 1. Therefore, since we have distance, d, available in the front view, we can lay off this distance to locate A_1.

Note that we were concerned with only three planes and the views on these three planes: the two that were given or assumed (H and F) and the view on plane 1, the view we wished to obtain.

Let us continue the analysis to locate A_2. Again, we observe that:

1. Planes H and 1 are mutually perpendicular.
2. Planes 1 and 2 are mutually perpendicular.
3. Therefore, planes H and 2 are perpendicular to plane 1.

Employing the *first fundamental principle*, views A_1 and A_2 lie on the perpendicular to the line marked 1/2. Employing the *second*

fundamental principle, we know that the distance, "d_1," that point A is from plane 1 will be seen twice, once in the top view, H, and again in the supplementary view on plane 2. Since distance, "d_1," is available in the H view, we can easily lay off this distance, as shown in the figure, to locate A_2. *We should note again that we were concerned with only three planes* (and the views on them): the two, H and 1, and the one we wished to obtain, the view on plane 2. We should recognize that the relationship among the views on planes H, 1, and 2 is basically the same as the relationship among the views on planes H, F, and 1.

To visualize the relationship of the views on planes H, 1, and 2, we can rotate the entire Figure 5.26 to a position that orients the line marked $H|1$ horizontally, and then regard the views H and 1 as the given views, and finally obtain the view on plane 2 (see Figure 5.27).

Visibility

We observe that some of the edges of the objects shown in Figures 5.24 and 5.26 are not visible in the supplementary views and that these edges are represented by dashed lines.

How shall we *rationally* determine which edges of an object are visible and which are hidden?

Let us consider the following examples.

EXAMPLE 1

The H and F views of lines m and n are shown in Figure 5.28. Which of the two lines is above the other and which is in front of the other?

Let us concentrate on the F view. The apparent intersection of m_F and n_F is the front view of two points, point 1 on line n and point 2 on line m. The top view of these

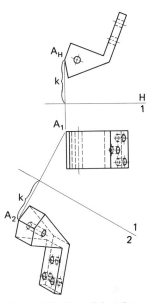

Figure 5.27 Visualizing the relationship of the views on planes H, 1, and 2 in Figure 5.26.

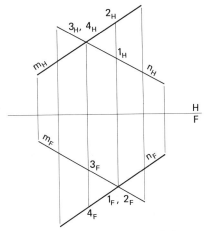

Figure 5.28 Determining the relative positions of skew lines m and n. Line m is above line n. Line n is in front of line m.

points is shown as 1_H and 2_H, respectively. As we observe the front view, we note that point 1 is in front of point 2. This is evident from the top view of the two points. Therefore, line n, which contains point 1, is in front of line m. We can similarly, analyze the top view of lines m and n. The apparent intersection of m_H and n_H is the top view of two points, point 3 on line m and point 4 on line n. The top view of these points is shown as 3_H and 4_H, respectively. As we observe the top view, we note that point 3 is above point 4. This is evident from the front view of the two points. Therefore, line m, which contains point 3, is above line n.

This analysis is the key to the determination of visibility. Applications of the analysis are discussed in the following examples.

EXAMPLE 2

The top and front views of pyramid $O-ABC$ are shown in Figure 5.29. The edges in question as to visibility are shown as light solid lines.

Let us examine the front view carefully (Figure 5.30). The apparent intersection of edges OA and BC is actually the front view of two points, one on edge BC and the other on edge OA. Let us designate these points as 1 and 2, respectively. The front view of these two points is shown as 1_F and 2_F. Now, note very carefully that 1_H is on $B_H C_H$ (since point 1 is on edge BC) and that 2_H is on $O_H A_H$. We see that point 1 is in front of point

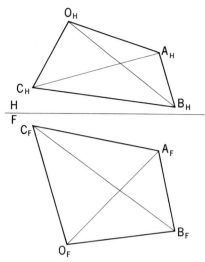

Figure 5.29 Pyramid *O-ABC* is shown with edges *OA* and *BC* drawn lightly. We need to determine which of the two edges is visible in the *F* view and in the *H* view.

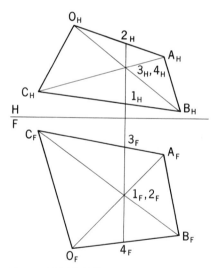

Figure 5.30 Visibility analysis of the pyramid in Figure 5.29.

2; therefore, *in observing the front view*, point 1 is closer to the observer than point 2 is. This mean that edge BC, which contains point 1, is the visible edge and that edge OA is invisible; hence, $B_F C_F$ is shown as a solid line and $O_F A_F$ is shown as a dashed line (Figure 5.31).

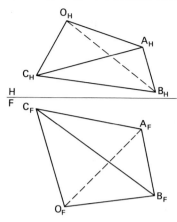

Figure 5.31 Results of the analysis (Figure 5.30) shows the correct visibility of edges *OA* and *BC*.

Now let us examine the top view (Figure 5.30). The apparent intersection of $A_H C_H$ and $O_H B_H$ is actually the top view of two points, 3 and 4, one on edge AC and the other on edge OB. From the top view we can easily locate 3_F (on $A_F C_F$, since point 3 is on edge AC) and 4_F (on $O_F B_F$, since point 4 is on edge OB). The front view now clearly shows that point 3 is above point 4; therefore, *in observing the top view*, point 3 is closer to the observer than point 4 is. This means that edge, AC, which contains point 3, is the visible edge and that edge OB is the invisible edge; hence, $A_H C_H$ is shown as a solid line and $O_H B_H$ is shown as a dashed line. Figure 5.31 shows the completed two views.

EXAMPLE 3

Let us consider the object shown in Figure 5.32. We will assume that the visibility has been determined in the top and front views. Our problem is to determine the visibility in supplementary views 1 and 2. We will first consider supplementary view 1. The apparent intersection of edges OA and BC in this view is actually the supplementary

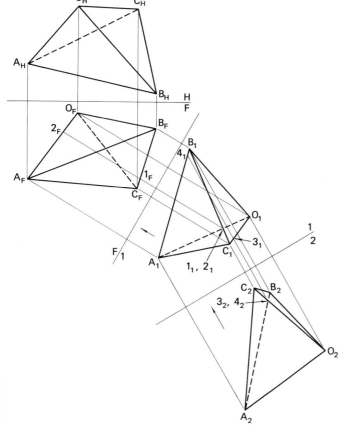

Figure 5.32 Visibility determination of pyramid *O-ABC* as seen in the views of planes *H*, *F*, 1, and 2.

direction of the arrow, edge *OC*, which contains point 3, is closer to us than is point 4; therefore, edge *OC* is visible and edge *AB* is invisible.

EXAMPLE 4

Let us consider the pyramid *O−ABC* shown in Figure 5.33. We know that the outside edges *AB*, *BC*, and *CA* are visible. What about the edges *OA*, *OB*, and *OC* as seen in the top view? Evidently there is no apparent intersection of edges in this view that might be compared with the previous examples. How shall we apply the method of analysis used in the previous examples? Suppose edge *OC* is extended so that its top view crosses the top view of edge *AB*.

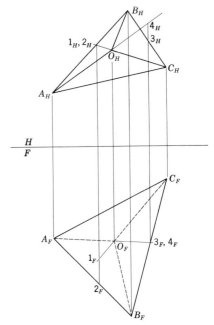

Figure 5.33 Visibility determination of pyramid *O-ABC*.

view of two points, one on edge *BC* (point 1) and the other (point 2) on edge *OA*. The front view of these two points, 1_F and 2_F, is easily located. When we look at the front view we see supplementary plane 1 as a line, and we observe that point 1 is closer to supplementary plane 1 than point 2 is. When we look at supplementary plane 1, in the direction of the arrow, we see that edge *BC*, which contains point 1, is closer to us than edge *OA*. This means that edge *BC* will be visible and that edge *OA* will be invisible. Note that the method of analysis used to determine the visibility in

the supplementary view is basically the same as the analysis used in the first two examples.

Now let us consider supplementary view 2. The apparent intersection of edges *OC* and *AB* is actually supplementary view 2 of two points, one (point 3) on edge *OC*, and the other (point 4) on edge *AB*. The views of these two points are shown as 3_2 and 4_2 on plane 2 and as 3_1 and 4_1 on plane 1. When we look at supplementary view 1, we see the edge view of plane 2 and we observe that point 3 is closer to plane 2 than point 4 is. This means that when we look at plane 2, in the

The *apparent intersection* of these two lines is the top view of two points, 1 and 2. The front view of the two points is shown as 1_F (on $O_F C_F$ extended) and 2_F (on $A_F B_F$). It is evident that the front view shows that point 1 is above point 2; therefore, in looking down on the pyramid, edge OC (extended) which contains point 1, will be visible in the top view, and certainly edges OB and OA will also be visible in the top view. We may similarly determine the visibility in the front view. You should have no difficulty in doing so.

EXAMPLE 5

Let us now consider triangle ABC and line m shown in Figure 5.34. We will assume that the triangle is opaque (not transparent) and that point P, the intersection of line m with the triangle, is known. Our problem is to determine which portion of m_H is visible and which portion of m_F is visible.

Consider the top view. The apparent intersection of line m with side BC is the top view of two points, one on line m (point 1) and the other on side BC (point 2). The front view of these two points, 1_F and 2_F, clearly shows that point 1 is above 2. Thus, when we look down (top view), line m, which contains point 1, is above side BC. Therefore, the portion of m_H from point 1 to point P is visible, and certainly the portion from P to side AC is invisible.

Now let us consider the front view. The apparent intersection of line m and side AB is the front view of two points, one on line m (point

3) and the other on side AB (point 4). The top view of these two points clearly shows that point 3 is in front of point 4. Thus, in looking at the front view, line m, which contains point 3, is in front of side AB. Therefore, the portion of m_F from point 3 to point P is visible and the portion from P to side AC is invisi-

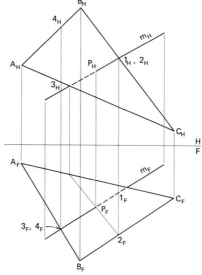

Figure 5.34 Visibility determination of line m, which intersects opaque triangle ABC.

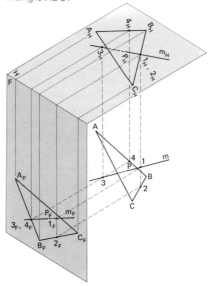

Figure 5.35 Pictorial study of visibility—line m and triangle ABC.

ble. A pictorial is shown in Figure 5.35.

EXAMPLE 6

Let us consider the sphere, center O, and line m shown in Figure 5.36. We will assume that the sphere is opaque and that the intersections, points P and Q, of the line with the sphere have been located. Our problem is to determine the visibility of line m.

Let us study the front view. The apparent intersections of line m and the circle are the front view of points 1, 2, 3, and 4. Points 1 and 2

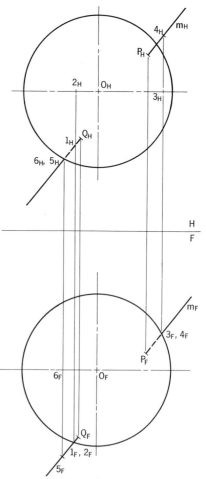

Figure 5.36 Visibility determination of line m, which intersects sphere O.

are, respectively, on line m and the great circle, which is parallel to the F plane. Point 1 is in front of point 2; therefore, line m, which contains point 1, is in front of the great circle. This means that the portion of m_F between 1_F and Q_F is visible. Points 3 and 4 are, respectively, on the great circle and line m. Point 3 is in front of point 4; therefore, the great circle, which contains point 3, is in front of the line. This means that the portion of m_F between 4_F and P_F is hidden and is, therefore, represented by a dashed line. The analysis for the visibility of m_H is the same. For example, points 5 and 6 are, respectively, on line m and the great circle, which is parallel to the H plane. Point 5 is below point 6; therefore, line m, which contains point 5, is below the great circle. This means that the portion of m_H between 5_H and Q_H is hidden and is, therefore, represented by a dashed line. You should experience no difficulty in verifying the visibility of the other portion of m_H.

INTERPRETING ORTHOGRAPHIC DRAWINGS

Design engineers must be capable of conveying their *design intent without ambiguity*, and those with whom they communicate—engineers, technicians, and production personnel—must be able to "read" the orthographic design drawings correctly.

We have had some experience in freehand sketching, the use of the fundamental principles of orthogonal projection, and the determination of visibility. Let us now apply this knowledge to enhance our ability to interpret orthographic views, which describe the shape of an object.

We try to form a mental image of the shape by "reading" the views,

and we then record the image by making a freehand pictorial sketch. Once that is done, we can verify the designer's intent as represented by the orthographic views. The pictorial sketch may be an isometric, an oblique, or a perspective. In Chapter 2, reference was made to the use of isometric and perspective grid sheets. Examples of orthographic, isometric, oblique, and perspective sketches were included.

For our purpose in interpreting orthographic views, we will use isometric grids and, in some cases, coordinate grid sheets for oblique sketches. Pads, ($8\frac{1}{2} \times 11$ in.), with 30° grid lines are available for freehand isometric sketching. Let us now study several examples of the interpretation of orthographic drawings.

EXAMPLE 1

Let us consider the object shown orthographically in Figure 5.37. At the outset, let us carefully "read" each of the views and then try to form a mental image of the object.

As an initial step in recording the mental image, it is a good proce-

dure to first sketch the "rectangular block," the dimensions of which are w, d, and h, or lengths that are proportional to those dimensions. See Figure 5.38a.

Second, sketch the lines of each orthographic view on the corresponding surfaces of the total block, as shown in Figure 5.38b.

Third, sketch the lines of intersection where surfaces meet, as shown in Figure 5.38c.

Finally, strengthen the edges of

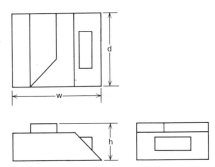

Figure 5.37 Orthographic views of given object.

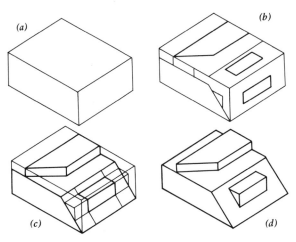

Figure 5.38 Steps in the interpretation of the object in Figure 5.37.

the object and the lines of intersection, previously located, as shown in Figure 5.38d. Now check the isometric sketch with the orthographic views to verify the interpretation. The light lines may be left in if they do not detract from the clearness of the final pictorial.

EXAMPLE 2

Consider the object shown orthographically in Figure 5.39. Again, study the three views and try to form a mental image of the object. You should encounter no difficulty in observing that the base of the object is a rectangular shape having dimensions w, d, and h.

Now suppose that the portion of the object above the base consists of the four sloping surfaces A, B, C, and D, and a horizontal top surface E, as shown in Figure 5.40a. We note, however, that there is a "cutout" portion, seen in the front view, that consists of two parallel vertical surfaces of height, k, and a horizontal bottom surface of

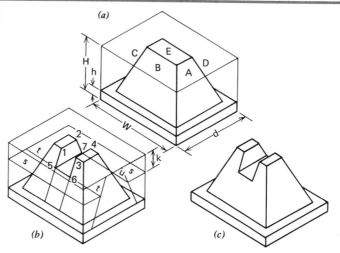

Figure 5.40 Steps in the interpretation of the object in Figure 5.39.

width, n. The profile view (right side) shows the other dimension, m, of this surface and also shows that the vertical surfaces are trapezoids.

Let us now locate the cutout portion in the isometric sketch (Figure 5.40b). In plane E we can easily establish the top edges 1–2 and 3–4 of the vertical surfaces. Since we know the height, k, of the cutout, we can locate the plane, S, which contains the bottom surface of the cutout. The lines of intersection, t and u, of plane S with sloping surfaces B and D are easily established from the profile view. Points 5 and 6, on line t, are readily located, since we know distances r and n, shown in the front view. Points 7 and 8 can be located in the same way. The final isometric sketch is shown in Figure 5.40c. It should be pointed out that what we did was to "read" the orthographic views first to form a mental image of the general shape of the object; then we "read" *again* to visualize the cutout portion. In the interpretation of design drawings of

more involved parts, or units, the "reader" can make good progress by using this "clinical" approach— read to get a general, overall impression of the object, and read again and again to fill in the details.

EXAMPLE 3

Let us consider the derrick shown orthographically in Figure 5.41.

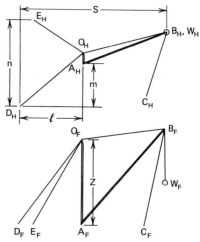

Figure 5.41 Orthographic views of a derrick.

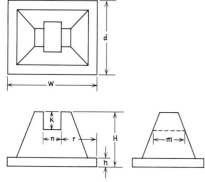

Figure 5.39 Freehand orthographic views of given object.

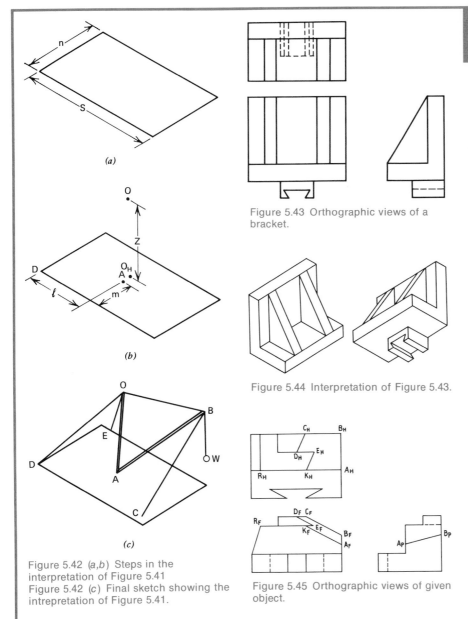

Figure 5.43 Orthographic views of a bracket.

Figure 5.44 Interpretation of Figure 5.43.

Figure 5.42 (*a,b*) Steps in the interpretation of Figure 5.41

Figure 5.42 (*c*) Final sketch showing the interpretation of Figure 5.41.

Figure 5.45 Orthographic views of given object.

metric sketch of the horizontal rectangle that contains the top view of all of the points, as shown in Figure 5.42*a*. Reading the front view, we observe that points, *D, E, A,* and *C* lie in the plane of the rectangle. Point *D* is easily located in the isometric sketch. Using *D* as an origin, we can locate points *E, A,* and *C.* For example, coordinates ℓ and m establish the location of point A. In a similar manner we can locate E and C. Now to locate point O, we first locate O_H, establish a vertical through O_H, and lay off distance, Z. Points B and W are established in the same way (see Figure 5.42*b*).

Finally, mast *OA,* boom *AB,* load *OW,* and cables *OD, OE, OB,* and *BC* are drawn as shown in the isometric sketch (Figure 5.42*c*).

EXAMPLE 4

Let us consider the *bracket* shown orthographically in Figure 5.43. It is not too difficult to form a mental image of the bracket once we have studied the three views carefully. An isometric sketch of the bracket, as shown in Figure 5.44*a*, does not reveal the dovetail feature. We can, however, *reverse the isometric axes* and make a sketch, as shown in Figure 5.44*b*, to show the dovetail feature.

EXAMPLE 5

Let us consider the object shown orthographically in Figure 5.45. Try to visualize (mental image) the object by reading the three views. We could start the interpretation by "blocking in" the entire object,

This derrick is the type that handles cargo aboard ships. Mast AO is slightly inclined so that the boom, AB, tends to swing toward its lowest position when loaded. Therefore, cable BC must be used to overcome this tendency and to position the boom. Careful study

of the two views will help us form a mental image of the mast, boom, cables, and load W. How shall we proceed to make an isometric sketch that satisfies the orthographic views? We can locate each of the points by the "coordinate" method. Let us first prepare an iso-as shown in Figure 5.46*a*, and then sketching in the lower portion, which contains the dovetail feature. Now we can read the views again and concentrate on the shape of the front vertical surface, defined by points *A*, *K*, and *R*. A sketch of the vertical surface is shown in Figure 5.46*b*. Now we can add the sloping line, *AB*, shown in the profile view. Since *BC* is parallel to *AK*, we can locate *BC*; and since *KE* is parallel to *CD*, we can locate *CD*. Finally, we can complete the isometric sketch as shown in Figure 5.46*c*. Some of you will be able to interpret the orthographic views without much difficulty.

EXAMPLE 6

Let us try to interpret the orthographic views of the object shown in Figure 5.47. Formulating a mental image of the object might be a bit frustrating. Perhaps we can start by simplifying the object. Suppose we eliminate (temporarily) the diagonal line of each view. Now we most likely can read the views and obtain an interpretation as represented in the isometric sketch shown in Figure 5.48. We note, however, that line *AB*, as shown in the sketch, is meaningless unless we realize that it represents the intersection of two plane surfaces. What two surfaces? Perhaps a surface such as *ABCD*, and the vertical surface that contains *AB* (see Figure 5.49). This solution, however, does not satisfy the three orthographic views. Moreover, we would have difficulty in justifying lines *BE* and *EC*, which

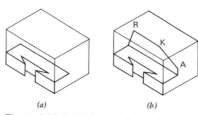
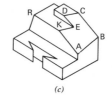
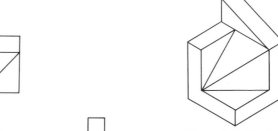

Figure 5.46 (*a*,*b*) Steps in the interpretation of Figure 5.45
Figure 5.46 (*c*) Final sketch showing the intrepretation of Figure 5.45.

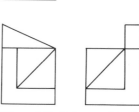

Figure 5.47 Orthographic views of given object.

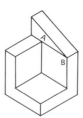

Figure 5.48 Interim intrepretation of Figure 5.47.

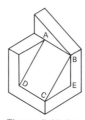

Figure 5.49 Another interim interpretation of Figure 5.47.

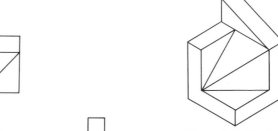

Figure 5.50 Final sketch showing the interpretation of Figure 5.47.

also represent intersections of plane surfaces. Now, if we study the orthographic views and concentrate on the possible interpretation of the diagonal line in each view, in light of the isometric sketch shown in Figure 5.49, we should arrive at the happy conclusion that the object, shown in Figure 5.50, is a valid solution.

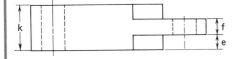

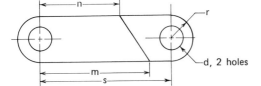

Figure 5.51 Link arm.

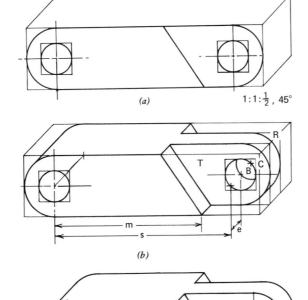

(a)   1:1:½ , 45°

(b)

(c)

Figure 5.52 Steps in the interpretation
of Figure 5.51.

EXAMPLE 7

Let us consider the *link arm* shown in Figure 5.51. Reading the front view, we observe that the ends of the link arm are semicircular, and that there are two circular holes, one near each end of the arm. This suggests that an *oblique* sketch is preferred over an isometric because the circular elements will appear circular. We should have no difficulty in forming a mental image of the link arm once we read the top and front views carefully.

To make the freehand, oblique sketch (1) we can block in the link arm as shown in Figure 5.52a. In the front face of the block, the front surface of the arm, to the left of the inclined portion, can be sketched; and (2) the plane that contains the front face of the tongue portion of the arm is located as shown in Figure 5.52b. In this plane the center, B, for the circle and for the semicircular end is located. In a similar manner, we can locate the plane that contains the rear face of the tongue and center, C, for the circle and circular end. The completed oblique sketch is shown in Figure 5.52c, which can be compared with the orthographic views to make certain that the interpretation is correct.

EXAMPLE 8

Let us study the orthographic views of the *gauge holder* shown in Figure 5.53. It should not be too difficult to form a mental image of the holder. The front view which shows its shape, suggests that an oblique sketch would portray the

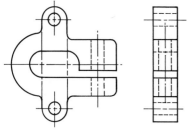

Figure 5.53 Gauge holder (half size).

interpretation of the orthographic views quite adequately.

Start by blocking in the entire piece. We should experience no difficulty in sketching the front face, since it is the same as shown in the front view (see Figure 5.54a). Now we can locate the center for the large semicircle of the back surface by locating point A, which is two units from center O. Similarly, centers B and C can be located, as shown in Figure 5.54b. The completion of the oblique sketch can now proceed without too much frustration (see Figure 5.54c).

Every effort should be made to develop your ability to read engineering drawings and record your design ideas. The following exercises will help you to strengthen your power of visualization and enhance your recording facility.

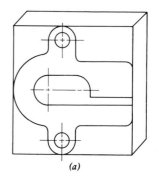

(a)

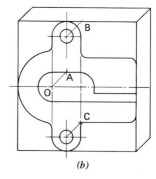

(b)

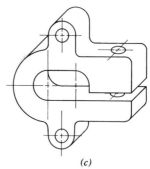

(c)

Figure 5.54 Steps in the interpretation of Figure 5.53.

Exercises

1. Locate the views of line segment *AB* on supplementary planes 1 and 2 as shown in Figure E-5.1. Place planes 1 and 2 as oriented in the figure.

2. Locate the views of surface *ABC* on supplementary planes 2 and 3 as shown in Figure E-5.2.

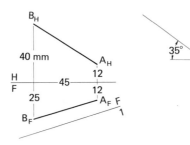

Figure E-5.1

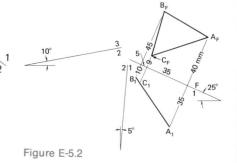

Figure E-5.2

3. Reproduce the views shown in Figure E-5.3. and then add the view on supplementary plane 1.
4. Reproduce the views shown in Figure E-5.4. Determine the correct visibility of the edges of the triangular pyramids in Figure E-5.4a, E-5.4b, E-5.4c, and E-5.4d.
5. Reproduce the views shown in Figure E-5.5. In Figure E-5.5a, E-5.5b, and E-5.5c, point P is the intersection of line t with the plane surface as shown.

Determine the correct visibility of line t, assuming that the plane surface is opaque.
6. Points P and Q are the intersections of line t with pyramid X-UVW as shown in Figure E-5.6. Determine (a) the correct visibility of the edges of the pyramid, and (b) of the line t.
7. In Figure E-5.7a and E-5.7b, determine the correct visibility of the intersecting plane surfaces.

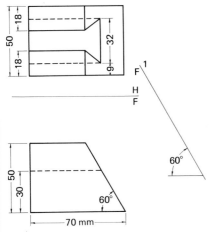

Figure E-5.3

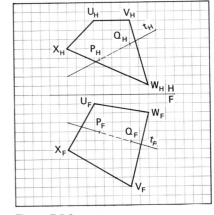

Figure E-5.6

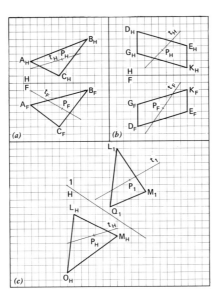

Figure E-5.5

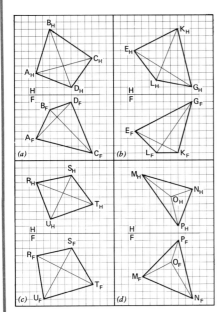

Figure E-5.4

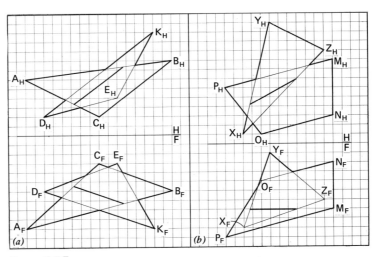

Figure E-5.7

10. Prepare *freehand* pictorial sketches of the pieces shown in Figures E-5.11 and E-5.12. Study the views in each case and then decide whether or not to make an isometric, oblique, or perspective sketch to shown the piece to the best advantage (ease of interpretation).

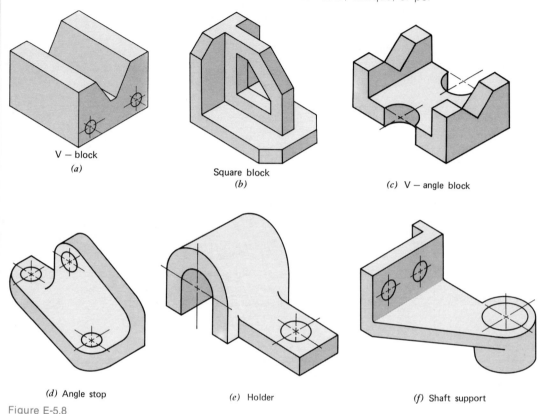

V — block
(a)

Square block
(b)

(c) V — angle block

(d) Angle stop

(e) Holder

(f) Shaft support

Figure E-5.8

8. Prepare *freehand* orthographic views of the objects shown in Figure E-5.8. Use $8\frac{1}{2} \times 11$-in., $\frac{1}{4}$-in. grid sheets (or metric grids if available). Count squares to obtain dimensions.

9. Some lines are missing in one or more views of the objects shown in Figures E-5.9 and E-5.10. Sketch the views of each object (approximate dimensions) on a separate $8\frac{1}{2} \times 11$-in., $\frac{1}{4}$-in. grid sheet, add the missing lines, and prepare *freehand* pictorial sketches of the objects. Use isometric, oblique, or perspective grid sheets in accordance with the choice you believe will best portray each piece. Verify the pictorial interpretation of the orthographic representation.

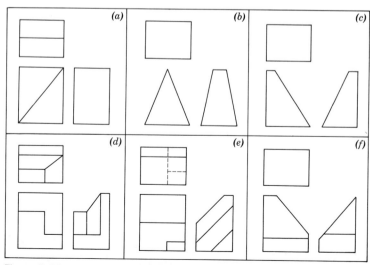

Figure E-5.9

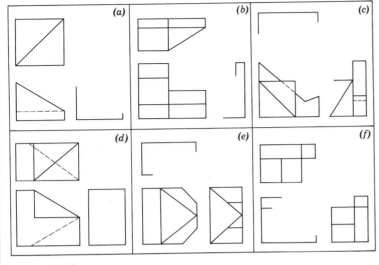

Figure E-5.10

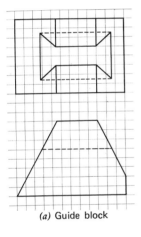

(a) Guide block

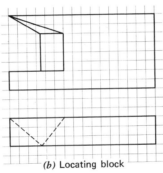

(b) Locating block

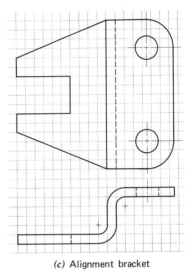

(c) Alignment bracket

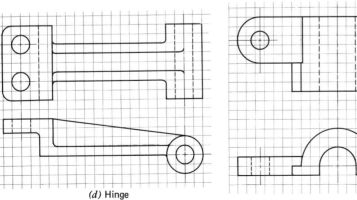

(d) Hinge

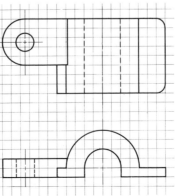

(e) Bearing cap

(f) Duck (the hook has been omitted)

Figure E-5.11

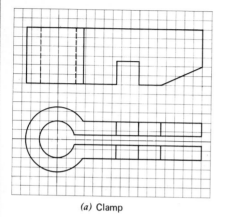

(a) Clamp

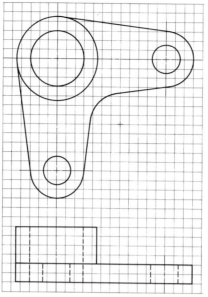

(d) Bell crank

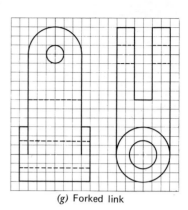

(g) Forked link

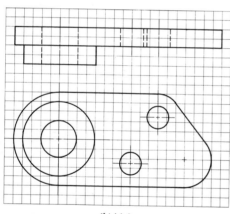

(b) Link

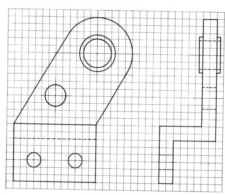

(e) Pulley holder

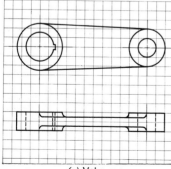

(c) Valve arm

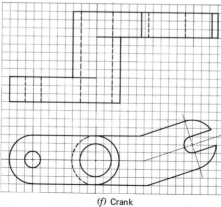

(f) Crank

Figure E-5.12

PART 2
Applied Graphics for Design, Analysis, and Communication

Chapter 6
Applications of the Fundamental Principles of Orthogonal Projection

INTRODUCTION

Now that you have a good grasp of the two fundamental principles of orthogonal projection and some experience in solving problems that strengthened your ability to visualize and to think in three dimensions, it is important to move ahead, to enhance further your analytic ability ("the mental process of thinking through the solution of a problem"), and to apply the fundamental principles to the graphical expression of the solution.

In Chapter 1 we learned that many design projects include technical problems that require the determination of the true lengths of members (e.g., rods, bars, pipes, and cables), the angles between plane surfaces, clearances between cables, wires, structural members, etc., the true shapes of surfaces, and many other similar problems.

The analysis and solution of such problems are not difficult once we thoroughly understand the solution of the *four basic problems*.

1. To find the true length of a line segment.
2. To find the point view of a line.
3. To find the edge view of a plane.
4. To find the true shape of a plane surface.

We now consider each of these problems.

TRUE LENGTH OF A LINE SEGMENT—FIRST BASIC PROBLEM

The true length of a line segment will be seen in the view on a plane that is parallel to the line segment.

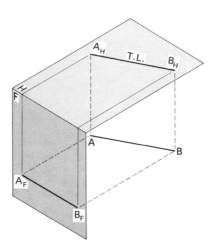

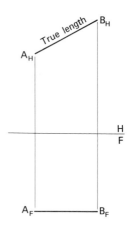

Figure 6.1 True length of line segment that is parallel to the *H* reference plane.

Case 1 The Line is Parallel to a Reference Plane

In Figure 6.1, line *AB* is parallel to the *H* reference plane. This is evident from the fact that the front view of line *AB* is horizontal, indicating that all the points on line *AB* are the same distance below the *H* plane. The top view, $A_H B_H$, therefore shows the true length of line *AB*.

A line that is parallel to the H *reference plane is called a horizontal line.*

In Figure 6.2, line *CD* is parallel to the *F* reference plane. The front view of the line shows its true length. Note that the top view of the line is seen as a line that is parallel to the top view (edge view) of the *F* plane. The top view of line *CD* clearly shows that all its points are the same distance behind the *F* plane.

A line that is parallel to the F *plane is called a frontal line.*

You should experience no difficulty in drawing the *H*, *F*, and *P* views of a line that is parallel to the *P* reference plane. *Try it freehand.*

The three lines, horizontal, frontal, and profile, are classified as *principal lines.*

Case 2 The Line is Not Parallel to a Reference Plane

Supplementary Plane Method Suppose that line segment *AB* is represented by the *H* and *F* views shown in Figure 6.3. It should be quite evident

that neither view shows the true length of the line, since the line is not parallel to either the *H* or *F* reference plane.

Let us introduce a supplementary plane that will be parallel to the line *and perpendicular* to one of the reference planes. Suppose we introduce plane *1*, parallel to line *AB* and perpendicular to the *H* plane (see Figure 6.4a). Careful study of this figure shows that the distances from the *H* plane to each of the points *A* and *B* are seen twice, once in the front view and again in the supplementary view. You will recall the second fundamental principle: *when two planes are perpendicular to a third, the distance that a point is from the third plane will be seen twice, once in each of the views on the other two planes.* In Figure 6.4a

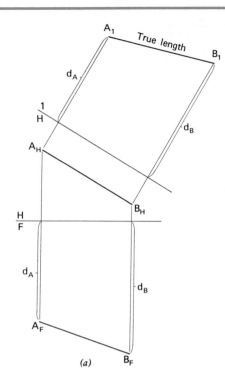

(a)

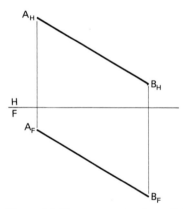

Figure 6.3 Line segment not parallel to either the *H* or *F* reference plane.

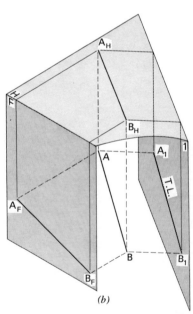

(b)

Figure 6.4 (*a*) True length of line segment is seen in the view on plane 1, which is parallel to the line and perpendicular to the *H* reference plane.
Figure 6.4 (*b*) Pictorial showing the line segment and its view on plane 1.

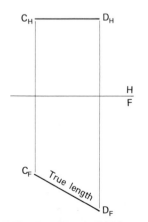

Figure 6.2 True length of line segment that is parallel to the *F* reference plane.

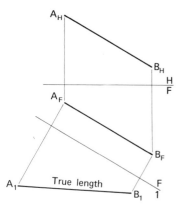

Figure 6.5 True length of line segment
is seen in the view on plane 1, which is
parallel to the line segment and perpendicular to the F reference plane.

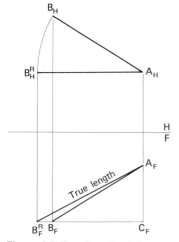

Figure 6.7 True length of line segment
using the method of rotation.

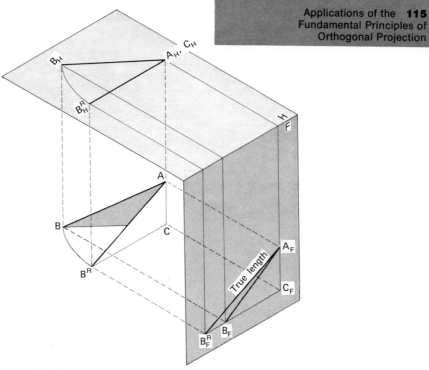

Figure 6.6 True length of line segment
using the method of rotation.

and 6.4*b*, we see that planes *F* and 1
are perpendicular to the *H* plane;
therefore the distances that points *A*
and *B* are below the *H* plane are seen
twice, once in the front view and again
in the supplementary view on plane 1.
We also see that since plane 1 is parallel to line *AB*, *its true length will be
seen in the view A_1B_1.*

It should be recognized that we
could have introduced a supplementary plane parallel to the line and perpendicular to the *F* plane. The view of
the line on the supplementary plane

would disclose the true length of the
line (see Figure 6.5).

Rotation Method. Let us consider the
pictorial shown in Figure 6.6. It is assumed that the given line segment is
AB. Note that neither the top nor front
views of the line shows the true length
of the line. Now let us introduce a vertical line through point *A* and a horizontal line through point *B* to intersect the
vertical line at point *C*, we have formed
the right triangle *ABC*. Let us rotate the
triangle about line *AC* until it is parallel

to the *F* plane. The front view of the rotated triangle will show its true shape
and, hence, the true length of line segment *AB*. The orthographic solution is
shown in Figure 6.7.

EXAMPLE

Suppose we are given the front
view and true length of line segment *AB* as 50 mm. How shall we
determine the top view of segment
AB?

Figure 6.8 shows the front view of line segment AB. In Figure 6.9, $A_F C_F B_F{}^R$ shows the rotated, or true shape, view of triangle ABC. This is similar to Figure 6.7. The location of A_H is arbitrary. The axis of rotation of triangle ABC is line AC. The true length of the base (CB) of the right triangle ABC is equal to the distance from C_F to $B_F{}^R$. Since line CB is horizontal, its top view will show its true length. Therefore, an arc of length C_F to $B_F{}^R$ and center C_H will cut the vertical through B_F in points B_H and B'_H.

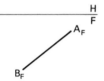

Figure 6.8 Front view of line segment that is 50 mm long.

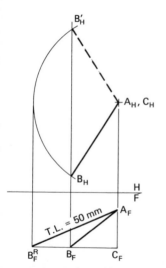

Figure 6.9 The two possible solutions to the problem of determining the top view of the line segment when the front view and the length of the segment are given.

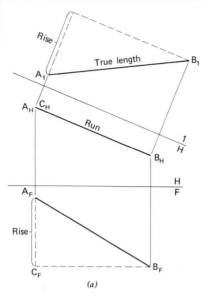

Figure 6.10 (a) Grade and slope of a line. The grade = Rise/Run × 100%. Slope is the angle θ with a horizontal plane.

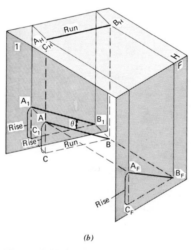

Figure 6.10 (b) Pictorial to show grade and slope of a line. Grade = Rise/Run × 100%. Slope angle $\theta°$.

Lines $A_H B_H$ and $A_H B'_H$ show the two solutions to the problem.

Grade of a line
The grade of a line is the ratio of the vertical displacement or "rise" of two of its points to the horizontal projected length, or "run" of the line segment. The percent of grade is this ratio multiplied by 100.

EXAMPLE 1
Let us consider Figure 6.10, which shows line AB and its views on the H and F planes. Now suppose we draw a vertical line through point A to intersect the horizontal line through point B to form the right triangle ABC.

The percent of grade of line AB is the ratio of AC (rise) to CB (run) times 100, or $AC/CB \times 100$, or $A_F C_F / A_H B_H \times 100$.

The grade of line AB is *negative*, since motion from A to B is downward. The grade of line BA, however, is *positive*, since motion from B to A is upward. The magnitude of the grade is the same in both cases.

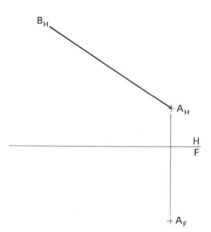

Figure 6.11 Top view of line segment AB and front view of point A are given. The grade of the line is −40%. *Problem:* Locate B_F and draw the front view of line segment AB.

EXAMPLE 2

Suppose we are given the top view of line segment AB, the front view of point A, and the grade of the line as -40% (see Figure 6.11).

We wish to locate the front view of point B.

At the outset we do know that B_F lies on the vertical line drawn through B_H, and that B_F is below A_F, since the grade of line AB is negative. How shall we locate B_F?

Let us refer back to either Figure 6.6 or Figure 6.10. We see that the percent of grade of the line is $AC/CB \times 100 = A_F C_F/A_H B_H \times 100$.

Now, in Figure 6.12, we introduce a vertical line $AD = 4$ units (of a convenient length). In the top view, we lay off a distance, $D_H K_H = 10$ units (same unit of measure), where point K is on line AB. Then we can establish the view K_F, and we can show the top and front views

of the right triangle AKD. The ratio AD/DK is the grade of the line. Finally, it is a simple matter to locate B_F. Since K is a point on line AB, $A_F K_F$ is extended to intersect the vertical through B_H to locate B_F.

Bearing of a Line

The bearing of a line is the angle, less than 90°, that its horizontal view makes with a north-south line. The bearing of a line is determined from its top view only.

EXAMPLE 1

The bearing of line AB is N $\theta°$ W, and the bearing of line CD is also N $\theta°$ W (see Figures 6.13 and 6.14).

The bearing of line EF is N $\phi°$ E; that of line GK is S $\phi°$ W (see Figures 6.15 and 6.16, respectively).

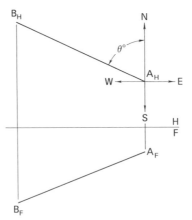

Figure 6.13 Bearing of line AB is N $\theta°$W.

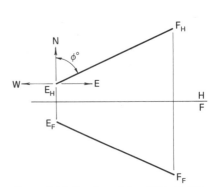

Figure 6.15 Bearing of line EF is N $\theta°$E.

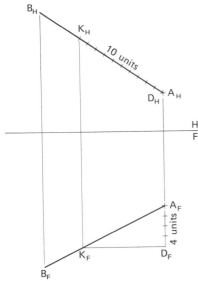

Figure 6.12 Solution to the problem in Figure 6.11.

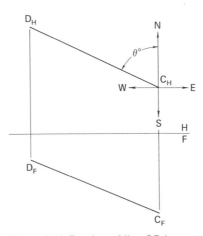

Figure 6.14 Bearing of line CD is N $\theta°$ W.

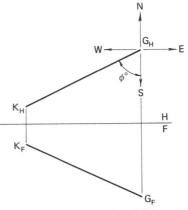

Figure 6.16 Bearing of line GK is S $\theta°$ W.

EXAMPLE 2

The front view and grade (-60%) of line segment AB are given (see Figure 6.17). It is required to find the bearing of line AB. We can readily construct the true shape of right triangle ABC. This is shown as $A_F C_F B_F{}^R$, where $A_F C_F = 6$ units and $C_F B_F{}^R = 10$ units. The grade, we recall, is rise over run or 6/10. Therefore, the distance from C_F to $B_F{}^R$ is the length of the run or the magnitude of $A_H B_H$ or $A_H B_H'$. There are these two solutions. The bearing is either S $\theta°$ W or N $\theta°$ W.

EXAMPLE 3

Suppose we have the following data: (1) Line segment AB has a bearing of N 60° W. (2) The grade of the line is 70%. (3) The true length of the line is 75 mm. (4) Point A is known.

It is required to establish the top and front views of line AB.

ANALYSIS AND SOLUTION

The bearing of line AB is seen in the top view; therefore, we can draw the top view of a portion of line AB (see Figure 6.18).

The view of the line on a supplementary plane that is parallel to the line and perpendicular to the H plane will show both the grade and the true length of the line segment.

Supplementary plane 1 is introduced parallel to the line and perpendicular to the H plane. The view on plane 1 shows the grade and true length of line segment AB (see Figure 6.19). Once we have determined the view $A_1 B_1$, it is a simple matter to locate B_H and B_F. The completed solution is shown in Figure 6.19.

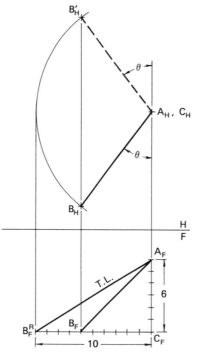

Figure 6.17 Bearing of line AB has two possible values, S $\theta°$ W and N $\theta°$ W, when the given data are the front view of AB and the grade of line AB.

EXAMPLE 4

Line segment AB has a bearing of N 50° W, a grade of -70%, and a true length of 50 mm. Assume that point A is known. Let us proceed to locate the top and front views of the line segment *employing the method of rotation*.

1. The top view of a portion of line AB is easily established, since the bearing is given (see Figure 6.20).

2. We recall the definition of grade of a line as "the ratio of rise to run of a segment of the line." Now, let us lay off 10 units (of convenient length) from A_H to K_H, where K is a point on line AB. The distance from A_H to K_H is the run of line segment AK.

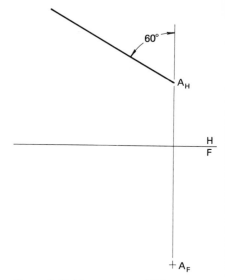

Figure 6.18 Line segment AB has a bearing N 60° W. A_F is known. *Problem:* Establish the H and F views of line AB when the grade is 70% and the true length of AB is 75 mm.

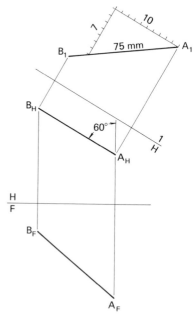

Figure 6.19 Solution to the problem in Figure 6.18.

Now it is a simple matter to locate K_F, since the stated grade is -70%. Note that the "rise" of 7 units establishes K_F.

Line AK has the correct bearing and grade. The length, AK, however, *is either longer or shorter than* the specified 50 mm for the line segment AB. By employing the rotation method, as discussed earlier, we can find the true length of segment AK. This is shown as length $A_F K_F{}^R$. Now we can lay off the prescribed length of segment AB on the true length line. This length is shown as $A_F B_F{}^R$. The locations of B_F and B_H are easily established and are shown in Figure 6.20.

Before we proceed to discuss the point view of a line and associated problems, we need to understand *the orthogonal representation of intersecting lines, skew lines, parallel lines, and perpendicular lines.*

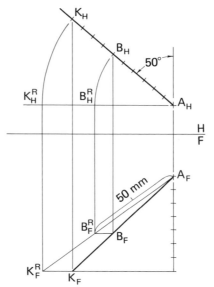

Figure 6.20 Line segment AB has a bearing N 50°W, a grade of -70%, and a length of 50 mm. Point A is known. *Problem:* Locate the H and F views of line segment AB.

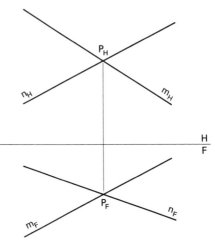

Figure 6.21 Intersecting lines shown orthographically.

Lines—Intersecting, Skew, Parallel, and Perpendicular

Intersecting Lines

Intersecting lines have a point in common. In Figure 6.21, the H and F views of lines m and n are shown. Note that their intersection, point P, is represented by the consistent views P_H and P_F.

In Figure 6.22, lines AB and CD seem to intersect. Since line AB is a profile line (parallel to the P plane), it is possible that the two lines do not intersect. A *profile view* of the two lines would clearly show whether or not there is an intersection. Obviously there is no intersection, since the F and P views of the "apparent" intersection do not lie on the same horizontal line.

When two lines are *general*, as shown in Figure 6.21, two adjacent views are sufficient to determine whether or not there is a point of intersection. What use can we make of intersecting lines?

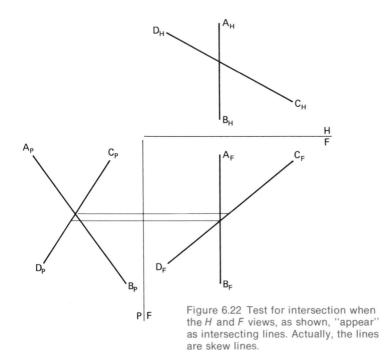

Figure 6.22 Test for intersection when the H and F views, as shown, "appear" as intersecting lines. Actually, the lines are skew lines.

EXAMPLE 1

Suppose we have a top view of four points and the front view of three of these points. Let us assume that the four points are in the same plane. How shall we locate the front view of the fourth point to satisfy the assumption?

In Figure 6.23, the top view of points A, B, C, and D and the front view of points A, B, and C are shown.

ANALYSIS AND SOLUTION

Since the four points lie in the same plane, the diagonals AC and DB must intersect. The top view of the point of intersection, P, is readily located.

The front view of diagonal AC may be drawn, since the front view of points A and C is known. Now, since point P is the intersection of the diagonals, we can easily establish the front view of point P on the front view of diagonal AC. Points B, P, and D lie on one line; therefore, we can establish the line through B_F and P_F that contains D_F, the location of which is now quite obvious.

Let us consider another example of a similar problem.

EXAMPLE 2

Figure 6.24 shows a top view of plane $ABCD$ and the corresponding front view of a portion of the plane. It is required to locate D_F and complete the front view.

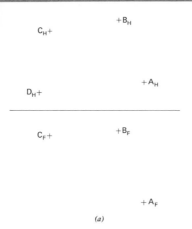

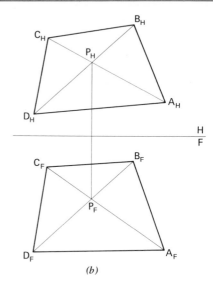

Figure 6.23 (a) Points A, B, C, and D lie in one plane. *Problem:* Locate D_F
Figure 6.23 (b) Solution to the problem in part (a).

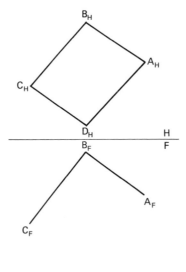

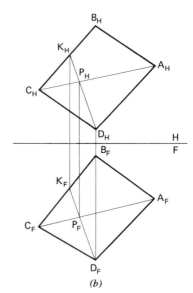

Figure 6.24 (a) The given data are shown. *Problem:* Locate D_F. Points A, B, C, and D lie in one plane
Figure 6.24 (b) The solution.

ANALYSIS AND SOLUTION

It is evident that we cannot use the point of intersection of the diagonals, because diagonal BD is a profile line and the profile view is not available.

We can, however, introduce a new line in the plane, such as line DK. Once the top view of DK is drawn, its intersection, P_H, with diagonal AC is readily located. Next P_F and K_F are located. Line PK contains point D. Therefore, the intersection of line $P_F K_F$, extended, with the vertical passing through B_F uniquely locates D_F. The front view of plane $ABCD$ is readily completed.

EXAMPLE 3

Let us consider Figure 6.25. It is assumed that the top view of point K is known and that point K is in plane ABC. Our problem is to determine the front view of point K.

ANALYSIS AND SOLUTION

Since point K is in plane ABC, there are many lines in plane ABC that pass through point K. One such line is AD.

The top view of line AD is easily established by connecting A_H and K_H and locating D_H, which is the top view of the intersection of line segment AK (extended) and BC.

Since lines AD and BC intersect at point D, we can easily locate D_F and then establish the front view of line AD.

Now K_F can be located on $A_F D_F$, since we know that "*when a point is on a line, the views of that point will lie on the corresponding views of the line.*"

Skew Lines

Skew lines do *not* have a common point. They are lines that are neither parallel nor intersecting. Figure 6.26 shows the H and F views of lines AB and CD. At first glance, we might conclude that the lines do intersect. However, careful study will show that the lines actually do *not* intersect. Note that the *apparent* intersection of the line $A_H B_H$ with $C_H D_H$ is the H view of two points, one on line AB and the other on CD. If we label these points E and F, respectively, we can locate both the H and F views. Now, since point E is above point F, we can see that line CD, which contains point E, is above line AB.

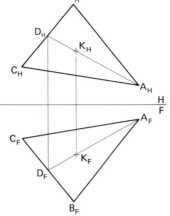

Figure 6.25 Plane ABC and K_H are given. Locate K_F, assuming that point K lies in the surface ABC. The solution is shown.

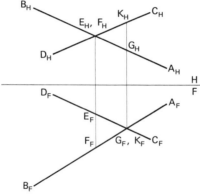

Figure 6.26 Skew lines AB and CD are shown orthographically.

In a similar manner, the apparent intersection of $A_F B_F$ with $C_F D_F$ is the F view of two points, one on line AB and the other on CD. If we label these points G and K, respectively, we can locate their H and F views. Now, since point G is in front of point K, line AB, which contains point K, is in front of line CD. We used this analysis when we discussed visibility in Chapter 5.

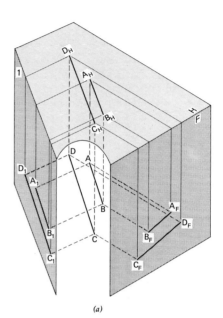

(a)

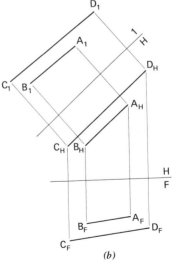

(b)

Figure 6.27 (a) Pictorial showing parallel lines AB and CD and their views on planes H, F, and 1.
Figure 6.27 (b) Orthographic views of parallel lines AB and CD on planes H, F, and 1.

Parallel Lines

Parallel lines appear as parallels in all views (see Figure 6.27). There are two special cases, however: (1) where one view of the two parallel lines will appear as points (see Figure 6.28), and (2) where one view will appear as a single line (see Figure 6.29).

Perpendicular Lines

Perpendicular lines are at right angles to each other.

A very important relationship to remember is: *When one of two perpendicular lines is seen in true length, the other line will make a 90° angle with the one in true length.*

EXAMPLE 1

Let us consider Figures 6.30 and 6.31, which show lines DC and DE perpendicular to line AB. Since line AB is parallel to the F plane, the true length of AB is seen in the front view as $A_F B_F$; and the front view of lines DC and DE forms a 90° angle with $A_F B_F$.

EXAMPLE 2

In Figure 6.32, we observe that both lines DC and DE are perpendicular to line AB. This is evident from the fact that the top view, $A_H B_H$, of line AB is true length, and the angle between $C_H D_H$ and $A_H B_H$ is 90°, and the angle between $E_H D_H$ and $A_H B_H$ is also 90°.

EXAMPLE 3

Suppose we are given line AB and point D (Figure 6.33) and that we wish to determine the top and front views of line DE, which is perpen-

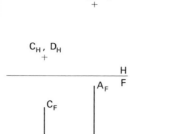

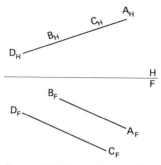

Figure 6.28 Parallel lines AB and CD are perpendicular to the H reference plane.

Figure 6.29 Parallel lines AB and CD lie in a plane perpendicular to the H reference plane.

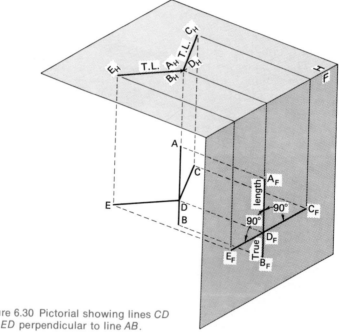

Figure 6.30 Pictorial showing lines CD and ED perpendicular to line AB.

dicular to line AB. It is also assumed that point E is on line AB. It is evident that neither the top view nor the front view of line AB shows the true length of line AB. We can determine the true length of line AB, however, by introducing a supplementary plane, 1, parallel to line AB and perpendicular to the F plane. The view on plane 1 shows

the true length of line AB and also the corresponding view of point D (shown as D_1). Now we can draw, through D_1, a line perpendicular to $A_1 B_1$ (based on what?). The point of intersection of this perpendicular with $A_1 B_1$ establishes E_1. Now it is a simple matter to locate E_H and E_F and the corresponding views of the perpendicular DE.

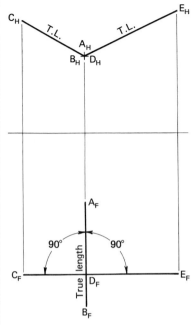

Figure 6.31 Orthographic views of Figure 6.30.

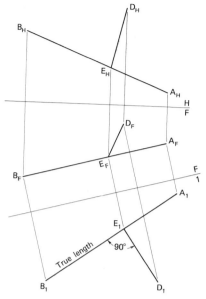

Figure 6.33 Line *AB* and point *D* are given. *Problem:* Line *DE* is ⊥ to *AB*. *E* lies on *AB*. Locate the *H* and *F* views of line *DE*. The solution is shown.

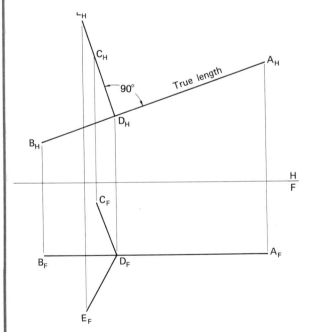

Figure 6.32 Lines *DC* and *DE* are perpendicular to line *AB*.

Problems Dealing with Parallels and Perpendiculars

We now consider several problems that can be solved by employing parallels and perpendiculars.

EXAMPLE 1

Suppose it is required to establish a plane through a given point, A, and parallel to two skew lines, *m* and *n*. Figure 6.34 is a graphical representation of the problem. Let us now analyze the problem and its solution.

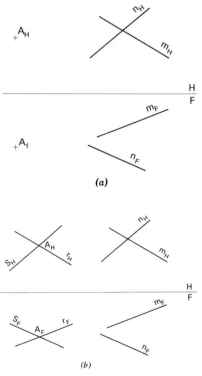

Figure 6.34 (a) The given data are shown. *Problem:* Establish the *H* and *F* views of a plane that passes through point *A* and is parallel to lines *m* and *n*. Figure 6.34 (b) The solution.

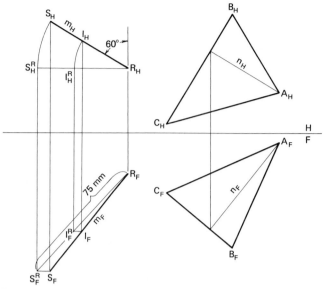

Figure 6.36 Solution to the problem in
Figure 6.35.

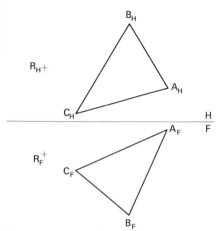

Figure 6.35 Point R and plane ABC are
given, as shown. *Problem:* Locate the H
and F views of line RS, when the
bearing of RS is N 60° W, when line RS
is parallel to plane ABC, and when line
RS is 75 mm long.

ANALYSIS AND SOLUTION
A plane may be determined by (1)
two intersecting lines, (2) two
parallel lines, (3) a point and a line
that does not contain the point, or
(4) three points that are not on one
line.

*When a line is parallel to a plane,
the line will be parallel to a line of
the plane.* We can now proceed to
establish (1) line, r, through point
A and parallel to line m, and (2) another line, s, also through point A
but parallel to line n. We now have
the required plane, determined by
intersecting lines r and s, passing
through point A and parallel to
both skew lines m and n. The solution is shown in Figure 6.34b.

EXAMPLE 2
Consider the following problem.
Establish the H and F views of line

RS, which fulfills the following
conditions.

1. The bearing of line RS is N 60° W.
2. Line RS is parallel to plane ABC.
3. Line RS is 75 mm long.

The information that is available is
shown in Figure 6.35. The solution
is shown in Figure 6.36.

ANALYSIS AND SOLUTION
To satisfy condition (1), we can
easily establish the bearing by
drawing a line m_H through R_H at N
60° W.

To satisfy condition 2, we know
that when a line is parallel to a
plane, the line (m in this case) is
parallel to a line of the plane; therefore, we can readily establish a
line, n, in plane ABC and parallel
to line m. The top view of line n is
drawn through A_H and parallel to
m_H. (Remember, parallel lines are
seen as parallels in the respective
views.) Now, since line n lies in
plane ABC, we can establish n_F.
The front view of line m passes

through R_F and parallel to n_F. Why
is this true?

Finally, to satisfy condition 3,
point S is located by first finding
the true length of a portion of line
m (such as R-1) and then by laying
off the specified length 75 mm.
You will recall that we solved a
similar problem in our discussion
of true length of a line segment.

EXAMPLE 3
Suppose we are given the H and F
views of angle ABC as shown in
Figure 6.37. Is the angle 90°? How
can we determine the answer?

ANALYSIS AND SOLUTION
One approach could be to find the
true length of each side of the
triangle ABC and then construct a
triangle with the true lengths previously found and, finally, measure
angle ABC. Although this approach yields an answer to the
question, it is a time-consuming
method (hence, costly).

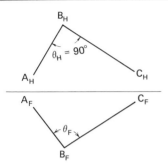

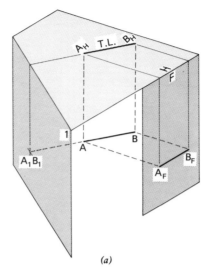

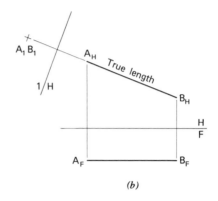

Figure 6.37 The H and F views of angle ABC; (θ), are shown. Assuming that the H view is fixed and that $B_F C_F$ is also fixed, where would you place A_F so that θ = 90°?

A *much simpler solution* requires only the recognition of the fact that, *"when two lines are perpendicular to each other, the view of the two lines on a plane that is parallel to one of them will reveal a right angle."*

Therefore, we may conclude at once that the angle ABC is *not* 90°, since the view marked 90° does not show the true length of either line AB or BC.

Question. If the top view, $A_H B_H C_H$ and the front view of BC remain as shown in Figure 6.37, where would you place A_F so that angle ABC = 90°?

POINT VIEW OF A LINE—SECOND BASIC PROBLEM

The point view of a line will be seen in the view on a plane that is perpendicular to the line.

Knowing how to obtain the point view of a line will enable us to analyze and solve space problems such as:

1. The clearance between skew cables or two skew pipelines, etc.
2. The shortest distance from a point to a line.
3. The distance between parallel lines.

Figure 6.38 (*a*) Pictorial showing point view of line *AB*
Figure 6.38 (*b*) Orthographic views showing line segment *AB* and its point view on plane 1.

4. The angle between plane surfaces.
5. The edge view of a plane surface.

A number of *practical applications* are included in the examples in this chapter and in the workbooks.

Now let us study Figure 6.38. We observe that line segment AB is parallel to the H plane, and that the *true length* of AB is seen in the top view.

Let us introduce supplementary plane 1, perpendicular to line AB. Plane 1 is therefore perpendicular to the H plane. The view of line AB on plane 1 is a point, shown as $A_1 B_1$.

If the given line were not parallel to either the H *or* F *plane, it would be impossible to introduce a supplementary plane that would be perpendicular to the given line and also to either the* H *or* F *plane.*

What would happen if we introduce a supplementary plane perpendicular to the H plane and *the*

top view of the line CD (see Figure 6.39)?

It should be evident that the view of line CD on plane 1 is *not* a point. This shows that plane 1 is *not* perpendicular to the line.

We recall that *adjacent planes must be at right angles to each other;* that is, H and F are adjacent planes, and H and 1 are also adjacent planes.

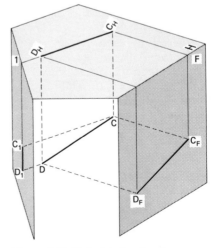

Figure 6.39 Pictorial showing line segment CD and its views on planes H, F, and 1, which is perpendicular to $C_H D_H$ and the H reference plane.

How shall we find the point view of line *CD*? We observed in the first case, Figure 6.38, that it was a simple matter to introduce plane 1 perpendicular to both the line and the *H* plane, since line *AB* was parallel to the *H* plane and the true length of line segment *AB* was seen in the *H* view. Therefore, if we can reduce the general problem to the simple case, we will encounter no difficulty in formulating the solution. *Actually, this means nothing more than first finding the true length view of the line segment and then determining the point view of the line.*

EXAMPLE

Let us consider the line segment *AB* shown in Figure 6.40. A true length view of *AB* can be obtained, quite readily, by introducing supplementary plane 1 parallel to the line and perpendicular to the *H* plane. The view, A_1B_1, on plane 1 shows the true length of line segment *AB*.

Now, *if we consider the views A_HB_H and A_1B_1 as the two given views* (forgetting the existence of the *F* view), we can readily place supplementary plane 2 perpendicular to both line *AB* and plane 1. The view, A_2B_2, on plane 2 is the point view of line *AB*.

Applications of the Second Basic Problem

There are many instances in engineering design where it is necessary to determine the clearance between cables, between structural members, between pipelines, etc.; or where it is essential to maintain specified distances between skew

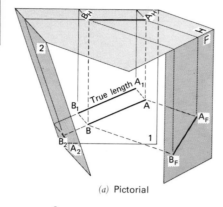

(a) Pictorial

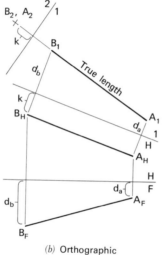

(b) Orthographic

Figure 6.40 Point view of line segment.

members (i.e., electrical currents, to prevent arcing).

The determination of the shortest distance between two skew members is based on the simple concept "point view of a line." When we obtain the point view of one of the skew lines (which could be taken as the axis of the member) and the corresponding view of the other line, we will be in position to see and measure the shortest (perpendicular) distance between the two skew lines.

EXAMPLE 1

Let us consider the skew lines *AB*

and *CD* shown in the *H* and *F* views of Figure 6.41. If we obtain a point view of either line, the perpendicular distance between the two lines will be apparent. Let us find the point view of line *AB*.

We know from our previous study of the second basic problem, "to find the point view of a line," that *it is first necessary to find the true length of the line*. Therefore, we will introduce supplementary plane 1 parallel to line *AB* and perpendicular to the *H* plane. The new view shows line *AB* in its true length and a foreshortened length of *CD*. Now the second supplementary plane 2 is introduced, perpendicular to both line *AB* and plane 1. The final view shows line *AB* as a point and line *CD* foreshortened. The shortest (perpendicular) distance between the skew lines can now be measured (shown as E_2F_2).

The other views of the common perpendicular, *EF*, can be readily established. The views of point *E*, which lies on line *CD*, are easily located, since the views of the point lie on the corresponding views of the line (i.e., E_1 lies on C_1D_1). Point *F*, however, especially the location of the view, F_1, requires some thought.

We recall that "when two lines are perpendicular to each other, the view of the two lines on a plane that is parallel to one of them will reveal a right angle"; therefore, the view E_1F_1 will be perpendicular to A_1B_1 (which is the true length of *AB*). Once we have located F_1, it is a simple matter to locate the other views of point *F* and of line *EF*.

EXAMPLE 2

Two circular pipes are shown orthographically in Figure 6.42. We are to determine the location of the centerline of the shortest connector pipe.

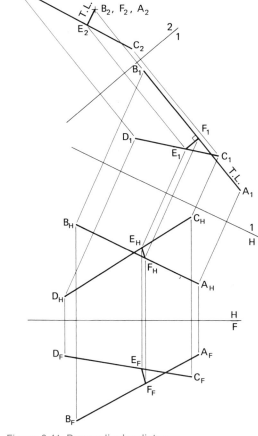

Figure 6.41 Perpendicular distance between skew lines.

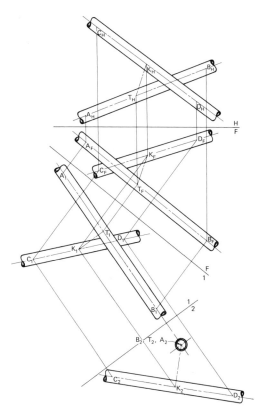

Figure 6.42 Shortest connector pipe between two skew pipes.

ANALYSIS AND SOLUTION

The shortest (perpendicular) distance between the two pipes will be seen in a view that shows the centerline of one of the pipes as a point.

We can obtain a point view of centerline AB by first introducing supplementary plane 1 parallel to AB and perpendicular to the F plane to obtain the true length view of AB and the corresponding

view of CD, and then introducing supplementary plane 2 perpendicular to AB to obtain its point view and the corresponding view of CD.

Now, in view 2, we can draw the perpendicular from B_2A_2 to C_2D_2. This is line K_2T_2, the centerline of the shortest connector pipe.

To locate the view of KT on plane 1, we observe that K_1 is readily located from point K_2. To locate T_1, we recall the statement: "When

one of two *perpendicular lines is seen in true length, the other line will make a 90° angle with the one in true length*"; therefore, we can draw a perpendicular from K_1 to line A_1B_1 (which is seen in true length) to locate point T_1. Finally, the F and H views of KT are located to show the front and top views of the centerline of the shortest connector pipe.

EXAMPLE 3

Two airplane cables KL and ST are shown orthographically in Figure 6.43. Two problems are to be solved.

1. The determination of the clearance, in millimeters, between the cables.
2. To establish the new location of cable ST to satisfy the following specification.
   (a) Points K, L, and S are fixed, as shown in the H and F views, and the length of ST cannot be changed.
   (b) Point T is to be moved, the shortest distance, in a path parallel to the F plane so that the clearance between the cables is twice the original distance.

ANALYSIS AND SOLUTION

The solution to the first problem is quite simple, since it is essentially a repeat of the previous example. We must, however, recognize that it is quite important to obtain a point view of *cable KL* and the corresponding view of cable ST to show the clearance and not a point view of cable ST and the corresponding view of cable KL in order to proceed conveniently, to the solution of the second problem. This is so because the specification states that the only moving point is T.

Once we obtain the point view of cable KL and the corresponding view of cable ST, the clearance can be measured, as shown in Figure 6.43.

With regard to the second problem, we can proceed in the following manner.

1. The new clearance between the cables is equal to the radius of the circle shown in supplementary view 2.
2. The new position of cable ST must be on a tangent from point S to the circle in order to satisfy the new clearance specification. Two tangents are possible; however, only one satisfies the constraint that "point T is to be moved the shortest distance" therefore the dashed tangent is the correct choice.
3. The new location of point T can be determined by finding the intersection of the solid tangent with the locus of point T as it moves in a plane parallel to the F plane. The top view of the locus appears as a horizontal line through T_H; the front view appears as a circular arc whose center is S_F and whose radius is $S_F T_F$. In supplementary view 1, the locus is the line

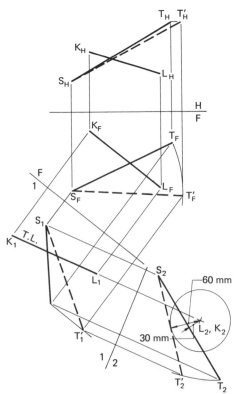

Figure 6.43 Clearance problem. Example 3.

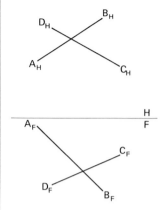

Figure 6.44 Clearance problem. Example 4.

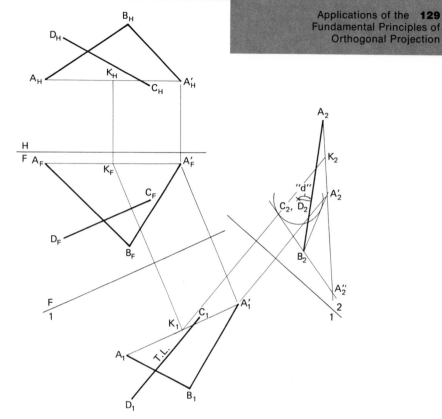

Figure 6.45 Solution to problem. Example 4.

that passes through T, and parallel to the F plane. To determine the locus in supplementary plane 2, we can select a few points such as A, B, C, . . . , and locate A_2, B_2, C_2, . . . , through which a smooth curve can be drawn. The intersection of this curve and the tangent locates T_2', the new position of point T, as seen in supplementary view 2. Once that is done we should encounter no difficulty in locating the other views of point T and the new position of cable ST, as shown in Figure 6.43.

EXAMPLE 4

Now let us try to solve another clearance problem. The specifications are:

1. First find the clearance between cables AB and CD, as shown in Figure 6.44.
2. Increase the clearance 12 mm.
3. Cable CD is to remain fixed.
4. Point B of cable AB is also fixed.
5. The new position of point A must be on a line that passes through the original position of A and parallel to both the H and F planes. The new position of A must be as close to its original position as possible.

ANALYSIS AND SOLUTION

The clearance between the cables is easily found, since this is the same problem as that in Example 1 (see Figure 6.41).

The new position of point A, represented by A', must lie somewhere on the tangents drawn through B_2 (shown in supplementary view 2) to the circle whose center is at C_2D_2 and whose radius is (d + 12 mm). This must be true if we are to satisfy the condition that the original clearance between the cables has been increased 12 mm.

The other condition, that point A moves on a line through the original position of A, parallel to both the H and F planes, defines the locus (path of motion) of point A. Graphically we can establish this

locus by drawing its top and front views (represented by A_HK_H and A_FK_F, respectively, where point K is selected arbitrarily), and then locating line AK in both supplementary views.

Point A must be somewhere on line AK (extended, if necessary).

In order to satisfy both conditions, as specified, point A' must be at the intersection of line AK and the tangents drawn through B. The possible locations of A' are shown in supplementary view 2 as A_2' and A_2''. Since the new position of A must be as close to its original position as possible, the correct location is at A_2'. The other views of point A' now are easily obtained. The complete solution is shown in Figure 6.45.

EDGE VIEW OF A PLANE SURFACE—THIRD BASIC PROBLEM

An edge view of a plane surface will be seen in the view on a plane that is perpendicular to a line that lies in the surface.

For example, an edge view of surface ABC, shown orthographically in Figure 6.46, can be obtained by finding the point view of a side of the triangle (i.e., BC). We know, from the second basic problem that a true length view of BC must be found before we can obtain the point view. In Figure 6.47, supplementary plane 1 is parallel to side BC and perpendicular to the F plane. The view on plane 1 shows side BC in true length. Supplementary plane 2 is perpendicular to both side BC and plane 1. The view on plane 2 shows the point view of side BC and the edge view of surface ABC.

We can shorten the solution by eliminating the need for two supplementary planes. In fact, it is a simple matter first to introduce a line, in surface ABC, so that its true length is available in either the top (H) or front (F) views. In Figure 6.48 line BK, a frontal line in surface ABC, is seen in true length in the front view, $B_F K_F$. Supplementary plane 1 has been introduced perpendicular to both line BK and the F reference plane. The view on plane 1 shows the point view of line BK and the edge view of surface ABC.

Now that we know how to find the edge view of a plane surface, let us consider a few problems that arise in engineering practice.

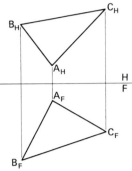

Figure 6.46 Plane surface ABC is known. *Problem:* Find an edge view of the surface.

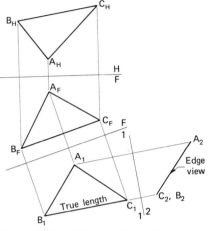

Figure 6.47 Solution to the problem in Figure 6.46.

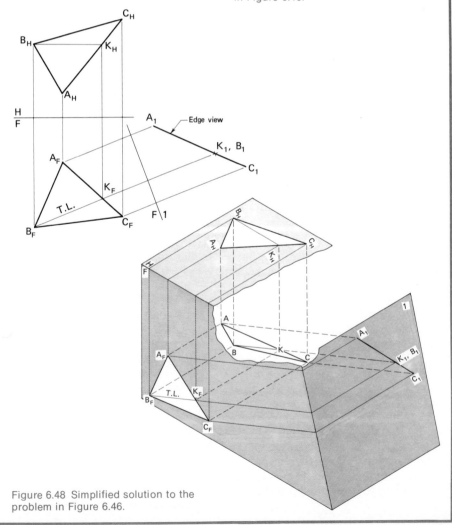

Figure 6.48 Simplified solution to the problem in Figure 6.46.

EXAMPLE 1

In the design of structures it is essential to know the safe bearing capacity of the material that will support, for example, a proposed office building. Generally, field tests are conducted to determine load-bearing values.

Let us assume that drilling operations defined the locations of points *A*, *B*, and *C* on the upper surface of a rock formation, and of point *D* on the lower parallel surface, as shown in Figure 6.49.

Our problem is to determine the thickness of the rock formation. This is very important in order to make certain that the thickness is sufficient to preclude the possible "punching through" (punching shear) of the load-supporting footings of the structure.

ANALYSIS AND SOLUTION

The perpendicular distance from point *D* to the surface defined by points *A*, *B*, and *C* is a measure of the thickness, "*t*," between the parallel surfaces of the rock formation.

Once we obtain an edge view of the surface and the corresponding view of point *D*, we can determine the thickness of the formation.

The edge view of surface *ABC* is seen on supplementary plane 1, which is perpendicular to line *CK* and, therefore, perpendicular to surface *ABC*. Figure 6.49 shows the edge view as $A_1B_1C_1$ and also the corresponding view, D_1, of point *D*. The thickness of the rock formation is shown as "*t*."

EXAMPLE 2

There are several "connector" problems that arise in engineering design projects. For instance, in the design of oil refineries, problems occur in piping systems that could include the need for determining the following.

1. The location of the *shortest pipe connector* between two skew pipelines.
2. The location of the *shortest horizontal* connector between two skew pipelines.
3. The location of the *shortest connector, at a given grade*, between two skew pipelines.

In a previous example we solved the first problem by finding the point view of one of the skew members. We can, however, employ an *alternative* method for solving the three problems. Essentially, the method makes use of the basic problem, "edge view of a plane."

Shortest Connector
Between Two Skew Members

Let us study the views of skew members (i.e., centerlines of pipes) *AB* and *CD* as shown in Figure 6.50. We are to determine the location of the shortest connector, *EF*, between *AB* and *CD*.

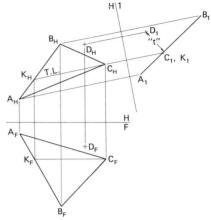

Figure 6.49 Determination of the thickness, *t*, of a rock formation.

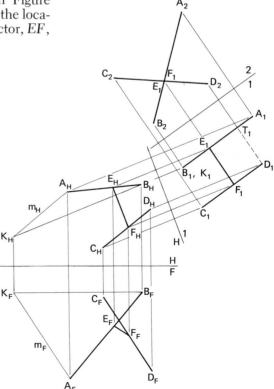

Figure 6.50 Shortest connector between skew lines: plane method.

ANALYSIS AND SOLUTION

We can introduce a plane that contains member AB and is parallel to member CD. A perpendicular from any point on CD to that plane is the length of the centerline of the shortest connector, EF, between members AB and CD. It remains, however, to locate the position of EF.

In Figure 6.50, it should be observed that a line m through point A and parallel to member CD has been introduced. (Recall that parallel lines appear as parallels in the respective views: m_F is parallel to $C_F D_F$, and m_H is parallel to $C_H D_H$.)

Now m and AB determine the plane that contains member AB and is parallel to member CD. We can easily obtain *an edge view* of that plane by finding the point view of the horizontal line BK. The view on supplementary plane 1 shows the point view of BK, the edge view of the plane determined by m and AB, and the corresponding view of member CD.

Now *the length* of the centerline of the shortest connector, EF, is the perpendicular distance, $D_1 T_1$, from any point (i.e., D_1) on member CD to the plane ABK, as seen in the view on supplementary plane 1.

The *location* of connector EF is easily found by introducing supplementary plane 2 perpendicular to the direction $D_1 T_1$. The view on plane 2 shows the connector, EF, as a point. The other views of EF are readily determined.

Shortest Horizontal Connector Between Two Skew Members

Let us assume the same two skew members AB and CD. The analysis and the steps that lead to the view of plane ABK and member CD on supplementary plane 1 are the same as in the previous problem (see Figure 6.51).

In plane 1, *the length* of the centerline of the shortest horizontal connector between members AB and CD is shown by the dashed line drawn from any point on member CD (i.e., G) to the plane ABK and parallel to the H reference plane. This is line $G_1 T_1$. *The location* of the shortest horizontal connector, EF, is obtained by introducing supplementary plane 2 perpendicular to the direction $G_1 T_1$. The view on plane 2 shows the connector, EF, as a point. The other views of EF are then easily located.

Shortest Connector, of a Given Grade, Between Two Skew Members

Again let us assume the same two skew members AB and CD as shown in Figure 6.52. Suppose it is required to locate the shortest connector, at a 25% grade, from member AB to CD. As in the first two parts of Example 2, the analysis and steps that lead to the view of plane ABK and member CD on supplementary plane 1 are the same.

We can establish the 25% grade in plane 1 by laying off the slope 2.5 to 10 (2.5:10), as shown in the figure. (One can visualize the shaded right triangle in space by forming a mental image of the three-dimensional representation of the planes H and 1.)

The *length* of the shortest connector, at the 25% grade, from any point on member CD to plane ABK is shown by the dashed line

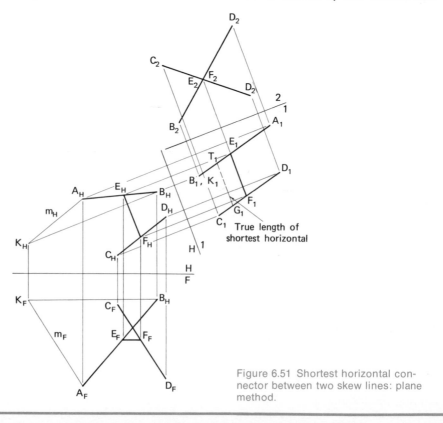

Figure 6.51 Shortest horizontal connector between two skew lines: plane method.

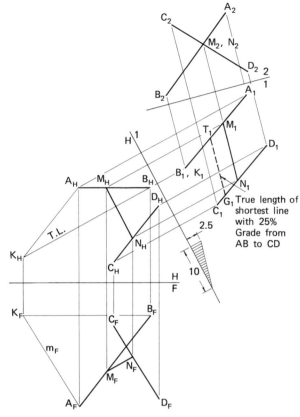

Figure 6.52 Shortest connector of a given grade, between two skew lines: plane method.

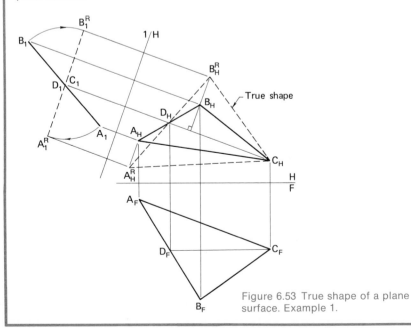

Figure 6.53 True shape of a plane surface. Example 1.

G_1T_1. The *location* of the shortest connector, MN, is obtained by introducing plane 2 perpendicular to the direction G_1T_1. The view on plane 2 shows connector MN as a point. The other views of MN are easily located.

TRUE SHAPE OF A PLANE SURFACE—FOURTH BASIC PROBLEM

A number of examples were cited in Chapter 1 to show the need for the determination of true shapes of plane surfaces.

How can we determine the true shape of a plane surface that is described by its orthographic views?

Let us consider the following examples.

EXAMPLE 1

We wish to determine the true shape of surface ABC shown orthographically in Figure 6.53.

ANALYSIS AND SOLUTION
We can establish a principal line in the surface (i.e., a horizontal line) and then rotate the surface about that line, as an axis, until it is parallel to the H reference plane. The H view would then reveal the true shape of the surface.

To implement the solution, we establish horizontal line CD as the axis of rotation. Supplementary plane 1 is then introduced perpendicular to both line CD and the H plane. The view on plane 1 shows the point view of line CD (the axis of rotation) and the edge view of the surface. As we rotate the surface about CD, parallel to the H plane, points A and B will move along the circular arcs shown in plane 1. Since the surface is rotated parallel to the H plane, the rotated position of the surface appears as the dashed line in the view on plane 1 and as the dashed surface shown in the H view. Surface $A_H^R B_H^R C_H$ is the true shape of the surface ABC.

EXAMPLE 2
An alternative method for finding the true shape of a plane surface is to obtain a view of the surface *on a plane to which it is parallel*. This is *a very useful method when it is necessary to dimension the surface and to locate components that may be attached to the surface.*

Let us now consider three orientations of the given plane surface.

**The Surface is Parallel
to a Reference Plane**
Suppose the given surface is triangle ABC as shown in Figure 6.54. Study of the pictorial and orthographic drawings leads to the conclusion that the surface ABC is parallel to the H reference plane. Therefore the top view, $A_H B_H C_H$, reveals the true shape of the surface.

**The Surface is Perpendicular
to a Reference Plane but not
Parallel to the Adjacent
Reference Plane**
Let us consider the plane surface ABC shown in Figure 6.55. We observe that the surface is perpendicular to the F plane, since it appears on edge in the front view. Moreover, the evidence is quite clear that the surface is *not* parallel to the H plane.

Since we have an edge view of

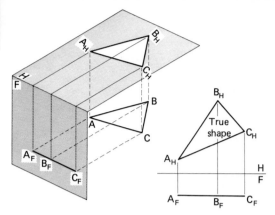

Figure 6.54 True shape of a plane surface that is parallel to the H reference plane.

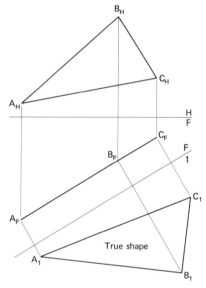

Figure 6.55 True shape of a plane surface that is perpendicular to the F reference plane.

the surface ABC, it is a simple matter to introduce a plane parallel to the surface. Supplementary plane 1 has been placed both parallel to the surface and perpendicular to the F plane. The view of surface ABC on plane 1 shows the true shape of the surface.

It should be stressed that the relationship between the front view and the supplementary view is basically the same as that between the top

and front views shown in the orthographic drawing of Figure 6.54.

**The Plane Surface is Not
Parallel or Perpendicular
to Any Reference Plane**
This is the most general orientation. Study of the top and front views of surface ABC, as shown in Figure 6.56, reveals that neither view shows the true shape of the surface. We can easily reduce this general orientation of the surface to that of the previous one (Figure 6.55) by finding an edge view of the surface. Such a view is seen in the view on supplementary plane 1 which is perpendicular to both line AD (a frontal line in surface ABC) and the F plane. Once we have obtained the edge view, shown as $A_1 B_1 C_1$, it is a simple matter to introduce supplementary plane 2 parallel to surface ABC and perpendicular to plane 1. The view, $A_2 B_2 C_2$, on plane 2 shows the true shape of the surface ABC.

Again, it should be stressed that the relationship among the front and supplementary views on plane 1 and plane 2, is basically the same as the relationship among the top, front, and supplementary views in Figure 6.55.

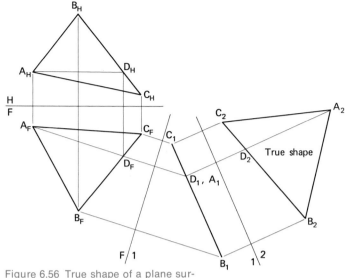

Figure 6.56 True shape of a plane surface that is neither parallel nor perpendicular to a reference plane.

Now let us consider a few additional examples.

EXAMPLE 3

Suppose we need to determine the area of the roof surfaces shown in Figure 6.57. (Why would we need to do this?)

ANALYSIS AND SOLUTION
Once we determine the true shapes of surfaces A and B, we can easily calculate their areas and then double the result to obtain the total roof area.

In Figure 6.57, we observe that an edge view of surface B is shown in the front view. Therefore, we can easily introduce supplementary plane 1, parallel to surface B and perpendicular to the F plane. The view of surface B on plane 1 is shown in true shape as triangle 1_1-2_1-3_1.

In order to obtain an edge view of surface A we can introduce supplementary plane 2, perpendicular to ridge 1-4, and the H plane. The edge view of surface A is shown in the view on plane 2. Now we can

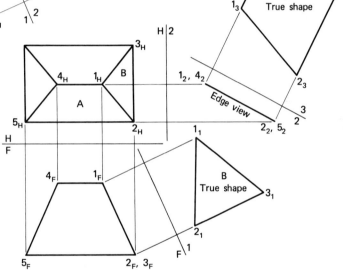

Figure 6.57 True shapes of a roof surface.

introduce supplementary plane 3, parallel to surface A and perpendicular to plane 2. The view of surface A on plane 3 is the true shape of the surface.

EXAMPLE 4

A partial front view and a complete profile view of a flat glass plate are shown in Figure 6.58(*a*). Let us determine the true shape of the plate and the complete front view.

ANALYSIS AND SOLUTION

In order to obtain the true shape of the plate it will be necessary to obtain an edge view. We observe that edge *AB* is seen in true length in the profile view; therefore, we can introduce supplementary plane 1, perpendicular to both edge *AB* and the profile reference plane, and obtain the edge view of the plate surface. See Figure 6.58(*b*). This is shown as $A_1B_1C_1D_1$. Now we can introduce supplementary plane 2 parallel to the plate surface and perpendicular to plane 1. The view on plane 2 is the true shape of the glass plate. We should experience no difficulty in completing the front view.

EXAMPLE 5

In a piping system design it was found necessary to connect a pipeline from point *C* to an existing pipeline *AB* at an angle of 60° with *AB* and closer to point *A* than to point *B* (Figure 6.59). How shall we locate the centerline of the pipe from *C* to *AB*?

ANALYSIS AND SOLUTION

A true shape view of triangle *ABC* will contain the centerline of the pipe from point *C* to pipeline *AB*.

We should have no difficulty in finding the true shape of triangle *ABC*. This is shown in the view on supplementary plane 2. In this view we can draw a centerline from point *C*, at an angle of 60° with centerline *AB*, to intersect *AB* at point *K*. Line *CK* is the required centerline of the pipeline from *C* to pipeline *AB*.

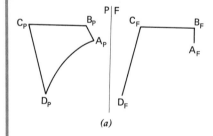

(a)

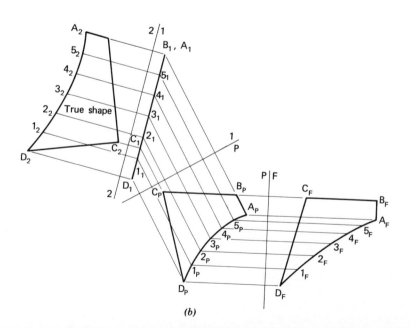

(b)

Figure 6.58 (*a*) Glass plate problem. Example 4
Figure 6.58 (*b*) Solution to the glass plate problem.

Figure 6.59 (a) Pipeline problem.
Example 5
Figure 6.59 (b) Solution to the pipeline
problem.

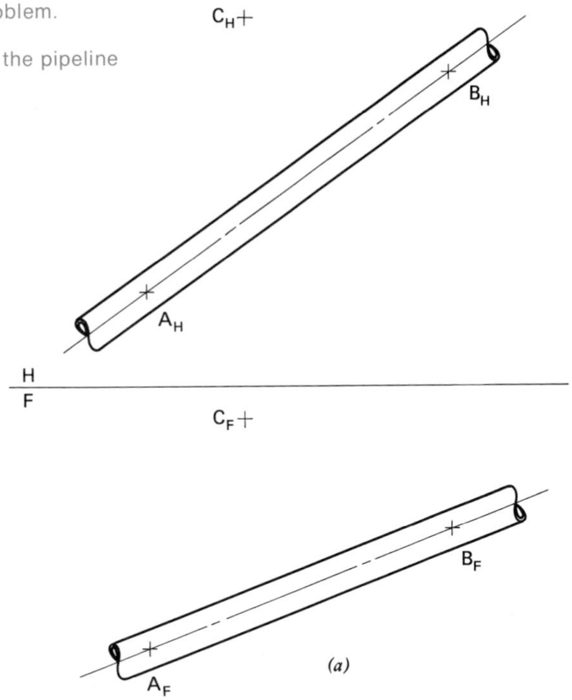

$C_H +$

B_H

A_H

$\dfrac{H}{F}$

$C_F +$

B_F

A_F

(a)

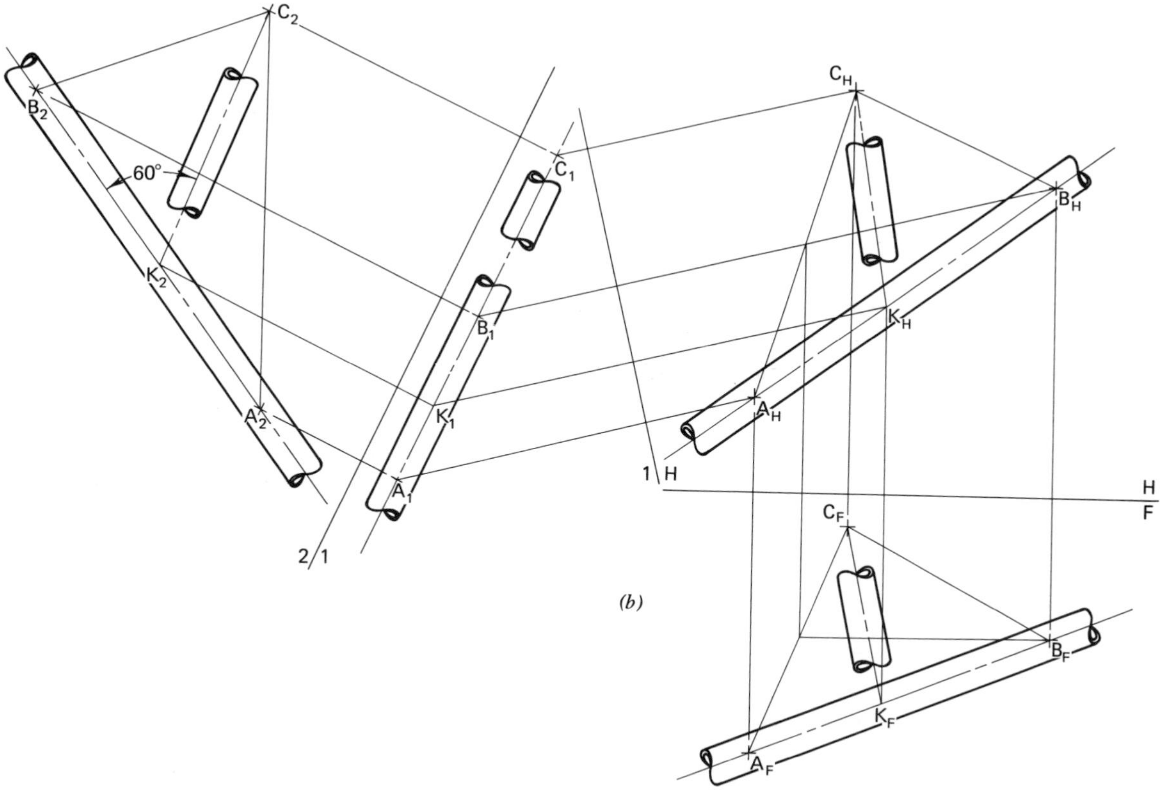

C_2

B_2

60°

K_2

A_2

C_1

B_1

K_1

A_1

2/1

1 \ H

C_H

B_H

K_H

A_H

$\dfrac{H}{F}$

C_F

B_F

K_F

A_F

(b)

Figure 6.60 Design of chute requires the determination of the angles between surfaces (i.e., *A* and *B*).

ANGLE PROBLEMS

Introduction

Many engineering design problems require the determination of the angle between plane surfaces. For example, in the design of the chute shown in Figure 6.60, it is necessary to determine the "bend angle" between surfaces *A* and *B*.

As another example observe the concrete footing shown in Figure 6.61. Assume that it is required to protect the sloping edges from possible damage. This can be done by providing "angle irons" as shown in the figure. The bend angle of the protection members must be determined. This is the angle between surfaces *A* and *B*.

In the fields of geology and mining there are problems that require the determination of the magnitudes of "dip angles" of plane surfaces of various formations. The dip angle of a plane surface is the angle between the surface and a horizontal plane. Examples will be given later.

There are additional "angle problems" that reduce to the determination of (1) the angle between lines, and (2) the angle between a line and a plane surface; examples of the former arise in the design of piping systems, electrical conduit systems, etc.

In a helicopter design it was necessary to determine the angle between the centerline of the Pitot tube and the surface of the plate shown in Figure 6.62.

Actually, this problem reduces to finding the angle between a line and a plane surface.

We now consider several practical examples.

Figure 6.61 Concrete footing with "angle irons" to protect edges from possible damage.

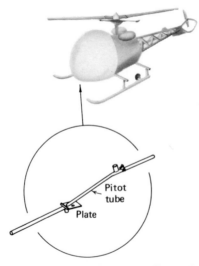

Figure 6.62 Angle between a line and a plane surface. Application in helicopter design.

EXAMPLE 1

The top and front views of a hip roof are shown in Figure 6.63. The design of a metal angle strip that will be placed along the horizontal ridge (1-5) and in contact with roof surfaces *A* and *C* requires the determination of the angle between these surfaces. Similarly, angle strips that are to be placed along the intersections of surfaces *A* and *B*, *B* and *C*, and *C* and *D* will require the determination of the angles between the corresponding roof surfaces.

ANALYSIS AND SOLUTION

The angle between two plane surfaces is seen in a view that shows the line of intersection of the two planes as a point. Due to the symmetry of the roof design, it is only necessary to determine the angles between surfaces *A* and *B*, and *A* and *C*. Let us first determine the angle between surfaces *A* and *B*. It is fairly obvious that line 1-2 is the intersection of surfaces *A* and *B*. A true length view of line 1-2 is seen in the view on supplementary plane 1 which is parallel to line 1-2, and perpendicular to the *F* plane. Surface *B* is seen as triangle 1_1-2_1-3_1, and a portion of surface *A* is seen as 1_1-2_1-4_1. Now we can introduce supplementary plane 2, per-

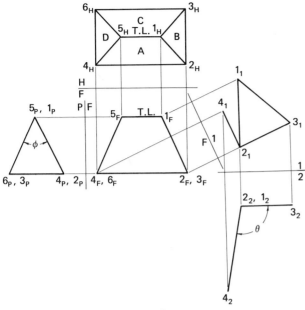

Figure 6.63 Angles between roof
surfaces. Example 1.

pendicular to both line 1-2 and
supplementary plane 1. The view
on plane 2 shows the point view of
line 1-2, the edge view of surfaces
A and B, and the angle, θ, between
them. Angle θ is the required bend
angle of the metal strip for the appropriate roof surfaces. It is a simple matter to find the angle between surfaces A and C, since we
have a true length view of their line
of intersection, 1-5. The profile
view shows the line 1-5 as a point,
the surfaces A and C on edge, and
the angle $\phi°$ between them. Angle
$\phi°$ is the required bend angle of the
metal strip for surfaces A and C.

EXAMPLE 2
In the introduction mention was
made of the term dip angle as used
in the fields of geology and mining.
Additional terms are "strike line"
and "direction of dip." A strike line
of a plane surface (i.e., of a rock
formation or of an ore deposit) is a
horizontal line that lies in the sur-

face. It is identified by its bearing.
Let us now consider the following
problem, the solution of which
should clarify the meaning of the
three terms.

We will assume that points A, B,
and C as shown in Figure 6.64 are
on an ore body surface. We will determine the following.

1. The bearing of a strike line.
2. The magnitude of the dip
angle.
3. The direction of dip.

ANALYSIS AND SOLUTION
The triangle ABC is a portion of
the surface. We know from our
previous work that the F view of a
line that is parallel to the H plane is
horizontal. A horizontal line, such
as AD, is a strike line and its bearing is N $\theta°$ E.

*The dip angle is the angle be-
tween surface* ABC *and the* H *reference plane.* Once we obtain the
point view of line AD, we will see
the edge view of surface ABC and

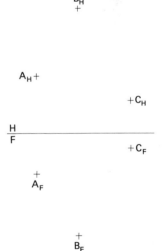

Figure 6.64 Points A, B, and C determine the top surface of an ore body.
Problem: Determine the bearing of a
"strike" line, the dip angle, and the
direction of dip.

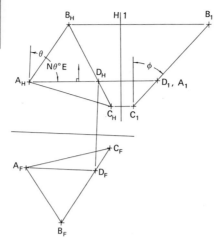

Figure 6.65 Solution to the problem
in Figure 6.64.

of the H reference plane. The view
on supplementary plane 1, which is
perpendicular to line AD and the
H plane, shows AD as a point, the
edge view of surface ABC and of
the H plane, and the dip angle, ϕ.
*Note that the dip angle is measured
below the H plane (see Figure 6.65).*

*The direction of dip is indicated
by the arrow that is perpendicular
to the strike line, as observed in the
top view. In this problem the direction of dip is northwesterly.*

ANGLE BETWEEN LINES

In addition to the examples cited in
the introduction, there are other
problems in design work that require the determination of the
"angle of bend" (i.e., of reinforcing
steel bars, pipe members, and rods
(i.e., that are used to connect an
accelerator pedal to a carburetor).
Basically, these problems reduce to
that of finding the angle between
lines.

EXAMPLE
Let us assume that the lines are
represented by the top and front
views shown in Figure 6.66. How
shall we determine the angle between them?

ANALYSIS AND SOLUTION
We can introduce a new line across
the given lines, thus forming a
triangle. Once we find the true
shape of the triangle, we can measure the magnitude of the angle between the lines.

Line AB, a frontal line, has been
introduced to form triangle ABC.

We can easily obtain an edge view
of triangle ABC, since we have a
true length view of line AB. The
view on supplementary plane 1
shows the point view of line AB
and the edge view of surface ABC.
Now it is a simple matter to introduce plane 2 parallel to surface
ABC and obtain the true shape of
triangle ABC. This is shown as
$A_2B_2C_2$. The magnitude of angle θ
is the required angle between lines
m and n.

In the event that the lines are
skew, the angle between them is
the same as the angle between intersecting lines that are respectively parallel to the skew lines.

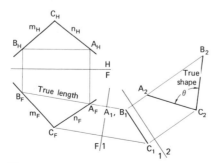

Figure 6.66 Angle between intersecting
lines.

ANGLE BETWEEN A LINE AND A PLANE

*The angle between a line and a
plane is defined as the angle between the line and its orthogonal
projection on the plane.* The orthogonal projection can be obtained
by (1) selecting two points on the
line; (2) establishing a line through
each point perpendicular to the
plane; and (3) finding the points in
which these lines intersect the
plane. The line that joins these
points is the required orthogonal
projection.

Much of this procedure can be
eliminated once we recognize that
*the angle between the given line
and one of the perpendiculars to
the plane is complementary to
the angle between the line and the
plane.* Therefore, we can find the
angle between a line and a plane by
(1) selecting a point on the line; (2)
introducing through that point a
new line perpendicular to the
plane; and (3) finding the angle between the two lines. That angle,
say α, is complementary to the required angle, θ, or $\theta = (90° - \alpha)$.

We know how to find the angle
between two lines, so the determination of angle α is quite simple.
But how shall we establish the new
line perpendicular to the plane? To
answer this question, let us consider plane ABC and point K
shown in Figure 6.67.

The perpendicular, n, through
point K to plane ABC must be perpendicular to two lines in the
plane. We can choose a horizontal
line, such as AD, as one of the two
lines. The true length of line AD is
seen as A_HD_H. The line, n, must
have its top view, n_H, perpendicular to the true length view of AD.
(We recall that "when two lines are
perpendicular to each other, the
view on a plane that is parallel to
one of them will reveal a right
angle.")

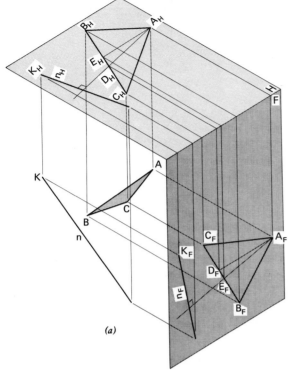

Figure 6.67 (*a*) Pictorial showing the
perpendicular, *n* through point *K*, to
plane of points *A, B,* and *C*
Figure 6.67 (*b*) Orthographic
representation part (*a*).

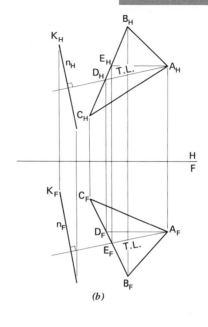

Figure 6.67 (*b*) Orthographic
representation of part (*a*).

Similarly, we can choose a frontal line such as *AE*. Its true length is seen in the front view as $A_F E_F$. Now line *n* must have its front view, n_F, perpendicular to the true length view of *AE*.

Since line *n* is perpendicular to both lines *AD* and *AE* of plane *ABC*, it is perpendicular to plane *ABC*.

EXAMPLE 1

The orthographic representation of line *m* and plane surface *ABCD* is shown in Figure 6.68. The angle, θ, between the line and plane will be determined by (1) establishing a perpendicular, *n*, through point *K* of line *m* to the plane; (2) finding the angle, α, between lines *m* and

n; and (3) determining angle θ from the relation, $\theta = (90° - \alpha)$.

In Figure 6.68, *BE*, a horizontal line, has been established in the plane *ABCD*. The true length of *BE* is seen as $B_H E_H$. We know that n_H must be perpendicular to $B_H E_H$. Now we can introduce line *AC*, a frontal line in plane *ABCD*. The true length of *AC* is seen as $A_F C_F$. The front view of *n* must pass through K_F and perpendicular to $A_F C_F$.

Now that we have established the views of line *n*, we can proceed to obtain the angle, α, between lines *m* and *n*. To do this we can easily introduce line *ST*, a frontal line, to intersect lines *m* and *n*, to form triangle *KST*. Since *ST* is a frontal line, its true length is seen

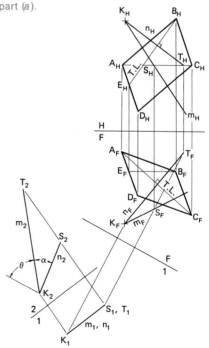

Figure 6.68 Angle, θ, between line *m*
and plane *ABCD*.

as $S_F T_F$. It is now a simple matter to establish supplementary plane 1 perpendicular to line ST and obtain a view on plane 1, which shows triangle KST on edge. Plane 2, which shows the true shape of triangle KST, is parallel to KST. In the true shape view we see angle α, and we can construct its complement, angle θ, as shown in Figure 6.68.

EXAMPLE 2

Suppose it is necessary to determine the angles that a line makes with the reference planes H and F.

Let us consider the problem of determining the angles θ and ϕ that line m makes with the H and F reference planes, respectively (see Figure 6.69). Angle θ is the angle between line AB and the H reference plane. We can determine θ by introducing line AC perpendicular to the H plane, finding the angle, α, between lines AB and AC, and then determining θ from the relation $\theta = (90° - \alpha)$. In the figure, the right triangle ABC has been rotated about line AC, as an axis, and parallel to the F reference plane. The view $A_F C_F B_F^R$ shows the true shape of the triangle ACB, the angle α between AB and AC, and the required angle, θ.

We can find the angle, ϕ, between line AB and the F reference plane, in a similar way. In Figure 6.70, we can introduce line AC perpendicular to the F plane. With AC as the axis of rotation, we can rotate triangle ACB until it is parallel to the H plane. The view $A_H C_H B_H^R$ shows the true shape of triangle ACB, the angle, β, between lines AC and AB, and ϕ, the

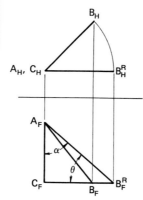

Figure 6.69 Angle θ, between line AB and the H reference plane.

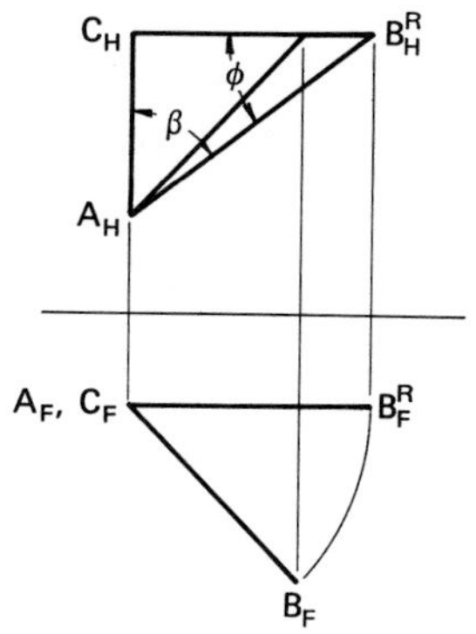

Figure 6.70 Angle ϕ, between line AB and the F reference plane.

required angle between line AB and the F reference plane.

EXAMPLE 3

The angle between a line and a plane is seen in a view that shows the line in true length and the plane on edge.

Let us consider the views of plane ABC and line DE as shown in Figure 6.71. It should be quite evident that the F view shows the true length of line DE and the plane ABC on edge. The angle, θ, is the angle between the line, DE, and the plane *"because it is the angle between the line and its orthogonal projection on the plane ABC."* Perpendiculars from points D and E to the plane ABC intersect that plane at points 1 and 2, respectively. The line joining those points is the orthogonal projection of line DE on plane ABC. The front view shows the true lengths of both lines DE and 1-2 and, therefore, the true angle, θ, between them.

EXAMPLE 4

Let us now consider a general case as shown in Figure 6.72. We should have no difficulty in finding two adjacent views that will disclose the information shown previously in Figure 6.71, that is one view similar to the top view and an adjacent view like the front view. The top view shows the true shape of surface ABC; therefore, let us proceed to obtain a true shape view of surface FGK and the corresponding view of line RS.

We can obtain an edge view of surface FGK by finding the point view of line KT, which is seen in true length in the front view. Supplementary plane 1 which is perpendicular to line KT, shows the edge view of surface FGK and the corresponding view of line RS. It is a simple matter to introduce plane 2 parallel to surface FGK and obtain the true shape view of that surface and the corresponding view of line RS. Finally, we can introduce plane 3 parallel to line RS and perpendicular to plane 2 to obtain a view that shows the true length of line RS, the surface FGK on edge, and the angle, θ, between the line and the surface.

From a practical point of view, this solution is ideal, because the views on planes 2 and 3 are best suited to the preparation of a working drawing that would be needed for the production, for example, of a plate (FGK) and a connecting member (centerline RS).

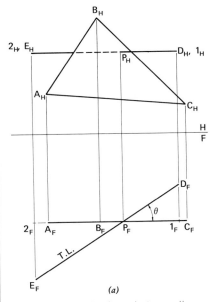

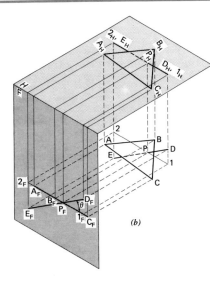

Figure 6.71 (a) Angle, θ, between line
DE and plane *ABC*, shown
orthographically
Figure 6.71 (b) Angle, θ, between line
DE and plane *ABC*, shown pictorially.

EXAMPLE 5

In some problems it is only necessary to determine the angle between a line (i.e., centerline of a cylindrical member) and a plane surface (i.e., steel plate). In such cases the solution can be simplified.

Let us consider plane *ABC* and line *DE* as shown in Figure 6.73. We note that the edge view of the plane is seen in the front view, but not the true length of line *DE*. We can, however, rotate the line, about the vertical, *DG*, until it is parallel to the *F* plane. The front view, $D_F E_F{}^R$, shows the true length of *DE* and the required angle, θ. **Remember that point E moved in a plane parallel to the surface ABC.** The importance of this statement will become evident in the following example.

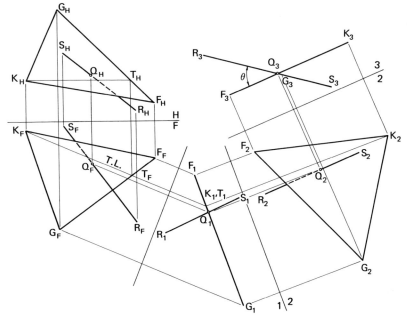

Figure 6.72 Angle, θ, between line *RS*
and plane *FGK*. General case.

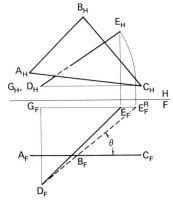

Figure 6.73 Angle, θ, between line *DE*
and plane *ABC*. Special case.

can introduce plane 2 parallel to the plane of the tibia and pin. The view on plane 2 shows the true shape of the tibia-pin plane, and the corresponding view of the femur, AB.

Let us concentrate on the views on supplementary planes 1 and 2. We can rotate line AB about BG as an axis until line AB is parallel to plane 1. The view on this plane now shows the true length of line AB as $A_1^R B_1$, and the required

angle, θ. Again, note that point A (i.e., A_1) moved in a plane parallel to the surface of the tibia-pin plane, which is shown on edge in the view on plane 1.

EXAMPLE 6

In one of the studies related to the design of artificial limbs it was necessary to find the angle between the femur (thigh bone) and the plane determined by the tibia (shin bone) and a 2.5 mm stainless steel pin that had been screwed into the tibia.

The graphical representation of the problem is portrayed in Figure 6.74. The solution is shown in Figure 6.75. In order to obtain an edge view of the plane (tibia and pin), supplementary plane 1 has been introduced perpendicular to line DF. The view on plane 1 shows line DF as a point, the edge view of the plane (tibia-pin), and the corresponding view of the femur, which is represented by line AB. Now we

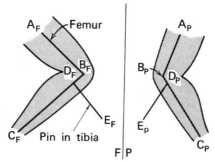

Figure 6.74 Angle between a femur (line AB) and a plane determined by a tibia (line BC) and pin, DE.

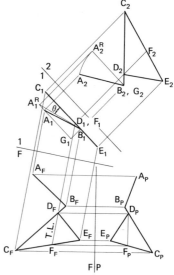

Figure 6.75 Orthographic solution to the problem in Figure 6.74.

Exercises

True Length, Grade, and Bearing of a Line Segment

1. Determine the true length, grade, and bearing of line segment AB. Dimensions are in millimeters (see Figure E-6.1).

2. Determine the true length, grade, and bearing of line segment CD. Dimensions are in millimeters (see Figure E-6.2).

3. Locate the H and F views of line segment AB, which is 60 mm long, bears N 45° W, and has a grade of 50%. Use the rotation method (see Figure E-6.3).

4. Locate the H and F views of line segment CD, which bears S 60° E and has a grade of −60%. Solve by the rotation method. Also find the true length of CD in millimeters (see Figure E-6.4).

5. A portion of a penstock runs from

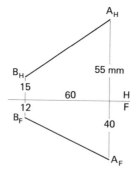

Figure E-6.1

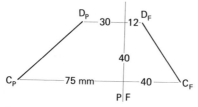

Figure E-6.2

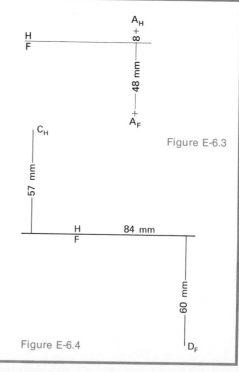

Figure E-6.3

Figure E-6.4

point A to point B, a distance of 120 m. The elevations of A and B are 685 m and 650 m, respectively. The bearing of the penstock is N 55° W. Determine (a) the H and F views of point B, and (b) the grade of AB (see Figure E-6.5).

6. Prepare two orthographic views of the space frame shown in Figure E-6.6 and then determine the true lengths of each member. Tabulate the results. Measure in millimeters.

7. *Review Exercise.* Locate the H, F, 1, and 2 views of line segment AB (see Figure E-6.7). Also determine the true length, grade, and bearing of the line. Show on drawing where measured. Give the true length measurement in millimeters.

Intersecting, Skew, Parallel, and Perpendicular Lines

1. Using instruments, reproduce the views shown in Figure E-6.8 and then determine if point O is or is not in the plane of which triangle ABC is a part.

2. Using instruments, reproduce the views shown in Figure E-6.9; then pass a plane (made up of two lines m and n) through point O and parallel to plane surface QRST. Also complete the H view of surface QRST.

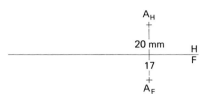

Figure E-6.5

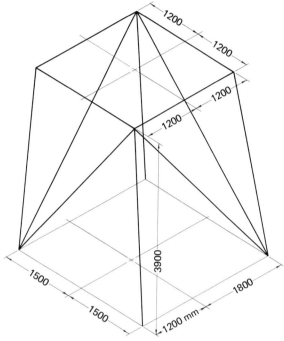

Figure E-6.6

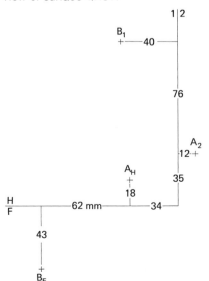

Figure E-6.7

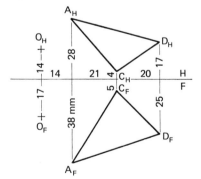

Figure E-6.8

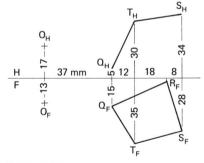

Figure E-6.9

3. Using instruments, reproduce the views shown in Figure E-6.10; then establish the H and F views of line m, which passes through point K and is parallel to both the H reference plane and plane $ABCD$.

4. Reproduce the views shown in Figure E-6.11 (a freehand sketch approximation is sufficient) and then determine the H and F views of line n, which passes through point P and is perpendicular to the plane of lines r and s.

5. Reproduce the views shown in Figure E-6.12 and then determine the H and F views of line m, which passes through point X and is perpendicular to plane ACD. Use the two given views only.

6. Reproduce the views shown in Figure E-6.13 on a $\frac{1}{4}$-in. grid sheet; then determine the H and F views of plane nk, which passes through point P, is parallel to line m and is perpendicular to plane rs.

7. *Review Exercise.* Using instruments, reproduce the views shown in Figure E-6.14 and then do the following.

(a) Complete the F view of the base of a right pyramid O-$RSTU$, whose axis extends down, forward, and to the left from the center of the base. The true length of the axis is 60 mm.

(b) Determine the correct visibility of the edges of the pyramid.

The Point View of a Line

1. Reproduce the views shown in Figure E-6.15 on a $\frac{1}{4}$-in. grid sheet; then locate the H and F views of the centerline of the shortest connector pipe (EF) between the pipelines AB and CD (point E is on AB). Show the true length of EF.

2. Reproduce the views shown in Fig-

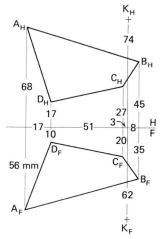

Figure E-6.10

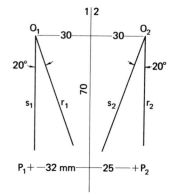

Figure E-6.11

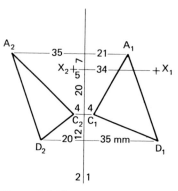

Figure E-6.12

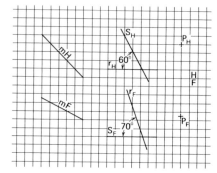

Figure E-6.13

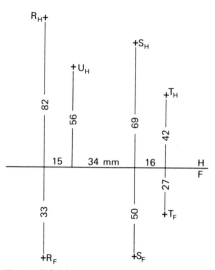

Figure E-6.14

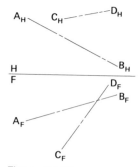

Figure E-6.15

ure E-6.16 on a ¼-in. grid sheet; then (a) determine the clearance between cables CD and EG; and (b) locate the new position of point E, which can be moved along a line through point E and perpendicular to the F plane, so that the new clearance between the cables will be twice the original clearance. How long is cable EG in the new position?

3. It is critical that the minimum clearance between electrical conductors AB and CD in Figure E-6.17 be at least 600 mm. Does the actual clearance exceed the minimum? If so, by how much? Reproduce the views shown in the figure and then solve the problem.

4. Determine the diameter of the circular cylinder in Figure E-6.18 whose axis is AB. Point C is a point on the surface of the cylinder. Measure the diameter to the nearest millimeter.

The Edge View of a Plane Surface

1. Points A, B, and C are (in Figure E-6.19) on the upper surface of an ore body. On the lower parallel surface is point D. How thick is the ore body? (Metric scale: 1:500.) Reproduce the views shown and then solve the problem.

2. Assume the H and F views of a plane surface (i.e., triangle ABC, which is *not* parallel to either the H or F reference plane). Determine the H and F views of triangle DEF, which is parallel to triangle ABC and 25 mm from ABC.

3. Determine the perpendicular distance from point P to plane surface ABC. Reproduce the views shown in Figure E-6.20 and then solve the problem. Measure the distance to the nearest millimeter.

4. Assume the H and F views of a line segment AB (which is neither parallel nor perpendicular to H or F). Now determine the H and F views of the plane, which is perpendicular to line segment AB and passes through its midpoint.

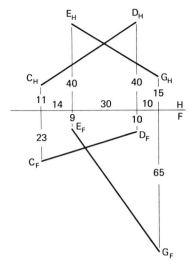

Figure E-6.16

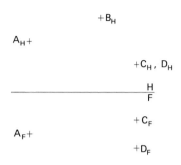

Figure E-6.19

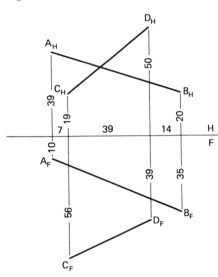

Figure E-6.17

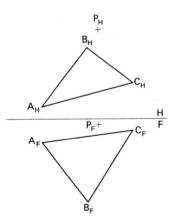

Figure E-6.20

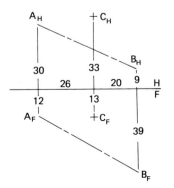

Figure E-6.18

The plane may be designated by two intersecting lines *m* and *n*.

Connectors Between Two Skew Members

1. Reproduce the *H* and *F* views shown in Figure E-6.21; then locate the views of the shortest horizontal connector between shafts *AB* and *CD*.

2. Reproduce the views shown in Figure E-6.22; then locate the *H* and *F* views of the shortest tunnel *EF*, from tunnel *AB* to tunnel *CD*, which is to have a −25% grade. Also find the true length of the connector tunnel. (Metric scale: 1:1000.)

True Shapes of Plane Surfaces

1. Reproduce the views shown in Figure E-6.23; then determine the true shapes of surfaces *A* and *B*. What are the areas of the surfaces in square meters?

2. Determine the diameter of the largest circular plate that can be cut from plate *DEF*. Reproduce the views shown in Figure E-6.24 and then proceed with the solution.

3. The centerline of a pipe from *P* to the centerline of pipe *A–B* forms an angle of 50°. The intersection of the centerline, point *K*, is closer to point *A* than it is to point *B*. Determine the location of point *K* and the true lengths of *AB* and *PK*. Reproduce the views shown in Figure E-6.25; then proceed with the solution.

4. Assume the *H* and *F* views of (*a*) a line *m* and an external point *P*; and (*b*) two intersecting lines *s* and *t*. Using the "fourth basic problem—true shape of a plane surface," find the perpendicular

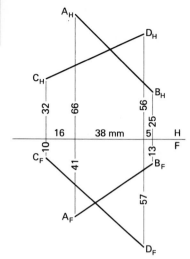

Figure E-6.21

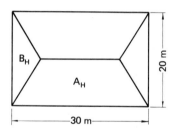

Figure E-6.23

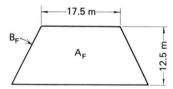

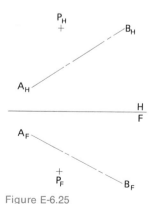

Figure E-6.25

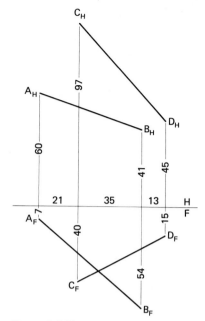

Figure E-6.22

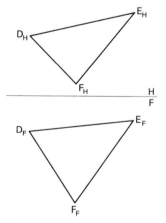

Figure E-6.24

distance from point P to line m, and the angles between lines s and t.

5. Reproduce the H and F views of the concrete footing shown in Figure E-6.26. Find the areas of the sloping surfaces and determine the board feet of lumber necessary to construct the form work for casting the concrete. (A board foot is 144 in.^{3}; allow 15% for spoilage.)

6. AC is the centerline of a mining shaft. The centerline of another shaft, BD, intersects AC at an angle of 90°, and is the same length as AC. Reproduce the views shown in Figure E-6.27 then locate the H and F views of centerline BD.

7. A portion $ABCD$ of a structural frame is shown in Figure E-6.28. In order to detail the steel members, it is necessary to find a true shape view. Reproduce the views shown in the figure on a $\frac{1}{4}$-in. grid sheet; then proceed to the solution of the problem.

Angles

1. Find the angles between the lateral surfaces of a portion of the package chute shown in Figure E-6.29. Show the angles where measured.

2. Points A, B, and C in Figure E-6.30 lie on the upper surface of an ore body. Point D lies on the lower parallel surface and is 25 m directly below point B. Determine the strike, dip, direction of dip, and thickness of the ore body. Also locate the front view of point D. (Metric scale: 1:600.)

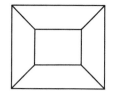

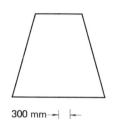

300 mm

Figure E-6.26

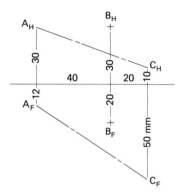

Figure E-6.27

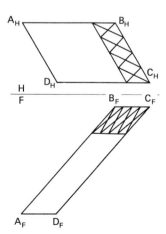

Figure E-6.28

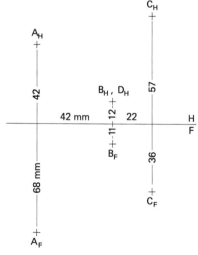

Floor

Figure E-6.29

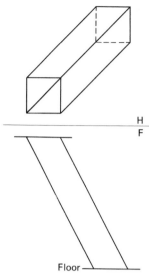

Figure E-6.30

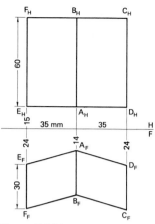

Figure E-6.31

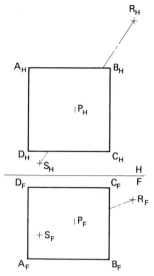

Figure E-6.32

3. The *H* and *F* views of a roof surface are shown in Figure E-6.31. It is planned to provide a sheet metal ridge along the intersection of surfaces *ABCD* and *AEFB*. Determine the angle of bend for the metal ridge. Reproduce the views and show the angle where measured.

4. Assume the *H* and *F* views of a line segment *AB*. Determine the angles θ, ϕ, and ψ that line *AB* makes with the *H*, *F*, and *P* reference planes, respectively.

5. The centerline of a control cable, *RS*, and a bulkhead (plane surface), *ABCD,* are shown in Figure E-6.32. The intersection, point *P*, of *RS* and the bulkhead is also shown. Reproduce the given views and then determine the magnitude of the angle, θ, between the cable and the bulkhead; show the correct visibility of *RS* in the *H* and *F* views.

6. Assume the *H* and *F* views of two intersecting lines *m* and *n*. Determine the *H* and *F* views of line *s* that makes an angle of 90° with the plane *mn* and passes through point *P,* which is the intersection of lines *m* and *n*.

7. *AE*, *BE*, and *CE* are guy wires fastened to an antenna support, only the centerline of which is shown in Figure E-6.33. Reproduce the views shown; then determine the true length of each guy wire and the angle each makes with the surface to which it is attached. Tabulate the true lengths and the angles.

8. Reproduce the views shown in Figure E-6.34, then determine the magnitude of the angle, θ, between rod *OC* and the plane of members *OA* and *OB*. Show the angle where measured.

9. Reproduce the views shown in Figure E-6.35; then determine the magnitude of the angle, θ, between member

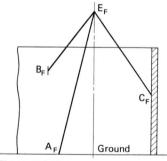

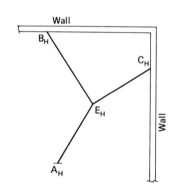

Figure E-6.33

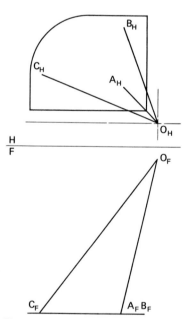

Figure E-6.34

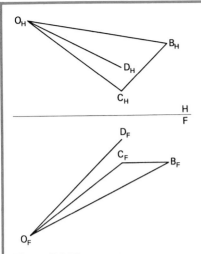

Figure E-6.35

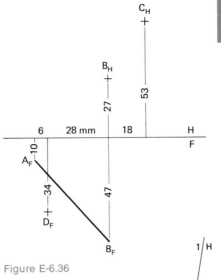

Figure E-6.36

DO and plate *OBC*. Show the angle where measured.

10. Rod *CD* has a bearing of S 60° W and a grade of −30%. Rod *AB* is perpendicular to *CD*. It is required to set up a plane *CDE* that makes an angle of 30° with rod *AB*. Point *E* lies on rod *AB* and is nearer to *B*. Show the *H* and *F* views of the plane *CDE* (see Figure E-6.36).

Faults

1. *Sample exercise.*
Given:
(a) The direction of movement.
(b) The offset of one vein.
  Required:
  The displacement of the fault.
  Known Data:
(a) Fault *Q* has a dip angle of 45°.
(b) Fault *Q* has a strike N 80° W.
(c) Vein *K* and its offset portion *R* have a dip angle of 30° and a strike N 30° E; *K* and *R* intersect *Q* at *E* and *D* respectively.
(d) Direction of the displacement on the fault is due south.
  Solution:
(a) Establish the strikes of *Q*, *K*, and *R* (see Figure E-6.37).
(b) Introduce supplementary plane 1, which is perpendicular to the *H* plane and the fault plane.

(c) Introduce supplementary plane 2, which is perpendicular to the plane of vein *R*.
(d) Find line *AD*, the intersection of planes *Q* and *R*.
(e) Establish a line through point *E*, due south and intersecting *AD* at *F*. *EF* is the *H* projection of the displacement on *Q*.
(f) Locate F_1. Now we have two views of the required line *EF*.
(g) Find the true length of *EF*. This is shown as EF^1.

Figure E-6.38 should aid you in forming a space conception of simple faulting.

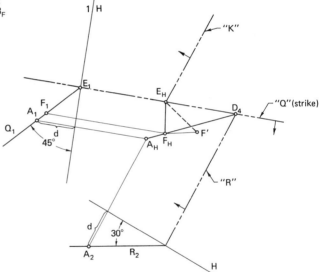

Figure E-6.37

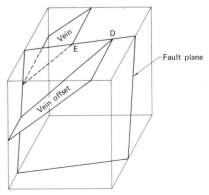

Figure E-6.38

2. *Given:*

(a) Fault Q and its dip angle (Figure E-6.39).

(b) Vein K and its offset portion R, with the dip angle as shown.

(c) Vein T and its offset portion W, with the dip angle as shown.

Required:

The displacement of the fault plane Q.

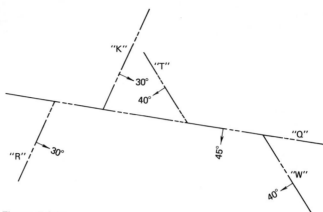

Figure E-6.39

Chapter 7

Developments and Intersections in Design

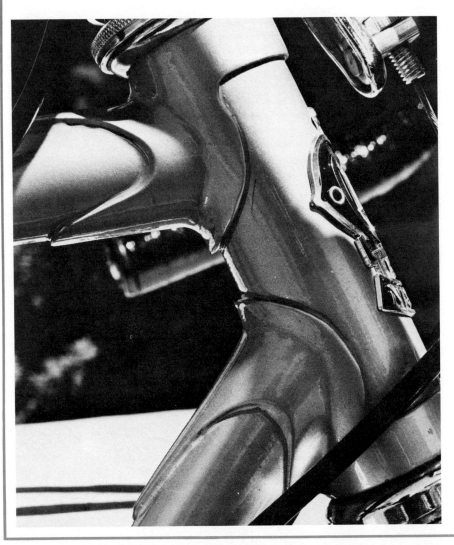

INTRODUCTION: DEVELOPMENTS

Developments, or patterns, are usually full-size layouts of various shaped surfaces.

Many engineering designs include parts and components that have plane, cylindrical, or conical surfaces. In a number of cases it is necessary to develop the surfaces to provide patterns for cutting and forming the material to the designed shapes. For example, in the fabrication of sheet-metal ducts for heating and ventilating systems, it is necessary to develop the lateral surfaces of the duct system to provide the patterns for cutting and bending the material to satisfy the requirements of the design. A partial duct system is shown in Figure 7.1. Development of surfaces A, B, C, D, . . ., would provide the patterns for cutting and bending the metal to the required shapes. Allowances for the seams and joints and the thickness of the material must be made to accommodate the angles of bend.

Developments of cylindrical and conical surfaces provide the patterns for cutting and forming the material for the fabrication of, for example, the "chute" shown in Figure 7.2.

In a number of designs double-curved surfaces occur, such as in the "tank" shown in Figure 7.3. Such surfaces can be developed, approximately, by assuming that they consist of small portions of developable surfaces, as cylinders or cones. An example will be given later. Also, some designs include warped surfaces, such as the "conoid surface" shown in Figure 7.4. Warped surfaces can be developed fairly accurately by assuming that they consist of relatively small triangles that are easily arranged to form the pattern.

The theory employed in making a development from the orthographic views of surfaces is quite simple. *Essentially it consists of determining the true lengths of the elements of the given surfaces (or of the substituted surfaces) and the correct placement of one element with respect to an adjacent element.*

Let us now consider the development of the surfaces of prisms, cylinders, pyramids, and cones.

DEVELOPMENT OF SURFACES OF PRISMS

Prisms are solids whose lateral surfaces are parallelograms and whose bases lie in parallel planes. Examples of right, oblique, and truncated prisms are shown in Figure. 7.5.

EXAMPLE 1
Let us develop the lateral surfaces of the truncated prism shown in Figure 7.6. The true lengths of the lateral edges are seen in the front view. The perpendicular distances

Figure 7.1 Duct system for ventilation; shows intake unit.

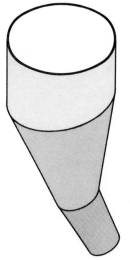

Figure 7.2 "Chute" made up from cylindrical and conical portions.

Figure 7.3 Spherical storage tank.

Figure 7.4 Conoid roof surfaces are used in building construction.

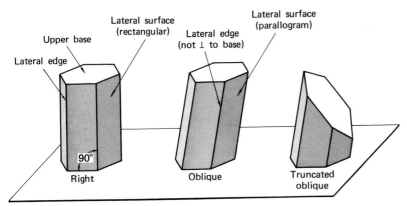

Figure 7.5 Types of prisms.

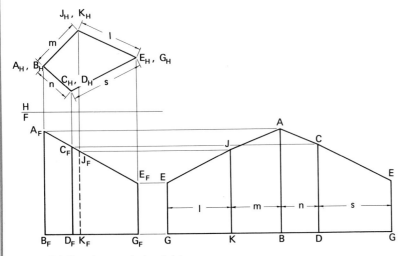

Figure 7.6 Development of a right truncated prism.

between the lateral edges are seen in the top view, since this view shows the edges as points. We can start the layout of the development, or pattern, by locating edge EG parallel to $E_F G_F$ and a convenient distance from it, as shown in the figure. The distances l, m, n, and s between edges EG and JK; JK and AB; AB and CD; and CD and EG, respectively, are laid off as shown. The true lengths JK, AB, CD, and EG are projected from the front view, and the lateral surfaces are completed as shown in the development.

EXAMPLE 2

Suppose the top and front views of the prism are oriented as shown in Figure 7.7. We observe that the true lengths of the lateral edges are available in the front view, since the edges are parallel to the F plane.

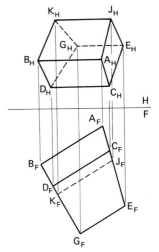

Figure 7.7 Orthographic views of a truncated prism. *Problem:* To develop its lateral surface.

In order to obtain the perpendicular distances between the edges. it will be necessary to obtain a view (Figure 7.8) that shows the edges as points. The view on supplementary plane 1, which is perpendicular to the edges, discloses the distances between them. We can now proceed to the layout of the development in a way similar to that employed in the previous example. It should be recognized that the orientation of the prism could be such that neither the top nor the front view would show the true lengths of the edges. In that case we could introduce one supplementary plane to obtain a view that would disclose the true lengths of the edges, and then another supplementary plane that would show the distances between the edges. The problem would then be reduced to the orientation shown in the previous Example 1.

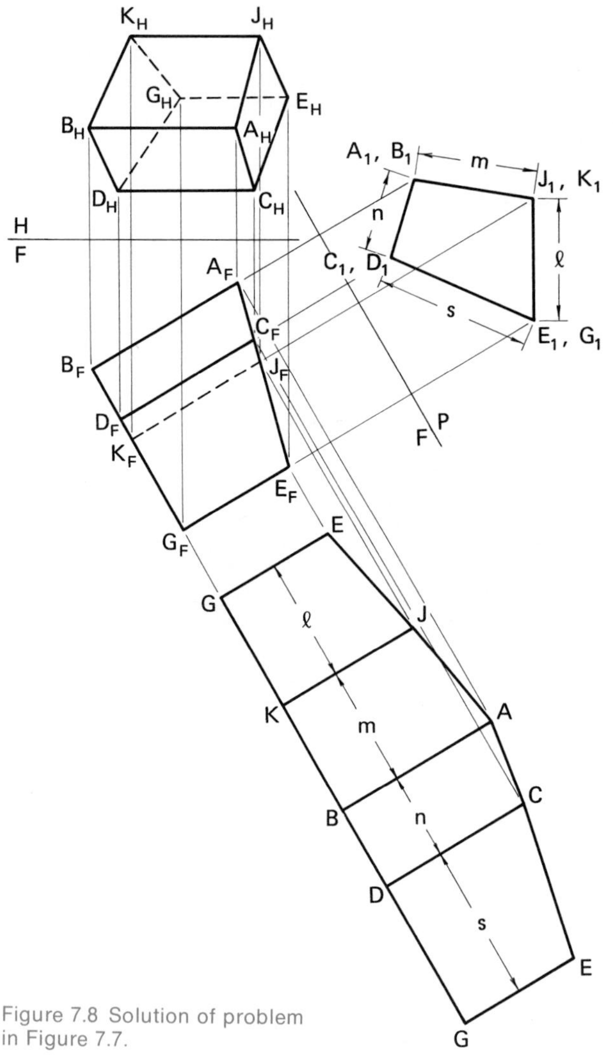

DEVELOPMENT OF SURFACES OF CYLINDERS

The development of the lateral surfaces of cylinders requires only the determination of "the true lengths of the elements and the correct placement of one element with respect to an adjacent element." Since we cannot, from a practical point of view, use all (the infinite number) the elements, we choose a sufficient number to define the development within the desired accuracy. By doing so we have, in effect, replaced the cylinders with prisms, so that the method employed in developing surfaces of cylinders is *basically* the same as that used in developing surfaces of

Figure 7.8 Solution of problem in Figure 7.7.

prisms. When a high degree of accuracy is required, the "stretch-out length" of the development can be determined by calculating the circumference of the cylinder.

EXAMPLE 1
Let us consider the circular cylinder shown in Figure 7.9. The top view shows the point views of the elements and, therefore, the distances between them. All the elements are shown in true length in

the front view. Now we have all the information necessary for the layout of the development. We note that the method is the same as that used in developing the lateral surface of the prism shown in Figure 7.6. The end points of the elements, shown in the development, are joined by smooth curves.

EXAMPLE 2
Now suppose the cylinder is represented by the front and profile

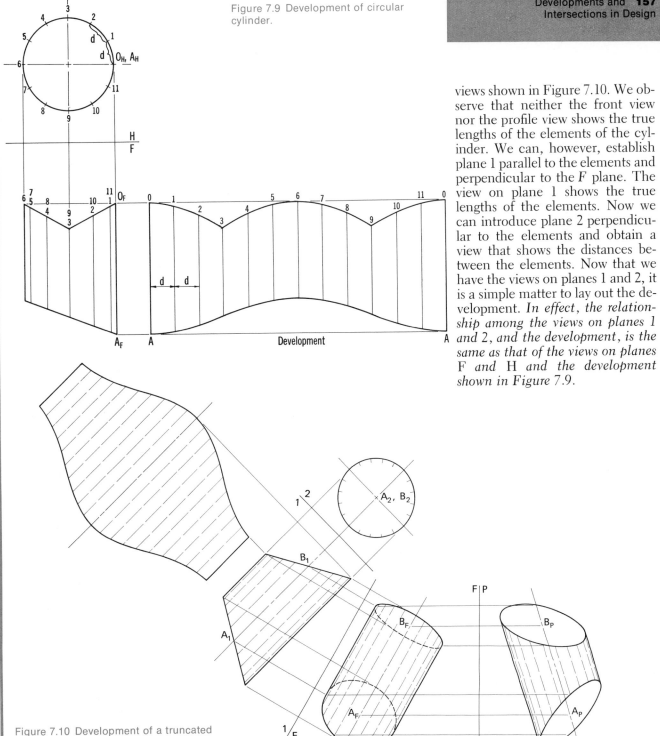

Figure 7.9 Development of circular cylinder.

views shown in Figure 7.10. We observe that neither the front view nor the profile view shows the true lengths of the elements of the cylinder. We can, however, establish plane 1 parallel to the elements and perpendicular to the F plane. The view on plane 1 shows the true lengths of the elements. Now we can introduce plane 2 perpendicular to the elements and obtain a view that shows the distances between the elements. Now that we have the views on planes 1 and 2, it is a simple matter to lay out the development. *In effect, the relationship among the views on planes 1 and 2, and the development, is the same as that of the views on planes F and H and the development shown in Figure 7.9.*

Figure 7.10 Development of a truncated circular cylinder described by its front and profile views.

EXAMPLE 3

The surface of an elliptical cylinder can be developed in the same way as in the previous examples. However, the elements are *not* equally spaced, since the curvature varies. In Figure 7.11, we observe that the elements are closer together where the curvature is sharper and further apart where the curvature is flatter. The development is made in the same manner employed in the previous Example 1.

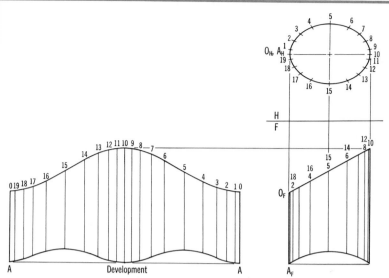

Figure 7.11 Development of an elliptical cylinder.

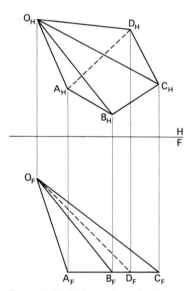

Figure 7.12 Oblique pyramid. *Problem:* Develop its lateral surface.

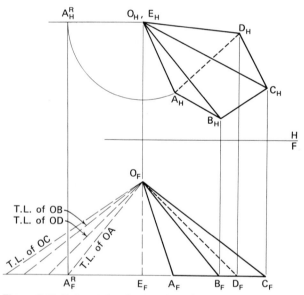

Figure 7.13 Determining the true length of the edges of the pyramid by the method of rotation.

DEVELOPMENT OF SURFACES OF PYRAMIDS

EXAMPLE 1

Let us consider the pyramid described by the two orthographic views shown in Figure 7.12. A pattern, or development, of the lateral surface of the pyramid can be made by joining the true shapes of triangles *OAB*, *OBC*, *OCD*, and *OAD*. The true shape of each triangle can be constructed once we have determined the true lengths of the sides of the triangle.

We can easily determine the true lengths of edges *OA*, *OB*, *OC*, and *OD* by the method of rotation, as shown in Figure 7.13. The true lengths of edges *AB*, *BC*, *CD*, and *DA* are available in the top view, since the plane *ABCDA* is parallel to the *H* plane.

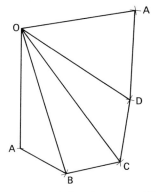

Figure 7.14 Development of the lateral surface of the pyramid in Figure 7.12.

We can lay out the patterns in the following manner.

1. Locate the true length of OA in an arbitrary position, as shown in Figure 7.14.

2. With point A as a center and the true length of AB as a radius, an arc is drawn; and with O as a center and the true length of OB as a radius, an arc is drawn to intersect the first arc at point B.

3. Join points A and B and points O and B to form the true shape of triangle OAB.

4. With points O and B as centers and the true lengths of OC and BC as radii, respectively, arcs are drawn to intersect at point C. Lines BC and OC are drawn to form the true shape of triangle OBC.

In the same way, we can construct the true shapes of triangles OCD and ODA to complete the development of the lateral surface of the pyramid.

EXAMPLE 2

Let us consider the truncated pyramid shown in Figure 7.15. The true lengths of edges OA, OB, and OC are easily obtained. The true length of OA is seen in the front view as $O_F A_F$. We can rotate edge OC into a position that is parallel to the F plane. The view $O_F C_F{}^R$

shows the true length of OC. Edge OB is the same length as edge OC. The development of the lateral surface of pyramid O-ABC is laid out in the same way as shown in the preceding Example 1.

On edge OA, we can lay off the true distance AD, which is seen as $A_F D_F$. (Why is this true?) The true lengths of BE and CF (which are the same) are found on the true length view of DC and are shown as $C_F{}^R F_F{}^R$. This length is laid off on edges OB and OC to locate points E and F. The completed development is shown in Figure 7.16.

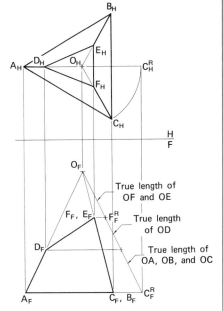

Figure 7.15 Truncated pyramid.
Problem: Develop its lateral surface.

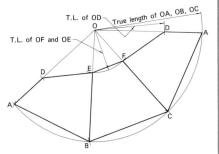

Figure 7.16 Development of the lateral surface of the pyramid in Figure 7.15.

DETERMINATION OF THE SURFACES OF A CONTAINER

Let us now consider the container shown in Fig. 7.17. First, let us determine the true lengths of the elements which, in this case, are regarded as the edges of the container. This can be done by the method of supplementary views or by the method of rotation.

The rotation method is more advantageous because the construction can be reduced to a minimum. Suppose we find the true length of edge DK. If we introduce a vertical line through point D and a horizontal line through point K intersecting the vertical line at point Q, we will have formed the

right triangle DQK. If we rotate this triangle about line DQ until it is parallel to the F-plane and then obtain a new front view, we will see the true length of DK as $D_F K_F{}^R$.

Now, if we observe that the altitudes of the right triangles formed in a similar manner for edges AE, BF, and CG will be the same length—equal to $D_F Q_F$ or $A_F S_F$—we can lay off the top view lengths of AE and BF at right angles to $A_F S_F$ through S_F and draw the hypotenuse for each right triangle. In a similar manner $C_H G_H$ is laid off at right angles to $C_F Q_F$ from Q_F. The hypotenuse of triangle $C_F Q_F G_F{}^R$ is then drawn. The hypotenuse lengths will be the true lengths of the edges. In this example they are shown as $A_F E_F{}^R$, $B_F F_F{}^R$, and $C_F G_F{}^R$, respectively.

The development, Figure 7.18, can now be started. Suppose we lay off a length equal to AD (true length is available in both the top

and front views). The true length of EK is also available in the top and front views. The location of edge EK with respect to edge AD cannot be established until we fix the location of either point E or point K. Let us agree to locate the position of point E. This can be done by drawing *a temporary line* from D to E. The true length of DE can be easily determined. It is shown as $D_F E_F{}^R$. Now, if we use A and D as centers and use radii respectively equal to the true lengths of AE and DE, we can draw arcs which intersect at point E. Line EK is now laid off parallel to AD, and line DK is drawn to complete the true shape of surface $ADKE$. If we introduce temporary line DG and find its true length, we can proceed in a similar manner to lay out the true shape of surface $DCGK$. The other two surfaces can be found similarly and laid out in proper sequence to form the complete development or pattern.

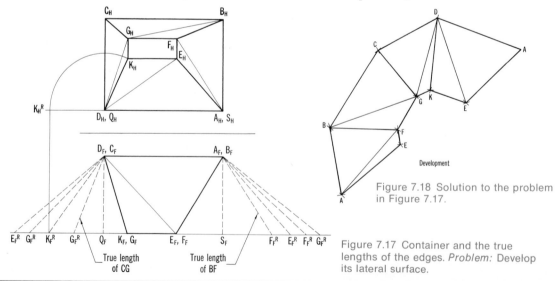

Development

Figure 7.18 Solution to the problem in Figure 7.17.

Figure 7.17 Container and the true lengths of the edges. *Problem:* Develop its lateral surface.

DEVELOPMENT OF THE SURFACES OF CONES

Patterns for the fabrication of designed conical elbows and of transition pieces often involve the de-

velopment of the surfaces of right circular and oblique cones. We will first consider the development of the surface of a right circular cone and its application to a conical

elbow design and we will then study an oblique conical surface and its application to a transition piece design.

EXAMPLE 1

We will consider the *right circular cone* shown in Figure 7.19. Since all of the elements are the same length, it is only necessary to find the true length of one element, such as OA, whose true length is seen in the front view, $O_F A_F$.

With O as the center and the true length of OA as the radius, an arc is drawn, as shown in the development. Point A is located, arbitrarily, on the arc, and element OA is drawn.

A simple way to complete the development is to calculate the magnitude of angle θ and then lay off this angle, with respect to element OA, to locate the boundary element of the pattern.

From the following relationship,

$$\frac{\theta°}{360°} = \frac{2\pi r}{2\pi S}$$

where r = radius of the circular base of the cone, and S = the slant height of the cone (equal to the true length of OA). It follows that

$$\theta° = \frac{r(360°)}{S}$$

Obviously, this solution obviates the need to introduce a number of elements and then to determine their proper locations in the development. Moreover, the method used is much more accurate.

EXAMPLE 2

Let us consider the three-piece, 90° turn, *conical elbow* shown in Figure 7.20. The dashed line outlines the frustum of the cone, which can be developed to provide the patterns for each piece of the elbow. It is a simple matter to determine the altitude of the full cone (in this case it is 1200 mm) and then proceed to lay out the surface of the full cone, as in the previous example. The patterns of the pieces of the elbow can now be laid out as shown in Figure 7.21.

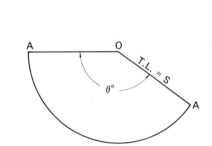
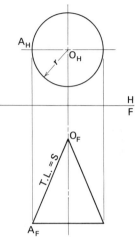

Figure 7.19 Right circular cone development of its lateral surface.

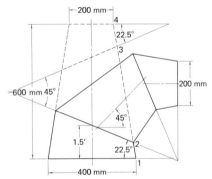

Figure 7.20 Three-piece conical elbow.
Problem: Develop its lateral surface.

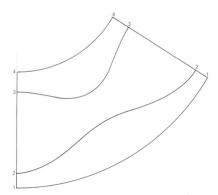

Figure 7.21 Development of the lateral surface of the three-piece elbow.

EXAMPLE 3

Suppose it is necessary to develop the surface of the *oblique cone* shown in Figure 7.22. The surface of the cone can be closely approximated by an inscribed pyramid whose apex is point A and whose base is the polygon 0, 1, 2, 3,- . . . . , 34, 35, 0. The true lengths of the lateral edges of the pyramid are found by the method of rotation. These are shown as $A_F 0_F{}^R$, $A_F 1_F{}^R$, . . . $A_F 18_F{}^R$. The sides of the base of the pyramid are shown in true length in the top view. Therefore, the true lengths of the sides of the triangles, A-0-1, A-1-2, A-2-3, . . . , are known and, hence, their true shapes can be constructed. When the triangles are joined as in Figure 7.23 and a smooth curve is drawn through points 0, 1, 2, 3, . . ., the pattern or development of the lateral surface of the cone is completed. Note that the cone is symmetrical about element A-0. In actual practice, therefore, it would only be necessary to lay out half the pattern.

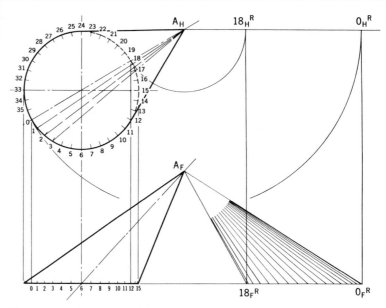

Figure 7.22 Oblique cone and true lengths of elements. *Problem:* Develop its lateral surface.

EXAMPLE 4

The design of ventilating systems quite often includes various types of ducts and transition units that are necessary to accommodate a change from a rectangular shape to one that is circular. For example, the photograph in Figure 7.24 shows such a transition unit that is part of the ventilating system in one of the library buildings on the Berkeley campus of the University of California.

The top and front views of a transition unit are shown in Figure 7.25a. The development consists of

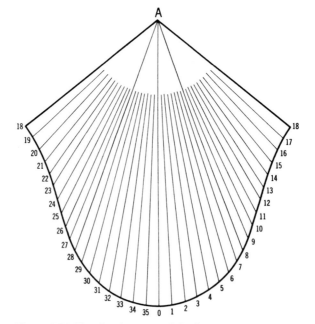

Figure 7.23 The development of the lateral surface of the oblique cone.

Figure 7.24 Transition piece in air-conditioning system. (University of California Library at Berkeley)

the true shapes of the triangles *DHC*, *AED*, *BAF*, and *BCG*, and the true shapes of smaller triangles such as *EKD*, joined together in proper sequence to form the patterns shown in Figure 7.25*b*.

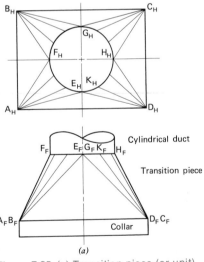

Figure 7.25 (*a*) Transition piece (or unit).
Figure 7.25 (*b*) Pattern (development) of transition piece.

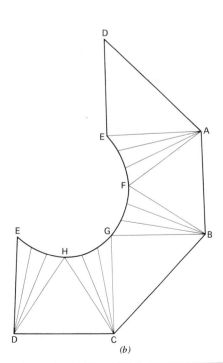

DEVELOPMENT OF A DOUBLE-CURVED SURFACE

In the introduction to this chapter mention was made of designs that include double-curved surfaces (Figure 7.3).

Let us now consider the problem of developing the spherical surface shown in Figure 7.26*a*.

Suppose the sphere is divided into an equal number of wedgelike segments formed by passing planes through the vertical axis *AB*. Now the surface of each segment can be approximated by a portion of a cylindrical surface. Since all the segments are the same, we will deal with the single segment *m* . . . A. The surface of this segment is approximated by the inscribed cylindrical surface with elements *m*, *n*, *r*, *s*, *o*, and point A (more elements could be used). Careful study of the top and front views shows that these elements are true length in the top view and that the distances between the elements are available in the front view, since they appear as points. The pattern or development can now be constructed. Element *m* can remain in its original position or be placed at a convenient distance from this position, as shown in Figure 7.26*b*. Element *n* is parallel to m and at a distance *d* from *m*, as seen in the front view. In a similar manner, elements *r*, *s*, *o*, and point A can be located. The

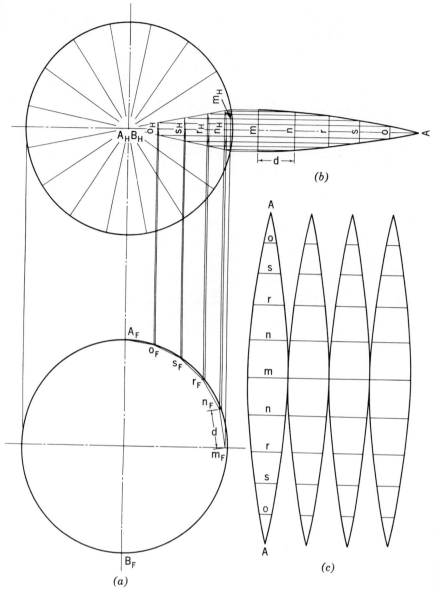

Figure 7.26 A sphere and its development (using wedges of the sphere).

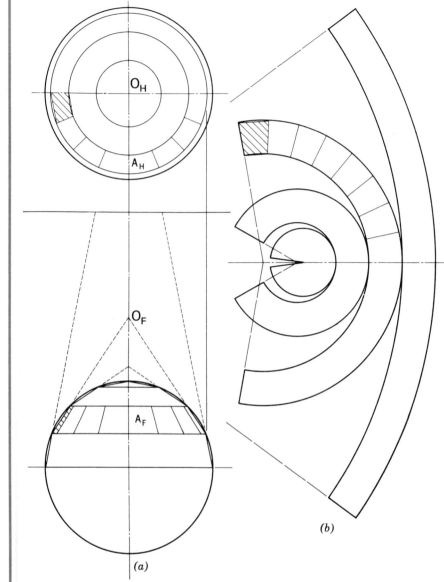

Figure 7.27 A sphere and its development (using zones of the sphere).

complete development of the sphere would consist of 16 areas, four of which are shown in Figure 7.26*c*.

Instead of using cylindrical surfaces to approximate the surface of the sphere, we can use conical surfaces. For example, in Figure 7.27*a*, the sphere has been divided into several horizontal zones whose surfaces can be closely approximated by the surfaces of frustums of right circular cones. For instance, the surface of the zone marked A is approximated by the surface of the frustum of the cone whose apex is point O. The bases of the frustum are seen in the front view. Now, the development of the surface of the frustum, which is shown in Figure 7.27*b*, is obtained, basically in the same way that was used in the development of the conical surface shown in Figure 7.19. In the same way we can obtain the developments of surfaces of the frustums of cones that closely approximate the spherical surfaces of the other zones. Developments of several zones for half the spherical surface are shown in Figure 7.27*b*. Problems dealing with cartography, surveying, and mapping make use of developments of spheres.

WARPED SURFACES

There are a number of warped surfaces, such as the *conoid*, *helicoid*, *cylindroid*, and *hyperbolic paraboloid* that are used in the design of roofs, package chutes, screw conveyors, etc.

Let us see how each of these surfaces is generated.

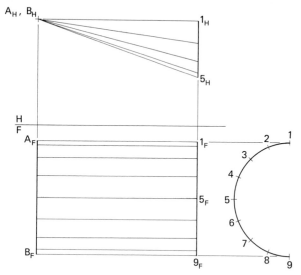

Conoid Surface

A *conoid surface* is generated by a straight line (the generatrix) that remains parallel to a plane (the plane director) while intersecting a straight line (the directrix) and a curved line (the other directrix). For example, in Figure 7.28 the front view shows several elements of the surface parallel to the *H* plane and intersecting line *AB* and the semicircle.

Figure 7.28 Conoid surface.

Helicoid Surface

A *helicoid surface* is generated by a straight line (the generatrix) that moves along two concentric helices (the directrices) while maintaining a constant angle with the axis of the helices. When the elements of the surface make an angle of 90° with the axis, the surface is identified as a *right* helicoid; otherwise, it is an *oblique* helicoid. Chutes and screw conveyors typify the right-helicoid surface. The sloping faces of thread forms are typical of the oblique-helicoid surface. The generation of a right-helicoid surface is shown in Figure 7.29, and an application of such a surface is shown in Figure 7.30.

Figure 7.30 (*a*) A helicoid surface.

Figure 7.30 (*b*) Screw conveyor at Mirassou Winery, San Jose, California, Fall 1977.

Cylindroid Surface

A *cylindroid surface* is generated by a line that remains parallel to a plane director while intersecting two curved line directrices that, generally, lie in nonparallel planes. In Figure 7.31, we observe that the elements of the surface are parallel to the *F* plane and intersect semicircle *AOB* and circular arc *CPD*.

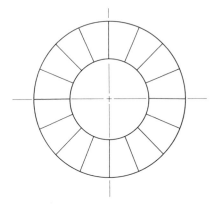

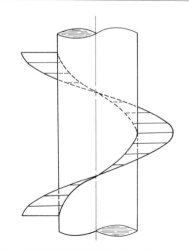

Figure 7.29 A right-helicoid surface.

Hyperbolic Paraboloid Surface

This surface is often used in the design of roof surfaces, concrete walls, culverts, etc. A *hyperbolic paraboloid* surface is generated by a straight line (the generatrix) that intersects two skew lines (directrices) and remains parallel to a

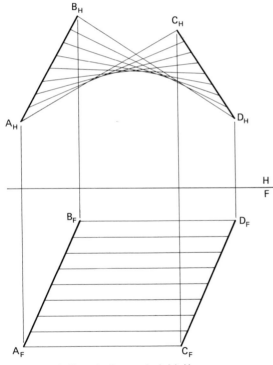

Figure 7.31 Cylindroid surface.

plane (director). As an example, let us assume skew lines AB and CD (Figure 7.32) as the directrices and the H plane as the director. The front view of elements of the surface will, of course, appear as horizontals, since the generatrix is parallel to the H plane. Several elements of the surface are shown in the two views.

The plane director need not be one of the reference planes. For example, in Figure 7.33, the plane director is perpendicular to the H plane. All of the elements of the

Figure 7.32 Hyperbolic paraboloid. H plane as director.

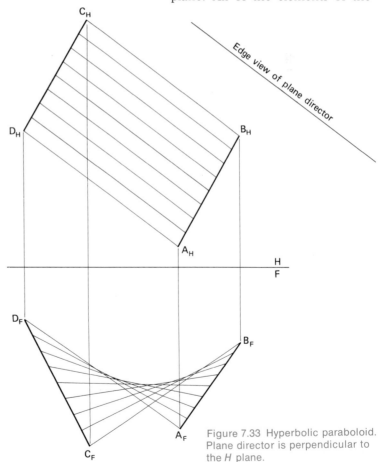

Figure 7.33 Hyperbolic paraboloid. Plane director is perpendicular to the H plane.

surface, in the top view, will be
parallel to the edge view of the
plane director. The top and front
views of nine elements of the surface are shown. An application of a
hyperbolic paraboloid surface is
shown in Figure 7.34.

DEVELOPMENT OF WARPED SURFACES

EXAMPLE 1

The warped transition piece shown
in Figure 7.35 is designed to connect a circular opening in the floor
with an elliptical opening in the
wall.

The warped surface is closely approximated by the triangles, as
shown in the top and front views.
The true lengths of the sides of the
triangles are easily found by using
the method of rotation. The development is made up of the true
shapes of the triangles arranged
in proper sequence to form the
pattern shown in Figure 7.36.

EXAMPLE 2

The transition piece shown in Figure 7.37 is designed to connect
two circular openings. Again, the
warped surface is approximated by
a number of relatively small triangles. The true shapes of the triangles have been constructed and arranged in proper sequence to form

Figure 7.34 Hyperbolic paraboloid
cooling tower. (Courtesy Ecodyne.)

the pattern (development) shown
in Figure 7.38. The true lengths of
sides such as 0'-1, 1-1', 1'-2, etc.,
have been found by using the
method of rotation. The true
lengths of the shorter sides of the
triangles are obtained in the manner shown in the top view of the
large circle and in the supplementary view, which shows the small
circle. The construction of one
of the triangles is shown in the
development. Smooth curves
through points 0 . . . . . 6 and
0' . . . . . 6' have been drawn to
compensate for the differences between chord distances (i.e., 0'-1',
1'-2', etc.) and the corresponding
circular arcs.

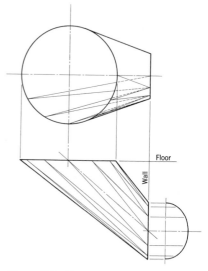

Figure 7.35 Warped transition piece.
Problem: Develop its lateral surface.

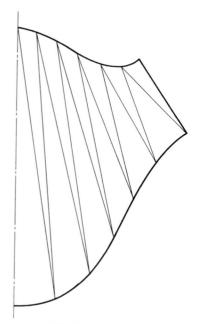

Figure 7.36 Solution to the problem
in Figure 7.35.

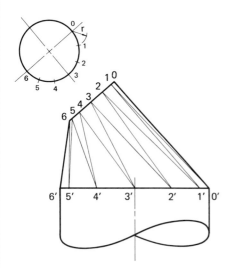

INTRODUCTION: INTERSECTIONS

In a number of design projects, problems occur that deal with the intersection of surfaces, such as the problem of determining the intersection of cylindrical surfaces arises in the design of piping systems. In some storage tank designs, the engineer is concerned with the determination of the intersection of a cylindrical surface with a spherical surface. In aircraft design intersection problems are quite common. And, certainly, there are many intersection problems that involve plane surfaces.

Several examples of designs that include the intersection of surfaces are shown in Figure 7.39 to 7.42.

Basic Problems

Once the student has a good grasp of the solutions of the four basic intersection problems, little difficulty will be encountered in solving problems such as:

1. The intersection of plane surfaces,

2. The intersection of solids bounded by plane surfaces,

Figure 7.37 Warped transition piece. *Problem:* Develop its lateral surface.

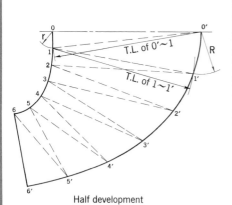

Figure 7.38 Solution to the problem in Figure 7.37.

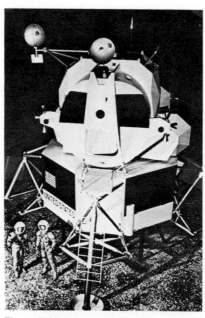

Figure 7.39 Intersection of plane surfaces.

Figure 7.40 Intersection of cylinders.

Figure 7.41 Intersection of cylinder with a spherical tank. (Courtesy Pittsburgh-Des Moines Steel Company)

Figure 7.42 Intersection of airplane wing and fuselage.

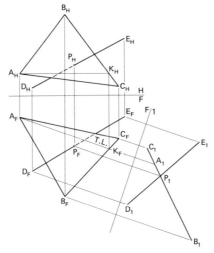

Figure 7.43 Intersection of line *CD* and opaque surface *ABC*.

3. The intersection of cylindrical surfaces, of conical surfaces, or of conical and cylindrical surfaces, etc.

4. The intersection of plane surfaces with topographic surfaces.

Now let us consider each of the four basic intersection problems.

1. The intersection of a line and a plane surface.
2. The intersection of a line and a conical surface.
3. The intersection of a line and a cylindrical surface.
4. The intersection of a line and a spherical surface.

Intersection of a Line and a Plane Surface

Let us assume that the H and F views of surface ABC and of line DE are given, as shown in Figure 7.43. To locate the intersection of line DE with surface ABC, we need only find an edge view of the surface and the corresponding view of line DE. The view on supplementary plane 1 shows the edge view of the surface as $A_1B_1C_1$, the line as D_1E_1, and the point of intersection as P_1. It is a simple matter to project from P_1 to locate P_F and then from P_F to locate P_H. If we assume that the surface is opaque, the visibility problem of line DE must also be solved. This is not difficult to do if we recall the analysis that was made when visibility was discussed in Chapter 5.

An *alternative* solution to finding the intersection of a line and a plane surface eliminates the need for a supplementary view.

As an example, consider Figure 7.44, which shows line m and the opaque surface ABC. Point P, the intersection of line m and plane ABC, can be located easily by the application of the following three steps.

1. *Pass a Plane Through the Line.* In Figure 7.45 we have introduced plane R, which contains line m and is perpendicular to the H plane (we could have introduced a plane through line m and perpendicular to the F plane).

2. *Find the Intersection of Plane R with Surface ABC.* The intersection is shown as n_H in the top view. It should be observed that the top view of line n and the top view of plane R appear as one line. Why is this true? Once we have n_H it is a simple matter to located n_F. Line n is designated by points 1 and 2.

3. *Find the Intersection of Line n with Line m.* This point P, whose front view is at the intersection of m_F and n_F. Now it is a simple matter to project from P_F to locate P_H on m_H as shown in Figure 7.44.

The latter method is the simpler of the two. Moreover, we will find that it is more convenient when we analyze and solve problems dealing with the intersection of solids.

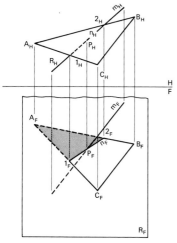

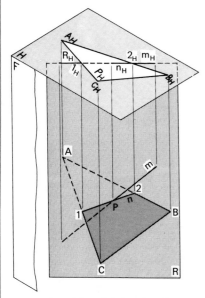

Figure 7.44 Intersection of line *m* and opaque surface *ABC*. Alternative solution.

Figure 7.45 Pictorial showing the method used in finding the intersection of line *m* and opaque surface *ABC*.

EXAMPLE 1
INTERSECTION OF
A LINE AND A PYRAMID

Suppose we wish to find the intersection of line *m* and pyramid *O-ABC* (see Figure 7.46).

Consider line *m* and triangular surface *OAC*. We can locate point *P* the intersection of line *m* and surface *OAC* in the same way that was used in solving the previous problem. Note the following steps.

1. Plane *R* is introduced through line *m* perpendicular to the *H* plane. *This is step 1*.
2. The intersection of planes *R* and surface *OAC* is line *n*, shown as 1_H2_H and 1_F2_F in the *H* and *F* views, respectively. *This is step 2*.
3. The intersection of line *n* with line *m* is point *P*, which is represented by P_H and P_F in the *H* and *F* views, respectively. *This is step 3*.

In a similar way we can find point *Q*, which is the intersection of line *m* with surface *OBC*. If the

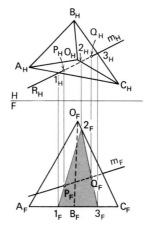

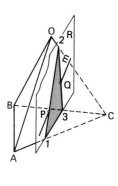

Figure 7.46 Intersection of line *m* and pyramid *O-ABC*.

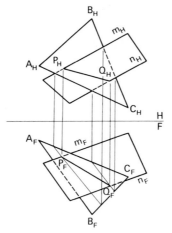

Figure 7.47 Intersection of two opaque
planes.

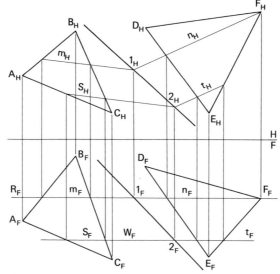

Figure 7.48 Intersection of two planes
defined by triangles *ABC* and *DEF*.

pyramid is regarded to be opaque, we can determine the visibility of line *m*. Satisfy yourself that the visibility of line *m* as shown in Figure 7.46 is correct.

EXAMPLE 2
INTERSECTION OF
PLANE SURFACES—CASE 1

Let us now consider Figure 7.47 which shows planes *ABC* and *mn*. We wish to determine the line of intersection of the two plane surfaces. The three steps used in Example 1 are applied to line *m* and surface *ABC*. Point *P* is the intersection of line *m* with plane *ABC*. We can locate the intersection of line *n* with surface *ABC* in the same way. This is point *Q*. Line *PQ* is the intersection of the plane surfaces *mn* and *ABC*. Satisfy yourself that the visibility shown in Figure 7.47 is correct.

EXAMPLE 3
INTERSECTION OF
PLANE SURFACES—CASE 2

Suppose the plane surfaces are defined by the limited surfaces *ABC* and *DEF* shown in Figure 7.48. It is quite obvious that there is no line of intersection of the limited surfaces. To determine the line of intersection of the *unlimited* surfaces, which are defined by *ABC* and *DEF*, we introduce a plane such as *R*, parallel to the *H* plane (it could be planed in any other convenient position) and crossing surfaces *ABC* and *DEF*. Plane *R* intersects surface *ABC* in line *m* and surface *DEF* in line *n*. Point 1, the intersection of lines *m* and *n*, is common to all three planes and,

therefore, is a point on the intersection of the unlimited surfaces *ABC* and *DEF*. To locate a second point, we introduce plane *W*, which is parallel to plane *R*. Plane *W* intersects surface *ABC* in line *s* and surface *DEF* in line *t*. Point 2, the intersection of lines *s* and *t*, is common to the three planes *ABC*, *DEF*, and *W*. The line that joins points 1 and 2 is the line of intersection of the unlimited surfaces *ABC* and *DEF*.

Intersection of a Line and a Cone

Let us consider Figure 7.49, which shows line *m* and a right circular cone. Let us see how the three steps are applied to this problem. We recall that step 1 was "to pass a plane through the line." If we pass a plane through line *m* cutting all of the elements of the cone, the resulting intersection would be an ellipse (step 2). The intersection of line *m* with the ellipse would locate the points in which line *m* inter-

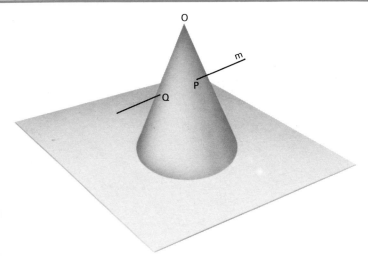

Figure 7.49 Intersection of line and cone.

EXAMPLE 1

The given line and cone are shown in the top and front views of Figure 7.50b. We can pass a plane through the apex, O, and line m by:

1. Selecting any convenient point, such as A, on line m. This is easily done by placing A_F on m_F and A_H on m_H. We know that "the views of the point lie on the corresponding views of the line."

2. Joining points O and A to establish a *new* line, n. Lines m and n determine the plane that contains line m and apex O. This completes *step 1*.

We should locate points B and C, which are the intersections of lines n and m, respectively, with *the plane that contains the base of the cone*. Since this plane appears

sects the cone (step 3). Other planes passed through line m could cut the cone to form a hyperbola or a parabola. In these cases the points in which line m intersects the hyperbola or the parabola would be the points in which line m intersects the cone. How can we avoid the need for one of the conic sections (ellipse, hyperbola, or parabola) in our solution?

Some thought about this question should lead us to conclude that if we pass a plane through the apex of the cone and line m, that plane will intersect the cone in a triangle such as 0-1-2, as shown in Figure 7.50(a). The intersection of line m with elements 0-1 and 0-2 locates points P and Q, which are the intersections of line m with the cone.

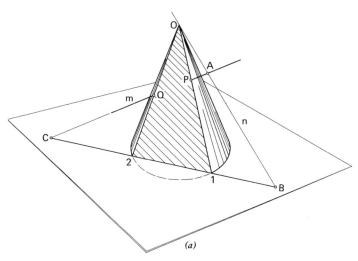

(a)

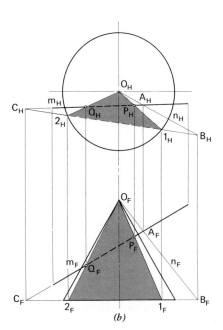

(b)

Figure 7.50 (a) Pictorial showing the intersections of line m and a cone, apex O. (b) Orthographic showing the intersections of line m and a cone, apex O.

on edge in the front view, we can locate B_F at the intersection of n_F with the base plane of the cone. Once B_F is determined, it is a simple matter to locate B_H. The top and front views of point C are determined similarly. The line joining points B and C cuts the base of the cone in chord 1-2 which, together with elements O-1 and O-2, forms triangle O-1-2. This completes *step* 2.

Finally, in *step* 3, the points of intersection, P and Q, are located by finding the points in which the respective views of line m intersect the corresponding views of the sides of the triangle, thereby locating P_H and P_F and Q_H and Q_F. A good check on the accuracy of the solutions is to see if P_H and P_F lie on a vertical line, and to do the same for Q_H and Q_F.

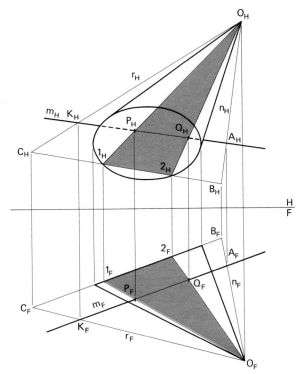

Figure 7.51 Intersections of line m and a cone, apex O.

EXAMPLE 2

Consider line m and the *oblique* cone shown in Figure 7.51

Again, we will employ the three steps that were used in Example 1. We can easily select a point, such as A, on line m, and establish line n, which joins points O and A. The plane that passes through the apex, O, and line m is defined by lines m and n. *This is step 1.*

The intersection of plane mn and the cone must be determined to complete step 2. As in the previous example, we can locate point B, which is the intersection of line n with the plane that contains the base of the cone. We observe, however, that line m is parallel to the base plane of the cone and, therefore, there is no intersection. How

shall we establish the line in which plane mn intersects the base plane of the cone? We can select a second point, K, on line m and establish line r, which joins points O and K. Now we can locate point C which is the intersection of line r and the base plane of the cone. Line BC cuts the base of the cone in chord 1-2 which, together with elements O-1 and O-2, forms triangle 0-1-2. The top and front views of this triangle are shown tinted.

The views of points P and Q, the points of intersection, are uniquely located by finding the points in which the *respective views of line m intersect the corresponding views of the sides of the triangle*. This completes step 3.

Intersection of a Line and a Cylinder

The three steps used in the previous examples can be applied to the problem of finding the points in which a line intersects a cylinder. The first step, "pass a plane through the line," is quite critical. Many planes can be passed through the given line, but which one should we select in order to cut the *simplest* shape from the cylinder? If we give some thought to this question, we will conclude that *a plane passed through the line parallel to the axis of the cylinder will cut elements of the cylinder*. The pictorial sketch in Figure 7.52 shows this. It should be observed that plane mn is parallel to axis S of the

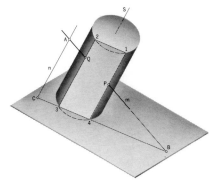

Figure 7.52 Pictorial showing the intersections of line *m* and a cylinder with axis *S*.

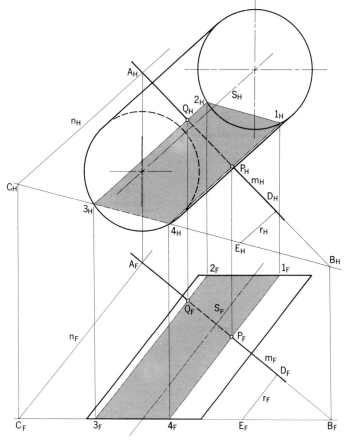

Figure 7.53 Orthographic showing intersections of line *m* and a cylinder with axis *S*.

cylinder. This is true because line *n*, which passes through point A (an arbitrary point on line *m*), is parallel to axis *S*. Plane *mn* intersects the cylinder in elements 1-4 and 2-3. The location of the elements is easily established, since we can find line *BC*, which is the intersection of plane *mn* and the plane *that contains the lower base of the cylinder*. Now the intersection of line *BC* with the base of the

cylinder establishes points 3 and 4, through which elements 2-3 *and* 1-4 *can be located. Finally, the intersection of line m with these elements are points P and Q, the* points in which line *m* intersects the surface of the cylinder.

The orthographic solution is shown in Figure 7.53. In some cases line *m* might not intersect the base plane of the cylinder within

the limits of the drawing area. In such cases we can introduce another line *r*, through point D (an arbitrary convenient point on line *m*) and parallel to the axis, *S*. Line *r* intersects the base plane of the cylinder at point E. Now line *CE* can be used to locate points 3 and 4.

Intersection of a Line and a Sphere
Let us consider Figure 7.54, which shows the top and front views of line m and sphere with center, O. Any plane that contains line m will intersect the sphere in a circle. The intersections of line m and the circle will locate the points in which line m intersects the sphere.

We will pass a plane, S, through line m and perpendicular to the F plane (we could have passed a plane through m and perpendicular to the H plane). Now the edge view (S_F) of plane S and the front

view of line m appear as one line. A supplementary (auxiliary) view on plane 1, which is parallel to the plane of the circle, shows the circle with center O_1, and the line as m_1 or as A_1B_1. The points in which line m intersects the sphere appear as P_1 and Q_1. It is a simple matter to project from P_1 and Q_1 to locate P_F and Q_F, and then P_H and Q_H. If we assume that the sphere is opaque, it will be necessary to determine the visibility of line m. Consider Figure 7.55, which only shows the top and front views of line m, the sphere, and the points of intersection P and Q.

We now consider the *apparent* intersection of m_H and the circle as seen in the top view. Point 1 is on the great circle, which is parallel to the H plane, and point 2 is on line m. As we look down (perpendicular to the H plane), point 1 is closer

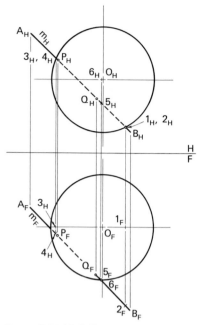

Figure 7.55 Visibility analysis—intersection of line AB and sphere O.

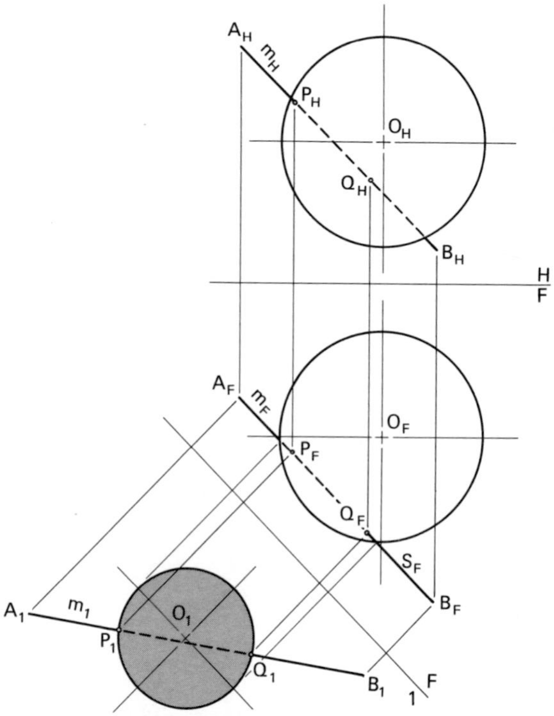

to us than point 2 is; therefore the great circle, which contains point 1 is above line m. This means that the portion of m_H from 2_H to Q_H is not visible and is represented by a dashed line. In a similar manner we can analyze the relative positions of points 3 and 4 and determine that the portion of m_H from 4_H to P_H is not visible.

Now we can consider point 5 on line m and point 6 on the great circle, which is parallel to the F plane. It should be quite evident that point 5 is in front of point 6; therefore, the portion of m_F from 5_F to Q_F is visible. You should experience no difficulty in determining the visibility of the other portion of m_F.

As we reflect on the four basic intersection problems, we should realize the analysis and the method of solution are basically the same.

We can now proceed to the analysis and solution of intersection problems concerned with:

Figure 7.54 Intersections of line AB and sphere O.

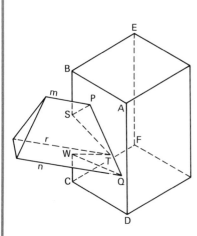

Figure 7.56 Intersections of solids bounded by plane surfaces.

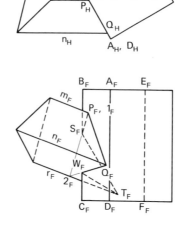

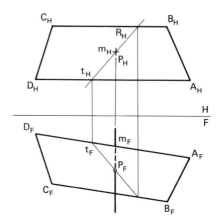

Figure 7.57 Intersection of vertical line *m* and plane surface *ABCD*.

1. Two solids bounded by plane surfaces.
2. A cone and a cylinder.
3. Two cones.
4. Any two solids.
5. A plane and a topographic surface.

Intersection of solids bounded by plane surfaces

The determination of the lines of intersection of two solids bounded by plane surfaces is basically the same as finding the intersection of plane surfaces.

Let us consider the solids shown in Figure 7.56. The top view of the large block shows surface *ABCD* on edge; therefore, we can easily locate points *P* and *Q*, which are, respectively, the intersections of edges *m* and *n* with surface *ABCD*. Similarly, we can locate point *T*, which is the intersection of line *r* with surface *BEFC*. Thus far we have found the points in which the edges of the triangular piece intersect the surfaces of the large block. The only edge of the large block that intersects surfaces of the triangular piece is *BC*. How shall we lo-

cate the points in which edge *BC* intersects surfaces *mr* and *nr*? Let us first investigate the problem shown in Figure 7.57. In this case we are given plane surface *ABCD* and line *m*. We wish to locate the point in which the line intersects the plane.

Recalling the three steps used in locating the point in which a line intersects a surface, let us introduce a plane through line *m*. It should be recognized that since line *m* is perpendicular to the *H* plane, all planes that contain line *m* will be perpendicular to the *H* plane. This means that the top view of each of these planes will appear as a line. Figure 7.57 shows one of these planes through line *m*. The top view of this plane, *R*, is marked R_H. Plane *R* intersects plane *ABCD* in a line, *t*, whose top view coincides with R_H. The front view of line *t* is easily determined from the top view. The intersection of lines *t* and *m* determines the location of point *P*, which is the intersection of line *m* and plane *ABCD*.

Returning to the previous prob-

lem, we see that the point in which edge *BC* intersects surface *mr* can be located in the manner just presented. In Figure 7.56, the point of intersection is point *S*. The plane, which is passed through edge *BC*, is an extension of surface *ABCD*. Note carefully that line 1-2 is the intersection of surface *mr* and surface *ABCD* extended.

We can find the intersection of edge *BC* with surface *nr* in the same way. This is point *W*.

Now the lines of intersection are drawn by joining points *P* and *Q*, *P* and *S*, *S* and *T*, and *T* and *W*. Each pair of points is common to two surfaces; therefore, the lines that join each pair are the lines of intersection of the two solids.

Satisfy yourself that the visibility problem is solved correctly.

Intersection of a cone and a cylinder

In our discussion of "the intersection of a line and cone," it was pointed out that a plane passed through the line and the apex of the cone would cut a triangle out of the cone if the plane intersected the base of the cone.

In the case of the "intersection of a line and cylinder," we found that a plane passed through the line and parallel to the axis of the cylinder would cut elements from the cylinder if the plane intersected the cylinder.

Therefore, to cut the simplest shapes out of both the cone and the cylinder, it is necessary to introduce planes through the apex of the cone and parallel to the axis of the cylinder.

This can be done by first introducing a line through the apex of the cone and parallel to the axis of the cylinder. Now all the planes that contain this line will cut elements from the both the cone and

the cylinder, if the plane intersects both surfaces.

Suppose we are given the cone and cylinder shown in Figure 7.58. Let us determine the location of points on the curve of intersection of the two solids.

Our first step is to pass a plane through the apex of the cone and parallel to the axis of the cylinder. In order to accomplish this, we first introduce line n through point O parallel to the axis of the cylinder. The top and front views of line n are easily determined, since we know that (1) the views of a line pass through the corresponding views of the point that lies on the line, and (2) views of parallel lines appear as parallel lines; in this case n_H is parallel to the top view of the axis of the cylinder and n_F is parallel to the front view of the axis of the cylinder.

All planes containing line n will intersect the cone and cylinder in elements if the planes cut the two solids.

Let us select one plane, S, for example, that contains line n and cuts both solids. How shall we proceed to do this? Before we attempt to answer this question, we should first investigate another one: How can we establish any plane through n? The question is easily answered. We should know that a plane may be determined by (1) two intersecting lines, (2) two parallel lines, (3) three noncollinear points, or (4) a line and an external point. Therefore, any plane through line n may be determined, for example, by line n and an intersecting line.

If we select the intersecting line m in the plane of the bases of the cone and cylinder, the point on n through which line m passes must also lie in the plane of the bases of the two solids. Thus, we see that this point, A, must be the point in which line n intersects the plane of the bases. This point is readily located, since an edge view of the plane of the bases is available in the front view.

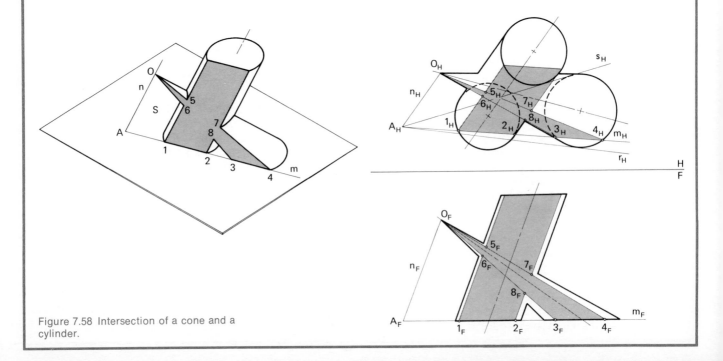

Figure 7.58 Intersection of a cone and a cylinder.

Now plane *S*, determined by lines *n* and *m*, intersects the cylinder in elements 1 and 2 and intersects the cone in elements 3 and 4. Since all four elements lie in plane *S*, the points in which the elements intersect will be common to the two surfaces. These four points are 5, 6, 7, and 8. In a similar manner we can establish other planes through line *n* and then determine additional points common to the two surfaces. The required intersection (curves in this case) is drawn through the points that lie in both surfaces.

It should be observed that planes *mr* and *ms* are the "limiting planes"; that is, the intersection of the two solids lies between these planes.

Intersection of two cones

We recall that "any plane that intersects a cone and passes through its apex will cut out elements of the cone."

It follows "that any plane that intersects two cones and contains the line that joins their apices will cut out elements of both cones."

The common points of the elements of the two cones are points on the curve of intersection of the two solids. A sufficient number of planes should be passed through the line joining the apices of the cones to determine an adequate number of points on the curve of intersection.

EXAMPLE 1

Consider the two cones shown in Figure 7.59. The planes that contain the line *n* (line *n* connects the apex *O* with apex *B*) and intersect the cones will cut triangles from the cones; that is, plane *mn* cuts triangle *OCD* from the cone with apex *O* and triangle *BGK* from the other cone. The points 1, 2, 3, and

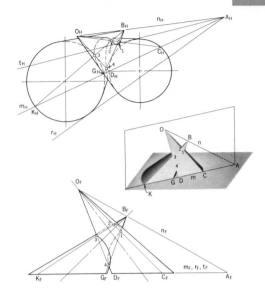

Figure 7.59 Intersection of two cones.

4 of the intersection of the sides (elements of the cones) of the triangles are points on the curve of intersection. Additional points can also be located by passing several planes through line *n*. The curve of intersection of the two cones will lie between the "limiting planes *nt* and *nr*."

EXAMPLE 2

Let us now consider a unique case where the line joining the apices is parallel to the plane that contains the bases of the cones. In Figure 7.60 we observe that line *n* is such a line. How shall we identify planes that contain line *n*?

Any plane that contains line n *will intersect the plane that contains the bases of the cones in a line that is parallel to* n. We know that the views of parallel lines will appear as parallels; therefore, we can easily establish planes through line

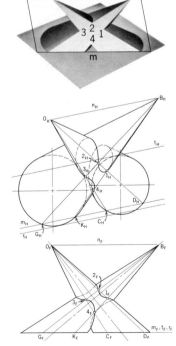

Figure 7.60 Intersection of two cones. Apices lie in a horizontal plane.

n. For instance, line *m* is drawn in the plane, which contains the bases, parallel to line *n*. This plane, *mn*, obviously contains line *n* and intersects the plane of the bases in line *m*. Moreover, plane *mn* intersects the cones in triangles *OCD* and *BGK*. The intersections of the sides (elements of the cones) of these triangles are points on the curve of intersection of the two cones, in this case, points 1, 2, 3, and 4. Additional planes may be passed through line *n* to determine more points on the curve of intersection. Again, it should be observed that the curve of intersection will lie between the limiting planes *nt* and *nr*.

Intersection of any two solids

The line (or curve) of intersection of any two solids generally can be determined by passing a number of planes across the two solids. The choice of the planes should be such as to cut the simplest shapes from the solids. Each plane will cut the solids in plane curves. Points that are common to the surfaces of the two solids lie at the intersections of the plane curves. Once a sufficient number of points has been established, the curve of intersection can be delineated.

In the problem of the intersection of a cone and a cylinder (Figure 7.58), we introduced planes through the apex of the cone and parallel to the axis of the cylinder. Consequently, *simple* shapes were cut from the two solids (triangles from the cone and parallelograms from the cylinder). We could also have used cutting planes parallel to the bases of the solids; such a plane

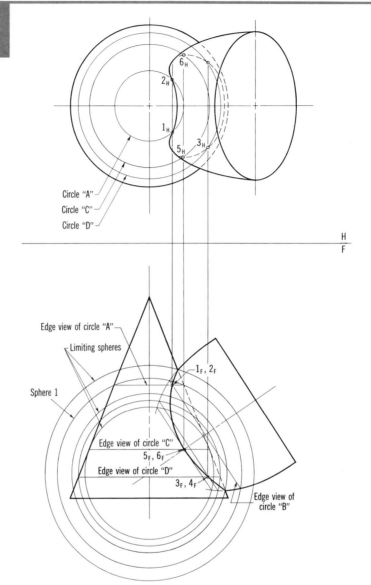

Figure 7.61 Intersection of two surfaces of revolution.

would intersect the cone in a circle and would also intersect the cylinder in a circle. Since both circles would lie in the same plane, their points of intersection would be points on the curve of intersection of the cone and cylinder.

EXAMPLE INTERSECTION OF TWO SURFACES OF REVOLUTION

Consider the two surfaces of revolution shown in Figure 7.61. We could find the intersection of the two surfaces by first finding the points in which elements of the cone intersect the ellipsoid and then joining the points of intersection by a smooth curve. This method is laborious, because the intersection of an element of the cone with the ellipsoid involves the three steps: (1) a plane through the element; (2) the intersection of the plane with the ellipsoid surface; and (3) the intersection of the element with the curve cut from the ellipsoid.

A simpler solution is possible by using the *sphere method*. This method is very useful in finding the intersection of surfaces of revolution, *if their axes intersect and are seen in true length in one view.* The sphere method is based on the fact that a sphere with its center at the intersection of the axes will cut circles from each surface of revolution. The circles will appear as lines in the view, which shows the axes in true length. The points of intersection of the edge views of the circles will be points on the curve of intersection of the surfaces of revolution.

In Figure 7.61, sphere 1 intersects the cone in circle A and the ellipsoid in circle B. The circles appear as straight lines in the front view. The intersections of the circles are points 1 and 2, shown in the front view as 1_F and 2_F. The top view of the points is shown as 1_H and 2_H, which are on the top view of circle A. Additional spheres may be introduced to obtain more points that, when properly connected, will establish the intersection of the cone and ellipsoid. The limiting spheres are shown in Figure 7.61. Other spheres will lie between the limiting spheres.

Intersection of a plane and a topographic surface

A practical intersection problem arises in highway design. The widths of "right-of-way" and the earth quantities of "cuts" and "fills" (in cubic yards or cubic meters) must be determined to provide needed information for land purchase and for earth-moving estimates.

EXAMPLE

Let us assume that line *AB* is the centerline of a proposed 50-ft (15.24 m) highway having a 10% grade (Figure 7.62). At station 1 + 00* the elevation of the finished subgrade is assumed to be 40 ft (12.2 m). Earth *cuts* are to be made at a slope of 1:1 (i.e., one unit horizontally to one unit vertically, as shown in Figure 7.63). Earth *fills* are to be made at a slope of $1\frac{1}{2}$:1 (see Figure 7.64).

Our problem is to determine the lines in which the sloping surfaces of the proposed highway intersect the topographic surface (earth surface). The line in which the 1:1 surface intersects the earth surface is called the "*top of the cut.*" The line in which the $1\frac{1}{2}$:1 surface intersects the earth surface is called the "*toe of the fill.*"

Let us start at station 1 + 00. If we lay off a distance of 10ft(3 m) from the upper edge of the road, the elevation of the end point, *C*, will be 50 ft (15.24 m). This is true because the contour† lines in the vicinity of stations 1 + 00, 2 + 00, etc., indicate that the ground is at a higher elevation than the elevation of the proposed subgrade; hence, cuts will be required. Since the slope of the cut is 1:1, a horizontal distance of 10 ft (3 m) will reflect an increase in elevation of 10 ft (3 m). If we lay off an additional 10 ft (3 m), from *C* to *D*, the elevation of *D* will be 60 ft (18.28 m), and so on for additional points.

Now let us move to station 2 + 00. The elevation of the subgrade at this station is 50 ft (15.24 m), because the roadway has a 10% grade. If we lay off a distance of 10 ft (3 m) from the upper edge of the road, the end point, *E*, will be at elevation 60 ft (18.28 m) and so on for additional 10-ft (3 m) increments. Points *E* and *D* are at the *same* elevation, 60 ft (18.28 m). *The line joining these points is a contour line on the sloping 1:1 surface. This contour line intersects the earth contour* [elevation 60 ft (18.28 m)] *at* F, *which is a point common to the earth surface and the plane surface, 1:1.* Point *F* is a point on the top-of-the-cut curve.

Now let us see how the three steps used in finding the intersection of a line with a plane were employed in this problem of finding the intersection of the contour line at, for example, elevation 60 ft (18.28 m) with the plane surface having the 1:1 slope.

You will recall that the first step in finding the intersection of a line with a plane is to *pass a plane through the line.* In this case the only plane that contains contour 60 is the horizontal plane in which the contour lies. The second step is to *find the line in which the horizontal plane intersects the 1:1 surface.* The line of intersection is *DE. The final step is to locate the point (or points) in which line DE intersects the contour 60. That*

---

* Notation used in surveying. Station 2 + 00 is 100 ft from station 1 + 00. A point halfway between these two stations would be marked 1 + 50.

† Lines all of whose points are at the same elevation.

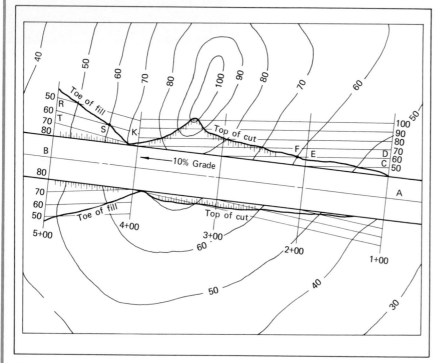

Figure 7.62 Intersection of planes with a ground surface. An application in highway design.

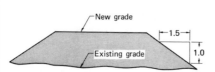

Figure 7.63 Typical cross section in a "cut."

New grade

Existing grade

Figure 7.64 Typical cross section in a "fill."

highway! At this station the subgrade elevation is 70 ft (21.34 m). If we lay off a distance of 15 ft (4.57 m) from the upper edge of the roadway (remember, the slope in a fill is $1\frac{1}{2}$:1, the elevation of the end point K, will be 60 ft (18.28 m). If we lay off another 15 ft (4.57 m) from point K, the elevation will be 50 ft (15.24 m), etc.

Now let us move to station 5 + 00. The subgrade elevation is 80 ft (24.38 m). Again let us lay off a distance of 15 ft (4.57 m) from the upper edge of the roadway. The elevation of the end point, T, is 70 ft (21.34 m). If we lay off another 15 ft (4.57 m) from point T to point R, the elevation of the latter will be 60 ft (18.28 m). The line joining points R and K is the 60-ft (18.28-m) contour on the sloping surface, $1\frac{1}{2}$:1. The intersection of this contour with the 60-foot (18.28-m) natural ground contour is the common point S. Additional points are similarly located. The line that joins the common points is the toe of the fill.

If the same procedure is followed in the determination of the top of the cut and the toe of the fill for the other side of the highway, it

point is F. We locate additional points in the same way. The line (usually irregular) that joins these points is the top of the cut.

As we approach station 4 + 00, we observe that the cut has nearly run out and that fill will be necessary for the continuation of the

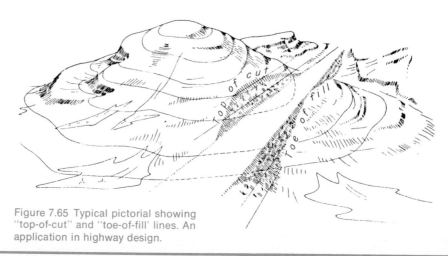

Figure 7.65 Typical pictorial showing "top-of-cut" and "toe-of-fill" lines. An application in highway design.

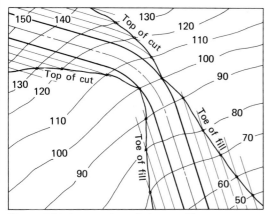

Figure 7.66 Intersection of planes with a ground surface. Another example of an application in highway design.

will be possible to draw cross sections of the road. The areas of the cross sections can be determined (usually by the use of a planimeter, shown in Figure 11.23). If the distances between the sections are known, the volumes of fill and cut can be calculated. The widths of right-of-way now can be established on the contour drawing, since the top-of-the-cut and toe-of-the-fill lines have been located. Figure 7.65 shows a typical pictorial of the top-of-the-cut and toe-of-the-fill lines.

Another example is shown in Figure 7.66. The same basic method used in the previous example is applied to the solution of this problem.

Exercises

The development of surfaces

1. Develop the lateral surface of the truncated prism shown in Figure E-7.1.

2. Develop the surface of the Hopper shown in Figure E-7.2. Assume the Hopper is fabricated from steel plate 1 cm thick. Calculate the mass of the Hopper based on 7500 k/m³.

3. Develop the lateral surface of the sheet metal chute shown in Figure E-7.3. Also determine the true shape of the opening in the wall.

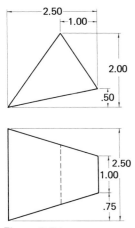

Figure E-7.1

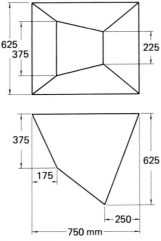

Figure E-7.2

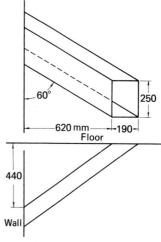

Figure E-7.3

4. Develop the lateral surface of the
truncated triangular pyramid shown in
Figure E-7.4. Include the base and the
upper face.

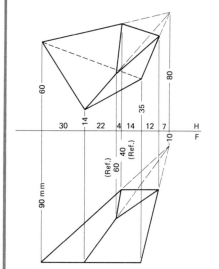

Figure E-7.4

5. Develop the surface of the oblique
cone shown in Figure E-7.5. Also show
the *H* and *F* views of the shortest path
on the surface of the cone between
points *A* and *B*.

6. Develop the surface of the three-
piece cylindrical elbow shown in Figure E-7.6.

7. Develop the surface of the conical
transition piece shown in Figure E-7.7.
It is only necessary to develop the front
half.

8. Design a transition piece to connect
the cylindrical portion to the rectangular opening shown in Figure E-7.8.
Also develop the transition piece.

9. Develop the sheet metal part shown
in Figure E-7.9. Provide for "bend allowance."

10. Design a pattern for the surface of
the conoid roof shown in Figure E-7.10.
In order to estimate the cost of shingles

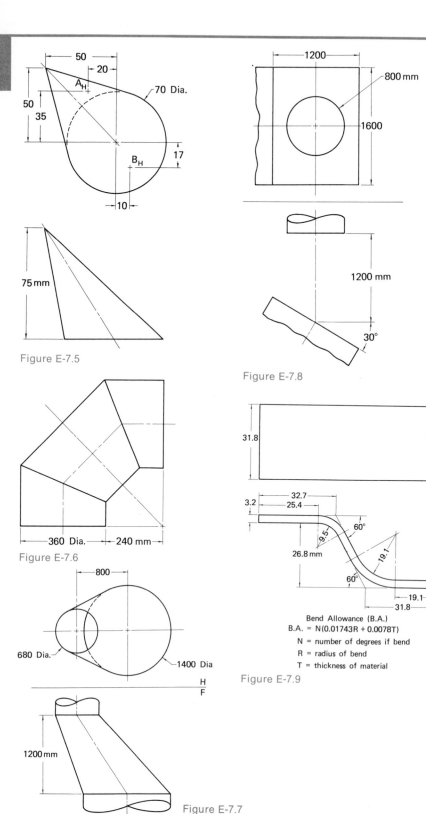

Figure E-7.5

Figure E-7.6

Figure E-7.7

Figure E-7.8

Bend Allowance (B.A.)
B.A. = N(0.01743R + 0.0078T)
N = number of degrees if bend
R = radius of bend
T = thickness of material

Figure E-7.9

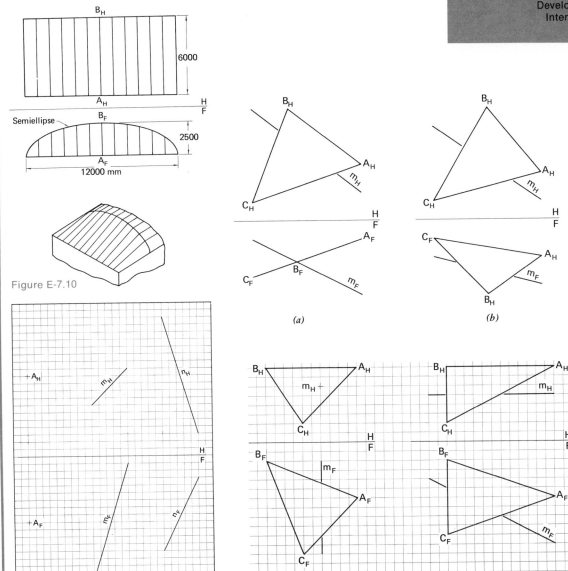

Figure E-7.10

(a)

(b)

(c)

(d)

Figure E-7.11

Figure E-7.12

to cover the surface, it is necessary to calculate the number of square meters in the roof surfaces. Assuming a 10% wastage, how many square meters will be required?

Intersections

1. Locate the H and F views of the intersection of line m and surface ABC as shown in Figure E-7.11. Assume that the plane surface is opaque and determine the correct visibility of the line.

2. In Figure E-7.12 lines m and n represent two tunnels. It is planned to construct a new tunnel AB that will intersect tunnels m and n. Locate the H and F views of tunnel AB, where B is on tunnel n.

3. Determine the *H* and *F* views of the intersections of line *m* and the cone shown in Figure E-7.13.

4. The *H* and *F* views of point *O* and line *n* are shown in Figure E-7.14. Locate the lines that pass through point *O*, intersect line *n*, and make an angle of 60° with the *F* reference plane.

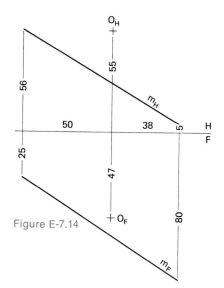

Figure E-7.14

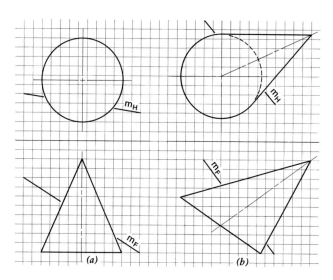

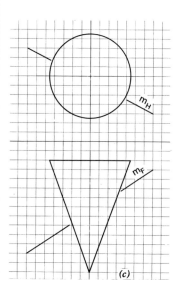

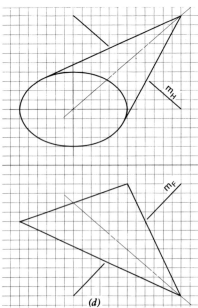

Figure E-7.13

5. Locate the *H* and *F* views of the intersections of line *m* and the cylinder shown in Figure E-7.15.

6. Determine the points of intersection of line *m* and the sphere shown in Figure E-7.16.

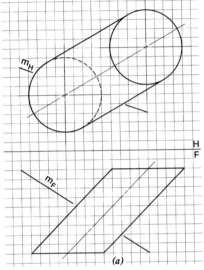

Figure E-7.15

(a)

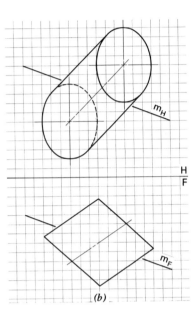

(b)

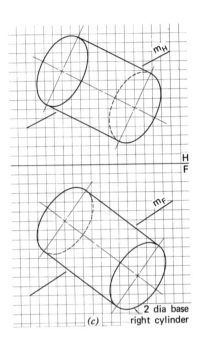

2 dia base
right cylinder

(c)

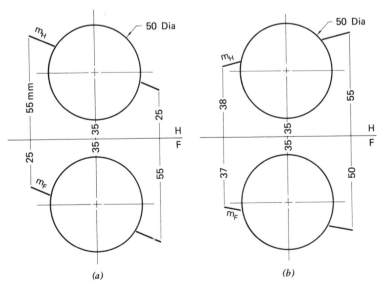

(a)

(b)

Figure E-7.16

7. Locate the *H* and *F* views of the intersection of the plane surfaces shown in Figure E-7.17.

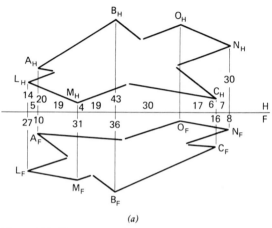

Figure E-7.17

8. Find the intersection of the plane surfaces shown in Figure E-7.18.

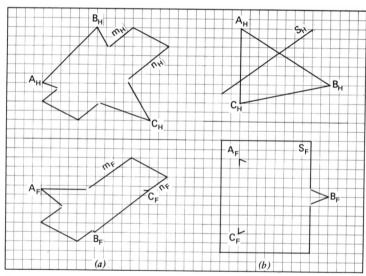

Figure E-7.18

9. Find the intersection of the solids shown in Figure E-7.19.

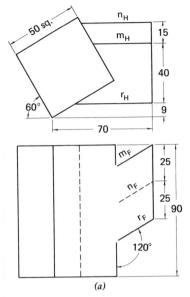

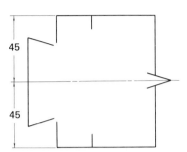

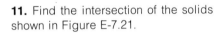

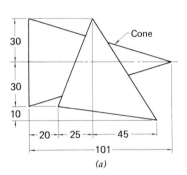

Figure E-7.19

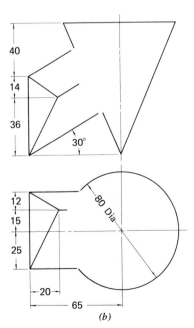

(a)

10. Find the intersection of the solids shown in Figure E-7.20.

11. Find the intersection of the solids shown in Figure E-7.21.

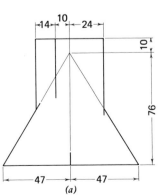

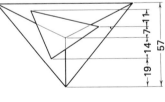

(a)

Figure E-7.20

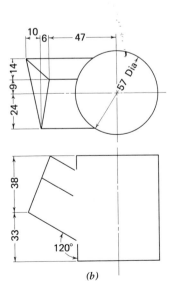

(b)

Figure E-7.21

12. Find the intersection of the cones shown in Figure E-7.22.

13. Find the intersection of the cone and cylinder shown in Figure E-7.23.

14. Find the intersection of the cylinders shown in Figure E-7.24.

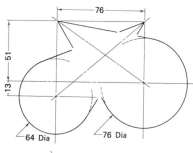

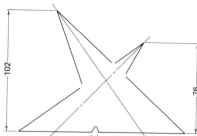

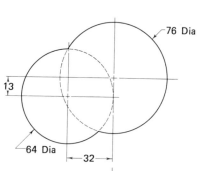

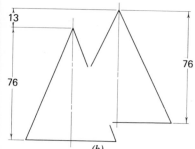

Figure E-7.22

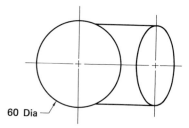

70 Dia

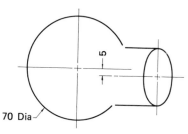

(a)

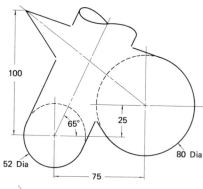

52 Dia

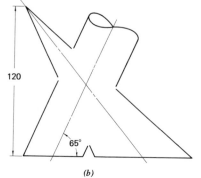

(b)

Figure E-7.23

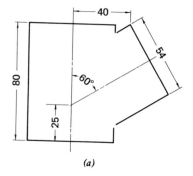

60 Dia

(a)

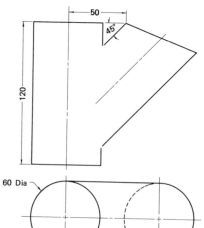

60 Dia

(b)

Figure E-7.24

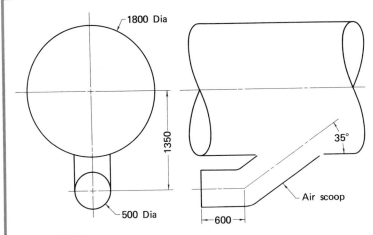

Figure E-7.25

15. Find the intersection between the air scoop and a portion of an engine macelle shown in Figure E-7.25.

16. Find the intersections of the center line of a pipe culvert and the embankments of the roadway shown in Figure E-7.26. What is the true length between the points of intersection?

17. In Figure E-7.27, find the intersection of the airplane windshield and the portion of the fuselage that is part of an ellipsoid.

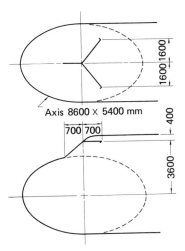

Figure E-7.27

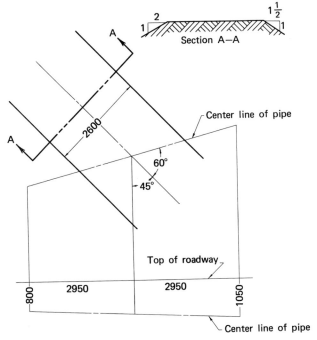

Figure E-7.26

18. In Figure E-7.28, determine the new elevation of the lower end of straight cable *AG* in order to maintain a minimum clearance of 30 m between the cable and the ground surface. What are the distances from points *B*, *C, D, E,* and *F* to the ground surface?

19. The upper surface of a rock formation is determined by points *A, B,* and *C* as shown in Figure E-7.29. Locate the intersection of the surface with the natural terrain.

20. The centerline of an access road and the top of an earth dam (10 m wide) runs N 30° E from point *X*, as shown in Figure E-7.30. The elevation of the road (the top of the dam) is 530 m. Show the toe of the fill on both sides of the dam (slope 3:1) and the top of the cut on both sides of the road (slope 2:1) Select scale.

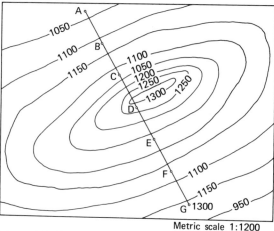

Metric scale 1:1200

Figure E-7.28

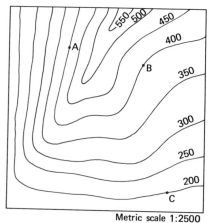

Metric scale 1:2500

Figure E-7.29

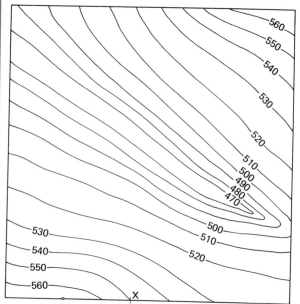

Figure E-7.30

Chapter 8   Vector Quantities and Vector Diagrams

Courtesy, T.Y. Lin Associates

INTRODUCTION

The determination of the sizes of components of a machine or structure depends on (1) the forces acting on them, (2) the environment to which the machine or structure is subjected, and (3) the selection of materials that are best suited to satisfy, economically, requirements 1 and 2. An example in bridge design was cited in the first chapter. Calculations of the forces acting in the members of the bridge structure are necessary in order to select the steel shapes that will safely carry the loads.

Also, there are a number of problems in design work that involve quantities such as (1) displacement (i.e., in cam design), (2) velocity (rate of change of displacement with respect to time), (3) acceleration (rate of change of velocity with respect to time), and (4) force (the product of mass and acceleration). These quantities have *magnitude*, *direction*, and *position* (line of action). They are *vector* quantities. Now let us represent these quantities graphically.

VECTOR DIAGRAMS

A vector quantity is represented graphically by a line segment that has a definite length (to some scale) and has an arrowhead at one end.

The length of the line segment represents the magnitude of the vector quantity, and the direction along the line segment from tail end to arrow end shows the direction of the vector quantity. The position, or line of action, is shown on a drawing.

EXAMPLE 1
Let us suppose that a point moves from A to B, as shown in Figure 8.1. When this has happened, the point has received a *displacement*, the magnitude of which is the length of line AB, and the direction of which is the direction of AB as shown by the arrow.

Figure 8.1 Vector *AB*.

Now suppose that the point that moved from A to B did not travel directly from A to B, but took a route through points E, D, C, and B, as shown in Figure 8.2. The final displacement of the point, however, is the same as in the first case, but the final displacement is the sum of the displacements, AE, ED, DC, and CB. Stated simply, vector AB is the resultant (or sum) of the other vectors, which are the components of vector AB. We should also recognize (Figure 8.3) that vector AD could replace vectors AE and ED, and that vector DB could replace vectors DC and CB without affecting the final displacement from A to B. In effect, vector AD is the resultant of vectors AE and ED, vector DB is the resultant of vectors DC and CB, and vector AB is the resultant of vectors AD and DB.

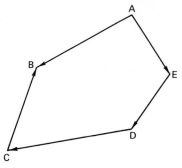

Figure 8.2 Displacement of point *A* to final position *B*.

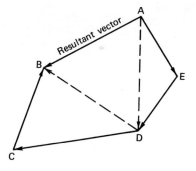

Figure 8.3 Vector *AB* is the sum of vectors *AE*, *ED*, *DC*, and *CB*.

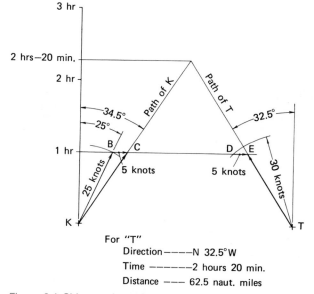

For "T"
Direction ————N 32.5°W
Time ———————2 hours 20 min.
Distance ——— 62.5 naut. miles

Figure 8.4 Ship travel problem.

EXAMPLE 2

Let us consider the following displacement problem. A ship travels at 25 knots on a course N 25° E from point K. A second ship travels at 30 knots from point T (Figure 8.4). There is a current of 5 knots due east.

What course should the second ship take to meet the first one?

SOLUTION

We can draw vector KB to show the travel (displacement) of the first ship during a 1-hour period, assuming no current. Vector BC shows the effect of the current, and vector KC (the resultant vector) shows the actual travel on the true course, N 34.5° E.

Starting with point T as the cen-

ter and a radius that represents 30 knots, we can locate point D. With no current the course of the second ship would be along vector TD. Adding the effect of the current, vector DE, we obtain vector TE, whose direction is the true course (N 32.5° W) of the second ship.

EXAMPLE 3

Let us assume that vectors OA and OB, as shown in Figure 8.5a represent the magnitude and direction of forces F_1 and F_2 acting on point O. The diagonal, OR, of the rect-

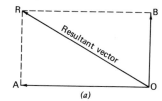

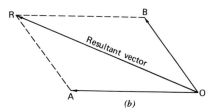

Figure 8.5 Obtaining resultant vectors.

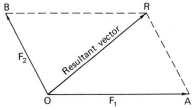

Figure 8.6 Resultant vector is less than vector OA.

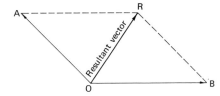

Figure 8.7 Resultant vector is less than either vector OA or OB.

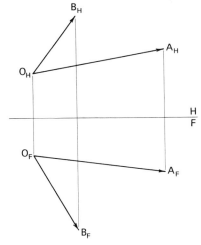

Figure 8.8 Vectors in space.

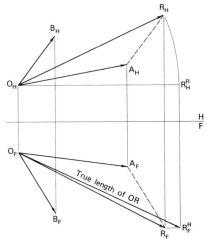

Figure 8.9 True length of resultant vector OR.

angle OBRA represents the magnitude and direction of the single force that produces the same effect on point O as do the forces F_1 and F_2. *Vector OR is the resultant of vectors OA and OB.*

In Figure 8.5b the diagonal, OR, of the parallelogram OBRA represents the magnitude and direction of the single force that has the same effect on point O as do the forces F_1 and F_2. Observe that the resultant vector could be located by drawing *either* AR or BR.

In some cases the resultant vector may have a magnitude that is less than one (Figure 8.6) or either of the given forces, as seen in Figure 8.7.

EXAMPLE 4

Figure 8.8 shows the top and front views of vectors OA and OB. Our problem is to determine the magnitude and direction of the resultant vector OR. We can easily establish the top and front views of line AR, which is equal to and parallel to OB, since we know that parallel lines have parallel views. This means simply that $A_H R_H$ is equal to and parallel to $O_H B_H$, and that $A_F R_F$ is equal to and parallel to $O_F B_F$. Now $O_H R_H$ and $O_F R_F$ can be established. The true length of OR has been obtained by using the method of rotation (see Figure 8.9).

GRAPHIC STATICS

We can apply our knowledge of the two fundamental principles of orthogonal projection, of the analysis and solution of the basic problems, and of vectors to the solution of problems that arise in the field of statics.

Graphic statics, as treated here, deals mainly with the graphical solution of elementary, two-dimensional and three-dimensional force problems. However, some additional problems are included to illustrate slightly more advanced applications of the simple principles; they will prove thought-provoking and will encourage further study.

Most problems in statics can be solved by either graphical or algebraic methods. In some cases the graphical solution is quicker and, therefore, more economical. In other cases the algebraic solution is more readily obtained. In a number of cases *computer solutions* may be well justified.

It is important for the student to become sufficiently familiar with several methods so that an intelligent choice of the one method that is best suited to a particular problem can be made. Regardless of which method is used, the basic conditions to be satisfied in the solution are the *principles of statics*. It is essential, therefore, that these few and simple principles be clearly understood and strictly applied to the solution of the problem at hand.

Statics is the area of mechanics that deals with balanced force systems (systems of forces in equilibrium). The term *statics* implies a static state or state of rest for the bodies on which the forces act, as in the case of roof trusses, walls, floors, columns of a building, the structural members of a highway or railroad bridge, storage tanks, etc. However, *all* bodies in equilibrium, including those in motion *with constant velocity* (e.g., constant-speed motors, shafts, pulleys, gears, and conveyor belts) may be analyzed by using the principles of statics. *Dynamics* is the branch of mechanics that deals in general with bodies in motion and the relations between their motions and the forces causing them. Whenever bodies move with varying velocities, their motions and the applied forces must be analyzed by using the principles of dynamics. The design and analysis of airplane structures, variable-speed mechanisms, and vibrating systems, for example, require careful study of their *dynamic behavior*.

Force is the action of one body on another; it changes, or tends to change, the state of rest or motion of the body acted on. Force is also equal to the product of mass and acceleration. Force is a vector quantity that has magnitude and direction. In addition, every force has a line of action on which its effect partly depends. Suppose, for example, that two forces, identical in magnitude and direction, are both applied perpendicular to the axis of a flywheel. Let us further suppose, however, that the line of action of one force intersects the flywheel axis, while the other force acts tangentially to the rim. The first force has no turning effect, but the second force tends to rotate the flywheel.

RESULTANT OF COPLANAR FORCES

EXAMPLE 1
Suppose that force F_1 is applied at point B and force F_2 at point A, as shown in Figure 8.10. Our problem is to find the resultant force of the coplanar (in the same plane) forces F_1 and F_2. We can do this *by sliding the vectors* along their respective lines of action to their point of intersection, O, completing the parallelogram and, finally, drawing the diagonal, R, which is the resultant force. It has the same effect on the body as the combined effect of the forces F_1 and F_2. Obviously, we could locate R by drawing a line through the arrow end of F_1', parallel to and equal to the length of F_2', to locate point P. Vector OP, then, is the resultant R.

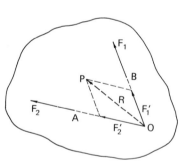

Figure 8.10 Resultant of coplanar forces.

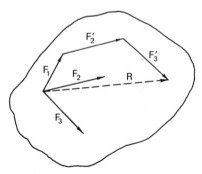

Figure 8.12 Simplified method for finding resultant, R.

EXAMPLE 2
Assume that a body is acted on by *three forces that have one point in common (concurrent forces) and that lie in one plane (coplanar forces)* (see Figure 8.11). If we use the *parallelogram construction*, we can first obtain R_1, the resultant of forces F_1 and F_2, and then combine R_1 with force F_3 to form a second parallelogram, whose diagonal is R, the resultant of the system of forces F_1, F_2, and F_3. Less construction is necessary if we use the *polygon method*. This is shown in Figure 8.12, where F_2' is parallel and equal to F_2 and where F_3' is parallel and equal to F_3.

EXAMPLE 3
Consider the body shown in Figure 8.13, acted on by forces F_1 and F_2

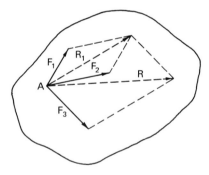

Figure 8.11 Resultant of three concurrent, coplanar forces.

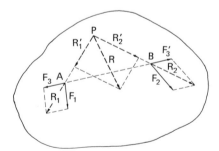

Figure 8.13 Determining the line of action, magnitude, and direction of resultant, R, when the intersection of the lines of action of the given forces is inaccessible.

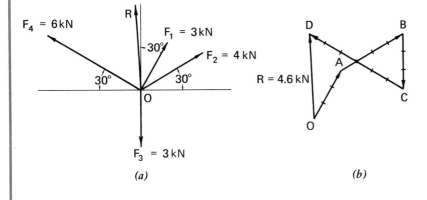

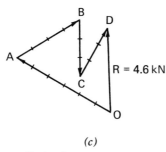

Figure 8.14 Direction and magnitude of the resultant, R, by constructing a force polygon.

sultant by constructing a force polygon in the following manner.

1. Through point O (Figure 8.14b) line OA is drawn parallel to force F_1 and is laid off to a length that represents 3.0 kN (thousands of Newtons)* (vector OA).

2. Through point A line AB is drawn parallel to force F_2 and to a length that represents 4.0 kN to the same scale as used to lay off OA. In a similar way we can draw vectors BC and CD.

3. Vector OD is the resultant, its direction is as shown in the figure, and its magnitude is 4.6 kN. In Figure 8.14a we can draw through point O a line parallel to OD of Figure 8.14b and show the resultant force of the system of forces.

The sequence of drawing the vectors of the force polygon need not follow the order F_1, F_2, F_3, and F_4. Any sequence may be used without affecting the direction and magnitude of the resultant. For instance, in Figure 8.14c, the sequence is F_4, F_2, F_3, and F_1.
We should carefully observe that:

1. *The arrow end of one vector is the beginning of the following vector.*

2. *The arrow end of the resultant vector touches the arrow end of the last vector.*

at points A and B, respectively. How shall we proceed to determine the line of action, magnitude, and direction of the resultant R, since the intersection of the lines of action of the given forces is inaccessible? Along line AB let us introduce force F_3 at point A and an equal, opposite, and collinear force F_3' at point B. The original force system is not affected by forces F_3 and F_3', since they cancel each other. Now two resultants, R_1 for forces F_1 and F_3, and R_2 for forces F_2 and F_3', are easily constructed. *Point O, the intersection of the lines of action of R_1 and R_2, is on the line of action of the resultant R of the forces F_1

and F_2. The direction and magnitude of R can now be established by drawing the diagonal of the parallelogram whose sides are R_1' and R_2'. Vector R_1' is obtained by sliding vector R_1 to point O; similarly, vector R_2' is obtained by sliding vector R_2 to point O. The magnitude of forces F_3 and F_3' will affect the location of point P without changing the line of action of the resultant force R.*

EXAMPLE 4
Let us consider the *coplanar, concurrent* force system shown in Figure 8.14a. We will determine the direction and magnitude of the re-

* The unit of force is the Newton, which is equal to mass (kilograms) times acceleration (meters per second squared).

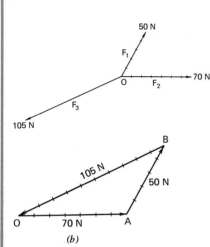

Figure 8.15 (a) Force system in equilibrium.
Figure 8.15 (b) Force triangle shows that the resultant of the forces shown in part (a) is zero.

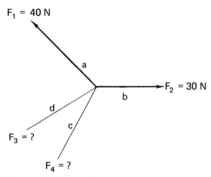

Figure 8.16 Two known forces and the lines of action of two other forces, all in a concurrent, coplanar system.

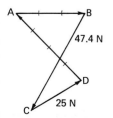

Figure 8.17 Determination of the magnitudes and directions of forces F_3 and F_4.

FORCES IN EQUILIBRIUM

If a body is in a state of equilibrium, the resultant of all forces applied to the body is zero. Suppose we have a system of concurrent forces, F_1, F_2, and F_3, as shown in Figure 8.15a. The *force triangle* shown in Figure 8.15b reveals that the resultant is zero, since the arrow end of the last force coincides with the beginning of the first one. Hence, *the system is in balance (i.e., it is in equilibrium). When the force system is in equilibrium, the arrowheads of the sides of the force polygon follow each other around the polygon.*

EXAMPLE 1

Consider Figure 8.16, which shows forces F_1 and F_2 and the lines of action of forces F_3 and F_4, which are concurrent with F_1 and F_2. If the system of forces is in equilibrium, what are the magnitudes of F_3 and F_4?

To simplify and shorten our further discussion, we introduce a system of lettering known as *Bow's notation*. In Figure 8.16, which shows the lines of action of the forces acting at a point, we write the lowercase letters *a*, *b*, *c*, and *d* in the spaces between the lines of action so that we can designate any line of action by the letters in the adjacent spaces; for example, the line of action of force F_2 is designated by the letters *ab*, and those of F_4, F_3, and F_1 are designated by *bc*, *cd*, and *da*, respectively. (*In concurrent force systems, the forces are conventionally designated by the letters as read in a clockwise progression around the point of concurrency.*) In the force polygon (Figure 8.17) capital letters are used to specify the corresponding vectors, the first letter denoting the tail end and the second denoting the arrow end; for example, force F_2 is lettered *AB*, F_4 lettered *BC*, etc.

With Bow's notation in mind, the analysis of the problem now proceeds as follows. Since the system is in equilibrium, the *vector sum of the four forces is zero*, so that the arrow end D of force F_3 must coincide with the beginning (also D) of F_1 when forces F_1, F_2, F_4, and F_3 are added head to tail in that order. Since forces F_1 and F_2 are known, the corresponding sides DA and AB may be constructed immediately. The direction and one end point of each of the remaining two sides are now known, so that the polygon can be completed and the unknown magnitudes determined, as shown in Figure 8.17.

EXAMPLE 2

A system of three weights held in equilibrium by the tensions in the connecting ropes is shown in Figure 8.18a. It is desired to determine the equilibrium position of point O and thus of the entire system. This may be done by finding the directions of the two inclined ropes. In Figure 8.18b, point O is *isolated* and is shown in equilibrium under the actions of forces F_1, F_2, and F_3. F_2 and F_3 are known in magnitude but have the *unknown* directions of the inclined ropes.

Equilibrium requires that the polygon of the three forces must close so that their resultant is equal to zero. In Figure 8.18c we first lay off the vector AB representing F_1, the only force that is *completely* known. Point C, the arrow end of the vector BC (force F_2), must now lie on a circular arc whose center is B and whose radius, to scale, is 220 N, the known magnitude of force F_2. Since point C is also the tail end of the closing vector CA (force F_3), C lies also on an arc whose center is A and whose ra-

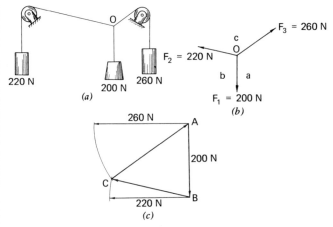

Figure 8.18 Determination of the equilibrium position of point O.

dius, to scale, is 260 N. The intersection of these two arcs locates point C and, when the arrowheads are placed in sequence, gives the directions of forces F_2 and F_3.

DETERMINATION OF FORCES IN TWO-DIMENSIONAL TRUSSES*

EXAMPLE 1

A simple truss is shown in Figure 8.19a. What are the forces acting in members *ca* and *ab*? We first *isolate* the joint to which the known load is applied and show all forces acting on this joint, as shown in Figure 8.19b. This isolation serves two very important purposes. First, it enables us to account, clearly and completely, for *all* forces that act on the joint. These forces, and only these forces, hold the joint in equilibrium. Second, it gives meaning to the distinction between the forces acting *on* the joint and those exerted *by* the joint on the members in contact with it. For ex-

---

* A truss is a structural framework that consists of members arranged and connected to form a system of triangles. The forces applied to any member are usually assumed to act in the direction of that member.

ample, the force that member *ab* exerts on the joint is accompanied by an equal, opposite, and collinear force having the same point of application but exerted by the joint on that member. *The isolation of the joint is an essential step in understanding the action of the forces.* The more complicated the structure and the more complex the force system, *the more vital this process of isolating the body considered becomes, since it is truly the key step in the solution of the problem.*

Why was the loaded joint isolated instead of one of the left-hand joints? The answer is that the loaded joint is the only one at which a known force acts and is therefore the only one for which the force polygon can be constructed.

The magnitude and direction of load *BC* are known; therefore, we can start the graphical solution by drawing a vertical line, *BC* (Figure 8.19c), to represent 10 kN. The lines of action of the forces in members *ca* and *ab* are known. Hence, we may draw through point *C* a line parallel to member *ca*, and we may draw through point *B* another line parallel to member

ab. The intersection of these lines locates point A. The magnitudes of the forces in members *ca* and *ab* are obtained by measuring line segments *CA* and *AB* with the unit of measure used in laying off *BC*. The directions of the forces in these members are established by placing arrowheads in sequence: at *A* (for *CA*) and at *B* (for *AB*). This we recall from our discussion of forces in equilibrium (Figures 8.15 and 8.19). We now observe that force *AB*, exerted by member *ab* on the isolated joint in Figure 8.19b, pulls away from the point of concurrency. This means that member *ab* is in *tension*. On the other hand, force *CA*, exerted by member *ca* on the isolated joint, pushes toward the joint. This means that member *ca* is in *compression*.

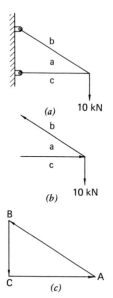

Figure 8.19 Determination of forces acting in the members of a simple truss.

EXAMPLE 2

Observe the truss shown in Figure 8.20. It is assumed that the reactions (the 2.5k N forces at the supports) have already been determined. What are the forces acting in the members of the truss?

Let us first consider the isolated portion of the truss shown in Figure 8.21*a*. The determination of the forces in members *b*-1 and 1-*a* is essentially the same as in the example just discussed. The force polygon is shown in Figure 8.21*b*. The magnitudes of vectors *B*-1 and 1-*A* are the forces in members *b*-1 and 1-*a*, respectively. Moreover, we observe that *member b*-1 is in *compression* and that *member* 1-*a* is in *tension*. Now let us analyze the forces acting at the joint that is common to members *c*-2, 2-1, and 1-*b* (Figure 8.21*c*). *We know the force in member b*-1 *from Figure* 8.21*b*. The force polygon for the forces acting at the joint is shown in Figure 8.21*d*. *BC* is laid off as a vertical line to represent 2.0 kN Through point *C*, a line is drawn parallel to member *c*-2, and through point 1 (known because we have both the magnitude and direction of the force in member 1-*b*) a line is drawn parallel to member 2-1. The intersection of these lines locates point 2. The magnitudes of the forces in members *c*-2 and 2-1 can be measured in the force polygon. The combined use of Figure 8.21*c* and 8.21*d* shows that both members *c*-2 and 2-1 are in compression. *The lines of action of all forces at a joint, including forces in the members, are identified by reading clockwise about the joint.* For example, at the peak joint (Figure 8.20), the reading

would be *c*-*d*, *d*-4, 4-3, 3-2, and 2-*c*. We will see that this same sequence, when applied to the force polygon, provides a simple method for determining the members that are in compression and also those that are in tension.

The analysis of the forces in the members of the peak joint and at the remaining joints can be made in a manner similar to that used with the first two joints. Note, particularly with reference to the peak joint, that the force in member 2-*c* is known from the analysis of the forces in the members of the previous joint and, likewise, the force in member *d*-4 is known, since the truss is symmetrical and is loaded symmetrically, so that the forces in both members 2-*c* and *d*-4 are the same. If this were not true, we could first determine the force in member 3-2 from an analysis of the forces acting at the joint common

to members *a*-1, 1-2, 2-3, and 3-*a* and then proceed to construct a force polygon that would contain the magnitudes of the forces in members 2-3 and 3-*a*. This will be done in the solution that follows.

MAXWELL DIAGRAM

Instead of constructing separate force polygons for all joints, it is most convenient to combine them into a single force diagram for the entire structure. Such a diagram is known as a *Maxwell diagram*, in recognition of the work of Clerk Maxwell, who presented this method in 1864.

Figure 8.22 shows the combined force diagram for the determination of the forces in the members of the truss. First the external loads —*BC, CD, DE, EA,* and *AB*—are laid off to a convenient scale. If we consider the first joint at the left

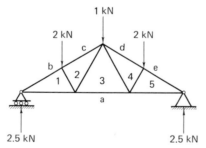

Figure 8.20 Truss and forces acting on it.

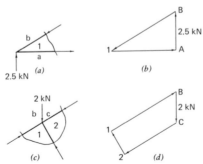

Figure 8.21 Analysis of forces acting in members of the truss shown in Figure 8.20.

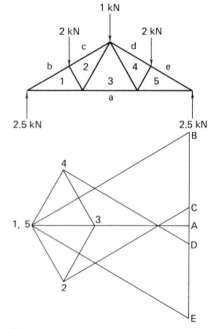

Figure 8.22 Maxwell diagram. A simple method to find the forces acting in the members of the truss.

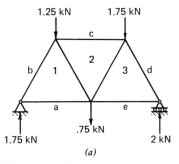

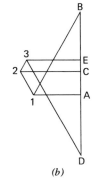

Figure 8.23 Use of Maxwell diagram to determine the forces acting in the members of the truss.

(same as shown in Figure 8.21*a*), we can locate point 1 by drawing through *B* a line parallel to member *b*-1 and another line through point *A* parallel to member *a*-1. The intersection of these lines locates point 1. Point 2 is located at the intersection of the line drawn through point *C* parallel to member *c*-2, with the line drawn through point 1 parallel to member 1-2. Point 3 is the intersection of the line drawn through point *A* parallel to member *a*-3, with the line drawn through point 2 parallel to member 2-3, and so forth for the other points 4 and 5.

Once the force diagram is completed, the magnitudes of all forces can be scaled directly. The determination of the kind of force—compressive or tensile—can also be made. For example, consider the peak joint. Starting with the known external load, *c*-*d*, and reading clockwise, the sequence is *c*-*d*, *d*-4, 4-3, 3-2, and 2-*c*. Using this sequence in the force diagram, we observe that force *D*-4 (arrow if placed would have been at 4) applied by the member *d*-4 to the isolated peak joint would push toward the joint; hence member *d*-4 is in compression. Reading force 4-3 (from point 4 to point 3) and applying it to the isolated joint shows that the force is pulling away from

the joint; hence, member 4-3 is in tension. Now, proceeding from point 3 to point 2 in the force diagram, applying force 3-2 to the isolated joint shows that the force is pulling away from the joint and, therefore, that member 3-2 is in tension. Finally, force 2-*C* on the peak joint acts toward the joint; hence, member 2-*c* is in compression.

EXAMPLE 3
Consider the truss shown in Figure 8.23*a*. Our problem is to find the forces acting in the members of the truss.

SOLUTION
First let us lay off, to a convenient scale, the "load line," which is the vector polygon—*AB*, *BC*, *CD*, *DE*, and *EA*—as shown in Figure 8.23*b*.

Consider the lower left joint. We can locate point 1 by drawing through *B* a line parallel to member *b*-1 and another line through *A* parallel to member *a*-1. The intersection of these lines is point 1.

Now consider the upper left joint. We can locate point 2 by drawing through *C* a line parallel to member *c*-2 and another line through point 1 parallel to member 2-1. The intersection of these two

lines is point 2. We can similarly locate point 3 and complete the Maxwell diagram.

Next let us find, for example, the forces acting in the members 1-*b*, *c*-2, and 2-1. The magnitudes of these forces are equal to vectors 1-*B*, *C*-2, and 2-1 (measured to the scale used in laying off the vectors *AB*, *BC*, . . .). Member 1-*B* is in compression. We can see that this is true by observing the action of the force in the direction from point 1 to point *B*. When this force is in the member 1-*b*, it is acting toward the joint that is common to members 1-*b*, *c*-2, and 2-1. Member *c*-2 is also in compression, and member 2-1 is in tension. We can determine in the same way the magnitudes of the forces in the other members of the truss and whether they are in compression or in tension.

NONCONCURRENT FORCES IN A PLANE

Before considering this subject, we should point out what is meant by the *moment of a force* such as *F* (Figure 8.24*a*) *about a point*, such as *P*, *is the product of the force and the perpendicular distance from the point to the line of action of the force.* The distance is known as the *moment arm*. If the length of the arm is in inches and the force is given in pounds, then the moment of the force about the point is given in inch-pounds. If the length of the arm is in meters and the force in Newtons, then the moment is in Newton-meters.

If we have *two equal and opposite forces whose lines of action are parallel* (Figure 8.24*b*), *the mo-*

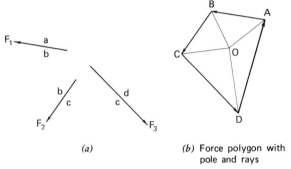

(a) Moment = F × d   (b) Moment = F × d

Figure 8.24 Moment of a force about a point and the moment of a couple.

(a)

(b) Force polygon with pole and rays

Figure 8.25 Determining the magnitude, direction, and line of action of the single additional force that will produce equilibrium in the system shown in part (a).

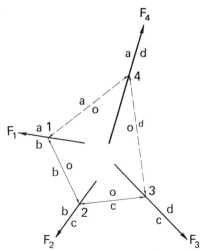

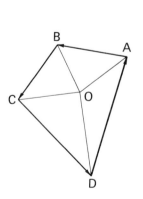

Figure 8.26 Determining a point on the line of action of the required force to produce equilibrium.

ment of the forces about any point is equal to Fd. This system of forces is known as a couple. A system of nonconcurrent forces in a plane is in equilibrium *when both the resultant force and the resultant moment are zero.*

Consider the force system shown in Figure 8.25a. It is required to determine the magnitude, direction, and line of action of the single additional force that will produce equilibrium. Again let us use Bow's notation, so that the line of action of any force is designated by adjacent lowercase letters and the vector representing the force in magnitude and direction is designated by the corresponding capital letters placed at its ends. The magnitude and direction of the required force can be easily determined from line DA, which closes the vector polygon (Figure 8.25b) and satisfies the requirement that the resultant force must be equal to zero.

It remains, however, to determine a point on the line of action *da* of the required force DA. Let us assume a pole O in (or outside) the vector polygon, and then draw lines OA, OB, OC, and OD. We may regard each of the four triangles OAB, OBC, OCD, and ODA as a vector triangle, one side of which in each case is the resultant of the forces represented by the other two sides. For example, consider triangle OAB. If vector AB is treated as the resultant of vectors AO and OB, the given force F_1 (AB) may be replaced by forces AO and OB.

Hence, if we select an arbitrary point 1 on the line of action *ab* of force AB (see Figure 8.26) and then draw lines *ao* and *ob* through point 1 parallel, respectively, to vectors AO and OB, we will have represented the lines of action of the two forces AO and OB by which force AB may be replaced. Now consider triangle OBC, in which force BC is the resultant of forces BO and OC. We observe that force BO is equal and opposite to OB, one of the two forces by which AB was replaced at point 1. If we extend the line of action *ob* until it intersects *bc* at point 2 and replace BC by the two forces BO and OC at this point, we see that the equal and opposite forces OB and BO have the common line of action *ob* and therefore cancel each other. The remaining forces AO and OC have now completely replaced the two forces AB and BC.

Similarly, we see that force CD may be replaced by the two forces CO and OD. If this replacement is made at point 3 on the line of action oc of force OC, the equal and opposite forces OC and CO will be collinear and will cancel each other. The forces AO and OD that remain are now completely equivalent to the original forces AB, BC, and CD. Finally, considering triangle DOA, we observe that the required force DA may be replaced by the two forces DO and OA. If this replacement is made at the point, 4, common to the lines of action of forces AO and OD, all forces cancel and the requirements for equilibrium are satisfied. Therefore, point 4 is a point on the line of action da of the required force DA, and this line of action may be drawn through point 4 parallel to vector DA.

The polygon 1-2-3-4 is known as a *string polygon or funicular polygon*. Each side of this polygon represents the common line of action of two equal and opposite forces and is quite analogous in this respect to a tension or compression member of a truss or, in the case of tension, to a string pulled tight under the action of the two forces.

We now see that the *force system* —F_1, F_2, F_3, F_4—*is in balance* (i.e., it is in equilibrium), *since both the force polygon and the string polygon are closed*. It is possible to have a closed force polygon and yet not have the system in equilibrium, if the string polygon does not close. This is clearly seen if we arbitrarily shift the line of action of the force F_4 and so replace it by the parallel and equal force F_4' (see Figure 8.27). We see that the system of forces F_1, F_2, F_3, F_4' is equivalent to the equal and opposite, but noncollinear, forces DO and OD, which are separated by the moment arm d and thus form a counterclockwise couple, which is the resultant of forces F_1, F_2, F_3, F_4'. Hence, these forces are *not* in equilibrium.

It should be quite clear that *a nonconcurrent coplanar force system is in equilibrium only when* both *the resultant moment and the resultant force of the system are zero; that the closed force polygon is the necessary condition for zero resultant force; and that the closed string polygon is the necessary condition for zero resultant moment.*

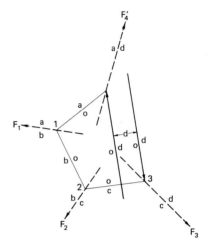

Figure 8.27 Arbitrary shift of force F_4 to a parallel and equal force F_4' shows that the force system is no longer in equilibrium.

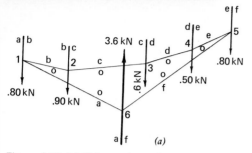

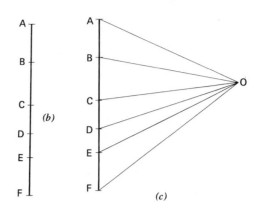

Figure 8.28 (a) String polygon locates a
point on the line of action of the required force to produce equilibrium.
Figure 8.28 (b) Force polygon
(a straight line, in this case).
Figure 8.28 (c) Basis for constructing
the string polygon in part (a).

EXAMPLE 1

Consider the force system shown
in Figure 8.28a. It is required to determine the magnitude, direction,
and line of action of the single additional force that will produce
equilibrium.

Start by constructing the force
polygon, *ABCDEF* (in this case a
single line), as shown in Figure
8.28b. The magnitude, 3.6 kN in
this case, and direction of the required force are represented by
vector *FA*. To locate a point on the
line of action of the required force,
we construct the string polygon as
shown in Figure 8.28c. First we assume a pole *O* (Figure 8.28c) and
then draw rays *OA*, *OB*, . . . ,
OF. Force *AB* is replaced by two
forces *AO* and *OB*, which have the
lines of action *ao* and *ob*, respectively (Figure 8.28c). Through
point 1, an arbitrarily selected
point on the line of action of force
AB, these two lines *ao* and *ob* are
drawn respectively parallel to *AO*
and *OB*. The intersection of line *ob*
with line *bc* locates point 2,
through which *oc* is drawn parallel

to *OC*. The intersection of line *oc*
with line *cd* locates point 3,
through which line *od* is drawn
parallel to *OD*. Points 4 and 5 are
similarly located. Finally, a line *of*
is drawn through point 5 parallel to
OF. The intersection of lines *oa*
and *of* locates point 6, a point on
the line of action of the required
force, which acts along a vertical
line (since *FA* of the force polygon
is vertical). The magnitude and direction of force *FA* are included.

EXAMPLE 2

Suppose we have a beam loaded as
shown in Figure 8.29a. It is required to determine the magnitudes of the support reactions
(forces *DE* and *EA*). In this problem we recognize that the direction
of the forces *DE* and *EA* is known
to be vertical, since the roller on
the right will support only a vertical
force. The force polygon (Figure
8.29b, is first drawn and then a pole
O is located arbitrarily. Rays (Figure 8.29b) *OA*, *OB*, *OC*, and *OD*
are then established. We know the
lines of action of the reactions *EA*
and *DE*; however, their magnitudes can be determined only by
locating point *E* in the force polygon. In order to do this we must
determine the direction of ray *OE*.
This ray is parallel to string *oe* of

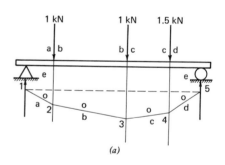

(a)

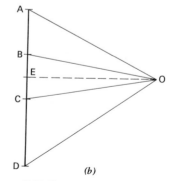

Figure 8.29 Determination of the beam-
support reactions, *DE* and *EA*.

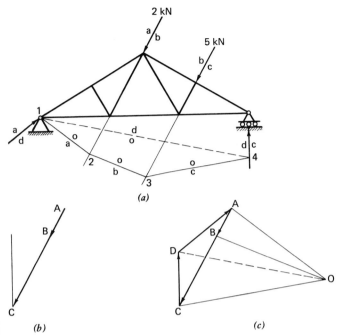

(a)

(b)

(c)

Figure 8.30 (a) Truss, forces, and string polygon.
Figure 8.30 (b) Partial vector polygon.
Figure 8.30 (c) Basis for constructing the string polygon in part (a) and completion of vector polygon.

the funicular polygon, whose direction may be established from the requirement that the funicular polygon must close. Starting from point 1 on the line of action *ea* of reaction *EA*, we construct the successive strings *ao*, *bo*, *co*, and *do* parallel respectively to the corresponding rays in the force polygon. Thus the points 2, 3, 4, and 5 are successively located. The string polygon is closed by the string *eo*, which joins points 1 and 5. Ray *OE* is drawn parallel to strong *oe*. The magnitude of vector *DE* determines the force (right-hand reaction) whose line of action is *de*, and the magnitude of vector *EA* is the left-hand force, whose line of action is *ea*. It should be clear from a study of Figure 8.29*a* and 8.29*b* that both the vector polygon,

ABCDEA, and the string polygon, 1-2-3-4-5-1, are closed polygons and, therefore, the force system is in equilibrium.

EXAMPLE 3
Consider the truss shown in Figure 8.30*a*. It is required to determine the magnitude of force *CD* and both the magnitude and direction of force *DA*.

A portion of the force polygon can be drawn by laying off vectors *AB* and *BC*, as shown in Figure 8.30*b*. A vertical line may then be drawn through point *C* because we know the direction of force *CD*, which must be vertical since the right support is on rollers. At this stage we do not know the location of point *D* because the magnitude

of force *CD* is unknown. We will employ a string polygon to locate point *D*.

In Figure 8.30*c*, pole *O* and rays *OA*, *OB*, and *OC* have been added to Figure 8.30*b*. The string polygon is started at point 1, the only known point on the line of action *da* of force *DA*. Lines *oa*, *ob*, and *oc* are respectively parallel to *OA*, *OB*, and *OC*. Closing line *od* (1-4) establishes the direction of ray *OD*. Now that point *D* is available, the force polygon is completed by drawing the closing line *DA*, thus establishing both the magnitude and the direction of force *DA*. The magnitude of force *CD* is the magnitude of vector *CD*. The supporting forces *CD* and *DA* are known as *reactions*.

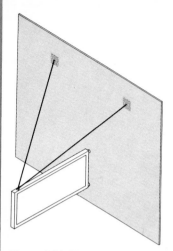

Figure 8.31 Three-dimensional space frame.

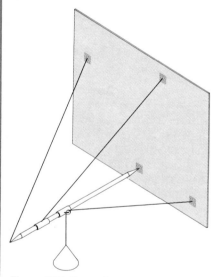

Figure 8.32 Space frame.

CONCURRENT NONCOPLANAR FORCE SYSTEMS

In the design of three-dimensional frames (space frames) it is necessary to determine the forces acting in the members of the frame in order to select the proper sizes of the supporting members. Simple space frames may consist of three concurrent noncoplanar members that support a load, such as the ones shown in Figures 8.31 and 8.32.

Before we consider problems that deal with the determination of forces acting in the members of space frames, let us see what is meant by a *space force polygon*.

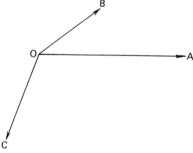

Figure 8.33 Three concurrent, noncoplanar forces.

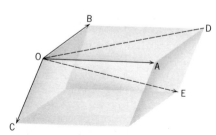

Figure 8.34 Vector *OE* is the resultant of forces *OA*, *OB*, and *OC*.

SPACE FORCE POLYGON

In Figure 8.33 three concurrent noncoplanar forces *OA*, *OB*, and *OC* are shown. Assume that we wish to find the resultant of these forces. To do this, we may employ the "parallelogram-of-forces" method. In Figure 8.34 we observe that force *OD* is the resultant of forces *OA* and *OB*. When we combine force *OD* with force *OC*, we will determine the resultant *OE* of the three forces *OA*, *OB*, and *OC*. *We note that OE is the body diagonal of the parallelepiped having edges OA, OB, and OC.*

It is now evident that it is not necessary to construct the entire parallelepiped to obtain the resultant force *OE*. Actually, it is sufficient to construct the space force polygon *OADEO* or the space force polygon *OAFEO*, shown in Figures 8.35 and 8.36. An orthographic solution is shown in Figure

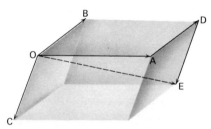

Figure 8.35 Resultant vector, *OE*, determined by constructing the space force polygon *OADEO*.

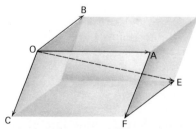

Figure 8.36 Resultant vector, *OE*, determined by constructing the space force polygon *OAFEO*.

8.37. In this figure only the space force polygon *OADEO* is shown. The magnitude of the resultant force *OE* can be determined by measuring the true length of *OE* in terms of the scale adopted in setting up the original data.

Figure 8.38 shows four concurrent forces, *OE*, *AO*, *BO*, and *CO*, which are in equilibrium. Therefore, their force polygon *OEDAO* closes. This figure suggests a solution to the following problem. *If three concurrent members of a space frame support a known load, what are the forces* [*] *in the members?* If we regard *OE* as the known load and *m*, *n*, and *s* as the concurrent

---

[*] It is assumed that the forces applied to any one of the members *m*, *n*, and *s* act only in the direction of that member. In other words, forces applied perpendicular to any member are neglected in comparison with the forces applied along that member. We should recognize that the same assumption is basic to the analysis of the two-dimensional truss; hence, the frame just analyzed is a simple example of a particular type of space frame, the space truss.

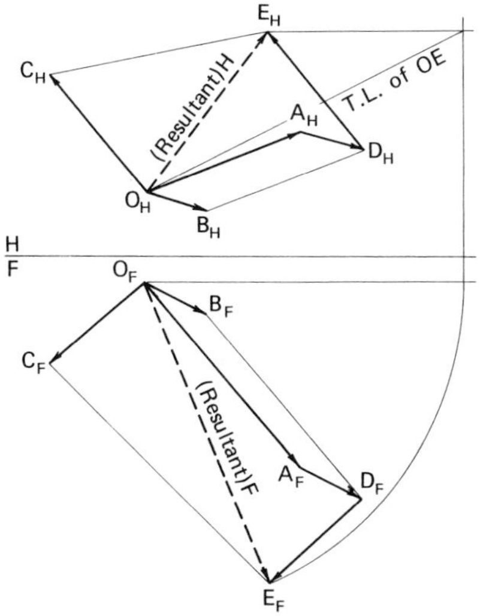

---

members, we can determine the force polygon, *OEDAO*. Since point *E* is known, we may introduce a line *t* through this point and parallel to member *n*. Line *t* intersects the plane of members *m* and *s* in point *D*. A line is drawn through point *D* parallel to member *m* and intersecting member *s* in point *A*. The force polygon is easily constructed, since it consists of sides *OE*, *ED*, *DA*, and *AO*. The true lengths of sides *AO*, *DA*, and *ED* determine the magnitudes of the forces in members *s*, *m*, and *n*, respectively. We now see that when force *ED* is applied to member *n*, to which it is parallel, *the force acts toward point O and, hence, the member is in compression.* Similarly, when force *DA* is applied to member *m*, this force also acts toward point *O* and, therefore, member *m* is in compression. Finally, we observe that force *AO* in member *s* also acts toward point *O*, so that member *s* is also in compression.

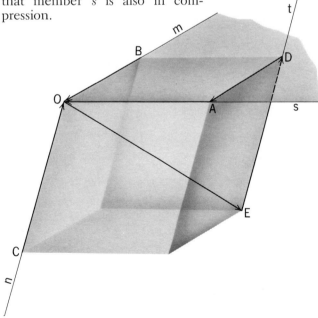

Figure 8.37 Obtaining the resultant of forces *OA*, *OB*, and *OC* by using force polygon *OADEO*.

Figure 8.38 Key to the solution of problems that require the determination of forces acting in three concurrent, noncoplanar members of a space frame that supports a known load.

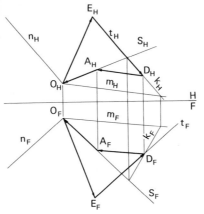

Figure 8.39 Space frame and force polygon. Example 1.

EXAMPLE 1
Now let us proceed to solve this type of problem described by the orthographic views shown in Figure 8.39. Members m, n, and s support a load that is represented by vector OE.

SOLUTION
Through point E (Figure 8.39) line t is introduced parallel to member n. The intersection of line t with the plane determined by members m and s is point D. *We may recall the three basic steps that are employed in locating the point in which a line intersects a plane.*

1. Pass a plane through the line. (In this example the plane containing line t is perpendicular to the horizontal reference plane.)
2. Find the intersection of this plane with the given plane. (This is line k in the example.)
3. Find the intersection of lines k and t. This is point D.

The line drawn through point D parallel to member m intersects member s in point A. The force polygon is $OEDAO$. The magnitudes of the forces in members m, n, and s are determined by measuring the true lengths (not shown) of vectors DA, ED, and AO, respectively. If the forces are applied to the members, it is evident that the members are in compression.

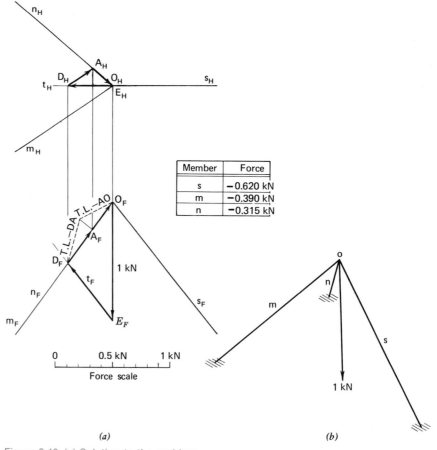

Member	Force
s	−0.620 kN
m	−0.390 kN
n	−0.315 kN

(a)

(b)

Figure 8.40 (a) Solution to the problem in part (b).
Figure 8.40 (b) Space frame and vertical load (force). *Problem:* To determine the forces acting in members, m, n, and s. Example 2.

EXAMPLE 2

Consider the space frame and the load shown in Figure 8.40*a* and pictorially in Figure 8.40*b*. Carefully observe that *the plane of members m and n is shown on edge in the front view*. Advantage should be taken of this because it enables us to locate very easily the intersection, point *D*, of line *t* with the plane of members *m* and *n*. Then we introduce a line through point *D* parallel to member *m* and intersecting member *n* in point *A*. (We could use a line through *D* parallel to member *n* and intersecting *m*.) The force polygon is *OEDAO*. The true lengths of the vectors are readily established, since the vector *ED* is in true length in the front view and a single auxiliary view shows the true lengths of vectors *DA* and *AO*. The true lengths, when measured in accordance with the scale shown in Figure 8.40*a*, result in the values given in the table. The minus signs are one way of indicating compression, since tensions in truss members are ordinarily considered positive.

EXAMPLE 3

Let us now consider the space frame and the load shown in Figure 8.41. The method employed in the previous Example 2 may be applied to this problem. It is necessary only to obtain a view that shows the plane of members *m* and *n* as a line. An edge view of plane *mn* is easily obtained by first establishing a point view of a line that lies in the plane. The horizontal line *k* is such a line. Supplementary plane 1 is perpendicular to the plane of members *m* and *n* and the *H* plane. Once the supplementary view of the space frame and the load is established, it is necessary only to regard the top and supplementary views as given and then to proceed as in Example 2 to determine the views of the force polygon.

In determining the forces in the members of the space frames just discussed, we have regarded the single joint as apart from the rest of the structure. We have thus *isolated the joint mentally* without actually drawing the views of the isolated joint and the forces acting on it. Exactly the same thing was done in constructing and using the Maxwell diagram for the two-dimensional truss. We can safely do this only because we are dealing with frames of the truss type, so that the unknown forces act along the members whose directions are already shown in the views of the frame. For almost all other types of structures and machines this is usually *not* true; therefore, the portion to be analyzed should be isolated by means of a clear diagram that shows all the forces that act on the body.

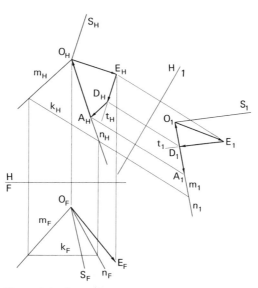

Figure 8.41 Space frame and force polygon. Example 3.

EXAMPLE 4
Carefully examine Figure 8.42. Members OA, OB, and OC of this space frame support a vertical load of 1.0 kN. Member OC is horizontal and is 3 m long. The force acting in member OA is 0.5 kN. Determine the views of member OC and the forces acting in members OB and OC.

SOLUTION
The solution is shown in Figure 8.43. We know the force acting in member OA is 0.5 kN; therefore, we can obtain a true length view of OA, shown as O_1A_1, and then lay off the vector $E_1O_1 = 0.5$ kN. The H and F views of point E are readily established. Through point E a parallel to member OB is drawn. The intersection of this parallel with the horizontal drawn through W_F establishes D_F (remember that member OC is horizontal). Once we have located D_F, the F view of the force polygon is fixed as $O_FW_FD_FE_FO_F$. The H view of the force polygon is readily established. The H and F views of member OC are now drawn; we know that OC is horizontal and is 3 m long. The forces in members OB and OC can be found by first obtaining the magnitudes of vectors DE and WD, respectively. Study of the force polygon shows that member OA is in compression and members OB and OC are in tension.

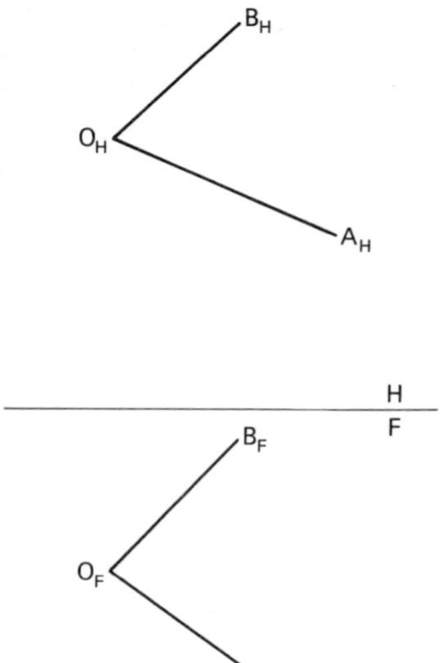

MORE INVOLVED SPACE FRAMES

The three examples that follow deal with the determination of the forces in space frames that are more complex than those previously discussed. These examples afford an opportunity to make more extensive use of the principles that have been employed in solving the problem of the simple tripod frame (Figure 8.39). The tripod problem is easily analyzed because there is only a single joint involved.

In the case of the more complex space frames, however, we again regard each joint as apart from the rest of the structure (isolated mentally) in analyzing the system of forces that acts at that joint. The force polygon for the force system at each joint will be drawn separately.

EXAMPLE 1
Consider the space frame in Figure 8.44a. We are to determine the forces acting in the boom AB, in member OB, and in cables OA, OC, and OD. Top and front views of the frame are shown in Figure 8.44b.

SOLUTION
Let us first isolate the joint at A (Figure 8.44c), since the known load, 2.5 kN is suspended from A. It should be observed that A-1 represents the cable that supports the load, line 1-2 is parallel to AB, the boom, and 2-A is parallel to cable OA. We can easily construct the vector triangle shown in Figure 8.44d, by laying off A-1 equal to 2.5 kN to a convenient scale and drawing a line through point 1 parallel to 1-2 and another line through point A parallel to 2-A to intersect the first line at point 2. The magnitude of the force acting in boom AB is represented by vector 1-2, measured to the scale that was used in laying off vector A-1. Vector 2-A represents the magnitude of the force acting in cable AO. We note that the direction of

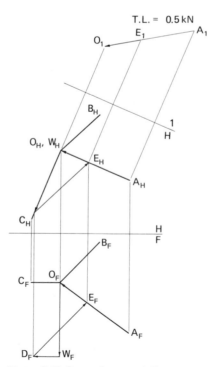

Figure 8.42 Space frame problem Example 4.

Figure 8.43 Space frame solution Example 4.

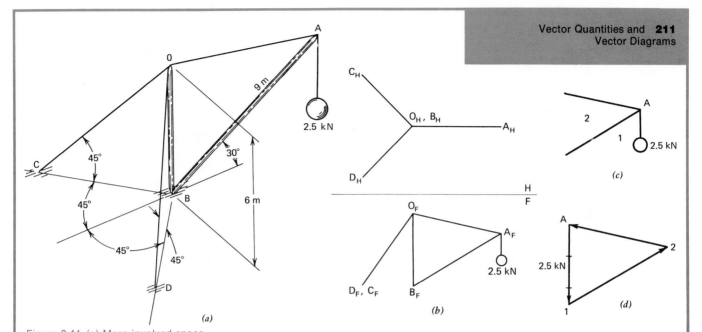

Figure 8.44 (a) More involved space
frame. Example 1.
Figure 8.44 (b) Top and front
views of the space frame shown in part
(a). Example 1.
Figure 8.44 (c) Joint A isolated
Example 1.
Figure 8.44 (d) Force triangle to determine
the magnitudes of the forces acting in
members 1-2 and 2-A. Example 1.

vector 1-2 is toward joint A; therefore boom AB is in compression. Vector 2-A clearly shows that cable OA is in tension.

Since we know the magnitude of the force acting in cable OA, we can determine the forces acting in member OB and cables OC and OD. In fact, we are back to a familiar problem, one similar to Example 1 (Figure 8.39). Consider Figure 8.45, which shows the top and front views of member OB, cables OC and OD, and vector OW, which represents the force acting in cable OA.

We can use the same steps that were used to solve the problem in the Example 1 (Figure 8.39).

1. Through point W, line t is introduced parallel to cable OD.
2. Line t intersects the plane of member OB and cable OC at point E.
3. Through point E a line is drawn parallel to OC to intersect OB (extended) at point F.
4. The space vector polygon is OW-WE-EF-FO.

Now we can determine the magnitudes of the forces acting in member OB and cables OC and OD by finding the true lengths of the vectors and their measurement to the scale that was used in laying off vector OW. We observe that member OB is in compression and that cables OC and OD are in tension.

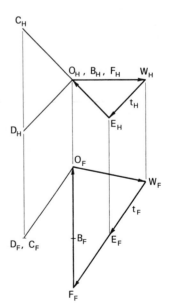

Figure 8.45 Top and front views of members OB, cables OC and OD, and vector OW, which represent the force acting in cable OA. Also, the vector polygon to determine the magnitude of the forces. Example 1.

EXAMPLE 2

Consider the space frame in Figure 8.46a. Before proceeding with the analysis, we should recognize that, for a balanced system of concurrent forces in space, the unknown magnitudes of three forces can be found if all the other forces are known. Thus, for the frame shown in Figure 8.46a, we consider first the joint at A, where the load W is known and the three forces in members AB, AC, and AD are unknown. We can complete the analysis by considering joint B, where the force in member AB is *now known* and the forces in members BC, BD, and BE are the only unknowns.

Figure 8.46b and 8.46c are the force polygons for joints A and B, respectively. The polygons are labeled with a modified Bow's notation so that the letter inside the polygon designates the joint on which the forces act, and the letter outside the polygon and adjacent to any force designates the other end of the member that exerts that force. Thus, in Figure 8.46b, the letter a inside the polygon identifies the forces as acting on joint A, and the adjacent letter b designates the force exerted on joint A by member AB. For example, the letters a_F and b_F identify the front view of the force AB that acts in member AB.

For each polygon the views of the vector representing the known force are first laid off to scale; then a line is drawn through one end of that vector parallel to one of the unknown forces, and a plane is determined by lines drawn through the other end parallel, respectively, to the remaining two unknown forces. The intersection of the line and the plane is found by the method reviewed in Example 1, and the views of the force polygon are completed. The true lengths of the vectors are not shown.

The force polygon for joint A shows that member AB is in tension. Hence, in the views of the force polygon for joint B (Figure 8.46c), the direction of force AB is such that it would pull away from that joint when applied by member AB. The completed force polygon clearly shows that members BD and BC are in compression, while member BE is in tension.

EXAMPLE 3

Consider the loaded frame in Figure 8.47a. Is the frame stable, or will it collapse? If it is not stable, there is no problem to solve; if it is stable, we can proceed to determine the forces in the members. We first observe that the location of point C is established by the members CD, CE, and CF, which are attached to the foundation at their lower ends. Point B is then uniquely located by the members connecting it to the fixed points C, E, and D. Finally, point A is fixed by members AB, AC, and AD.

Next, forces AB, AC, and AD are found from the force polygon for joint A (Figure 8.47b) in the same manner as previously discussed in Example 2 (Figure 8.40). It is seen that the plane of forces AB and AC is shown as a line in the front view. This enables us to locate immediately the front view of the point of intersection of the line of force AD with the plane of forces AB and AC. The views of the force polygon are then completed. Since force AB is now known, only three unknown forces remain at joint B. The force polygon for joint B is shown in Figure 8.47c. The two known forces, AB and T, are first drawn. Since member BC appears as a point in the front view of the frame, two planes of unknown forces, the plane of forces BC and BE and the plane of forces BC and BD, will appear as lines in the front view of the force polygon. We may thus regard the completion of the front view of the force polygon as locating the front

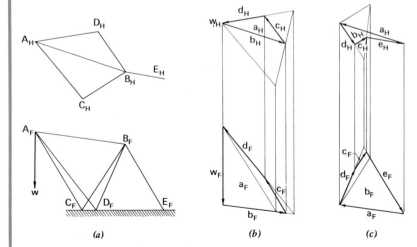

Figure 8.46 Analysis of forces acting in the members of the space frame shown in part (a). Example 2.

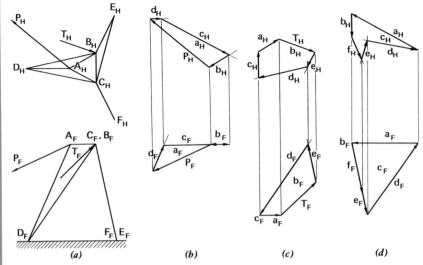

Figure 8.47 Analysis of forces acting in the members of the space frame shown in part (a). Example 3.

labeling *and the top view* would be different. The other two are BC, BE, and BD, and BE, BD, and BC.

We may now construct the views of the force polygon for joint C (Figure 8.47*d*) where forces AC and BC are now known and the three forces CD, CE, and CF are unknown. The plane of forces CE and CF appears as a line in the front view, and the intersection of the line of force CD with this plane is easily found in that view.

In this example, several planes appear as lines in the front view. This is a convenience in solving the part of the problem that deals with the intersection of a line and a plane. However, if this were not true, we could still find the intersection by employing the three basic steps that have been referred to several times.

view of either of two points (i.e., either the intersection of force *BD* with the plane of forces *BC* and *BE*, or the intersection of force *BE* with the plane of forces *BC* and *BD*). We then complete the top

view of the force polygon, which shows the sequence *BE*, *BC*, and *BD* for the three unknown forces. This sequence is only one of three for which the front view would have the same shape, although the

Exercises

Solve all problems graphically. Select suitable scales.

1. Determine the resultant force of the coplanar concurrent force system shown in Figure E-8.1.

2. Determine the resultant force of the coplanar concurrent force system shown in Figure E-8.2.

3. Determine the resultant force of the coplanar concurrent force system shown in Figure E-8.3.

4. Determine the magnitude of force *F* of the force system shown in Figure E-8.4.

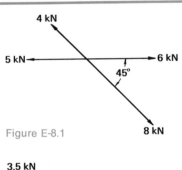

Figure E-8.1

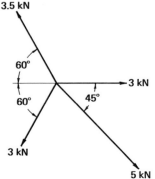

Figure E-8.2

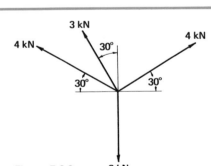

Figure E-8.3

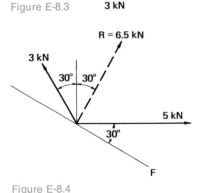

Figure E-8.4

5. Determine the magnitudes of the vertical resultant force R and of the force F of the force system shown in Figure E-8.5.

6. Point O is in equilibrium when acted on by the forces shown in Figure E-8.6. What are the magnitudes and directions of the unknown forces F_1 and F_2?

7. A 500-N force, inclined at an angle of 40° with the horizontal, acts on a body that rests on the floor. What is the magnitude of the force that tends to lift the body?

8. Determine the magnitudes of the forces acting in members OA and OB of the frame shown in Figure E-8.7. Indicate whether each force is in compression or tension.

9. Determine the magnitudes of the forces acting in the members of the truss shown in Figure E-8.8. Use a minus sign to designate compression.

10. Determine the forces acting in the members of the truss shown in Figure E-8.9. Use Bow's notation and the Maxwell diagram.

11. Determine the forces acting in the members of the truss shown in Figure E-8.10.

12. Determine the magnitude, direction, and line of action of the resultant of the force system shown in Figure E-8.11.

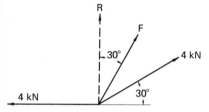

Figure E-8.5

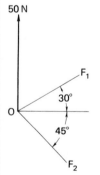

Figure E-8.6

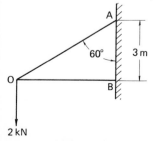

Figure E-8.7

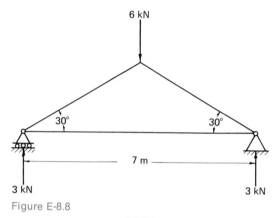

Figure E-8.8

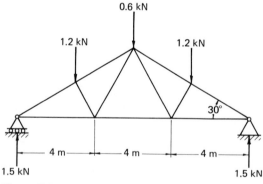

Figure E-8.9

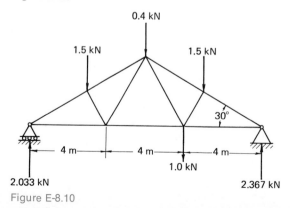

Figure E-8.10

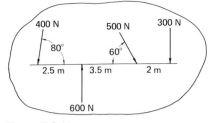

Figure E-8.11

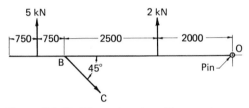

Figure E-8.12

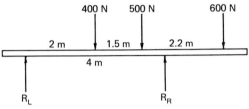

Figure E-8.13 (Dimensions in millimeters.)

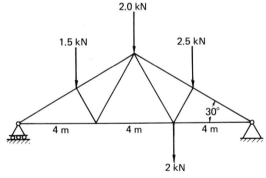

Figure E-8.14

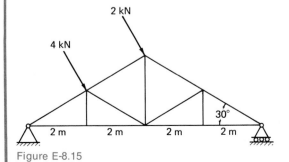

Figure E-8.15

13. Determine the magnitudes of the reactions R_L and R_R on the beam shown in Figure E-8.12.

14. Determine the magnitude of force *BC* acting on the beam shown in Figure E-8.13. (*Hint.* Start the funicular polygon at point *O*.)

15. Determine the reactions and the forces acting in the members of the truss shown in Figure E-8.14.

16. Determine the reactions and the forces acting in the members of the truss shown in Figure E-8.15.

17. Determine the reactions and the forces acting in the members of the truss shown in Figure E-8.16.

18. Determine the reactions and the forces acting in the members of the truss shown in Figure E-8.17.

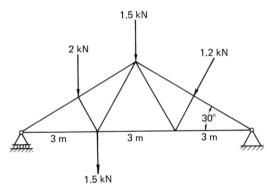

Figure E-8.16

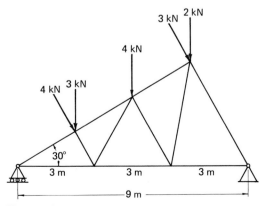

Figure E-8.17

22. Determine the magnitudes of the forces acting in the members of the space frame shown in Figure E-8.21. Load *OW* = 2 kN.

23. Determine the magnitudes of the forces acting in the members of the space frame shown in Figure E-8.22. Load *OW* = 2.5 kN.

19. A simplied version of one of the supporting towers of the San Francisco Bay Bridge is shown in Figure E-8.18. Possible wind and traffic loads are as indicated. Determine the forces acting in each of the members of the structure. Tabulate the results.

20. Determine the resultant of the concurrent noncoplanar force system shown in Figure E-8.19. Assume the magnitudes of the forces.

21. Determine the resultant of the concurrent noncoplanar force system shown in Figure E-8.20. Assume the magnitudes of the forces.

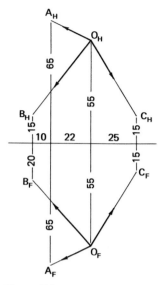

Figure E-8.19

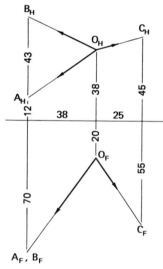

Figure E-8.20

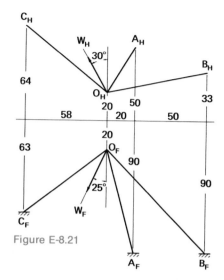

Figure E-8.21

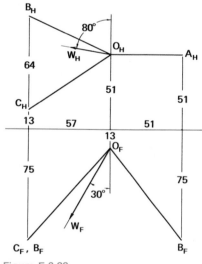

Figure E-8.22

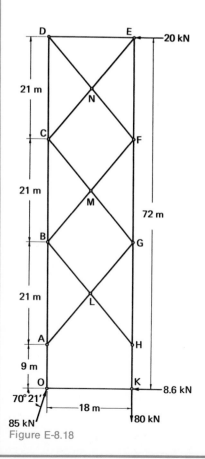

Figure E-8.18

24. Find the forces acting in the members of the space frame shown in Figure E-8.23. A vertical load of 3 kN is suspended from point O.

25. Find the forces acting in the members of the space frame shown in Figure E-8.24. Load $OW = 3.5$ kN.

26. Find the forces acting in the members of the space frame shown in Figure E-8.25. Load $OW = 2.5$ kN.

27. Find the forces acting in the members of the space frame shown in Figure E-8.26. Load $OW = 1.0$ kN.

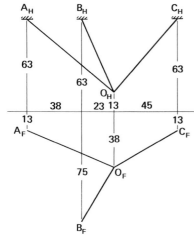

Figure E-8.23

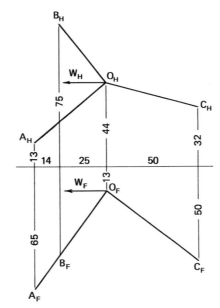

Figure E-8.25

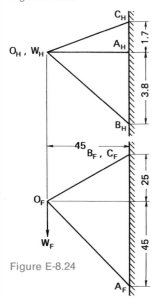

Figure E-8.24

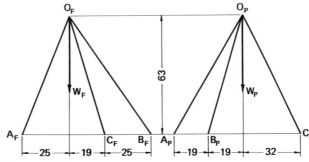

Figure E-8.26

28. Find the forces acting in the members of the space frame shown in Figure E-8.27. Load $OW = 3.0$ kN.

29. Find the forces acting in the members of the space frame shown in Figure E-8.28. Load $L = 1.0$ kN.

30. Find the forces acting in the members of the space frame shown in Figure E-8.29. Load $L = 1.5$ kN.

31. Find the forces acting in the members of the space frame shown in Figure E-8.30. Load $W = 5.0$ kN.

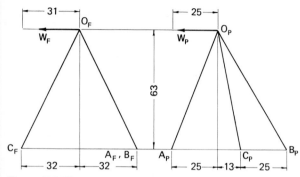

Figure E-8.27

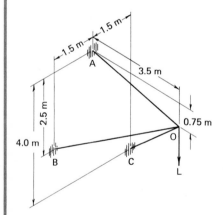

Figure E-8.28

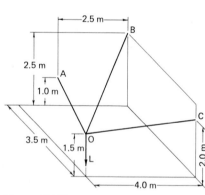

Figure E-8.29

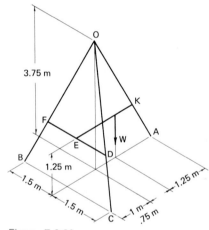

Figure E-8.30

32. Find the forces acting in the members of the space frame shown in Figure E-8.31. Points A, B, C, and D, E, G are the corners of two horizontal equilateral triangles whose centers lie on the same vertical line and whose sides are respectively parallel. Length $\overline{AB} = 2$ m, and $\overline{DE} = 5$ m.

33. Find the forces acting in the members of the space frame shown in Figure E-8.32.

34. Find the forces acting in the members of the space frame shown in Figure E-8.33.

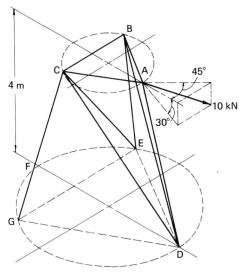

Figure E-8.31

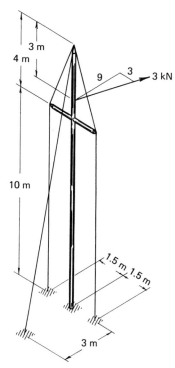

Figure E-8.32

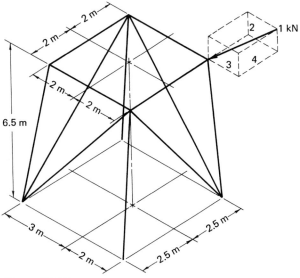

Figure E-8.33

Chapter 9 Pictorial Representation

INTRODUCTION

It has been said that a picture is worth 1000 words. In a technological setting it is difficult to carry on effective communication among engineers, designers, drafters, technicians, and production personnel without the use of pictorial drawings (freehand isometrics, oblique or perspective sketches, conceptual designs, etc.). You may recall the emphasis placed on freehand sketching in Chapter 2 and its use in interpreting orthographic drawings as discussed in Chapter 5.

The use of pictorials greatly enhances the effectiveness of technical reports and oral presentations. In the preparation of various technical manuals (i.e., service, parts, and installation), many pictorials are included to portray clearly what is intended.

Training Programs

Another important use of pictorials is in technical training programs. An excellent example is the Instructional Systems Approach (ISA) program developed by the Product Support Training Department at the Douglas Aircraft Company Division of McDonnell Douglas Cor-

poration at Long Beach, California. This department has developed a variety of teaching programs that call on many types of audiovisual (a.v.) projectuals and other technical visual aids within a multimedia instructional framework. More than 50 instructors supported by 22 technical experts annually teach an average of 1600 executives and supervisory and maintenance personnel sent to McDonnell Douglas Long Beach classrooms from 53 major airlines all over the world. The visitors are taught every aspect of maintenance of the latest Douglas commercial transport, the DC-10 wide-cabin trijet, and the smaller DC-9 jetliner. In addition, a parallel training program instructs U.S. and foreign users on Douglas military aircraft.

The keystone of ISA is technical graphics, and no effort is spared to provide technical drawings, diagrams, illustrations, charts, and graphs to illustrate every step and phase of instruction.

Examples of some of the material that has been used in the ISA program are shown in Figures 9.1 to 9.6.

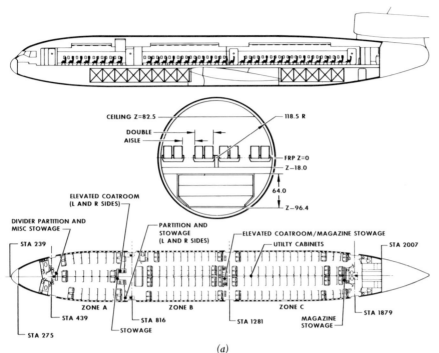

Figure 9.1 (*a*) Interior Arrangement, DC-10.

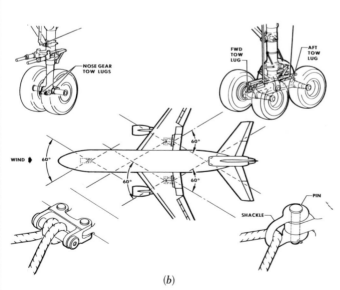

(*b*)

Figure 9.1 (*b*) Mooring attach points, DC-10, (Courtesy Douglas Aircraft Company, Long Beach, California).

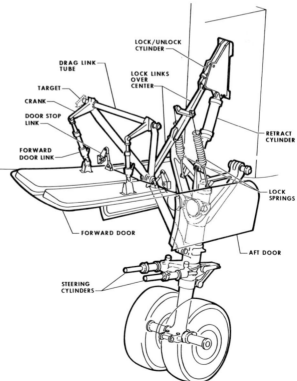

Figure 9.2 Nose landing gear and doors, DC-10. (Courtesy Douglas Aircraft Company, Long Beach, California).

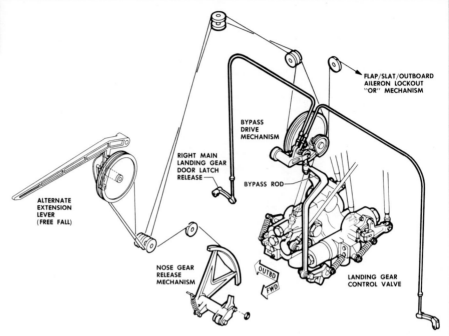

Figure 9.3 Landing gear alternate mechanical control diagram, DC-10. (Courtesy Douglas Aircraft Company, Long Beach, California)

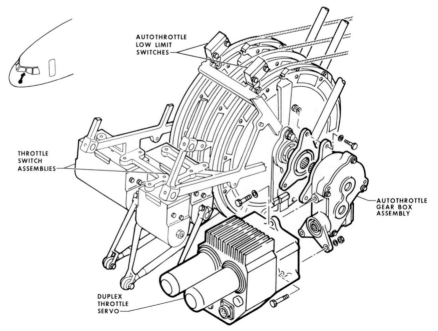

Figure 9.4 Autothrottle servo drive components, DC-10 (Courtesy Douglas Aircraft Company, Long Beach, California)

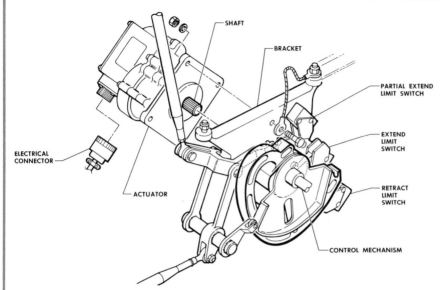

Figure 9.5 Spoiler control electrical actuator removal/installation, DC-10. (Courtesy Douglas Aircraft Company, Long Beach, California)

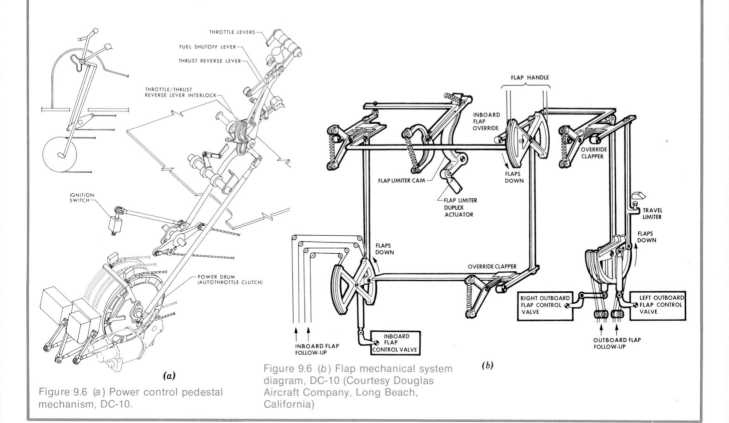

(a)

(b)

Figure 9.6 (*a*) Power control pedestal mechanism, DC-10.

Figure 9.6 (*b*) Flap mechanical system diagram, DC-10 (Courtesy Douglas Aircraft Company, Long Beach, California)

TYPES OF PICTORIALS

To further our knowledge and use of pictorials, we now consider the *types of pictorials*.

1. **Views**
   (*a*) Axonometric: isometric, dimetric, and trimetric views.
   (*b*) Oblique views.
   (*c*) Perspective views.
2. **Drawings.**

(*a*) Isometric, dimetric, and trimetric drawings.
(*b*) Oblique drawings.
(*c*) Perspective drawings.

Isometric Views

Consider the two views, *F* and *P*, of a cube as shown in Figure 9.7. It should be noted that the front view is a pictorial view of the cube; that the body diagonal *AB* appears as a point ($A_F B_F$); that the three concurrent edges, at point *A*, make equal angles with the *F* plane; and that the edges make angles of 120° with each other, as observed in the front view. These three edges are regarded as the *isometric axes*.

Lines parallel to the axes are identified as isometric lines. *The front view is the isometric view of the cube*. The edges in this view are foreshortened by an amount, $\dfrac{\sqrt{2}}{\sqrt{3}}$, times the original length.

EXAMPLE 1

Let us consider the object in Figure 9.8. How shall we proceed to obtain an isometric *view* of the object?

First, we introduce an *inscribed cube* with a body diagonal *AB*, as shown in Figure 9.9. Second, we can obtain a point view of *AB* and the corresponding view of the object.

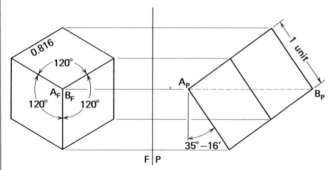

Figure 9.7 Isometric view (front view) of a cube.

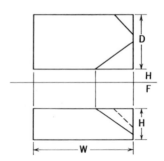

Figure 9.8 Orthographic views of cut block.

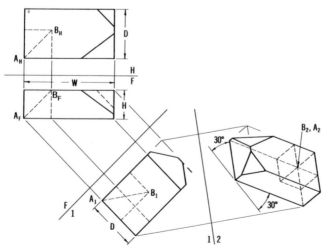

Figure 9.9 Isometric view, using inscribed cube.

SOLUTION

To obtain the point view of *AB* we must first obtain a true length view. This is easily done by introducing supplementary plane 1, parallel to line *AB* and perpendicular to the *F* plane. The view on plane 1 includes the true length of *AB*, shown as A_1B_1. Now we can introduce supplementary plane 2 perpendicular to line *AB* and plane 1. The view on plane 2 shows the point view of the body diagonal of the inscribed cube and the *isometric view* of the object.

EXAMPLE 2

When the given object does not accommodate the introduction of an inscribed cube, we can set up the views of a small cube adjacent to the views of the object, as shown in Figure 9.10. The placements of the supplementary planes 1 and 2 are related to the choice of a body diagonal of the cube. In this example the choice was made to show the object to the best advantage (fewest number of hidden surfaces).

It should be quite clear that the method of attack in *obtaining an isometric view is basically an application of the fundamental principles of orthogonal projection previously employed* (Chapter 6) in *finding the point view of a line*.

Dimetric Views

Let us consider the *H*, *F*, and 1 views of the cube in Figure 9.11. We observe that the true length of body diagonal *AB* is seen in the view on supplementary plane 1.

Now we can establish supplementary plane 2 perpendicular to plane 1, but *not* perpendicular to line *AB*. The view on plane 2 is a *dimetric view* of the object. In this view, *it should be noted that only two of the three concurrent edges (AF and AG) of the cube are equally foreshortened*. Edge *AE*, however,

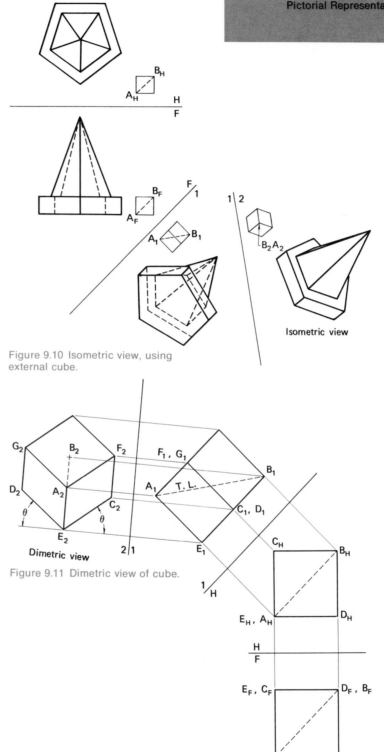

Figure 9.10 Isometric view, using external cube.

Figure 9.11 Dimetric view of cube.

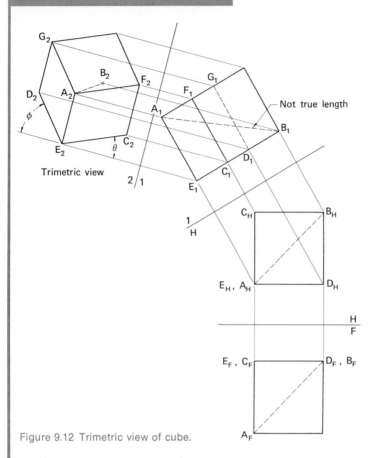

Figure 9.12 Trimetric view of cube.

A trimetric view of an object that is described by an orthographic drawing can be obtained by employing the procedure previously used to obtain an isometric view.

Isometric, Dimetric, and Trimetric Drawings

Isometric, dimetric, and trimetric *drawings* are similar to isometric, dimetric, and trimetric *views*, respectively. These pictorial drawings are made without resorting to the use of supplementary planes. You may recall (Chapter 5) the direct use of pictorials in the "reading" of orthographic drawings.

Isometric Drawing

In our study of the isometric view of a cube (Figure 9.7) we observed that (1) the isometric axes make angles of 120° with each other, and (2) that the edges are equally foreshortened by an amount $\frac{\sqrt{2}}{\sqrt{3}}$ times the original length of an edge.

In an isometric *drawing* of a cube the axes are the same as in the isometric view, but the length of the edges are drawn *full* size (or to a convenient scale).

EXAMPLE 1

An isometric drawing of a cube with an edge dimension of 20 mm is shown in Figure 9.13. Edge AB is

is not foreshortened the same amount. By varying the position of plane 2, angle θ can be changed, thereby changing the ratio of the lengths of the edges AF and AE. The three concurrent edges AE, AF, and AG as seen in the view on plane 2 may be regarded as the *dimetric axes*.

For a given object that is described by two (or more) orthogonal views, the procedure used previously in obtaining an isometric view can be applied to obtain a dimetric view.

Trimetric Views

Once again let us consider the cube described by the H and F views of Figure 9.12. Suppose we introduce supplementary plane 1 perpendicular to the H plane but *not* parallel to body diagonal AB. The view on plane 1 contains diagonal AB, which is not seen in true length. Now we can introduce supplementary plane 2 perpendicular to plane 1 and oriented in any desired position. The view on plane 2 is a *trimetric view*. Note that the three concurrent edges at point A, *the trimetric axes, are foreshortened unequally*. By changing the positions of supplementary planes 1 and 2, a variety of trimetric views can be obtained.

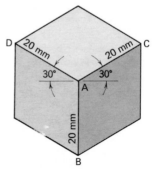

Figure 9.13 Isometric drawing of 20-mm cube.

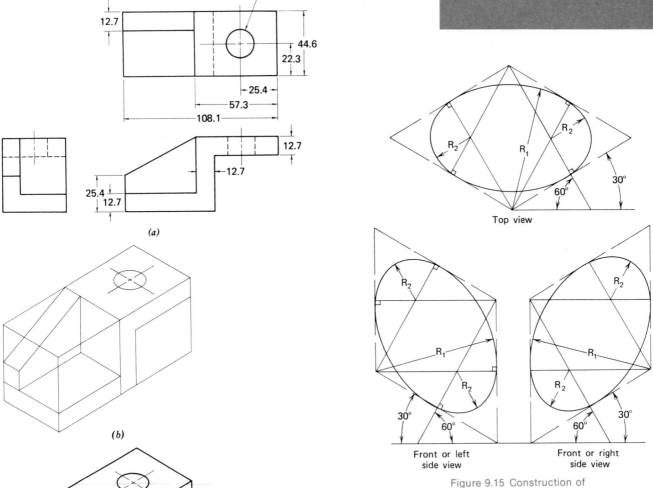

Figure 9.15 Construction of isometric circles.

Figure 9.14 (a) Orthographic views of support bracket.
Figure 9.14 (b) Isometric drawing of support bracket, "blocking-in" step.
Figure 9.14 (c) Isometric drawing of support bracket.

laid off as a vertical line 20 mm long. Edges AC and AD, each 20 mm long, form 30° angles with the horizontal that passes through point A. The three edges AB, AC, and AD are regarded as the *isometric axes*. All measurements that are parallel to the axes are true lengths. Again, note that the use of isometric grid sheets greatly simplifies the preparation of isometric drawings.

EXAMPLE 2
Let us consider the support bracket in Figure 9.14a. How shall we pro-

ceed to make an isometric drawing of the bracket? At the outset we can make an isometric drawing (using light lines) of the "block" that would enclose the bracket and then add features of the bracket, as shown in Figure 9.14b. Finally, we can strengthen the lines that are essential to the completion of the isometric drawing, as shown in Figure 9.14c. A convenient construction for drawing the isometric circle that represents the hole is shown in Figure 9.15; however, ellipse templates may be used for drawing isometric circles.

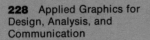

Figure 9.16 NACA 6321 airfoil from NACA TR 460.

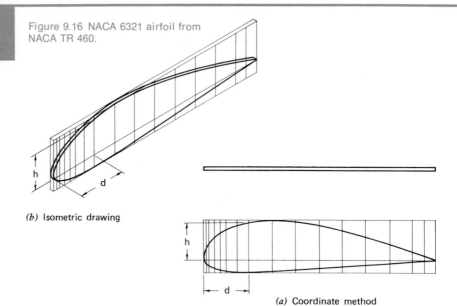

(b) Isometric drawing

(a) Coordinate method

Figure 9.17 Isometric drawing of meat deboning system. (Courtesy Richard Price, Beehive Machinery, Inc.)

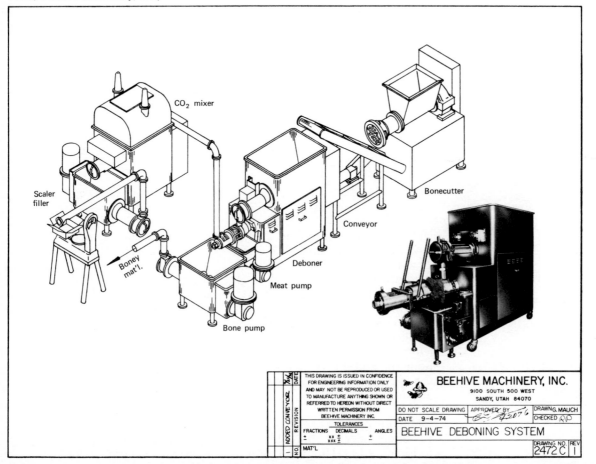

EXAMPLE 3

An airfoil with an irregular-shaped surface is shown in Figure 9.16a. The blocking-in procedure and a construction to locate points on the irregular-shaped surface are shown in the isometric drawing in Figure 9.16b.

EXAMPLE 4

Isometric drawings can be quite helpful to the layperson. Consumers of processed meat have been concerned with the question "Just how big are those bone chips when meat is mechanically deboned?" The USDA Animal and Plant Health Inspection Service found that the chips varied from about 400 to 450 μ (about the size of a grain of salt). Opponents argued that there had been no studies that assured the safety of ingesting bone particles no matter how small. The Food and Drug Administration assured the USDA that there was no problem with the presence of the size of bone particles resulting from the deboning process. The American Institute of Nutrition felt that the increased calcium would be beneficial. The USDA regulations specify a *maximum* of 1% in certain deboned products and 0.7% in others.

To help the consumer understand the mechanical deboning process, the isometric drawing in Figure 9.17 was prepared.

Dimetric Drawing

Basically there is little difference between the preparation of an isometric drawing and a dimetric drawing. It is only necessary to establish the dimetric axes and the scale ratios. The angles between the oblique axes and a horizontal line will vary for different positions of the cubes. Four different combinations of angles and scale ratios are shown in Figure 9.18.

EXAMPLE 1

Suppose we wish to make a dimetric drawing of a rectangular block 100 mm long, 50 mm high, and 75 mm wide. Let us use the scale ratios 1:1:1/2, as shown in Figure 9.18b. The 100-mm and 50-mm dimensions would be laid off full size, but the 75-mm dimension would be laid off as 37.5 mm. If the drawing were not made full size, the scale ratios would nevertheless be maintained (see Figure 9.19).

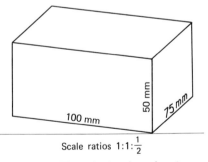

Scale ratios 1:1:$\frac{1}{2}$

Figure 9.19 Dimetric drawing of rectangular block using Figure 9.18b ratios and angles.

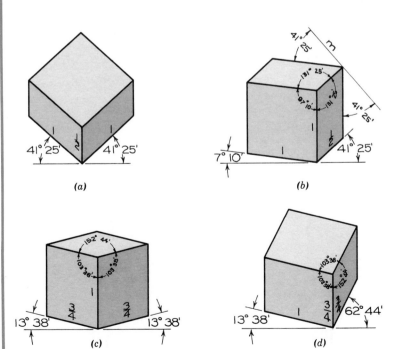

Figure 9.18 Dimetric drawings; scale ratios and axes angles.

EXAMPLE 2

Study the orthographic views of the bracket in Figure 9.20a. In order to show the dovetail portion on the bottom surface of the bracket, it is good practice to reverse the dimetric axes. Use the axes angles and scale ratios in Figure 9.18d with reversed axes. The dimetric drawing is shown in Figure 9.20b.

Trimetric Drawing

A trimetric drawing can be made just as easily as an isometric or dimetric drawing once the axes angles and scale ratios have been selected. Four different combinations of axes angles and scale ratios are shown in Figure 9.21.

EXAMPLE 1

Let us make a trimetric drawing of a rectangular block 100 mm long, 50 mm high, and 75 mm wide. We will assume the axes angles and scale ratios, $\frac{7}{8}:1:\frac{2}{3}$, as shown in Figure 9.21c. The 100-mm dimension is laid off $\frac{7}{8}$ size, the 50-mm dimension is laid off full size, and the 75-mm dimension is laid off $\frac{2}{3}$ size. The completed trimetric drawing is shown in Figure 9.22.

EXAMPLE 2

Consider the bracket shown previously in Figure 9.20a. A trimetric drawing of the bracket is shown in Figure 9.23. The axes angles and scale ratios have been based on Figure 9.21b. Again, reversed axes have been used to show the dovetail feature clearly.

Comment:
There are cases where a dimetric or a trimetric drawing may have some advantages over an isometric drawing. In engineering practice, however, much greater use is made

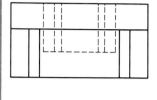
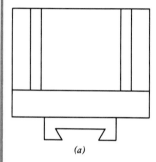

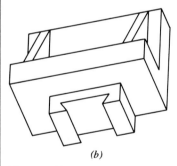

Figure 9.20 (a) Orthographic drawing of bracket.
Figure 9.20 (b) Dimetric drawing of bracket, using scale ratios and axes angles shown in Figure 9.18d.

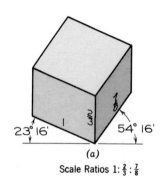

(a)
Scale Ratios $1:\frac{2}{3}:\frac{7}{8}$

23° 16'   54° 16'

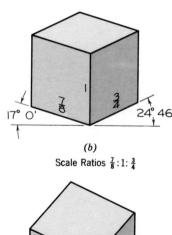

(b)
Scale Ratios $\frac{7}{8}:1:\frac{3}{4}$

17° 0'   24° 46'

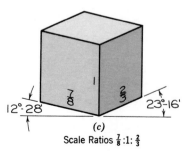

(c)
Scale Ratios $\frac{7}{8}:1:\frac{2}{3}$

12° 28'   23° 16'

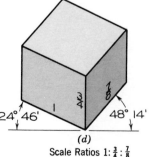

(d)
Scale Ratios $1:\frac{3}{4}:\frac{7}{8}$

24° 46'   48° 14'

Figure 9.21 Trimetric drawings; scale ratios and axes angles.

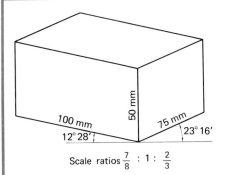

Figure 9.22 Trimetric drawing of a rectangular block.

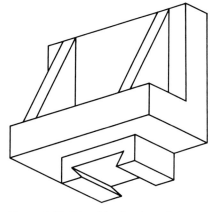

Figure 9.23 Trimetric drawing of bracket, using scale ratios and axes angles in Figure 9.21*b*.

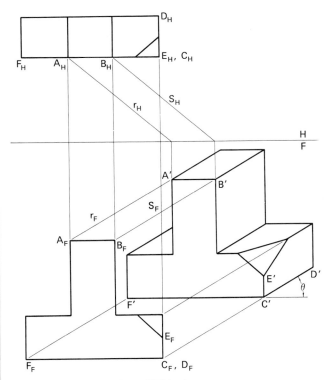

Figure 9.24 Oblique view of T block.

of isometric and oblique drawings. *We now study the methods used to make* (1) *an oblique* view, *and* (2) *an oblique* drawing.

Oblique Views

Consider the *T* block in Figure 9.24. An oblique view of the block can be obtained in the following way. (1) Through point A let us introduce projector (ray) *r*, which is *not* perpendicular to the reference planes. (2) Locate the intersection of *r* with the *F* plane. This is the point marked A'. In a similar manner we introduce a *parallel* projector (ray) *s* through point *B* and then locate *B'*, which is the intersection of *s* with the *F* plane. (3) Join points A' and B'. Line A'B' is the oblique view of edge AB of the *T* block. (4) In a like manner additional *parallel* projectors are introduced through salient points of the object; the intersections of the projectors with the *F* plane are found and are properly connected to form the oblique view of the object, as shown in Figure 9.24.

The angle θ will vary with the selection of the desired direction of the parallel projectors. The three concurrent edges C'F', C'E', and C'D' may be regarded as the *oblique axes*.

Oblique Drawing

An oblique *drawing* is similar to an oblique *view*. Once we have established the three concurrent axes—one horizontal, one vertical, and one at an angle $\theta°$, usually 30° or 45°, we can proceed to construct an oblique drawing quite easily. The scale ratios are usually 1:1:1 or 1:1:$\frac{1}{2}$. The latter ratio is helpful in overcoming the pictorial distortion caused by the nonconvergence of the lines that are parallel to the oblique axis.

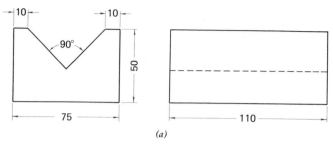

Figure 9.25 (*a*) Orthographic views of a V block.

Drawings made with scale ratios 1:1:1 and with the oblique axis at 45° are known as "Cavalier" oblique drawings; those with scale ratios 1:1:½ and with the oblique axis also at 45° are known as "Cabinet" oblique drawings (see Figure 9.25).

Advantages in the Use of Oblique Drawings

In many cases an oblique drawing has certain advantages over iso-metric, dimetric, and trimetric drawings. For example, irregular-shaped surfaces will appear true (without distortion) when such shapes lie in a plane that is parallel to the plane of the horizontal and vertical axes. This is also true of

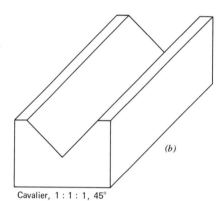

Cavalier, 1 : 1 : 1, 45°

Figure 9.25 (*b*) "Cavalier" oblique drawing of the V block.

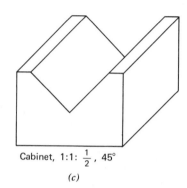

Cabinet, 1:1: $\frac{1}{2}$, 45°

Figure 9.25 (*c*) "Cabinet" oblique drawing of the V block.

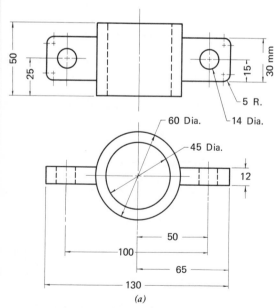

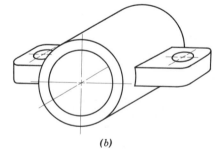

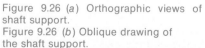

Figure 9.26 (*a*) Orthographic views of shaft support.
Figure 9.26 (*b*) Oblique drawing of the shaft support.

circular holes, slots with rounded ends, etc.

It is good engineering practice to place the object (i.e., part, component, or unit) in a position that orients the irregular-shaped plane surfaces in the plane of the horizontal and vertical axes.

EXAMPLE 1

Consider the orthographic views of the shaft support in Figure 9.26a. The circles shown in the front view will be retained as circles in the oblique drawing (Figure 9.26b.) The circles in the top view, however, will appear as ellipses in the oblique drawing because they do not lie in a plane parallel to the F plane. This is also true of the rounded corners of the support tabs. A simple construction for drawing circles in oblique is shown in Figure 9.27.

EXAMPLE 2

Orthographic views of a swivel washer are shown in Figure 9.28a. The oblique drawing (Figure 9.28b) retains the circular shapes as circles because they have been positioned in planes parallel to the plane of the horizontal and vertical axes. The oblique axis is at an angle of 30° with the horizontal axis. The scales ratios are 1:1:1.

The effect of changing the scale ratios to $1:1:\frac{1}{2}$ is shown in Figure 9.28c.

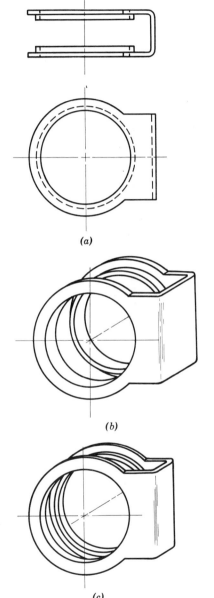

(a)

(b)

(c)

Figure 9.28 (a) Orthographic drawing of a swivel washer.
Figure 9.28 (b) Oblique drawing of swivel washer 1:1:1, 30°.
Figure 9.28 (c) Oblique drawing of swivel washer $1:1:\frac{1}{2}$, 30°.

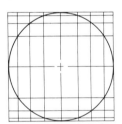

True shape of circle as it would appear in front plane, showing grid. (See below)

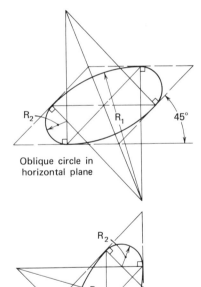

R_2 R_1 45°

Oblique circle in horizontal plane

One—half depth

30°

30° cabinet oblique circle obtained by location of points. (See view above)

R_2 R_1 45°

Oblique circle in profile plane

Figure 9.27 Construction of oblique circles.

EXAMPLE 3

The orthographic views of a spur gear sector are shown in Figure 9.29a. The oblique drawing is seen in Figure 9.29b. The advantages of the oblique drawing should be quite evident. Virtually all of the circular features are retained as circles in the pictorial drawing.

Perspective Views—Ray Method

In Chapter 2 we discussed perspective sketching in a general manner. Now we will consider the simple theory that is used in making perspective views.

The top and front views of a cut block are shown in Figure 9.30. Assume point S as a *station* point or point of sight through which projectors (or rays) are drawn to salient points of the block. When we find the points in which the projectors intersect the F plane (the "picture" plane), we will have located the perspective view of the salient points of the object. The lines that

properly join these points will form the perspective view of the block.

The location of the points of intersection of the projectors with the picture plane is *not* a new problem. Basically it is no different from any problem that deals with *the intersection of a line with a plane surface* (see Chapter 6).

For instance consider ray r which joins point A of the block with the station point S. Note that we have an edge view of the picture plane; therefore, the top view of the point of intersection of ray r with the picture plane is readily located and is shown as A'_H. The front view, of course, is on r_F, and is shown as A'_F, which is the perspective view of point A. The perspective views of the other points can also be located in this way. Finally, the lines that properly join these points will form the perspective view of the block, as shown in Figure 9.30.

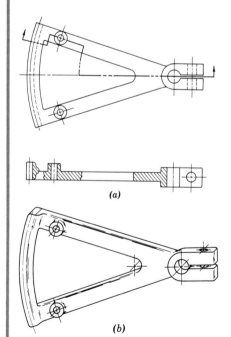

Figure 9.29 (a) Orthographic views of a spur gear sector.
Figure 9.29 (b) Oblique drawing of spur gear sector.

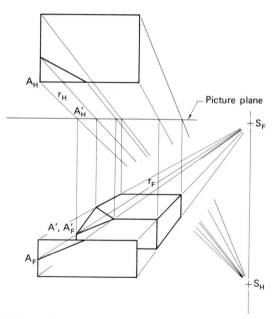

Figure 9.30 Perspective view of cut block. Ray method.

Eliminating Overlap

We can eliminate the overlap of a portion of the perspective view with the front view by using the *top and profile* views of the object, as shown in Figure 9.31. For instance, the perspective view of point A is found by locating the intersection of ray r with the F plane, using the top and profile views of the ray. The top view, A_H', is located in the same way as previously discussed. The profile view, A_P', is located at the intersection of r_P with the profile (edge view) of the F plane (the picture plane).

Now that we have located the top and profile views of point A', it is a simple matter to locate the front view, A_F', which is the perspective view of point A. The perspectives of the other points of the object can be similarly obtained, and the perspective view of the entire object can be completed, as shown in Figure 9.31.

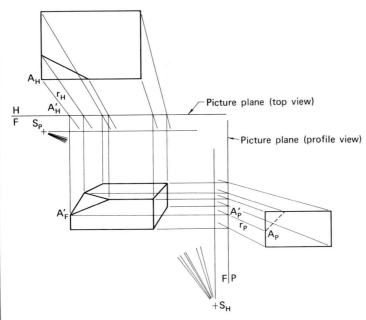

Figure 9.31 Perspective view of cut block, using the H and P views. Ray method.

Perspective Views—Line Method

Consider the object in Figure 9.32. Its top view is oriented so that the vertical surface $ACDEB$ makes an angle of $\theta°$ with the F plane, the picture plane. Usually angle $\theta°$ is taken as 30°, although it could be any desired angle. The view on the right shows the vertical distances between points on the upper and lower surfaces of the object. We can regard the view as a profile view when $\theta = 0°$. Station point, S, is selected so that the perspective view nicely discloses the features of the object.

Careful study of the top view shows that edge AB is in the picture plane; consequently, the front view and the perspective view are one and the same. Now locate the perspective view of point D by first locating the perspective view of edges CD and KD. Their intersec-

tion will locate D', the perspective view of point D.

Assume that edge CD is extended to infinity. The perspective view of the point at infinity can be found by the ray method. Ray r is drawn through the station point, S, parallel to edge CD. We know that parallel lines have parallel projections; therefore, r_H is drawn through S_H parallel to $C_H D_H$, and r_F is drawn through S_F in a horizontal direction, since edge CD is a horizontal line. Ray r intersects the F plane (picture plane) at the point marked V.P.R. This point is known as the vanishing point for all lines parallel to edge CD. Now we can extend edge DC to the picture plane at point O'. The line that joins O' and V.P.R. is the perspective view of the line from point O to point V.P.R. CD is a segment of that line.

Now we can locate the perspective view of edge KD extended from the picture plane to infinity. Edge KD extended to the picture plane is point T'.

The perspective view of the point at infinity on DK extended can be found by the ray method as was done previously in locating point V.P.R.

Ray m is drawn through point S parallel to edge DK; therefore m_H passes through S_H parallel to $D_H K_H$, and m_F passes through S_F in a horizontal direction, since edge DK is a horizontal line. Now line m intersects the picture plane at point V.P.L., which is the perspective view of the point at infinity on DK extended. This point is the vanishing point for all lines parallel to edge DK. The line that joins points T' and V.P.L. is the perspective view of the extended line KD. Finally, the intersection of the perspective views of lines CD (infinitely long) and KD (infinitely long) is the perspective view of point D, shown as D'.

The line method has many advantages over the ray method. This will become evident after you have prepared a few perspective views.

Basically, the method used in preparing an oblique view, or a perspective view, is no different from finding the intersection of a line with a plane. The preparation of perspective views can be facilitated by the use of available printed perspective grid sheets.

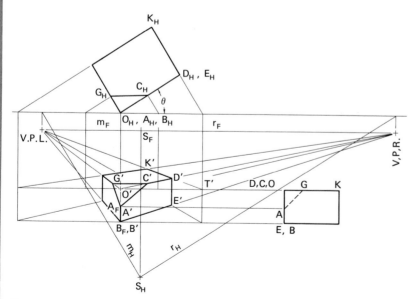

Figure 9.32 Perspective of cut block. Line method.

Modern Applications

Several pictorials of modern applications are shown in Figures 9.33 to 9.40.

Some computer programs are available for the preparation of perspective drawings; a few examples are shown in Chapter 12.

Engineering designers (and, for that matter, engineers, scientists, and technicians) should understand the basic principles used in preparing pictorial views and pictorial drawings; they should be capable of making intelligible pictorial sketches and, thereby, greatly enhancing their power to express themselves effectively in the graphical language. Surely it will help clear the lines of communication between themselves and their co-workers—scientists, engineers, technicians, production personnel, and management.

Figure 9.33 Viking lander capsule, aerodecelerator. (Source: Report NSAI-9000, June 1976, Martin Marietta Corp., Denver Division)

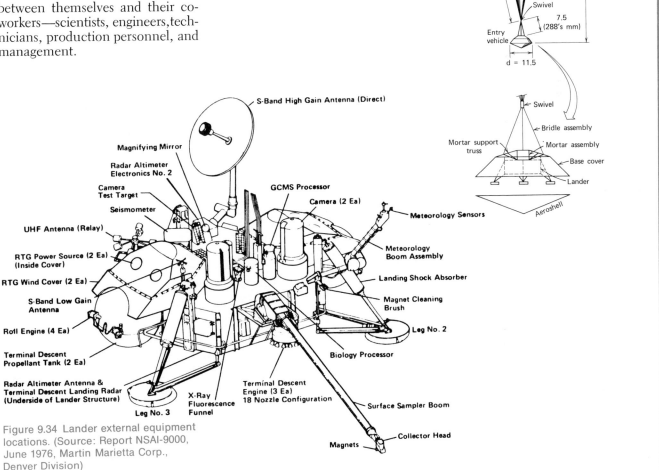

Figure 9.34 Lander external equipment locations. (Source: Report NSAI-9000, June 1976, Martin Marietta Corp., Denver Division)

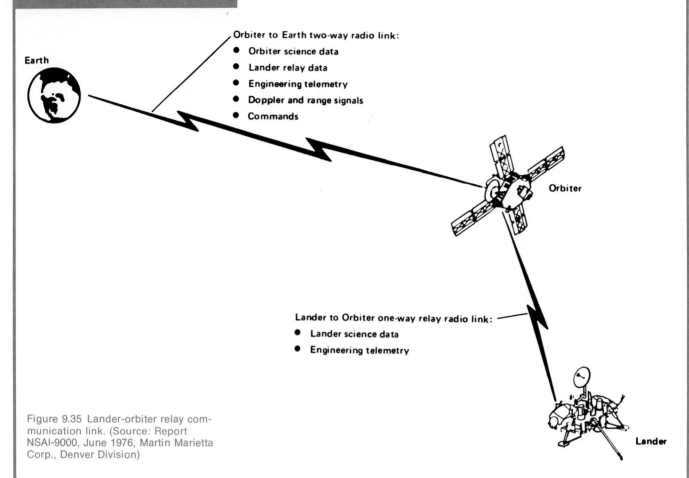

Orbiter to Earth two-way radio link:
- Orbiter science data
- Lander relay data
- Engineering telemetry
- Doppler and range signals
- Commands

Lander to Orbiter one-way relay radio link:
- Lander science data
- Engineering telemetry

Earth

Orbiter

Lander

Figure 9.35 Lander-orbiter relay communication link. (Source: Report NSAI-9000, June 1976, Martin Marietta Corp., Denver Division)

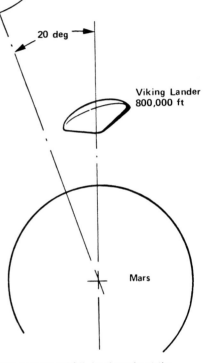

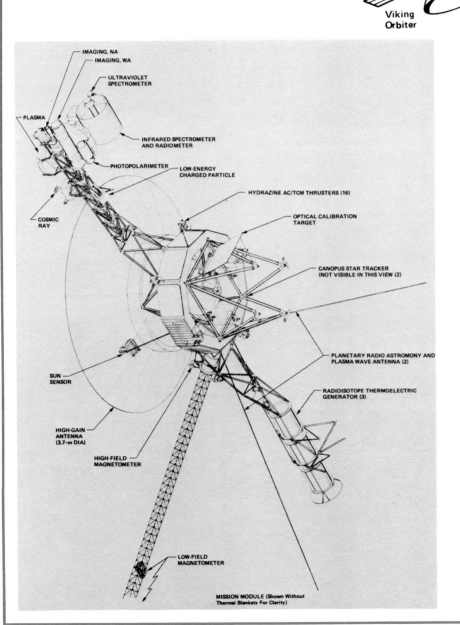

Figure 9.36 VO/VL lead angle at the entry altitude. (Source Report NSAI-9000, June 1976, Martin Marietta Corp., Denver Division)

Figure 9.37 Voyager spacecraft, showing components. Jupiter-Saturn missions. Arrivals in March and July 1979 for Jupiter missions and November 1980 and August 1981 for Saturn mission. (Courtesy Jet Propulsion Laboratory, California Institute of Technology)

Figure 9.38 Voyager spacecraft. Eleven scientific experiments, including photography, are planned in interplanetary space and near the planets and their satellites. (Courtesy Jet Propulsion Laboratory. California Institute of Technology)

Figure 9.39 Voyager spacecraft with its propulsion module, which provides an acceleration of 2 km (4475 mph) per second to escape from Earth orbit at the start of the craft's 39-month journey to Jupiter and Saturn. Small liquid rockets on outriggers stabilize the spacecraft during the 43-s firing of the 15,300-lb thrust, solid rocket motor. After the booms, which hold science instruments and electric power generators, are deployed the module is jettisoned (black struts remain). The diameter of the dish antenna is 3.66 m (12 ft). (Courtesy Jet Propulsion Laboratory, California Institute of Technology).

Figure 9.40 Mission to the outer planets. Painting shows voyager spacecraft as it speeds past Saturn and its rings. The trajectory line shows the spacecraft path from Earth and past Jupiter. The inner planets of the solar system (Mercury, Venus, Earth, and Mars) are shown orbiting the sun. (Courtesy Jet Propulsion Laboratory, California Institute of Technology).

Exercises

1. Draw the necessary orthographic views to describe the objects in Figure E-9.1. Add the supplementary views to obtain an *isometric view*. *Omit* the hidden edges in the isometric view.

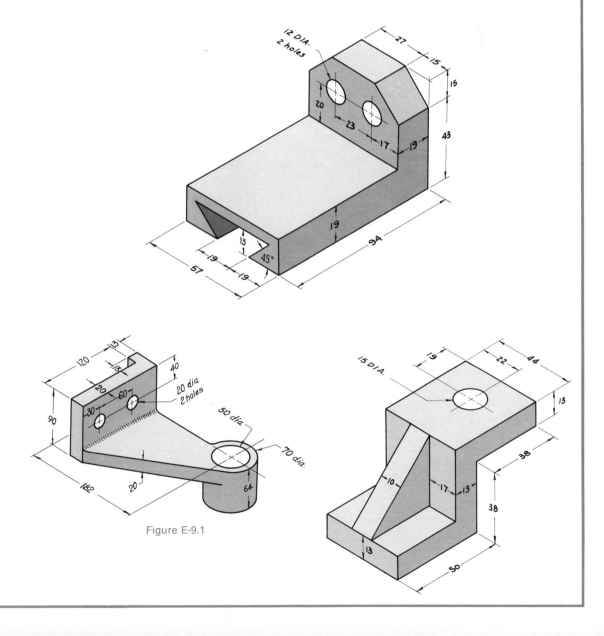

Figure E-9.1

2. Prepare isometric, dimetric, and trimetric *drawings* of each of the parts in Figure E-9.2. Select the direction of the axes that will best show the parts.

3. Prepare *freehand* isometric *drawings* of the parts in Figure E-9.3. Select the direction of the axes to show each part to the best advantage.

4. Prepare *freehand* dimetric or trimetric drawings of the parts in Figure E-9.4. Select the axes angles and scale ratios to show each part to the best advantage.

5. Reproduce the views in Figure E-9.5 and add an *oblique view*. Select the appropriate scale and angle for the oblique view to show the part to the best advantage. Omit the hidden edges in the oblique view.

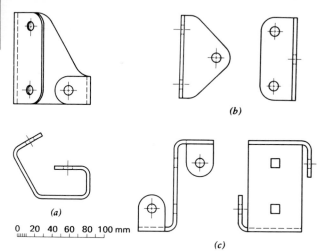

Figure E-9.2

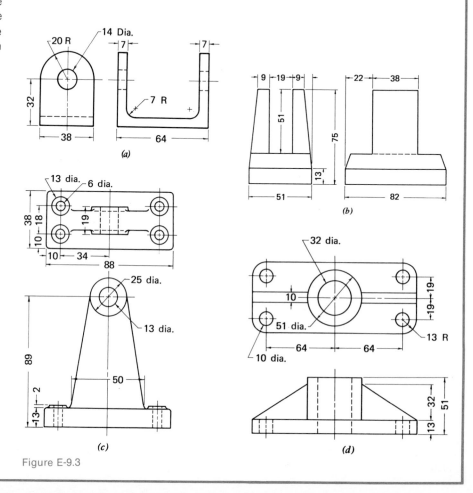

Figure E-9.3

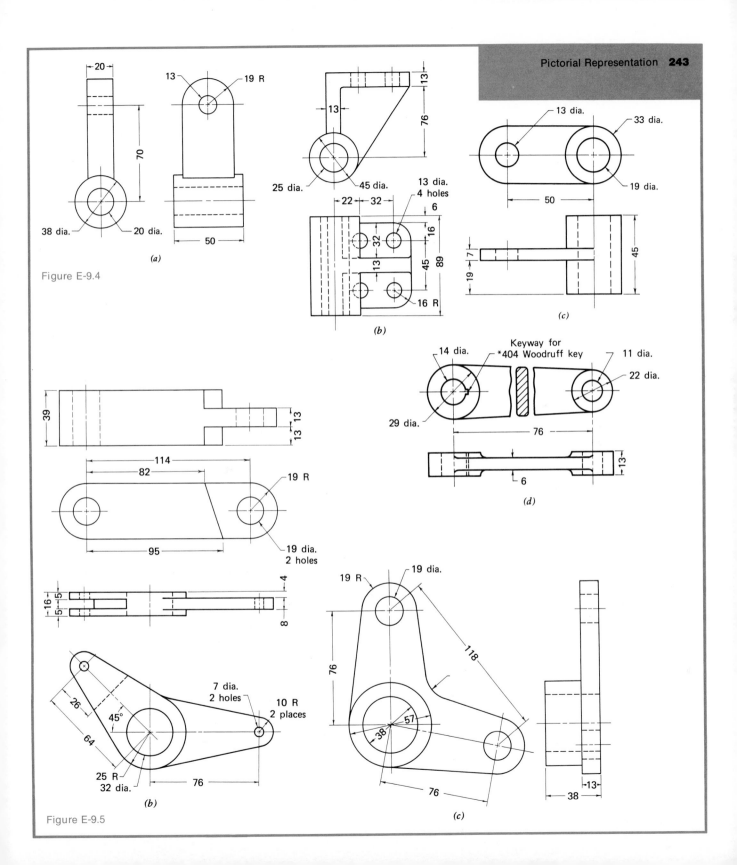

Figure E-9.4

Figure E-9.5

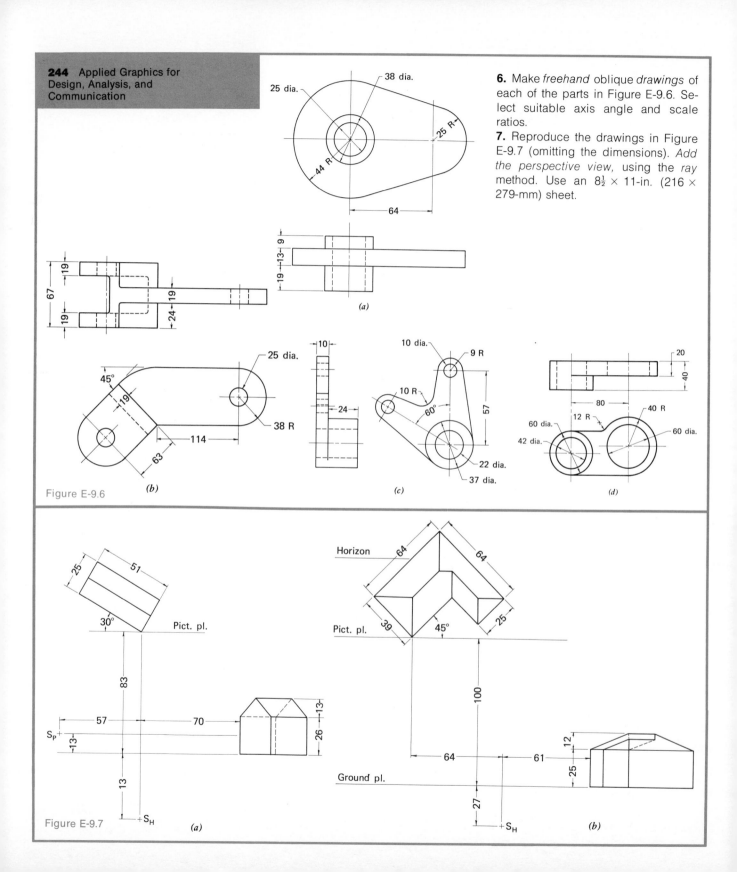

6. Make *freehand* oblique *drawings* of each of the parts in Figure E-9.6. Select suitable axis angle and scale ratios.

7. Reproduce the drawings in Figure E-9.7 (omitting the dimensions). *Add the perspective view,* using the *ray* method. Use an $8\frac{1}{2} \times 11$-in. (216×279-mm) sheet.

Figure E-9.6

Figure E-9.7

8. Reproduce the drawing in Figure E-9.8; then add the perspective view using the *line* method.

9. Reproduce the drawing in Figure E-9.9; then add the perspective view using the *line* method.

10. Reproduce the drawing in Figure E-9.10; then add the perspective view using the *line* method.

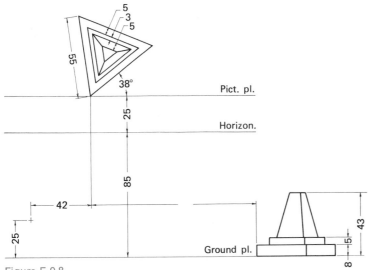

Figure E-9.8

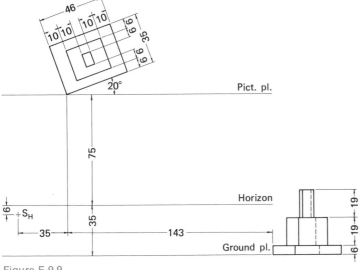

Figure E-9.9

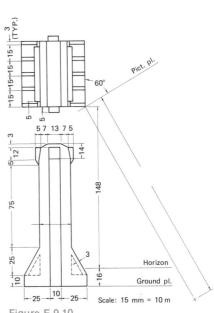

Figure E-9.10

Chapter 10 Graphical Representation and Significance of Design Data

INTRODUCTION

The graphical presentation of technical and nontechnical data is an effective means of conveying facts and their significance to laypersons, executives, production personnel, scientists, engineers, technicians, and other professionals.

Our daily newspapers, magazines, annual corporation reports, technical reports, and technical journals contain various types of charts and graphs to simplify and enhance communication with the interested reader.

The weighing and evaluating of alternative solutions of a design project usually include the collection and interpretation of data, such as the performance of components (pumps, engines, motors, electrical systems, etc.). Of course, cost factors and the degree to which the design requirements are satisfied are of paramount importance.

In some design situations data such as traffic flow, population trends, weather information, etc., are necessary.

Much of the required data is in numerical form as tables. Interpretation of numerical data is greatly facilitated by the presentation of the information in graphical form.

The selection of the graphical form (pie charts, bar charts, polar charts, rectangular coordinate charts, etc.) depends on the kind of data to be presented and interpreted.

PIE CHARTS

Circular diagrams subdivided on the basis of percent are known as pie charts. Examples are shown in Figure 10.1a and 10.1b. Figure 10.1a shows where the U.S. budget dollar comes from and how it is disbursed (for a specified fiscal year). Figure 10.1b shows water-analysis data. The segments of the circle are indicative of the percentage composition of the total equivalents per million.

BAR CHARTS

The bar chart in Figure 10.2a shows the relative proportions of dissolved material present in natural water. Several studies and analyses were made, as indicated by the numbers at the head of the bars. Figure 10.2b shows the average monthly mean temperature, over a four-year period, in Berkeley, California.

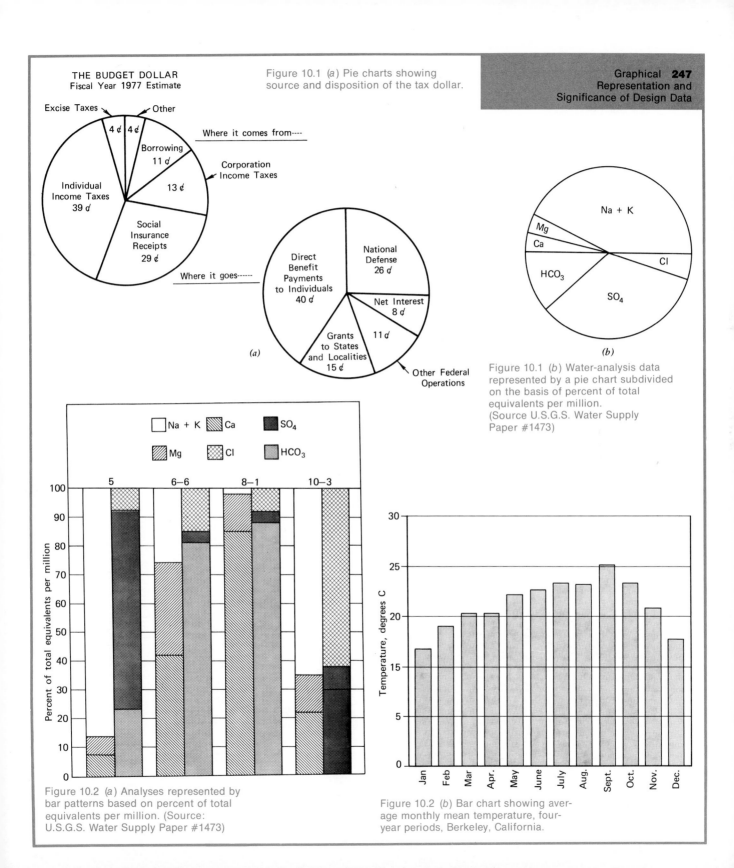

THE BUDGET DOLLAR
Fiscal Year 1977 Estimate

Figure 10.1 (*a*) Pie charts showing
source and disposition of the tax dollar.

Excise Taxes

Other

4 ¢ 4 ¢

Where it comes from----

Borrowing
11 ¢

Corporation
Income Taxes

13 ¢

Individual
Income Taxes
39 ¢

Social
Insurance
Receipts
29 ¢

Where it goes------

Direct
Benefit
Payments
to Individuals
40 ¢

National
Defense
26 ¢

Net Interest
8 ¢

11 ¢

Grants
to States
and Localities
15 ¢

Other Federal
Operations

(*a*)

Na + K

Mg

Ca

Cl

HCO₃

SO₄

(*b*)

Figure 10.1 (*b*) Water-analysis data
represented by a pie chart subdivided
on the basis of percent of total
equivalents per million.
(Source U.S.G.S. Water Supply
Paper #1473)

Figure 10.2 (*a*) Analyses represented by
bar patterns based on percent of total
equivalents per million. (Source:
U.S.G.S. Water Supply Paper #1473)

Figure 10.2 (*b*) Bar chart showing average monthly mean temperature, four-
year periods, Berkeley, California.

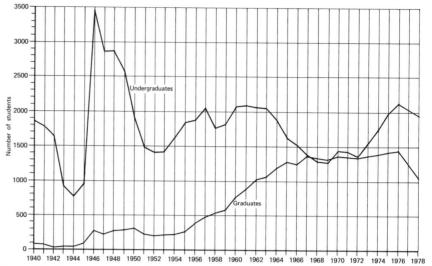

Figure 10.3 Rectangular coordinate chart—fall enrollment 1940 to 1978, College of Engineering, University of California at Berkeley.

Having a higher starting torque than the split-phase type, this motor is useful for applications requiring more torque, such as air compressors, pumps, etc. Sizes range to 7½ hp. Integral-horsepower versions are made in sizes up to 10 hp at speeds from 3,600 to 900 rpm.

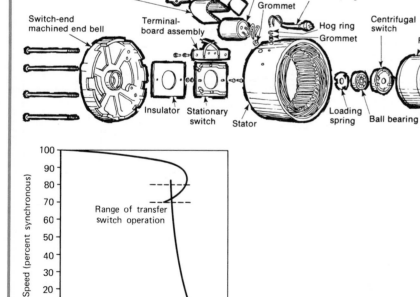

Figure 10.4 (a) Single-phase induction motors: Capacitor-start motors.

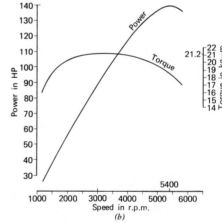

Figure 10.4 (b) Performance curves of Gamma engine.
(Courtesy *Automotive Engineering*, October 1976)

RECTANGULAR COORDINATE CHARTS

Examples of *rectangular coordinate charts* are shown in Figures 10.3 to 10.5. The engineering student registration data (Figure 10.3) shows the large increase in enrollment that followed the end of World War II and the return to a more normal enrollment that started in 1952. The significant increase in the number of graduate students from 1952 to the present time reflects the tremendous growth in sponsored research and the need for graduate study to satisfy the requirements of education and industry.

Analysis of enrollment data is quite important in planning for new facilities (laboratory space, new equipment, some of which will require *in-house design and fabrication*, study space for students, etc.).

Design engineers often need data concerning the performance, for example, of motors and engines in order to make a selection that best satisfies specified requirements. Examples are shown in Figure 10.4.

Figure 10.5 shows the relationship between runoff coefficients and rainfall intensity for subdivision and small watershed design in Santa Barbara County.

The designer must have access to up-to-date information in order to compete successfully. Practicing design engineers usually maintain a good library of various commercial catalogs, for example, on structural shapes, motors, refrigeration and air-conditioning equipment, conveyors, fasteners, properties of materials, pertinent reports, various standards, and cost data.

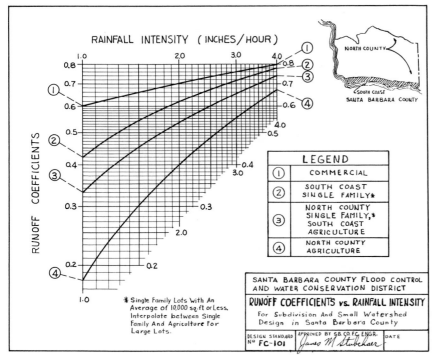

Figure 10.5 Runoff coefficients versus rainfall intensity. (Courtesy Lompoc City Engineering Dept.)

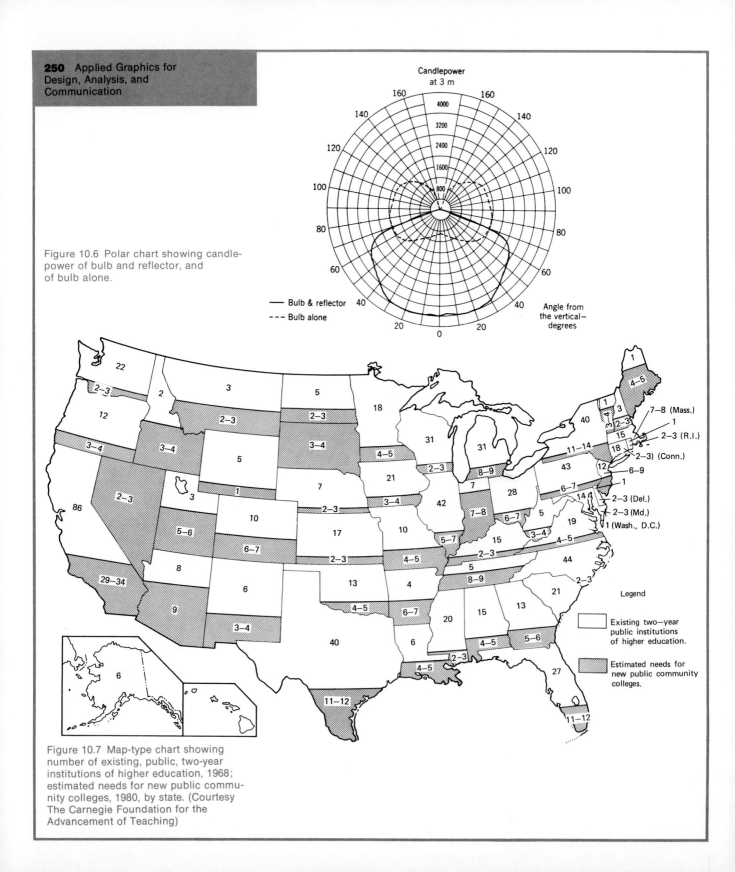

Figure 10.6 Polar chart showing candle-power of bulb and reflector, and of bulb alone.

Figure 10.7 Map-type chart showing number of existing, public, two-year institutions of higher education, 1968; estimated needs for new public community colleges, 1980, by state. (Courtesy The Carnegie Foundation for the Advancement of Teaching)

OTHER TYPES OF CHARTS AND GRAPHS

Other types of charts and graphs include the following.

1. *Polar Charts*. Polar coordinate sheets are used to plot continuous curves, each point of which represents the simultaneous measurement of a linear distance from the pole (center of concentric circles) and the corresponding angle expressed in degrees or radians. A polar coordinate sheet usually consists of equally spaced concentric circles and radii. Recording devices are generally used to plot the curves. Examples are shown in Figure 10.6.

2. *Map-type Charts*. These are used to show distribution systems such as power systems, city zoning (areas designated for various types of living units, such as single dwellings, apartments, commercial buildings, etc.), and earthquake magnitudes and locations. Map-type charts are important in planning the locations of distribution centers (warehouses for published materials—books, magazines, etc.) in order to minimize transportation costs. Examples of map-type charts are shown in Figure 10.7 and Figure 10.8a and 10.8b.

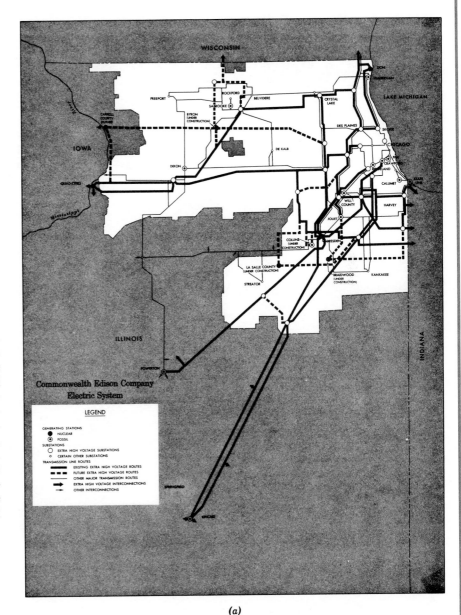

(a)

Figure 10.8 (*a*) Map-type chart showing an electric system. (Courtesy Commonwealth Edison Co.)

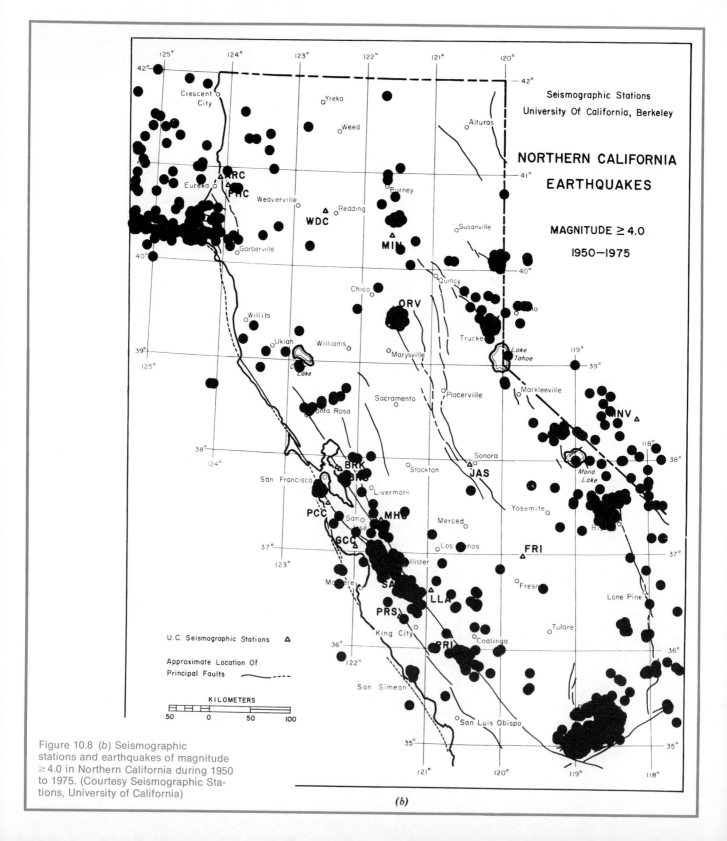

Figure 10.8 (*b*) Seismographic
stations and earthquakes of magnitude
≥ 4.0 in Northern California during 1950
to 1975. (Courtesy Seismographic Stations, University of California)

(b)

3. *Trilinear Charts*. These charts are in the form of an equilateral triangle. The sum of the perpendiculars from a point within the triangle to the sides of the triangle is equal to the altitude of the triangle.

When we graduate each side of the triangle in equal divisions from 0 to 100%, as shown in Figure 10.9, and regard each side as representing a variable, we have a means for determining the percentages of each of the variables that will make their sum equal 100%.

We observe in Figure 10.9 that for any point such as *P*, the perpendicular distance, *PD*, repre-sents the percentage of substance *A* 30% in this case); the perpendicular distance, *PE*, represents the percentage of substance *C* (20%); and the perpendicular distance, *PF*, represents the percentage of substance *B* (50%). Their sum, *PD* + *PE* + *PF*, equals 100%, which is represented by an altitude of the triangle.

When various combinations of the three substances exhibit a common characteristic [i.e., that their combination weighs 1000 g (1 Kg)], we could establish the 1000-g contour on the chart. Once that is done, perpendiculars from any point on the contour to the sides of the triangle would determine the percentage of each substance. Choices of the location of point *P* could be made to *minimize* the cost of the mixture. Of course, additional contours could be established for the desired range of weights.

4. *Flowcharts*. This category includes organization charts, progress charts, process or operations charts, and computer flowcharts.

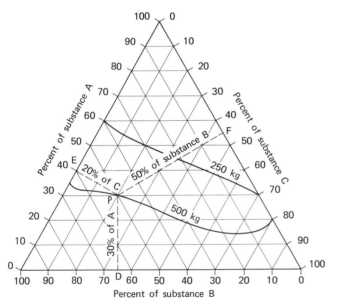

Figure 10.9 Trilinear chart showing percentages of three substances (totaling 100%) to make a mixture of 500 Kg.

(a) *Organization charts* usually show the structure of a company, laboratory, or industry and the relationship of its components. Examples are shown in Figure 10.10 and 10.10*a*.

(b) *Progress charts* show the extent to which scheduled tasks are completed as planned. These charts are helpful in proposing and implementing steps to complete a project on time or in negotiating for an extension of time if justified. An example of a progress chart is shown in Figure 10.10*b*.

(c) *Process or operations charts* show the sequence of steps required to achieve a desired end product. Examples are shown in Figure 10.10*c*, 10.10*d*, and 10.10*e*.

(d) *Computer flowcharts* are shown in Figure 10.10*f* and 10.10*g*.

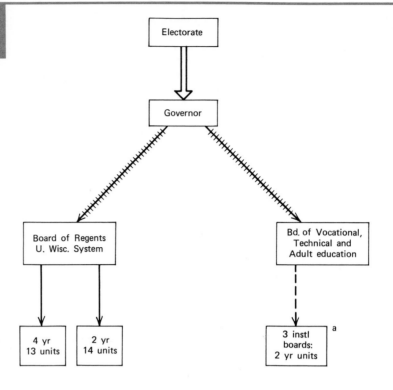

[a] Board members are appointed by a local appointment committee headed in two cases by the county board chairman and, in the third case, by the local school board president.

Note: The appointing processes illustrated above for the University of Wisconsin board and the Board of Vocational, Technical and Adult account for the majority of members of these boards, but two ex officio members also serve on each board.

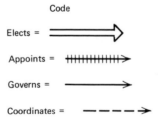

Figure 10.10 Organizational chart of public higher education in Wisconsin, 1976. (Courtesy The Carnegie Foundation for the Advancement of Teaching)

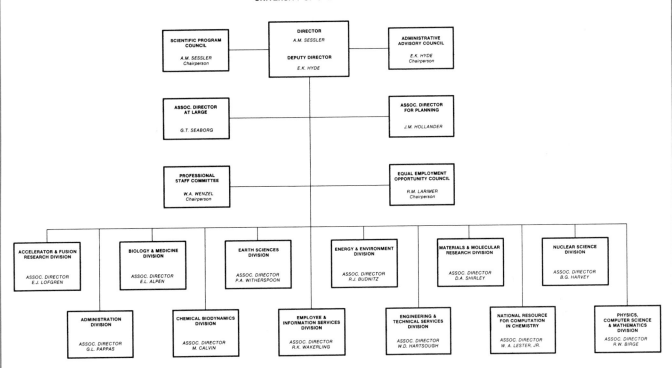

Figure 10.10 (*a*) Organization chart of
Lawrence Berkeley Laboratory.
(Courtesy of the Lawrence Berkeley
Laboratory, Berkeley, California)

(a)

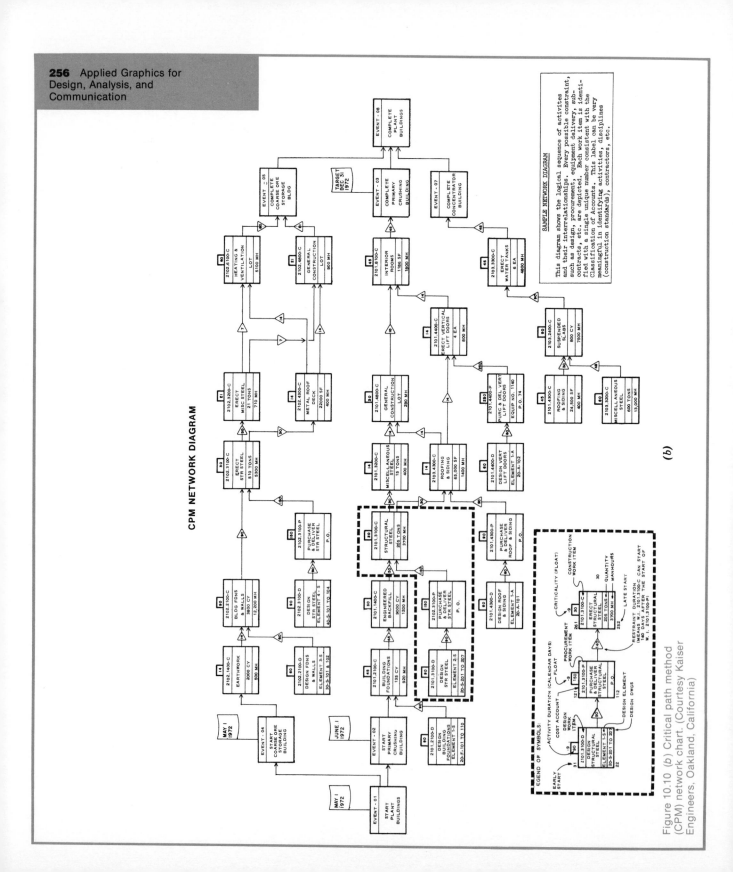

Figure 10.10 (b) Critical path method
(CPM) network chart. (Courtesy Kaiser
Engineers, Oakland, California)

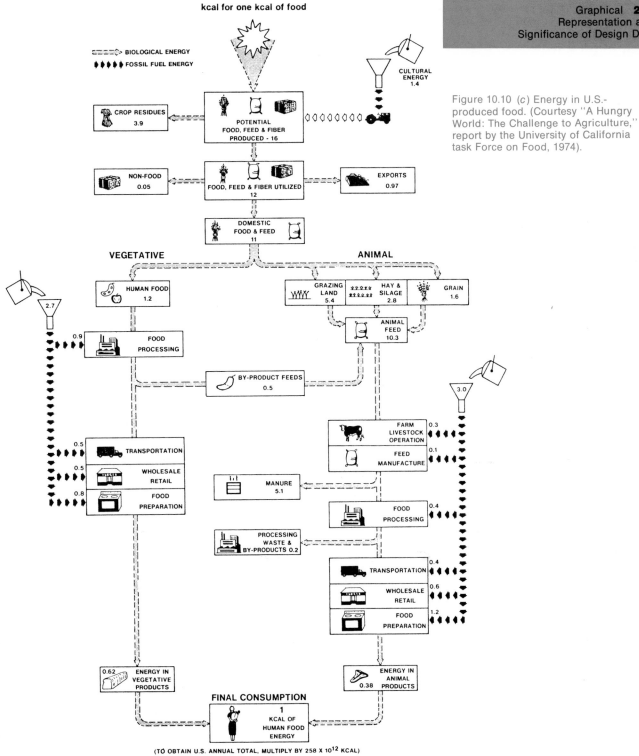

kcal for one kcal of food

Figure 10.10 (*c*) Energy in U.S.-produced food. (Courtesy "A Hungry World: The Challenge to Agriculture," report by the University of California task Force on Food, 1974).

(TO OBTAIN U.S. ANNUAL TOTAL, MULTIPLY BY 258 X 10^{12} KCAL)

(*c*)

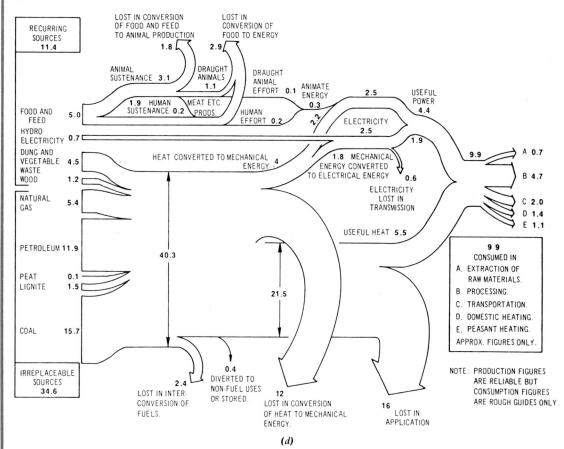

Figure 10.10 (*d*) Origin and utilization
of the world's energy resources.
(Courtesy "A Hungry World: The
Challenge to Agriculture," report by the
University of California Task Force on
Food, 1974)

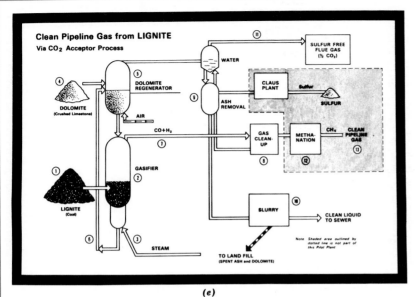

(e)

Figure 10.10 (e) Process chart for
coal gasification. (Courtesy
National Coal Association)

Figure 10.10 (f) Computer flow-chart
to solve $f(x) = ax^2 + bx + c$ for

$$ax^2 + bx + c \quad \text{if } x < d$$
$$0 \quad \text{if } x = d$$
$$-ax^2 + bx - c \quad \text{if } x > d$$

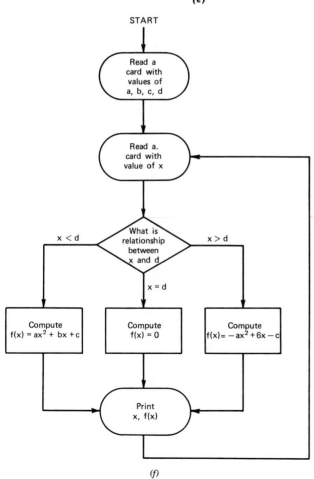

(f)

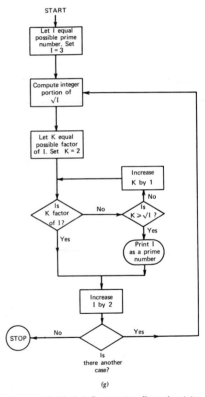

(g)

Figure 10.10 (g) Computer flowchart to
find all of the prime numbers between
1 and 1000.

PREPARATION OF GRAPHS

The basic steps in preparing graphs follow.

1. *Examine the tabular data* to determine the ranges of each variable.

2. *Select a suitable commercial grid sheet* or, if necessary, prepare a grid to accommodate the ranges of each variable (including the zero lines, if desired).

3. *Select a scale* that will convey the information to the best advantage.

4. *Place the origin* in the lower left portion of the sheet when the ranges of the variables are positive. Otherwise, locate the origin to accommodate both positive and negative values. It is customary to use the x-axis for the independent variable. There are, however, some exceptions (e.g., when a dependent variable, such as time, is usually plotted on the x-axis).

5. *The plotted data points should be identified by a symbol*, such as a small open circle. When more than one set of data is plotted on the same grid sheet, different symbols should be used to identify each set.

6. *The drawn curve* connects the data points by straight lines when the data are discontinuous or do not follow an implicit relationship; otherwise, a smooth curve is drawn to balance the plotted points, so that some points are on the curve and others are near the curve. *Do not draw the curve through the symbols, but stop at the boundary of each symbol* (see Figure 10.11). If more than one curve is needed, different line designs, such as dashed lines, may be used.

7. *The curve, or curves, should be properly identified by* suitable labels placed to advantage—in some cases on the curve, in others, near the curve.

8. *Appropriate titles and notes* should accompany the charts. The lettering should be placed so it is readable from the bottom and right side of the chart.

9. *If the chart is to be presented in a technical publication or used as a slide* in the oral presentation of a technical paper, it is suggested that reference be made to ANSI Y15.1M-1979, *Illustration for Publication and Projection*, for many valuable suggestions concerning the preparation of charts for each purpose.

CHOICE OF SCALES

The choice of scales is the most important and at the same time the most difficult step in chart construction.

Scale selection does more to shape the picture than all other steps combined. It can make the difference between an accurate picture; between a revealing presentation and a misleading one.

Scale selection is a subtle problem: the proper scale under one set of conditions may not be proper under other conditions; the proper scale for a chart shown alone may be improper for the same chart shown with other charts. Scales should never be chosen carelessly or left to chance: only rarely will you get a satisfactory scale by accident.*

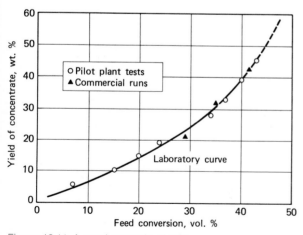

Figure 10.11 A good example of data curve, symbols, and notes. (Courtesy ANSI Y15.2 1960)

* From Section 4, *Scale Selection*, ANSI Y15.2M-1979. This standard, "Time-Series Charts," is highly recommended for good practice in the design of various types of charts.

NOMOGRAMS

There are two types of nomograms: (1) *the concurrency chart* (rectangular coordinate system), and (2) *the alignment chart* (parallel coordinate system). Two examples of concurrency charts are shown in Figures 10.12 and 10.13. Point A of Figure 10.12 is the point of concurrency of the lines $u = 3$, $v = 3$, and $w = 6$. The alignment chart in Figure 10.14 represents the same equation, $u + v = w$, shown in Figure 10.12. It is evident that the alignment-chart nomogram is easy to construct and is much simpler to use. Of course, nomograms can be

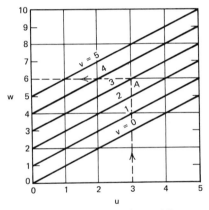

Figure 10.12 Cartesian chart of the equation $u + v = w$.

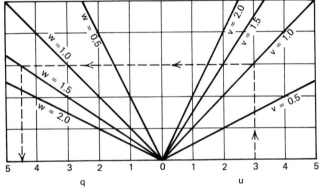

Figure 10.13 Cartesian coordinate representation of the equation $uvw = q$. *Example:* When $u = 3$, $v = 1.0$, and $w = 1.5$, then $q = 4.5$.

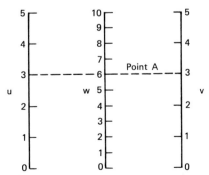

Figure 10.14 Alignment chart of the equation $u + v = w$.

designed to solve more involved problems. Additional examples are shown in Figures 10.15 to 10.19.

Nomograms are very useful, not only as time-savers in making repetitive calculations of design formulas, but also *as a means for quickly analyzing the interrelationship among the variables*. This is important in the designer's consideration of "trade-offs" among de-

sign parameters. Applications of nomography are not restricted to engineering and physical sciences; they also embrace fields such as the biological sciences, medicine, statistics, food technology, and business.

The subject of nomography is discussed more fully in the next chapter.

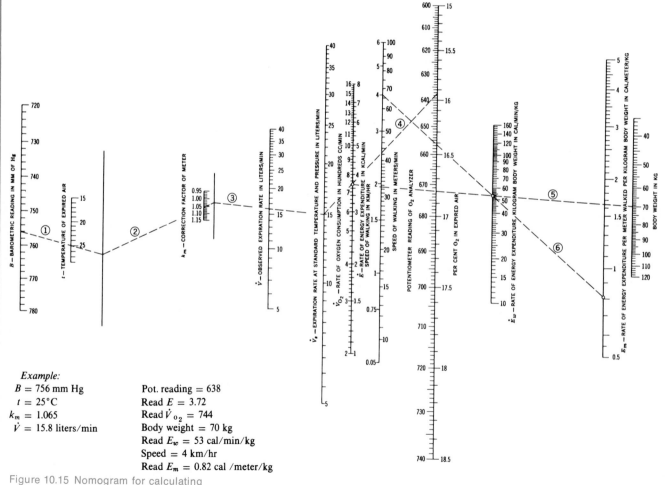

Example:
$B = 756$ mm Hg
$t = 25°C$
$k_m = 1.065$
$\dot{V} = 15.8$ liters/min

Pot. reading = 638
Read $E = 3.72$
Read $\dot{V}_{O_2} = 744$
Body weight = 70 kg
Read $E_w = 53$ cal/min/kg
Speed = 4 km/hr
Read $E_m = 0.82$ cal /meter/kg

Figure 10.15 Nomogram for calculating rates of energy expenditure of normal subjects, amputees, and hemiplegics for various rates of walking.

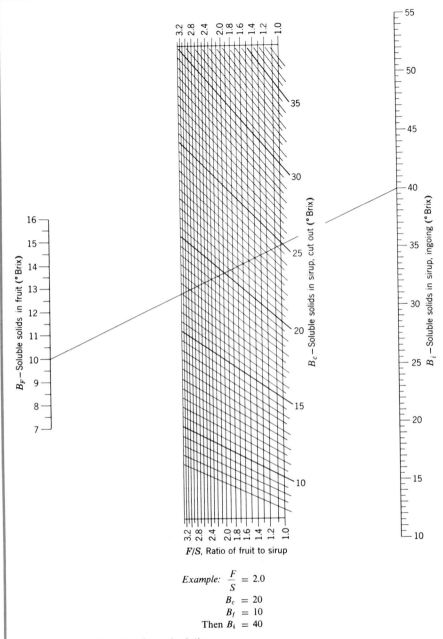

Example: $\dfrac{F}{S} = 2.0$

$B_c = 20$

$B_f = 10$

Then $B_i = 40$

Figure 10.16 Nomogram for calculating
relation between syrup strength and
fruit solids content for various ratios of
weight of fruit to syrup.

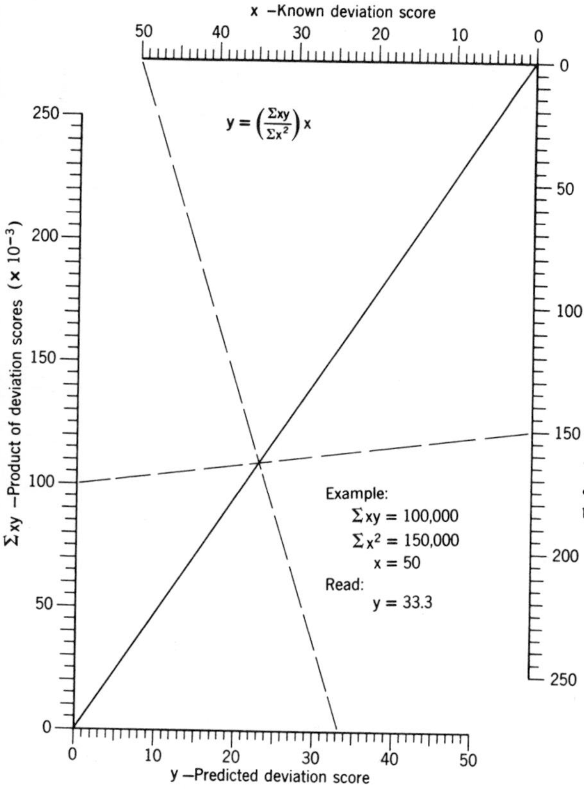

x – Known deviation score

$$y = \left(\frac{\Sigma xy}{\Sigma x^2}\right) x$$

Σxy – Product of deviation scores ($\times 10^{-3}$)

Σx^2 – Square of known deviation score ($\times 10^{-3}$)

Example:
$\Sigma xy = 100{,}000$
$\Sigma x^2 = 150{,}000$
$x = 50$

Read:
$y = 33.3$

y – Predicted deviation score

Figure 10.17 Regression line for predicting scores.

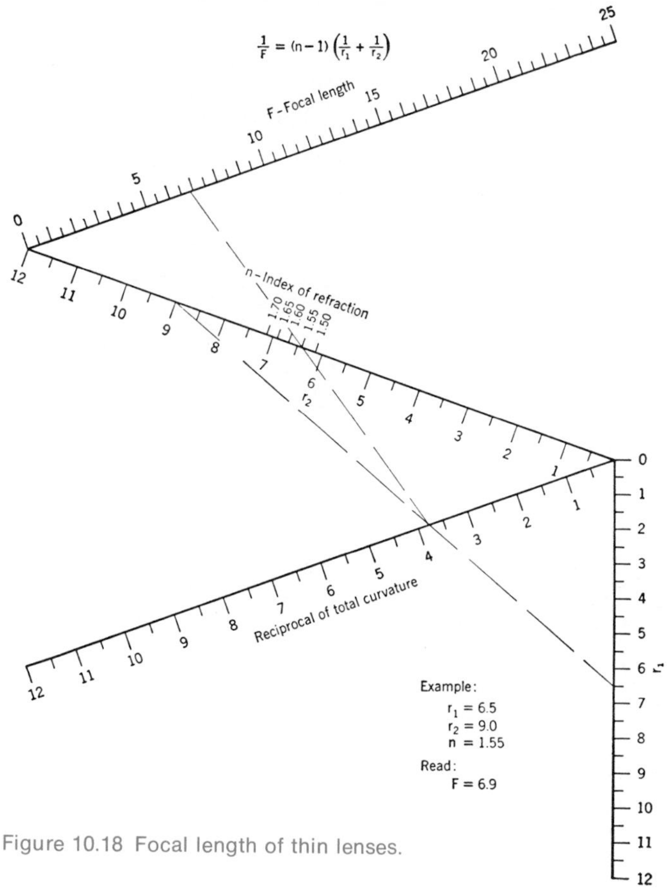

$$\frac{1}{F} = (n-1)\left(\frac{1}{r_1} + \frac{1}{r_2}\right)$$

F – Focal length

n – Index of refraction

1.70 1.65 1.60 1.55 1.50

r_2

Reciprocal of total curvature

Example:
$r_1 = 6.5$
$r_2 = 9.0$
$n = 1.55$

Read:
$F = 6.9$

r_1

Figure 10.18 Focal length of thin lenses.

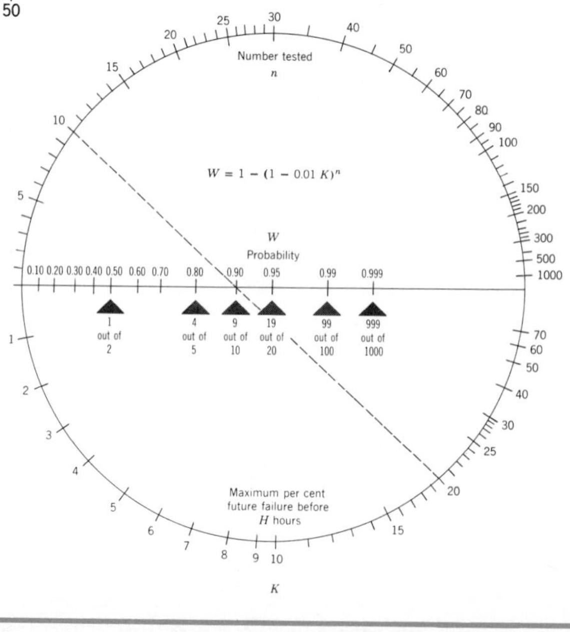

Number tested
n

$$W = 1 - (1 - 0.01 K)^n$$

W
Probability

0.10 0.20 0.30 0.40 0.50 0.60 0.70   0.80   0.90 0.95   0.99   0.999

1
out of
2

4
out of
5

9
out of
10

19
out of
20

99
out of
100

999
out of
1000

Maximum per cent
future failure before
H hours

K

Figure 10.19 Life test nomogram. (Courtesy E. C. Varnum, Barber-Colman Co.)

GRAPHS AND CHARTS IN ENERGY REPORTS

The mounting dependence on foreign energy, primarily oil, is a serious problem that faces the United States and other countries. We have witnessed the effects of the 1973 to 1974 oil embargo by the OPEC cartel on the economy of our country and, much more seriously, on the countries that have little or virtually no oil resources.

In order to increase the energy self-sufficiency of our country we must substantially increase our domestic energy production over the next several decades. Our success in doing so will depend considerably on a good working relationship between our government and industry. It is essential to develop and implement government policies that are conducive to the fullest and most economical development of our energy resources—oil, gas, coal, shale, geothermal, solar, etc. Oil and natural gas in the United States are in limited supply; however, there is an abundance of coal and oil shale resources. Most likely these two sources will be used extensively before economical use, on a large scale, can be made of breeder reactors, fusion, solar, geothermal, wind, and tidal wave energies. Figure 10.20 shows an estimated time scale for the first commercial application and significant uses of various forms of energy; it tells us that for the next decade or two we will need to depend primarily on oil, natural gas, and coal.

Hundreds of articles, technical papers, books, and reports have been written about our energy problems. Most of them contain many nomographs and charts in order to "tell the story" in a manner that can be comprehended by laypersons, interested groups concerned with the environmental effects of energy development, and legislative committees that are charged with the responsibility for recommending legislation that can have a direct bearing on the success of our efforts to meet the energy needs of our growing nation.

Instead of getting into the lengthy discussions of the arguments, pro and con, concerning the "best" choice of priorities, we will limit ourselves to the "pictorial" representations of several salient energy projections.

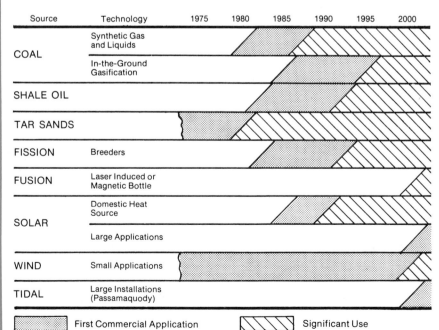

Figure 10.20 Alternate fuels. (Courtesy Standard Oil Co. of California)

Forecast of Energy Demand to the Year 2000

Estimates have been made by (a) the National Petroleum Council; (b) the U.S. Department of the Interior; (c) the U.S. Office of Science and Technology; and (d) the Shell Oil Company. These estimates are shown in Figure 10.21.

Supply and Demand Data to the Year 2000

These data curves (Figure 10.22) show the increased use of coal, the expected additional inputs of energy from geothermal sources, Alaska oil, shale, and solar uses. Dramatically shown is the importance of nuclear fission (breeder reactors, most likely) and fusion reactors.

Energy Demand and Uses, Projected to 1985

The amount of energy expressed in millions of barrels per day oil equivalent (MBPDOE) needed for transportation, generation of electricity, residential and commercial uses, industrial operations, and other uses is shown in Figure 10.23.

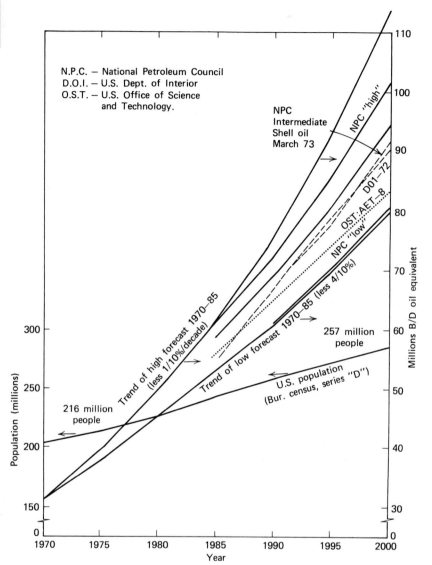

Figure 10.21 Forecast of energy demand to year 2000. (Courtesy Joint Committee on Atomic Energy)

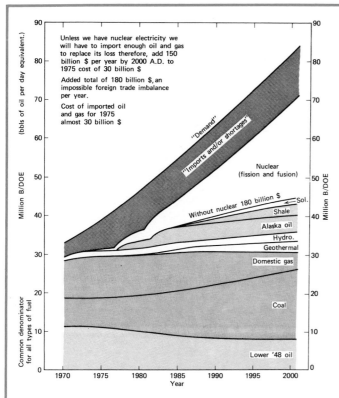

Unless we have nuclear electricity we
will have to import enough oil and gas
to replace its loss therefore, add 150
billion $ per year by 2000 A.D. to
1975 cost of 30 billion $

Added total of 180 billion $, an
impossible foreign trade imbalance
per year.

Cost of imported oil
and gas for 1975
almost 30 billion $

"Demand"

"Imports and/or shortages"

Nuclear
(fission and fusion)

Without nuclear 180 billion $ — Sol.
Shale
Alaska oil
Hydro.
Geothermal
Domestic gas

Coal

Lower '48 oil

Figure 10.22 Supply and demand energy
data to the year 2000. (Courtesy Joint
Committee on Atomic Energy)

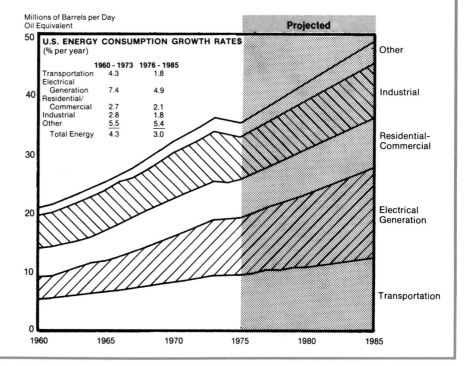

Figure 10.23 How much energy do we
need? (Courtesy Standard Oil Co., of
California)

Millions of Barrels per Day
Oil Equivalent

Projected

Other
Industrial
Residential-
Commercial
Electrical
Generation
Transportation

U.S. ENERGY CONSUMPTION GROWTH RATES (% per year)	1960 - 1973	1976 - 1985
Transportation	4.3	1.8
Electrical Generation	7.4	4.9
Residential/ Commercial	2.7	2.1
Industrial	2.8	1.8
Other	5.5	5.4
Total Energy	4.3	3.0

Principal Fuels Used in the United States, Projected to 1985

It has been estimated that the growth rates of U.S. energy sources from 1976 to 1985 will be 2.4% per year for oil; a drop of about 1% for natural gas; an increase of 4.4% for coal; a smaller growth for hydro-electric power, approximately 3%; and a substantial increase in nuclear and geothermal power of 19.4% and 14.6%, respectively (see Figure 10.24).

It has been estimated that coal consumption will increase by about 65% from 1975 to 1985. It is believed that if coal production could be doubled during this period, oil imports could be decreased 2 million barrels daily by 1985.

Natural Gas Problem—Forecast

Natural gas production in the United States peaked in 1973 and has been declining ever since. Shortages are likely to occur, with unfavorable weather conditions a real threat to the supply. It is estimated that by 1980 the United States will receive only about 20% of its energy from natural gas (it was about 33⅓% in the early 1970s). By 1985 about 40% of the total natural gas production will come from developed reserves. Synthetic gas, although costly, will become a most important source of clean-burning fuel. As projected to 1985, gas production will come from several sources, as shown in Figure 10.25.

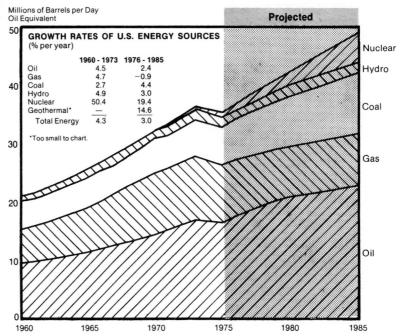

Figure 10.24 What fuels will we use?
(Courtesy Standard Oil Co. of California)

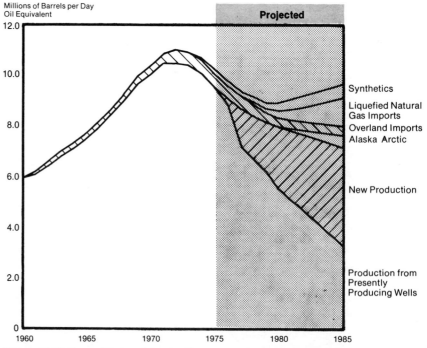

Figure 10.25 How will we get our gas?
(Courtesy Standard Oil Co. of California)

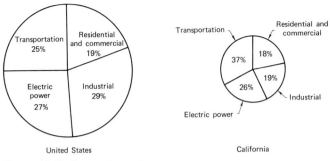

Figure 10.26 Comparison of energy uses in California and the United States (1974). California used $8\frac{1}{2}$% of total energy consumed in the United States (Courtesy Bechtel Power Corp.)

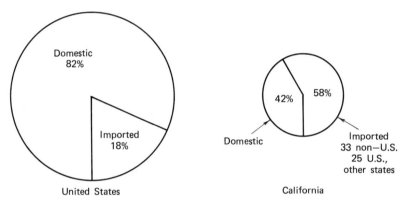

Figure 10.27 Energy sources for California and the United States, 1974. (Courtesy Bechtel Power Corp.)

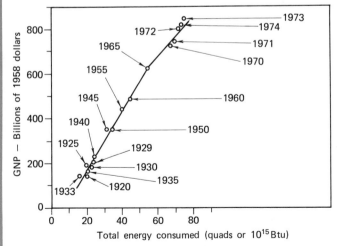

Figure 10.28 Relationship of GNP to total energy consumed in the United States, 1920–1974. (Courtesy Bechtel Power Corp.)

Nuclear Power—Its Significance in the Overall Energy Situation in the United States.

The figures included here have been selected from a presentation* made to the Committee on Resources, Land Use, and Energy of the California State Assembly. The presentation focused on the range of energy possibilities for 1975 to 1990.

For background information, Figure 10.26 shows 1974 energy uses in California and the United States. The areas of the circles (pie charts) are proportional to the relative energy consumption. California used $8\frac{1}{2}$% of the total energy consumed in the United States.

Energy sources for California and the United States are shown in Figure 10.27. California is a major importer of energy—about 83% natural gas, 30% oil, 10% electricity, and 100% uranium and coal.

Growth in total energy demand, correlated with GNP (gross national product) in 1958 dollars is shown in Figure 10.28. Obviously there is a close relationship between energy demand and GNP.

Most projections for 1985 have predicted demands ranging from 54 to 60 MBPDOE. The National Academy of Engineering (NAE) Energy Task Force forecast, for planning purposes, 58 MBPDOE as most probable. It also considered a reduced growth in energy consumption through a major em-

* W. Kenneth Davis, "The United States Energy Outlook," October 1975. W. Kenneth Davis is Vice-President of the Bechtel Power Corp and Chairman of the Energy Task Force of the National Academy of Engineering.

phasis on the potential for *energy conservation* by *more efficient equipment design, more efficient automobiles and other transportation facilities, better building insulation, kitchen appliances designed to use less energy*, etc. It has been estimated that by 1985 the savings could be 8 MBPDOE, or about 14% less than the level that would result from the continuation of historical growth. To meet the demand for about 50 MBPDOE, expected by the middle of the 1980s, it will be necessary to rely almost completely on oil, gas, coal, nuclear fuels, and a small contribution from existing hydroelectric power. Deficiencies will require oil imports (or shortages). It is hoped that when we reach the 1990s and the 2000s, other energy sources [coal, synthetic gas, shale, solar, fission (breeder), fusion, wind, tides, etc.] will have been sufficiently developed to fill our needs economically.

Figure 10.29 shows the recoverable energy from domestic sources, the largest of which is coal. Although there is an abundance of coal, uncertainties concerning the cost of competing energy forms, questions on environmental constraints, mine safety, problems in the methods of transporting increased volume of coal to various parts of the country, the effects of burning coal in many areas, etc., may have produced a situation that is not favorable to huge investments.

The effort required to meet the 1985 demand, and what the Task Force has believed possible, includes increasing coal production

by a factor of two, increasing nuclear power capacity to one-third of total electrical capacity, and duplicating present domestic oil and gas capacity to get a net small increase in output.

The task force considered new technologies to produce gas and oil from coal and oil shale and believed that by 1985, 1.1 MBPD of synthetic gas from coal, 0.6 MBPD of synthetic liquids from coal, and 0.5 MBPD of oil from shale could be produced. Modest increases in hydroelectric and geothermal power could be expected.

The various energy sources and

requirements in 1985 are shown in Figure 10.30. Note the projected increase (from 1974 to 1985) in the need for nuclear power from 0.6 to 8.3 MBPDOE.

Another estimate of the 1985 energy supply situation was made (1975) by the U.S. Department of Commerce's Technical Advisory Board (CTAB). This is shown in Figure 10.31. The projections were based on rapid implementation of new energy programs.

Both studies (NAE and CTAB) include an estimated "conservation savings" of 8.0 MBPDOE.

Figure 10.29 Available energy from domestic resources (10^{15} BTU). (Courtesy Bechtel Power Corp.)

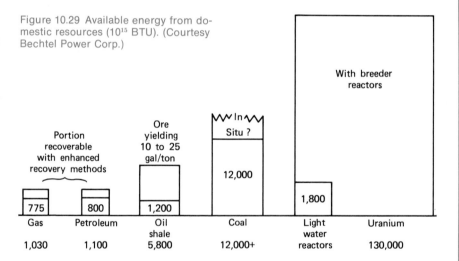

Figure 10.30 U.S. energy sources and supply in 1985. (Courtesy Bechtel Power Corp.)

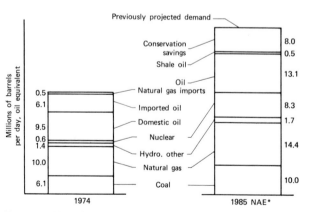

CONSERVATION SAVINGS

Conservation of energy is a must in order to reduce the pressure for energy, maintain reasonable prices, and allow more time to develop the new technologies more fully and economically.

Efficiency in energy use is essential to energy savings. For example, the choice of transportation systems can make a big difference in the use of energy; note, in Figure 10.32, the savings that could be made by effective choices of the transportation modes in each category. Also, the use of good insulation material in residences would reduce the use of fuels for home

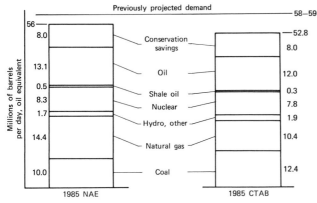

Figure 10.31 U.S. energy sources and supply in 1985. Comparison of estimates made by NAE and CTAB. (Courtesy Bechtel Power Corp.)

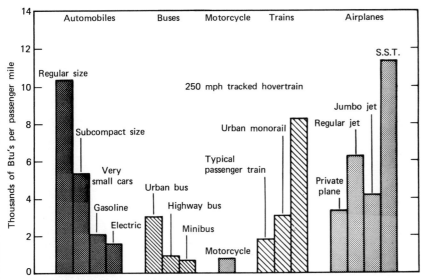

Figure 10.32 Energy consumption of various transportation modes. (Courtesy "Explaining Energy," LBL-4458, ERG 71 to 4, University of California, Berkeley)

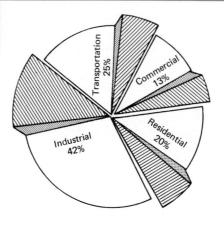

Percentages given are todays breakdown of energy consumption. Shaded areas give savings.

Figure 10.33 U.S. energy use and potential savings in various sectors. (Courtesy "Explaining Energy," LBL-4458, ERG 71 to 4, University of California, Berkeley)

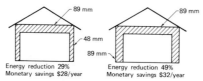

Figure 10.34 Energy savings by installing various amounts of insulation (N.Y. climate 1972). (Courtesy "Explaining Energy," LBL-4458, ERG 71 to 4, University of California, Berkeley)

heating; in turn, more efficient equipment and processes could reduce the amount of energy required to produce the insulation.

There are many ways* to save energy, including *plugging leaks*; *mixing modes* (changing the mix of transportation to reduce energy requirements per passenger, or ton mile); *thrifty technologies* (heat pumps for industrial heat, electric ignition of gas water heaters); *juggling inputs* [more efficient machines, solar energy, recycling, no throwaways (beverage bottles), etc.]; *juggling outputs* (smaller cars, longer lifetime of consumer goods, changes in recreation or travel patterns); *belt tightening* (turning off lights, changes in thermostat settings, car pooling, driving at reduced speeds, more use of public transportation, etc.); and curtailment (rationing of fuels, driving less, shutdown of factories, etc.); these methods are all summarized in Figure 10.33. Examples of money and energy savings are shown in Figures 10.34 to 10.37. Much more is discussed in the report LBL-3299.†

* Lee Schipper, "Explaining Energy," Lawrence Berkeley Laboratory, University of California, Berkeley, January 1976, pp. 43–44.

† Lee Schipper, "Towards More Productive Energy Utilization," Energy and Environment Division, Lawrence Berkeley Laboratory, University of California, Berkeley, October 1975.

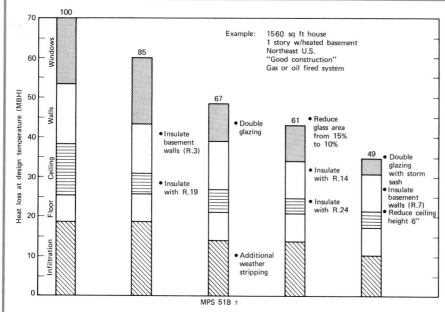

Figure 10.35 Leak plugging and input juggling. Effect of more elaborate modifications of design and construction of a house on energy use for heating. Note the progressive savings. (Courtesy "Towards More Productive Energy Utilization," LBL-3244, University of California, Berkeley)

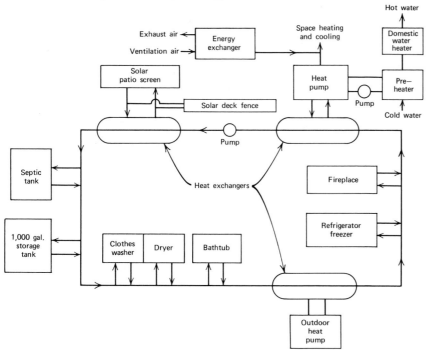

Figure 10.36 Input juggling and thrifty technology. Schematic of the house of the future. This system is installed in the experimental low-energy house built by Pennsylvania Power and Light Company in Allentown, Pennsylvania. Heat exchangers recapture as much heat as possible. (Courtesy "Towards More Productive Energy Utilization," LBL-3244, University of California, Berkeley)

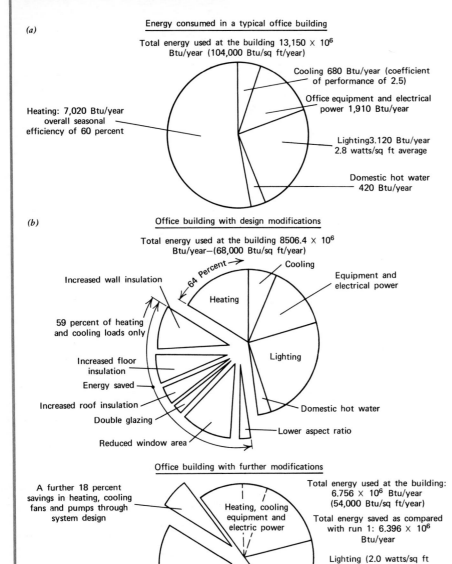

(a)

Energy consumed in a typical office building

Total energy used at the building $13,150 \times 10^6$ Btu/year (104,000 Btu/sq ft/year)

Cooling 680 Btu/year (coefficient of performance of 2.5)

Office equipment and electrical power 1,910 Btu/year

Heating: 7,020 Btu/year overall seasonal efficiency of 60 percent

Lighting 3.120 Btu/year 2.8 watts/sq ft average

Domestic hot water 420 Btu/year

(b)

Office building with design modifications

Total energy used at the building 8506.4×10^6 Btu/year—(68,000 Btu/sq ft/year)

64 Percent

Cooling

Heating

Equipment and electrical power

Increased wall insulation

59 percent of heating and cooling loads only

Lighting

Increased floor insulation

Energy saved

Increased roof insulation

Double glazing

Domestic hot water

Reduced window area

Lower aspect ratio

Office building with further modifications

A further 18 percent savings in heating, cooling fans and pumps through system design

Heating, cooling equipment and electric power

Total energy used at the building: 6.756×10^6 Btu/year (54,000 Btu/sq ft/year)

Total energy saved as compared with run 1: 6.396×10^6 Btu/year

Lighting (2.0 watts/sq ft average)

59 percent of heating and cooling saved through modification of original building (run 1)

30 percent of lighting saved through better lighting design (2.8 w/sq ft reduced to 2 w/sq ft)

100 percent of domestic hot water saved through heat recovery

Figure 10.37 All strategies—effect of progressive design modifications of an office building. Actual predictions of building being built for Government Services Administration in Manchester, New Hampshire. Figures show equivalent en-energy units of 10^6 Btu year.
Figure 10.27 (a) In New England; 126,000 sq. ft.; design based on "typical" New England design criteria; weather data from Manchester, New Hampshire: wall U value = 0.3 Btu/° F-hr-ft; floor U value = 0.25; roof U value = 0.2; single glazing, 50% window/wall area ratio; shading coefficient = 0.5 (year round); 6 stories tall; 2:1 aspect ratio (length:width); long axis, north-south.

Figure 10.37 (b) Wall, floor, roof U values = 0.06; double glazing, 10% window/wall area ratio; shading coefficient = 0.5 (year round). (Courtesy "Towards More Productive Energy Utilization," LBL-3244, University of California, Berkeley)

WORLD FOOD SITUATION

A very significant report‡ has evaluated expected food supply and demand conditions in 1985 and beyond. A number of graphs and charts that highlighted important findings and projections are presented in this report.

The world food system is depicted in Figure 10.38. The system, although quite complex, is delicately balanced. There are two major components of world food demand, population and per capita consumption. They are shown in

‡ "A Hungry World: The Challenge to Agriculture," General Report by the University of California Food Task Force, 1974.

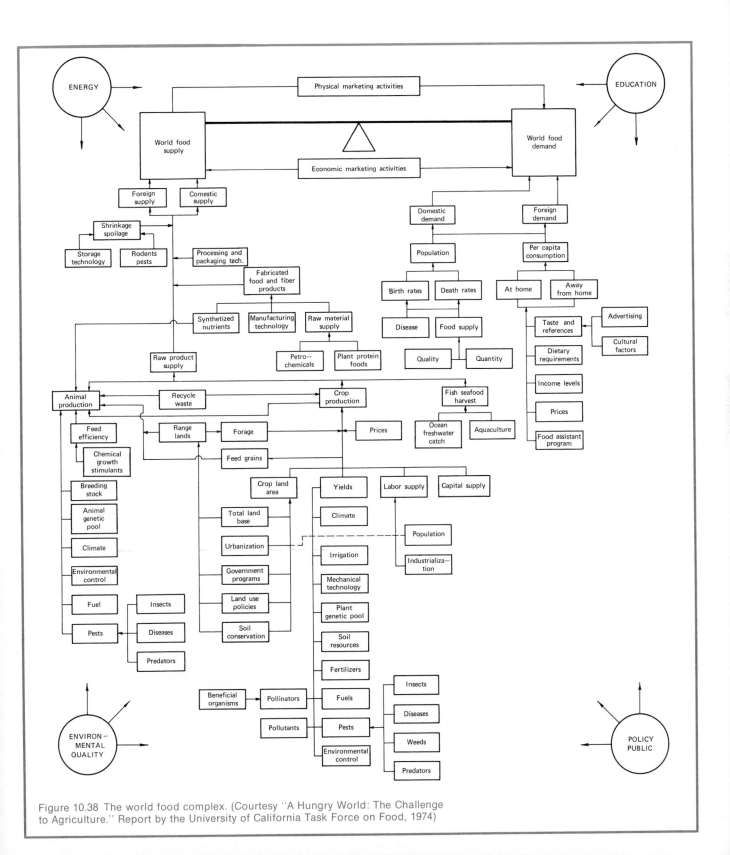

Figure 10.38 The world food complex. (Courtesy ''A Hungry World: The Challenge to Agriculture.'' Report by the University of California Task Force on Food, 1974)

Figures 10.39 and 10.40. A projection of world net food consumption of major commodities for 1985 and 2000 is shown in Figure 10.41. (The letters *L, M,* and *H* refer to low, medium, and high estimates for the year 2000.)

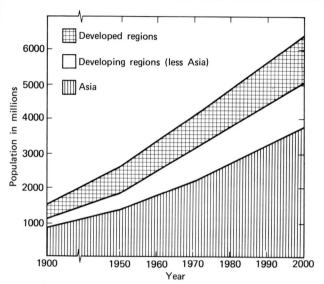

Figure 10.39 World population growth, 1900–2000. (Courtesy "A Hungry World: The Challenge to Agriculture." Report by the University of California Task Force on Food, 1974)

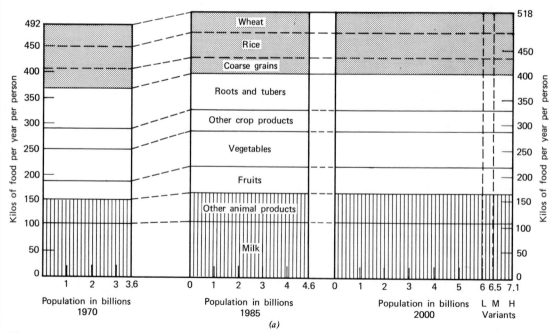

(a)

Figure 10.40 (*a*) and (*b*) Average per capita consumption of all food products by world and selected regions. (Courtesy "A Hungry World: The Challenge to Agriculture." Report by the University of California Task Force on Food, 1974)

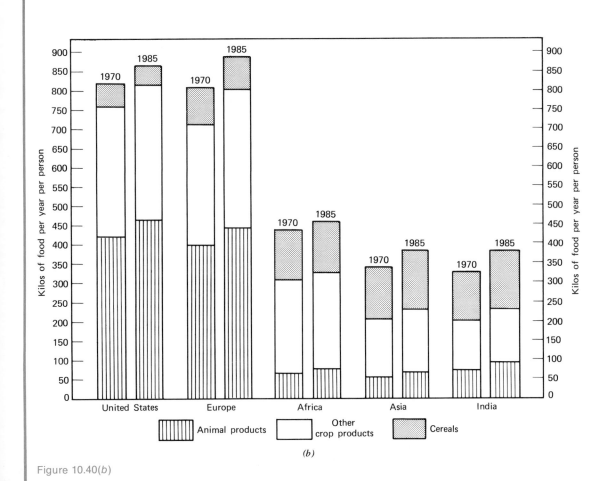

Figure 10.40(*b*)

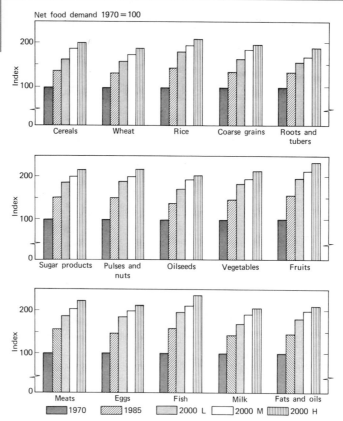

Figure 10.41 Projection of world total net food consumption of major commodities 1970-1985-2000. (Courtesy "A Hungry World: The Challenge to Agriculture." Report by the University of California Task Force on Food, 1974)

ENERGY USE— U.S. AGRICULTURE*

The energy picture for agriculture (Figure 10.42) shows that on-highway vehicles used in agriculture consume almost as much fuel as off-highway vehicles. The latter's fuel use is only one-half gasoline. The other half is diesel fuel, with a small portion liquefied petroleum gas (LPG). Chemical manufacture is another major energy input to agriculture.

It has been estimated that the total of all agricultural energy inputs is about 2.5% of the nation's total energy.

EFFORTS TO SAVE ENERGY IN AGRICULTURAL PRODUCTION

Corn has the highest energy input, per acre, of the U.S. major crops (see Figure 10.43). Except for corn, few energy-saving alternatives are available for crop production. Reduced tillage systems offer a way to use less energy. Comparisons of the energy requirements, per acre, for corn production employing various cultural practices is shown in Figure 10.44. It has been shown

* L. F. Nelson and W. C. Burrows, "Putting the U.S. Agricultural Energy Picture Into Focus," American Society of Agricultural Engineers, Paper No. 7A-104D, 1974.

Figure 10.42 U.S. agricultural energy use. (Courtesy "Putting the U.S. Agricultural Energy Picture into Focus," by L. F. Nelson and W. C. Burrows, Deere & Co.)

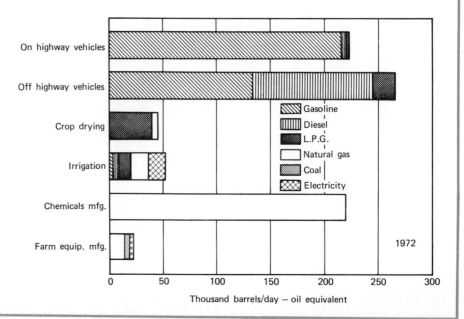

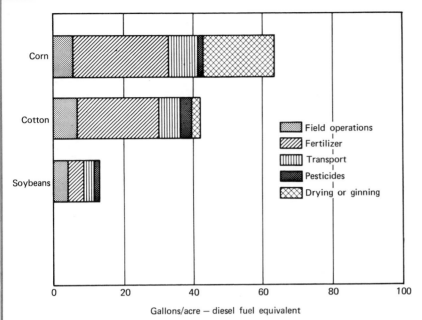

Figure 10.43 Energy use for three crops.
(Courtesy "Putting the U.S. Agricultural
Energy Picture into Focus," by L. F.
Nelson and W. C. Burrows, Deere & Co.)

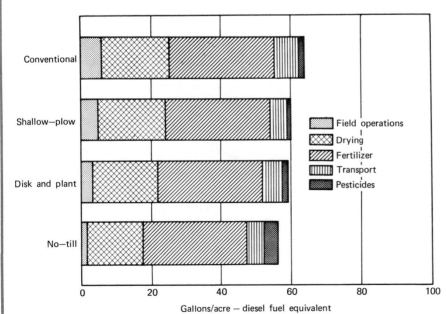

Figure 10.44 Energy use for corn with
four cultural practices. (Courtesy "Putting the U.S. Agricultural Energy Picture
into Focus," by L. F. Nelson and W. C.
Burrows, Deere & Co.)

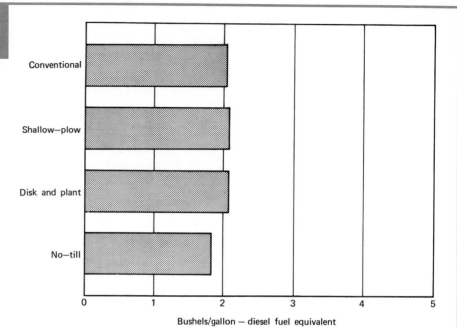

that each of the practices produces about two bushels of corn per gallon of diesel fuel equivalent total requirement (see Figure 10.45). Shallow plowing, however, produces slightly more than "disc-and-plant" or conventional methods. Obviously, the no-till method produces the least output. It has been estimated that the energy content of the harvested grain is about six times greater than the energy input of the fossil fuel. To satisfy domestic needs for corn and to provide some margin for export, it is necessary to maintain nearly normal fertilizer application rates and to use conventional, shallow plowing or disc-and-plant tillage methods (see Figure 10.46).

Efficient designs of field equipment and the appropriate selection of equipment to accomplish field tasks can result in fuel economy ranging from 10 to 30%.

Figure 10.45 Energy use efficiency for corn. (Courtesy "Putting the U.S. Agricultural Energy Picture into Focus," by L. F. Nelson and W. C. Burrows, Deere & Co.)

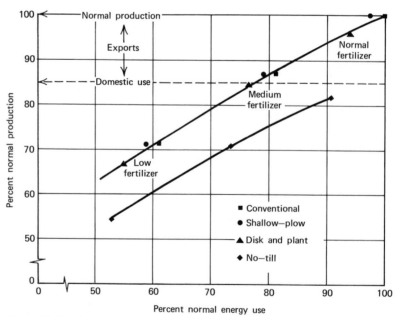

Figure 10.46 Effect of energy input on U.S. corn production. (Courtesy "Putting the U.S. Agricultural Energy Picture into Focus," by L. F. Nelson and W. C. Burrows, Deere & Co.)

Exercises

In the following problems present the data graphically. Study the data carefully and then make a judgment as to the "best" form for effective presentation.

1. The following data show energy consumption in the United States for 1960, 1973, 1976, 1980, and 1985.

Energy consumption[a]	Millions of barrels per day oil equivalent				
	1960	1973	1976	1980	1985
Oil	9.8	17.3	17.2	19.9	21.3
Gas	5.9	10.9	9.6	8.7	9.0
Coal[b]	4.5	6.4	6.8	8.8	10.3
Hydro[c]	0.7	1.3	1.5	1.7	1.8
Nuclear	—	0.4	1.0	2.0	4.9
Imports of electricity	—	0.1	0.1	0.1	0.1
Total	21.0	36.4	36.1	41.2	47.3

[a] Some totals may not add due to rounding.

[b] Includes coal used for synthetic gas production.

[c] Includes small amount of geothermal in later years.

2. The following data show oil demand in the USA for 1960, 1973, 1976, 1980, and 1985.

Demand[a]	Millions of barrels per day				
	1960	1973	1976	1980	1985
Domestic	9.8	17.3	17.2	19.9	21.3
Export	0.2	0.2	0.2	0.2	0.2
	10.0	17.5	17.4	20.1	21.5

Source. Courtesy of Standard Oil Company of California.

[a] Some totals may not add due to rounding.

3. Data for oil supply are shown in the following table.

Supply[a]	Millions of barrels per day oil equivalent				
	1960	1973	1976	1980	1985
Production					
Crude[b]	7.0	9.2	8.1	9.4	9.7
MGL	0.9	1.7	1.6	1.5	1.4
Total	8.0	10.9	9.7	10.9	11.0
Process gain, inventory change, etc.	0.2	0.4	0.6	0.3[c]	0.7
Imports, crude and products	1.8	6.3	7.1	8.9	9.8
Total supply	10.0	17.6	17.4	20.1	21.5

Source. Courtesy of Standard Oil Company of California.

[a] Some totals may not add due to rounding.

[b] Includes 50,000 barrels per day of shale oil in 1985.

[c] Includes small amount of geothermal in later years.

5. Typical property tax data for an Oakland home are shown in the following table for 1976.

4. The data in the following table show ownership of common stock.

Owners	Percent of stock
Women	39.6
Joint and common tenancies	25.4
Men	25.2
Individuals as fiduciaries	8.0
Corporations	0.9
Bank nominees, brokerage firms, etc.	0.8
Other	0.1

Source. Courtesy of Con Edison, 1976.

Tax Rates Per $100 Assessed Valuation. Rates and Amounts Levied by Taxing Agencies

Taxing agency	Rate	amount
County tax	3.1100	280.66
City of Oakland 1	2.9403	265.36
School unified	6.1270	552.97
School comm coll	.8880	80.14
Co sup sch spec ed	.0830	7.49
Co flood control	.0140	1.26
Air poll control	.0180	1.63
Mosquito abatemnt	.0100	.90
Ac transit sv 1	.4480	40.43
Bay area rapid tra	.4220	38.09
East bay regional	.1950	17.60
Ebmud	.1240	11.19
Ebmud sewage dist	.0690	6.22
* Total a.v.tax *	14.4483	1303.94
12 flood zone	.2400	
		25.86
* Total tax		1329.80

6. Data concerning projected average enrollment in two-year public institutions of higher education and estimates of needs for *new* community colleges by 1980 are shown in the following table.

State	Number of public two-year institutions, 1968[a]	Average enrollment, 1968 in thousands	Projected average enrollment. 1980		Estimated needs for new public community colleges, 1980
			Based on trends, 1960–1968 and 1964–1968	Averaged and adjusted for criteria relating to size	
United States	781	2201			226–280
Alabama	15	1215	3000–4500	3750	4–5
Alaska	6	91	500–1500	1500	
Arizona	8	4010	4500–5000	4750	9
Arkansas	4	653	1500–2500	2000	6–7
California	86	6982	9000–11000	8500	29–34
Colorado	10	1317	2500–3000	2750	6–7
Connecticut	18	974	2500–3000	2750	2–3
Delaware	1	913	3000–4000	3500	2–3
District of Columbia	1	1178	3000–6000	4500	
Florida	27	3439	4500–7000	5750	11–12
Georgia	13	1277	2500–3000	2750	5–6
Hawaii	6	1029	2500–3500	3000	
Idaho	2	1196	1500–2500	2000	3–4
Illinois	42	2386	3000–4000	4000	5–7
Indiana	7	2600	4000–5000	4500	7–8
Iowa	21	725	1300–1800	1800	3–4

Source. Courtesy of the Carnegie Foundation for the Advancement of Teaching. Similar data for the other states are available.

[a] Carnegie Commission

7. The following table shows data on Guggenheim Fellowship Awards and Renewals, 1964 to 1975, for ranks through 25.

State	Number of awards in public institutions	Rank	Number of awards in private institutions	Rank	Total number of awards	Rank (all)
California	521	1	151	3	672	1
New York	134	2	425	1	559	2
Massachusetts	21	16	329	2	350	3
Illinois	94	3	128	5	222	4
Pennsylvania	44	7	136	4	180	5
Connecticut	7	21	124	6	131	6
New Jersey	20	17	81	7	101	7
Michigan	78	4	3	18	81	8
Wisconsin	76	5	3	18	79	9
Indiana	69	6	2	21	71	10
Texas	40	9	17	12	57	11
Maryland	19	18	35	9	54	12
North Carolina	28	11	24	10	52	13
Ohio	25	13	24	10	49	14
Rhode Island	2	31	43	8	45	15
Washington	44	7	0	—	44	16
Minnesota	35	10	3	18	38	17
Oregon	26	12	2	21	28	18
Iowa	25	13	2	21	27	19
Virginia	25	13	2	21	27	19
Missouri	2	31	16	13	18	21
New Hampshire	5	26	13	14	18	21
Kansas	12	19	0	—	12	23
Georgia	7	21	4	17	11	24
Arizona	8	20	0	—	8	25
Florida	7	21	1	26	8	25
Tennessee	2	31	6	15	8	25

Source. Courtesy of The Carnegie Foundation for the Advancement of Teaching.

8. Data showing the percent of federal research and development funds received going to private institutions within each state, for the fiscal year 1974, are shown in the following table.

State	Percent	State	Percent
Massachusetts	93.9	Wisconsin	7.0
Connecticut	82.0	Alabama	6.4
New York	78.3	Maine	4.6
Missouri	72.6	Utah	3.3
New Hampshire	65.7	Oregon	3.0
Pennsylvania	64.5	South Dakota	1.4
Maryland	63.3	Mississippi	1.3
Rhode Island	58.7	South Carolina	1.2
Illinois	57.4	Oklahoma	1.1
New Jersey	51.3	Arkansas	0.8
Tennessee	51.0	Minnesota	0.6
Louisiana	48.4	Arizona	0.5
North Carolina	46.2	Kentucky	0.4
Ohio	43.8	Michigan	0.4
Florida	43.7	Virginia	0.4
Georgia	36.3	New Mexico	0.2
California	31.2	Iowa	0.1
Texas	24.4	(All others)	0.0
Indiana	11.5	(Wyoming has no	
Nebraska	10.7	private institution)	
Colorado	9.5		

Source. Courtesy of The Carnegie Foundation for the Advancement of Teaching.

9. The following table lists some familiar items and the *total* energy required for *manufacture*.

Energy Cost of Things: Common and Uncommon Figures in parentheses give equivalent energy in gallons (or liters) of gasoline

Car, standard size	123 million Btu (980 gallons) (or 3,710 liters)	(Uses about 130 million Btu/year)
Color TV $400	20,000,000 Btu (160 gallons) (or 606 liters)	
Refrigerator, average size $400	21,000,000 Btu (168.0 gallons) (or 636 liters)	
Men's suit $200	4,000,000 Btu (32.0 gallons) (or 121 liters)	Note that $200 suit requires as much energy as $80 bicycle. However, bicycle replaces much energy use
Bicycle $80 average cost	4,000,000 Btu (32.0 gallons) (or 121 liters)	
Three sources of protein	Energy required to produce 1 gram of protein	
Fish	450 Btu	Beef is one of the most *expensive* sources of protein
Cheese	475 Btu	
Meat	700 Btu	
Electric can opener	690,000 Btu (6 gallons) (or 2.3 liters)	Uses as much energy in 10 years as was needed for manufacture
$100 Worth of food consumed in the average home	4,100,000 Btu (33 gallons) (or 125 liters)	Average energy to *prepare* food is about half as much
Beverage containers	Energy required per 12 ounce filling	
Throwaway bottle	5,800 Btu	Throwaway bottles are the worst and most expensive, recycled cans are better, but returnable bottles are cheaper, use less energy and create more employment than the other two. (Hannon, 1972; Herendeen, private communication)
Aluminum can		
No recycling	7,800 Btu	
16% recycled (national average)	6,800 Btu	
50% recycled	4,900 Btu	
87.5% recycled	3,000 Btu	
Returnable bottles		
5 returns	3,800 Btu	
15 returns	1,900 Btu	

Source. Courtesy of the Laurence Berkeley Laboratory, University of California, Berkeley, Report UCID-3707.

10. The following table lists the Btu content and energy value content of selected goods and services (partial list).

Product	Energy content, Btu/$	Energy value content, ¢/$
Plastics	218,097	13.2
Man-made fibers	202,641	7.4
Paper mills	177,567	7.9
Air transport	152,363	12.0
Metal cans	136,961	7.3
Water, sanitary services	116,644	11.6
Metal doors	109,875	6.7
Cooking oils	94,195	7.1
Fabricated metal products	91,977	5.8
Metal household	91,314	5.9
Knit fabric mills	88,991	6.5
Toilet preparations	85,671	5.1
Blinds, shades	81,472	6.3
Floor coverings	79,323	5.8
House furnishings	75,853	5.3
Poultry, eggs	75,156	7.3
Electric housewares	74,042	5.6
Canned fruit, vegetables	72,240	5.2
Motor vehicles and parts	70,003	5.9
Photographic equipment	64,718	3.8
Mattresses	63,446	4.5
New residential construction	60,218	4.5
Boat building	60,076	4.9
Food preparation	58,690	4.8
Soft drinks	55,142	4.5
Upholstered house-hold furniture	51,331	4.1
Cutlery	50,021	4.0
Apparel, purchased materials	45,905	4.0
Alcoholic beverages	43,084	3.0
Hotels	40,326	5.4
Hospitals	38,364	5.4
Retail trade	32,710	4.4
Insurance carriers	31,423	4.4
Misc. professional services	26,548	4.3
Banking	19,202	2.5
Doctors, dentists	15,477	1.9

These values are for producer's prices, and do not take into account mark up to retail price, about 66%.

Source. Courtesy of the Laurence Berkeley Laboratory, University of California, Berkeley, Report UCID-3707.

11. The data in the following table show the per capita demand for fish by country for the year 1970 and projected for the year 1985.

	Level of per capita demand, kg/year, roundweight basis	
	1970	1985
World	11.8	14.0
Europe	17.6	22.6
Western Europe	20.3	25.9
Eastern Europe	8.7	11.6
EEC-9	17.4	21.5
EEC-6	16.1	20.8
U.K.	21.1	23.6
USSR	23.9	32.8
North America	15.4	17.4
Canada	16.2	18.3
United States	15.3	17.3
Oceania	12.4	14.2
Australia	11.3	13.2
New Zealand	17.3	18.4
Asia	8.0	10.2
Japan	56.6	64.9
China	7.9	10.0
India	2.9	4.2
Indonesia	10.5	11.5
Latin America	6.5	8.3
Mexico	3.6	4.7
Brazil	5.8	7.8
Africa	6.7	9.0

Source. General Report by the University of California Food Task Force, 1974.

		Numbers				Production		Demand
		1950	1960	1970	1985[a]	1970[b]	1985[c]	1985[d]
		Numbers × 10⁶				M.T. × 10⁶		
World	Beef cattle	763.3	920.6	1250.7	1936.8	40.29	58.06	60.18
	Sheep and goats	1012.0	1218.7	1457.3	2155.7	7.09	10.27	11.35
	Swine	227.7	343.1	626.9	1075.4	37.14	54.22	50.81
	Poultry	—	—	5560.2	9668.7	17.67	30.41	28.12
	Milk eq	—	—	—	—	398.50	511.67	565.86
	Eggs	—	—	—	—	17.68	36.13	27.13
Europe	Beef cattle	100.6	117.5	124.3	182.3	8.93	11.84	13.49
	Sheep and goats	144.2	150.9	140.9	196.2	1.03	1.24	2.01
	Swine	69.4	109.8	130.9	177.6	12.59	16.87	15.45
	Poultry	—	—	1226.2	2218.6	4.29	7.59	6.39
	Milk eq	—	—	—	—	149.88	176.40	175.24
	Eggs	—	—	—	—	5.93	10.42	7.10
USSR	Beef cattle	56.0	76.0	95.6	154.6	5.93	7.38	7.98
	Sheep and goats	92.6	140.3	135.8	206.1	1.00	1.49	1.55
	Swine	19.7	58.7	56.0	83.2	4.54	6.56	5.34
	Poultry	—	—	590.3	953.7	1.07	1.75	1.49
	Milk eq	—	—	—	—	82.90	123.52	101.36
	Eggs	—	—	—	—	2.22	5.34	4.29
N. America	Beef cattle	88.5	106.2	124.1	186.9	10.95	15.11	17.88
	Sheep and goats	35.4	36.7	23.5	28.8	0.26	0.19	0.42
	Swine	63.7	61.8	63.1	70.9	6.69	7.38	7.39
	Poultry	—	—	541.4	920.5	6.77	10.77	11.44
	Milk eq	—	—	—	—	61.57	60.42	74.69
	Eggs	—	—	—	—	4.49	7.14	4.79
Oceania	Beef cattle	19.7	23.5	31.4	46.9	1.46	2.20	1.78
	Sheep and goats	145.6	201.2	240.6	305.7	1.37	1.98	1.19
	Swine	1.9	2.4	3.3	5.2	0.23	0.34	0.72
	Poultry	—	—	31.5	67.9	0.13	0.25	0.36
	Milk eq	—	—	—	—	13.54	15.78	10.77
	Eggs	—	—	—	—	0.24	0.46	0.36
Asia	Beef cattle	275.9	330.2	473.6	814.8	4.05	5.98	7.18
	Sheep and goats	229.0	295.3	484.6	902.4	2.02	3.09	3.74
	Swine	19.2	38.0	269.4	517.3	11.04	19.28	17.82
	Poultry	—	—	2102.3	5877.3	3.93	7.11	6.60
	Milk eq	—	—	—	—	53.94	77.73	132.34
	Eggs	—	—	—	—	6.16	5.65	6.61
Latin America	Beef cattle	162.9	204.0	242.8	373.2	7.01	11.23	9.22
	Sheep and goats	156.8	163.4	169.9	313.2	0.44	0.67	0.78
	Swine	47.6	77.0	97.5	256.6	1.76	2.96	3.48
	Poultry	—	—	657.0	1427.7	1.00	1.84	1.78
	Milk eq	—	—	—	—	23.44	38.27	44.62
	Eggs	—	—	—	—	1.38	2.54	2.45
Africa	Beef cattle	86.2	107.4	158.8	289.9	2.04	3.58	4.24
	Sheep and goats	193.4	220.8	261.9	491.3	0.97	1.61	1.80
	Swine	4.1	4.8	6.6	12.2	0.28	0.47	0.48
	Poultry	—	—	411.4	792.3	0.47	0.88	0.90
	Milk eq	—	—	—	—	9.63	14.67	26.80
	Eggs	—	—	—	—	0.49	0.92	0.79

Source. General Report by the University of California Food Task Force, 1974.

[a] Projected on the basis of 1970 published values adjusted for population increases and change in anticipated per capita consumption.

[b] Actual production values from FAO.

[c] Extension of FAO projection production to 1980.

[d] Economic demand calculated as projected per capital consumption x population.

12. The data on page 286 record Historical trends in livestock numbers, Production in 1970, and Projected production and economic demand for 1985.

13. The following data record world Production and Gross Utilization for Food Commodities in 1970 and Projected to 1985.

Commodities	Gross utilization[a]				Production			
	Metric tons × 10⁶ 1970	Metric tons × 10⁶ 1985	Calories mcal × 10⁹ 1985	Protein M.T. × 10⁶ 1985	Metric tons × 10⁶ 1970	Metric tons × 10⁶ 1985	Calories mcal × 10⁹ 1985	Protein M.T. × 10⁶ 1985
Wheat	317.5	447.0	1492.9	54.5	318.0	454.7	1518.8	55.5
Rice	307.6	452.8	1630.2	30.3	307.5	443.7	1597.4	29.7
Coarse grains	602.0	877.4	3071.2	78.9	583.3	878.2	3073.7	79.0
Roots and tubers	529.7	608.1	547.3	9.1	529.7	800.2	720.2	12.0
Sugar	840.9	1314.5	460.1	—	816.6	1219.3	426.7	—
Pulses, nuts	42.6	62.7	216.3	13.9	42.3	60.9	210.2	13.5
Oilseeds	101.0	189.2	700.1	66.2	103.3	151.3	559.8	53.0
Vegetables	212.2	316.3	69.6	4.4	212.2	316.3	69.6	4.4
Fruits	201.7	322.7	119.4	1.9	202.1	316.1	117.0	1.9
Crop total	—	—	8307.1	259.2	—	—	8293.2	249.0
Animal feed[b]	—	—	3052.0	147.2	—	—	3052.0	147.2
Beef	39.4	61.8	135.9	9.1	39.8	57.3	126.1	8.4
Mutton	7.2	11.5	27.6	1.4	7.1	10.3	24.6	1.2
Pork	32.8	50.7	192.5	5.1	37.1	53.9	204.7	5.4
Poultry	15.9	29.0	37.7	3.5	17.7	30.2	39.3	3.6
Total meat	95.3	152.9	393.6	19.1	101.7	151.7	394.7	18.6
Eggs	18.4	26.4	38.3	2.9	20.9	32.5	47.1	3.6
Fish	64.0	99.8	69.9	11.5	69.5	94.8	66.4	10.9
Milk	386.8	544.0	217.6	19.0	394.9	506.8	202.7	17.7
Animal total[c]	—	—	800.5	52.5	—	—	800.7	50.9
Nutrient total[d]	—	—	6055.6	164.5	—	—	6041.9	152.7

Source. General Report by the University of California Food Task Force, 1974.

[a] Gross includes waste, feed, seed, and manufacturing in addition to food uses.

[b] Nutrient content of cereals and oilseeds fed to livestock within this region.

[c] Includes animal fats and oil.

[d] Crop and animal totals less crops used for domestic animal feed.

14. The following table contains data for a summary of California harvested averages of field crops, vegetables, tree fruits, nuts, and grapes.

| Crop | Harvested acreage[a] | | | | | Percent change |
	1968–1972 average actual	1970 actual	1971 actual	1972 actual	1985 projected	1968–1972 average to 1985
	Acres					Percent
Field crops						
Rice	362,800	331,000	331,000	331,000	356,890	−1.63
Other grains	2,221,200	2,363,000	2,241,000	2,003,000	2,432,529	9.51
Sugar beets	310,440	320,500	346,500	326,000	329,175	6.04
Cotton	730,480	662,400	860,400	851,007	16.50	
Dry beans	179,400	174,000	148,000	157,000	179,450	0.03
Potatoes	84,080	87,500	82,600	67.200	83,431	−0.77
Safflower	211,800	201,000	242,000	235,000	211,800	0.00
Alfalfa seed	92,800	104,000	91,000	70,000	92,800	0.00
Alfalfa hay	1,168,200	1,152,000	1,210,000	1,198,000	1,213,463	3.87
Other field crops	661,573	670,800	685,900	689,000	661,573	0.00
Subtotal	6,022,773	6,066,200	6,119,600	5,936,600	6,412,118	6.46
Vegetables						
Tomatoes (all)	205,520	117,000	192,800	208,900	231,931	11.30
Dark green and deep yellow	78,480	74,700	80,700	87,200	87,523	11.52
Other vegetables	345,920	341,500	336,900	352,800	388,309	12.25
Melons	79,610	80,800	73,900	75,000	88,622	11.32
Strawberries	8,320	8,500	8,300	7,800	6,719	−19,24
Subtotal[b]	768,198	732,200	741,370	785,300	853,452	11.10
Fruits and nuts						
Citrus fruits	292,302	293,220	298,795	303,276	299,718	2.54
Semitropical	82,235	82,150	88,750	89,450	95,057	15.59
Deciduous	375,551	383,280	358,373	351,099	315,441	−16.01
Tree nuts	414,203	401,990	433,557	474,509	632,691	52.75
Grapes	495,590	475,050	500,473	547,927	481,000	−2.94
Subtotal	1,659,881	1,635,690	1,679,948	1,766,261	1,823,907	9.88
Total	8,400,504	8,384,390	8,492,148	8,434,561	9,089,477	8.20

Source. General Report by the University of California Food Task Force, 1974.

[a] Includes bearing plus nonbearing acreages for tree fruits, nuts, and grapes.

[b] Includes acreage for minor and exotic crops.

15. The following table shows data of identified energy saving potential of eight industrial plants.

Plant type	Total annual energy bill	Identified savings	Percentage
Basic chemicals	5.5	2.39	43.4
Textiles	0.9	0.29	32.0
Agricultural chemicals	1.7	0.28	16.7
Oil refinery	10.3	1.12	10.8
Chemical intermediates	13.2	1.87	14.2
Food processing	1.1	0.33	30.1
Pump and paper	5.3	1.70	31.5
Rubber and tires	2.9	0.47	16.4
Average	5.1	1.05	20.6

Source. Courtesy of the Laurence Berkeley Laboratory, University of California, Berkeley, Report LBL-3299, p. 56.

16. Plot the following data; then draw the "best" curve to represent the data.

E, m	0	300	915	1580	2160	3530
P_1 mm of mercury	750	725	675	625	575	500

What is the value of P when $E = 600$ m?

17. Plot the following data; then draw the "best" curve to represent the data. What is the value of t when $r = 50$ mm?

r, mm	23.1	30.5	35.8	42.6	48.5	54.9	61.2
t, degrees Celsius (C°)	222.2	188.3	167.8	150.0	133.3	118.3	99.4

18. Plot the data shown in the following table and draw the "best" curve to represent the data.

h, differential head, mm of water	7500	6000	3100	2800	2080	1200	900
Q, flow rate, 3 mm/s	6.2×10^6	5.5×10^6	4.2×10^6	3.7×10^6	3.2×10^6	2.7×10^6	2.2×10^6

20. Research and development (R&D) data for R&D in industry are shown in the following table.

Type of industry	R&D/Sales ratios			R&D, $ million	
	1967	1971	1985	1974	1985
Manufacturing (all other)	0.8	0.7	0.75	2,091	3,530
Electrical equipment	3.5	3.6	3.6	2,503	3,140
Machinery	3.2	3.2	3.3	1,836	2,660
Motor vehicles	2.5	2.6	3.0	1,814	2,420
Industrial chemicals	3.9	3.3	3.5	965	1.900
Optical and other instruments	3.9	4.9	5.1	622	920
Drugs and medicines	7.9	7.4	7.3	600	900
Aircraft and missiles	4.0	3.3	3.5	1,008	830
Other chemicals	2.1	1.8	1.8	285	610
Nonmanufacturing	—	—	—	271	510
Scientific instruments	2.5	2.6	2.5	87	170
All-industry total	2.1	2.1	2.0	12,082	17,590

Source. 1985 R&D Funding Projections, NSF 76-314.

19. Research and Development (R&D) accounts for the employment of about one-third of U.S. engineers and scientists. The following data show R&D funds and classification for 1974 and projected in 1985.

Classification	R&D funds, $ million	
	1974	1985
Fund sources		
Federal government	14,541	19,290
Industry	12,266	17,830
Universities and colleges	579	640
Other nonprofit institutions	429	480
R&D performer		
Industry	19,250	27,330
Federal government	4,144	5,400
Universities and colleges	2,600	3,330
Other nonprofit institutions	1,077	1,215
Federally funded R&D centers	744	965
Total	27,815	38,240

Source. 1985 R&D Funding Projections, NSF 76-314.

Chapter 11 Graphical Solutions and Computations

INTRODUCTION

In many design projects it is often necessary to do some research in order to achieve answers to a number of subproblems before progress can be made in the overall design.

An example dealing with the design of artificial legs was cited in Chapter 1. In this design project, for instance, a number of experiments were conducted to determine the relationship between power supply by muscles and energy level of the shank during level walking, as a function of time (see Figure 11.1). The solid curve shows the power supplied by the muscles as a function of time. This curve

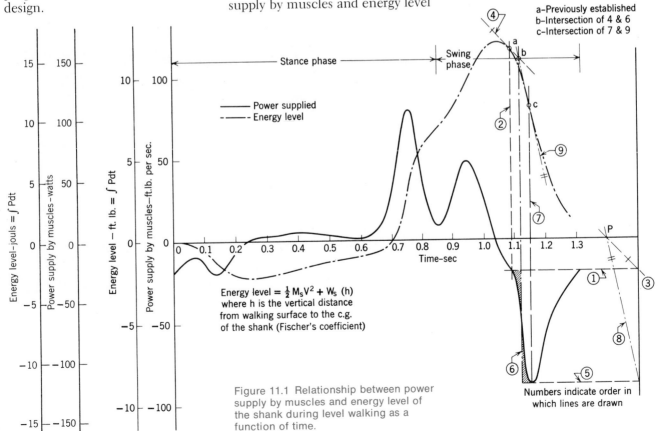

Figure 11.1 Relationship between power supply by muscles and energy level of the shank during level walking as a function of time.

was *integrated graphically* to establish the dashed-line curve, which shows the energy level as a function of time.

In another investigation a study of the variation of ankle angle and angular velocity as a function of time was undertaken. A plot of the data relating ankle angle as a function of time is shown in Figure 11.2. The curve of the data points was *differentiated graphically* to obtain the velocity-time curve.

Studies of gait (in human locomotion), both in normal subjects and in amputees, were conducted in order to understand their walking patterns. These studies were helpful in determiming normal or average gaits and deviations from the normal.

Displacements in the vertical plane of progression of salient points of the leg were determined by an "interrupted light" technique. Small light bulbs were placed at selected positions of the subject's hip, leg, and shoe. The subject walked in front of the open lens of a camera, whose field was interrupted by a slotted rotating disc, so that the displacement pattern seemed to consist of small lights moving along the paths of motion. Another camera was used to obtain an exposure of the subject at midfield for purposes of identification. Figure 11.3 shows the results of this technique, with the points joined to delineate the pattern of walking. Differentiation of the various curves established velocity curves, which were differentiated to establish the acceleration curves.

Other problems relating to this project and a variety of additional

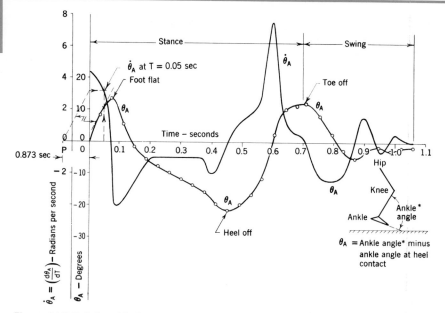

Figure 11.2 Relationship between ankle angle and angular velocity as a function of time.

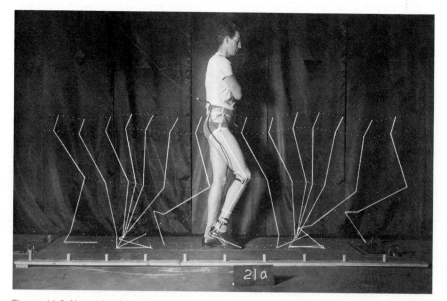

Figure 11.3 Normal subject, level walking. Locomotion study, using the interrupted light technique.

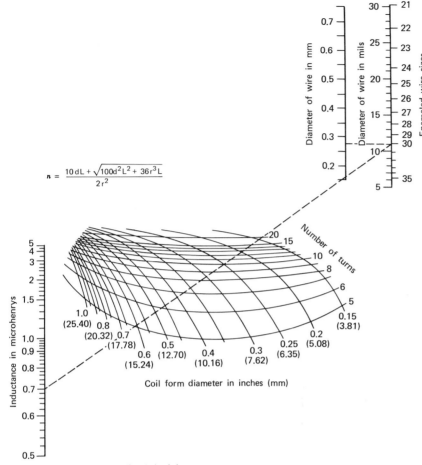

$$n = \frac{10\,dL + \sqrt{100d^2L^2 + 36\,r^3L}}{2\,r^2}$$

Figure 11.4 Nomogram for television
intermediate-frequency coil design
(Courtesy J. H. Felker; *Electronics*)

of empirical equations is limited to three forms that occur quite frequently in engineering practice.

This chapter also includes an introduction to *nomography*, which is the graphical representation and solution of mathematical expressions that contain three or more variables. Studies of the interrelationships among the variables are greatly enhanced by the use of properly designed nomograms. *This is quite important in evaluating trade-offs* among the variables, thereby enabling the design engineer to choose combinations of parameter values that will best satisfy the specified constraints of the problem. A nomogram for this purpose is much more effective than a computer printout. For example, consider the nomogram in Figure 11.4.

Various nomograms, charts, and calculators are available for calculating the inductance of coils; however, most of them do not cover the range of values that are of interest to the designer. The one shown in Figure 11.4 does fulfill this need. It gives in one operation the number of closely wound turns required to obtain a desired inductance. Suppose we wish to determine how many turns of #30 wire are required on a 0.25-in. (6.35-mm) diameter coil form to obtain 0.7 μH? Place a straightedge between 0.7 on the left-hand vertical scale and 30 on the right side of the right-hand vertical scale, as shown by the dashed line. Now trace upward along the 0.25 curve to the dashed line and read 10 turns as the value of the other curve that passes through this intersection.

If the designer cannot use the

examples are included in this chapter to show the advantages of using graphical calculus to solve problems that arise in engineering design.

Also, in dealing with physical data that result from tests to determine the relationship, for example, between two related variables, the designer can gain an invaluable understanding of the test data by determining the *empirical* equation that best represents the data. Laboratory tests are often conducted to observe the behavior of related variables. For example, in

the area of materials, tests to determine the relation between tensile loads applied to a newly composed metal bar and the resulting corresponding elongations of the bar are necessary to apprise designers of the elastic and plastic properties of the material. The data gathered from such tests, from prototype tests (i.e., wing deflection tests of the Boeing 707), from vehicle road tests, etc., are known as empirical data. A "best" curve drawn through the plotted data points usually reveals the form of the empirical equation. In this chapter our study

0.25-in. (6.35-mm) coil form but can use a 0.20-in. (5.08-mm) size, the number of turns (13) required can be quickly determined. The nomogram enables the designer to make trade-offs that are consistent with the constraints that may be imposed.

We now discuss *graphical calculus* and its application to a variety of problems that arise in engineering design.

GRAPHICAL CALCULUS

There are many technical problems in engineering design that require the determination of areas, volumes, accelerations, velocities, displacements, etc. We can solve such problems graphically, numerically, mechanically, mathematically, or by a combination of these methods. The choice of method depends on the manner in which the problem can be best defined. If, for example, it is most convenient to define the problem by an algebraic expression, we can solve the problem mathematically and, if it is desirable, a computer program can be prepared to obtain a printout of the results. When the problem, however, is defined by experimental data or by field data (i.e., ground elevations from which a contour map can be generated), it will most likely be more economical to solve the problem graphically, or by a combination of graphical and mechanical methods. We first consider graphical integration.

Graphical Integration

In a physical sense integration is a summation process of many small

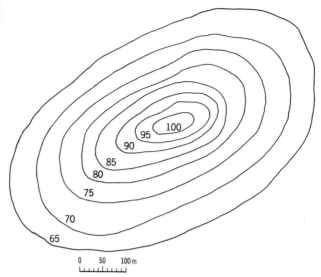

Figure 11.5 Topographic map of proposed building site.

quantities. In the algebraic form of the calculus the quantities are regarded as infinitely small, whereas in practical, technical applications that *cannot be expressed conveniently in algebraic form*, the quantities are taken sufficiently large to effect a good, workable solution.

It would not be economically feasible, for example, to find the volume of an irregularly shaped hill by the formal calculus methods of integration, because a *precise* mathematical expression to define the hill surface would be quite difficult to determine, if it is at all feasible. Problems of this kind are best solved by graphical, mechanical, or numerical methods of integra-

tion or by a combination of these methods.

EXAMPLE

A building site for a proposed manufacturing plant requires the leveling of the hill shown topographically in Figure 11.5. It has been decided that the hill should be cut to contour line 70 m. As part of the cost estimate, it is necessary to determine the number of cubic meters of earth to be removed. The following method is used.

1. The area bounded by each contour line is determined mechanically by using a *planimeter*. The values thus obtained are shown in the following table.

Contour, m	70	75	80	85	90	95	100
Area, m²	164,500	95,000	59,000	36,200	18,800	9,400	2,000

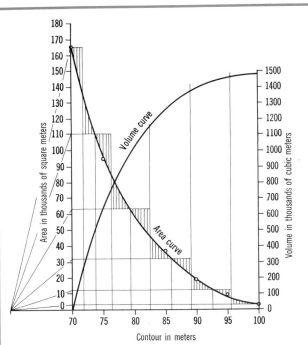

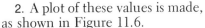

Figure 11.6 Graphical integration to determine volume of earth to be removed for building site.

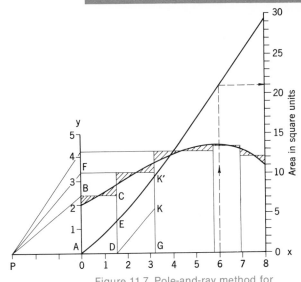

Figure 11.7 Pole-and-ray method for graphical integration.

2. A plot of these values is made, as shown in Figure 11.6.

3. The volume, in cubic meters, is obtained by integrating the area curve. This is done graphically using the "pole-and-ray" method, which will be discussed shortly. Observe that the volume scale is shown in cubic meters. The amount of earth to be removed is 1,480,000 m³.

Question. How would you use Fig. 11.6 to determine the number of cubic meters to be removed if it had been decided to level the hill to contour line 80 m?

The Pole-and-Ray Method

This method of graphical integration is based on the geometric relationship "corresponding sides of similar triangles are in the same ratio."

Consider Figure 11.7. First, we divide the area under the curve into strips whose widths vary in accordance with the sharpness or flatness of the curve (i.e., wider strips for the flatter portions of the curve, and narrower strips for the sharper portions).

The area of the first strip is closely approximated by that of the rectangle *ABCD,* where line *BC* is placed so that the shaded areas are in balance. This is done quite accurately by eye.

Second, lay off a convenient "pole distance," *AP,* measured in terms of the unit on the horizontal scale. In Figure 11.7 we show that $AP = 3$. A larger value would *decrease* the length of the area scale shown on the right, and a smaller value would *increase* the length of the area scale.

Through point *P,* rays such as *PB* are drawn. Through point *A,* line *AE* is drawn parallel to ray *PB.* The area of the first strip is equal to $\overline{DE} \times \overline{AP}$. This is easily shown in the following manner.

1. Triangles *BAP* and *EDA* are similar by construction.

2. Therefore, $ED/DA = BA/AP$ (corresponding sides of similar triangles are in the same ratio).

3. $ED \times AP = DA \times BA$.

4. $DA \times BA$ is the area of the rectangle *ABCD,* which is equivalent to the area of the first strip.

5. Therefore, $ED \times AP$ is equal to the area of the first strip.

The area of the second strip is found in a similar manner. Line *DK* is drawn parallel to ray *PF.* The area of this strip is equal to $\overline{KG} \times \overline{AP}$. Now, if a line is drawn through point *E* parallel to *DK,* distance *K'G* equals $\overline{DE} + \overline{KG}$ and, therefore, the sum of the areas of the first two strips is equal to $\overline{K'G} \times$

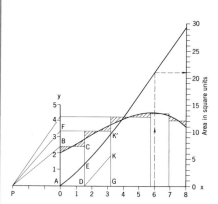

Figure 11.7 (*repeated*).

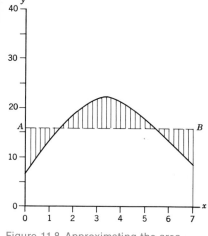

Figure 11.8 Approximating the area
under the curve.

corresponding y-scale values multi-plied by the pole distance. In Figure 11.7 the pole distance is 3; therefore, the location of a gradua-tion, say, 15, on the area scale is on the horizontal passing through $y = 5$. Once a convenient graduation, such as the one marked 15, has been located, other graduations are readily established without further calculation, since the distance from the 0 to the 15 is known.

Choice of Pole Distance

A judicious choice of the pole dis-tance can be made by first esti-mating the area under the curve and then selecting a desired length for the area scale. *The pole dis-tance, then, is very close to the esti-mated area divided by the desired length of the area scale.*

Suppose, for example, that the given curve is that shown in Figure 11.8. The horizontal dashed line has been located (by eye) so that the shaded areas below the line AB balance (approximately) the area above the line. The rectangular area ($112 \pm$ square units) is approx-imately equal to the area under the

$\overline{AP}$. It is evident that it is not neces-sary to draw DK, since EK' can be drawn directly. When this process is repeated for the other strips and a smooth curve is drawn through points A, E, K', etc., we will have established the *integral curve*— the summation curve—which will enable us to determine the total area under the given curve or the area of any portion of the total area. It is readily seen that the area under the given curve between $x = 0$ and $x = 6$ is found by following the vertical dashed line drawn through $x = 6$ to the *integral curve* and then the horizontal dashed line to the area scale shown on the right. *The values of the graduations on the area scale are equal to the*

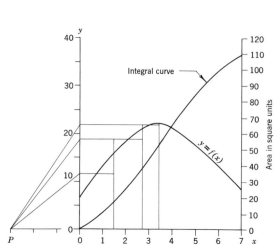

Figure 11.9 Graphical calculus; integra-tion, pole-and-ray method.

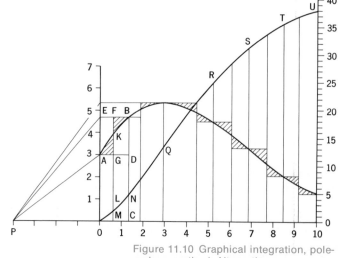

Figure 11.10 Graphical integration, pole-and-ray method. Alternative
construction.

curve. If we assume that the area-scale length corresponds to the length from $y = 0$ to $y = 40$, the approximate value of the pole distance is $(112 \pm /40) \cong 3$. Figure 11.9 shows the use of the pole-and-ray method to determine the integral curve when the pole distance is 3.

We should always remember that the pole distance is laid off in terms of the units on the x-axis.

The pole-and-ray method as used in Figures 11.7 and 11.9 establishes chords of the integral curve. Consequently, the path of the integral curve may vary some. To improve the accuracy of the integral curve, an alternative construction is used.

Pole-and-Ray Method, Alternative Construction

Let us study strip *OABCO* in Figure 11.10. Through points *A* and *B* horizontals *AD* and *BE* are drawn. Vertical line *FG* is drawn so that areas *KGA* and *KFB* are approximately equal. Line *OL* is drawn parallel to ray *PA*. We know from our previous discussion that area $OAGMO = \overline{LM} \times \overline{OP}$. Now line *LN* is drawn parallel to ray *PE*. It follows that $NC \times OP = $ area *OAKBCO*. The integral curve passes through points *O* and *N* tangentially to *OL* and *NL*. Additional points and tangents are similarly established. The integral curve passes through points *O*, *N*, *Q*, *R*, *S*, *T*, and *U* tangentially to the line segments adjacent to these points. *Q* is the inflection point. An enlargement of a portion of Figure 11.10 is shown in Figure 11.11, which clearly shows the integral curve passing through points *O* and *N* tangentially to line segments *OL* and *LN*.

Several examples employing the pole-and-ray method of integration follow.

EXAMPLE 1

Part of a building project required the subdivision of the plot of ground, shown in Figure 11.12, into three lots of equal area by property lines perpendicular to the street line *OD*.

The pole-and-ray method afforded a simple graphical solution of this problem.

SOLUTION

The integral curve is established in the same way as shown in Figure 11.10. Distance *DG* is a measure of the total area of the plot of ground. We divide line *DG* into three equal segments *DE*, *EF*, and *FG*, since we wish to subdivide the plot into three lots of equal area. The horizontal lines drawn through points *E* and *F* intersect the *integral* curve at points 1 and 2, respectively. The verticals *KL* and *MN* drawn through points 1 and 2, respectively, locate the required boundary lines.

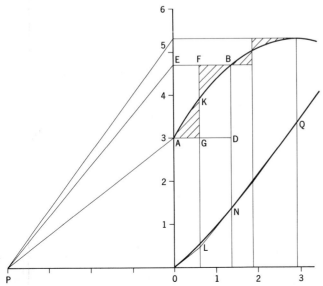

Figure 11.11 Enlargement of a portion of Figure 11.10.

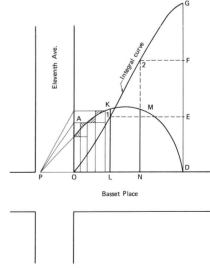

Figure 11.12 Plot subdivision, employing the pole-and-ray method of graphical integration.

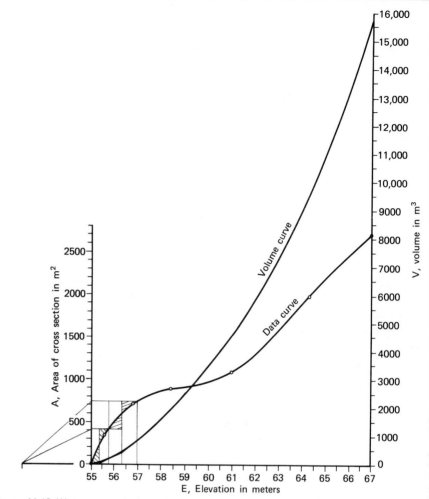

Figure 11.13 Water reservoir capacity determination by graphical integration.

EXAMPLE 2

One of the subproblems in a water project required the determination of the capacity, in cubic meters, of a reservoir for which the following data were available.

E, elevation, m	55.0	55.6	56.8	58.4	61.0	64.3	67.0
A, area of section, m²	0	344.4	722.2	888.9	1088.9	1988.9	2711.1

A plot of the data is shown in Figure 11.13. The total volume is obtained by graphically integrating the area under the curve. Again, the pole-and-ray method is used (employing the alternative construction).

The volume scale shows that the reservoir, when full, contains 15,750 m³ of water.

EXAMPLE 3

Test data of the relationship between velocity, V, in meters per second and time, t, in seconds are shown in the table.

t, s	0	0.1	0.2	0.3	0.5	0.55	0.7	0.85	1.0
V, velocity, m/s	0	3.0	10.0	16.5	23.5	24.5	22.5	18.3	14.1

A plot of these data is shown in Figure 11.14.

It is required to determine the curve that will show the relationship between displacement and time. This is easily accomplished by integrating the velocity curve. The pole-and-ray method has been used to do this. *It should be carefully observed that the pole distance is 0.3 units, not 3.*

At $t = 0.36$ s, for example, the displacement is 3 m. (Follow the vertical dashed line from $t = 0.36$ to the displacement curve and then horizontally to the displacement scale.)

EXAMPLE 4

For the design of beams it is necessary to know shear, bending moment, and deflection values. The shear curve discloses locations in the beam where shear forces (forces that tend to cut the beam) are critical. The moment curve shows the bending moment values at any point of the beam, and the deflection curve shows values of deflection at any point of the beam. In your study of mechanics of materials (usually in your junior year) you will learn beam theory and much more (torsion, thin shells, design criteria, etc.).

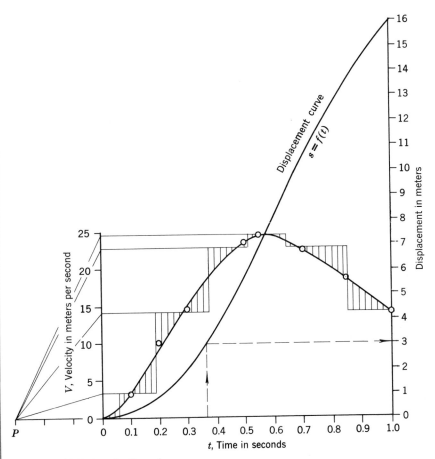

Figure 11.14 Determination of displacement-time curve from a velocity-time curve by graphical integration (alternative construction).

For our purposes (an application of graphical integration), consider the simple beam that supports the concentrated loads 5 kN and 8 kN, as shown in Figure 11.15a. The support reaction R_L = 6.55 kN and R_R = 6.45 kN can be *determined graphically* (see Chapter 8), or by taking moments about either R_L or R_R; that is, $20 \times R_R$ = 5(5 kN) + 13(8 kN), from which R_R = 6.45 kN.

The shear curve, shown in Figure 11.15b, is determined by integrating the loads. In symbolic form

$$V(\text{shear}) = \int l(\text{loads})dx.$$ In this

case, since the loads are concentrated, we can easily establish the shear curve by first plotting the magnitude of R_L = 6.55 kN (segment OA) *upwardly*, as shown in the figure. As we move to the right of R_L, we see that there is no change until we reach the downward load, 5 kN. This is represented by segment AB. At point B, we subtract the load 5 kN. This is shown as segment BC. We proceed to the right to obtain segments CD, DE, EF, and FG. Now that we have completed the shear curve, we can proceed to determine the bending moment curve by integrating the shear curve. In symbolic form M (bending moment) = $\int V(\text{shear})$. The pole-and-ray method has been used to integrate the shear curve to obtain the bending

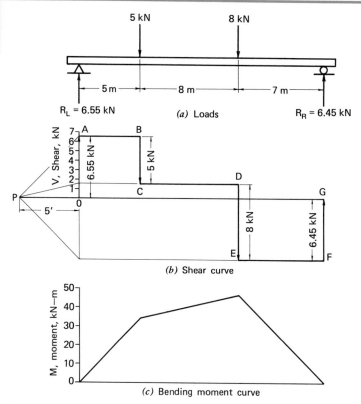

Figure 11.15 Beam problem; graphical integration.

moment curve shown in Figure 11.15c.

EXAMPLE 5
The centerline section of an arched dam is shown in Figure 11.16. Part of a model study of this dam, the Ross High Dam in the state of Washington, was concerned with a test to determine the effect of adding 1 ft of concrete, at various elevations, on the strains recorded by a particular strain gage. Lead weights were placed on a number of horizontal sections of the physical model (see Figure 11.17). By using similitude factors, the applied loading was related to the effect of 1 ft (304.8 mm) of concrete, at the various sections, on the stresses computed from strains recorded by the gage, which was placed at elevation 1200 ft (365.76 m). The vertical stress increments in pounds per square inch per foot of concrete and the corresponding elevations were plotted from the following tabular values.

Load elevation, ft[a]	1736	1650	1570	1500	1495	1475	1470	1400	1300	1200
Vertical stress increment, psi per foot of concrete	−0.445	−0.663	−0.808	−0.811	−0.721	−0.746	−0.668	−0.738	−0.738	−0.738

[a] Corresponding values in meters are shown in Figure 11.16.

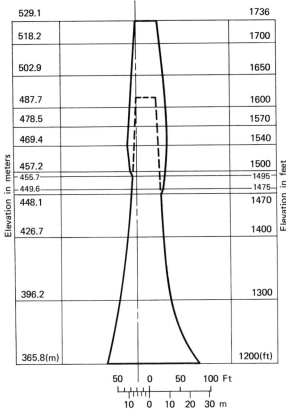

Figure 11.16 Cantilever section of dam.

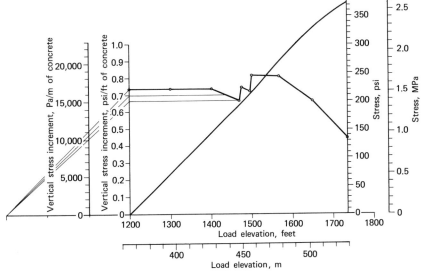

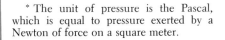

Figure 11.17 Ross High Dam model.

The plotted curve (Figure 11.18) was integrated graphically using the pole-and-ray method to obtain the vertical stress [374 psi or 2.58 MPa (thousands of Pascals)]* on the upstream face of the dam, at elevation 1200 ft (365.76 m).

Other Methods of Integration

There are several methods of integration that are quite good, especially when the integral curve is not needed. Among them are (1) the method of parabolas, (2) a numerical method, using Simpson's rule, and (3) a mechanical method, using a planimeter. We now consider each of these methods.

* The unit of pressure is the Pascal, which is equal to pressure exerted by a Newton of force on a square meter.

Figure 11.18 Determination of the vertical stress on upstream force of the dam at elevation 1200 (365.8 m) by graphical integration.

Method of Parabolas

This method is quite easy and provides very close approximations of the actual areas. Find the area $OABCDO$ in Figure 11.19. First, consider Figure 11.19a. We will assume that curve $A'B'C'$ is a parabola and that $\overline{B'E'}$ is equal to $\overline{BE}$. Now area $A'B'C' = hd$, where $h = (\frac{2}{3})B'E'$. If all the vertical lines, such as $B'E'$, of area $A'B'C'$ are approximately equal to corresponding lines, such as BE, of area ABC, then the two areas are approximately equal to each other.

Therefore, area $OABCDO$ is equal to Hd, where $H = EF + (\frac{2}{3})BE$. This is true, since the total area is made up of trapezoid $OACDO$ whose area is $EF \times d$, and area ABC, whose area is $(\frac{2}{3})BE \times d$.

Now let us apply this method to a larger area, such as that in Figure 11.20. Suppose we divide area $OABC$ into two strips of equal width and then determine H_1 and H_2 for the two strips in a manner similar to that used in Figure 11.19. The line that joins points a and b (the upper end points of H_1 and H_2, respectively) intersects line DE at point c.

The total area is equal to $H \times 2d$, where $H = cE$. Obviously, this method may be extended to include larger areas.

Numerical Method, Employing Simpson's Rule

Consider area $OABCDO$ in Figure 11.21. We divide the area into two strips of *equal* width. Curve ABC may be regarded as part of a parabola. Using the previous method, the area under the curve is

$$A = \frac{2}{3}\left(y_1 - \frac{y_0 + y_2}{2}\right)2d$$
$$+ \left(\frac{y_0 + y_2}{2}\right)2d$$
$$= \frac{d}{3}(y_0 + 4y_1 + y_2)$$

Now let us determine the area under the curve in Figure 11.22.

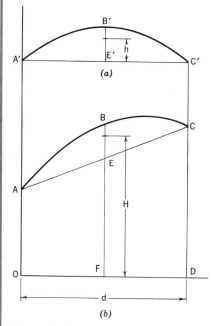

Figure 11.19 Area under a curve determined by the method of parabolas.

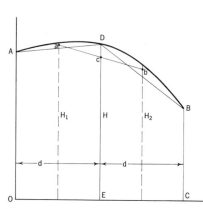

Figure 11.20 Method of parabolas applied to larger areas.

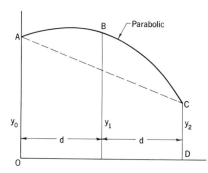

Figure 11.21 Area under a curve employing Simpson's rule.

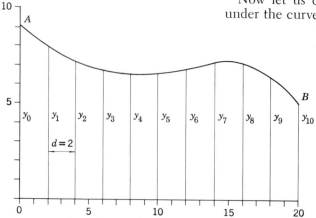

Figure 11.22 *Example:* Area under a data curve, using Simpson's rule.

First, we divide the area into an *even* number of strips—10 in this example.

Second, we measure the distances y_0, y_1, . . . y_{10}, which are recorded in the following table.

y_0	y_1	y_2	y_3	y_4	y_5	y_6	y_7	y_8	y_9	y_{10}
9.1	8.0	7.1	6.8	6.5	6.5	6.8	7.1	7.1	6.5	5.0

From our previous discussion we recall that the area of the first two strips can be determined from the expression $(d/3)(y_0 + 4y_1 + y_2)$.

Now the area of the third and fourth strips can be similarly determined from the expression $(d/3)(y_2 + 4y_3 + y_4)$. Therefore, the sum of the areas of the first four strips is equal to $(d/3)(y_0 + 4y_1 + 2y_2 + 4y_3 + y_4)$.

It should be clear that the single expression (Simpson's rule) $(d/3)(y_0 + 4y_1 + 2y_2 + 4y_3 + 2y_4 + 4y_5 + 2y_6 + 4y_7 + 2y_8 + 4y_9 + y_{10})$, sums the areas of all of the strips.

The total, therefore, is $\frac{2}{3}(9.1 + 32 + 14.2 + 27.2 + 13 + 26 + 13.6 + 28.4 + 14.2 + 26 + 5) = 139.1$ square units.

A hand calculator could be used to make the calculation.

Mechanical Method, Using a Planimeter

This very practical method uses an instrument known as a planimeter,* which is a mechanical integrator. A photograph of a planimeter set up for the determination of the area of a roadway cross section is shown in Figure 11.23. The tracer is started at point A (any other point could be chosen) and is moved to follow along the irregular curve to point B, then to point C along line BC, then along line CD and, finally, back to the starting point A along line DA. The number of square units in the cross-section area is read on the graduated wheel, whose axis is horizontal. When the areas of consecutive cross sections are determined, the number of cubic yards (or cubic meters) of earth required for "cuts" (earth removed) and for "fills" (earth added) can be computed, since the distance between the cross sections would be known (see Figure 11.24). *This information is necessary when considering costs* of highway construction.

Additional uses of the planimeter are discussed in the following examples.

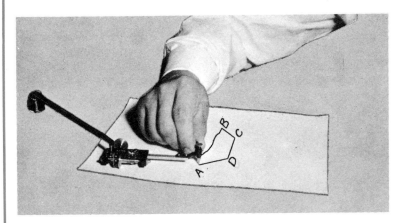

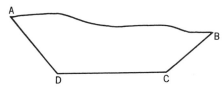

Figure 11.23 Photo showing use of a planimeter to obtain the area of an earth cross-section.

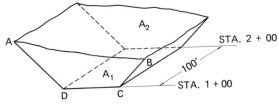

Figure 11.24 Pictorial to show earth sections A_1 and A_2 for end-area method in computing volume between stations $1 + 00$ and $2 + 00$.

* A mathematical development of the principle on which the design of a planimeter is based is usually available in calculus books.

EXAMPLE 1 LOCATING THE CENTROID OF AN AREA

We will consider the area bounded by the curve in Figure 11.25. It is required to determine the location of its centroid.

First we draw parallel lines m and n an arbitrary distance (H) apart. Now we introduce line BC parallel to m and then project points B and C to line n by perpendiculars to line n. The projected points are B_1 and C_1, respectively. Lines PB_1 and PC_1 are drawn to intersect line BC at points B_2 and C_2. Note that point P is any point on line m.

From similar triangles PB_1C_1 and PB_2C_2 it follows that

$$\frac{B_2C_2}{B_1C_1} = \frac{y}{H} \qquad \text{or} \qquad \frac{dA_1}{dA} = \frac{y}{H}$$

if we let $dA_1 = B_2C_2 \times dy$, and $dA = BC \times dy$. Therefore,

$$A_1 = \int dA_1 = \frac{1}{H} \int y \, dA$$

$$\text{and} \qquad \bar{y} = \frac{A_1H}{A}$$

$$\text{since} \qquad \bar{y} = \frac{\int y \, dA}{A}$$

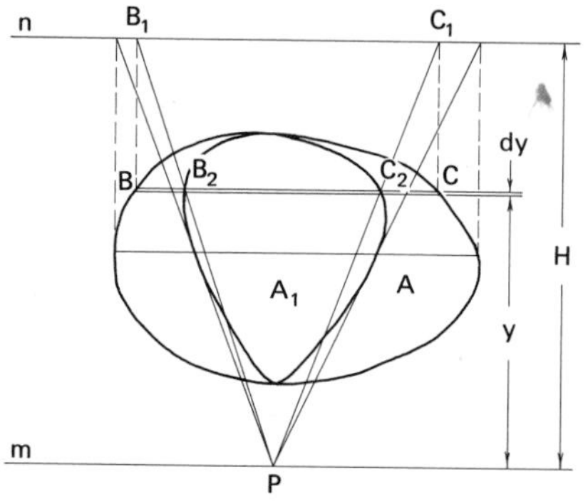

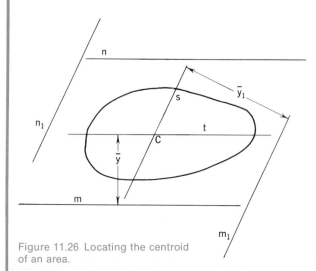

Figure 11.25 Graphomechanical method to locate the centroid of an area.

When the construction shown in Figure 11.25 is repeated for additional parallels to line m, the area, A_1, within the curve that passes through such points as B_2 and C_2 can be determined quite easily by the use of a planimeter. Of course, this is also true for area A. Now that areas A_1 and A and distance H are known, y is calculated from $\bar{y} = A_1H/A$.

To locate the centroid, we can determine the value of y_1 (see Figure 11.26) by repeating the construction in Figure 11.25. It should be clear that the parallels m_1 and n_1 were oriented arbitrarily. The intersection point, C, of lines t and s is the centroid of the area.

EXAMPLE 2 MOMENT OF INERTIA OF AREAS

The moment of inertia of area A about the axis, m, as shown in Figure 11.27, is the summation of all of the products formed by multiplying every elementary area by the square of its distance from the axis. We will find that by adding one step to the previous construction (Figure 11.25) we can find the moment of inertia.

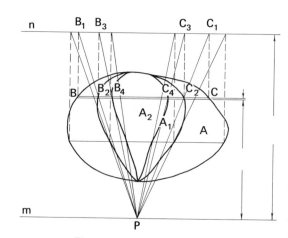

Figure 11.27 Moment of inertia of an area about axis m. Graphomechanical method.

Figure 11.26 Locating the centroid of an area.

First, we will repeat the construction shown in Figure 11.25 to locate points such as B_2 and C_2. Now we project these points to line n to locate points B_3 and C_3, respectively. Lines PB_3 and PC_3 intersect line BC at points B_4 and C_4, which are on the curve that encloses area A_2.

Again, from similar triangles PB_3C_3 and PB_4C_4, it follows that

$$\frac{B_4C_4}{B_3C_3} = \frac{y}{H} \quad \text{or} \quad \frac{dA_2}{dA_1} = \frac{y}{H}$$

where $dA_2 = B_4C_4 \times dy$ and $dA_1 = B_3C_3 \times dy$.

Therefore, the moment of inertia, I, about axis m is A_2H^2, since

$$I = \int y^2 \, dA$$

and

$$A_2 = (1/H^2) \int y^2 \, dA$$

Area A_2 is easily determined by the use of a planimeter.

EXAMPLE 3 CENTROID OF A SYMMETRICAL SECTION

The cross section of a structural steel member is shown in Figure 11.28. It is required to determine the location of its centroid, C. Since the cross section is symmetrical about the vertical axis, PK, of the web, it is only necessary to determine the value of $\bar{y}$, which is A_1H/A. The construction used in the previous examples defines the shaded area, A_1, which can be determined by using a planimeter.

Lines 1-2 and 3-4 are not actually straight lines, but are so drawn because they do not appreciably deviate from the correct curve. The enlargement in Figure 11.29 shows the slight difference between line 1-2 and correct curve 1-2.

To find the moment of inertia about line m, we can regard area A_1 as the given section and then repeat the construction to determine area A_2. Then the moment of inertia about axis m would be A_2H^2.

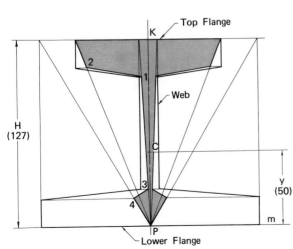

Figure 11.28 Location of the centroid of a structural steel section.

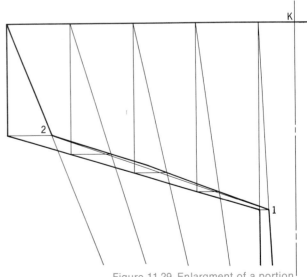

Figure 11.29 Enlargment of a portion of Figure 11.28.

GRAPHICAL DIFFERENTIATION

We may, in a physical sense, regard differentiation as meaning *the determination of the rate of change of one variable with respect to a related variable.*

Consider this statement carefully. Assume that the velocity, V, of a body as measured at various distances, S, from a fixed location is as shown in the following table.

S, m	0	0.10	0.20	0.30	0.40	0.50	0.60	0.70	0.80	0.90	1.00
V, m/s	0	0.02	0.03	0.12	0.32	0.43	0.57	0.64	0.77	0.95	1.10

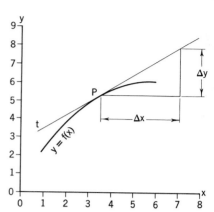

Figure 11.30 Slope of a tangent to a curve.

The values of the average rate of change of velocity with respect to distance for the intervals shown in the table can be calculated. For example, the change in velocity corresponding to the displacement from 0 to 0.10 is shown as:

$$\frac{\text{Change in V} (\Delta_1 V)}{\text{Change in S} (\Delta_1 S)} = \frac{0.02 - 0}{0.10 - 0}$$
$$= 0.20 \, (\text{m/s})/\text{m}$$

This value is the *average* rate of change of velocity with respect to the corresponding distance in the first interval.

In the second interval it is

$$\frac{\Delta_2 V}{\Delta_2 S} = \frac{0.03 - 0.02}{0.20 - 0.10} = \frac{0.01}{0.10}$$
$$= 0.10 \, (\text{m/s})/\text{m}$$

We can compute the average values for the other intervals similarly.

If the distance interval, ΔS, had been made smaller, we still would have obtained an *average* value for each interval. If, however, ΔS were made infinitely small (ΔS approaching zero as a limit), the *average* values would become *exact* values at the respective values of S. The *exact* values are denoted by the expression dV/dS, where this ratio is the rate of change of V with respect to S. In the algebraic form of the calculus the relation between the variables is usually ex-pressed algebraically (i.e., $V = S^2$) and then differentiated. (That is, $dV/dS = 2S$ in this case. The rate of change of V with respect to S can be obtained for any value of S from this equation.)

If the tabular data had been plotted and a curve drawn through the data points, the *average* rate of change, $\Delta V/\Delta S$, would be determined from the slopes of chords (formed by joining consecutive points) of the curve. Then, as ΔS approached zero, each chord would approach a tangent position to the curve.

Graphically, differentiation is the determination of the slope of a tangent to a curve. If, for example, a curve is defined by the equation $y = f(x)$, then dy/dx is the slope of the tangent to the curve at any point of the curve.

Observe, in Figure 11.30, that at point P of the curve, dy/dx is equal to the slope, $\Delta y/\Delta x$, of tangent, t, at point P.

Note that Δy must be measured in terms of the scale on the y-axis, and Δx must be measured in terms of the scale on the x-axis.

The location of a tangent to a curve can be closely approximated by drawing two parallel chords and then locating the point at which the line joining the midpoints of the chords intersects the curve. The tangent to the curve at that

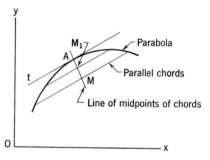

(a) Geometric method

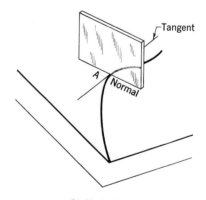

(b) Mechanical method

Figure 11.31 Location of a tangent to a curve.

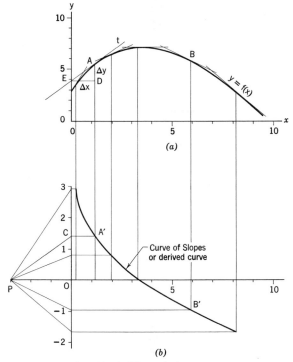

Figure 11.32 Graphical differentiation;
pole-and-ray method.

follows that $\Delta y/\Delta x = AD/DE = CO/OP$.

Moreover, if distance CO, which is a measure of the ordinate of point A', is to represent the correct magnitude of the slope of tangent t, the y-axis scale must be determined accordingly. Since the x-axis graduations are the same for both figures (usual and convenient practice), it should be fairly evident that the values on the y-axis of Figure 11.32b are equal to the values on the y-axis of Figure 11.32a *divided by the pole distance* (measured in terms of the horizontal scale). In this case the location of point 1 on the *vertical axis of the derived curve* is the same distance from the origin as point 3 of the given curve, because the pole distance is 3.

The curve of slopes, or the derived curve, passes through points such as A'.

The construction shown in Figure 11.32b locates additional points through which the derived curve is drawn.

Several examples showing the application of graphical differentiation to a variety of problems follows.

EXAMPLE 1 DETERMINING ACCELERATION FROM VELOCITY DATA

The relation between velocity, V, in kilometers per hour and time, t, in seconds is shown in the following table.

point is parallel to the chords, as shown in Figure 11.31.

If necessary, a more precise method can be used to locate the tangent. It consists of placing a polished surface or a mirror perpendicular to the paper and across the given curve. If the reflected portion of the curve shows no break in the curve (i.e., if it appears to be a smooth continuation of the portion of the curve in front of the polished surface), the line of intersection of the planes of the paper and of the polished surface is the normal (perpendicular) to the curve at the point of intersection. The required tangent is perpendicular to this normal (see Figure 11.31b).

Next we employ a simple graphical method, the pole-and-ray method, to differentiate a given curve.

Pole-and-Ray Method

Assume that the given curve is that shown in Figure 11.32a. A convenient pole distance $OP = 3$ units of x has been chosen, as shown in Figure 11.32b. Ray PC is parallel to the tangent, t, which passes through point A of the curve, $y = f(x)$. The vertical line drawn through point A intersects the horizontal line drawn through point C at the point labeled A', which is a point on the "curve of slopes" or the "derived curve." The ordinate value of A' is a measure of the magnitude of the slope of line t. Why is this true? First, we observe that $AD/\overline{DE} = \Delta y/\Delta x$ is a measure of the slope of line t. Since triangles ADE and COP are similar, it

t, s	0	5	10	15	20	25	30
V, km/hr	0	13.8	26.5	36.8	42.3	45.3	47.2

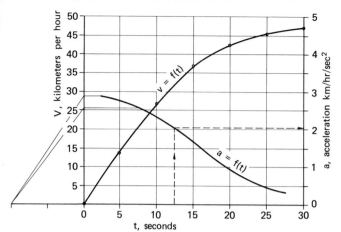

Figure 11.33 Graphical differentiation to obtain $a = f(t)$ from $v = f(t)$. *Example:* When $t = 12.5$ s, $a = 2.05$ km/hr/s².

How shall we determine the curve that will show the acceleration, a, as a function of time, t?

Since acceleration is the rate of change of velocity with respect to time, we should recognize that the acceleration-time curve can be obtained by differentiating the velocity-time curve. A plot of the given data is shown in Figure 11.33. The curve $V = f(t)$ through the data points has been differentiated, using the pole-and-ray method, to obtain the acceleration-time curve, $a = f(t)$. Note that the graduations on the acceleration scale are equal to the graduations on the velocity scale *divided* by the pole distance, which is equal to 10 units of the t scale.

EXAMPLE 2 RATE OF HEATING WATER DETERMINATION

Let us consider the relation of temperature, C, in degrees Celsius of heating water to time, t, in minutes, as shown in the following table.

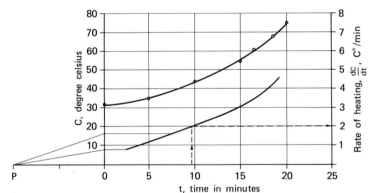

Figure 11.34 Determination of the rate of heating water as a function of time by the graphical differentiation method. *Example:* $t = 9.3$ min, then the rate of heating is 2°C/min.

t, min	0	5	10	15	17	19	21
C, degrees Celsius (C°)	32.2	36.6	44.3	56.7	62.1	68.4	75.4

Our problem is to determine the rate of heating at any time, t, for the range shown in the table. Again, we should recognize that differentiation is involved.

A plot of the data and a "fair" curve drawn through, or quite near, the plotted data points are presented in Figure 11.34. This curve has been differentiated graphically, using the pole-and-ray method, to obtain the curve, which shows the rate of heating at any time, t. The scale for the rate-of-heating curve has been obtained by dividing the values on the C scale by the pole distance. The example in Figure 11.34 shows that when $t = 9.3$ min, the rate of heating, $dC/dt = 2°$/min.

EXAMPLE 3 CAM DISPLACEMENT, VELOCITY, AND ACCELERATION

A *cam* is a machine part, usually a plate or a cylinder, whose surface is shaped in a manner to impart a desired motion to another part or unit with which it is in contact. The latter part is known as the *follower* element. Several cams with various followers are shown in Figure 11.35.

The cam shown in Figure 11.35*a* is typical of plate cams with various types of reciprocating followers. As the shaft rotates at a desired uniform velocity the follower is displaced in a vertical direction; that is, it rises and falls in accordance with the contour of the plate. If the weight of the follower unit is not sufficient to maintain adequate contact with the cam surface, it may be necessary to use springs. Figure 11.35*a* to 11.35*e* shows a flat follower, a knife-edge offset follower, an oscillating arm roller follower, and a cylindrical cam with a cone follower. These are just a few examples of a great variety of cams that are used in modern design.

Geometric Shape of Cam

The geometric shape of a cam will depend on the desired displacement pattern of the follower. Suppose we wish to make a cam layout to satisfy the displacement data in the following table, assuming counterclockwise rotation with uniform velocity, a knife-edge follower, a base circle whose radius, 48 mm, is equal to the shortest radial distance from the center of rotation of the cam to its surface, and the axis of the follower passes through the center of rotation.

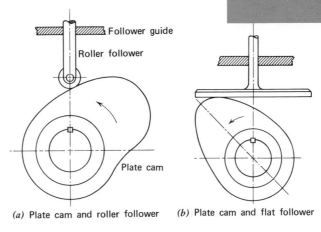

(a) Plate cam and roller follower (b) Plate cam and flat follower

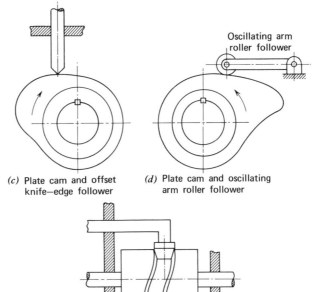

(c) Plate cam and offset knife—edge follower

(d) Plate cam and oscillating arm roller follower

(e) Cyllindrical cam and cone follower

Figure 11.35 Examples of cams and cam followers.

Follower, mm	0	6	15	24	33	42	48	42	33	24	15	6	0
Cam	0°	30°	60°	90°	120°	150°	180°	210°	240°	270°	300°	330°	360°
Point on cam	0	1′	2′	3′	4′	5′	6′	7′	8′	9′	10′	11′	0

Layout of Cam Shape

First let us establish the 48-mm base circle with center C (Figure 11.36) and then assume the vertical position of the follower as the point of zero displacement. Other points can be located in the following manner. First, we lay off on the follower axis distances 0-1, 0-2, 0-3, 0-4, 0-5, and 0-6, respectively, equal to 6, 15, 24, 33, 42, and 48 mm. Now we can assume that the cam is fixed and that the follower rotates *clockwise* to positions 1′, 2′, 3′, 4′, 5′, and 6′. It is quite evident, for example, that point 1′ is the intersection of the arc of radius C-1, and the radial line that makes an angle of 30° with the axis of the follower; that point 2′, similarly, is the intersection of the arc of radius C-2 and the adjacent 30° radial line, and so forth, for the successive points 3′, 4′, 5′, and 6′. The table of displacement data shows that when point 6′ has been located, the follower has reached its maximum upward displacement (frequently identified as the *rise* of the follower).

The remaining points 7′, 8′, 9′, 10′, and 11′ are located in the same manner as the others. Distances 0-7, 0-8, 0-9, 0-10, and 0-11 on the follower axis are, of course, respectively equal to 42, 33, 24, 15, and 6 mm. This portion of the cam from point 6′ to 7′ and, finally, back to the starting point, accommodates the downward displacement (or *fall*) of the follower.

The smooth curve drawn through points 0, 1′, 2′, . . . 11′, 0 is the cam shape, or profile.

The pattern of motion of the follower during the rise phase is the same as the motion during the fall

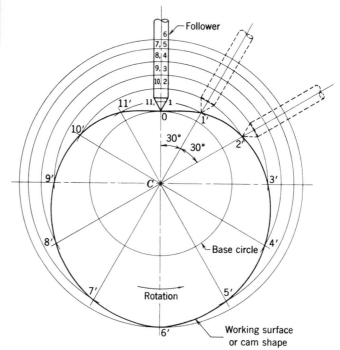

Figure 11.36 Cam shape layout.

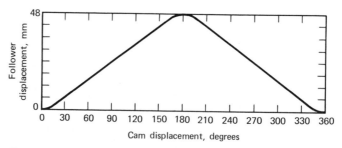

Figure 11.37 Graphical representation of cam displacement data.

phase. This is evident from the displacement data, whose graphical representation is shown in Figure 11.37.

Other types of follower motion —straight line, modified straight line, simple harmonic, and parabolic (constant acceleration and deceleration)—can be represented graphically. The latter two are described as follows.

Simple Harmonic Motion. If a point that moves with uniform speed on

the circumference of a circle is projected on a diameter, the projection is said to have simple harmonic motion.

Let us consider point P, which moves on the circumference of the circle (Figure 11.38) with uniform speed. For a general position of point P, its projection on the vertical diameter is in position P′, and similarly for other positions of point P, such as position 1. The displacement chart is easily established by simply locating points

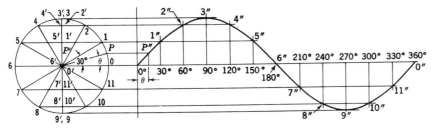

Figure 11.38 Simple harmonic motion.

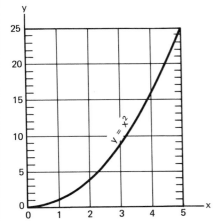

Figure 11.39 Parabolic motion.

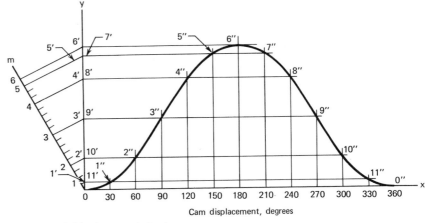

Figure 11.40 Displacement chart—
constant acceleration and deceleration.

such as 1″, 2″, etc., and then drawing a smooth curve through these points.

Parabolic Motion—Constant Acceleration or Deceleration. This motion is described by the equation $y = x^2$, where y represents the follower displacement and x the cam displacement. Figure 11.39 shows the parabola or constant acceleration curve for an arbitrary range of x from 0 to 5 and the corresponding displacements when $x = 1, 2, 3, 4$, and 5.

Now suppose that a cam follower is to move upward a distance of 2 units, with constantly accelerated and decelerated motion from 0 to 180° of cam rotation, and then downward the same distance, with the same motion from 180 to 360° of cam rotation. The displacement

chart can be constructed in the following manner. In Figure 11.40, line m is drawn at a convenient angle with the y-axis. Points 1, 2, and 3 are located on line m by laying off 1, 4, and 9 units, respectively, from the origin, 0. Since accelerated motion is desired for half of the upward displacement, point 3′ is at the midpoint of the 2-units vertical line. If a line is drawn between points 3 and 3′, parallels drawn through points 2 and 1 will intersect the vertical in corresponding points 2′ and 1′.

Points 1″, 2″, and 3″ are easily located by simply finding the intersection of the horizontals drawn through 1′, 2′, and 3′ with the corresponding verticals drawn through 30°, 60°, and 90°. In a similar manner we may locate points 4″, 5″,

and 6″ for the decelerated motion from 90 to 180°. The construction for the portion of the curve from 180 to 360° is fairly obvious.

We now proceed to lay out the cam shape for a roller follower to satisfy the following design requirements.

Base circle diameter = 72 mm.
Roller follower diameter = 18 mm.
No offset.
Rise of follower = 36 mm, with constantly accelerated and decelerated motion in first half of revolution of the cam.
Fall of follower = 36 mm, with simple harmonic motion in second half of revolution of the cam.
Cam rotation—counterclockwise.

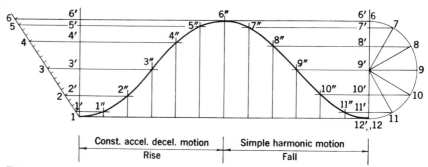

Figure 11.41 Displacement chart.

The displacement chart is shown in Figure 11.41. It is advantageous to orient this chart in a convenient position in relation to the center of the roller at its lowest point, as shown in Figure 11.42.

On the vertical line drawn through the center of the roller, point C (Figure 11.42), the center of rotation of the cam is located at a distance of 45 mm (the sum of the radii of the base circle and the roller) below the center of the roller. Points $1''$, $2''$, etc., are projected horizontally to the axis of the follower. Now an arc is drawn with radius C-1 and center C, intersecting radial line n and locating point 1^R, which is the center of the roller if we suppose that the cam remains stationary and the follower moves clockwise (the cam actually moves counterclockwise and the follower remains in a vertical position). In a similar manner, points 2^R, 3^R, etc., are located. The smooth curve passing through these points is the "theoretical curve," whereas the curve that is tangent to the roller positions is the "working surface," or the actual cam shape layout.

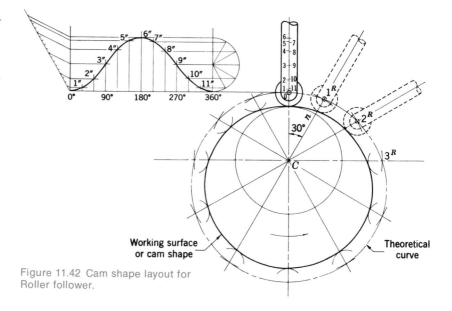

Figure 11.42 Cam shape layout for Roller follower.

Now consider the determination of velocity and acceleration from displacement data. We will assume that the cam-and-follower displacement data in the following table are given and that it is necessary to determine (1) the velocity-time relationship, and (2) the acceleration-time relationship. The rotational speed of the cam is 125 rpm.

S, follower displacement, m	0	0.002	0.01	0.02	0.03	0.038	0.04	0.04	0.04	0.04	0.03	0.01	0
θ, cam displacement, degrees	0	30	60	90	120	150	180	210	240	270	300	330	360

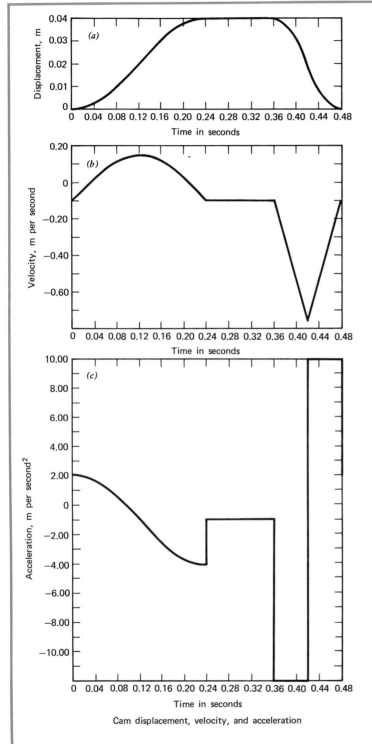

Figure 11.43 Velocity and acceleration
curves determined from displacement
curve by graphical differentiation.

A plot of the data (with θ converted to time in seconds) is shown in Figure 11.43*a*. Observe that since the rotational speed is 125 rpm, one revolution of the cam takes place in 0.48 s. The time scale, *t*, is based on this value.

The displacement-time curve is differentiated graphically to obtain the velocity-time curve shown in Figure 11.43*b*. The latter curve is then differentiated to obtain the acceleration-time curve shown in Figure 11.43*c*.

The pole-and-ray method was used to differentiate the curves.

EXAMPLE 4 HUMAN LOCOMOTION—PELVIS STUDY

In a recent study* of human locomotion, data were recorded on the anterior (forward) and posterior (backward) displacements of the pelvis during a walking cycle on a

* L. W. Lamoreux, "Kinematic Measurements in the Study of Human Locomotion," Biomechanics Laboratory, University of California, San Francisco-Berkeley. Published in the *Bulletin of Prosthetics Research*, BPR10-15, Spring 1971.

treadmill. A typical curve plot is shown in Figure 11.44a.

This curve was differentiated graphically, using the pole-and-ray method, to obtain the velocity-time curve shown in Figure 11.44b.

EXAMPLE 5 HUMAN LOCOMOTION—ANKLE FLEXION STUDY

Another study* conducted in the Biomechanics Laboratory dealt with ankle flexion in degrees as a function of time in seconds. A typical curve is shown in Figure 11.45a. The velocity-time curve was obtained by differentiating the former curve. The velocity-time curve is shown in Figure 11.45b. The pole-and-ray method is quite appropriate, economically, when the number of data curves is relatively small; however, in the actual study, the large number of investigations required the use of computer programs and ancillary equipment to produce results within a reasonable amount of time.

More material on the subject is presented in Chapter 12.

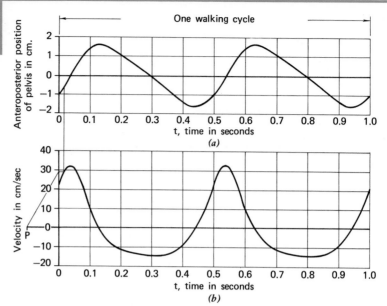

Figure 11.44 Human locomotion. Displacement and velocity of the pelvis as a function of time. (The displacement curve is based on data supplied by the Biomechanics Laboratory, University of California, Berkeley.) The velocity curve was obtained by graphical differentiation of the displacement curve.

EXAMPLE 6 ENERGY EXPENDITURE AND WALKING SPEED.

Another interesting study† in biomechanics dealt with the relation-

* L. M. Lamoreus, "Kinematic Measurements in the Study of Human Locomotion," Biomechanics Laboratory, University of California, San Francisco-Berkeley. Published in the *Bulletin of Prosthetics Research*, BPR10-15, Spring 1971.

† In-house report, No. 44, published by the Biomechanics Laboratory, University of California, San Francisco-Berkeley, February, 1963.

ship between energy expenditure and walking speed of above-the-knee amputees using a University of California Biomechanics Laboratory polycentric hydraulic prosthesis. Figure 11.46 shows the data points and curve for one amputee. The open circles represent values for the morning tests and the solid circles represent values for the afternoon tests.

The problem was to determine the minimum energy value and the corresponding walking speed. This was done by differentiating the data curve. The derived curve is shown as a dashed line. The minimum energy expenditure, 0.86 cal/m/kg, corresponds to the walking speed, 3.5 km/hr. It should be observed that since the derived curve is the "curve of slopes" of tangents to the data curve, the point of intersection of the derived curve with the x-axis has the value zero and therefore locates the x-coordinate of the horizontal tangent to the data curve. Once we know the x value, it is a simple matter to obtain the y value.

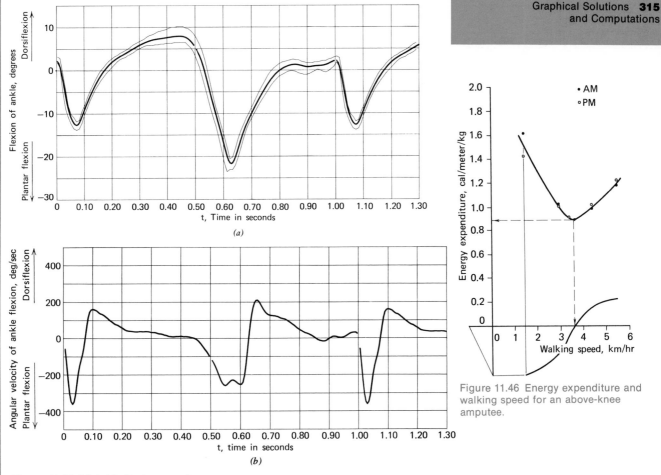

Figure 11.45 (a) Ankle flexion as a function of time. (b) Angular velocity of ankle flexion as a function of time.

Figure 11.46 Energy expenditure and walking speed for an above-knee amputee.

GRAPHONUMERICAL METHOD OF DIFFERENTIATION

In a number of cases it has been observed that relatively large errors can occur in the *second* derived curve due to some variations in the first derived curve. For example, variations (small errors) in a velocity-time curve, which was obtained by graphical differentiation of a displacement-time curve, can reflect larger errors in the acceleration-time curve, which was

obtained by differentiating the velocity-time curve.

To overcome this difficulty, a *graphonumerical method* has been developed and successfully used in the prosthetic devices research project. The method is generally applicable.

It essentially consists of (1) plotting recorded data, (2) drawing a "best" smooth curve to represent the data graphically, and (3) reading the *y* values of points *on*

the curve corresponding to *equal* increments, *h*, along the *x*-axis.

The ordinates of these points are then used in formulas that yield ordinate values for the first and second derived curves. The formulas can be developed in the following manner.

Let us assume that a portion of a parabola passes through points A, B, and C whose ordinates are y_{n-1}, y_n, and y_{n+1}, respectively, as shown in Figure 11.47. When the

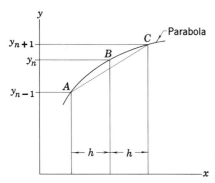

Figure 11.47 Basis for the graphonumerical method of differentiation.

increment, h, is made small, the slope of chord AC closely approximates the slope of the tangent to the curve at point B. This slope may be expressed as

$$\frac{y_{n+1} - y_{n-1}}{2h} = \text{Slope at point } B$$

Now, when we replace n with $n + 1$, the slope of the tangent to the curve at point C is

$$\frac{y_{n+2} - y_n}{2h} = \text{Slope at point } C$$

Similarly, when we replace n with $n - 1$, the slope at point A is

$$\frac{y_n - y_{n-2}}{2h} = \text{Slope at point } A$$

When, for example, experimental data express the relationship between displacement and time, we now have available, in the preceding expressions, a simple means for computing velocities (first derived curve) at points on the curve corresponding to y_{n+1} and y_{n-1}. But, more important, since acceleration is the rate of change of velocity with respect to time, we can easily develop a formula for acceleration from the preceding expressions.

Since,

$$V_{n+1} = \frac{y_{n+2} - y_n}{2h} \quad \text{(velocity at point } C)$$

and

$$V_{n-1} = \frac{y_n - y_{n-2}}{2h} \quad \text{(velocity at point } A)$$

then

$$a_n = \frac{(y_{n+2} - y_n) - (y_n - y_{n-2})}{4h^2}$$

(acceleration at point B)

or

$$a_n = \frac{y_{n+2} + y_{n-2} - 2y_n}{4h^2}$$

Now we see that by using the last equation, we may compute accelerations (second derived curve) *without actually first determining the velocity curve.*

EXAMPLE 1

Assume that the following tabular data are given, where t represents time in seconds and S represents displacement in meters.

t, s	0	1	2	3	4	5
S, m	10	11	18	37	74	135

A plot of these data points and a "fair" curve through (or near) them are shown in Figure 11.48.

Now we measure and record the ordinate values of points *on the curve* $S = f(t)$, assuming $h = 0.5$ s. Corresponding values of t and S are shown in the first two columns of the accompanying table. Additional columns for $2S_n$, S_{n+2}, and S_{n-2} are included to facilitate the calculations for acceleration, a_n.

t, s	S, m	$2S_n$	S_{n+2}	S_{n-2}	$a_n = \dfrac{S_{n+2} + S_{n-2} - 2Sn}{4h^2}$	$h = 0.5$
0.0	10.0	20.0	11.0			
0.5	10.125	20.250	13.375			
1.0	11.0	22.0	18.0	10.0	$a = 18 + 10 - 22$	$= 6$ m/s²
1.5	13.375	26.750	25.625	10.125	$a = 25.625 + 10.125 - 26.750$	$= 9$ m/s²
2.0	18.0	36.0	37.0	11.0	$a = 37 + 11 - 36$	$= 12$ m/s²
2.5	25.625	51.250	52.875	13.375	$a = 52.87 + 13.375 - 51.250$	$= 15$ m/s²
3.0	37.0	74.0	74.0	18.0	$a = 74 + 18 - 74$	$= 18$ m/s²
3.5	52.875	105.750	101.125	25.625	$a = 101.125 + 25.625 - 105.750$	$= 21$ m/s²
4.0	74.0	148.0	135.0	37.0	$a = 135 + 37 - 148$	$= 24$ m/s²
4.5	101.125	202.250		52.875		
5.0	135.0	270.0		74.0		

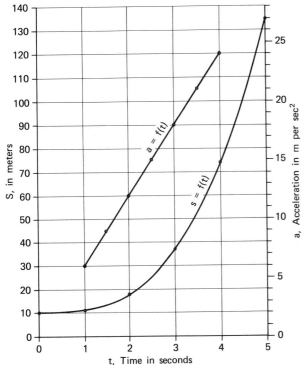

Figure 11.48 Plot of displacement and
acceleration data curves.

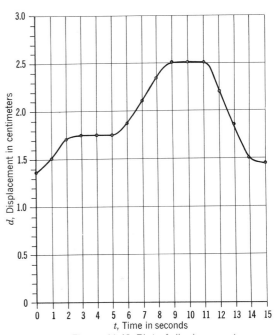

Figure 11.49 Plot of displacement
values.

A plot of the acceleration values is shown in Figure 11.48. Note that, in this case, the curve through these points is a straight line. In an experiment that involves many data points, the calculations for acceleration can be obtained from a computer program output.

EXAMPLE 2
The following tabular data record the relation of time, t, in seconds to displacements, d, in centimeters of the center of a cam follower from the center of the rotating cam.

The graphical representation of the data points and the curve through them are shown in Figure 11.49. Repeating the steps used in the previous example, we arrive at the values of acceleration as calculated from the equation for a_n, as

t, s	0	1	2	3	4	5	6	7	8	9	10	11	12	13	14	15
d, cm	1.37	1.51	1.72	1.75	1.75	1.75	1.87	2.11	2.35	2.51	2.51	2.51	2.22	1.87	1.52	1.46

shown in the accompanying tabulation.

t, sec	d, cm	$2d_n$	d_{n+2}	d_{n-2}	$a_n = \dfrac{d_{n+2} + d_{n-2} - 2d_n}{4h^2}$ $\quad h = 0.50$
0.0	1.37	2.74	1.51		
0.5	1.43	2.86	1.61		
1.0	1.51	3.02	1.72	1.37	$a = 1.72 + 1.37 - 3.02 = 0.07$ cm/s^2
1.5	1.61	3.22	1.73	1.43	$a = 1.73 + 1.43 - 3.22 = -0.06$ cm/s^2
2.0	1.72	3.44	1.75	1.51	$a = 1.75 + 1.51 - 3.44 = -0.18$ cm/s^2
2.5	1.73	3.46	1.75	1.61	$a = 1.75 + 1.61 - 3.46 = -0.10$ cm/s^2
3.0	1.75	3.50	1.75	1.72	$a = 1.75 + 1.72 - 3.50 = -0.03$ cm/s^2
3.5	1.75	3.50	1.75	1.73	$a = 1.75 + 1.73 - 3.50 = -0.02$ cm/s^2
4.0	1.75	3.50	1.75	1.75	$a = 1.75 + 1.75 - 3.50 = 0$
4.5	1.75	3.50	1.78	1.75	$a = 1.78 + 1.75 - 3.50 = 0.03$ cm/s^2
5.0	1.75	3.50	1.87	1.75	$a = 1.87 + 1.75 - 3.50 = 0.12$ cm/s^2
5.5	1.78	3.56	1.99	1.75	$a = 1.99 + 1.75 - 3.56 = 0.18$ cm/s^2
6.0	1.87	3.74	2.11	1.75	$a = 2.11 + 1.75 - 3.74 = 0.12$ cm/s^2
6.5	1.99	3.98	2.24	1.78	$a = 2.24 + 1.78 - 3.98 = 0.04$ cm/s^2
7.0	2.11	4.22	2.35	1.87	$a = 2.35 + 1.87 - 4.22 = 0$
7.5	2.24	4.48	2.46	1.99	$a = 2.46 + 1.99 - 4.48 = -0.03$ cm/s^2
8.0	2.35	4.70	2.51	2.11	$a = 2.51 + 2.11 - 4.79 = -0.08$ cm/s^2
8.5	2.46	4.92	2.51	2.24	$a = 2.51 + 2.24 - 4.92 = -0.17$ cm/s^2
9.0	2.51	5.02	2.51	2.35	$a = 2.51 + 2.35 - 5.02 = -0.16$ cm/s^2
9.5	2.51	5.02	2.51	2.46	$a = 2.51 + 2.46 - 5.02 = -0.05$ cm/s^2
10.0	2.51	5.02	2.51	2.51	$a = 2.51 + 2.51 - 5.02 = 0$
10.5	2.51	5.02	2.41	2.51	$a = 2.41 + 2.51 - 5.02 = -0.10$ cm/s^2
11.0	2.51	5.02	2.22	2.51	$a = 2.22 + 2.51 - 5.02 = -0.29$ cm/s^2
11.5	2.41	4.82	2.00	2.51	$a = 2.00 + 2.51 - 4.82 = -0.31$ cm/s^2
12.0	2.22	4.44	1.87	2.51	$a = 1.87 + 2.51 - 4.44 = -0.06$ cm/s^2
12.5	2.00	4.00	1.65	2.41	$a = 1.65 + 2.41 - 4.00 = 0.06$ cm/s^2
13.0	1.87	3.74	1.52	2.22	$a = 1.52 + 2.22 - 3.74 = 0$
13.5	1.65	3.30	1.47	2.00	$a = 1.47 + 2.00 - 3.30 = 0.17$ cm/s^2
14.0	1.52	3.04	1.46	1.87	$a = 1.46 + 1.87 - 3.04 = 0.29$ cm/s^2
14.5	1.47	2.94		1.65	
15.0	1.46	2.92		1.52	

If it should be necessary to increase the accuracy of values of acceleration in some regions (i.e., between $t = 9.00$ and $t = 11.00$), the value of h can be reduced (e.g., from $h = 0.50$ to $h = 0.25$), and a new set of values can be calculated, as shown in the following table.

t	d_n	$2d_n$	d_{n+2}	d_{n-2}	$a_n = \dfrac{d_{n+2} + d_{n-2} - 2d_n}{4h^2}$ $\quad h = 0.25$
8.50	2.46	4.92	2.51	2.35	$a_n = (2.51 + 2.35 - 4.92)4 = -0.24$ cm/s^2
8.75	2.50	5.00	2.51	2.41	$a_n = (2.51 + 2.41 - 5.00)4 = -0.32$ cm/s^2
9.00	2.51	5.02	2.51	2.46	$a_n = (2.51 + 2.46 - 5.02)4 = -0.20$ cm/s^2
9.25	2.51	5.02	2.51	2.50	$a_n = (2.51 + 2.50 - 5.02)4 = -0.04$ cm/s^2
9.50	2.51	5.02	2.51	2.51	$a_n = (2.51 + 2.51 - 5.02)4 = 0$
9.75	2.51	5.02	2.51	2.51	$a_n = (2.51 + 2.51 - 5.02)4 = 0$
10.00	2.51	5.02	2.51	2.51	$a_n = (2.51 + 2.51 - 5.02)4 = 0$
10.25	2.51	5.02	2.51	2.51	$a_n = (2.51 + 2.51 - 5.02)4 = 0$
10.50	2.51	5.02	2.51	2.51	$a_n = (2.51 + 2.51 - 5.02)4 = 0$
10.75	2.51	5.02	2.49	2.51	$a_n = (2.49 + 2.51 - 5.02)4 = -0.08$ cm/s^2
11.00	2.51	5.02	2.41	2.51	$a_n = (2.41 + 2.51 - 5.02)4 = -0.40$ cm/s^2
11.25	2.49	4.98	2.31	2.51	$a_n = (2.31 + 2.51 - 4.98)4 = -0.64$ cm/s^2

EMPIRICAL EQUATIONS

Our study of empirical equations, as mentioned in the introduction to this chapter, will be limited to the three forms that arise most frequently in engineering.

The three forms are:

1. $y = mx + b$ (straight line on a rectangular uniform grid).
2. $y = bx^m$ (straight line on a rectangular log-log grid).
3. $y = bm^x$ (straight line on a rectangular semilog grid).

Consider the first form, $y = mx + b$.

EXAMPLE 1

The relationship between R, resistance in ohms, and C, temperature in degrees Celsius, of a coil of wire resulting from a test yielded the following data.

R, Ω	11.42	11.94	12.32	12.80	13.25	13.67
C, degrees Celsius (C°)	10.50	30.11	43.00	60.41	75.60	91.25

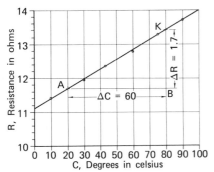

Figure 11.50 Data plot shows that the empirical equation is of the form $y = mx + b$.

The relationship between the quantities is portrayed graphically on a rectangular uniform grid whose x-axis is used (generally) for values of the *independent variable*, and whose y-axis is used for values of the *dependent variable*. A plot of the data points and a "fair" curve through (or very near) them to represent the relation between the variables graphically are shown in Figure 11.50.

In most cases, the observations of the magnitudes of the variables are based on readings of instruments (voltmeters, ammeters, gages, thermometers, etc.) that have graduated scales. The *readings are therefore graphical and approximately correct*. Once this is recognized, it will help us to understand that the equation that symbolizes the relation between the two variables is a close approximation. *Such an equation is known as an empirical equation.*

OBTAINING THE EMPIRICAL EQUATION

The form $y = mx + b$ contains constants m and b, where m is the slope of the line and b is the y-intercept (the value of y when $x = 0$).

The values of the constants m and b can be determined by one of the following methods: (1) *graphical*, (2) *selected points*, (3) *averages*, and (4) *least squares*. Methods (1) and (2) are adequate for data accurate to two significant figures. Method 3 is quite accurate for most cases, and method 4 is generally more precise than the others, but it is also more time consuming. We will apply the four methods to the given example. In subsequent examples we will consider methods 1 and 2 only.

Graphical Method

This method is quite simple. Once we determine the slope of the line that best represents the data and determine the y-intercept of the line, we will have the values of the constraints m and b, respectively.

EXAMPLE 1

First, we assume that the "best" straight line (Figure 11.50) has been drawn to represent the relation between the two quantities. This means that the line passes through some of the plotted points and approximately balances the others so that nearly the same number of points are on each side of the line.

The slope, m, of the line is determined in the following manner.

1. Select two points such as A and K *on the line* and at a reasonable distance apart.

2. Form the right triangle ABK. Now $m = \Delta R/\Delta C = 1.7/60 = 0.028$. Note carefully that ΔC is the length of the horizontal side AB *measured in terms of the units of the horizontal C scale*, and that ΔR is the length of the vertical side BK *measured in terms of the units of the vertical (R) scale*.

3. The R-intercept, b, is the value of R when $C = 0$. In this case $b = 11.1$, which is read directly on the R-axis.

4. The empirical equation is $R = 0.028C + 11.1$. Let us now check the accuracy of the equation by computing R values corresponding to the given C values and then comparing the computed values with the observed values of R. These are shown in the following table.

C	$R_{(obs)}$	$R_{(comp)}$	$R_o - R_c$	Percent deviation
10.50	11.42	11.39	0.03	0.26
30.11	11.94	11.94	0.00	0.00
43.00	12.32	12.30	0.02	0.16
60.41	12.80	12.79	0.01	0.08
75.60	13.25	13.22	0.03	0.23
91.25	13.67	13.66	0.01	0.07

From these values and the percentage deviations we note good agreement between the observed and the calculated values of the R quantities.

Method of Selected Points

In this method we first select two points on the line. Let us use points A and K. The coordinates of point A are $C = 20$ and $R = 11.7$; the coordinates of point K are $C = 80$ and $R = 13.4$. Two points are selected because there are two constants, m and b, that must be determined. Since points A and K are on the line, the coordinates of these points must satisfy the equation $y = mx + b$; therefore

$$11.7 = 20m + b$$
$$\text{and} \quad 13.4 = 80m + b$$

The simultaneous solution of these equations yields

$$m = 0.028 \qquad \text{and} \qquad b = 11.14$$

Therefore, the empirical equation of the line is

$$R = 0.028C + 11.14$$

Method of Averages

The best line is based on the assumption that its location is such as to make the algebraic sum of the differences of the observed and calculated values equal to zero. Algebraically, this means

$$\Sigma(y - mx - b) = 0$$

where y represents any observed value and $(mx + b)$ the corresponding calculated value.

We know that two constants, m and b, must be determined; therefore, the given data are divided into two nearly equal groups. The sum of the differences (r) of the observed and calculated values of each group is placed equal to zero; that is:

$$11.42 - 10.50m - b = r_1$$
$$11.94 - 30.11m - b = r_2$$
$$\underline{12.32 - 43.00m - b = r_3}$$
$$35.68 - 83.61m - 3b = 0$$
(where $\Sigma r_i = 0$)

$$12.80 - 60.41m - b = 4_4$$
$$13.25 - 75.60m - b = r_5$$
$$\underline{13.67 - 91.25m - b = r_6}$$
$$39.72 - 227.26m - 3b = 0$$
(where $\Sigma r_i = 0$)

The simultaneous solution of the equations

$$35.68 = 83.61m + 3b$$
$$\text{and} \quad 39.72 = 227.26m + 3b$$

results in $m = 0.028$ and $b = 11.11$. Therefore, the empirical equation is $R = 0.028C + 11.11$.

Method of Least Squares

The best line is based on the assumption that its location is such as to make the sum of the squares of the differences of the observed and calculated values a *minimum*. Algebraically, this means that $\Sigma(y - mx - b)^2 = $ minimum; therefore, the derivatives of this expression with respect to m and b must equal zero. Hence,

$$\Sigma[2(y - mx - b)(-x)] = 0$$

and

$$\Sigma[2(y - mx - b)(-1)] = 0$$

or

$$\Sigma xy = b\Sigma x + m\Sigma x^2 \tag{1}$$

and

$$\Sigma y = bn + m\Sigma x \tag{2}$$

where n represents the number of observations.

The simultaneous solution of these two equations will determine the values of constants b and m.

First, let us compute the values of the terms in Equations 1 and 2.

This is conveniently done in the accompanying tabular form.

C	R	RC	C²
10.50	11.42	119.91	110.25
30.11	11.94	359.51	906.61
43.00	12.32	529.76	1,849.00
60.41	12.80	773.25	3,649.37
75.60	13.25	1,001.70	5,715.36
91.25	13.67	1,247.39	8,326.56
Σ = 310.87	75.40	4,031.52	20,557.15

The two equations are:

$$4031.52 = 310.87b + 20557.15m$$
$$75.40 = 6.00b + 310.87m$$

The simultaneous solution of the equations yields

$$b = 11.113 \quad \text{and} \quad m = 0.028$$

Therefore, the empirical equation is $R = 0.028C + 11.113$.

EXAMPLE 2

Suppose we have the following data.

x	0	10	20	30	35	40	50	60	70	80	90
y	32.5	29.8	26.1	23.1	22.0	20.0	17.1	13.9	10.3	8.0	5.0

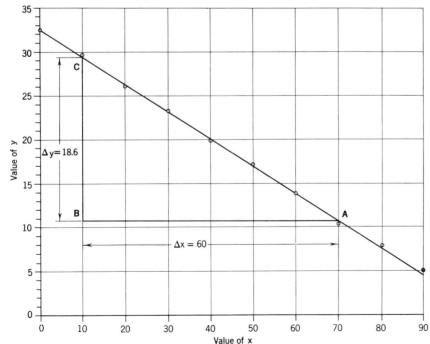

Figure 11.51 Plot of data and "fair" curve for empirical equation of the form $y = mx + b$. Example 2.

A plot of the data is shown in Figure 11.51. The relation between the two variables is linear, and the empirical equation, therefore, is of the form $y = mx + b$. Determine the values of the constants m and b by employing the graphical method and then checking the values by the method of selected points.

Graphical Method

The value of the constant b is readily available, since $y = b$ when $x = 0$. We observe that $b = 32.5$.

The value of the constant m is the slope of the line, or $m = \Delta y / \Delta x = 18.6/-60 = -0.31$. Why is $\Delta x = -60$? Note that since points $C(x_1 y_1)$ and $A(x_2 y_2)$ are on the line, as shown in Figure 11.51, $\Delta y = (y_1 - y_2)$ and $\Delta x = (x_1 - x_2)$, or $\Delta y = (29.4 - 10.8) = 18.6$ and $\Delta x = (10 - 70) = -60$. The slope of the line, therefore, is negative, because

$$m = \frac{(y_1 - y_2)}{(x_1 - x_2)} = \frac{18.6}{-60} = -0.31$$

Method of Selected Points. Points $A(70,10.8)$ and $C(10,29.4)$ are on the line; therefore, their coordinates must satisfy the equation of the line. The equation $y = mx + b$ can be expressed as $y = (y_1 - y_2/x_1 - x_2)x + b$, since $m = y_1 - y_2/x_1 - x_2$. Therefore,

$$y = \left(\frac{29.4 - 10.8}{10 - 70} \right) x + b$$

or

$$y = \frac{18.6}{-60} x + b = -0.31x + b$$

Now, since $y = -0.31x + b$, we can substitute the x- and y-coordinates of either point A or point C to evaluate b. When we use the x- and y-coordinates of point A, we obtain

$10.8 = (-0.31)(70) + b$

from which

$b = 32.5$

The empirical equation is $y = -0.31x + 32.5$, which checks the equation previously obtained from the graphical method.

EXAMPLE 3
Let us assume that the plot of known data and the "best" straight-line representation of those data are as shown in Figure 11.52. Using the graphical method, we observe that

$$m = \frac{\Delta y}{\Delta x} = \frac{BC}{CA} = \frac{60 - 40}{60 - 20} = \frac{1}{2}$$

The intercept b, however, is *not* the y value of point A, because the x value of A is not zero. How can we obtain the value of b without actually extending line AB to $x = 0$? The method is shown in Figure 11.53. We note that it is only necessary to shift point A to position A', a distance of 20 units of x, and then draw A'K parallel to line AB. Now the y value of K is the intercept, b, because point K is on the line $x = 0$. The value of intercept b is 30, and the empirical equation is $y = 0.5x + 30$.

Using the selected point method, we can select point A, whose coordinates are $x = 20$ and $y = 40$, and point B, whose coordinates are $x = 60$ and $y = 60$. These coordinates must satisfy the equa-

tion $y = mx + b$; therefore,

$$40 = 20m + b$$
$$\text{and} \quad 60 = 60m + b$$

The simultaneous solution of these two equations yields the values $m = 0.5$ and $b = 30$. The empirical equation is $y = 0.5x + 30$.

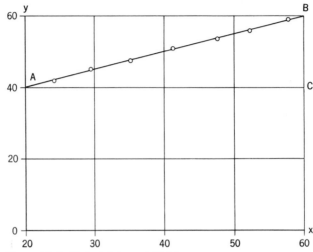

Figure 11.52 Plot of data and "fair" curve for an empirical equation of the form $Y = mx + b$. Example 3.

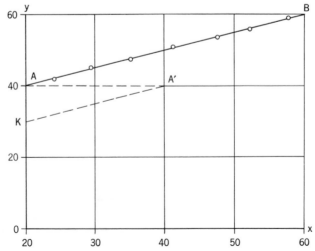

Figure 11.53 Graphical method to determine the y-intercept value when it is inaccessible in the original plot.

EQUATIONS OF THE FORM $y = bx^m$

Equations of this form, when plotted on a Cartesian chart having uniform scales, are curves; they are either parabolas when m is positive, or hyperbolas when m is negative.

When a plot of experimental data, on a Cartesian chart having uniform scales, reveals a curve that belongs to the family of curves $y = bx^m$, *we can verify this fact by plotting the data on a chart having logarithmic scales for the x- and y-axes.* In this case the plot of the data should lie approximately on a straight line. This is true because the equation $y = bx^m$, when expressed logarithmically, is $\log y = m\log x + \log b$, which is of the form $y' = mx' + b'$ where

$y' = \log y$,
$x' = \log x$,   and $b' = \log b$

EXAMPLE 1

Consider the following data. We will assume that the data were plotted on a chart having uniform scales and that the curve that best represented the data belonged to the family of curves, $y = bx^m$.

x	1	2	3	4	5	6
y	7.0	15.5	25.3	39.6	50.1	60.2

Now, when we plot the data on a chart having logarithmic scales, we observe that the points are very nearly on a straight line, as shown in Figure 11.54.

Using the graphical method, we can readily determine the values of m and b.

$$m = \frac{\Delta y}{\Delta x} = \frac{\log 60 - \log 7}{\log 6 - \log 1}$$
$$= \frac{1.778 - 0.845}{0.778 - 0} = 1.2$$

In the equation $\log y = m\log x + \log b$ we observe that when $x = 1$, $m\log 1 = 0$ and, therefore, $\log y = \log b$ or $y = b$. Now that we know that the value of $b = 7$, the empirical equation is $y = 7x^{1.2}$.

As a check on this equation, let us employ the method of selected points to compute the values of b and m. Let us select points $A(1,7)$ and $B(6,60)$. The coordinates of these points must satisfy the equation $\log y = m\log x + \log b$; therefore,

$$\log 60 = m\log 6 + \log b \qquad (1)$$

and

$$\log 7 = m\log 1 + \log b \qquad (2)$$

From Equation 2, $b = 7$; substituting this value into Equation 1, we obtain

$$\log 60 = m\log 6 + \log 7$$

from which

$$m = \frac{\log 60 - \log 7}{\log 6} = 1.2$$

The empirical equation is $y = 7x^{1.2}$.

EXAMPLE 2

Determine the empirical equation for the relation between I, current in amperes, and E, terminal voltage of a 120-V tungsten lamp. The following test data are available.

Figure 11.54 Data plot and "fair" curve for empirical equation of the form $y = bx^m$. Example 1.

E, V	4	8	16	25	32	50
I, A	0.039	0.059	0.086	0.113	0.130	0.172

A preliminary plot of these data, on a Cartesian chart having uniform scales, revealed the fact that the curve through (or near) the data points belonged to the family of curves $y = bx^m$. A plot of the data on a log-log coordinate sheet shows that the plotted values lie very nearly on a straight line, as shown in Figure 11.55. In terms of the given variables, the equation is $I = bE^m$. The constants b and m can be determined in the following manner.

$$m = \frac{\Delta I}{\Delta E} = \frac{\log 0.15 - \log 0.03}{\log 40 - \log 2.5}$$

$$= \frac{\log 5.0}{\log 16} = 0.581$$

$b = 0.0175$, which is the value of the intercept when $E = 1$. The empirical equation is

$$I = 0.0175 \, E^{0.581}$$

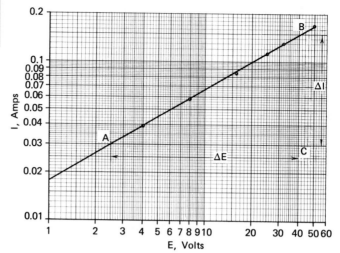

Figure 11.55 Data plot and "fair" curve for an empirical equation of the form $y = bx^m$. Example 2.

EXAMPLE 3 FLAME-SPREAD PHENOMENON

To understand flame-spread phenomenon, thin sheets of plexiglass were burned in a specially designed wind tunnel. The purpose of the experiment was to measure the spread rate as a function of the flow velocity. The specimen was ignited at its downstream end by a heated wire. The flame spread at a constant velocity upstream. It was observed that the flame-spread velocity increased as the flow velocity increased because of more heat transfer to the plastic. Specimens of two different thicknesses were used in the experiment. A sketch of

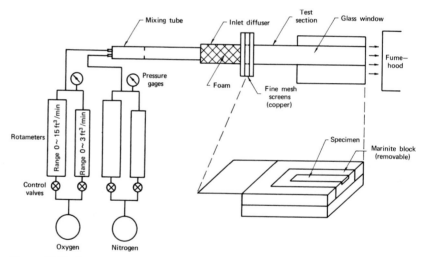

Figure 11.56 Schematic of the apparatus for the flame spread in opposed flow experiment.

the experimental setup is shown in Figure 11.56. The cross-sectional area of the wind tunnel was 12.7 × 127 mm, the specimen size was 25.4 to 152.4 mm thickness, and the gas composition was 50% O_2 and 50% N_2.

The following data were recorded for thickness 0.051 cm. (Similar data were recorded for the other specimen, of thickness 0.089 cm.)

A plot of the data on a log-log grid is shown in Figure 11.57. Ob-

u_∞, stream velocity, cm/s	80	90	100	120	150	170	200
V_n, flame-spread velocity, cm/s	0.122	0.132	0.136	0.145	0.155	0.165	0.174

serve that the data are well represented by a straight line, which indicates that the empirical equation belongs to the form $y = bx^m$.

The values of constants b and m will be determined graphically and by the method of selected points.

Graphical Method

The value of the constant, b, is the intercept value of the line shown in Figure 11.57. Note the graphical construction to obtain this value, which corresponds to the stream velocity value of 1 cm/s. (Recall that since $y = bx^m$ and $\log y = m\log x + \log b$, $\log y = \log b$, or $y = b$ when $x = 1$.)

Graphically, the value of b equals 0.03. Now the value m is the value of the slope of the line, or $m = \Delta y/\Delta x = 0.34$. Therefore, the equation is

$$y = 0.03x^{0.34}$$

or, in terms of the specified variables, $V_{fl} = 0.03\ U_\infty^{0.34}$.

Method of Selected Points

We will select points $K(100, 0.135)$ and $T(200, 0.170)$ on the line. The coordinates of these points must satisfy the equation $y = bx^m$; therefore,

$$\log 0.170 = m\log 200 + \log b$$

and

$$\log 0.135 = m\log 100 + \log b$$

from which $m = 0.334$ and $b = 0.0291$. The empirical equation is $V_{fl} = 0.0291\ U_\infty^{0.334}$, which is a very good check on the equation obtained previously.

EQUATIONS OF THE FORM $y = be^{mx}$

Curves of this form, when plotted on a Cartesian chart having uniform scales, are generally flatter than the curves that belong to the form $y = bx^m$. Figure 11.58 shows several curves of the form $y = be^{mx}$, where $b = 2$ and $m = 1, 2, -1$, and -2.

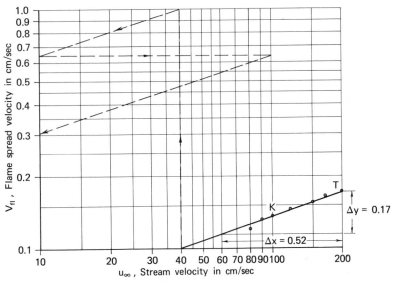

Figure 11.57 Data plot and "fair" curve for the flame spread study. Empirical equation of the form $y = bx^m$. Example 3.

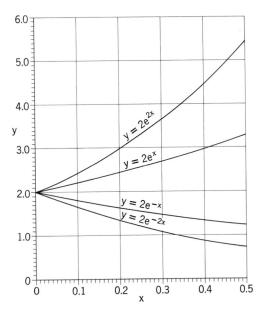

Figure 11.58 Equations of the form $y = be^{mx}$.

EXAMPLE 1

Consider the following data.

t, s	1	2	3	4	5	6
S, m	0.9	1.9	4.1	8.2	15.9	32.3

A preliminary plot of these data on a Cartesian chart with uniform scales showed that the "best" curve, which passed through (or near) the data points, belonged to the family of curves $y = be^{mx}$. A plot of the data on a semilogarithmic grid shows that the points lie very nearly on a straight line (see Figure 11.59).

The empirical equation is $S = be^{mt}$. We can proceed to evaluate the constants b and m. Since log $S = (m\log e)t + \log b$, it is seen that when $t = 0$, log $S = \log b$, or $S = b$. Graphically, $b = 0.46$. We also observe that

$$m = \frac{\log S - \log b}{(0.434)(t)}$$

where $\log e = 0.434$.

Let us select point $B(5,16)$ on the line; then

$$m = \frac{\log 16 - \log 0.46}{(0.434)(5)} = 0.710$$

Therefore, the empirical equation is $S = 0.46e^{0.710t}$. Of course, the value of m could be determined by

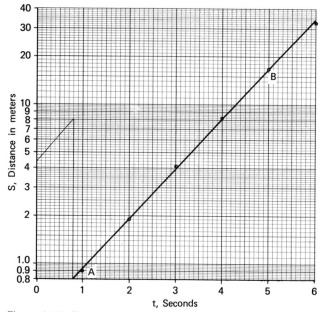

Figure 11.59 Data plot and "fair" curve for an empirical equation of the form $y = be^{mx}$.

using two points on the line [i.e., $B(5,16)$ and $A(1,0.93)$].

$$(m\log e) = \frac{\Delta S}{\Delta t}$$

$$= \frac{\log 16 - \log 0.93}{5 - 1}$$

$$= 0.309$$

and

$$m = \frac{0.309}{0.434} = 0.712$$

which is a bit different from the previous value.

EXAMPLE 2

A plot of the data, recorded as follows, on a semilogarithmic grid shows that the data points lie very nearly on a straight line (see Figure 11.60).

t, min	11	13	15	18	20	21
S, km	4.3	5.0	5.9	7.3	8.6	9.2

The empirical equation is $S = be^{mt}$. Let us evaluate the constants b and m. We know that log $S = (m\log e)t + \log b$ and that $S = b$ when $t = 0$. *Note, however, that $t = 0$ is not available in the original plot of the data.* Nevertheless, we can shift point A to the right a distance of 10 units to point A',

through which a parallel to the original line can be drawn to locate point B'. Now it is quite evident that the value of $b = 1.88$ is the intercept of the line.

We recall that

$$m = \frac{\log S - \log b}{(0.434)(t)}$$

When we choose point $C(18,7.3)$ on the line and substitute the values $S = 7.3$ and $t = 18$ in the preceding equation, we obtain

$$m = \frac{\log 7.3 - \log 1.88}{(0.434)(18)} = 0.0754$$

Now the empirical equation is $S = 1.88e^{0.0754t}$.

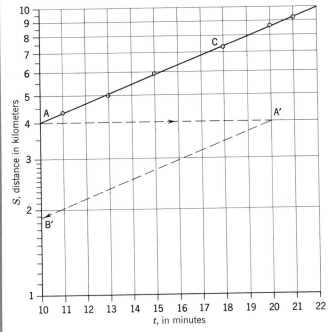

Figure 11.60 Graphical method to obtain the intercept value when the original "fair" curve does *not* show the value.

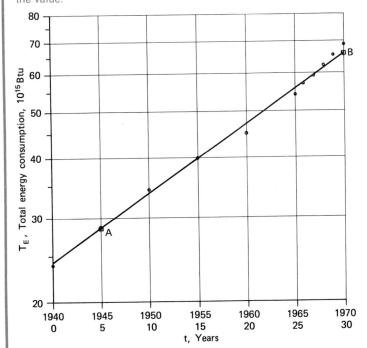

Figure 11.61 Total energy consumption 1940 to 1970. Plot of data and "fair" curve for an empirical equation of the form $y = be^{mx}$. Example 3.

EXAMPLE 3   U.S. TOTAL ENERGY CONSUMPTION— 1940–1970*

The total energy consumption, in trillions of British thermal units, for each year from 1940 to 1970 is shown in the following table. The sources of the energy included coal (anthracite, bituminous, and lignite), crude petroleum, natural gas, and electricity.

Year	Total
1940	23,908
1950	34,153
1955	39,956
1960	44,816
1965	53,969
1966	57,130
1967	59,156
1968	62,448
1969	65,773
1970	68,810

A plot of these data is shown in Figure 11.61. The data are represented fairly well by the line on the semilogathmic grid. The empirical equation, therefore, is of the form, $y = be^{mx}$. We can determine the values of the constants b and m by employing the method of selected points, using the coordinates of points A and B, or by the graphical method.

Let us use the method of selected points. The coordinates of points A and B are A(5,28.5) and B(30,66.0). †

From $y = be^{mx}$ it follows that

$$\log y = (m\log e)x + \log b$$

* Mineral Yearbook, Department of the Interior, Bureau of Mines, 1972.
† For calculation purposes we regard 1940 = 0, 1945 = 5, 1950 = 10, etc; we also omit 10^{15}, since it has no effect on the solution.

Substituting the preceding values, we obtain

$$\log 66.0 = (0.434m)30 + \log b \quad (1)$$

and

$$\log 28.5 = (0.434m)5 + \log b \quad (2)$$

The simultaneous solution of Equations 1 and 2 yields $b = 24.1$ and $m = 0.0336$. The empirical equation is $y = (24.1)10^{15}e^{0.0336}$ or $T_E = (24.1)(10^{15})e^{0.0336t}$, where T_E represents total energy and t represents time in years. The reader should have no difficulty in checking the values of the constants by using the graphical method.

Other useful forms of empirical equations include the following.

1. $y = bx^m + c$

A plot of $\log (y - c)$ and $\log x$ approximates a straight line. The value of c is obtained by selecting two points x_1y_1 and x_2y_2 on the plotted curve and then measuring the value of y_3 that corresponds to $x_3 = \sqrt{x_1x_2}$. It can be shown that

$$c = \frac{y_1y_2 - y_3^2}{y_1 + y_2 - 2y_3}$$

2. $y = bm^x + c$

Log $(y - c) = x \log m + \log b$. A plot of $\log (y - c)$ against x approximates a straight line. To determine a value of c, select points x_1y_1 and x_2y_2 on the curve (the points should be quite far apart), and then find y_3 for $x_3 = (x_1 + x_2)/2$. It can be shown that c has the same value as in the preceding form.

3. $y = nx^2 + mx + b$

Select a point x_1y_1 on the "best"

curve of the plotted data. Then

$$y_1 = nx_1^2 + mx_1 + b$$

and $(y - y_1) = n(x^2 - x_1^2) + m(x - x_1)$,

or $(y - y_1)/(x - x_1) = n(x + x_1) + m$.

Therefore, a plot of

$$(y - y_1)/(x - x_1)$$

against x approximates a straight line, since

$$(y - y_1)/(x - x_1) = nx + (nx_1 + m).$$

4. $y = px^3 + nx^2 + mx + b$

Select four points x_1y_1, x_2y_2, x_3y_3, and x_4y_4 on the curve and then solve simultaneously the four equations:

$$y_1 = px_1^3 + nx_1^2 + mx_1 + b$$
$$y_2 = px_2^3 + nx_2^2 + mx_2 + b$$
$$y_3 = px_3^3 + nx_3^2 + mx_3 + b$$
$$y_4 = px_4^3 + nx_4^2 + mx_4 + b$$

for the values of the constants p, n, m, and b.

NOMOGRAPHY

Nomography deals with the graphical representation and solution of mathematical expressions that contain three or more variables. Studies of the interrelationships among the variables are greatly enhanced by the use of a properly designed nomogram. This is quite important in evaluating trade-offs among the values of the variables, thereby enabling the designer to choose combinations of values that will best satisfy specified constraints of the problem. A nomogram for this purpose is much more effective than a computer printout. For example, the nomogram in Figure 11.62 cites a problem that deals with the determination of a size of a shop weld ($\frac{5}{16}$ in. or 7.94 mm) to satisfy specified values of the leg size (3 in. or 76.2 mm) of an I beam connection angle, its length ($10\frac{5}{8}$ in. or 269.9 mm), and

the end reaction (R) value 50,000 lb (50 kips, or 222.4 kN). It is assumed that the weld size must not exceed $\frac{5}{16}$ in. (7.94 mm).

1. Could we use a smaller leg size and still satisfy the value of R and the weld value?
2. Could we use a shorter angle?

The answer to both questions is yes. We can establish a dashed line through $w = \frac{5}{16}$ in. (7.94 mm) and $R = 50$ kips (222.4 kN). This line intersects the "net" chart at point K, whose x-coordinate is $2\frac{1}{2}$ in. (63.5 mm) for the leg size of the connection angle, and whose y-coordinate is $L = 9$ in. (228.6 mm), the length of the connection angle.

The use of nomograms is a great time-saver in making repetitive calculations, for example, of design formulas. Once the nomogram is designed, all of the useful solutions are available and can be obtained very quickly.

Nomography has a wide application, not only in the various fields of engineering, but also in areas such as biomechanics, ballistics, food science and technology, medicine, physical and biological sciences, production, and statistics.

Engineering and science students will find many opportunities to design and use nomograms in other areas of study and certainly later on in the professional practice of engineering. A number of professional journals (*Design News*, *Machine Design*, *Product Engineering*, *Petroleum Refiner*, *Civil Engineering*, *Electronics*, *American Journal of Clinical Pathology*, *Radiological Health Handbook*, etc.) continue to publish technical articles that show the application of nomograms. You will encounter no difficulty in understanding the simple mathematical theory employed in the design of nomograms.

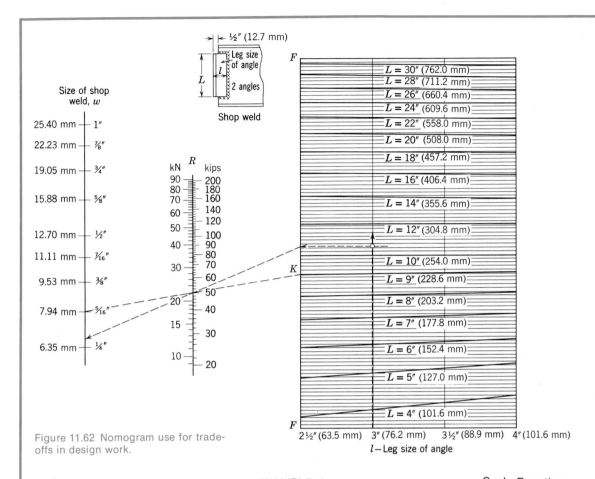

Figure 11.62 Nomogram use for trade-offs in design work.

TYPES OF NOMOGRAMS

There are, in general, two types of nomograms; one is recognized as a rectangular Cartesian coordinate chart (a concurrency chart), and the other as an alignment chart. Both types will be discussed; however, it is essential first to learn of "functional scales" and their use in the design of nomograms.

Functional Scales

A functional scale is one on which the graduations are marked with *the values of the variable* and on which the distances to the graduations are laid off in proportion to the corresponding *values of the function of the variable*.

EXAMPLE 1

Observe the logarithmic scale in Figure 11.63. Note that the graduations are marked 1, 2, 3, 4, . . . 10 and that the distances between consecutive values of the variable are proportional to the logarithums of the numbers.

Scale Equation

Let us assume that the distances from graduation 1 to the other graduations is X and that the variable is u; then we can write the *scale equation* as $X = \log u$. Thus the distance from $u = 1$ to $u = 10$ is $X = (\log 10 - \log 1) = 1$ unit.

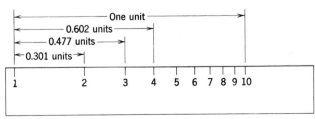

Figure 11.63
Functional scale—log scale.

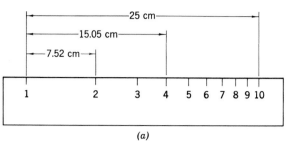

(a)

The distance from $u = 1$ to $u = 2$ is $X = (\log\ 2 - \log\ 1) = 0.301$ units, etc. Now suppose we wish to fix the length of the scale; it would be necessary to introduce a scale factor, or *scale modulus*. This means that the scale equation would be $X = m\log\ u$ or, *in general*, $X_u = mf_u(u)$, where m is the scale modulus.

EXAMPLE 2
Suppose that the $f(u) = \log u$, that u varies from 1 to 10, and that the scale length, l, is 25 cm. Now the scale modulus $m = 25/(\log\ 10 - \log 1) = 25$, and the scale equation is $X = 25 \log u$. As seen in Figure 11.64a, the distance from $u = 1$ to $u = 10$ is $X = 25(\log 10 - \log 1) = 25$ cm; the distance from $u = 1$ to $u = 4$ is $X = 25(\log 4 - \log 1) = (25)(0.602) = 15.05$ cm, etc.

EXAMPLE 3
Now we will use the same function of u, a range from $u = 2$ to $u = 8$, and a scale length of approximately 25 cm. We can calculate the value of the scale modulus from

$$m = \frac{25\pm}{\log 8 - \log 2} = \frac{25\pm}{0.602} = 41.5$$

Since the length of the scale is approximately 25 cm, we will use $m = 40$. Now the scale equation is $X_u = 40(\log\ u - \log\ 2)$. The distance from $u = 2$ to $u = 8$ is then

$$X_u = 40(\log 8 - \log 2)$$
$$= 40(0.602) = 24.08 \text{ cm}$$

In general, the *scale equation* is written as

$$X_u = m_u[f(u) - f(u_1)]$$

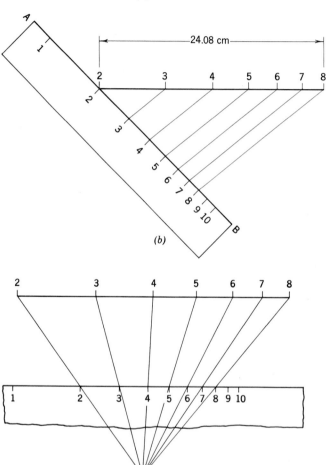

Figure 11.64 (a) Functional scale for $f(u) = \log(u)$; u varies from 1 to 10. (b) Functional scale for $f(u) = \log u$. u varies from 2 to 8. (Projection method.) (c) Alternative projection method for graduating a log scale.

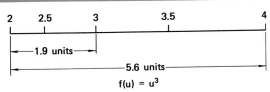

Figure 11.65 Functional scale for $f(u) = u^3$ when u varies from 2 to 4.

where $f(u_1)$ is the minimum value of the function. The subscript identifies the variable.

It should be noted that since logarithmic scales are available, it is not necessary to compute the values of the distances from $u = 2$ to the other graduations of the scale. The simple construction in Figure 11.64b shows how to do it. We lay off a line 24.08 cm long and mark the ends 2 and 8. Through point 2 we introduce a line, AB, making a convenient angle with the first line. On line AB we mark points 2, 3, 4, 5, 6, 7, and 8 from a printed log scale. Lines drawn through these points, parallel to the line that joins the 8s, will intersect the 24.08-cm scale in points that have corresponding values.

An alternative method that is more accurate is shown in Figure 11.64c. This is often preferred to the one shown previously, because it usually overcomes the difficulty of accurately locating the points of intersection with the given line. This is especially true when the length of the printed scale is much shorter than the line to be graduated. In such a case the angle of intersection of the parallels with the given line is very small, making it difficult to locate the points of intersection.

Often the functional scales are not logarithmic. In such cases it is necessary to compute distances from the end values, using the scale equation. Whenever printed forms (for functions other than logarithmic) are available, it is recommended that they be used if appropriate.

EXAMPLE 4
Suppose the $f(u) = U^3$, that u varies from 2 to 4, and that the scale length, L, is approximately 6 units. The scale equation is $X = m(u^3 - 2^3)$. The scale modulus, m, is computed from the relation

$$m = \frac{L}{f(u)_{\text{max}} - f(u)_{\text{min}}}$$

or

$$m = \frac{6\pm}{4^3 - 2^3} = \frac{6\pm}{64 - 8}$$

$$= \frac{6\pm}{56} = 0.1$$

This means that the scale length is actually 5.6 units. If we had decided to use $m = 0.11$, the scale length would be $(0.11)(56) = 6.16$ units. In either case, *the method for locating the graduations would be the same*. Let us agree to use $m = 0.1$; then the scale equation is $X = 0.1(u^3 - 2^3)$. Distances from the end point, $u = 2$, to several other values of u are shown in the following table. The scale is shown in Figure 11.65.

Observe that:

1. The distance between any two graduations, u_1 and u_2, is equal to $X = m[f(u_2) - f(u_1)]$.

2. Any convenient unit of length can be adopted as the unit of measure.

3. The distance between any two graduations is equal to the product of the scale modulus and the difference in the *values of the function of the variable. The graduations, however, are marked with the values of the variable*.

u	2	2.5	3.0	3.5	4
$f(u) = u^3$	8	15.63	27	42.88	64
$X = 0.1(u^3 - 2^3)$	0	0.76	1.9	3.49	5.6

EXAMPLE 5
Let us consider the function $f(u) = 1/u$. We will assume that u varies from 1 to 10 and that the scale length is 9 units. The scale modulus is

$$m = \frac{9}{(1/1 - 1/10)} = \frac{9}{9/10} = 10$$

The scale equation is $X_u = 10[(1/u) - (1/10)]$. We note that when $u = 10$, $X_u = 0$. Values of X_u for other values of u are shown in the following table.

When the value of u is increased arbitrarily to 100 and to 1000, we note that the distances are negative, which means that they are laid off to the left of the *starting*

u	1	2	3	4	5	10	100	1000
$X_u = 10(1/u - 1/10)$	9	4	$2\frac{1}{3}$	$1\frac{1}{2}$	1	0	$-\frac{9}{10}$	$-\frac{99}{100}$

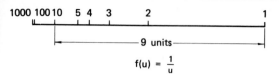

Figure 11.66 Functional scale for $f(u) = 1/u$.

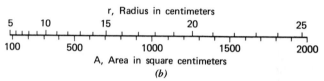

Figure 11.67 (a) Relation between degrees Celsius and degrees Fahrenheit: $C = \frac{5}{9}(F - 32)$.

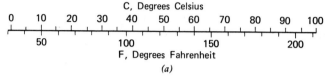

Figure 11.67 (b) $A = \pi r^2$, area of a circle.

point $u = 10$ (see Figure 11.66). Also, it is quite apparent that as u increases in value, the scale becomes more congested; therefore, it is recommended that *scales for reciprocal functions be restricted to short ranges of the variable*.

Adjacent Scales for Equations of the Form $f_1(u) = f_2(v)$

Equations of this form are easily solved by graduating both sides of one line in such a manner that any point on the line will yield values which satisfy the given equation.

The scale equations are:

$$X_u = m_u f_1(u) \qquad [\text{from } f_1(u) = 0]$$

and

$$X_v = m_v f_2(v) \qquad [\text{from } f_2(v) = 0]$$

For any position on the scale,

$$X_u = X_v$$

or

$$m_u f_1(u) = m_v f_2(v)$$

Since $f_1(u) = f_2(v)$, it follows that $m_u = m_v$

EXAMPLE 1

Consider the relationship between Fahrenheit and Celsius temperatures, as expressed by the equation

$$C = \frac{5}{9}(F - 32) \qquad (1)$$

Assume that C varies from 0 to 100° and that the scale length is 10 units.

The scale equation for the variable, C, is

$$X_c = m_c C \qquad (2)$$

from which

$$m_c = \frac{10}{100 - 0} = 0.1$$

Therefore, the scale equation is

$$X_c = 0.1C \qquad (3)$$

Since the moduli must be the same for both functions of the variables, the scale equation for the function of F is

$$X_F = (0.1)\left(\frac{5}{9}\right)[(F - 32) - (F_1 - 32)] \qquad (4)$$

or

$$X_F = \frac{1}{18}[F - F_1] \qquad (5)$$

It should be pointed out that F_1 is a value that corresponds to a selected value of C; that is, when we select $C = 0$, $F_1 = 32$ in order to satisfy Equation 1. Therefore, Equation 5 now becomes,

$$X_F = \frac{1}{18}(F - 32) \qquad (6)$$

The adjacent scales are shown in Figure 11.67a.

EXAMPLE 2

Consider the relation $A = \pi r^2$, where r (5 to 25 cm) is the radius of a circle, A is the area in square centimeters, and the scale is approximately 10 units long.

It is suggested that the equation be written as $r^2 = A/\pi$, since the range of r is *specified*, thereby freeing variable r of the constant π.

Now we can calculate m_r from the equation,

$$m_r = \frac{10\pm}{25^2 - 5^2} = \frac{10\pm}{600} = 0.017$$

The scale equation is $X_r = 0.017(r^2 - 5^2)$. Graduations on the r scale can be located from this equation or can be projected from an available squared scale once the end points $r = 5$ and $r = 25$ are located.

Now we can write the scale equation for variable A as

$$X_A = \frac{0.017}{\pi}(A - A_1)$$

$$= 0.0054(A - A_1)$$

where A_1 is a value that corresponds to a selected value of r. For example, when we select $r = 5$, $A = \pi 5^2 = 78.6$.

When we use this value for A_1, the scale equation is $X_A = 0.0054(A - 78.6)$. From the equation we can compute the distances to $A = 500$ and $A = 1000$. Once the graduations for these two values are located, no further computations are necessary to complete the graduation of the A scale, because the function of A is linear. Additional graduations are easily located by subdivision (see Figure 11.67b).

EXAMPLE 3

Consider the relation between inches, I, and centimeters, C. We know there are 2.54 cm in 1 in. When we let I represent the *number* of inches and C the corresponding number of centimeters, the relation is

$$(2.54)I = C \quad \text{or} \quad I = \frac{C}{2.54}$$

Suppose I varies from 0 to 20 in., and that the scale length is 10 units. Now,

$$X_I = m_I I \quad \text{from which}$$

$$m_I = \frac{10}{20} = 0.5$$

The scale equation is

$$X_I = 0.5I$$

The scale equation for the $f(C) = C/2.54$ is

$$X_C = 0.5 \left(\frac{C}{2.54} \right) = \frac{C}{5.08}$$

The solution is shown in Figure 11.68.

Nonadjacent Scales for Equations of the Form $f_1(u) = f_2(v)$

In our previous discussion of adjacent scales, it was shown that the *same* modulus must be used in both of the scale equations. The value of the modulus was based on the desired length of scale and the range of *one of the variables*. As a consequence, the adjacent scale for the other variable was fixed. This means that the scale increment between successive values of the variable *could be* too small for the desired accuracy in reading. To help overcome this difficulty, the scales for the variables can be separated so that *different* moduli can be used. In this manner we can select scale lengths that are consistent with the desired accuracy of reading the values of each variable.

Design of Nonadjacent Scales

Let us now proceed to the design of nonadjacent scales.

From the geometry (similar triangles Au_1K and Bv_1K) in Figure 11.69, we observe that

$$\frac{X_u}{X_v} = \frac{AK}{KB} = \frac{a}{b}$$

Now, since $X_u = m_u f_1(u)$ and $X_v = m_v f_2(v)$, we obtain the relation

$$\frac{m_u f_1(u)}{m_v f_2(v)} = \frac{a}{b} \quad \text{or} \quad \frac{m_u}{m_v} = \frac{a}{b}$$

since $f_1(u) = f_2(v)$. Now point K can be located on the diagonal AB by dividing it into the ratio

$$\frac{AK}{KB} = \frac{a}{b} = \frac{m_u}{m_v}$$

EXAMPLE 1

Suppose the given equation is $C = \pi D$, where D (diameter) varies from 0 to 25 cm and the scale

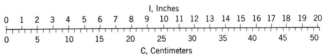

Figure 11.68 Adjacent scales for the relation 2.54$I = C$.

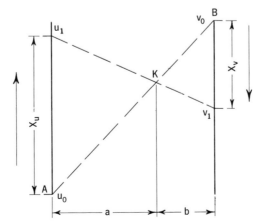

Figure 11.69 Geometry for separated scales when $f_1(u) = f_2(v)$.

length is 10 units. Since the range of variable D is specified, we will rewrite the equation as $D = C/\pi$. The scale equation is

$$X_D = m_D D \quad \text{and} \quad m_D = \frac{10}{25} = 0.4$$

Therefore,

$$X_D = 0.4D \qquad (1)$$

The scale equation for variable C is

$$X_C = m_c \frac{C}{\pi}$$

Let us assume that the desired scale length is 20 units; then

$$m_c = \frac{20}{C/\pi} = \frac{20}{25\pi/\pi} = 0.8$$

$(25\pi$ is the maximum value of $C)$ and

$$X_c = \frac{0.8C}{\pi} = \frac{14}{55} C, \qquad (2)$$

where $\left(\pi = \frac{22}{7} \right)$

Scales D and C have been graduated from Equations 1 and 2 respectively. The distance between the scales is arbitrary (see Figure 11.70).

Obviously, it is only necessary to establish the 0 and the 25 on the D scale, since the scale is uniform. Intermediate points can be easily located graphically. As to the C scale, it is only necessary to locate the 0 and one additional point, such as the 75. Since the C scale is also uniform, other values of C can be located graphically without any difficulty. Point K, which lies on the line joining the zero values of the functions of C and D, is located by dividing the diagonal into the ratio

$$\frac{a}{b} = \frac{m_D}{m_C} = \frac{0.4}{0.8} = \frac{1}{2}$$

All lines through point K intersect the C and D scales in values that satisfy the given equation. Note the directions for plotting increasing values of the variables and that the length of the unit on the diagonal is independent of the unit used to graduate the C and D scales.

EXAMPLE 2

Let us consider the equation $F = \frac{9}{5}C + 32$, where F varies from -40 to 212°, and its scale length is approximately 25 units.

$$m_F = \frac{25\pm}{[212 - (-40)]}$$

$$= \frac{25\pm}{252} = 0.1$$

and

$$X_F = 0.1[F - (-40)]$$
$$= 0.1(F + 40) \qquad (1)$$

The scale modulus for the function of C, for a scale length of approximately 10 units, is

$$m_C = \frac{10\pm}{\frac{9}{5}[(100) - (-40)]} = \frac{10\pm}{\frac{9}{5}(140)}$$

$$= \frac{10\pm}{252} = 0.04$$

Note that C varies from -40 to 100° when F varies from -40 to 212°.)

The scale equation for the variable C is

$$X_C = 0.04\{(\tfrac{9}{5}C + 32) - [\tfrac{9}{5}C(-40) + 32]\}$$

or

$$X_C = 0.04[\tfrac{9}{5}(C + 40)] \qquad (2)$$

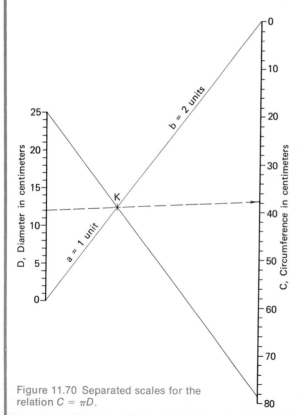

Figure 11.70 Separated scales for the relation $C = \pi D$.

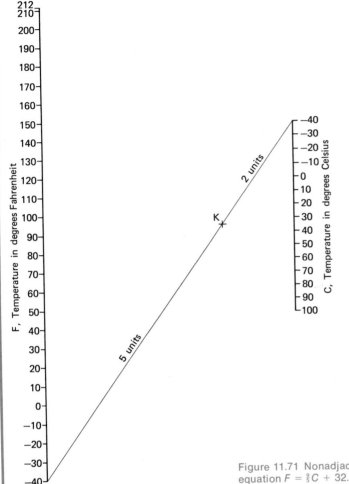

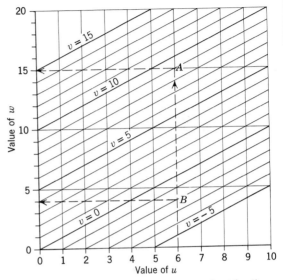

Figure 11.72 Concurrency chart for the relation $u + v = w$. *Example 1:* $u = 6$; $v = 9$; then $w = 15$. *Example 2:* $u = 6$; $v = -2$; then $w = 4$.

Figure 11.71 Nonadjacent scales for the equation $F = \frac{9}{5}C + 32$.

We can locate point K on the diagonal from the ratio

$$\frac{m_F}{m_C} = \frac{0.1}{0.04} = \frac{5}{2}$$

A quick check on the location of point K can be made quite easily by connecting $F = 212$ with $C = 100$. That line should pass through point K (see Figure 11.71).

NOMOGRAMS— CONCURRENCY TYPE

The graphical solution of a relationship among three or more variables as presented in a Cartesian coordinate system is known as a concurrency type nomogram or, simply, as a concurrency chart.

Concurrency Charts for Equations of the Form $f_1(u) + f_2(v) = f_3(w)$

EXAMPLE 1

Assume the relation $u + v = w$, where u and v each ranges from 0 to 10.

When we let $x = u$ and $y = w$, we obtain

$$y = x + v \tag{1}$$

This equation is one of the straight line family,

$$y = mx + b \tag{2}$$

in which m is the slope of the line and b is the value of the y-intercept.

In Equation 1 we observe that the slope is one and the intercept value is v. As we assign different values to v, we will obtain a family of parallel lines, as shown in Figure 11.72. The concurrency chart serves as an addition chart, since it represents the equation $u + v = w$.

EXAMPLE 2

Let us consider the relation $I = E/R$ (Ohm's law). I, the current in amperes, varies from 1 to 100 A, R, the resistance in ohms, varies from 1 to 10 Ω, and E is the electromotive force in volts.

We can reduce the equation to the form $y = mx + b$ by writing

$$\log I = -\log R + \log E$$

Thus, $y = \log I$ and $x = \log R$; therefore

$$y = -x + \log E$$

For each value of E we will obtain a straight line that has a slope of -1 and an intercept value that corresponds to the value of E. The graphical solution is shown in Figure 11.73.

Concurrency Charts for Equations of the Form $f_1(u) \cdot f_2(v) = f_3(w)$

EXAMPLE 1

Let us assume that the given equation is $u \cdot v = w$. It is fairly obvious that we can reduce this equation to the previous form by writing $\log u + \log v = \log w$ and then proceeding to a graphical solution in a manner similar to that used in the previous Example 2. In many cases, when the ranges of the variables are suitable, it is not necessary to use logarithmic scales for the graphical representation of equations belonging to the family $f_1(u) \cdot f_2(v) = f_3(w)$.

The given expression $u \cdot v = w$ can be treated in the following manner: Let

$$x = u \tag{1}$$

and

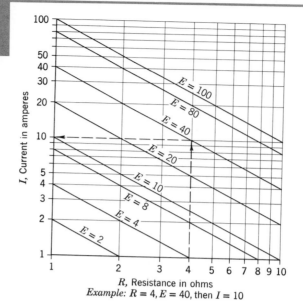

Figure 11.73 Concurrency chart for the equation $I = E/R$.

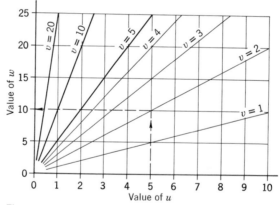

Figure 11.74 Concurrency chart for the relation $uv = w$. *Example:* $u = 5$; $v = 2$. Read $w = 10$.

$$y = w \tag{2}$$

then

$$y = vx \tag{3}$$

This equation is of the form $y = mx + b$. In Equation 3 v is the slope and $b = 0$. This means that all of the lines pass through the origin and have slopes equal to values of v. The graphical solution is shown in Figure 11.74.

EXAMPLE 2

Suppose the given equation is $uv = wq$ and that ranges of the variables are as follows: u from 0 to 8; v from 0 to 20; w from 0 to 30; and q from 0 to 10.

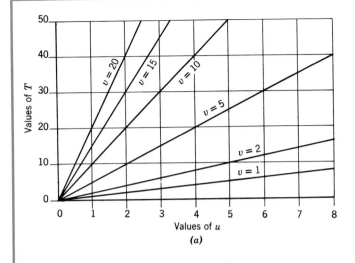

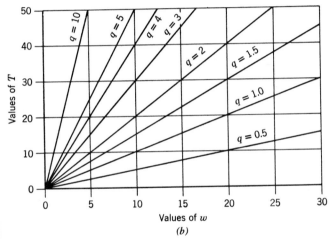

Now we can let

$$uv = T \quad\quad\quad (1)$$

and

$$wq = T \quad\quad\quad (2)$$

In Equation 1, when

$$x = u \quad \text{and} \quad y = T$$

then

$$y = vx$$

The graphical representation and solution of Equation 1 is shown in Figure 11.75a. This is similar to the solution shown in Figure 11.74. The solution of Equation 2 is shown in Figure 11.75b.

It should be stressed that the T scales in Figure 11.75a and 11.75b must be identical in order to combine them to form the concurrency chart in Figure 11.75c.

Figure 11.75 (a) Concurrency chart for the relation, $T = uv$. (b) Concurrency chart for the relation, $T = wq$. (c) Concurrency chart for the relation, $uv = wq$.

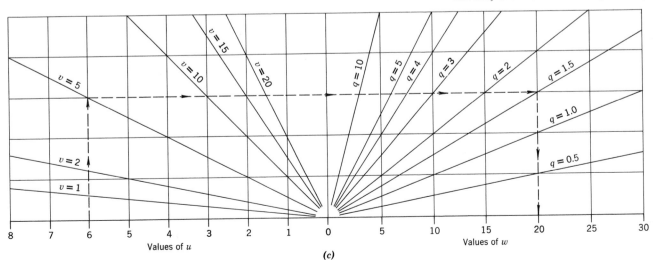

EXAMPLE 3

Consider the equation

$$S = \frac{(rpm)D}{5308R} \qquad (1)$$

where

S = vehicle road speed (15 to 90 km/hr)

D = tire size, outside diameter (350 to 750 mm)

R = gear ratio (2:1 to 6:1)

rpm = engine revolutions per minute (0 to 4000)

Equation 1 can be divided into two parts.

$$5308\,RS = T \qquad (2)$$

and

$$(rpm)(D) = T \qquad (3)$$

The graphical representation and solution of Equations 2 and 3 are shown in Figure 11.76a and 11.76b, respectively. These two figures have been combined to form the concurrency chart shown in Figure 11.76c. It should be clear that this method could be extended to include additional variables.

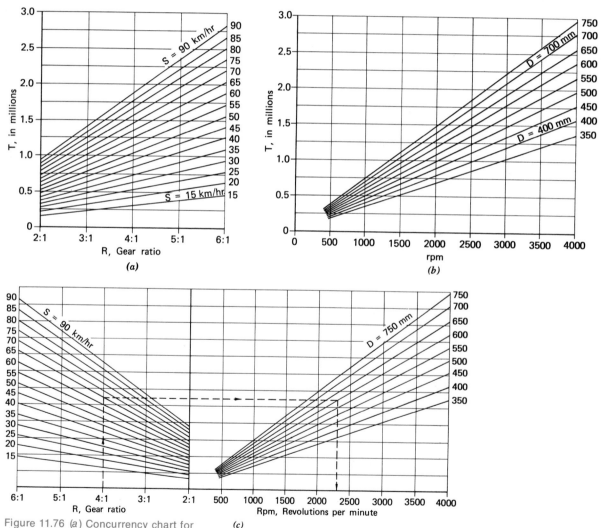

Figure 11.76 (a) Concurrency chart for $T = 5308RS$. (b) Concurrency chart for $T = (rpm)\,D$. (c) Concurrency chart for $S = (rpm)D/5308\,R$.

NOMOGRAMS—ALIGNMENT TYPE OF THE FORM
$f_1(u) + f_2(v) = f_3(w)$

Alignment nomograms are known generally as alignment charts. In their simplest form they consist of three parallel scales, each graduated so that a straight line that joins values on two of the scales will intersect the third scale in a value that satisfies the relation $f_1(u) + f_2(v) = f_3(w)$.

Design of Alignment Nomograms of the Form $f_1(u) + f_2(v) = f_3(w)$

In order to design an alignment nomogram for this form we need to know how to graduate each of the scales and how to position the scales with respect to each other.

Let us assume that the three parallel scales, shown in Figure 11.77a, have been so graduated that lines (known as isopleths) 1 and 2 intersect the scales in values that satisfy the equation $f_1(u) + f_2(v) = f_3(w)$.

Now we observe from Figure 11.77a that

$$X_u = m_u[f_1(u_1) - f_1(u_0)] \qquad (1)$$
$$X_v = m_v[f_2(v_1) - f_2(v_0)] \qquad (2)$$
$$X_w = m_w[f_3(w_1) - f_3(w_0)] \qquad (3)$$

These equations are the scale equations that enable us to determine the distance from the initial graduations u_0, v_0, and w_0 to other graduations, such u_1, v_1, and w_1 on each of the three scales, respectively.

If u_0, v_0, and w_0 represent the *zero values of the functions*, and if line 2 is any line, we may drop the subscripts and simply write

$$X_u = m_u f_1(u)$$
$$X_v = m_v f_2(v)$$
$$X_w = m_w f_3(w)$$

where m_u, m_v, and m_w are the scale moduli (or scale factors) for the u, v, and w scales, respectively. Once we graduate the scales for $f_1(u)$ and $f_2(v)$ from their respective scale equations, it will be necessary to determine the value of the modulus, m_w, for the scale equation of $f_3(w)$, and also the ratio a/b to locate the position of the w scale.

Further study of Figure 11.77a shows that

$$\frac{X_u - X_w}{X_w - X_v} = \frac{a}{b}$$

(since the shaded triangles are similar by construction)

From this equation we obtain the relation

$$X_u b + X_v a = X_w(a + b)$$

and now

$$\frac{X_u}{a} + \frac{X_v}{b} = \frac{X_w}{ab/(a + b)}$$

Since

$$X_u = m_u f_1(u) \qquad (1)$$
$$X_v = m_v f_2(v) \qquad (2)$$
$$X_w = m_w f_3(w) \qquad (3)$$

it follows that

$$\frac{m_u f_1(u)}{a} + \frac{m_v f_2(v)}{b} = \frac{m_w f_3(w)}{ab/(a + b)}$$

This equation can be made identical to the desired form

$$f_1(u) + f_2(v) = f_3(w)$$

when we set

$$\frac{a}{b} = \frac{m_u}{m_v} \qquad \text{and}$$

$$m_w = \frac{ab}{a + b} = \frac{m_u m_v}{m_u + m_v}$$

Design Summary

To construct an alignment chart for an equation of the form $f_1(u) + f_2(v) = f_3(w)$:

1. Place the parallel scales for u and v a *convenient* distance apart.
2. Graduate them in accordance with their scale equations $X_u = m_u f_1(u)$ and $X_v = m_v f_2(v)$.
3. Locate the scale for w so that its distance from the u scale is to its distance from the v scale as $m_u/m_v = a/b$.
4. Graduate the w scale from its scale equation

$$X_w = m_u m_v/(m_u + m_v)f_3(w).$$

Let us now design several nomograms of the form $f_1(u) + f_2(v) = f_3(w)$.

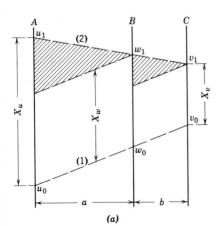

(a)

Figure 11.77 (a) Geometry for the determination of the scale equations and the spacing of the scales in the type form $f_1(u) + f_2(v) = f_3(w)$.

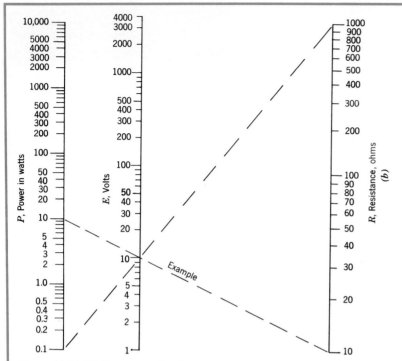

Figure 11.77 (b) Nomogram for the equation $P = E^2/R$, with the E scale between the P and R scales. *Example:* $E = 10, R = 10$; then $P = 10$.

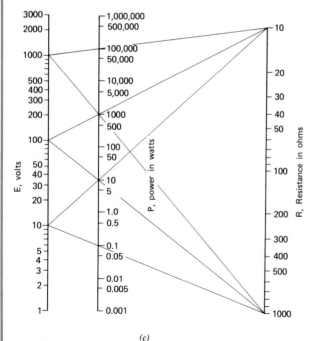

(c)

Figure 11.77 (c) Nomogram for $P = E^2/R$, with the P scale between the E and R scales.

EXAMPLE 1

Let us consider the relation

$$P = \frac{E^2}{R}$$

where

P = power in watts (0.1 to 10,000)
E = voltage
R = resistance in ohms (10 to 1000)

The equation is put in type form by taking logarithms; thus we write

$$\log P + \log R = 2 \log E$$

Now the moduli for the scales are computed as follows.

From $X_P = m_P (\log P - \log P_1)$ we can calculate m_P by assuming a suitable length, X_P, for the P scale. Suppose we let $X_P = 10\pm$ units.

$$m_P = \frac{10\pm}{\log 10,000 - \log 0.1} = 2$$

The scale equation is $X_P = 2(\log P - \log 0.1) = 2 \log 10P$. For the R scale we obtain,

$$m_R = \frac{10\pm}{\log 1000 - \log 10} = 5$$

and

$$X_R = 5(\log R - \log 10)$$
$$= 5 \log \frac{R}{10}$$

The modulus of the E scale equation is

$$m_E = \frac{(m_P)(m_R)}{m_P + m_R} = \frac{(2)(5)}{2 + 5} = \frac{10}{7}$$

Therefore,

$$X_E = \frac{10}{7} (2 \log E - 2 \log 1)$$

(where $E_{\min}$ is 1)

$$= \frac{20}{7} \log E$$

The ratio a/b, which controls the spacing of the scales, is $m_P/m_R = 2/5$. The completed nomogram is shown in Figure 11.77b.

EXAMPLE 2

Again consider the equation $P = E^2/R$ and plan the design of the nomogram so that the P scale is located between the E and R scales. We can do this by writing

$$2 \log E - \log R = \log P \qquad (1)$$

Assume that E varies from 1 to 3000 V and that R varies from 10 to 1000 Ω.

$$m_E = \frac{10\pm}{2 \log 3000 - 2 \log 1}$$

$$= \frac{10\pm}{6.95} = 1.5$$

where the scale length is approximately 10 units. The scale equation is

$$X_E = 1.5(2 \log E) = 3 \log E \qquad (2)$$

The scale equation for the R scale is

$$X_R = 5(\log R - \log 10)$$

$$= 5 \log \frac{R}{10} \qquad (3)$$

where

$$m_R = \frac{10\pm}{\log 1000 - \log 10} = 5$$

The negative sign in Equation 1 *indicates that increasing values of R are plotted downwardly*, since we usually plot upwardly for plus signs.

As in the previous example, the spacing of the scales is based on the ratio

$$\frac{a}{b} = \frac{m_E}{m_R} = \frac{1.5}{5} = \frac{3}{10}$$

Now the modulus for the P scale is

$$m_P = \frac{(1.5)(5)}{1.5 + 5} = \frac{7.5}{6.5} = \frac{15}{13}$$

and the scale equation is $X_P = 15/13 \ (\log P - \log P_1) = 15/13 \log P/P_1$, where P_1 is a value that sat-

isfies the relation $P = E^2/R$ for selected values of E and R; that is, when $E = 10$ and $R = 1000$, $P_1 = 0.1$. Therefore the line which joins $E = 10$ with $R = 1000$, intersects the P scale at $P_1 = 0.1$. Now we can graduate the P scale with reference to the initial point $P_1 = 0.1$. The scale equation, therefore, is

$$X_P = \frac{15}{13} \log \frac{P}{0.1}$$

$$= \frac{15}{13} \log 10P \qquad (4)$$

Design of the Nomogram

Graduations on the E scale are plotted from Equation 2. It is only necessary to locate the logarithmic cycles (1 to 10, 10 to 1000, etc.). Intermediate graduations can be projected from a printed log scale. In a similar manner the R scale can be graduated from Equation 3, noting that *increasing values of R are plotted downwardly*. The P scale is graduated from Equation 4, starting with the previously located point $P_1 = 0.1$. The completed nomogram is shown in Figure 11.77c. The designer's choice of the nomographic solution of the equation, Figure 11.77b or 11.77c, may depend on the ranges of the variables and the desired accuracy in reading values of P, the power in watts.

Practical Shortcut Method

When the designer is very familiar with the theory involved in alignment-type nomograms and fully understands the algebraic steps that are necessary to reduce a given equation to a desirable form [i.e., $f_1(u) + f_2(v) = f_3(w)$], the designer can frequently short-cut the actual construction of the nomogram.

EXAMPLE 1

Consider the nomogram shown in Figure 11.77c for the equation $P = E^2/R$. Once the equation is reduced to the form $f_1(u) + f_2(v) = f_3(w)$ as $2 \log E - \log R = \log P$, and the ranges of the variables $E(1$ to 3000 V) and $R(10$ to 1000 Ω) are stated, it is only necessary to do the following.

1. Draw the E and R scales a convenient distance apart and assume their lengths.

2. Mark the E scale from 1 (lower value) to 3000 (the upper value) and geometrically project the intermediate graduations from a four-cycle log sheet.

3. In a similar manner graduate the R scale, starting with the 10 at the upper end of the scale and the 1000 at the lower end. A two-cycle log sheet can be used to locate the intermediate graduations geometrically.

4. Locate the P scale by, for example, letting $E = 100$ and $R = 1000$. The line that connects these points will contain a point whose value is $P = 10$ (from $P = E^2/R$). Similarly, we can let $R = 10$ and $E = 10$. The line that joins these values also contains a point whose value is $P = 10$. Therefore, the intersection of the two lines is $P = 10$ and the vertical line through $P = 10$ locates the position of the P scale. Another point, such as $P = 1000$, is readily located by joining $E = 100$ and $R = 10$. This line intersects the P scale at $P = 1000$. Additional values of P can be easily located by projecting geometrically from a four-cycle log sheet.

EXAMPLE 2

Consider the equation $Q = 3.33bH^{3/2}$ (Francis' Weir formula), where

Q = discharge in cubic feet per second and in cubic meters per second
b = width of the weir (3 to 30 ft or 1 to 9 m)
H = head above the crest of the weir (0.5 to 1.5 ft or 0.15 to 0.45 m)

If a nomogram consisting of three parallel scales is desired, the designer recognizes that the equation must be converted to the form

$$\log b + \tfrac{3}{2} \log H = \log Q - \log 3.33$$

The nomogram can now be constructed without making further calculations. The following procedure is suggested.

1. Draw two vertical lines a convenient distance apart. Lable the left vertical b and the right H (see Figure 11.78).
2. Graduate the b scale by simply marking the lower end 3 and the upper end 30. Other graduations may be located by projecting from a log scale (two-cycle slide rule scale or commercial log sheet having two cycles). (Corresponding metric values are shown.)
3. Mark the lower end of the H scale 0.5 and the upper end 1.5. Again, locate additional graduations by projecting from a log scale. (Metric values are also shown.)
4. To locate the Q scale and two graduations on it, make calculations as follows.

   (a) Let $b = 3$ and $H = 1$. From the equation $Q = 3.33bH^{3/2}$, we now get $Q = 10$, which lies some-

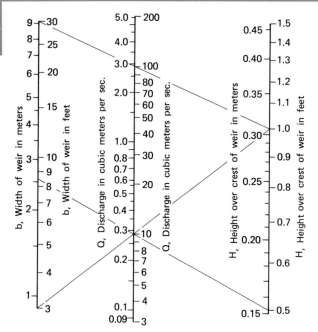

Figure 11.78 Nomogram for the equation $Q = bH^{\frac{3}{2}}$.

where on the line joining $b = 3$ and $H = 1$.

(b) Now let $H = 0.5$ and $Q = 10$; then $b = 8.45$. The line joining $H = 0.5$ and $b = 8.45$ contains $Q = 10$. Therefore, the intersection of this line with the line in part a uniquely locates $Q = 10$. The vertical through $Q = 10$ is the Q scale.

(c) A second graduation on the Q scale can now be located by letting $H = 1$ and $Q = 100$, from which $b = 30$. The line joining $b = 30$ and $H = 1$ intersects the Q scale at $Q = 100$. Other graduations can be easily located by projecting from a log scale. The completed nomogram is shown in Figure 11.78. (Metric values are included.)

Alignment Charts for Equations of the Form $f_1(u) + f_2(v) + f_3(w) = f_4(q)$

EXAMPLE 1

Let us first analyze the alternative

solutions that are possible. For this purpose we will use the simple relation

$$u + v + w = q \qquad (1)$$

Case 1

Let

$$u + v = T \qquad (2)$$

and

$$T + w = q \qquad (3)$$

Each of Equations 2 and 3 is of the form $f_1(u) + f_2(v) = f_3(w)$. A sketch of the solution is shown in Figure 11.79a.

Case 2

Let

$$u + v = T \qquad (4)$$

and

$$q - w = T \qquad (5)$$

A sketch of the solution is shown in Figure 11.79b.

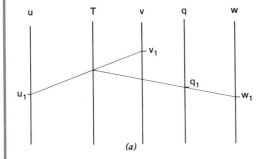

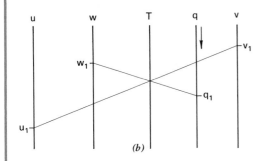

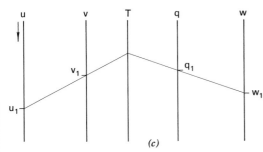

Figure 11.79 (a) Nomogram for $u + v + w = q$. Case 1. (b) Nomogram for $u + v + w = q$. Case 2. (c) Nomogram for $u + v + w = q$. Case 3.

Case 3

Let

$$-u + T = v \tag{6}$$

and

$$T + w = q \tag{7}$$

A sketch of the nomographic solution of Equation 1 is shown in Figure 11.79c.

EXAMPLE 2

Now consider the equation

$$D^3 = \frac{16KT}{\pi S}$$

where

D = shaft diameter (1/2 to 6 in. or 12.7 to 152.4 mm)
T = torque (1000 to 100,000 in-lbs or 113 to 1,129,848 N-m)
S = design stress (5,000 to 30,000 psi or 34.5 to 206.8 MPa)
K = shock factor (1 to 3)

The equation can be converted to the type form by writing

$$3 \log D = \log \frac{16}{\pi} + \log K$$
$$+ \log T - \log S$$

This equation can be replaced by the following two equations.

$$\log S + \log R = \log T \tag{1}$$

and

$$\log R + \log K + \log \frac{16}{\pi}$$
$$= 3 \log D \tag{2}$$

where R is the ungraduated turning axis.

Although several combinations of variables could be taken to form

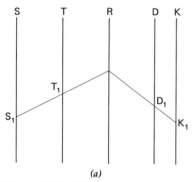

Figure 11.80 (a) Sketch layout of scales.
Example 2.

equations such as Equations 1 and 2, the variables S and T were placed in Equation 1 *because the ranges of these two variables are the large ones.*

A sketch layout of Equations 1 and 2 is shown in Figure 11.80a.

Calculations for Equation 1 follow.

$$m_S = \frac{10\pm}{\log 30{,}000 - \log 5000} = 12$$

$$X_S = 12(\log S - \log 5000)$$

$$= 12 \log \frac{S}{5000}$$

$$m_T = \frac{10\pm}{\log 100{,}000 - \log 1000} = 5$$

(but use 6 for convenience in further calculations)

$$X_T = 6 \log \frac{T}{1000}$$

$$m_T = \frac{m_S \, m_R}{m_S + m_R} \quad \text{or}$$

$$6 = \frac{12 m_R}{12 + m_R}$$

from which $m_R = 12$. Now,

$$\frac{m_S}{m_R} = \frac{12}{12} = \frac{1}{1}$$

Thus we see that the T scale is half way between the S and R scales.

From Equation 2,

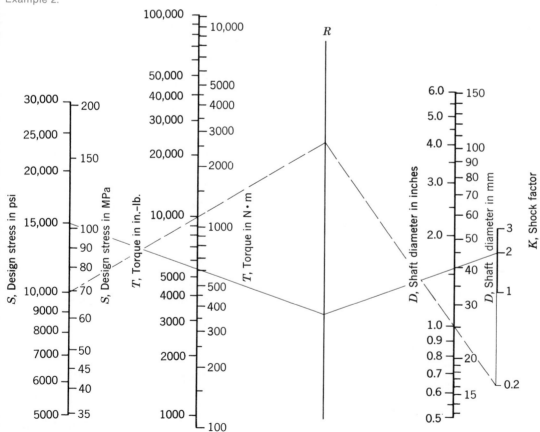

Figure 11.80 (b) Nomogram for the equation $D^3 = \dfrac{16KT}{\pi S}$.

(b)

$$m_D = \frac{10\pm}{3(\log 6 - \log 0.5)} = 3$$

and

$$X_D = 3(3 \log D - 3 \log 0.5)$$
$$= 9 \log 2D$$

Now

$$m_D = \frac{m_R m_K}{m_R + m_K}$$

or

$$3 = \frac{12 \, m_K}{12 + m_K}$$

from which $m_K = 4$. Therefore,

$$X_K = 4(\log K - \log K_1)$$

K_1 is a value on the K scale and is determined by letting $S = 10,000$, $T = 10,000$, and $D = 1$. From the given equation we obtain $K_1 = 0.2$. Once we have located $K_1 = 0.2$, we can locate values of K from the expression

$$X_K = 4 \log \frac{K}{0.2} = 4 \log 5K$$

We also note that

$$\frac{m_R}{m_K} = \frac{12}{4} = \frac{3}{1}$$

from which we can position the D scale.

The completed nomogram is shown in Figure 11.80b. (Metric values are included.) Now the designer can easily and quickly determine shaft sizes for various combinations of values of stress, torque, and shock factor.

z-Type Nomograms for Equations of the Form $f_1(u) = f_2(v) \cdot f_3(w)$

An equation of this form can be readily reduced to the previous form of parallel scales by writing $\log f_2(v) + \log f_3(w) = \log f_1(u)$. In a number of cases, however, a better nomographic solution is the z-type nomogram, especially when one or more of the functions are linear and the ranges of the variables are not too large to affect the desired accuracy in scale readings.

Let us now study the geometry of the z-type nomogram in Figure 11.81. The parallel scales u and v are graduated from their scale equation $X_u = m_u f_1(u)$ and $X_v = m_v f_2(v)$, respectively. The w scale connects $f_1(u_0)$ and $f_2(v_0)$, which are the *zero values of the functions of u and v, respectively.*

How shall we determine the scale equation for $f_3(w)$ so that a line joining u_1 and v_1 will intersect the w scale in w_1, a value that satisfies the equation $f_1(u) = f_2(v) f_3(w)$?

From similar triangles $u_1 w_1 u_0$ and $v_1 w_1 v_0$ we obtain the following relations.

$$\frac{X_u}{X_v} = \frac{K - X_w}{X_w}$$

From which

$$X_u = X_v \frac{K - X_w}{X_w}$$

Now

$$m_u f_1(u) = m_v f_2(v) \frac{K - X_w}{X_w}$$

When

$$\frac{K - X_w}{X_w} = \frac{m_u}{m_v} f_3(w) \qquad (1)$$

then

$$f_1(u) = f_2(v) f_3(w)$$

From Equation 1 we obtain

$$X_w = \frac{K m_v}{m_u f_3(w) + m_v}$$

or

$$X_w = \frac{K}{(m_u/m_v) f_3(w) + 1} \qquad (2)$$

Equation 2 is the scale equation for $f_3(w)$.

To construct a nomogram of the Z type:

1. Place the parallel scales, u and v, a convenient distance apart.
2. Graduate the u and v scales from $X_u = m_u f_1(u)$ and $X_v = m_v f_2(v)$, respectively, *plotting positive values of v downward*.
3. Graduate the w scale from the upper end of the diagonal scale using the equation

$$X_w = \frac{K}{(m_u/m_v) f_3(w) + 1}$$

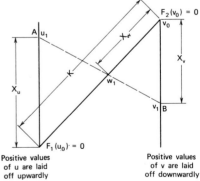

Positive values of u are laid off upwardly

Positive values of v are laid off downwardly

Figure 11.81 Nomographic layout for z-type nomograms of the form $f_1(u) = f_2(v) f_3(w)$.

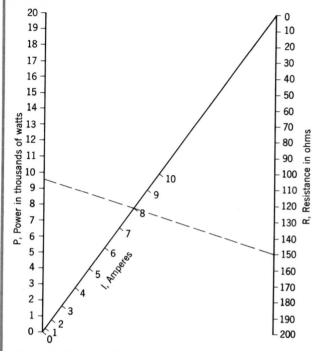

Figure 11.82 $P = I^2R$. Example: $I = 8$, $R = 1500\Omega$; then $P = 9600$ W.

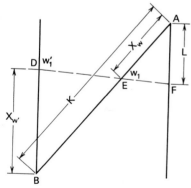

Figure 11.83 Nomographic layout for shortcut method to graduate the w scale of the equation $f_1(u) = f_2(v)f_3(w)$.

EXAMPLE

Consider the expression $P = I^2R$, where $P = $ power in watts (0 to 20,000), $I = $ current in amperes (0 to 10), and $R = $ resistance in ohms (0 to 200). We will assume that the parallel scales will be 10 units long.

Calculations

$$m_P = \frac{10}{20,000} = 0.0005$$

and

$$X_P = 0.0005P \tag{1}$$

$$m_R = \frac{10}{200} = 0.05$$

and

$$X_R = 0.05R \tag{2}$$

The diagonal scale, for variable I, will be graduated from the scale equation

$$X_I = \frac{K}{(0.0005/0.05)I^2 + 1}$$

$$= \frac{K}{(0.01)I^2 + 1}$$

The length of the diagonal scale is K. We can assign any convenient value, such as 1, to K.

The scale equation is now

$$X_I = \frac{1}{(0.01)I^2 + 1} \tag{3}$$

Values of X_I for various values of I are shown in the following table.

I	0	. . .	5	. . .	10
X_I	1		0.8		0.5

Since the functions of P and R are linear, it is only necessary to locate the end values. Other graduations can be easily located geometrically by proportion, thereby eliminating the need to use scale Equations 1 and 2.

The completed nomogram is shown in Figure 11.82.

Simplified Method for Graduating the Diagonal Scale

Study Figure 11.83. Point F is a fixed point on the right vertical scale at a distance L from point A.

The *right side* of the left vertical scale carries a *temporary* w scale. From similar triangles DEB and FEA we obtain the relation

$$\frac{X_{w'}}{L} = \frac{K - X_w}{X_w}$$

From this equation we obtain

$$X_{w'} = L\left(\frac{K - X_w}{X_w}\right)$$

And, from Equation (1) (p. 345), we obtained the relation

$$\frac{K - X_w}{X_w} = \frac{m_u}{m_v} f_3(w)$$

Therefore, it follows that

$$X_{w'} = L\frac{m_u}{m_v} f_3(w)$$

Using this equation we can graduate the *temporary* w scale. The lines that connect point F with the graduations on the temporary scale will intersect the diagonal in points which have the corresponding values of the variable w. This method of graduating the w scale has two advantages over the previous one: (1) when the $f_3(w)$ is linear, a *uniform* scale can be easily graduated on the *temporary* scale; and (2) the length, K, of the diagonal need not be known.

EXAMPLE 1

$$R = \frac{S^2}{A} \quad \text{(aspect ratio of airplane wing)}$$

where

S = span of airplane wing (5000 to 30,000 mm)
R = aspect ratio (4 to 7)
A = area of wing (0 to 230 million mm²)

The equation may be written in type form as $S^2 = AR$. A sketch of the nomogram would look like Figure 11.84a. Note that the S scale is plotted from S = 0. This is done because the distance from S = 0 to S = 5000 is a very small part of the scale from S = 0 to S = 30,000. This is evident from the scale equa-

tion for f(S) in Equation 1, which follows.

$$m_s = \frac{10\pm}{30,000^2 - 5000^2}$$
$$= 0.000000011$$

and

$$X_s = 0.000000011S^2 \qquad (1)$$

For the R scale,

$$m_R = \frac{7.5\pm}{7 - 4} = 2.5$$

and

$$X_R = 2.5(R - 4) \qquad (2)$$

From Equations 1 and 2 we can graduate the S and R scales, respectively. This is shown in Figure 11.84b.

Please note that the *same unit of measure* was used on both scales. Now suppose we place the fixed point F at the 4 on the R scale (actually, F can be placed anywhere on the R scale).

The scale equation for the temporary A' scale is

$$X_{A'} = L\frac{m_S}{m_R} A'$$

or

$$X_{A'} = (10)\frac{0.000000011}{2.5} A'$$
$$= 0.000000044A'$$
$$\text{(for F)} \quad (3)$$

Note that L is the distance from F to the zero value of the function of R, measured in terms of the same unit that was used for the S and R scales.

Since the function of A is linear, the temporary A scale, A', is uniform. Once the maximum value of A is located on the A' scale, it is a simple matter to subdivide the A' scale as is deemed necessary. The maximum value of A is

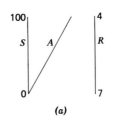

Figure 11.84 (a) Sketch layout of z chart for $R = S^2/A$.

$$A_{max} = \frac{S^2}{R} = \frac{30,000^2}{4}$$

$$= 225 \text{ million}$$

(use given value of 230 million)

Therefore,

$$X_{A'max} = (0.000000044)(230 \text{ million})$$
$$= 10.12 \text{ units}$$

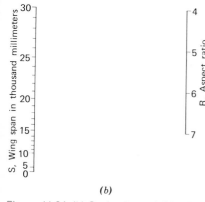

(b)

Figure 11.84 *(b)* Scales for variables S
and R of equation $R = S^2/A$.

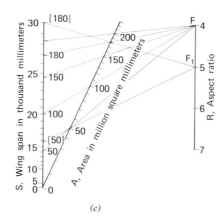

(c)

Figure 11.84 *(c)* Shortcut method for
location and graduation of diagonal
scale A in the equation $R = S^2/A$.

Having located this temporary
graduation, we can locate others.
The lines that pass through points
F and $A' = 0, 50, \ldots, 230$ million mm² contain these values of
A, respectively.

Now we introduce fixed point
F_1. The temporary-scale equation
becomes

$$X_{A'} = (12.5)\frac{0.000000011}{2.5}$$

$$= 0.000000055A'$$

(for F_1)

Values of A' are shown in brackets
in Figure 11.84c. The line that connects F_1 and 180 million contains
$A = 180$ million. Therefore, the
intersection of this line with the
line joining F and $A' = 180$ million
uniquely locates $A = 180$ million.
The diagonal scale, A, is established by joining $A = 180$ million
and $A = 0$.

Finally, the intersections of the
rays through F with the diagonal
establish the locations of the other
graduations of the A scale.

There is no difficulty in using the
simplified method when the zero
values of both functions $f_1(u)$ and
$f_2(v)$ are inaccessible.

z-Type Nomograms for Equations of the Form

$$f_1(u) + f_2(v) = \frac{f_2(v)}{f_3(w)}$$

Study the nomographic layout in
Figure 11.85. It is assumed that the
u and v scales have been graduated
from their scale equations

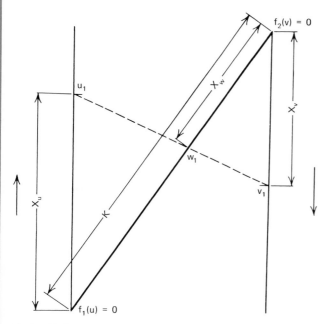

Figure 11.85 Nomographic layout for
equations of the form $f_1(u) + f_2(v) =$
$f_2(v)/f_3(w)$.

$$X_u = m_u f_1(u)$$

and

$$X_v = m_v f_2(v)$$

We should also note that positive values of the variable u are plotted upward and those of v downward. *The diagonal scale passes through the zero values of the functions of u and v.*

From the similar triangles (which are easily recognized in Figure 11.85) we obtain the relation

$$\frac{X_u}{X_v} = \frac{K - X_w}{X_w}$$

or

$$\frac{m_u f_1(u)}{m_v f_2(v)} = \frac{K - m_w f_3(w)}{X_w} \qquad (1)$$

By adding 1 to each side of Equation 1, we obtain

$$\frac{m_u f_1(u) + m_v f_2(v)}{m_v f_2(v)} = \frac{K}{m_w f_3(w)}$$

or

$$m_u f_1(u) + m_v f_2(v) = \frac{K m_v f_2(v)}{m_w f_3(w)}$$

Now, when $m_u = m_v$ and $K = m_w$,

$$f_1(u) + f_2(v) = \frac{f_2(v)}{f_3(w)}$$

Therefore, to construct a nomogram for the preceding form, graduate the scales in accordance with these scale equations.

$$X_u = m_u f_1(u)$$
$$X_v = m_v f_2(u); \; m_u = m_v$$

and

$$X_w = K f_3(w)$$

EXAMPLE
Consider the equation

$$C = \frac{87}{0.552 + m/\sqrt{R}}$$

where

m = coefficient of roughness (0.1 to 2)
R = hydraulic radius (0.5 to 30 ft;
C = Bazin's coefficient for velocity in open channel flow

The equation is placed in the form

$$0.552\sqrt{R} + m = \frac{87\sqrt{R}}{C}$$

from which we obtain the correct form:

$$1.81 m + \sqrt{R} = \frac{\sqrt{R}}{0.0063 C}. \qquad (1)$$

For the m scale,

$$m_m = \frac{10\pm}{1.81(2 - 0.1)} = 3$$

and

$$X_m = 3(1.81 m) = 5.43 m$$
$$\text{(from } m = 0) \quad (2)$$

$$m_R = m_m = 3$$

and

$$X_R = 3\sqrt{R} \qquad \text{(from } R = 0) \quad (3)$$

(*Note.* Scales m and R are plotted from zero because the distances

from $m = 0$ to $m = 0.1$ and from $R = 0$ to $R = 0.5$ are small portions of the m and R scales, respectively.)

Using scale Equations 2 and 3, we plot the graduations for the variables m and R.

For the C scale, we let the length of the diagonal scale equal 10 units (could be 20, 50, or any assigned number). Now

$$X_C = (10)(0.0063 C)$$
$$= 0.063 C$$

The completed nomogram is shown in Figure 11.86.

Additional Forms of Nomograms
Several additional forms of nomograms are used to solve a variety of problems that occur in the area of research, design, manufacturing, etc. Among these are the following, which are used frequently.

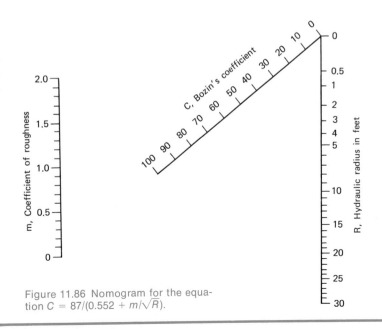

Figure 11.86 Nomogram for the equation $C = 87/(0.552 + m/\sqrt{R})$.

1. $\dfrac{1}{f_1(u)} + \dfrac{1}{f_2(v)} = \dfrac{1}{f_3(w)}$ (see Figure 11.87a).

$$Y_w = \frac{m_u m_v f_4(w)}{m_u f_3(w) + m_v}$$

The scale equations are:

$$X_u = m_u f_1(u)$$

$$X_v = m_v f_2(v)$$

Location of the w scale is determined from

$$\frac{R}{S} = \frac{m_u}{m_v}$$

The w scale is graduated by using one of the following.

(a) $R = m_u f_3(w)$ and parallels to the B axis.

(b) $S = m_v f_3(w)$ and parallels to the A axis.

(c) $X_w = (m_u{}^2 + m_v{}^2 + 2 m_u m_v \cos \theta)^{1/2} f_3(w)$.

2. $f_1(u) + f_2(v) \cdot f_3(w) = f_4(w)$ (see Figure 11.87b. The scale equations are:

$$X_u = m_u f_1(u)$$
$$X_v = m_v f_2(v)$$
$$X_w = \frac{K m_u f_3(w)}{m_u f_3(w) + m_v}$$

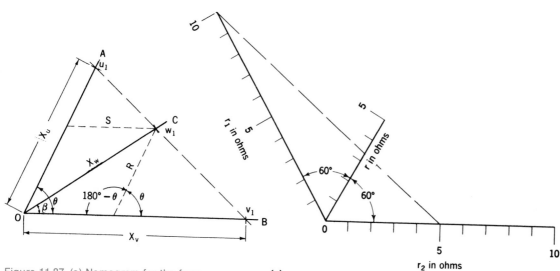

Figure 11.87 (a) Nomogram for the form
$[1/f_1(u)] + [1/f_2(v)] = [1/f_3(w)]$ and nomogram for the equation $1/r_1 + 1/r_2 = 1/r$.

(a)

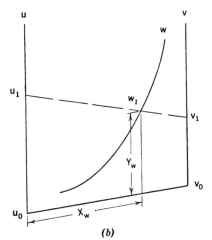

Figure 11.87 (b) Nomogram for the form
$f_1(u) + f_2(v)f_3(w) = f_4(w)$ nomogram
for visibility of signal lights—daylight.
(Courtesy Civil Aeronautics Administration).

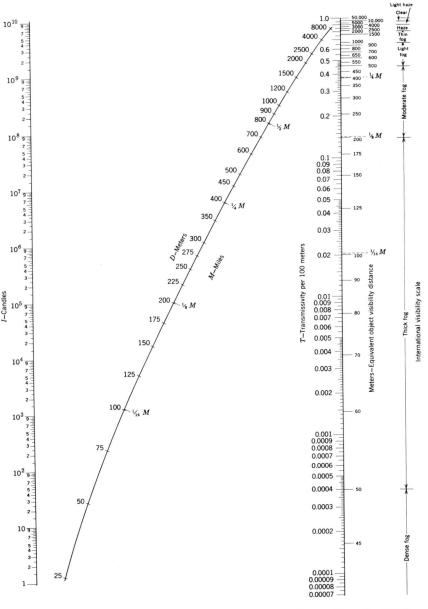

Visibility of signal lights for various candlepowers and transmissivities as calculated by Allard's
law $I = E_0 D^2 T^{-D}$

where I = candle powers of signal light;
E_0 = threshold illumination = 2 hm C (day);
D = distance, hectometers;
T = transmissivity per 100 meters (hectometers).

Equivalent object visibility distance calculated by formula $T^u = 0.02$.

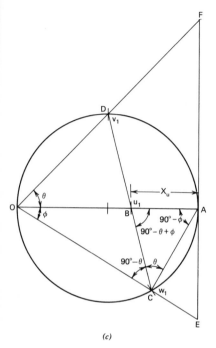

(c)

Figure 11.87 (c) Circular nomogram for
the form $f_1(u) = f_2(v)f_3(w)$ and nomogram
to determine percent of original toxic
substance remaining after several
exchange transfusions. (Courtesy
E. C. Varnum, Barber-Colman Co.)

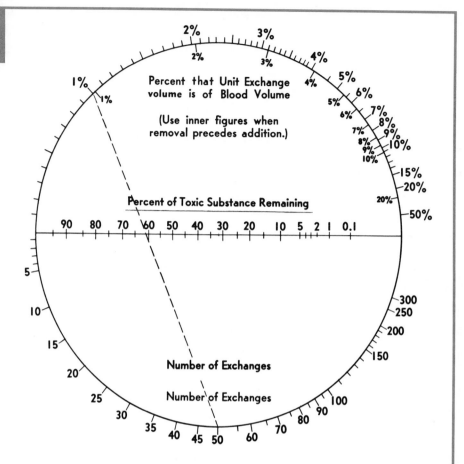

Example: Assume a blood volume of 1000cc.
and that 10cc. is added and with-
drawn. Then the unit exchange
volume is 1% of the blood volume.

After 50 exchanges, there remains 61%
of an original content of toxic substance.

(c)

3. $f_1(u) = f_2(v) \cdot f_3(w)$ (see Figure 11.87c. For a circular nomogram,

$$\tan \theta = m_v f_2(v)$$

$$\tan \theta = m_w f_3(w)$$

$$m_u = m_v \cdot m_w$$

$$\frac{X_u}{2R - X_u} = m_u f_1(w), \text{ where } R \text{ is}$$
the radius of the circle

$$X_u = \frac{2R[m_u f_1(u)]}{1 + [m_u f_1(u)]}$$

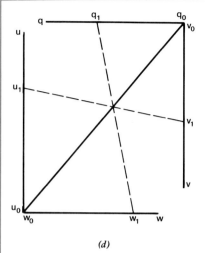

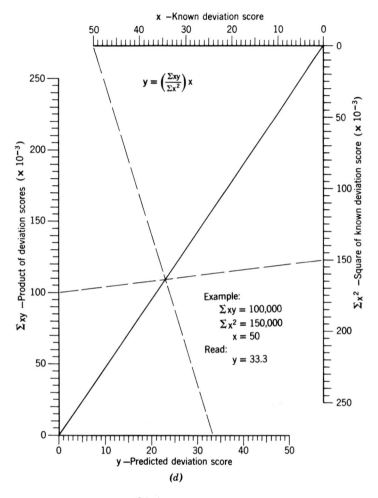

(d)

Figure 11.87 (*d*) Nomogram for the form $f_1(u)/f_2(v) = f_3(w)/f_4(q)$ and nomogram for the equation $y = \dfrac{\Sigma xy}{\Sigma x^2}$.

$$y = \left(\frac{\Sigma xy}{\Sigma x^2}\right) x$$

Example:
$\Sigma xy = 100{,}000$
$\Sigma x^2 = 150{,}000$
$x = 50$
Read:
$y = 33.3$

4. $\dfrac{f_1(u)}{f_2(v)} = \dfrac{f_3(w)}{f_4(q)}$ (see Figure 11.87*d*).

Scale equations:

$X_u = m_u f_1(u)$

$X_v = m_v f_2(v)$

$X_w = m_w f_3(w)$

$X_q = m_q f_4(q)$

$\dfrac{m_u}{m_v} = \dfrac{m_w}{m_q}$

5. $f_1(u) + f_2(v) = \dfrac{f_3(w)}{f_4(q)}$ (see Figure 11.87*e*).

Scale equations:

$X_u = m_u f_1(u)$

$X_v = m_v f_2(v)$

$X_w = m_w f_3(w)$

$m_u = m_v$

$X_q = m_q f_4(q)$ where

$m_q = \dfrac{K m_w}{m_u}$

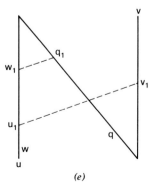

(e)

Figure 11.87 (*e*) Nomogram for the form $f_1(u) + f_2(v) = f_3(w)/f_4(q)$.

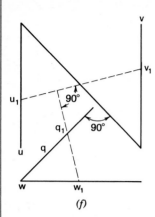

(f)

Figure 11.87 (f) Nomogram for the form $f_1(u) + f_2(v) = f_3(w)/f_4(q)$. Alternative design is preferred. Nomogram for the optimum direction angle, ϕ, from horizontal of first stage (launch direction) for the case $\rho_c \leqq \rho_r$.

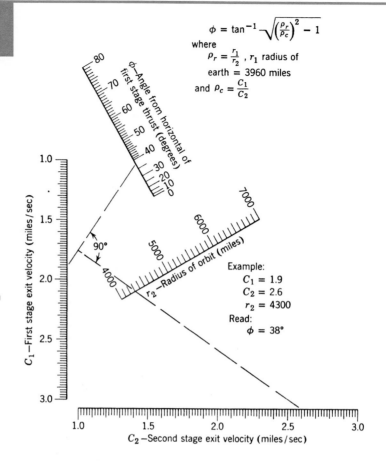

$$\phi = \tan^{-1} \sqrt{\left(\frac{\rho_r}{\rho_c}\right)^2 - 1}$$

where

$\rho_r = \dfrac{r_1}{r_2}$, r_1 radius of earth = 3960 miles

and $\rho_c = \dfrac{c_1}{c_2}$

ϕ—Angle from horizontal of first stage thrust (degrees)

C_1—First stage exit velocity (miles/sec)

r_2—Radius of orbit (miles)

Example:
$C_1 = 1.9$
$C_2 = 2.6$
$r_2 = 4300$

Read:
$\phi = 38°$

C_2—Second stage exit velocity (miles/sec)

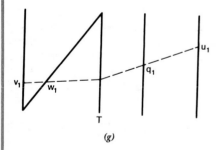

(g)

Figure 11.87 (g) Nomogram for an equation of the form $f_1(u) + f_2(v)f_3(w) = f_4(q)$.

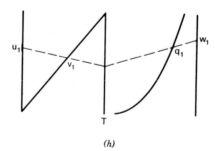

(h)

Figure 11.87 (h) Nomogram for an equation of the form $f_1(u)f_2(v) + f_3(w)f_4(q) = f_5(q)$.

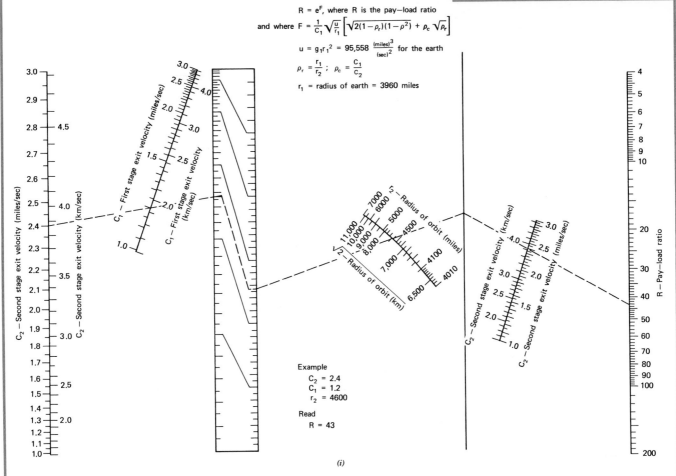

$R = e^F$, where R is the pay—load ratio

and where $F = \frac{1}{C_1}\sqrt{\frac{u}{r_1}}\left[\sqrt{2(1-\rho_r)(1-\rho^2)} + \rho_c\sqrt{\rho_r}\right]$

$u = g_1 r_1^2 = 95,558 \frac{(miles)^3}{(sec)^2}$ for the earth

$\rho_r = \frac{r_1}{r_2}$; $\rho_c = \frac{C_1}{C_2}$

r_1 = radius of earth = 3960 miles

Example
$C_2 = 2.4$
$C_1 = 1.2$
$r_2 = 4600$

Read
$R = 43$

(i)

Figure 11.87 (*i*) Example of combinations of nomographic forms. Nomogram for optimum payload ratio, *R*, for the case $\rho_c \leqq \rho_r$.

6. $f_1(u) + f_2(v) = \dfrac{f_3(w)}{f_4(q)}$ (see Figure 11.87*f*.

Same as in step 5 except for alternative form of nomogram.

7. *Combinations of Type Forms* (see Figure 11.87*g*).

$f_1(u) + f_2(v) \cdot f_3(w) = f_4(q)$

Let

$f_2(v) \cdot f_3(w) = T$

and

$f_1(u) + T = f_4(q)$

8. *Combinations of Type Forms* (see Figure 11.87*h*).

$f_1(u) \cdot f_2(v) + f_3(w) \cdot f_4(q) = f_5(q)$

Let

$f_1(u) \cdot f_2(v) = T$ (1)

and

$T + f_3(w) \cdot f_4(q) = f_5(q)$ (2)

Equations 1 and 2 can be combined to form the nomogram shown in the figure.

An example of a combination type is shown in Figure 11.87*i*.

Exercises

Graphical Calculus

1. It is planned to use an irregularly shaped ground depression for a water reservoir. Field data have been obtained to determine the contours from which the cross-sectional areas were obtained. Tabular values of the areas and the distances to the cross sections are shown below.

Make a plot of these data, draw a "fair" curve through (or near) the data points, and then integrate the curve, using the pole-and-ray method to determine (a) the total volume in cubic meters, and (b) the volume that would be required to bring the reservoir up to full capacity after the level had dropped 3 m.

A, area, m²	5,000	10,000	30,000	51,500	74,600	91,000	84,500	75,000	59,800
D, distance, m	0	2.0	3.0	3.6	5.0	6.1	7.2	8.4	10.2

2. The ampere-time record of a power station is shown in the following table. What was the total number of ampere-hours supplied by this station?

Plot the data and then draw a curve that fits the data well. Use the pole-and-ray and planimeter methods and Simpson's rule to solve the problem. Tabulate the results.

t, time	12 noon	1	3	3	4	5	5:30	6
A	4,700	3,100	3,000	2,700	3,100	7,900	15,000	19,000
t, time	6:30	7	8	9	10	11	12	
A	22,500	23,000	20,000	17,000	11,000	9,500	8,200	

3. A relation between time, t, in minutes and speed, S, in kilometers per hour is shown in the following table. Plot the curve $S = f(t)$ and then use the pole-and-ray method to integrate the curve. Include a scale to read distance traveled in kilometers during any interval of time. What is the total distance traveled?

t, min	0	0.5	1	2	3	4	5	6	7	8	9	10	11	12
S, km/hr	0	11	19	25	29	32	39	45	42	36	31	21	15	10

4. Determine the flow, in cubic meters per minute, of a river whose average cross section can be plotted from the following field data.

The average velocity of flow is 1.2 m/s. Use (a) the pole-and-ray method, and (b) Simpson's rule.

S, horizontal distance, m	0	3	6	9	12	15	18	21	24	27	30
d, depth, m	0	4	4.5	5.0	5.5	6.0	5.2	5.0	4.7	4.2	0

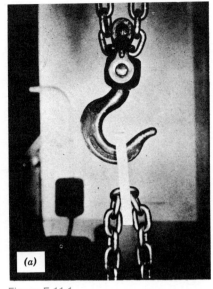

Figure E-11.1

5. Figure E-11.1*a* shows the hoist hook unloaded; Figure E-11.1*b* shows the hoist hook with maximum load. The hook has a rated capacity of 2 metric tons. The table shows the test results of elongation of the hook by the various loads.

Plot the data and draw the curve that best represents the plotted points. Determine the work done in opening the hook when loads varying from 35.64 to 93.54 kN were applied. Solve graphically and also solve by using Simpson's rule. Compare the results.

Elongation, mm	0	8.13	9.90	14.98	18.80	22.35	33.00	43.18	55.88
Load, kN	35.64	53.46	66.82	77.95	89.00	93.54	98.00	98.00	93.54

6. Subdivide the plot shown in Figure E-11.2 into three lots of equal area. The dividing lines are parallel to edge *OA*.

7. The profile of a solid concrete playground slide is shown in Figure E-11.3. The width of the slide is 1 m. What is the mass of the concrete in kilograms? Assume that 1 m³ weighs 2400 kg.

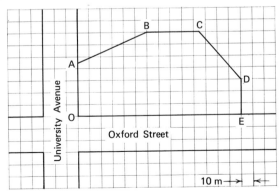

Figure E-11.2

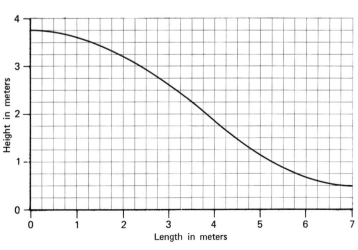

Figure E-11.3

8. The velocity, V, of a moving body at various times, t, is shown in the following table. Graphically determine the total distance traveled in 10 s. $S = 0$ when $t = 0$.

t, s	0	1.4	2.6	3.4	5.0	6.2	7.7	8.9	10.0
V, m/s	9.8	8.0	6.8	6.1	5.2	4.6	4.0	3.6	3.0

9. A simple beam 7 m long supports a uniform load of 150 N/m. Graphically determine (a) the shear diagram, and (b) the bending-moment diagram.

10. A vertical rectangular plate 2 m wide and 3 m long is immersed in fresh water. One of the shorter edges is in contact with the water surface. Locate the center of pressure (C.P.) on the plate and determine the total force on one side of the plate. (*Note.* The C.P. is the point at which the total force is assumed to act.)

11. Determine the $\bar{X}$- and $\bar{Y}$-coordinates of the centroid of the area shown in Figure E-11.4. Use the *graphomechanical* method in the solution of the problem.

12. Determine the moment of inertia of the structural steel section shown in Figure E-11.5 about axis AB. Use the *graphomechanical* method.

13. Find the moment of inertia of a 20×30 cm rectangle about an axis

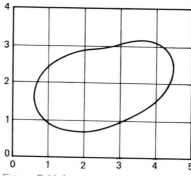

Figure E-11.4

that contains a 30-cm edge. Use the graphomechanical method.

14. The relation of temperature, C, of cooling water to time, t, is shown in the

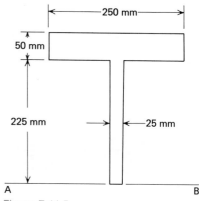

Figure E-11.5

following table. Graphically, determine the rate-of-cooling curve and then find the rate at $t = 8$ min.

t, min	0	2	4	6	10	14	20	25
C, degrees Celsius (C°)	79.8	68.0	60.5	55.2	45.1	37.2	31.8	25.6

15. The slider-crank mechanism shown in Figure E-11.6 is composed of crank $CA = 10$ cm and arm $AB = 30$ cm; arm AB is attached to the crank at A and to the sliding unit at B. The crank rotates about C at 600 rpm. Plot the displacement-time curve for point B and then determine the velocity-time

and acceleration-time curves. Include scales for velocity and acceleration.

16. The tangential cam shown in Figure E-11.7 rotates counterclockwise about point C at 300 rpm. Plot the displacement-time curve of the follower and then graphically determine the velocity-time and acceleration-

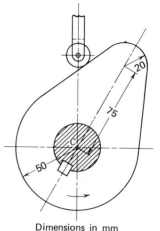

Dimensions in mm
Figure E-11.7

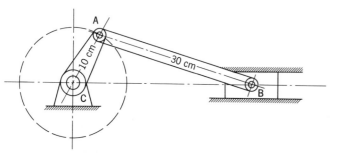

Figure E-11.6

time curves. What is the maximum velocity and maximum acceleration of the follower?

17. The approximate speeds of a moving van on a 30-min run are shown in the following table. Plot a curve to represent the data and then determine (a) the displacement-time curve, and (b) the acceleration-time curve.

t, min	0	0.5	1	3	5	7	10	13	19	20	22	23	25	26	27	30
S, km/hr	0	16	32	49	56	61	67	72	72	74	85	88	88	83	67	0

18. A simply supported beam 600 cm long carries a concentrated load at a distance of 360 cm from the left end and has the following bending moments.

l, distance from left end	0	360	0
M, bending moment, cm-kN	0	2880	0

(*Note.* The bending-moment diagram is linear from the largest value to the zero values.) Plot the bending-moment diagram and then graphically determine the shear diagram. What is the magnitude of the load?

19. Plot the curve $y = \sin \theta$, θ varies from 0 to 2π rad. Graphically determine the curve $y = \cos \theta$.

20. Plot a "fair" curve to represent the following data.

Determine the acceleration-time curve. What is the acceleration when $t = 9$ s? Check the result by using the graphonumerical method.

t, s	0	2	4	6	8	10	12	14
V, velocity, m/s	0	7.4	15.0	21.3	26.1	29.8	32.4	35.0

Empirical Equations

1. The data in the following table relate values of R, resistance in ohms and the corresponding values volts, resulting from a lamp test. Determine the empirical equation using the graphical method; then check by the method of selected points.

V, V	65	70	75	80	85	90	95	100	105	110	115
R, Ω	138	142	146	150	154	157	161	166	170	173	177

2. The following data relate values of L, load in Newtons, and the corresponding effort, E, also in Newtons, to lift the load. Determine the empirical equation using the graphical and selected points methods.

L, N	60	80	120	160	200	240	280	320	360
E, N	5.0	6.0	8.4	10.8	13.2	15.2	18.0	20.4	22.5

3. Determine the equations of lines m and n as shown in Figure E-11.8.

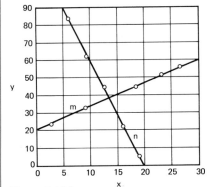

Figure E-11.8

4. Determine the equations of lines r and s as shown in Figure E-11.9.

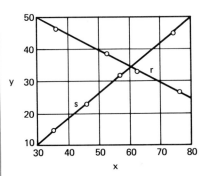

Figure E-11.9

5. Determine the relation between H, the head of water over a weir, and Q, the quantity of water flowing over the weir. Plot the following data and find the equation graphically and by using the selected-points method.

H, cm	5.70	7.20	8.10	8.73	9.36	10.53	12.33
Q, kg/min	64.6	114.2	143.5	189.5	221.0	279.1	403.0

6. The water displacement, V, of a vessel varies with h, the distance from the keel to the water level, as shown.

Plot the data as a straight line on the appropriate coordinate grid; then determine the relation between the variables. Solve graphically and check by the method of selected points.

h, m	2.9	3.3	3.9	5.4	6.2
V, m³	1161	1440	1832	3020	4400

7. Determine the relation between V, the speed of a vessel and the horsepower required to maintain the speed. The data follow. Solve graphically and by the method of selected points.

V, knots	11.6	13.8	16.0	17.7	20.1	21.8
hp	1,880	3,150	4,960	7,355	10,525	13,100

8. Determine the equations of lines A and B in Figure E-11.10. Solve graphically and check by using the method of selected points.

9. Determine the equations of lines C and D in Figure E-11.11.

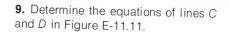

Figure E-11.11

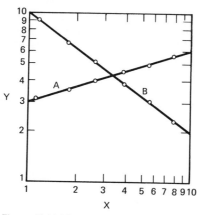

Figure E-11.10

10. Determine the relation between barometric pressure, P, and elevation, E, above sea level. The data follow.

Solve graphically and check by the method of selected points.

E, m	0	273	838	1450	2110	3240
P, cm mercury	75.0	72.5	67.5	62.5	57.5	50.0

11. Determine the relation between, μ, coefficient of friction, and T, temperature of bearings operating at a constant speed. The data follow. Solve graphically and check the result by the method of selected points.

t, degrees Celsius (C°)	48.9	43.3	37.8	32.2	26.7	21.1	15.6
μ, coefficient of friction	0.005	0.006	0.007	0.009	0.010	0.021	0.075

12. Determine the relation between the compressive strength of concrete, S, and the water-cement ratio, Cr. The data follow. Solve graphically and check by using the method of selected points.

S, kN/m²	31,000	25,500	20,700	16,900	13,800	11,700	9,650
Cr, liters/kg	0.60	0.70	0.80	0.90	1.00	1.10	1.20

13. Determine the relation between speed, S, in revolutions per minute and capacity, C, in liters per minute, of a pump. The data follow. Solve graphically and check by using the method of selected points.

S, rpm	600	700	800	900	1000	1100	1200
C, liters/min	5,678	7,380	8,720	9,850	11,000	12,100	13,250

14. Determine the relation between, A, the amplitude in centimeters of a long pendulum, and t, the swinging time in minutes, since it was set in motion. The data follow. Solve graphically.

t, min	0	1	2	3	4	5	6	7
A, cm	26.7	13.0	6.4	3.2	1.7	1.0	0.9	0.4

15. Determine the equations of lines C and D as shown in Figure E-11.12. Solve graphically and check by the method of selected points.

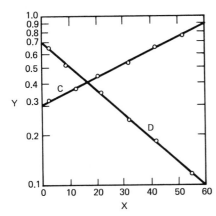

Figure E-11.12

16. Determine the equations of lines A and B as shown in Figure E-11.13. Solve graphically and check by the method of selected points.

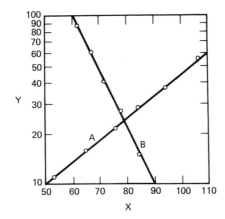

Figure E-11.13

Functional Scales

1. Construct a scale for the function $f(u) = u^2$. u varies from 0 to 12. The scale length is approximately 20 cm. Show the scale equation.

2. Construct a scale for the function $f(u) = (u^2 + 3u)$. u varies from 3 to 12. The scale length is approximately 20 cm. Show the scale equation.

3. Construct a scale for the function $f(u) = 3 \log u$. u varies from 100 to 500. The scale length is approximately 15 cm. Show the scale equation.

4. Construct a scale for the function $f(u) = 1/u$. u varies from 5 to 10. The scale length is approximately 15 cm. Show the scale equation.

5. Construct a scale for the function $f(u) = (u - 2)$. u varies from 0 to 12. The scale length is approximately 20 cm. Show the scale equation.

6. Construct a scale for the function $f(u) = \cos u$. u varies from 0 to 90°. Select a suitable scale length in centimeters. Show the scale equation.

7. Construct adjacent scales for the relation $A = \pi R^2$. R varies from 5 to 25 cm.

8. Construct adjacent scales for the relation $V = \sqrt{2gh}$, where V is the velocity in centimeters per second, and h (60 to 360 cm) is the velocity head in centimeters.

9. Construct adjacent scales for the relation $S = 4\pi R^2$, where S is the area of a spherical surface in square centimeters, and R (10 to 30 cm) is the radius of the sphere in centimeters.

10. Construct *nonadjacent* scales for the relation $C = \frac{5}{9}(F - 32)$. Temperature in degrees Celsius (°C) varies from -40 to $100°$.

11. Construct *nonadjacent* scales for the relation $V = \frac{4}{3}\pi R^3$, where V is the volume of a sphere in cubic centimeters and R (10 to 30 cm) is the radius of the sphere in centimeters.

12. Construct *nonadjacent* scales for the relation $A = 0.433S^2$, where A is the area of an equilateral triangle in square centimeters and S (5 to 30 cm) is the length of a side of the triangle in centimeters.

Nomography

CONCURRENCY CHARTS

1. Construct a concurrency chart for the equation $I = bd^3/12$, where I is the moment of inertia of a rectangle; its axis is parallel to b, which is the width of the rectangle, and d, its height. The ranges of the variables are b (1 to 5 m) and d (2 to 10 m). I has the dimension m^4.

2. Construct a concurrency chart for the equation $S = \pi DN/12$, where S is the cutting speed in millimeters per minute, D is the diameter of the work (10 to 300 mm), and N equals 150 to 2500 rpm.

3. Construct a concurrency chart for the equation $P = 3EI \cos\theta$, where P is the power in watts, E is the line voltage (50 to 220 V), and I is the line current (5 to 60 A).

ALIGNMENT CHARTS

1. Design an alignment chart for the equation $kpl = d/l$, where kpl equals kilometers per liter, d is the distance traveled in kilometers (40 to 400 km), and l is the number of liters of fuel (10 to 50 liters).

2. Design an alignment chart for the equation $V = (\pi r^2 h)10^{-6}$, where V is the volume of a right circular cylinder in cubic meters, r is the radius of the circular base (10 to 25 cm), and h is the height of the cylinder (25 to 100 cm).

3. Design an alignment chart for the equation $M = wl^2/8$, where M is the bending moment of a simply supported and uniformly loaded beam, w is the load (50 to 500 N/m), and l is the length of beam (2 to 10 m).

4. Design an alignment chart for the equation $P_1/P_2 = (V_2/V_1)^{1.41}$ (adiabatic expansion of air), where P_1 is the initial pressure (35 to 2000 kN/m²), P_2 is the final pressure (20 to 1700 kN/m²), V_1 is the initial volume (0.3 to 3.0 m³), and V_2 is the final volume (0.3 to 3.5 m³).

5. Design an alignment chart for the equation $N_s = NQ/H^{3/4}$, where N_s is the specific speed of a centrifugal pump, Q is the flow rate (4000 to 30,000 liters per minute), N is the speed (100 to 3000 rpm), and H is the head (6 to 60 m).

6. Design a z-type alignment chart for the equation $B = d^2 n/0.215$, where B is the brake horsepower in watts (0 to 75,000 W), d is the diameter of cylinder (0 to 127 mm), and n is the number of cylinders (2 to 12).

7. Design a z-type alignment chart for the equation $f = 1 + 20,000/(144l^2/9000r^2)$ (Gordon column formula), where f is the fiber stress in thousands of pounds per square inch (0 to 20), l is the length of the column in feet (0 to 50), and r is the radius of gyration in inches. *Also, graduate each scale with the appropriate SI units.*

8. Design a z-type alignment chart for the equation $S = (N_s - N_R/N_s)100$ (induction motor slip), where S is the percent of slip, N_s is the synchronous speed (0 to 1800 rpm), and N_R is the rotor speed (0 to 1800 rpm).

9. Design an alignment chart for the equation $X_L = 2\pi fL$, where X_L is the reactance in ohms, f is the frequency in cycles per second (20 to 130), and L is the inductance in henrys (0 to 30).

10. Design an alignment chart for the equation $1/u + 1/v = 1/f$ (lens formula), where u is the object distance (0 to 100), v is the image distance (0 to 80), and f is the focal distance (0 to 50). All distances are in the same unit of measure.

11. Design an alignment chart for the equation $N = 0.41QP + 75P - M$, where N is the profit ($1000 to $15,000), Q is the quantity of books sold (1000 to 10,000), P is the list price ($5 to $20), and M is the manufacturing cost ($4000 to $25,000).

12. Design an alignment chart for the equation $A = d(b_1 + b_2)/2$, where A is the area of a trapezoid, b_1 is the length of one of the parallel sides (0 to 500 mm), b_2 is the length of the other parallel side (0 to 450 mm), and d is the altitude of the trapezoid (0 to 300 mm).

Chapter 12 Computer Graphics

INTRODUCTION

Computer use has greatly enhanced the work of engineers. Much time is saved in the solution of complex mathematical problems, in analyzing data, in storing, and retrieving design information, in modifying design data, in making design changes, etc.

The design engineer who is relieved of time-consuming repetitive work can devote much more talent and experience to professional creative endeavors.

The availability of reasonably priced minicomputers and peripheral equipment such as x-y plotters, digitizers, cathode-ray tubes (CRT) with light pen operation, and alphanumeric and function keyboards for design displays (and their modification when necessary) makes it economically possible to input graphical information and to obtain graphical outputs. This person-machine relationship enables the engineer-designer to carry on effective communication through the use of *interactive computer-graphics systems*.

Many companies, such as The Boeing Company, General Motors, and McDonnell-Douglas Corporation, successfully use interactive computer-graphics systems.

Highlights of the system used by the McDonnell Aircraft Company follow.

McDONNELL AIRCRAFT COMPANY

In the early 1970s the company organized the computer-aided technology (CAT) project, which pulled together all of the engineering disciplines under one project management and provided for complete interface coordination with computer-aided manufacturing (CAM) activities.

The interactive graphics capability ranges from advanced design sizing and performance analysis to quality assurance inspection data (see Figure 12.1).

Primary Graphics Module

"The primary graphics module in this system is Computer-Aided Design Drafting (CADD) after which all the others are patterned. All are resident on the central computer complex and hence are accessible to central data base files."

The files are created by each discipline and accessed, as needed, by the other disciplines. For instance, the "loft" information is directly accessible by the structures analysts, the design engineers, the numerical control programmers, and the part inspectors. Output in the form of drawings or numerical control milling instructions is sent by direct numerical control wire (DNC) to plotters and milling and inspection machines. All of the interactive graphic display devices are of the IBM 2250 type, as shown in Figure 12.2.

Figure 12.2 Cadd display console arrangement. (Courtesy McDonnell Aircraft Co.)

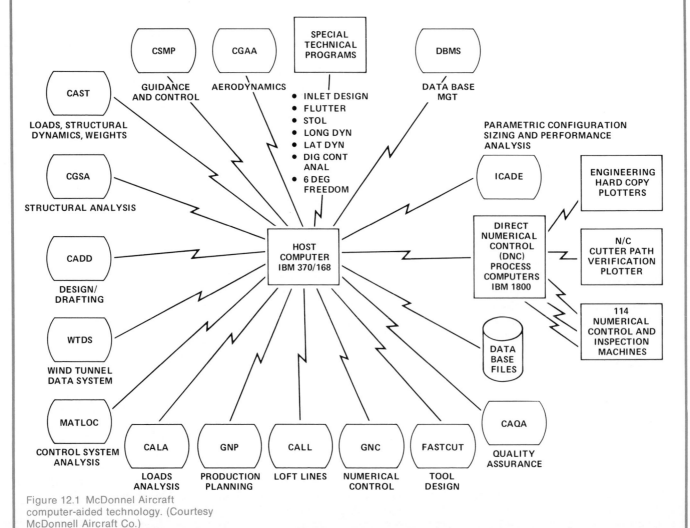

Figure 12.1 McDonnel Aircraft computer-aided technology. (Courtesy McDonnell Aircraft Co.)

The term CADD (pronounced caddy) denotes various computer techniques and applications where data are either presented or accepted by a computer in a geometric form as opposed to alphanumerics only. CADD is "interactive," which implies that there is an efficient, real time interplay of actions between the console operator and the system hardware devices. CADD therefore describes an interactive and conversational mode of operation, utilizing a display console where the engineer may describe his design, perform analysis procedures, and make changes to the design if he so chooses.

A designer is normally concerned with creating a geometric representation of a physical object. In the drawing board mode, these are lines drawn on paper or mylar in two dimensions. This representation, when the CADD package is used, is in terms of an exact mathematical description in the computer's memory, either as a two-dimensional line drawing, a three-dimensional wireframe drawing, or as a completely surfaced definition of the model in three dimensions. In addition, the graphic representation is displayed on the cathode-ray tube (CRT) (Figure 12.3). Hardware and software features enable the operator to converse with the computer by detecting, with a light pen, elements displayed on the CRT screen and by inputting specific instructions with the alphanumeric and functional keyboards.*

The designer is in command of the computer while the essence of the normal working environment is retained. Consequently, the designer is able to address and solve problems in a comfortable geometric language *without having to comprehend the intricacies of the computer or programming languages.*

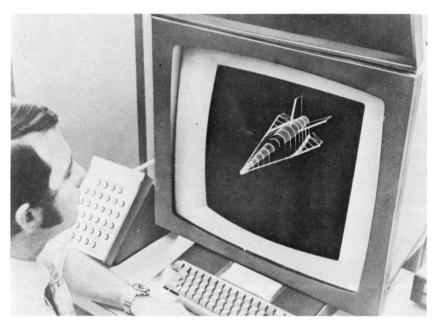

Figure 12.3 Graphic representation displayed on the cathode-ray tube (CRT). (Courtesy McDonnell Aircraft Co.)

* From a paper by C. H. English (McDonnell Aircraft Co.), "Interactive Computer Aided Technology, Evolution in the Design/Manufacturing Process." October 1978.

How the CADD System Operates

The CADD module is a very generalized, versatile system for creating, interfacing, and storing geometric data in mathematical form in the computer. Hence, the CADD system is the focal point for a wide range of development and fabrication capability at MCAIR. The engineer interfaces with the computer using the graphics console in three ways. He can use the fiber optics *light pen* to address the face of the display tube with the computer programmed to respond to the function lo-cated at the tube face addressed by the light pen. This is the prime interface used in CADD. The *alphanumerical keyboard* below the display surface is used to key in specific data or text information. The third interface is the *function keyboard* to the left of the screen. The keyboard (Figure 12.2) consists of 32 keys which initiate computer commands for frequently used functions. Three types of functions are on this keyboard: create type function (i.e., point, line, arc, conic, etc.); manipulate functions (e.g., translate/rotate, flip, depth, intersection); and administrative type functions such as reject, delete, display type, etc.

By these interfaces the engineer can access and create data in a very generalized, versatile manner permitting real time solutions of geometry problems which are fast, accurate and cost effective."

The CADD system is effective as a design tool in Advanced Design. The aircraft configurationist is able to rapidly and accurately construct the vehicle, using time-saving CADD capabilities such as the automated wing planform routine, the kinematics routine for defining the landing gear and controls spatial mechanism geometry, the vision plot routine for ascertaining the limits of the pilot's vision from the cockpit, etc.*

A few applications of the CADD system follow.

Landing Gear Kinematics.

Among the early uses of the CADD was the synthesizing of three-dimensional spatial geometry and path of travel of the main landing gear mechanism of the F-15 to make certain that the required precision of motion was obtained (see Figure 12.4).

Crew Station Design

The cockpit layout is based on requirements such as number of crew; method of crew egress; pilot vision limits; instruments needed; and cockpit clearance.

By positioning the pilot's eye at the desired location, the designer can construct the canopy and establish the fore-aft and over-the-

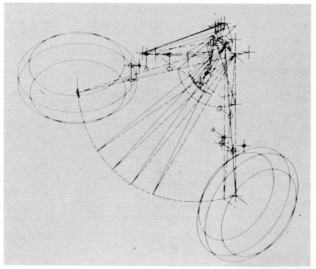

Figure 12.4 F-15 landing gear kinematics. (Courtesy McDonnell Aircraft Co.)

side vision constraints (see Figure 12.5). With this information the designer can construct console width, side wall thickness, sill width, etc.

Sheet Metal Design

Conventional sheet metal parts are readily defined with CADD. A CADD operator can input the contour and flange limits of the part by light pen detection. In turn, the computer asks questions that may be answered by selecting menu items displayed on the CRT screen or on the alphanumeric keyboard.

The variables that require answers are, for example, flange width, flange bend radius, thickness of material, and rivet size and spacing. From these inputs, the program develops a completed flat pattern of the part showing joggles, developed flanges, location of rivets, bend angles, etc. A typical flat pattern sheet metal part is shown in Figure 12.6.

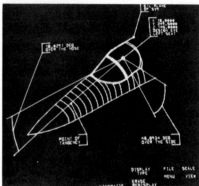

Figure 12.5 Crew station design. (Courtesy McDonnell Aircraft Co.)

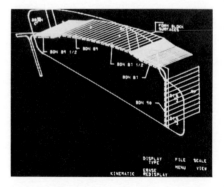

Figure 12.6 Sheet metal design. (Courtesy McDonnell Aircraft Co.)

* From a paper by S. A. LaFavor and J. H. Schulz (McDonnell Aircraft Co.), "Integrating the Design and Manufacturing Process Through Computer Aided Technology, November 1976.

Other Types of Design Tasks

Among the design tasks that have been accomplished by using the CADD system are: complex spatial kinematic synthesis problems; determination of clearance problems of moving elements with respect to structure; routing of hydraulic and fuel lines; and generally any type of layout work (see Figure 12.7).

These tasks are accomplished economically, through faster and more reliable outputs than was ever possible prior to the advent of interactive computer graphics.

Manufacturing Interface

The manufacturing programmer using CADD retrieves a copy of the CADD part model and then can add tooling lugs, clamps, and other manufacturing requirements (see Figure 12.8).

The N/C (numerical control) parts programmer, using GNC (graphics numerical control), accesses the revised CADD drawing and, using the graphics display interactively, creates an N/C source program that describes the manner for a cutter to shape the part. Included are the N/C milling machine to be used, cutter size, feeds, and speeds.

Once the programmer is satisfied with the graphically depicted cutter motion, the source program is processed through the APT (automatically programmed tool), which retrieves the engineering geometry mathematics and generates the cutter path (see Figure 12.9).

"The parts programmer need only be concerned with the geometry labels and need not be involved with the complex mathematics which define the part." If changes in cutter sizes or part geometry are necessary, the source program can be easily changed from an alpha-numerical terminal, and the source program can be reprocessed through APT for revised cutting in-

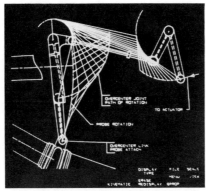

Figure 12.7 Mechanism equivalent positions. (Courtesy McDonnell Aircraft Co.)

Figure 12.8 Manufacturing interface. (Courtesy McDonnell Aircraft Co.)

Figure 12.9 Generating the cutter path. (Courtesy McDonnell Aircraft Co.)

structions. The part is then machined on one of the many N/C milling machines"* (see Figure 12.10).

Pictorial Schematic of CAD and CAM

An overall view of the flow of steps in computer-aided drafting, design, and computer-aided manufacturing using the APT language is

* Ibid.

Figure 12.10 Numerical control milling machines. (Courtesy McDonnell Aircraft Co.)

shown in Figure 12.11. Careful study of this figure will help you to understand the flow of operations that are used in a fairly comprehensive system.

Even though the computer and its ancillary hardware relieve the engineer and drafter of many repetitive tasks and make it easy to retrieve data and drawings, *it is still the engineer's creativity and experience that bring about the new designs—not the computer*. The engineers and design drafters must have a thorough understanding of the fundamental principles of engineering graphics in all of its aspects and a good working knowledge of their application to the various problems that arise in design. This is essential to effective computer programming and to the design engineer's efforts in carrying on design work using interactive computer graphics systems.

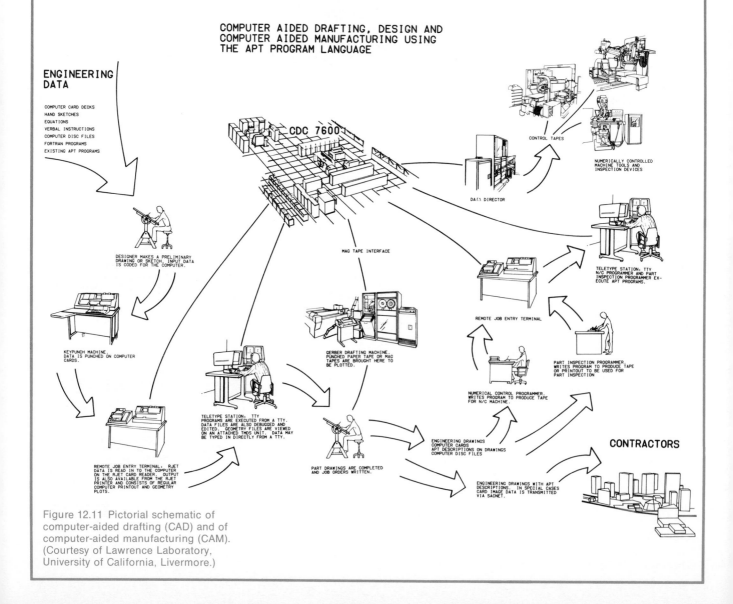

COMPUTER AIDED DRAFTING, DESIGN AND COMPUTER AIDED MANUFACTURING USING THE APT PROGRAM LANGUAGE

ENGINEERING DATA

COMPUTER CARD DECKS
HAND SKETCHES
EQUATIONS
VERBAL INSTRUCTIONS
COMPUTER DISC FILES
FORTRAN PROGRAMS
EXISTING APT PROGRAMS

CDC 7600

CONTROL TAPES

NUMERICALLY CONTROLLED MACHINE TOOLS AND INSPECTION DEVICES

DATA DIRECTOR

DESIGNER MAKES A PRELIMINARY DRAWING OR SKETCH. INPUT DATA IS CODED FOR THE COMPUTER.

MAG TAPE INTERFACE

TELETYPE STATION; TTY N/C PROGRAMMER AND PART INSPECTION PROGRAMMER EXECUTE APT PROGRAMS.

REMOTE JOB ENTRY TERMINAL

KEYPUNCH MACHINE. DATA IS PUNCHED ON COMPUTER CARDS.

GERBER DRAFTING MACHINE. PUNCHED PAPER TAPE OR MAG TAPES ARE BROUGHT HERE TO BE PLOTTED.

PART INSPECTION PROGRAMMER WRITES PROGRAM TO PRODUCE TAPE OR PRINTOUT TO BE USED FOR PART INSPECTION

NUMERICAL CONTROL PROGRAMMER. WRITES PROGRAM TO PRODUCE TAPE FOR N/C MACHINE.

TELETYPE STATION; TTY PROGRAMS ARE EXECUTED FROM A TTY. DATA FILES ARE ALSO DEBUGGED AND EDITED. GEOMETRY FILES ARE VIEWED ON AN ATTACHED TMDS UNIT. DATA MAY BE TYPED IN DIRECTLY FROM A TTY.

CONTRACTORS

REMOTE JOB ENTRY TERMINAL. RJET DATA IS READ IN TO THE COMPUTER ON THE RJET CARD READER. OUTPUT IS ALSO AVAILABLE FROM THE RJET PRINTER AND CONSISTS OF REGULAR COMPUTER PRINTOUT AND GEOMETRY PLOTS.

PART DRAWINGS ARE COMPLETED AND JOB ORDERS WRITTEN.

ENGINEERING DRAWINGS COMPUTER CARDS APT DESCRIPTIONS ON DRAWINGS COMPUTER DISC FILES

ENGINEERING DRAWINGS WITH APT DESCRIPTIONS. IN SPECIAL CASES CARD IMAGE DATA IS TRANSMITTED VIA SACNET.

Figure 12.11 Pictorial schematic of computer-aided drafting (CAD) and of computer-aided manufacturing (CAM). (Courtesy of Lawrence Laboratory, University of California, Livermore.)

Selected Applications of Computer Graphics

Several interesting applications of computer graphics occur in the following areas.

1. Biomechanics research—human locomotion.
2. Surface display—mapping.
3. Generating mathematical surfaces.
4. Computer-aided design of nomograms.
5. Solutions to Space Problems, such as true length of a line segment, perpendicular distance from a point to a line, perpendicular distance from a point to a plane, and distance between two skew lines.
6. Isometric drawings.

Biomechanics Research—Human Locomotion

In the previous chapter we discussed graphical differentiation and cited several examples; among them were two dealing with human locomotion—the Pelvis Study (Example 4) and the Ankle Flexion Study (Example 5). It was pointed out that in the actual studies the many investigations required the use of computer programs and ancillary equipment to produce results economically within a reasonable amount of time.

EXAMPLE: COMPUTED DERIVATIVES OF ANKLE FLEXION ANGLES

Most of the data curves were plotted, as they were recorded, as a function of time. Once the measurements are in the computer, it is a relatively straightforward step to compute derivatives of the curves. A simple procedure was used to compute first and second derivatives of hip, knee, and ankle flexion. "The derivative at each point on the curve is defined as the slope of a least-square-error straight line through that point and an arbitrary but equal number of adjacent preceding and succeeding points. Points near the ends of the curve are treated as if the ends of the curve met each other. The more points that are used in each slope computation, the smoother will be the resulting derivative curve. Multiple derivatives can be obtained by simply repeating the derivative computation. Much more sophisticated techniques could be used, but even this elementary method produces useful results with the relatively low-frequency content and high sample rate of the data collected here (see Figure 12.12)."* The zero reference on the vertical scale of each plot is the value of the variable recorded when the subject was standing still in a relaxed position. Standing reference runs were made both before and after each

* L. W. Lamoreux, "Kinematic Measurements in the Study of Human Walking," *Bulletin of Prosthetics Research*, BPR10-15; Spring 1971, pp. 3–84.

series of data runs. To insure that the standing position was as repeatable as possible for all of these reference runs, a footprint tracing was made and attached to the side of the treadmill, as can be seen in Figures 12.13 and 12.14. Note that the derivatives of the ankle angle curves yield the ankle angular velocity curves, and the derivatives of the latter yield the ankle angular acceleration curves, as shown in Figure 12.12.

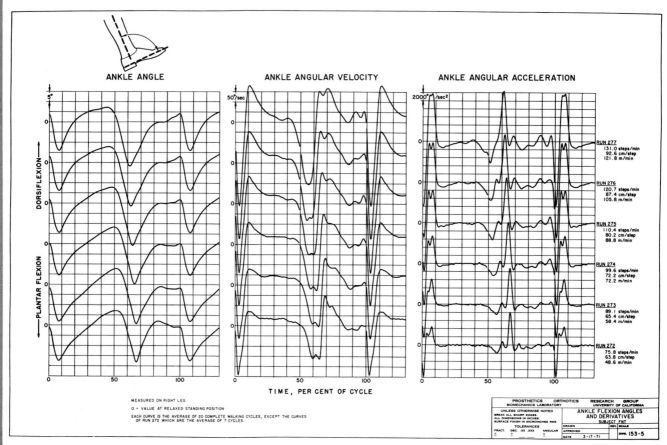

Figure 12.12 Ankle angle, angular velocity, and angular acceleration curves. (Courtesy Biomechanics Laboratory, University of California, Berkeley.)

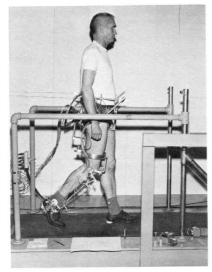

Figure 12.13 Footprint tracing attached to side of treadmill. (Courtesy Biomechanics Laboratory, University of California, Berkeley.)

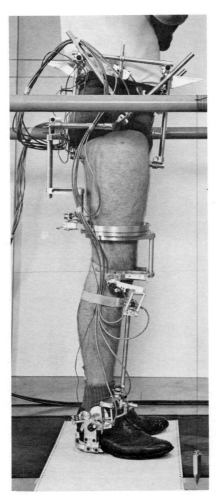 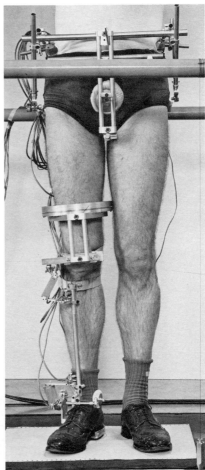

Figure 12.14 Exoskeletal linkages fitted for test runs and data collections. (Courtesy of Biomechanics Laboratory, University of California, Berkeley.)

Data Processing

The selection of data processing
procedures was influenced by the
availability of a seven-track instrumentation tape recorder, a digital
drafting machine (plotter), an electronics engineer (part time), and a
skilled programmer of the PDP-7
computer; the result was a workable data collection and data processing system, a block diagram of
which is shown in Figure 12.15.

A new and more versatile data
acquisition system is based on a
minicomputer that permits direct
digital recording of 16 or more analog data channels on industry-
compatible magnetic tape. Simple
computations are accomplished
immediately in the minicomputer,
while bulk data can be transferred
on magnetic tape to any large computer center for more complex
analysis. A block diagram of the
new data acquisition system is
shown in Figure 12.16.

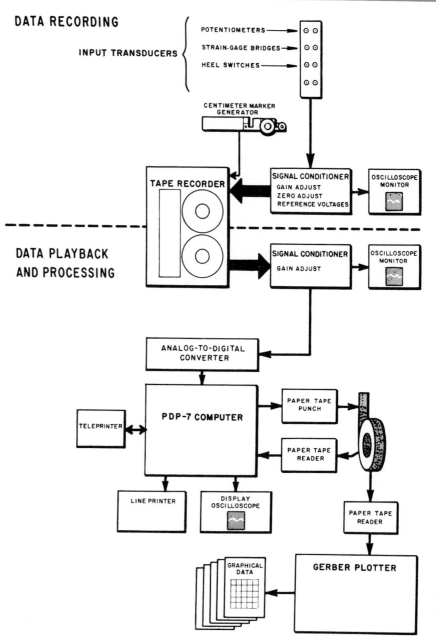

Figure 12.15 Block diagram of data col-
lection and data processing system.
(Courtesy of Biomechanics Laboratory,
University of California, Berkeley)

UC-BL DIGITAL DATA ACQUISITION SYSTEM

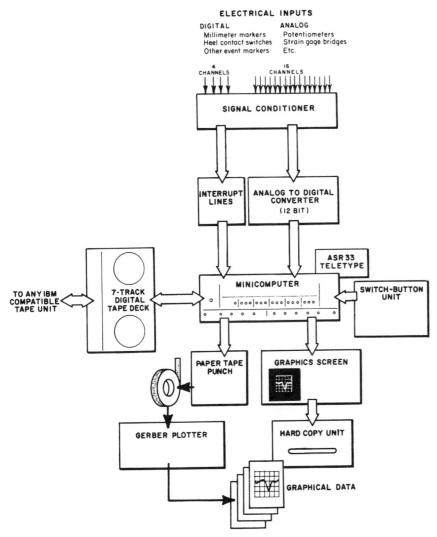

ELECTRICAL INPUTS

DIGITAL	ANALOG
Millimeter markers	Potentiometers
Heel contact switches	Strain gage bridges
Other event markers	Etc.

4 CHANNELS   16 CHANNELS

SIGNAL CONDITIONER

INTERRUPT LINES

ANALOG TO DIGITAL CONVERTER (12 BIT)

7-TRACK DIGITAL TAPE DECK

TO ANY IBM COMPATIBLE TAPE UNIT

ASR 33 TELETYPE

MINICOMPUTER

SWITCH-BUTTON UNIT

PAPER TAPE PUNCH

GRAPHICS SCREEN

GERBER PLOTTER

HARD COPY UNIT

GRAPHICAL DATA

Figure 12.16 Block diagram. UC-BL data acquisition system. (Courtesy Biomechanics Laboratory, University of California, Berkeley)

Surface Display—Mapping

The most common graphical presentation of three-dimensional data which are concerned with topography, subsurface structure (i.e., deposits of minerals, gas, and oil), temperature and pressure data, geophysical data, etc., is the two-dimensional contour map. Such maps are frequently prepared manually; this is a slow and costly process.

To the trained technical person, the two-dimensional contour map is very useful in the computation of quantities such as the number of cubic meters of earth fills and cuts that result from the modification of ground surfaces in highway design, the volume of a water or gas reservoir, calculations of drainage slopes, etc.

A unique program that generates three-dimensional mesh surfaces and three-dimensional contour maps that greatly enhance the interpretation of two-dimensional contour presentation and are an important aid to persons whose ability to visualize a surface from a two-dimensional contour map may be limited is currently available.

The two-dimensional contour map shown in Figure 12.17 is not as readily visualized as the three-dimensional mesh shown in Figure 12.18 or the three-dimensional contour surface shown in Figure 12.19. The latter enables the reader to obtain a reliable picture quickly. The combination of the two-dimensional and three-dimensional contour maps is invaluable to the engineers and to the management team.

The Computer Graphics Program
Figures 12.17, 12.18, and 12.19 and the other graphic displays were produced by a set of computer programs and a Calcomp plotter. The figures were produced from *the same data base* that represented the input information. The three-dimensional views are perspectives drawn as they appear to an observer stationed at some point in space and looking toward the center of the data area. The station point (viewer location) is arbitrary and may be specified by the user.

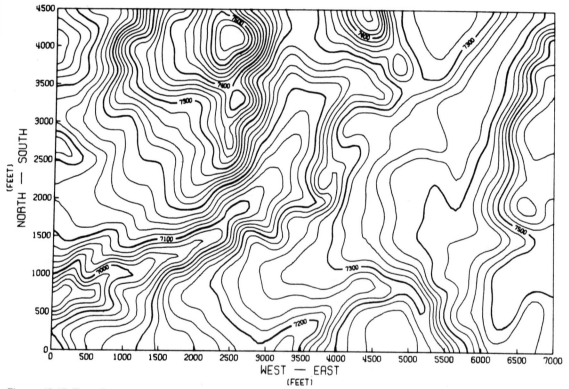

Figure 12.17 Two-dimensional Contour map of a watershed study area produced by the Surface Display Library. (Courtesy Dynamic Graphics, Berkeley, California)

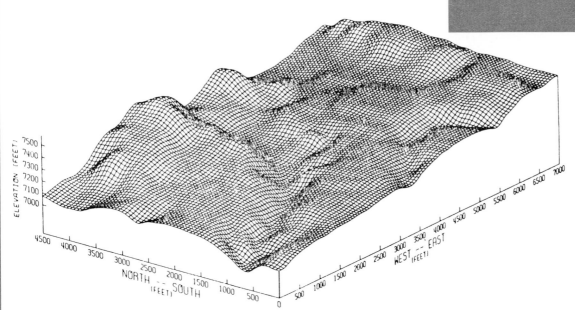

Figure 12.18 A *mesh* perspective view of Figure 12.17. (Courtesy Dynamic Graphics, Berkeley, California)

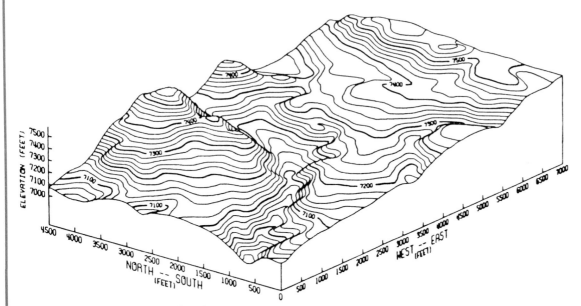

Figure 12.19 A *contour* perspective view of Figure 12.17. (Courtesy Dynamic Graphics, Berkeley, California)

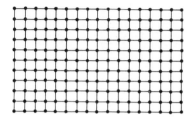

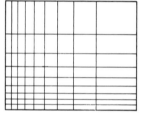
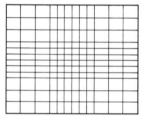

Input Information

To utilize the computer programs, it is necessary to have the information in "gridded" form. The difference between random and gridded data is shown in Figure 12.20. Before random data can be used, a gridded set of data must be interpolated from the random points. Gridding can be done manually, but a computer program is usually used.

Once the data are in gridded form, it is possible to perform many operations ranging from area, slope, and volume calculations to graphical presentations of two-dimensional contour maps, three-dimensional perspective mesh and contour surfaces, cross sections, etc.

Data for the graphical display programs may be supplied in either the "random" or "gridded" form. The data must be supplied as a set of triplet (x, y, z) values covering the areas of interest. Detailed information defining data for surface displays is available in the *Users Manual for the Surface Display Library*, which is available from Dynamic Graphics, Inc., Berkeley, California.

Figure 12.20 Random and gridded data forms. (*Top*) Three examples of random data that must be reduced to a rectangular grid by a gridding program before it can be used by the Surface Display Library. Left, elevations taken along contour lines on an existing contour map; center, spot data from geologic borings; right, aeromagnetic data recorded along airplane flight lines. (*Middle*) Gridded data, uniform. (*Bottom*) Gridded data, non—uniform.

Applications of the Surface Display Packages

GENERATING TOPOGRAPHIC MAPS.

Photogrammetry methods are here compared with the computer graphics surface display packages. In photogrammetry, a three-dimensional model of a portion of the ground surface can be created by orienting a pair of overlapping aerial photographs, printed on glass plates, in a Wild B9 Aviograph Stereoplot-ting instrument, in the same orientation that existed between the photographs at the time of the exposures during flight.

The model is then related to the ground survey control through identifiable images of ground points in the model. A measuring mark, which is seen in three dimensions, is moved about through the model to locate lines of equal elevations. This is performed by successively setting the measuring mark to predetermined elevations and then moving the mark through the model in a manner that results in apparent contact with the surface of the model at these elevations. The movement of the measuring mark is reproduced on a drawing table by means of a linear pantograph. This tracing generates the contours on the map sheet.

The contour map in Figure 12.21 was produced by this process, which required approximately 20 hours (work done by a graduate student).

The contour map in Figure 12.22, which is a good replica of Figure 12.21 was produced by the surface display computer packages in conjunction with an *x-y* plotter.

Input Information

The input information for this contour map consisted of the *x*, *y*, and *z* coordinate values of the elevations at the intersections of a gridded system over the same terrain.

The *x*, *y*, and *z* values were obtained in a high-order stereoscopic

plotting instrument known as the Sopelem Presa 226 RC. The function of the Presa is the same as that of the B9 Aviograph: the formation of a three-dimensional model for measuring purposes. The *x*, *y*, and *z* positions of the measuring mark in the model can be recorded automatically to the nearest 5 μm by means of linear encoders read by photoelectric transducers. The operator traversed the model area in the *y* direction, and at 10-mm intervals measured the elevation of the model point, whose coordinates were recorded on perforated tape. After the completion of readings on the first line, the operator traversed the next line, which was 10 mm in the *x* direction from the initial orientation, and repeated the operation. This procedure was carried on for each line until the entire stereoscopic model was covered. The net result was a model that had been digitized 10 mm in both the *x* and *y* directions. This took about 4 hours.

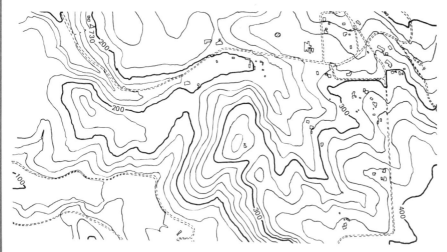

Figure 12.21 Contour map using photogrammetry method.

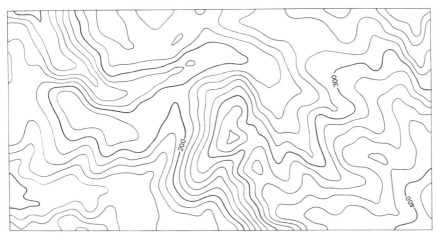

Figure 12.22 Computer graphics contour map based on the data used to produce Figure 12.21. (Courtesy Dynamic Graphics, Berkeley, California)

The same input information was used to generate the three-dimensional contour map shown in Figure 12.23. The three-dimensional representation was plotted in 15 min. The input information can be used by other programs to produce slope maps, drainage maps, cross-sections and cut and fill calculations.

In summary, the computer program, in conjunction with a plotter, can produce both two-dimensional and three-dimensional contour maps quickly and economically.

Generating Mathematical Surfaces

The programs in the surface display library enabled a mathematician with no previous experience to plot a variety of graphs that represented specific equations.

After a few days with a computer and a plotter, the mathematician was able to produce elegant results; the following are among them.

1. *Simple Graphs*. These include:

(*a*) The paraboloid; $z = x^2 + y^2$ (see Figure 12.24).

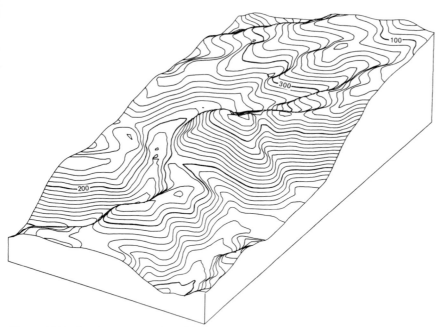

Figure 12.23 Contour perspective of Figure 12.22. (Courtesy Dynamic Graphics, Berkeley, California)

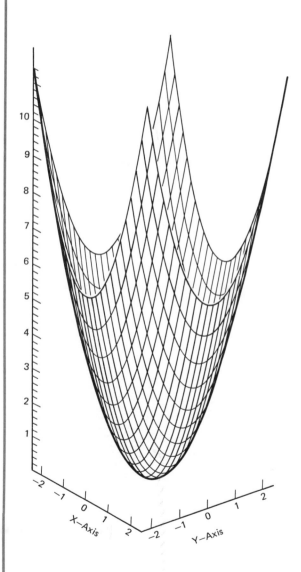

(a) Mesh surface

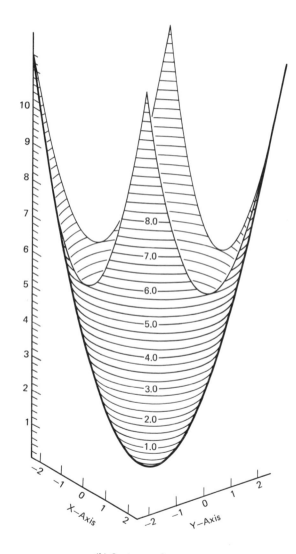

(b) Contour surface

Figure 12.24 Paraboloid $Z = X^2 + Y^2$.
(Courtesy Professor J.L. Kazdan, University of Pennsylvania, and Dynamic Graphics, Berkeley, California)

(b) The hyperboloid; $z = x^2 - y^2$ (see Figure 12.25).

(c) The four-legged saddle; $z = 4x^3y - 4xy^3 =$ imaginary part of $(x + iy)^4$ (see Figure 12.26).

2. *More Involved Graphs.* These include:

(a) The volcano; $z = (x^2 + y^2)e^{-(x^2+y^2)}$ (see Figure 12.27).

(b) The surface; $z = y \sin \pi x$, which has infinitely many saddles, at $(n, 0)$ for $n = \pm 1, \pm 2, \ldots$ (see Figure 12.28).

(c) The sombrero surface; $z = (\sin \pi r)/\pi r$, where $r^2 = (x^2 + y^2)$ (see Figure 12.29).

(d) The surface; $z = xy(x^2 - y^2)/(x^2 + y^2)$ (see Figure 12.30).

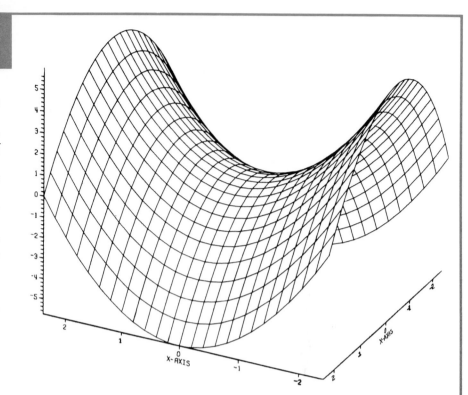

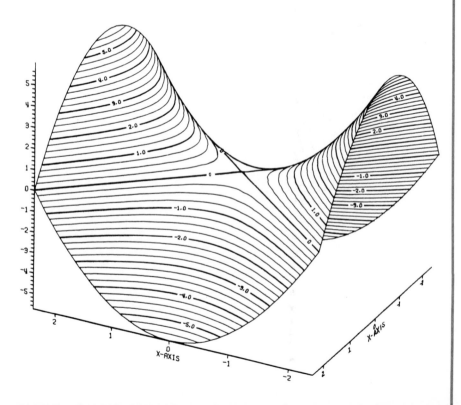

Figure 12.25 Hyperboloid $Z = X^2 - Y^2$. (Courtesy Professor J.L. Kazdan, University of Pennsylvania, and Dynamic Graphics, Berkeley, California)

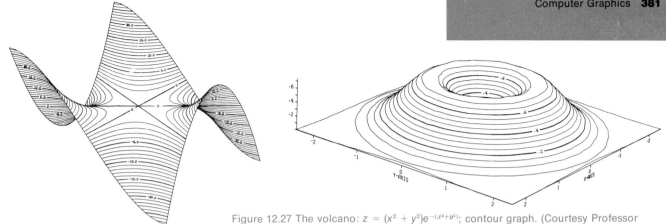

Figure 12.26 The monkey saddle $z = x^3 - 3xy^2 = $ real part of $(x + iy)^3$. (Courtesy Professor J.L. Kazdan, University of Pennsylvania, and Dynamic Graphics, Berkeley, California)

Figure 12.27 The volcano: $z = (x^2 + y^2)e^{-(x^2+y^2)}$; contour graph. (Courtesy Professor J.L. Kazdan, University of Pennsylvania, and Dynamic Graphics, Berkeley, California)

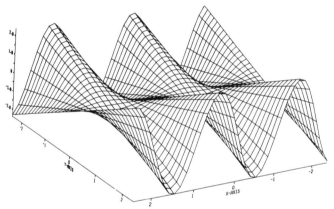

Figure 12.28 The surface: $z = y \sin \pi x$, which has infinitely many saddles, at $(n, 0)$ for $n = \pm 1, \pm 2, \ldots$. (Courtesy Professor J.L. Kazdan, University of Pennsylvania, and Dynamic Graphics, Berkeley, California)

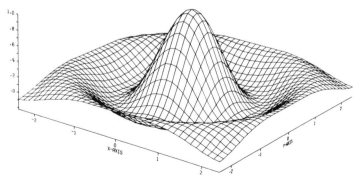

Figure 12.29 The sombrero surface: $z = (\sin \pi r)/\pi r$, where $r^2 = (x^2 + y^2)$. (Courtesy Professor J.L. Kazdan, University of Pennsylvania and Dynamic Graphics, Berkeley, California)

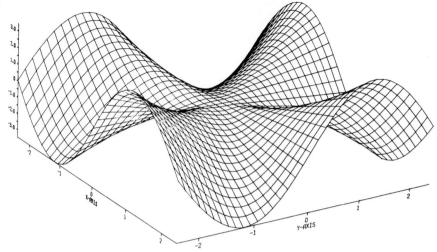

Figure 12.30 The surface: $z = xy(x^2 - y^2)/(x^2 + y^2)$. (Courtesy J.L. Kazdan, University
of Pennsylvania, and Dynamic Graphics, Berkeley, California)

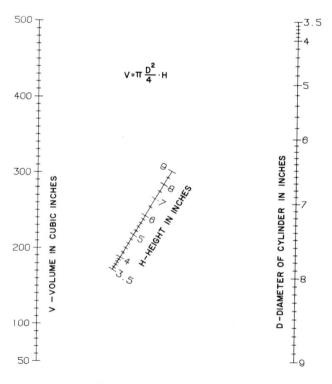

NOMOGRAM FOR V=PI×D××2/4.×H

Figure 12.31 *z*-Type nomogram.

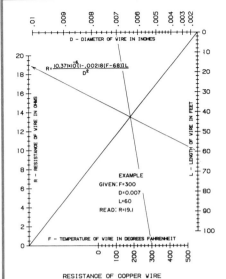

Figure 12.32 Proportional-type nomogram.

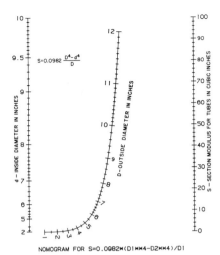

Figure 12.33 Nomogram for the type form $f(u) + f_2(v) \times f_3(w) = f_4(w)$.

Computer-Aided Design of Nomograms

Over the past several years we have developed programs to produce nomograms that consist of parallel scales for equations of the form $f_1(u) + f_2(u) + f_3(u) + \ldots = f_n(q)$.

In addition, programs have been written for the Z-type nomogram for equations of the form $f_1(u) = f_2(v) \cdot f_3(w)$ (see Figure 12.31). Other programs for equations of the following forms have been developed (see Figures 12.32 to 12.34).

1. $\dfrac{f_1(u)}{f_2(v)} = \dfrac{f_3(w)}{f_4(q)}$ (the proportional type)

2. $f_1(u) + f_2(v) \cdot f_3(w) = f_4(w)$ (two parallel scales and one curve)

3. $f_1(u) \{g_2(v) - g_3(w)\} + f_2(v) \{g_3(w) - g_1(u)\} + f_3(w) \{g_1(u) - g_2(v)\} = 0$ (three curves)

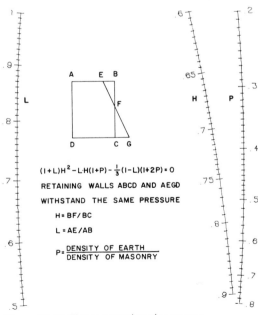

Figure 12.34 Three curved scales nomogram for the equation, $(1 + L)H^{**}2 - L^{*}H^{*}(1 + P) - (1 - L)^{*}(1 + 2P)/3 = D$, or $(1 + L) H^2 - LH (1 + P) - 1/3(1 - L)(1 + 2P) = D$.

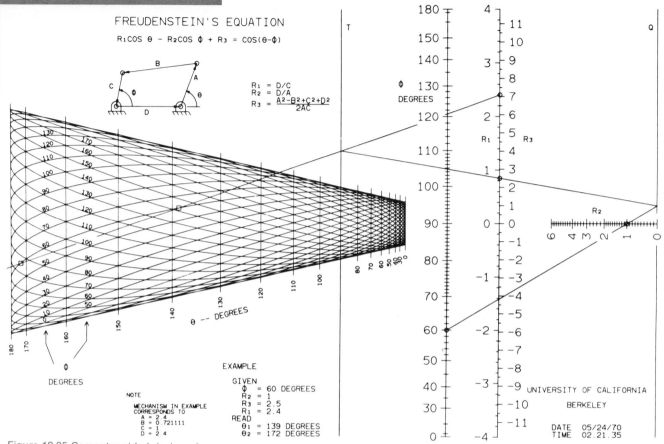

Figure 12.35 Computer-aided design of a net-type nomogram for Freudenstein's equation.

A *net-type nomogram* for the equation $R_1 \cos \theta - R_2 \cos \phi + R_3 = \cos (\theta - \phi)$ (Freudenstein's equation) was also developed. Its nomographic solution is shown in Figure 12.35.

Extensive Use of Parallel-Scale Nomograms

Since a large proportion of nomograms consist of parallel scales, it was found desirable to develop a new procedure that could be used to produce economically such no-mograms for the solution of equations that contain three or more variables. A *single* main computer program, which would be augmented by the few short subroutines required to identify the functions of the variables, their ranges, and the width and height of the desired nomogram, was developed and operated quite successfully.

ADVANTAGES OF THE PROGRAM. The most significant advantages are the speed, accuracy, economy, and ease with which the nomogram can be produced.

The program requires very little input information and can be operated by students who have had introductory computer courses.

OPERATING THE PROGRAM. The person who operates the program need perform only the simple steps outlined on page 385.

1. Convert the given equation, if necessary, to the form $f_1(u) + f_2(v) + f_3(w) + \ldots = f_n(q)$. For example 1, the equation $S = u + v^2 + 3w$ is in the required form and needs no change, but the equation $P = E^2/R$ is not in the proper form; however, it can be put in the correct form by writing $\log P = 2 \log E - \log R$.

2. Read into the main program the functions of each variable (subprogram).

3. Read into the main program the following data.

(*a*) Number of variables in the equation.

(*b*) Desired height and width of the nomogram.

(*c*) Maximum and minimum values of each variable.

(*d*) Scale and nomogram titles.

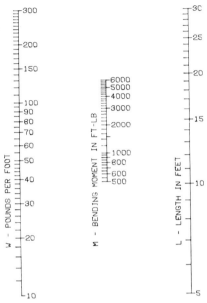

Figure 12.36 Nomogram for the equation $M = WL^2/8$.

With these inputs the program automatically selects the spacing between the scales and the "best" width for the nomogram, graduates and labels the functional scales, and titles the nomogram.

The entire job deck is placed in the following sequence.

1. Control cards.
2. Main program.
3. Subprogram function.
4. Data cards.

Examples of plotted (Calcomp) nomograms are shown in Figures 12.36, 12.37, and 12.38.

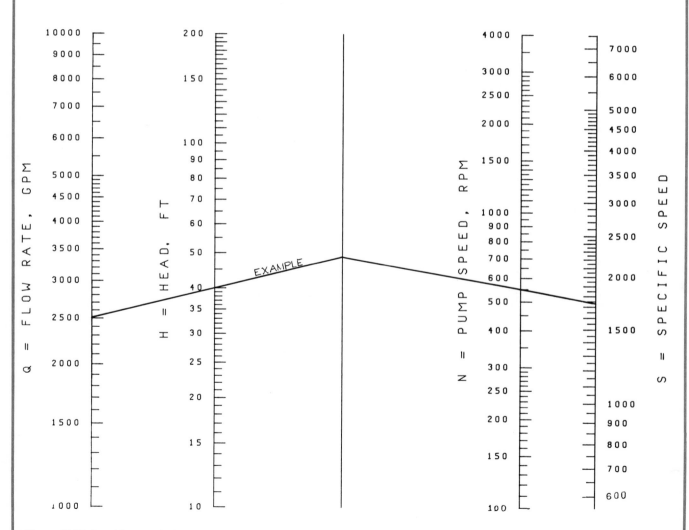

Figure 12.37 Specific speed of a centrifugal pump.

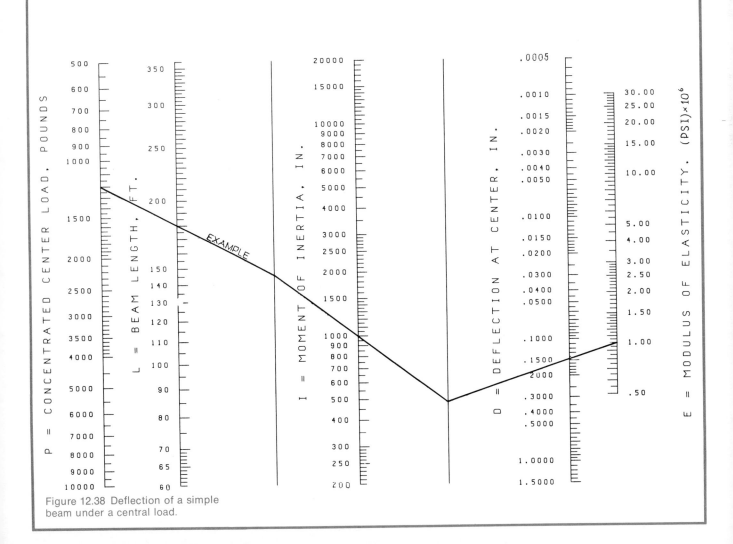

Figure 12.38 Deflection of a simple beam under a central load.

Solutions to Space Problems.
Programs are available (developed by Dynamic Graphics, Inc., Berkeley, California) to solve the following three-dimensional problems both mathematically and graphically.

1. True length of a line segment.
2. Perpendicular distance from a point to a line.
3. Perpendicular distance from a point to a plane.
4. Distance between two skew lines.

The hardware that is used is the Tektronix 4051 graphics system. See Figure 12.39 for a hard-copy listing of the problems.

Inputs to the Program
The inputs to the program are x, y, and z-coordinates of points and the constants in the equations that describe the plane or the line.

For example, for the mathematical solution to problem 3 in Figure 12.39, the input for the point and the constants in the equation are shown in Figure 12.40.

The inputs for the graphical solution shown in Figure 12.41 are point P (7.0, 1.0, 6.0), and for the plane, A(1.5, 1.5, 2.5), B(4.5, 4.5, 7.0), and C(7.5, 2.5, 4.5). The distance, d, is shown in the view on supplementary plane 1. Photographs of the operator using the Tektronix 4051 are shown in Figure 12.42.

Isometric Drawings
There are several software packages that produce isometric drawings in conjunction with plotters.

```
          PROBLEM TYPES

1: Length of a line segment
2: Perpendicular distance from a point to a line
3: Distance from a point to a plane
4: Distance between two skew lines
5: No more problems

Select problem type (1 to 5), then hit RETURN

Problem type =
```

Figure 12.39 Types of space problems. This is hard copy using the Tektronix 4051.

```
   3
Calculate distance from a point (X1,Y1,Z1) to

a plane (AX+BY+CZ+D=0)




ENTER X1 =6.5
ENTER Y1 =1.0
ENTER Z1 =4.0

ENTER A =4
ENTER B =5
ENTER C =3
ENTER D =0



DISTANCE =
  6.0811183182




would you like to see this plotted?
if yes, press 2; if no, press 1, then return
```

Figure 12.40 Inputs and solution to Problem 3. This is hard copy using the Tektronix 4051.

The example shown in Figure 12.43 is an isometric drawing of a piping system designed by the Bechtel Corporation. Input information is displayed on a CRT (cathode-ray tube) and then, if necessary, is modified by the designer. A Univac computer is employed to produce the tape that is used in the plotting operation that produces the hard copy.

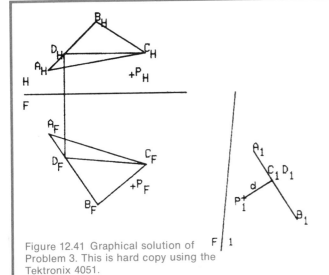

Figure 12.41 Graphical solution of Problem 3. This is hard copy using the Tektronix 4051.

Figure 12.42 The operator using the Tektronix 4051.

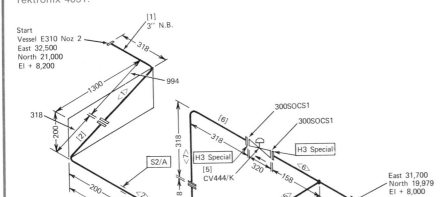

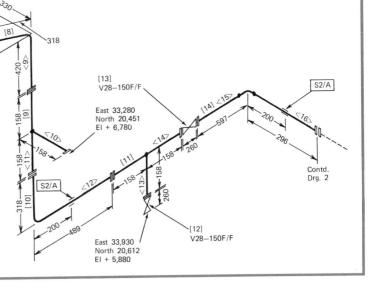

Figure 12.43 Isometric piping system. (Courtesy Bechtel Incorporated.)

	A	B	C	D	E	F	G	H	I	J	K	L	M	N	O	P
16	393	390	388	391	392	393	391	386	383	379	370	360	350	328	320	309
15	380	374	373	377	384	390	387	380	377	377	374	370	363	358	350	342
14	379	367	357	360	376	382	379	374	368	368	370	370	365	362	357	351
13	378	368	359	350	362	370	370	368	361	356	361	364	359	353	350	352
12	378	370	359	340	340	352	351	360	350	337	346	361	350	339	340	347
11	380	367	351	338	334	330	327	320	330	322	334	344	337	323	320	336
10	376	365	359	347	340	333	325	315	309	314	325	332	331	320	310	318
9	375	372	366	360	354	350	340	322	302	300	307	320	325	320	308	307
8	372	372	370	368	365	362	361	339	312	294	283	290	310	317	306	298
7	366	365	365	366	364	362	361	347	319	300	285	276	290	307	300	289
6	363	353	350	360	362	359	350	347	328	303	288	270	273	290	287	279
5	361	340	335	343	352	350	338	334	328	307	290	273	259	264	270	269
4	360	344	329	328	337	339	325	321	321	310	292	274	260	250	253	258
3	360	343	320	317	318	318	313	312	312	307	293	271	257	247	239	242
2	358	343	323	316	310	308	303	302	303	300	283	264	251	241	234	236
1	353	340	326	318	310	303	297	292	291	282	268	255	241	233	226	230

Problem: Plot a grid to a scale of 1 in. = 200 ft and draw in 10-ft contour lines.

Figure E-12.1

Exercises

1. Visit the computer center or computer facilities at your college and learn about the hardware and software that are available for the solution of three-dimensional space problems, for drawing two-dimensional and three-dimensional contour maps, for drawing nomograms, etc. Prepare an adequate report (including charts if needed) to describe the available equipments and their capability in so far as *computer graphics* is concerned. Some output examples could be included.

2. Manually locate the contour lines (10-ft interval) based on the data in Figure E-12.1).

3. Assuming that you have access to a contour software package, produce a two-dimensional contour map, using the same data as in Figure E-12.1. Compare the results with the contour map produced manually in the second exercise.

4. The elevations for the grid in Figure E-12.2 are as follows.

Point P = 750; point Q = 740

Construct a 1-in. grid and manually sketch in the 5-ft contour lines. Label and broaden every fifth contour.

5. The requirement is the same as in Exercise 2, except that the data shown in Exercise 4 apply.

6. Prepare (or, if available, use) a program to solve the following space problems mathematically.

(*a*) To determine the angle between two intersecting planes.

(*b*) To determine the angle between a line and a plane.

(*c*) To determine the angle between two intersecting lines.

	1	2	3	4	5	6	7	8
A	691	710	725	731	734	739	742	746
B	695	716	728	735	739	736	738	739
C	701	723	732	741	736	736	744	736
D	706	727	731	731	728	739	739	729
E	708	726	722	721	729	737	730	726
F	710	722	714	714	731	729	718	721

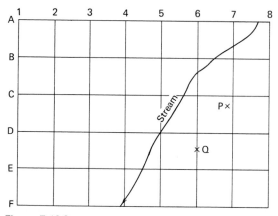

Figure E-12.2

7. Prepare (or, if available, use) a program to produce graphical displays (and hard copy if facilities are available) of the solutions to the same problems as listed in Exercise 6. The input data, for example, for Exercise 6a could be the x, y, and z values of three points to define each plane, or the equations of the planes (see Appendix E, p. 673–679).

8. Prepare a program (or use a program that may be available at your institution) to produce parallel-scale nomograms for the following problems.

(a) $S = \pi DN$

where

S = cutting speed in millimeters per minute in a lathe or boring mill
D = diameter of work (10 to 250 mm)
N = (10 to 100 rpm)

(b) $E = 0.67\ \dfrac{V^2}{R}$

where

E = maximum superelevation in feet per foot of width of roadway
V = velocity of vehicles (10 to 65 mph).
R = radius of curve (100 to 6000 ft)

(c) $P = \dfrac{E^2}{R}$

where

P = power in watts used in passing a current through a resistance, R
E = voltage (10 to 220 V)
R = resistance (10 to 1000 Ω)

(d) $D = 68.5\ \sqrt{\dfrac{P}{NS}}$

where

D = shaft diameter (1 to 6 in.)
P = horsepower (10 to 1500)
N = speed (100 to 1000 rpm)
S = design stress (2000 to 8000 psi)

(e) $V = C\sqrt{RS}$ (Chezy formula for velocity in an open channel, in feet per second)

where

R = hydraulic radius (5 to 30 ft)
S = slope of channel (0 to 0.02)
C = Bazins coefficient (60 to 150)

(f) $T = \dfrac{W}{Nf}$ (shaper operation)

where

T = time in minutes (1 to 180)
W = width of the piece (2 to 20 in.)
f = feed inches per stroke (0.005 to 0.030)
N = number of strokes per minute (10 to 100)

(g) $I = PRT$ (simple interest law)

where

I = interest (\$10 to \$500)
P = principal (\$5 to \$2000)
R = rate of interest per year (5 to 10%)
T = time or period in years (1 to 5) (subdivide the time scale into months)

(h) $Q = \dfrac{(A_1 + A_2)L}{54}$ (roadway earth quantities)

where

A_1 = one end area (500 to 2500 ft²)
A_2 = adjacent end area (300 to 3000 ft²)
L = perpendicular distance between the end areas (20 to 100 ft)
Q = earth quantity in cubic yards (also graduate the scale to read cubic meters)

PART 3
Graphics for Design Implementation

Chapter 13 Technical Drawings: Sections and Conventional Practices; Fasteners and Joining; Machine Elements

In "Part A: Sections and Conventional Practices" we discuss sectioning techniques for illustrating shapes that are difficult to visualize and conventions for portraying relationships of components in assembled products. "Part B: Fasteners and Joining" describes methods of holding assemblies together; related topics such as keys, retaining rings, and set screws are included. To conclude the chapter, in "Part C: Machine Elements" we comment on several common machine elements and manufacturers' catalogs.

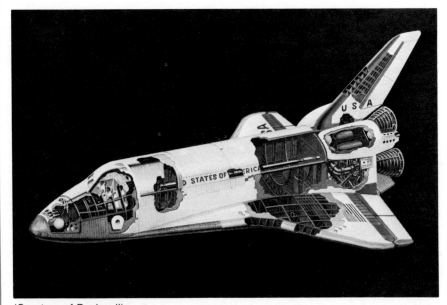

(Courtesy of Rockwell)

Part A   Sections and Conventional Practices

A student in one of our classes volunteered the following response when asked to prepare a poster to illustrate how a hydraulic cylinder worked: "Why not get a real one and cut it in half? Then you could mount it on a display board." Basically, this is what sectioning is all about. The only difference is that you only *imagine* cutting an object, and you draw what you would expect to see. Of course, your ultimate objective in sectioning is effective communication, so that those who are concerned with "reading" your design drawings can do so with a correct interpretation of your *design intent*.

SECTIONS

"A section is drawn to show how the object would appear if an imaginary cutting plane were passed through the object perpendicular to the direction of sight and the portion of the object between the observer and the cutting plane were removed or broken away. The exposed cut surface of the material is indicated by section lining or crosshatching."*

* From ANSI (ASA) Y14.2-1973. ANSI is the American National Standards Institute.

Figure 13.1*a* shows a sectioned drawing of a moment-of-inertia device that contains several types of sections; it also illustrates the use and need of a general-purpose symbol for sectioning. Figure 13.1*b* is a photograph of the actual device and of an object supported by the turntable. All portions of the device that are sectioned are "full sections" except the forelegs, which are "broken-out sections." These two types of sectioning and others will be discussed in succeeding paragraphs. The conventional sectioning symbol* used for the turntable tells us that it may be made of brass, bronze, or copper. The symbol utilized for the center hub represents babbit, lead, zinc, or alloys. The remaining symbol, equally spaced lines, depicts cast iron. The spacing may vary to differentiate between different cast iron parts. Equally spaced lines may also be used for *any material* as a general-purpose symbol. The use of the general-purpose symbol on design drawings is economical and, moreover, obviates the necessity of redrawing sectioned portions of the design when the material is changed. For example, if the turntable in Figure 13.1*a* was changed to aluminum, a different conventional sectioning symbol would be required.

Types of Sections

Various types of sections are employed to enhance the interpretation of the designer's intent. The location and size of the cutting plane determine the type of section, as seen in the following list.

* Ibid.

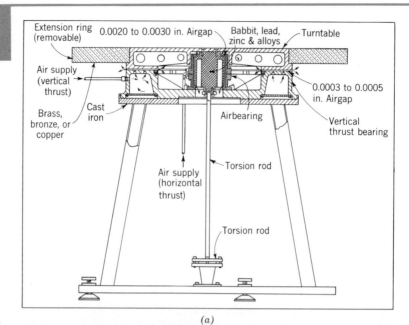

(a)

(b)

Figure 13.1 (*a*) Moment of inertia device: full section. (*b*) Photograph of moment of inertia device. (Courtesy *Product Engineering*)

1. *Full Sections.* The cutting plane extends through the entire object, as in Figure 13.2. A full section in pictorial is shown in Figure 13.3. Figure 13.4*a* shows a commercial application. Note the use of the general-purpose symbol in Figure 13.4*b*, which shows full sectioning utilized by a student design team.

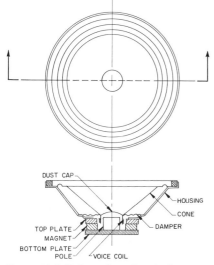

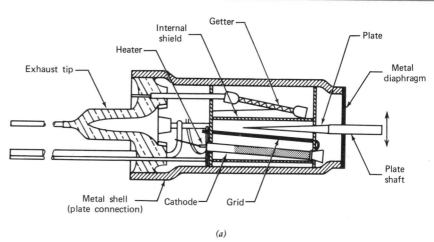

Figure 13.2 Full section of typical
hi-fi speaker.

Figure 13.4 (*a*) A mechanoelectronic transducer which is a specially constructed
vacuum tube, used for measuring vibration. When the plate shaft connected to a
flexible metal diaphragm moves the current through the tube changes proportionately.
(Courtesy TEKTRONIX)

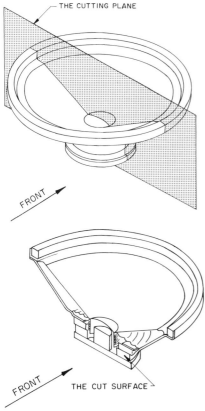

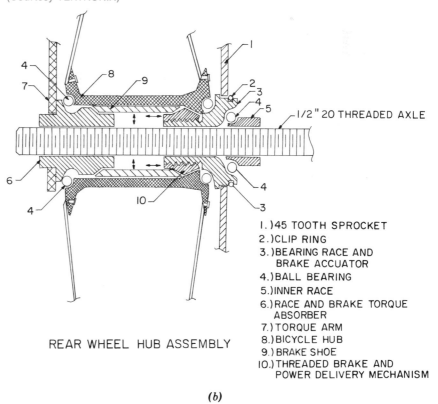

REAR WHEEL HUB ASSEMBLY

1.) 45 TOOTH SPROCKET
2.) CLIP RING
3.) BEARING RACE AND
    BRAKE ACCUATOR
4.) BALL BEARING
5.) INNER RACE
6.) RACE AND BRAKE TORQUE
    ABSORBER
7.) TORQUE ARM
8.) BICYCLE HUB
9.) BRAKE SHOE
10.) THREADED BRAKE AND
    POWER DELIVERY MECHANISM

(*b*)

Figure 13.3 Pictorial of full section.

Figure 13.4 (*b*) Full sectioning showing a modified wheelchair hub, to be used by a
student design team.

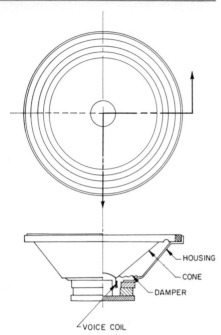

Figure 13.5 Half section.

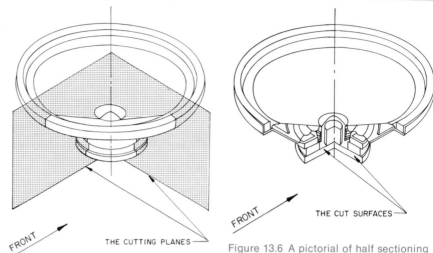

Figure 13.6 A pictorial of half sectioning showing cutting planes at 90° to each other.

FFTF REACTOR

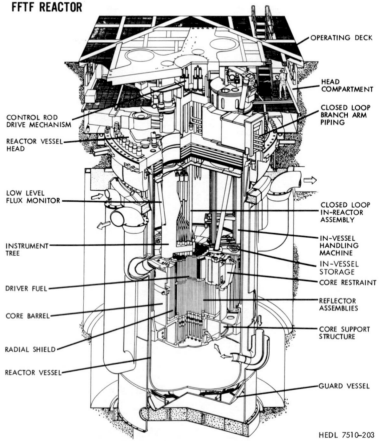

Figure 13.7 Commercial application of half sectioning. (Courtesy Westinghouse)

2. *Half Sections*. The cutting plane appears to extend halfway through the object, as in Figure 13.5. A half section shown in a pictorial would show the true cutting plane, which is, as you would expect, really two cutting planes at 90° to each other, as in Figure 13.6. Figure 13.7 is a commercial application of half sectioning.

3. *Broken-out Sections*. We recently overheard two students conversing about the construction details of a new automobile that is to appear on the market soon. One student asked the other, "How'd you find that out?" The response was, "You know, those cutaways in ads show everything." The cutting plane for broken-out cutaways is more like a torn surface than a clean slice. Figure 13.8 shows a typical use of broken-out sections to expose internal details of a large

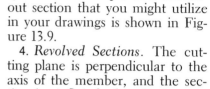

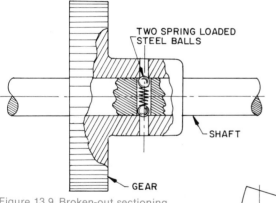

Figure 13.8 Commercial application of
broken-out sectioning. (Courtesy
Westinghouse)

system. An example of a broken-
out section that you might utilize
in your drawings is shown in Figure 13.9.

4. *Revolved Sections.* The cutting plane is perpendicular to the
axis of the member, and the section (an infinitely thin slice) as cut
is revolved 90° about a line that is
perpendicular to the axis. Figure
13.10 contains two revolved sections, one superimposed on the object (on the spoke), and one in
place of a portion of the rim.

5. *Removed Sections.* These are
usually the same as revolved sections but are placed outside the object (see Figures 13.11 and 13.12).

TWO SPRING LOADED
STEEL BALLS

SHAFT

GEAR

Figure 13.9 Broken-out sectioning
showing an overload protection
concept.

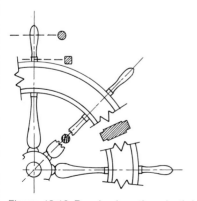

Figure 13.10 Revolved sections both in
the figure and alongside.

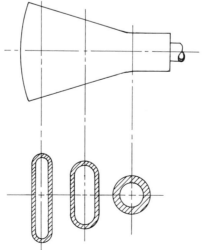

Figure 13.11 Removed sections of a
blowtorch attachment.

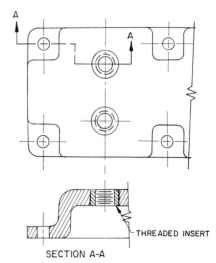

A

A

THREADED INSERT

SECTION A-A

Figure 13.12 Removed section showing
a threaded insert.

If two or more sections from the same object are on the same sheet they should, if possible, be arranged in a consistent sequence, as illustrated in Figure 13.13.

6. *Offset Section*. The imaginary plane is offset to include features that are desired but are not in a line. The section is drawn, however, as though the features were in line. An example of offset sectioning is shown in Figure 13.14.

7. *Auxiliary (Supplementary) Sections*. Figure 13.15 shows logical placement of section views. Section A-A is essentially an orthographic view. Section B-B is drawn as an auxiliary view projected in a direction perpendicular to a portion of the offset cutting plane B-B.

8. *Thin Materials in Section*. Solid lines are employed to depict materials too thin to use sectioning effectively (see Figure 13.16).

9. *Assembly Drawings in Section*. Refer again to Figure 13.1a, which is an example of an assembly drawing in section. Note that the general-purpose symbol could have been used throughout and thus simplify the drawing. Figure 13.17 is an additional example of assembly sectioning.

CONVENTIONAL PRACTICES

A reasonable approach for communicating effectively through engineering design drawings seems to involve three aspects: (1) use a common language (words and symbols); (2) employ accepted standards and conventions; and (3) occasionally deviate from the language or conventions for clarity, even though the modification may violate some

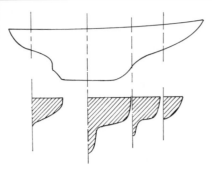

Figure 13.13 Several revolved half sections of a sailboat hull.

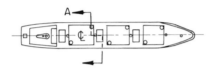

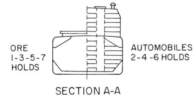

Figure 13.14 Offset section showing cargo holds in a commercial ship.

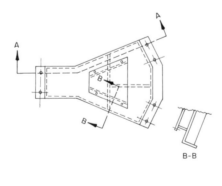

Figure 13.15 Auxiliary section utilized for a base plate.

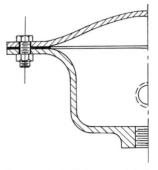

Figure 13.16 Thin materials in section are solid lines.

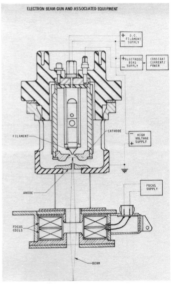

Figure 13.17 Assembly sectioning showing an electron beam gun and associated equipment. (Courtesy Boeing Company)

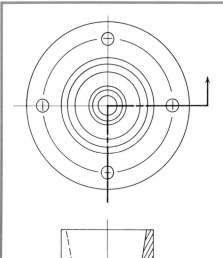

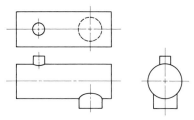

Figure 13.18 Theoretical lines of intersection are shown in the top view.

Figure 13.20 Intersections: small and
large circular shapes.

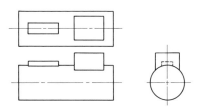

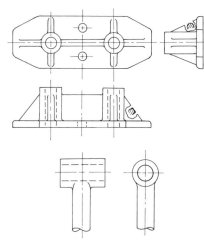

Figure 13.21 Intersections: small and
large rectangular shapes.

Figure 13.19 Filleted intersections. Note
the use of revolved sections.

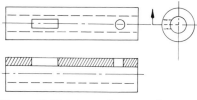

Figure 13.22 Intersections: small rectangular and circular shapes.

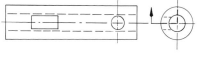

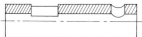

Figure 13.23 Intersections: large rectangular and circular openings.

basic principles. The common language of drawing has been incorporated throughout this textbook.
Conventional practices for sectioning are discussed in the preceding
pages. Further conventional practices that include some violations
follow.

Intersections: Unfinished
Surfaces and Fillets

Although it is true that the intersections of two unfinished surfaces
theoretically show no line, it is
considered good practice to include the theoretical line of intersection. An example is shown in
Figure 13.18. A closely related situation occurs in filleted intersections (see Figure 13.19).

Intersections: Small Circular and
Rectangular Shapes

The intersections of small circular
and rectangular shapes are conventionalized in the manner shown in
Figures 13.20 and 13.21.

Intersections: Small
and Large Curves

When a section is drawn through
an intersection in which the exact
figure or curve of intersection is
small or is of little significance, the
figure or curve may be shown in
simplified form, as in Figure 13.22.
Larger figures of intersection may
be shown as in Figure 13.23.

Violations of Orthogonal Projection Theory

In the interest of clarity in interpreting the designer's intent, strict adherence to theory may be waived. A few selected samples of the violation of orthogonal projection theory are shown in Figures 13.15, 13.24, and 13.25.

In Figure 13.15 the treatment of section A-A leads to a view that is easy to understand, whereas a true, theoretically correct, front view would be confusing. In Figure 13.24 one arm of the bell crank has been revolved to show the actual relationship of the portions of the piece. A true front view would not add to the clarity of the drawing. In most cases it would, in fact, make the reading of the drawing more difficult. In Figure 13.25 the section shows the ribs and the holes as though they were revolved into the cutting plane. Note the sectional treatment of the ribs. A true orthogonal projection certainly would hamper the ease of interpreting the views.

Treatment of Shafts, Bolts, Nuts, Webs, Keys, etc., in Section

Section lines are omitted when the section passes through shafts, bolts, nuts, webs, keys, rivets, and similar parts whose axes lie in the cutting plane. Figures 13.26 and 13.27 are typical of this treatment.

When the cutting plane cuts across the elements, they should be section lined (see Figure 13.28).

In cases where the presence of a flat element is not clear without section lining, alternate section lines may be used to clarify the designer's intent. Figure 13.29 shows the sectioning, which clearly indicates the presence of the ribs.

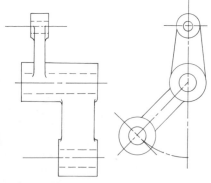

Figure 13.24 An example of violations of orthogonal projection theory to clarify a view of the bell crank.

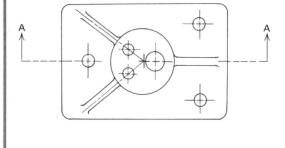

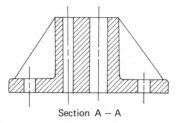

Section A — A

Figure 13.25 An example of violations of orthogonal projection theory to clarify a sectioned view of a cast base.

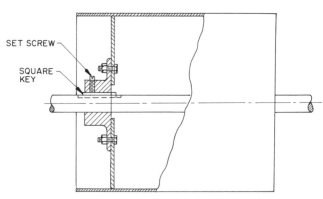

SET SCREW

SQUARE KEY

Figure 13.26 A welded steel belt conveyor pulley showing a shaft, a key, and bolts in section.

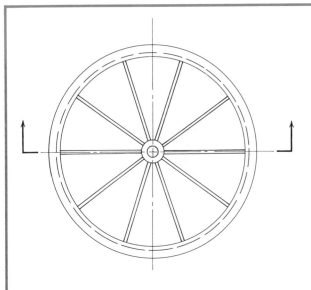

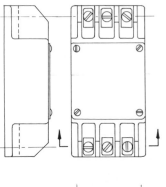

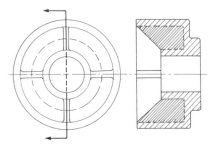

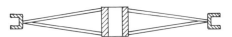

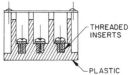

Figure 13.27 Spokes in section are shown without sectioning lines. Note the revolved spokes for clarity.

Figure 13.29 Alternate section lines for clarity.

THREADED INSERTS

PLASTIC

Figure 13.28 Section lines are shown in the ribs of the junction box shown.

Part B Fasteners and Joining

The variety of fastening techniques for fastening two or more materials together matches the variety of needs demanded by products. "Threaded fasteners remain the basic assembly method of industry despite the advances of welding, adhesives and other joining techniques. Over 500,000 fasteners can be classified as standard. That is, their physical characteristics are covered by published standards. At least as many specials are also made. These fasteners are engineered to meet some specific product need."*

Fundamentals

You might ask, "How can I learn all about half a million fasteners? How about other methods of assembly: welding, brazing, and adhesives? And, aren't there metric fasteners, too?" Since we cannot cover a half-million fasteners in one chapter, we have provided you with the fundamentals of both the inch-based fasteners and metric fasteners so that you can study them in the future on your own. Other methods of joining are also briefly discussed.

Furthermore, we have geared our discussion of fastening and joining techniques to assist you directly in your design problems now. That is, we have presented only a few of the more common fasteners and joining techniques, knowing that as you gain experience in school and, later, "on the job," you will obtain

* *Machine Design*, November 20, 1975, *Reference Issue*; Fastening and Joining.

the information you need to solve specific fasteners and joining problems as you encounter them.

We also acknowledge that many of you have already had experience with fastening and joining techniques, primarily at home on appliances, mowers, and plumbing, in hobbies on glued models, bicycles, and boats, and in shops that use fasteners, welding, brazing, and soldering. A majority of the students in our own classes have had some experience with fasteners on automobiles! Therefore our examples deal mostly with these familiar products. (*Note*. "A typical passenger car uses 3500 fasteners of 500 different types and sizes."*)

General Applications

Many designs require the use of both *removable fasteners* such as *screws*, *bolts*, *keys*, or *pins* and *permanent fasteners* such as *rivets*, *welds*, or *adhesives*. Removable-type fasteners are employed in the assembly of components of automobiles, hi-fi equipment, household appliances, health care components, switches, electric bulbs and sockets, belt-driven components on shafts, and a myriad of machine tools. Permanent fastening and joining are utilized for riveted or welded structures such as steel bridges, buildings, space vehicles, trains, trucks, honeycombs in aircraft, boilers, automobile frames, health care components, and agricultural machines. Conscientious designers will accumulate "data sheets" of fasteners and their uses, especially those that pertain to the

* *Machine Design*, November 18, 1976.

design areas of their interest. (See Exercise 13.8 in "Machine Elements" regarding reader service cards.)

As a beginning designer you should become familiar with the variety of fastening techniques employed on a range of products, from relatively rugged machinery such as earth-moving equipment to sophisticated items such as cameras, health care units, and optical apparatus. Then you will have a broad basis for selecting fastening techniques for your specific designs. (See Exercise 13.1 in "Fasteners and Joining.")

Moreover, you should know how to use recognized standards for graphic representation and identification of fasteners and joining techniques. The most frequently used fasteners use screw threads.

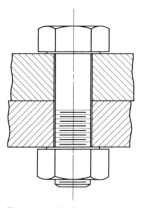

Figure 13.30 Through bolt connection.

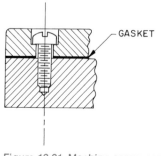

Figure 13.31 Machine screw connection.

The most frequently used joining technique is welding. Therefore we must study these two major areas.

Metric versus Inch-Based Systems

When you enter industry, you will be associating with individuals who learned technology within the inch-based system of units, and you will be working with documents made in that system. You will have to be able to be conversant in metric units and terms and in inch-based units as well. Therefore, we will note comparisons of both types of fasteners where appropriate.

FASTENERS: BOLTS AND MACHINE SCREWS

If you were to do a survey of a variety of machines and instruments, from rugged machines to delicate instruments, as to the types of threaded fasteners employed in them, you would most likely find two fastening techniques utilized for connections more than others: *through bolts* and *machine screws*. Figure 13.30 shows a typical through-bolt connection. Observe that the clearance holes are not threaded. Also, through-bolt connections are used more often for heavier machinery and frames. Figure 13.31 shows a typical machine-screw connection. One member has a threaded hole, and one is a plain clearance hole. Machine-screw connections are utilized more often for thin materials, in instruments generally and in light machinery applications.

To continue our discussion of fasteners, we will establish some *definitions* first and then note some *standard practices* for both inch-based and metric threaded fasteners.

Definitions and Nomenclature;
Screw Threads: Inch Based*

Abstracted from the American National Standards Institute's ANSI B1.7 (formerly ASA) are the following definitions, correlated with the graphical representation shown in Figure 13.32.

Screw Thread. A screw thread (hereafter referred to as a thread) is a ridge of uniform section in the form of a helix on the external or internal surface of a cylinder, or in the form of a conical spiral on the external or internal surface of a cone or frustum of a cone. A thread formed on a cylinder is known as a "straight" or "parallel" thread to distinguish it from a "taper" thread, which is formed on a cone or frustum of a cone. (M)

External Thread. An external thread is a thread on the external surface of a cylinder or cone. (M)

Internal Thread. An internal thread is a thread on the internal surface of a hollow cylinder or cone. (M)

Major Diameter. The major diameter is the largest diameter of a screw thread (Figure 13.32). (M)

Pitch Diameter. Pitch diameter is the diameter of the imaginary coaxial cylinder whose surface would cut the threads so as to make the widths of the threads equal to the widths of the spaces between the threads (Figure 13.32).

Minor Diameter. The minor diameter is the smallest diameter of a screw thread (Figure 13.32).

Pitch. Pitch is the distance measured parallel to the axis from a point on one thread to the corresponding point on an adjacent thread (Figure 13.32). (M, but measured along the major diameter.)

Lead. The lead is the distance a threaded part moves axially with respect to a fixed mating part in one complete rotation. (M)

* When applicable to metric threads, the letter M (M) will be added in parentheses.

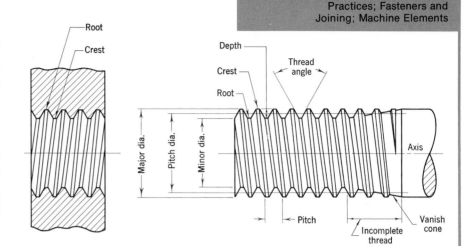

Figure 13.32 Screw thread nomenclature.

Crest. The crest is the top surface that joins the sides of a thread (Figure 13.32). (M)

Root. The root is the bottom surface that joins the sides of a thread (Figure 13.32). (M)

Depth of Thread. The depth of thread is the distance, measured perpendicular to the axis, between the crest and root surfaces. (M)

Thread Angle. The thread angle is the angle between the flanks (sides) of the thread measured in an axial plane (Figure 13.32). (M)

Single Thread. A single (single-start) thread is one having *lead* equal to pitch (Figure 13.33). (M)

Double Thread. A double (double-start) thread is one in which the *lead is twice the pitch* (Figure 13.33). (M)

Multiple Thread. A multiple (multiple-start) thread is one in which the lead is an integral multiple of the pitch (Figure 13.33). (M)

Right-Hand Thread. A thread is a right-hand thread if, when viewed axially, it recedes when turned clockwise. (M)

Left-Hand Thread. A thread is a left-hand thread if, when viewed axially, it

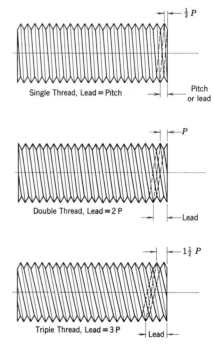

Figure 13.33 Relationship between lead and pitch.

recedes when turned counterclockwise. All left-hand threads are designated "L.H." (M)

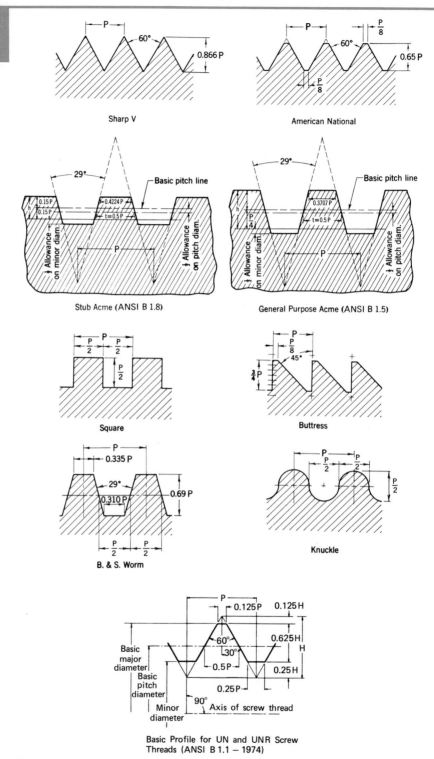

A simple way to verify the "hand" of an externally threaded member, metric or inch based, is to place the member horizontally on a table in front of you so that the axis of the member is perpendicular to your line of sight. Position your arms slightly away from your body and pointing toward the floor. The arm that the threads parallel is the "hand" of the threads.

Thread Profiles; Designer's Choice: Inch Based

The designer's choice of a thread profile (cross-section shape) is directly related to the function that is to be served. Several thread forms have been produced for specific purposes (i.e., for making adjustments and for transmitting motion or power). Among these are threads such as the *sharp V*, *American National*, *acme*, *square*, *buttress*, *worm*, and *knuckle* threads (see Figure 13.34).

The *sharp* V is used to some extent for adjustment. Usually the sharp V is employed in the small diameter range. The *Unified Inch Screw Threads*, which is a modification of the sharp V, is used in much of the screw thread work in this country. The *acme* and *square* thread forms are used for transmitting power. The *buttress* thread is also used for transmitting power, but in one direction only. The *worm* thread is another form for transmitting power and is used in mechanisms that involve the transmission of power to worm wheels. *Knuckle* thread forms are commonly used for fuses, electric bulbs, and lamps.

Figure 13.34 Thread profiles.

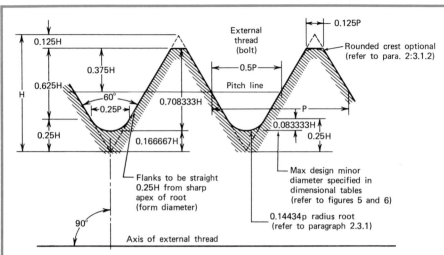

Design profile for external UNR screw thread

The unified thread form is shown in Figure 13.35.

Thread Series and Suggested Applications: Inch Based

Thread series are groups of diameter-pitch combinations that are distinguished from each other by the number of threads per inch applied to a specific diameter (see Appendix B, Table 1, p. 544). The following are included in the thread series.

Coarse-Thread Series. This series is utilized for the bulk production of bolts, screws, and nuts. It is used in general applications for threading into lower-tensile strength materials (cast iron, mild steel, bronze, brass, plastics, etc.) to obtain the optimum resistance to stripping of the internal thread. It is applicable to rapid assembly and disassembly. Coarse-thread series are designated UNC (unified coarse) or UNRC (rounded root).

Fine-Thread Series. This series is used where the length of engagement is short or the wall thickness demands a fine pitch. It is recommended for general use in automotive and aircraft work and where special conditions require a fine thread. The designation is UNF (unified fine) or UNRF (rounded root).

Extrafine-Thread Series. This series is applicable where even finer pitches of threads are needed for short lengths of engagements and for thin-walled tubes, nuts, ferrules, or couplings. It is used especially in space vehicles and auxiliary equipment. The designations for this series are UNEF (unified extrafine) or UNREF (rounded root).

Eight-Thread Series. The 8-thread series is a uniform-pitch series for large diameters or a compromise be-

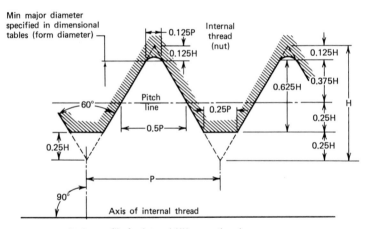

Design profile for internal UN screw thread

Figure 13.35 Unified internal and external thread design forms. (Courtesy ANSI B1.1-1974)

Unified Screw Threads: Inch Based

As a consequence of agreements reached by the standards associations of the United States, Great Britain, and Canada on November 18, 1948, a new *Unified and American Screw Threads Standard*, ASA (American Standards Association) B1.1-1949, was made available through the sponsorship of the Society of Automotive Engineers and the American Society of Mechanical Engineers. Further developments and revisions since that time culminated in the *Unified Screw Threads Standard*, ANSI B1.1-1960 and, more recently, ANSI B1.1-1974.

tween coarse- and fine-thread series. It is widely used as a substitute for the coarse-thread series for diameters larger than 1 in. The designation is 8 UN or 8 UNR (rounded root).

Twelve-Thread Series. The 12-thread series is a uniform-pitch series for large diameters requiring threads of medium-fine pitch. It is used as a continuation of the fine-thread series for diameters larger than $1\frac{1}{2}$ in. It is designated as 12 UN or 12 UNR.

Sixteen-Thread Series. The 16-thread series is a uniform-pitch series for large diameters requiring fine-pitch threads. It is used as a continuation of the extrafine-thread series for diameters over $1\frac{11}{16}$ in. It is also used for adjusting collars and retaining nuts.

Although there are additional constant-pitch series with 4, 6, 20, 28, and 32 threads per inch that may be used when the threads in the coarse, fine, and extrafine series do not meet design requirements, preference should be given wherever possible to the 8, 12, or 16-thread series. In some cases design requirements may dictate the use of selected combinations designated UNS or UNRS.

Screw Thread Classes* and Their Uses; Unified and American: Inch Based

"Classes of thread are distinguished from each other by the amount of tolerance or the amount of tolerance and allowance as applied to pitch diameter."*

The *tolerance†* on a dimension is the total permissible variation in its size.

An *allowance†* is an intentional difference in correlated dimensions of mating parts. It is the minimum clearance (positive allowance) or maximum interference (negative allowance) between such parts.

The *fit†* between two mating parts is the relationship existing between them with respect to the amount of clearance or interference that is present when they are assembled.

Basic size‡ is the theoretical size from which the limits of size for that dimension are derived by the application of the allowances and tolerances.

Classes 1A, 2A, and 3A apply to *external* threads only. *Classes 1B, 2B, and 3B* apply to *internal* threads only.

Classes 1A and 1B are intended for ordnance and other special uses. They are used on components that require easy and quick assembly. These classes replace American National Class 1 for new designs.

Classes 2A and 2B are the *most commonly used* standards for general applications, including production of bolts, nuts, screws, and similar fasteners.

Classes 3A and 3B are used when tolerances closer than those provided by Classes 2A and 2B are required.

Classes 2 and 3, because of their long-established use, are still retained in the American Standard. They apply to both internal and external threads. Unified classes, however, are rapidly replacing the American Standard classes in new designs.

Specification and Designation of Screw Threads: Inch Based

A screw thread is designated on a drawing by a note with leader and arrow pointing to the thread. The minimum of information required in all notes is the specification, in sequence, of the nominal size (or screw number), number of threads per inch, thread series symbol, and the thread class number or symbol, supplemented optionally by the pitch diameter limits. Unless otherwise specified, threads are right-hand and single lead; left-hand threads are designated by the letters L.H. following the class symbol; and double- or triple-lead threads are designated by the words DOUBLE or TRIPLE preceding the pitch diameter limits.*

The following examples demonstrate applications of this specification.

EXAMPLE 1

$\frac{1}{4}$-20 UNC-2A (see Figure 13.36).

PD 0.2164—0.2127 (optional).

$\frac{1}{4}$ = nominal diameter; 20 = number of threads per inch.

UNC = unified coarse series; 2A = class of fit (external thread).

PD = pitch diameter limits (optional).

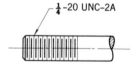

Figure 13.36 External threads.

---

* ANSI B1.1-1974.

† ANSI (ASA) B1.7-1977.
‡ ANSI (ASA) Y14.6-1978.

* ANSI (ASA) Y14.6-1978.

EXAMPLE 2

$\frac{3}{8}$-16 UNC—2A-L.H. (see Figure 13.37).

$\frac{3}{8}$ = nominal diameter; 16 = number of threads per inch.

UNC = unified coarse series; 2A = class of fit.

L.H. = left hand.

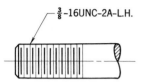

Figure 13.37 External threads.

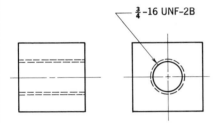

Figure 13.38 Internal threads.

EXAMPLE 3

$\frac{3}{4}$-16 UNF-2B (see Figure 13.38).

$\frac{3}{4}$ = nominal diameter; 16 = number of threads per inch.

UNF = unified fine series; 2B = class of fit.

Bolts: Inch Based

ANSI (ASA) B18.2.1-1972, *Square and Hex Bolts and Screws*, and ANSI B18.2.2-1972, *Square and Hex Nuts*, include complete general and dimensional data for the various types of square and hexagon bolts, nuts, and screws. Typical details of a modern bolt are shown in Figure 13.39. Several definitions follow.

Washer Face. The washer face is a circular boss on the bearing surface of a nut, bolt, or screw head.

Height of Head. The height of head is the overall distance from the top of the head to the bearing surface, including the thickness of the washer face where provided.

Thread Length. For purposes of this standard, thread length is the distance from the extreme end of the bolt or screw to and including the last complete (full-form) thread.

Bolt or Screw Length. Bolt or screw length is the distance from the bearing surface of the head to the extreme end of the bolt or screw, including point, if product is pointed.

Thickness of Nut. The thickness of nut is the overall distance from the top of the nut to the bearing surface, including the thickness of the washer face where provided.

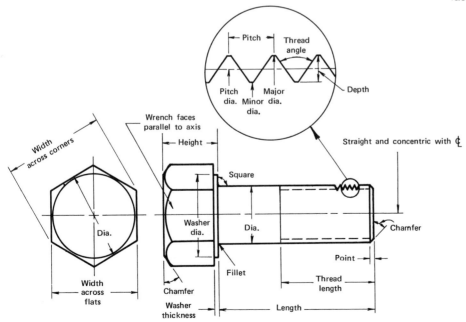

Figure 13.39 Details that make up the modern bolt. (Courtesy ANSI)

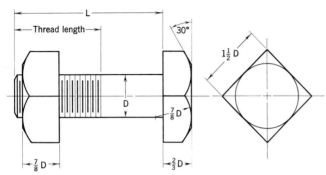

Figure 13.40 Square bolt and nut.

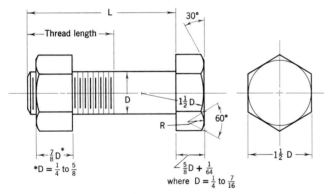

Figure 13.41 Hexagon bolt and nut.

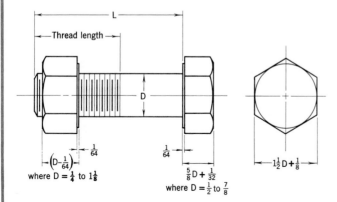

Figure 13.42 Heavy hexagon bolt
and nut.

Bolt Dimensions: Inch Based

Dimensions for hexagonal and square-head bolts and nuts are given in Appendix B, Tables 5 to 8, pp. 554–557. Formulas for widths across flats and heights of head are given in the American standards. For example, in the case of a square bolt, the width across flats, W, is one and one-half times the nominal bolt size, D; the height of the head is two-thirds D; and the height of the nut is seven-eighths D. Variations from these dimensions for different body sizes, semifinished, and finished bolts and nuts are given in the standards. Typical examples are shown in Figures 13.40, 13.41, and 13.42.

Bolt Specifications: Inch Based

Detail drawings of standard bolts and nuts are not necessary. Commercial templates are available for drawing bolt heads and nuts. Bolts and nuts are specified by giving the following information: nominal size (diameter of bolt); length (from bearing surface of head to end of shank); thread specification; material (if other than steel); series and finish; shape of head and nut (if different from head); and name (i.e., bolt).

EXAMPLES

1. $\frac{3}{4} \times 2$, 10 UNC-2A, semifinish hex head bolt.
2. $\frac{7}{8} \times 2\frac{1}{2}$, 14 UNF-2A, brass fin. hex head bolt.
3. $\frac{3}{4} \times 2$, 10 NC-2, heavy unfin. sq. head bolt.
4. $1\frac{5}{8}$-11 UNC-2B semifin. hex nut.

Machine Screws: Inch Based

A typical machine screw connection is shown in Figure 13.31. The callout for that fastener is similar to a through-bolt callout.

EXAMPLE

$\frac{1}{4}$-20 NC −3 ×1 oval head (the 1 is length of screw). The callout for the *threaded hole* to receive the screw would be $\frac{1}{4}$-20 NC −3 × (hole depth to accommodate the fastener). The *clearance hole* callout could read $\frac{9}{32}$ diameter or 0.278/0.288 diameter with a leader and arrow to the hole.

Locking Devices: Inch Based

There are several types of standard locking devices, such as jam nuts, which tighten against the gripping nut; slotted and castle nuts, which are prevented from turning by the use of cotter pins that pass through a hole in the bolt and are placed in slots of the nuts;* and cotter pins, which pass through bolt holes just above the nuts.

Standard spring lock washers† are intended for automotive and general industrial application. These devices compensate for developed looseness and the loss of tension between component parts of an assembly.

Today there are locknuts available for nearly all known applications. It is primarily a matter of proper selection to meet given and anticipated conditions.

Locknuts

Locknuts that depend on pressure contact against the piece to maintain the locking effect are very efficient on bolts in shear; however, where some bolt stretch occurs, as in the case of bolts in tension, pressure on the piece may relax so that the locking effect is lost. Temperatures affect the operation of locknuts. Some are excellent for use at low and medium temperatures, but lose their grip at high temperatures. Others are most efficient, ranging upward in size from $\frac{1}{2}$ in., but would be quite useless in a smaller size because of production difficulties that often result in the manufacture of locknuts of nonuniform holding power.

---

* ANSI (ASA) B18.2.2-1972.
† See Appendix B, Tables 21 and 22, pp. 572–573.

BETHLEHEM ANCO LOCK NUT

Description: A precision-made, single unit, self-locking nut in which the locking element is a special alloy steel locking pin which has extra high tensile strength. The nut can be reused repeatedly without appreciable loss of locking effectiveness. In sizes from ¼-in. to 3-in. diameter, the ANCO Lock Nut is made from a full range of metals, including carbon steel, stainless steel, aluminum, brass, bronze, and silicon-bronze.

Manufacturer: Bethlehem Steel Company, Bethlehem, Pennsylvania.

BI-WAY LOKUT

Description: An all-metal, prevailing-torque type self-locking nut designed for free starting from either end. Center threads are altered into an oval configuration to give a stiff spring locking action. Furnished in American Standard Finished Hexagon Series with American Coarse and Fine Threads of Class 2B tolerance.

Manufacturer: Shakeproof Division, Illinois Tool Works Inc., Elgin, Illinois.

CONELOK

Description: The Conelok is a one-piece, reusable, prevailing-torque lock-nut with the locking characteristics obtained by accurately preforming the threads in the locking section.

Manufacturers: Automatic Products Company, Detroit, Michigan; The National Screw & Manufacturing Co., Cleveland, Ohio.

LAMSON LOCK NUT

Description: The Lamson Lock Nut is a one-piece, spring-action collar style prevailing-torque type lock nut. It is characterized by ability to maintain locking action. The Lamson Lock Nut is available in a light (regular) and a heavy series. Both plain and plated nuts are standard.

Manufacturer: The Lamson & Sessions Co., Cleveland, Ohio.

LOKUT

Description: An all-metal, prevailing-torque type, self-locking nut designed for free starting. Unique three point shear depression assures positive locking action. Several top threads are projected inward providing efficient distribution of the locking load. Manufactured in Machine Screw American Standard Light and Regular Series.

Manufacturer: Shakeproof Division, Illinois Tool Works Inc., Elgin, Illinois.

TWO LOCKING PRINCIPLES

'M-F' TWO-WAY LOCK NUT

Description: The 'M-F' Two-Way Lock Nut is an automatable, reusable, one-piece, all-metal, prevailing-torque type lock nut available in low and high-carbon steels as well as non-ferrous materials. The Two-Way locking feature can be added to other fasteners including cap nuts and weld nuts.

Manufacturers: MacLean-Fogg Lock Nut Co., Chicago, Illinois; Russell, Burdsall & Ward Bolt and Nut Co., Port Chester, New York.

It is difficult to estimate the man-hours that have been saved in the speeding up of assembly operations and in the reduction of maintenance services as a result of using modern locknuts. A few of the modern designs are shown in Figure 13.43.

Setscrews

Setscrews are used to prevent relative motion between parts by entering the threaded hole of one part and setting the point against the other part (i.e., a pulley hub fastened to a shaft) (see Figure 13.44). There are several standard setscrews with a variety of points (cup, flat, oval, cone, full-dog, and half-dog). Dimensions are given in Appendix B, Table 4, p. 551.

Keys

Keys are used to prevent relative motion between shafts and wheels, pulleys, etc. There are several commonly used keys, such as:

1. *Square and flat*, plain, parallel stock keys.
2. *Woodruff* keys, which are semicircular.
3. *Taper* stock keys, both flat and square.
4. *Pratt and Whitney* keys.

Square, flat, taper, and gib-head taper stock keys are specified by giving the width, height, and length. Sizes should whenever possible, conform to ANSI B17.1. American standard Woodruff keys are generally referred to by number. Appendix B, Tables 26 to 29, pp. 577–579 give key details and show figures.

Taper Pins

In designs for relatively light loading, pins may be used as holding fasteners. Appendix B, Table 30, p. 581, contains dimensions for the commonly used taper pins.

Retaining Rings

Retaining rings are used to prevent axial movement of components mounted on shafts or in bored holes. Appendix B, Table 32, p. 583, contains figures and dimensions of commonly utilized retaining rings.

Cotter Pins

Cotter pins are also a device for holding assemblies together and are removable. Appendix B, Table 31, p. 582, contains dimensions of commonly used cotter pins and several figures showing uses.

Figure 13.44 Set screw, cone point.

Figure 13.43 Modern locking designs.
(Courtesy Industrial Fasteners Institute)

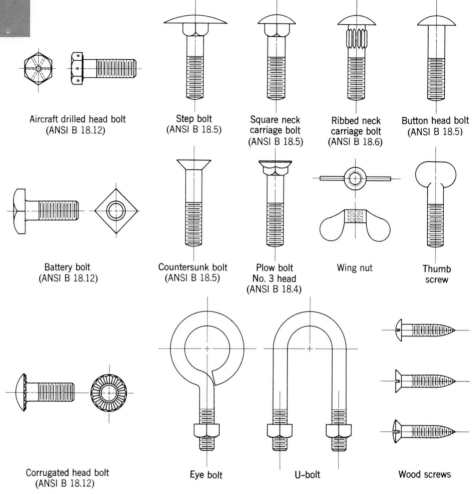

Aircraft drilled head bolt
(ANSI B 18.12)

Step bolt
(ANSI B 18.5)

Square neck
carriage bolt
(ANSI B 18.5)

Ribbed neck
carriage bolt
(ANSI B 18.6)

Button head bolt
(ANSI B 18.5)

Battery bolt
(ANSI B 18.12)

Countersunk bolt
(ANSI B 18.5)

Plow bolt
No. 3 head
(ANSI B 18.4)

Wing nut

Thumb
screw

Corrugated head bolt
(ANSI B 18.12)

Eye bolt

U–bolt

Wood screws

Figure 13.45 Miscellaneous screws
and bolts.

**Miscellaneous Hardware:
Locking Devices, Screws, and Bolts:
Inch Based and Metric**
Most of the locking devices in the
preceding section and the miscellaneous hardware in this section
are available in metric sizes. Many
types of screws and bolts other
than those presented earlier are
used in commercial practice. A few
of the more common ones, which
you may need for your design
specifications, are shown in Figure
13.45.

**Graphic Representation of Threads:
Inch Based or Metric**
Screw thread representation has
been greatly simplified by the use
of standards* that eliminate the laborious and time-consuming task
of drawing true helical curves.

There are three conventions generally used for the representation
of threads on design drawings.

* ANSI Y14.6-1978.

1. The detailed representation
(see Figure 13.46).

2. The schematic representation
(see Figure 13.47).

3. The simplified representation
(see Figure 13.48).

The detailed representation is a
good approximation of the actual
appearance of screw threads. The
helices are conventionalized as
slanting straight lines, and the
thread contour is shown as a sharp
V at 60°.

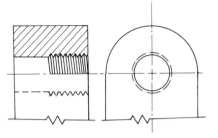

Figure 13.46 Detailed representation.

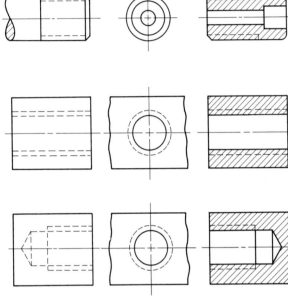

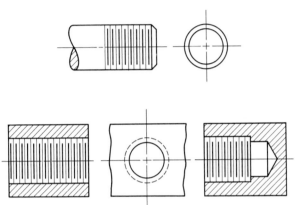

Figure 13.47 Schematic representation.

Figure 13.48 Simplified representation.

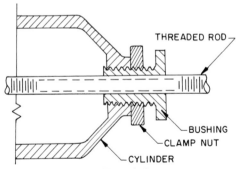

THREADED ROD

BUSHING
CLAMP NUT
CYLINDER

Figure 13.49 Threads in
section assembly.

In the schematic representation the staggered lines, symbolic of the thread crests and roots, are usually drawn perpendicular to the axis. The short lines are usually drawn heavier than the long lines. The spacing of the lines is independent of the actual pitch of the thread, as long as the distances appear reasonable. The construction shown should not be used for external threads in section or for hidden internal threads. In the interest of economy, the simplified representation is quite justified and is recommended for extensive use. In some cases, clarity may dictate the use of all three conventions (see Figure 13.49).

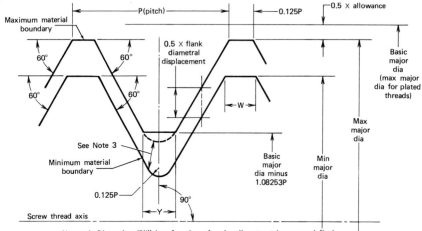

Notes: 1. Dimension "W" is a function of major diameter tolerance and flank
diametral displacement.
2. Dimension "Y" is the crest width at the minor diameter of a GO
thread ring gage with flank set for tolerance position "g", and with
minor diameter at basic.
3. The thread root of property class 8.8 and higher strength externally
threaded fasteners shall have a non—reversing curvature, no portion
of which shall have a radius less than 0.125P, and blend tangen—
tially into the flanks and any flat portion if present. The maximum
root radius is limited by the boundary profiles. The thread root of
lower strength externally threaded fasteners shall preferably have a
non—reversing curvature, no portion of which shall have a radius
less than 0.125P, however, a flat root is optional if permitted by
the purchaser.

Boundary profiles for gaging external threads

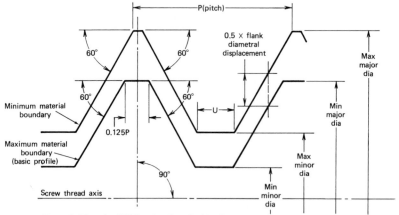

Notes: 1. Dimension "U" is a function of minor diameter tolerance and flank
diametral displacement.

Boundary profiles for gaging internal threads

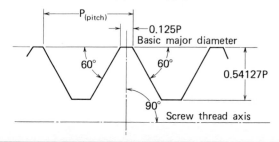

Figure 13.50 Metric basic thread profiles
for internal and external threads. (Courtesy Industrial Fasteners Institute)

Unified-Inch and IFI Metric Compared

Unified inch				IFI-500 Screw Thread standard
Coarse		Fine		
Nom. size	Threads/ (in.)	Nom. size	Threads/ (in.)	Major thread dia. pitch (mm) (mm)
—	—	0	−80	M1.6x0.35
1	−64	1	−72	—
2	−56	2	−64	M2x0.4
3	−48	3	−56	M2.5x0.45
4	−40	4	−48	—
5	−40	5	−44	M3x0.5
6	−32	6	−40	M3.5x0.6
8	−32	8	−36	M4x0.7
10	−24	10	−32	M5x0.8
12	−24	12	−28	—
$\frac{1}{4}$	−20	$\frac{1}{4}$	−28	M6.3x1
$\frac{5}{16}$	−18	$\frac{5}{16}$	−24	M8x1.26
$\frac{3}{8}$	−16	$\frac{3}{8}$	−24	M10x1.5
$\frac{7}{16}$	−14	$\frac{7}{16}$	−20	—
$\frac{1}{2}$	−13	$\frac{1}{2}$	−20	M12x1.75
$\frac{9}{15}$	−12	$\frac{9}{16}$	−18	M14x2
$\frac{5}{8}$	−11	$\frac{5}{8}$	−18	M16x2
$\frac{3}{4}$	−10	$\frac{3}{4}$	−16	—
$\frac{7}{8}$	− 9	$\frac{7}{8}$	−14	M20x2.5
1	− 8	1	−12	M24x3
$1\frac{1}{8}$	− 7	$1\frac{1}{8}$	−12	—
$1\frac{1}{4}$	− 7	$1\frac{1}{4}$	−12	M30x2.5
$1\frac{3}{8}$	− 6	$1\frac{3}{8}$	−12	M
$1\frac{1}{2}$	− 6	$1\frac{1}{2}$	−12	M36x4
—	—	—	—	—
$1\frac{3}{4}$	− 5	—	—	M42x4.5
—	—	—	—	—
2	− 4$\frac{1}{2}$	—	—	M48x4
$2\frac{1}{4}$	− 4$\frac{1}{2}$	—	—	M56x5.5
$2\frac{1}{2}$	− 4	—	—	M64x6
$2\frac{3}{4}$	− 4	—	—	—
3	− 4	—	—	M72x6
$3\frac{1}{4}$	− 4	—	—	M80x6
$3\frac{1}{2}$	− 4	—	—	—
$3\frac{3}{4}$	− 4	—	—	M90x6
4	− 4	—	—	M100x6

Figure 13.51 Metric screw thread series.
Unified inch and IFI metric compared.
(Courtesy Iron Age)

Screw Threads Profile: Metric

Information we include here is taken primarily from the *Metric Fasteners Standard*† which was prepared for the fasteners industry. The foreward to the publication stated its purpose: "The purpose was to make available to American Industry metric fastener standards which could be used with confidence until their replacement with national standards developed through the procedures of the American National Standards Institute" (ANSI). Many of the standards in the IFI publication are referenced to ISO (International Standardization Organization) standards.

The metric basic thread profiles are shown in Figure 13.50.

Screw Thread Series: Metric

The new IFI standard consists of a single series of diameter-pitch combinations ranging from 1.6 to 100 mm in diameter. Figure 13.51 lists the 25 combinations with the nearly equivalent unified thread series (English). Note the striking difference; metric gives the actual *pitch distance*, unified gives the *threads per inch*.

Screw Thread Classes of Fit: Metric

Two classes of fits are specified; one for general purpose applications, 6H/6g, and one where closer thread fits are required, 6H/5g6g.

The two classes are tolerance classes that closely approximate the unified thread series Classes 2A and 2B. Capital letters indicate internal threads, lowercase letters indicate

† Compiled by IFI (Industrial Fastener Institute) 1976.

external threads. The letters also indicate the amount of allowance between the mating parts (e.g., the allowance between a nut and a bolt fastener). The ranges of allowances are:

1. *External Threads*. e—large allowance; g—small allowance; h—no allowance.
2. *Internal Threads*. G—small allowance; H—no allowance.

The table in Figure 13.52 shows the ranges of tolerances and the relative position of the two classes. The numbers are indicative of the actual amount of tolerance; that is, a 4 is less tolerance than a 6.

Specification and Designation of Screw Threads: Metric

A typical callout for metric threads is shown in Figure 13.53. Note the

similarities to the inch-based system, such as nominal diameter, thread information, class, and fit.

Bolt Dimensions and Specifications: Metric

IFI recommended hex bolts and dimensions appear in Appendix B, Table 35, p. 590. Examples of metric bolt "callouts" follow. (Note the capital M for metric.)

M6.3 × 1 × 125 hex head bolt (length = 125 mm).
M100 × 6 × 200 brass finish hex head bolt.

Examples of clearance hole callouts for a through-bolt connection relative to the preceding two examples might be:

Drill 7.1 diameter for the M6.3 fastener.
Drill 107 diameter for the M100 fastener.

Refer to Appendix B, Table 40, p. 601, for recommended clearance holes for metric fasteners.

Machine Screws and Spline Flange Screws: Metric

The callouts for machine screws would be similar to bolts; for example, M6.3 × 1 × 125 socket head machine screw. The threaded hole callout might be: M6.3 × 1 − 5g 6g × 30 deep. The clearance hole callout would be: Drill 7.1 diameter.

A new product, the 12-spline-flange screw, offers increased underhead bearing area, improved resistance to fatigue, and less material content. See Appendix B, Table 37, p. 594.

JOINING

Joining refers to permanent connections such as welding, brazing, soldering, riveting, and adhesives. Each of these topics could comprise a complete course; therefore, we will present basic ideas to assist you now and to provide you with a basis for future learning. The discussion is limited primarily to welding and riveting.

Table 5 Preferred Tolerance Classes

Quality	External threads (bolts)									Internal threads (nuts)					
	Tolerance position e (large allowance)			Tolerance position g (small allowance)			Tolerance position h (no allowance)			Tolerance position g (small allowance)			Tolerance position h (no allowance)		
	Length of engagement			Length of engagement			Length of engagement			Length of engagement			Length of engagement		
	Group S	Group N	Group L	Group S	Group N	Group L	Group S	Group N	Group L	Group S	Group N	Group L	Group S	Group N	Group L
Fine							3h4h	4h	5h4h				4H	5H	6H
Medium		6e	7e6e	5g6g	6g	7g6g	5h6h	6h	7h6h	5G	6G	7G	5H	6H	7H
Coarse					8g	9g8g					7G	8G		7H	8H

Figure 13.52 Metric screw thread classes and tolerances. (Courtesy American Society of Mechanical Engineers; ISO Metric Screw Threads)

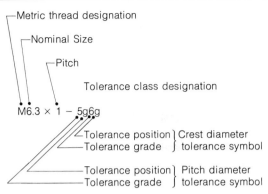

Figure 13.53 An example of specification of metric screw threads.

Welded Joints

Approximately 80% of welds in machines and structures are fillet welds, 15% are butt welds, and the rest are special welds. Refer to Appendix C, pp. 663–668, for commonly used graphical symbols for the representation of types of weld and associated dimensions. In addition, there are examples of the use of welding symbols on a machine drawing or a structural drawing.

A few basic guidelines for your use in specifying and selecting welds are as follows.

1. Generally, if the throat thickness of a weld is at least as thick as the thinnest member at a joint, the welded joint will be as strong as the thinnest member. The throat thickness is the smallest dimension of the cross section of the weld.

2. The arrangement of the parts to be welded in a frame or structure should provide the welder with access to do the welding.

3. It is usually less expensive to use standard shapes and forms than to make up shapes and forms by welding.

4. The members of a welded joint should be of the same material.

Riveted Joints

Riveted joints are employed primarily for the following conditions.

1. To join dissimilar materials.
2. For use in automatic fabricating processes.
3. For relatively thin members.
4. For odd shapes.
5. For "blind" connections.

Rivets consist of a cylindrical portion known as the body, or "shank," and a "head" that is integral with the body. Rivets are designated by giving the diameter, length and type of head. Additional information would be the series and the rivet number. Figure 13.54 shows the nomenclature applied to a button-head rivet as manufactured and after being driven.

There are a number of head shapes such as acorn, cone, flat-top, etc. A few are shown in Figure 13.55. Special rivets can also be used where only one side of the work is exposed. Head shapes and a few typical examples of "blind rivets" are shown in Figure 13.56.

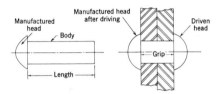

Figure 13.54 Rivet nomenclature: button-head rivet.

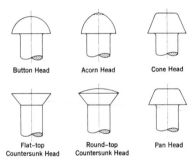

Figure 13.55 Rivet head shapes.

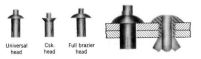

Figure 13.56 Special purpose rivets.

Part C Machine Elements

Machine elements for our design purposes include the components in a wide variety of electrical-mechanical devices. The devices are used in industry and by the general public to convert electrical or chemical energy to a useful mechanical output, such as the internal combustion engine automobiles and electrical appliances. Additional devices are farm and construction machinery and production and shop tools.

How do designers select the appropriate machine elements for these devices to do the required jobs? It is beyond the scope of our text to pursue this topic in depth; however, we can give you some help in getting started in learning the procedures for selecting some of the more common machine elements for your designs.

MANUFACTURER'S CATALOGS

Imagine for this discussion that you are a manufacturer of V belts. It would be too expensive for you to publish tables to cover all possible applications of the belts, but you could do what most manufacturers of machine elements do. You could publish "conversion" tables and charts that would provide a means for the potential buyer of your product to relate his or her application needs to your basic product line. Let us see how this works using a typical catalog.

1. Your catalog would have a table that lists typical V-belt applications and the severity of service for these applications. For example, V belts utilized on mining machinery would probably have to meet more demanding service requirements than belts used in a food-processing plant. For simplicity you may only list three catego-

Selection of A, B, C, D, and E Drives
Service Factors

	Driver					
	ac Motors: Normal torque, squirrel cage synchronous, split phase. dc Motors: Shunt wound. Engines: Multiple-cylinder internal combustion.			ac Motors: High torque, high slip, repulsion-induction, single phase, series wound, slip ring. dc Motors: Series wound, compound wound. Engines: Single-cylinder internal combustion. Line shafts Clutches		
	Intermittent service	Normal service	Continuous service	Intermittent service	Normal service	Continuous service
	3 to 5 hr daily or seasonal	8 to 10 hr daily	16 to 24 hr daily	3 to 5 hr daily or seasonal	8 to 10 hr daily	16 to 24 hr daily
Blowers and exhausters Centrifugal pumps and compressors Fans up to 10 hp Belt conveyors for sand, grain, etc.	1.0	1.1	1.2	1.1	1.2	1.3
Fans over 10 hp Laundry machinery Printing machinery Bucket elevators Piston compressors	1.1	1.2	1.3	1.2	1.3	1.4
Piston pumps Textile machinery	1.2	1.3	1.4	1.4	1.5	1.6
Crushers (gyratory-jaw-roll) Hoists	1.3	1.4	1.5	1.5	1.6	1.8

Figure 13.57 Tables for selection of V-belts; annotated from manufacturer's catalog. (Courtesy of Dodge Division of Reliance Electric.)

(a)

ries of service: *intermittent, normal service*, and *continuous service*.

2. The table would contain service factors (factors to multiply times basic horsepower requirements) related to the type of power utilized, such as an internal combustion engine or various types of electric motors. Service factors for V belts range between 1.0 and 2.0; for other machine elements such as roller chains, service factors range between 1.0 and 3.0. The result of the service factor calculation would give the buyer an *equivalent horsepower*, sometimes called the *design horsepower*.

3. A chart would specify *design horsepower* versus *output speed* of the power source. This chart would help the buyer find an appropriate V-belt size. Finally, the size, such as A, B, C, D, or E, could be used further to select sheave diameters. An example selection of V belts is given in annotated tables in Figure 13.57.

EXAMPLE

Select a V-belt drive for a continuous-service, two-cylinder compressor to run at about 275 rpm and to be driven by a 10-hp, 1160-rpm normal torque squirrel cage motor. Centers between the drive shaft and driven shaft are to be about 3 ft.

1. The service factor in Figure 13.57*a* is 1.4. Then *design horsepower* is equal to $1.4 \times 10 = 14$.

2. Figure 13.57*b* gives the type B belt (14 on horizontal axis and 1160 on vertical axis).

3. One line of information from a *V-Belt Drive Selection Table—"B" Section Belts* is shown in Figure 13.57*c*. Note that 278 rpm is close to the desired 275 rpm. Note also that the sheave diameters are 6.0 and 25.0 in., and that a B120 belt will give a 35.3 in. center distance.

4. A further factor, 0.95, is obtained from the same table; it is used to modify the listed horsepower rating for the belt as follows: $0.95 \times 4.31 = 4.1$.

5. Finally, dividing the *design horsepower* by the rated belt *horsepower* will give the recommended number of belts: $(14.0/4.1) = 3.41$, or 4 belts.

Nominal Belt Cross Sections

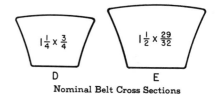

(b)

V-Belt Drive Selection Table—"B" Section Belts

Driven speeds for motor speeds of				Pitch diameter of sheaves		Grooves	Horsepower per belt for motor speeds of			Center distance and combined arc-length correction factor; center distances are close approximations and provision should be made for installing belts and for belt take up				
1750	1160	870	Ratio	Driver	Driven	Stock sheaves	1750	1160	870	B68	B90	B120	B180	B300
501	332	249	3.49	8.6	30.0	2 to 10▲	9.44	7.20	5.76	. . .	. . .	28.6	59.6	119.4
								Factors	→	.84	.83	.92	1.13	1.25
420	278	209	4.17	6.0	25.0	2 to 10▲	5.78	4.31	3.43	. . .	19.2	35.3	65.9	125.4
								Factors	→	. . .	.82	.95	1.02	1.24
303	201	151	5.77	5.2	30.0	2 to 6	4.49	3.36	2.69	. . .	. . .	30.8	62.0	121.9
								Factors	→	. . .	. . .	.92	1.07	1.22

(c)

Your catalog would also provide the potential buyer with basic formulas for determining a tentative basic horsepower requirement for his or her application. The catalog would also include tables of dimensions of the various sizes of your product to help the designer fit the element to the machine.

This general approach of converting anticipated loads and service requirements to an *"equivalent"* so you can select an appropriate machine element is utilized for a number of machine elements, such as:

1. Ball and roller bearings.
2. Roller chain.
3. Clutches and brakes.
4. Couplings.
5. Gearmotors (motor and gear box combined).
6. Gear drives.
7. Shafting.
8. Motors (AC and DC).
9. Controls.

Appropriate dimensions for these elements would also be included in the manufacturer's catalogs.

SPRINGS

A spring is a unique element that appears in a wide variety of machines, controls, and products but is neither a fastener nor a typical power transmission element. Springs exert forces and absorb energy. We include them because of their omnipresence in devices and machines.

Springs are elastic units made from wire or strip material; they stretch, compress, or twist under applied forces. Wire springs may be helical or spiral and may be made from round, square, or special-section wire. They are classified as (1) compression, (2) extension, and (3) torsion types.

Compression springs are open-coiled helical springs that resist compression. The most common form has the same diameter throughout its entire length and is known as a straight spring. Extensive use, however, is made of tapered and cone-shaped compression springs.

Extension springs are close-coiled springs that resist pulling forces. They are closely wound, in contact with each other, and made from round or square wire. The coils may be wound so tight that they require an effort to pull them apart. This coiling load is known as the initial tension.

Torsion springs exert force along a circular path, thus providing a twist or a torque. Although compression and extension springs are subjected to torsional forces, a torsion spring is subjected to bending forces. The ordinary spring hinge is typical of one of the common uses of torsion springs.

Spring specifications usually include material; outside diameter, inside diameter; free length, number of coils; type of ends; position of loops; winding, left or right; and length of coils. Figure 13.58 shows the graphical representation and dimensioning of compression, extension (tension), and torsion springs.

Two common approaches are used for selecting coil springs.

1. Appropriate information is transmitted to a spring manufacturer, who makes the spring to your specifications.
2. A standard spring is selected from existing stocks (i.e., it is an off-the-shelf item).

Other types of springs, such as flat springs, are designed to fit specific installations such as the springs under the keys in a hand calculator, the springs in an automobile suspension system, etc.

CONCLUSION

This chapter combined several topics to help you to design and to prepare effective design drawings. Part A gave you guidelines for sectioning. Part B described techniques for assembling parts of devices and machines, both temporary and permanent connections, plus guidelines for selecting and specifying the components of the connection on drawings. Part C introduced you to manufacturers' catalogs and the information that you would find for preparing drawings of the machine elements in the catalogs.

The initial exercises that follow are grouped to coincide with the major parts of this chapter. Later exercises combine several of the topics, such as fasteners and machine elements.

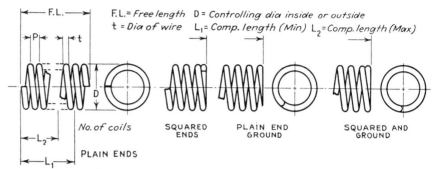

REPRESENTING AND DIMENSIONING COMPRESSION SPRINGS

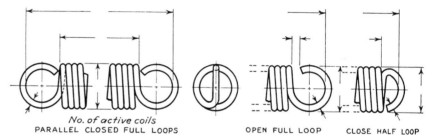

REPRESENTING AND DIMENSIONING TENSION SPRINGS

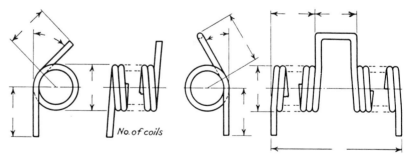

REPRESENTING AND DIMENSIONING TORSION SPRINGS

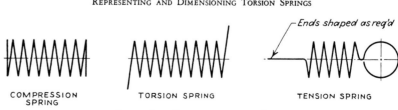

SINGLE LINE REPRESENTATION OF SPRINGS

Figure 13.58 Representation and dimensioning of springs. (Courtesy of ANSI)

Exercises

Sections and Conventional Practices

1. Select one of your favorite modes of transportation—an airplane, a hang glider, a sailboat, a car, a motorcycle, a bicycle, or a skateboard—and prepare freehand sketches of selected portions of your choice to illustrate one, several, or all of the following types of sectioning (include conventional practices where appropriate).

(*a*) Full.
(*b*) Half.
(*c*) Broken-out.
(*d*) Revolved.
(*e*) Removed.
(*f*) Offset.
(*g*) Auxiliary.
(*h*) Thin.
(*i*) Assembly.

Manufacturers' owners' manuals or repair manuals may be consulted for help.

2. Refer to Exercise 1 and prepare scaled sectional drawings. (*Suggestion.* Visit a repair shop relating to your choice and request the loan of a part or several parts of an assembly so you can take measurements.)

3. Refer to Exercises 1 and 2 and select other devices of your interest, such as:

(*a*) Hi-fi components.
(*b*) Appliances.
(*c*) Hobby equipment.
   (*1*) Trains.
   (*2*) Clocks.
   (*3*) Radio-controlled models.
   (*4*) Mountain climbing apparatus.
   (*5*) Scuba components.
   (*6*) Shop tools.

4. Draw the casting shown in Figure E-13.1 using the circular view and a full section view on cutting plane *A-A*. *Scale* full size or as directed by your instructor.

5. Redraw the part as shown in Figure E-13.2 using the circular view plus a half section view on cutting plane *A-A*. Scale four times size or as directed by your instructor.

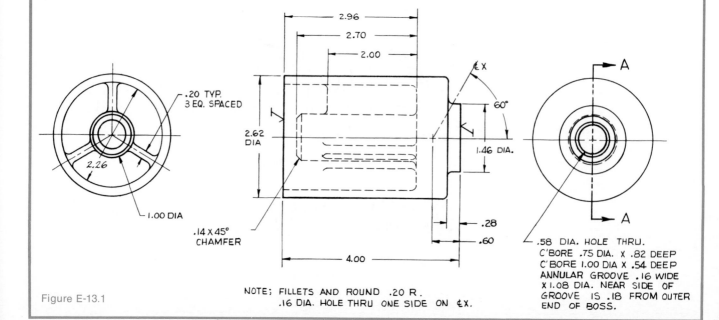

NOTE; FILLETS AND ROUND .20 R.
.16 DIA. HOLE THRU ONE SIDE ON ₵X.

Figure E-13.1

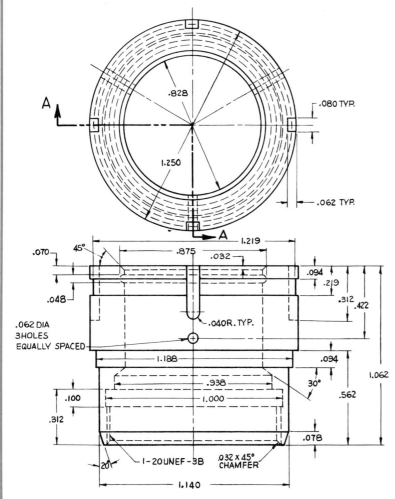

Figure E-13.2

6. Make a drawing of a *hook* that consists of a front view (Figure E-13.3), a full-section view on cutting plane *D-D,* a removed section on cutting plane *B-B,* a revolved section on line *A-A,* and a revolved section on line *C-C.* Variations of the types of sections may be specified by your instructor. (*Note.* The 1 9/16-in. radius contour at the outer edge of the *hook* as shown in the side view is uniform along the *hook* except for the eye at the upper end.)

Fasteners and Joining

1. Select one of your favorite means of transportation—an airplane, a glider, a sailboat, a motor-scooter, an all terrain vehicle, a bicycle, or a skateboard—and prepare freehand sketches of typical fastening and joining techniques utilized.

2. Refer to Exercise 1. Do the exercise and add standard callouts for the fastener(s) and standard symbol(s) for the joining technique.

3. On centerlines in Figure E-13.4 show appropriate-size hexagon head bolts and nuts (four required). (*Note.* If you are required to select fasteners you should consider that construction, shop, and field-assembly personnel are often husky individuals and can easily overstress fasteners smaller than M12.) On centerline *A,* place a safety setscrew, and on centerline *C,* place a hexagon socket setscrew. Use cone-pointed setscrews. Dimension and specify (callout) the fasteners and indicate the clearance hole sizes.

4. Prepare a *freehand* section assembly of the hanger shown in Figure E-13.5. Assume needed dimensions. Call out the fasteners and include the threaded hole and the clearance hole details.

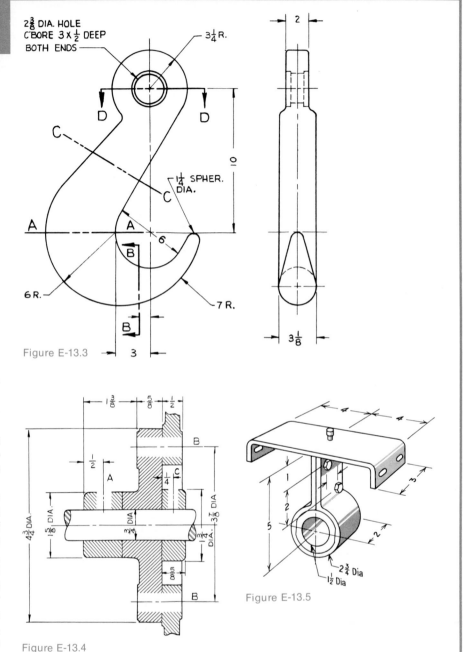

Figure E-13.3

Figure E-13.4

Figure E-13.5

5. The tower frame shown in Figure E-13.6 was sketched by a research engineer who needs a means to support air sampling apparatus in the field. The tower frame is to be made of aluminum structural shapes held together with hexagon bolts and nuts. Use the guidelines in Figure E-13.6 to establish hole locations and do one or all of the following, as assigned by your instructor.

(a) Lay out to scale
  (1) Joint A.
  (2) Joint B.
  (3) Joint C.
  (4) Joint D.
  (5) Joint E.

(b) Refer to part a and determine the required lengths of members
  (1) AB.
  (2) AC.
  (3) CD.
  (4) DE.

(c) Design a tower top that weighs approximately 800 N to support air sampling equipment. Utilize the hole connections specified at the top. Employ a combination of weldments and fasteners.

(d) Design a ladder to bolt to the tower for easy access to the top. Employ a combination of weldments and fasteners.

Machine Elements

1. Limit switches are often employed to indicate positions of objects or to act as interlocks; for example, limit switches are utilized to prevent an elevator from moving while the doors are open. Use the information in Figure E-13.7 to determine the dimensions and location of an actuating shoe attached to the vertical antenna shaft to operate the limit switch within the range given. Roller diameter is 15 mm; arm length is 50 mm.

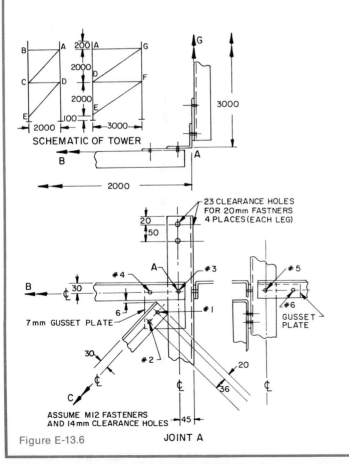

Figure E-13.6

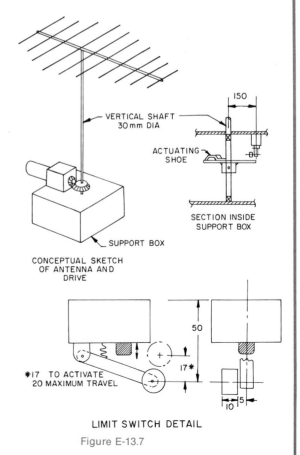

Figure E-13.7

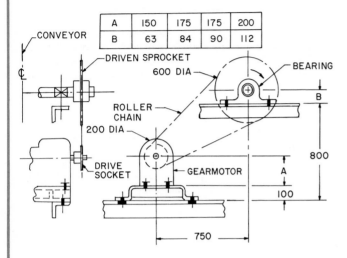

A	150	175	175	200
B	63	84	90	112

CONVEYOR

DRIVEN SPROCKET
600 DIA

BEARING

ROLLER CHAIN

200 DIA

DRIVE SOCKET

GEARMOTOR

800

A

100

750

B

DRIVE MACHINERY FOR CONVEYOR

Figure E-13.8

2. Use of manufacturers' catalogs.

(*a*) Figure E-13.8 shows a possible assembly of power transmission equipment at the head of a conveyor. Prepare a layout drawing of the front view so you can measure the approximate chain length required for the given dimensions *A* and *B* (selected by your instructor). The table of *A* and *B* dimensions in Figure 13.8 was made up of typical dimensions given in manufacturers' catalogs.

(*b*) Obtain manufacturers' catalogs and propose your own arrangements of the power transmission equipment, including some or all of the following components.

   (*1*) Gearmotor (power and output speed).

   (*2*) Chain drive plus sprockets.

   (*3*) Shaft size and antifriction bearings (ball or roller).

   Investigate the yellow pages of your phone book to determine if there are manufacturers' representatives in your area. They are usually willing to help students and provide design information and literature that relates to their products.

3. Refer to Exercise 2 and investigate similar layouts for V-belt drives for driving equipment such as compressors or large ventilating fans for buildings.

4. Refer to the general approach of Exercise 2 and propose hi-fi speaker housing dimensions for speakers of your choice. Then prepare a layout drawing of your proposed housing.

5. Refer to Exercise 1, 2, 3, or 4 and specify appropriate fasteners with locking devices for the assemblies.

6. *Springs.* One successful use of springs is to employ them in antibacklash assemblies. Typical applications are often found in instruments where continued accuracy is critical while adjusting settings in either direction. Examples are optical equipment, electronic apparatus, telescopes, and laser assemblies. One clever use of springs and fasteners for minute, mirror-angle adjustments is found in an optical instrument. The mirror-plate assembly is shown in Figure E-13.9. Note that the springs always provide tension on the mirror plate, so the mirror plate always bears on the same side of the threads on the fasteners. Lay out to double size the front view of the mirror assembly and measure the change in the mirror angle for four complete rotations of fastener *A*.

7. Design an adjustable base for a home telescope (or for a mount that holds binoculars) based on the principles shown in Exercise 6. Prepare the following as assigned by your instructor.

(*a*) Freehand sketches of your concept with approximate dimensions.

(*b*) Layout drawings with your concepts drawn to scale.

(*c*) Technical drawings including appropriate callouts and parts lists.

8. *Reader Service Card* (*RSC*). Locate an RSC in a technical journal of your choice and ask the owner (may be in a library) if you can send in the RSC in his or her name. Usually the RSC contains reference numbers to specific machine elements. Circle the numbers that interest you and mail the RSC. You should receive a response in two to four weeks.

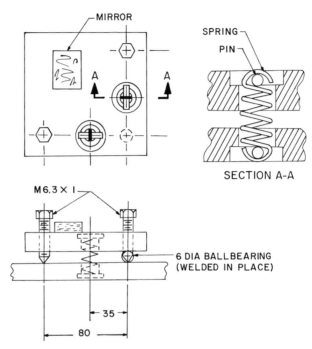

Figure E-13.9

Chapter 14 Technical Drawings: Dimensions and Specifications; Detail and Assembly Drawings

GENERAL DIMENSIONING

Introduction

The definition of the end product, Phase 3 of the design process, must include properly dimensioned detail and assembly drawings and specifications for the implementation of the project.

The orthographic (or pictorial) views of a part, or unit, including necessary sections, tell the *shape story* of the object. The production of machines, structures, devices, etc., requires much more than "shape descriptions of their components." Dimensions must be added to the views to specify *size*. Also, specifications regarding, for example, the kind of materials to be used, treatment and finish of the materials, and special instructions *if necessary* about methods of manufacture and verification, shipping of subassembly units, methods of assembly at the site, etc., must be prepared. *The design drawings alone are useless for practical purposes without dimensions and specifications!*

The well-qualified professional engineer should have a good working knowledge of materials, manufacturing processes, construction methods, and product design in order to produce devices, machines, and structures that are functional, reliable, and economical.

The engineering student should make every effort to learn as much as possible about materials, manufacturing processes, and construction methods. A good deal can be gained from course work, plant visitations, and relevant material that can easily be found in technical periodicals and journals (e.g., *Product Engineering*, *Machine Design*, and *Plant Engineering*). The engineering student should be encouraged to seek summer employment that will enhance knowledge and understanding of manufacturing processes and construction methods.

A properly prepared engineering drawing tells the true and complete story about the design. There is no room for ambiguity that could result in costly errors.

In dimensioning, as in any form of communication, the message must be carefully thought out. If the communique is to be effective it must be clear, complete, and understandable. To achieve a properly dimensioned design drawing, the designer must anticipate possible misrepresentations and then establish specifications in such a manner that the reader must accept the information *as it is intended*.

Before we explore what can be done to help the student in the effort to understand what is involved in the preparation of carefully dimensioned and specified drawings, we must first become acquainted with recommended dimensioning practices and techniques.

Dimensioning Practices and Techniques

Dimensions are applied by use of dimension lines, extension lines, and leader lines from a dimension, note, or specification directed to the appropriate features. General notes are used to convey additional information.

1. The selected *unit of measurement* should be compatible with the uses of the drawing. Although the commonly used U.S. customary linear unit employed on engineering drawings has been the inch (and may continue as such for several years), the *SI* (International System of Units) *metric* linear unit, *the millimeter* (mm), has been adopted. Quite a number of American corporations have converted to the SI system. Within a

reasonable period of time (perhaps by 1990) the United States will have completed the conversion to the SI system and thereby greatly enhance our industrial relationships with countries throughout the world. *In this treatment of dimensioning we will use the SI unit, millimeters.*

2. *Dimension lines*, with their arrowheads, show the direction and extent of dimensions. Dimension lines are usually broken for the insertion of numerals to indicate the number of units of the measurements, as shown in Figures 14.1, 14.2, and 14.3. Dimension lines should be aligned if practicable, grouped for uniform appearance, and drawn parallel to the direction of measurement (see Figure 14.4). Be sure to allow adequate space between dimension lines. Recommended distances are shown in Figure 14.5. When there are several parallel dimension lines, the numerals should be staggered (displaced), as shown in Figure 14.6.

3. *Extension lines* are used to indicate the distance measured when the dimension is placed outside of the view. Extension lines are thin, full lines that start with a short visible gap from the outline of the part and extend beyond the outermost related dimension line, as shown in Figure 14.5. Where space is limited, extension lines may be drawn at an oblique angle to show clearly where they apply (see Figure 14.7). This is a good technique to use when dimensioning in crowded spaces. *Clarity is of prime importance.*

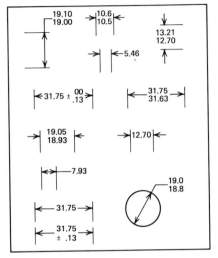

Figure 14.1 Decimal millimeter dimensions. (ANSI Y14.5)

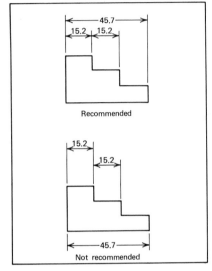

Figure 14.4 Grouping of dimensions. (ANSI Y14.5)

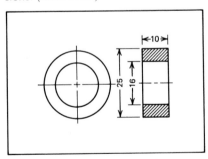

Figure 14.2 Millimeter dimensions. (ANSI Y14.5)

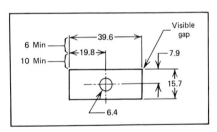

Figure 14.5 Spacing of dimensions. (ANSI Y14.5)

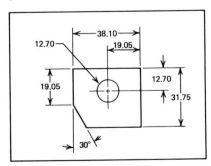

Figure 14.3 Placement of dimensions. (ANSI Y14.5)

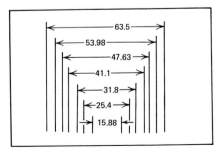

Figure 14.6 Staggered dimensions. (ANSI Y14.5)

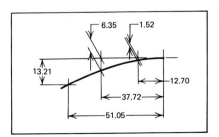

Figure 14.7 Oblique extensions lines. (ANSI Y14.5)

4. *Leaders* are thin, straight lines terminated by arrowheads that point to elements of the part. Leaders are used to direct dimensions, notes, or symbols to the intended places on the design drawing. The plain ends of the leaders are followed by descriptive notes, as shown in Figure 14.8. Where a leader is directed to a circle, its direction should be radial, as shown in Figure 14.9.

5. *Reading directions* of dimensions on technical drawings are either *aligned* or *unidirectional*. *The latter is preferred*. Dimension lines and their corresponding numbers are placed so that they may be read from the bottom edge of the drawing. The aligned system for orienting dimensions is shown in Figure 14.10, and that of the unidirectional system is shown in Figure 14.11.

6. *Reference dimensions* are enclosed in parentheses or may be followed by the abbreviation REF (see Figures 14.12 and 14.13).

7. *Angular dimensions* are expressed in degrees (°) and decimals of a degree (25.50°). Still in use is the form 23″ 14′ 20″ (degrees, minutes, and seconds). Examples are shown in Figure 14.14.

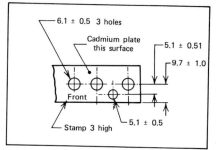

Figure 14.8 Leaders—general use. (ANSI Y14.5)

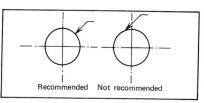

Figure 14.9 Leaders to circles. (ANSI Y14.5)

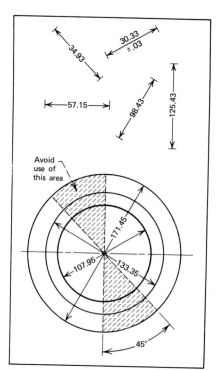

Figure 14.10 Aligned dimensions. (ANSI Y14.5)

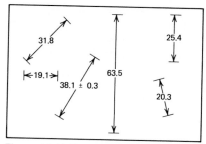

Figure 14.11 Unidirectional dimensions.

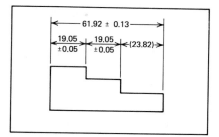

Figure 14.12 Reference dimensions using parentheses.

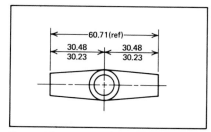

Figure 14.13 Reference dimensions using abbreviation reference.

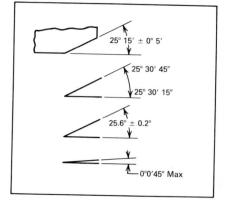

Figure 14.14 Angular dimensions.

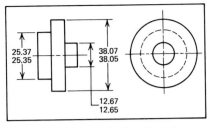

Figure 14.15 Dimensioning diameters.
(ANSI-Y14.5)

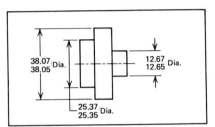

Figure 14.16 Dimensioning diameters of
concentric cylinders. (ANSI-Y14.5)

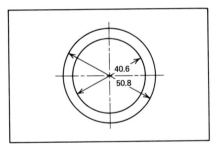

Figure 14.17 Dimensioning diameters in
circular view.

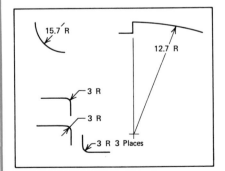

Figure 14.18 Dimensioning radii of
circular arcs.

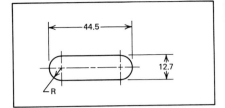

Figure 14.19 Dimensioning a part
having rounded ends.

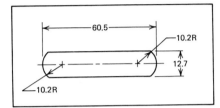

Figure 14.20 Dimensioning a part with
partially rounded ends.

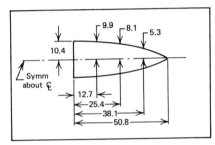

Figure 14.21 Dimensioning symmetrical
outlines.

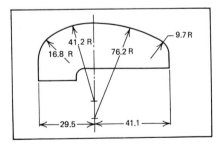

Figure 14.22 Dimensioning outlines
consisting of arcs.

Dimensioning Features

Various characteristics and features of parts are uniquely dimensioned. Consider the following.

1. *Diameters*. Where the diameters of a number of concentric cylindrical features are specified, the dimensions are shown in a longitudinal view, as shown in Figure 14.15. Other treatments are shown in Figures 14.16 and 14.17.

2. *Radii*. A circular arc is dimensioned by giving its radius. Each numeral should be followed by "R" (see Figure 14.18).

3. *Rounded Ends*. Overall dimensions should be used for parts having rounded ends. Examples of fully rounded and partially rounded ends are shown in Figures 14.19 and 14.20.

4. *Symmetrical outlines*. Symmetrical outlines are dimensioned on one side of the axis of symmetry (see Figure 14.21). Note that this method is an exception to general practice, which states that centerlines, extension lines, and lines of the views should *not* be used as dimension lines.

5. *Outlines Consisting of Arcs*. A curved outline consisting of two or more circular arcs is dimensioned by giving the radii and locating their centers (see Figure 14.22).

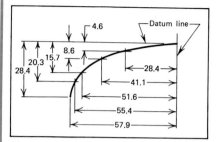

Figure 14.23 Dimensioning irregular
outlines.

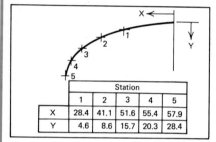

Figure 14.24 Tabulated outlines' use of
tabulated coordinates to delineate an
irregular outline.

Station					
	1	2	3	4	5
X	28.4	41.1	51.6	55.4	57.9
Y	4.6	8.6	15.7	20.3	28.4

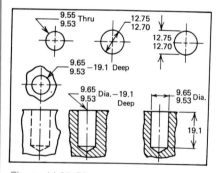

Figure 14.25 Dimensioning round holes.

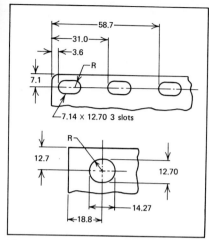

Figure 14.26 Dimensioning slotted
holes.

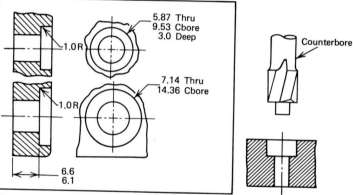

Figure 14.27 Specifying counterbored
holes.

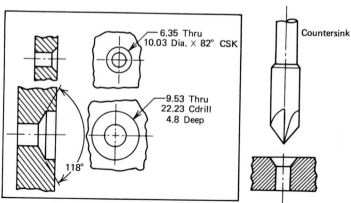

Figure 14.28 Specifying countersink and
counterdrilled holes.

6. *Irregular Outlines*. Irregular
outlines may be dimensioned as
shown in Figures 14.23 and 14.24.

7. *Round Holes*. Round holes
are specified by giving their diameters and depths. When the hole
does not go through the material, it
is known as a *blind* hole, and the
depth dimension is only to the
shoulder (see Figure 14.25).

8. *Slotted Holes*. A regular-shaped slot is dimensioned by giving its width and length. Its location is given by a dimension to its
center plane and one end, as
shown in Figure 14.26.

9. *Counterbored Holes*. Counter-boring provides a space for receiving a bolt head or the head of a screw. Counterbored holes are specified by a note giving the diameter, depth, and corner radius (see Figure 14.27).

10. *Countersunk and Counter-drilled Holes*. For countersunk holes, the diameter and the included angle are specified. For counterdrilled holes, the diameter, depth, and included angle are specified. The depth dimension is the depth of the full diameter of the counterdrill (see Figure 14.28).

11. *Spotfaces*. In order to provide a good bearing surface for a bolt head or nut, the area around the hole through which the bolt passes must be made smooth. This is done by the operation known as spot-facing, which is called out by a note, as shown in Figure 14.29.

12. *Chamfers*. The recommended practice is to give an angle and a length. A chamfer is a beveled edge (see Figure 14.30).

13. *Keyseats*. Keyseats are dimensioned by giving the width, depth, location and, if required, the length (see Figure 14.31a and 14.31b).

14. *Knurling*. Knurls provide a rough surface for gripping. Knurling is specified in terms of type, pitch, diameter before and after knurling, and the axial length of the knurled area, as shown in Figures 14.32 and 14.33.

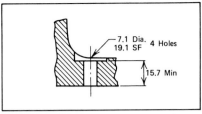

Figure 14.29 Specifying spot-faced holes.

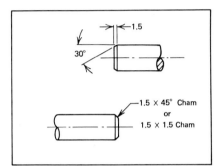

Figure 14.30 Dimensioning chamfers.

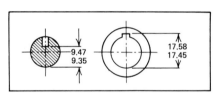

Figure 14.31 (a) Dimensioning Key seats.

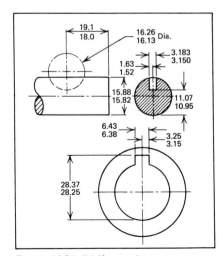

Figure 14.31 (b) Key seats.

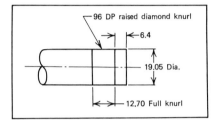

Figure 14.32 Specifying knurls.

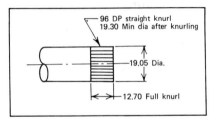

Figure 14.33 Specifying knurls for press fits.

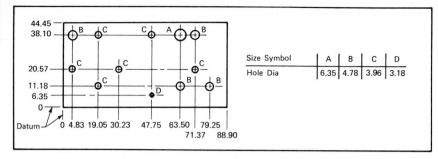

Figure 14.34 Rectangular coordinate dimensions without dimension lines.

Size Symbol	A	B	C	D
Hole Dia	6.35	4.78	3.96	3.18

Coordinate Dimensioning

Coordinate dimensions originate from datum planes indicated as zero coordinates. From these, dimensions are shown on extension lines *without the use of dimension lines or arrowheads*. An example is shown in Figure 14.34.

Tabular Dimensioning

Dimensions from mutually perpendicular datum planes are listed in a table that accompanies the drawing. The method is used on drawings that require the location of a fairly large number of similarly shaped features. Examples are shown in Figures 14.35 and 14.36.

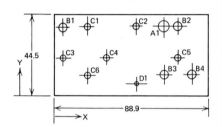

Figure 14.35 Tabular dimensioning.

Reqd		1	4	6	1
Hole dia.		.250	.188	.156	.125
Position		Hole symbol			
X →	Y ↑	A	B	C	D
63.5	38.1	A1			
4.8	38.1		B1		
7.4	38.1		B2		
63.5	11.2		B3		
76.2	11.2		B4		
19.1	38.1			C1	
47.8	38.1			C2	
4.8	20.6			C3	
30.2	20.6			C4	
71.4	20.6			C5	
19.1	11.2			C6	
47.6	6.4				DL

Figure 14.36 Drawings for numerical control production.

Point	X	Y
1	0	0
2	15.88	21.59
3	15.88	105.41
4	22.23	111.76
5	180.48	116.76
6	187.33	105.41
7	187.35	21.54
8	180.48	15.24
9	69.85	50.80
10	22.23	15.24

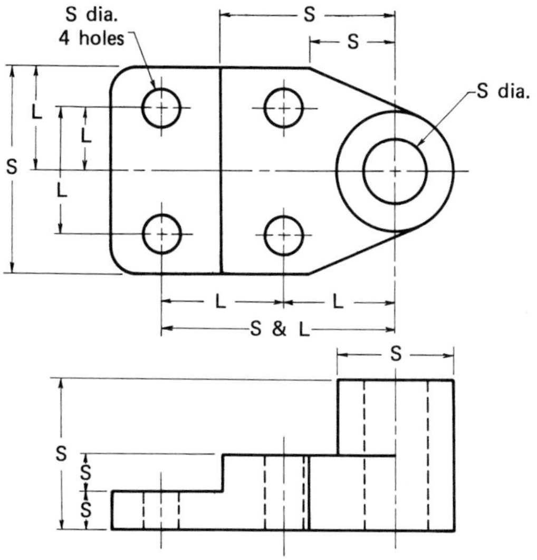

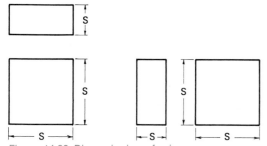

Figure 14.37 Size and location dimensions.

Size and Location Dimensions

Designs of components, machines, devices, and structures usually consist of a combination of simple geometric shapes such as prisms, pyramids, cylinders, cones, and spheres. To produce a part, machine, device, or structure, it is obvious that it is necessary to know the *size* of the object and its components, and also the *location* of the components with respect to each other. In many cases, dimensioning a single unit will require both size and location dimensions. An example of size (S) and location (L) dimensions is shown in Figure 14.37.

1. *Prisms*. The prism, which is the shape that occurs most often in design work, is dimensioned by giving its height and width on one view and its depth on an adjacent view (see Figure 14.38).

2. *Pyramids*. The base dimensions are given on the appropriate view and the altitude on an adjacent view, as shown in Figure 14.39.

Figure 14.38 Dimensioning of prisms

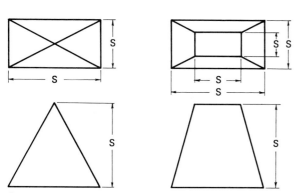

Figure 14.39 Dimensioning of pyramids.

3. *Cones*. The height and diameter of the base are shown on the *expressive* view, as shown in Figure 14.40.

4. *Cylinders*. A cylinder is dimensioned by giving its diameter and length in the view that shows the cylinder as a rectangle. For holes (negative cylinders) a specification by note is sufficient (see Figure 14.41).

5. *Spheres*. A sphere is dimensioned by giving its diameter.

6. *Specific Features*. Specific features of a part should be dimensioned in the view that most clearly delineates the features. For example, in Figure 14.42, the holes of the shaft support are best described in the top view. In Figure 14.43, however, the holes in the bell crank are best dimensioned in the expressive view, the front view in this case.

Surface Texture Quality
The design engineer should be familiar with surface texture characteristics, their significance, and

where certain textures are required for functional reasons. In a number of cases, for example, the engineer is concerned with the reduction of friction in moving parts that

may be required to operate at high speed. The degree or extent of irregularities of the surfaces of solid materials, usually caused by machining operations, must be con-

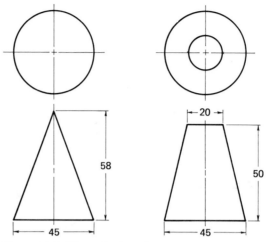

Figure 14.40 Dimensioning of cones.

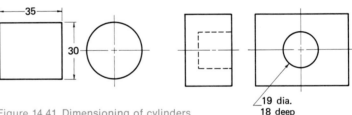

Figure 14.41 Dimensioning of cylinders

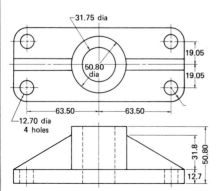

Figure 14.42 Shaft support; specific features, such as the holes, should be noted (in the top view in this case).

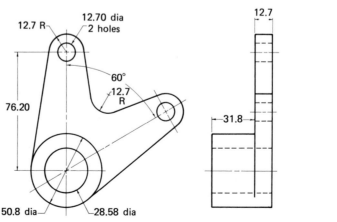

Figure 14.43 Bell crank; specific features, such as the holes, are best dimensioned in the expressive view (the front view in this case).

sidered in the design of the product. Surface quality can be specified by the use of well-recognized notes and symbols to indicate surface *roughness*, *waviness*, and *lay*. The following definitions of these terms are based on ANSI B46.1.

Roughness is defined as the predominant surface pattern resulting from the finely spaced surface irregularities produced by cutting edges of machine tools. The *height* of the irregularities is rated in micrometers (one millionth of a meter). *Waviness* refers to irregularities that result from factors such as machine or work deflections, vibration, heat treatment, or warping strains. These irregularities are of greater spacing than roughness. *The height is expressed in millimeters.* *Lay* is the direction of the predominant surface pattern, produced by tool marks or grains of the methods of production.

Specifying Surface Quality by Use of Surface Symbols

Where it is only necessary to specify surface *roughness height*, the simplest form of symbol is used, as shown in Figure 14.44*a*. The height, represented by the rating, is in micrometers and indicates the maximum value.

A specification of maximum and minimum values is shown in Figure 14.44*b*. The numerals indicate the permissible range of roughness height.

Where it is necessary to include *waviness* height in addition to roughness height, the symbol shown in Figure 14.44*c* is used. The numerical value of the height of waviness is placed above the horizontal line. When it is necessary to include the maximum waviness *width*, the rating is placed above the horizontal line and to the right of the waviness height rating, as shown in Figure 14.44*d*.

To specify *lay* in addition to roughness and waviness, another symbol is included, as shown in Figure 14.44*e*. The parallel lines indicate that the dominant lines of the surface are parallel to the boundary line of the surface in contact with the symbol. Other *lay* symbols are shown in Figure 14.45.

The relation of symbols to surface characteristics is quite clearly conveyed in Figure 14.44*g*.

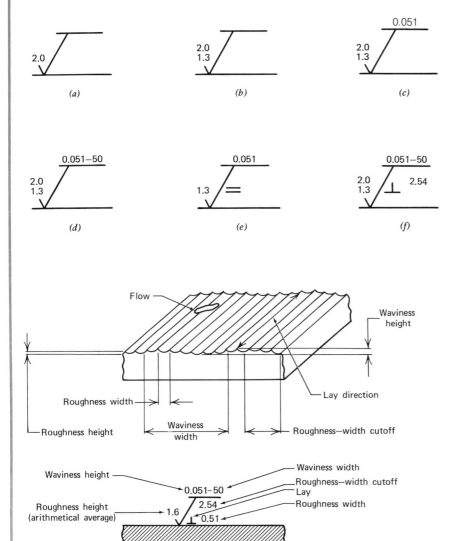

Figure 14.44 Surface quality symbols and their relation to surface characteristics.

Lay Sym-bol	Meaning	Example Showing Direction of Tool Marks
=	Lay approximately parallel to the line representing the surface to which the symbol is applied.	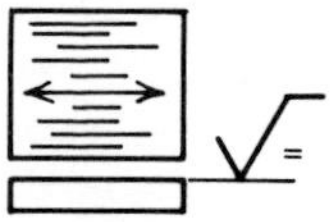
⊥	Lay approximately per-pendicular to the line representing the surface to which the symbol is applied.	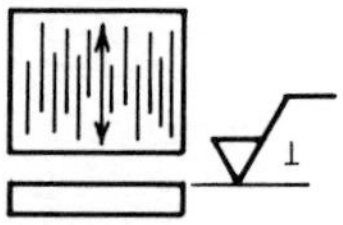
X	Lay angular in both direc-tions to line representing the surface to which the symbol is applied.	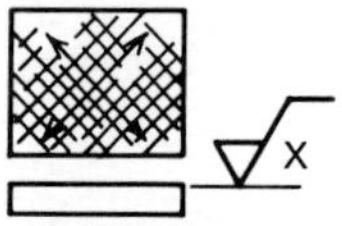
M	Lay multidirectional.	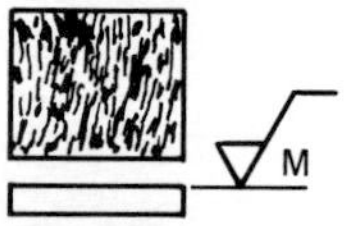
C	Lay approximately circular relative to the center of the surface to which the symbol is applied.	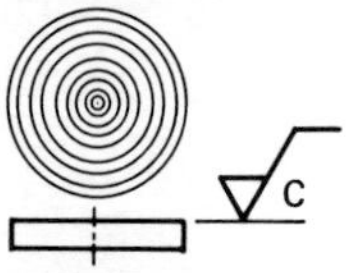
R	Lay approximately radial relative to the center of the surface to which the symbols is applied.	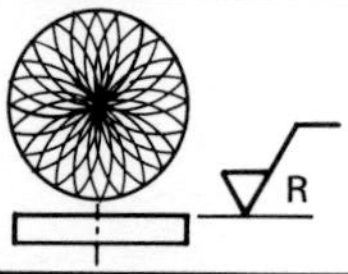
P[3]	Lay particulate, non-di-rectional, or protuberant.	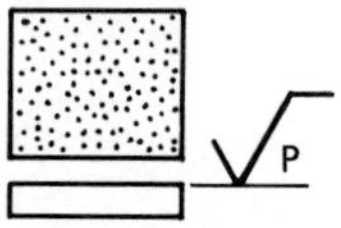

(ANSI Y 14.36—1978)

[3] The "P" symbol is not currently shown in ISO Standards. American National Standards Committee B46 (Surface Tex-ture) has proposed its inclusion in ISO 1302 "Methods of indicating surface texture on drawings."

Figure 14.45 Lay symbols (ANSI Y14.36-1978).

Surface Roughness Produced by Common Production Processes

Typical ranges of surface roughness height values that may be obtained by various production processes are shown in Figure 14.46. Note that because of the many variables in processing operations, the given ranges of values are for *general* guidance only. The engineer must be aware of *costs* and should not specify operations beyond those necessary to fulfill the functional requirements of the components of the product.

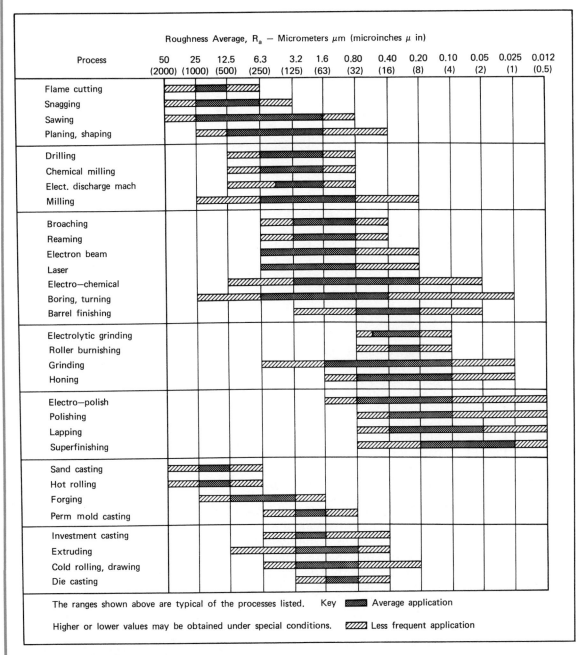

Figure 14.46 Surface roughness produced by common production processes. (Courtesy ANSI B46.1-1962)

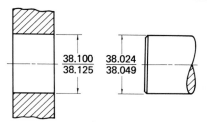

Figure 14.47 Hole and shaft dimensions. Tolerance is 0.025 mm.

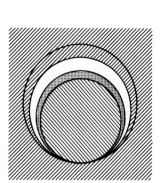

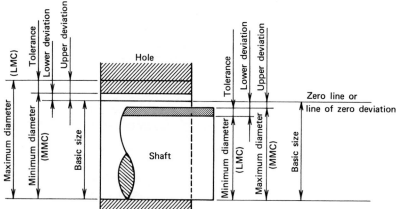

Figure 14.48 Limits and fits conventional and related terms. (Courtesy AS 1654, 1974)

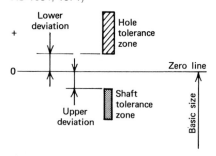

Figure 14.49 Conventional diagram showing tolerance zones and basic size. (Courtesy AS 1654, 1974)

GENERAL TOLERANCING

Introduction

Mass production of various types of machines often involves components that consist of mating parts. For example, in the design of an automobile crankshaft and its supporting bearings, dimensions of these components must reflect the *designer's intent* in providing the desired functions of the parts.

We must recognize that mating components cannot be produced to meet exact sizes. It is essential, therefore, to provide "tolerances" in the parts so that production is *feasible and economical. The parts must be held to the tolerances necessary for proper functioning.*

For example, in Figure 14.47, the dimensions of the shaft and of the part with which it mates show that the shaft diameter can vary from 38.024 mm to 38.049 mm and that the hole diameter can vary

from 38.100 mm to 38.125 mm. The tolerance for the shaft is 0.025 mm (38.049 mm − 38.024 mm), and for the hole it is also 0.025 mm (38.125 mm − 38.100 mm).

Specifying "limit" dimensions of the parts may be based on the manufacturing experience of the company; however, in general it is good practice to follow the recommendations that are contained in the *special metric* publication of the ANSI, adapted from ISO Recommendation R286-1962 and the standard AS-1694-1974, "Limits and Fits for Engineering (Metric Units)."

Dimensioning for mating parts introduces terms and definitions with which we should become familiar.

Terms and Definitions

1. *Tolerances of parts* occur since a part cannot be made precisely to

a stated dimension, primarily because of the inaccuracies of manufacturing methods. It is generally sufficient, however, to produce the part so as to lie within two possible limits of size (as previously shown in Figure 14.47). *The difference between the limit dimensions is the tolerance.*

2. As a convenience, a *basic size* ascribed to the part and each of the two limits is defined by its *deviation* from this basic size. The magnitude and sign of the deviation are obtained by subtracting the basic size from the limit in question. These definitions are illustrated in Figure 14.48. In practice, however, the illustration is replaced by a schematic diagram similar to Figure 14.49. In this simplified diagram the axis of the part, which is not shown, always lies, by convention, below the diagram. In Figure 14.49 the two deviations of the hole are positive, and those of the shaft are negative.

3. *Limits of size* are the maximum and minimum sizes permitted for a feature.

4. *Maximum material condition* (MMC) denotes the condition in which the greatest amount of material is present at the surface of a feature (i.e., the upper limit of size of a shaft or the lower limit of size of a hole).

5. *Least material condition* (LMC) denotes the condition in which the least permissible amount of material is present at the surface of a feature (i.e., the upper limit of size of a hole or the lower limit of size of a shaft). An example of an exception is an exterior corner radius where the maximum radius is the least material condition and the minimum radius is the maximum material condition.

6. *Upper deviation* is the algebraic difference between the maximum limit of size and the corresponding basic size (see Figure 14.49).

7. *Lower deviation* is the algebraic difference between the minimum limit of size and the corresponding basic size (see Figure 14.49).

8. *Zero line* is the line of zero deviation and represents the basic size (see Figures 14.48 and 14.49). By convention, when the zero line is drawn horizontally, positive de-

viations are shown above it and negative deviations are shown below it.

9. *Tolerance zone* is the zone between the two lines representing the upper and lower deviations and is defined by its magnitude (tolerance) and by its position in relation to the zero line. Conventionally, *hole tolerance zones* are shown *crosshatched* and *shaft tolerance zones* are fully shaded (see Figure 14.49 or, as shown in Figures 14.50 and 14.51).

10. *Tolerance grade* is a group of tolerances considered as corresponding to the same level of accuracy for all basic sizes.

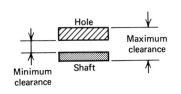

(a) Clearance fit extremes

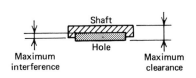

(b) Transition fit extremes

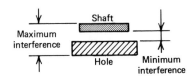

(c) Interference fit extremes

Figure 14.50 Schematic diagram for various fits—extremes. (Courtesy AS 1654, 1974)

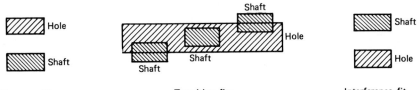

Clearance fit   Transition fits   Interference fit

Figure 14.51 Schematic diagram of various fits. (Courtesy ANSI R2.86-1962)

11. *Fundamental deviation* is the one of the two deviations (the one nearer to the zero line) that is conventionally chosen to define the position of the tolerance zone in relation to the zero line (see Figure 14.52).

12. *Tolerance class* is the association of a fundamental deviation with a tolerance grade.

13. *Basic hole* is a hole whose *lower deviation* is zero. More generally, it is the hole selected as a basis for a hole-basis system of fits.

14. *Basic shaft* is a shaft whose *upper deviation* is zero. More generally, it is the shaft selected as a basis for a shaft-basis system of fits.

Fits*

The relation that results from the difference, before assembly, between the sizes of the two parts that are to be assembled is called a fit. The fit may be one of the following.

1. A *clearance fit*, which always provides a clearance, as shown in Figure 14.48. Note that the tolerance zone of the hole is entirely *above* that of the shaft, as shown in Figure 14.49. Clearance fit extremes are shown in Figures 14.50*a* and 14.51.

2. A *transition fit*, which may provide either a clearance or an interference. The tolerance zones of the hole and the shaft overlay, as shown in Figure 14.51. Transition fit extremes are shown in Figure 14.50*b*.

3. An *interference fit*, which always provides an interference. In this case the tolerance zone of the hole is entirely below that of the shaft (see Figure 14.51). Interference fit extremes are shown in Figure 14.50*c*.

Applications of the System of Limits and Fits

The system primarily consists of (1) a series of grades of tolerances associated with a range of basic sizes, and (2) a range of deviations for holes and shafts that defines the position of the tolerance zones with respect to the basic size. (Although "holes" and "shafts" are referred to explicitly, the recommendations apply equally well to the space contained by or containing two parallel faces or tangent planes of any part, such as the width of a slot or tongue, key thickness, etc.).

Tolerance Grades

There are 18 grades of tolerance to provide for different classes of work. The tolerance grades are numbered 01, 0, 1, 2, 3, . . . 16. The standardized numerical values for grades 6 to 12 are given in Appendix B, Table 3, p. 630, and also grades 01 to 5 and 13 to 16 are given in Appendix B, Table 3, p. 631. (In general engineering practice, grades 6 to 12 are used for fits, 01 to 5 are used for gauges, and 13 to 16, are used for general dimensioning).

Deviations

There are 27 different fundamental deviations for holes and shafts for sizes up to and including 3150 mm.

Each deviation is designated by a letter symbol—*uppercase* letters *for holes* and *lowercase* for *shafts*. The designations are:

Holes. A, B, C, CD, D, E, EF, F, FG, G, H, J, K, M, N, P, R, S, T, U, V, X, Y, Z, ZA, ZB, and ZC.
Shafts. a, b, c, cd, d, e, ef, f, fg, g, h, j, k, m, n, p, r, s, t, u, v, x, y, z, za, zb, and zc.

Tolerance Classes

A tolerance grade with a fundamental deviation constitutes a tolerance class (i.e., H7, a hole with deviation H and tolerance grade 7, or k6, a shaft with deviation k and tolerance grade 6). Figure 14.51 shows the relative positions of tolerance zones that result from a combination of a fundamental deviation and a tolerance grade for holes and shafts with respect to the zero line representing the basic size.

Limits of Size

The upper and lower limits of size are determined by the algebraic addition of relevant upper and lower deviations, respectively, to the basic size.

The maximum material condition (MMC) concept is adopted and use is made of the fundamental deviations together with the tolerance, where applicable, to calculate the lower deviations for holes (MMC holes) and the upper deviations for shafts (MMC shafts).

Selection of Fits

The number of possible combinations is quite large because of the availability of 27 fundamental deviations for holes and shafts and 18 tolerance grades.

Most of the engineering requirements, however, can be satisfied by a restricted selection of fits by using the hole-basis system or the shaft-basis system of Fits.

Hole-Basis System

The different fits (clearance, transition, and interference) are obtained by associating various shafts with a single hole (or possibly with holes of different tolerance grades) but always having the same fundamental deviation.

---

* Descriptions of fits using *inch units* are given in Appendix B, Tables 42 to 46, pp. 604–610.

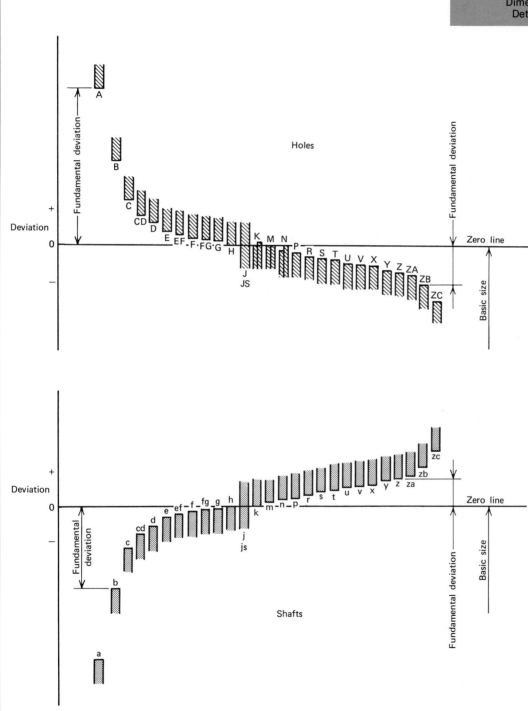

Figure 14.52 Relative positions of the
various tolerance zones. (Courtesy
AS 1654, 1974)

The H hole-basis fits are recommended for most general applications. The lower deviation of the H hole is zero (see Figure 14.51).

The types of fits that generally result when shafts are assembled with the H hole are:

1. Clearance fits—shafts a to h.
2. Transition fits—shafts j, k, and m.
3. Interference fits—shafts n to zc.

Tables 2 and 3 in Appendix B, pp. 621–630, are used for hole-basis fits. The upper deviation for the shaft from Table 2, regardless of its sign, gives the tightest fit (i.e., minimum clearance for shafts a to h and maximum interference for shafts k to zc) when the shaft is used with the H hole.

Shaft-Basis System
The different fits are obtained by associating various holes with a single shaft (or possibly with shafts of different tolerance grades) but always having the same fundamental deviation.

The h shaft-basis fits are recommended for most general applications. The upper deviation of the h shaft is zero (see Figure 14.51).

The types of fits that generally result when holes are assembled with the h shaft are:

1. Clearance fits—holes A to H.
2. Transition fits—holes J, K and M.
3. Interference fits—holes N to ZC.

Tables 1 and 3, in Appendix B, pp. 612–630, are used for shaft-basis fits. The lower deviation for the hole from Table 1, irrespective of its sign, gives the tightest fit (i.e., minimum clearance for holes A to H, and maximum interference for holes K to ZC) when the hole is used with the h shaft.

Examples of Calculations of Limits and Fits

1. Limits of Size of a Hole Designated 25D7
   Basic size   25.000 mm
   From Table 1, lower deviation for MMC   + 0.065
        25.065 Lower limit of size
   From Table 3, tolerance grade 7   + 0.021
        25.086 Upper limit of size

2. Limits of Size for a Hole Designated 20M8
   Basic Size   20.000 mm
   From Table 1, lower deviation for MMC   − 0.029
        19.971 Lower limit
   From Table 3, tolerance grade 8   + 0.033
        20.004 Upper limit

3. Limits of Size for a Shaft Designated 150f9
   Basic size   150.000 mm
   From Table 2, upper deviation for MMC   − 0.043
        149.957 Upper limit
   From Table 3, tolerance grade 9   − 0.100
        149.857 Lower limit

4. Limits of Size for a Shaft Designated 75m7
   Basic size   75.000 mm
   From Table 2, upper deviation for MMC   + 0.041
        75.041 Upper limit
   From Table 3, tolerance grade 7   − 0.030
        75.011 Lower limit

5. Limits of Size for a Shaft Designated 50h6
   Basic size   50.000 mm
   From Table 2, upper deviation for MMC   0
   _____
   50.000 Upper limit

   From Table 3, tolerance grade 6   − 0.016
   _____
   49.984 Lower limit

6. Limits of Size for a Hole Designated 50H6
   Basic size   50.000 mm
   From Table 1, lower deviation for MMC   0
   _____
   50.000 Lower limit

   From Table 3, tolerance grade 6   + 0.016
   _____
   50.016 Upper limit

7. Limits of Size for a Hole Designated $100J_s8$
   From Table 3, tolerance grade 8   0.054 or ±0.027
   Basic size   100.000 mm
   $\dfrac{\text{Tolerance}}{2}$ + 0.027
   _____
   100.027 Upper limit
   Basic size   100.000 mm
   $\dfrac{\text{Tolerance}}{2}$ − 0.027
   _____
   99.973 Lower limit

Note. The limits of size for a shaft designated $100j_s8$ are the same as the limits shown above.

Choice of Fits

In the choice of fits for a specific application (e.g., shaft and bearing design) the engineer considers factors such as bearing load, speed, length of engagement, lubrication, temperature, humidity, and materials. The engineer also recognizes that *manufacturing costs increase as closer tolerances are specified*; therefore, it is essential to prescribe the most liberal tolerances consistent with *function*.

For a given application the design engineer chooses an appropriate hole and shaft combination to satisfy the design requirements. It is assumed that the engineer has chosen the hole basis or the shaft basis as appropriate for the specific application.

Examples of How Various Design Requirements are Satisfied

1. *Hole Basis, Clearance Fit*
Design requirements:
Basic size: 75.000 mm
Fit desired: 0.060 mm minimum clearance to 0.160 mm maximum clearance
Solution:
From Table 2, shaft e, for the basic size 75.000 mm gives the minimum clearance of 0.060 mm. The variation of fit is 0.100 mm (0.160 mm − 0.060 mm), which is *the sum* of the tolerances on the hole and the shaft. Now, from Table 3, the tolerance combination grade 7 plus grade 9 on the basic size 75.000 mm gives the sum of the tolerances 0.030 mm + 0.074 mm = 0.104 mm

which is close to the design requirement; therefore the fit H9/e7 is selected. In terms of limits of size, for hole 75H9, the dimensions are 75.074 mm and 75.000 mm, and for Shaft 75e7, the dimensions are 74.940 mm and 74.910 mm (see Figure 14.53).

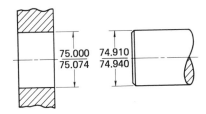

75.000
75.074   74.910
74.940

Figure 14.53 Dimensions of hole and shaft for fit H9/e7 (hole basis, clearance fit).

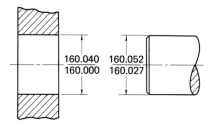

Figure 14.54 Dimensions of hole and shaft for fit H7/n6 (hole basis, transition fit).

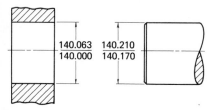

Figure 14.55 Dimensions of hole and shaft for fit H8/u7 (hole basis, interference fit).

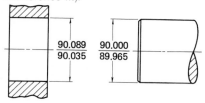

Figure 14.56 Dimensions of hole and shaft for fit F8/h7 (shaft basis, clearance fit).

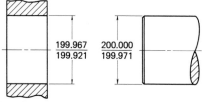

Figure 14.57 Dimensions of hole and shaft for fit P7/h6 (shaft basis, interference fit).

2. *Hole Basis, Transition Fit*
Design requirements:
Basic size: 160.000 mm
Fit desired: 0.052 mm maximum intereference to 0.013 mm maximum clearance
Solution:
From Table 2, shaft n6 gives the maximum interference of 0.052 mm. The variation of fit is 0.065 mm, which is *the sum* of the tolerances on the hole and the shaft. Now, from Table 3, the tolerance combination of grade 6 plus grade 7 gives the sum of the tolerances 0.025 mm + 0.040 mm = 0.065 mm, which satisfies the design requirement. The fit, therefore, is H7/n6. The limits of size are shown in Figure 14.54

3. *Hole Basis, Interference Fit*
Design requirements:
Basic size: 140.000 mm
Fit desired: 0.210 mm maximum interference to 0.105 mm minimum interference
Solution:
From Table 2, shaft u7 gives the desired maximum interference 0.210 mm. Variation of fit is 0.105 mm, which represents *the sum* of the tolerances on the hole and the shaft. Now, from Table 3, the tolerance combination of grade 7 plus grade 8 gives the sum of the tolerances 0.040 mm + 0.063 mm = 0.103 mm, which is close to the desired value 0.105 mm. The fit, therefore, is 140H8/u7. The limits of size are shown in Figure 14.55.

4. *Shaft Basis, Clearance Fit*
Design requirements:
Basic size: 90.000 mm
Fit desired: 0.035 mm minimum clearance to 0.124 mm maximum clearance
Solution:
From Table 1, hole F gives the desired minimum clearance of 0.036 mm. Variation of fit is 0.089 mm (0.124 mm − 0.035 mm), which represents *the sum* of the tolerances on the hole and the shaft. Now, from Table 3, the tolerance combination of grade 7 plus grade 8 gives the tolerances 0.035 mm + 0.054 mm = 0.089 mm, which is the desired value 0.089 mm. The fit, therefore, is 90F8/h7. The limits of size are shown in Figure 14.56.

5. *Shaft Basis, Interference Fit*
Design requirements:
Basic size: 200.000 mm
Fit desired: 0.079 mm maximum interference to 0.004 mm minimum interference
Solution:
From Table 1, the hole P7 gives the desired maximum interference of 0.079 mm. Variation of fit is 0.075 mm, which represents *the sum* of the tolerances on the hole and the shaft. Now, from Table 3, the tolerance combination of grade 6 plus grade 7 gives the sum of the tolerances 0.029 mm + 0.046 mm = 0.075 mm. The fit, therefore, is P7/h6. The limits of size are shown in Figure 14.57.

Selected Fits for General Engineering Application
The majority of fit conditions for general engineering applications can be accommodated by a limited selection of deviations and tolerances, as follows.

Selected Hole-Basis Fits
The following holes and shafts are most commonly used.

Holes. H7, H8, H9, and H11.
Shafts. c11, d10, e9, f7, g6, h6, k6, n6, p6, and s6.

Combinations of these holes and shafts, shown in Appendix B, Table 4, p. 632, form a selected series of fits. Some manufacturers may need to adopt a different range.

Assuming that a manufacturer has decided on a selected range and has installed the necessary tooling and gauging facilities, it is possible to combine the holes and shafts in different ways to provide a wider range of fit conditions without any additional investment in tools and equipment. For example, given that the range of fits shown in Table 4 has been introduced and that, for a particular application, the fit H8/f7 is appropriate but provides too much variation, the hole H7 could equally well be associated with the shaft f7 and may provide exactly what is required without additional tooling. This facility may be extended to include any combinations of the selected holes and selected shafts.

EXAMPLE (USING TABLE 4)
Given:

Basic size = 80.000 mm
Fit H9/e9

From Table 4 the hole dimensions are

80.000 mm + 0 mm = 80.000 mm
(smallest diameter)

and

80.000 mm + 0.074 mm
= 80.074 mm (largest diameter)

For the shaft the dimensions are

80.000 mm − 0.060 mm
= 79.940 mm (largest diameter)

and

80.000 mm − 0.134 mm =
79.866 mm (smallest diameter)

Selected Shaft-Basis Fits
A shaft-based system is desirable, for example, where a single shaft is to accommodate bearings, collars, and couplings. In such a case a constant diameter shaft can be maintained, and the bore of the ac-cessories can vary in accordance with the desired fits.

Table 5 in Appendix B, p. 633, features the shaft-basis equivalents of the hole-basis fits shown in Table 4. These are all direct conversions except that the fit H9/d10, instead of being converted to D9/h10, is adjusted to D10/h9, thereby avoiding the introduction of the additional shaft h10.

EXAMPLE (USING TABLE 5)
Given:

Basic size = 80.000 mm
Fit = E9/h9

From Table 5 the hole dimensions are

80.000 mm + 0.134 mm
= 80.134 mm (largest diameter)

and

80.000 mm + 0.060 mm
= 80.060 mm (smallest diameter)

For the shaft the dimensions are

80.000 mm − 0 mm = 80.000 mm
(largest diameter)

and

80.000 mm − 0.074 mm
= 79.926 mm (smallest diameter)

Unilateral Tolerance
A tolerance in which variation is permitted only in one direction from the specified dimension (see Figure 14.58*a*).

Bilateral Tolerance
A tolerance in which variation is permitted in both directions from the specified dimensions (see Figure 14.58*b*).

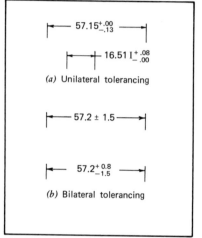

$57.15^{+.00}_{-.13}$

$16.51\ I^{+.08}_{-.00}$

(*a*) Unilateral tolerancing

57.2 ± 1.5

$57.2^{+0.8}_{-1.5}$

(*b*) Bilateral tolerancing

Figure 14.58 Unilateral and bilateral tolerancing.

Tolerance Accumulation

Where it is important to maintain close tolerancing, it is imperative to avoid errors caused by the accumulation of tolerances. For example, the type of "chain dimensioning" shown in Figure 14.59a is not good practice. It is seen that the distance between X and Y could be a maximum of 69.95 mm or a minimum of 69.77 mm, or of a dimension between these two values. In order to reduce the accumulation of tolerance between X and Y, the

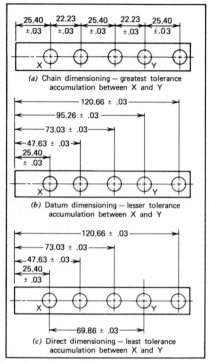

Figure 14.59 Control of tolerance accumulation.

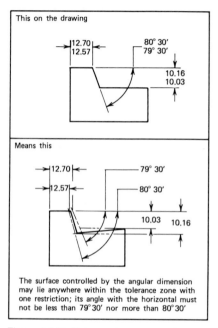

Figure 14.60 Tolerancing an angular surface using a combination of linear and angular dimensions.

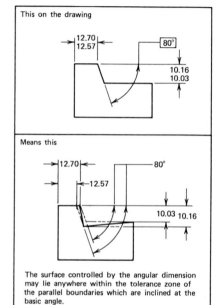

Figure 14.61 Tolerancing an angular surface with a basic angle.

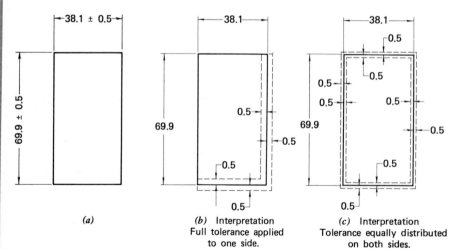
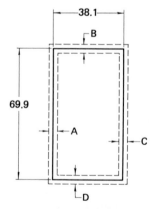

Figure 14.62 Dimensioned part and possible interpretations.

dimensions are placed as shown in Figure 14.59*b*. This method provides for a simple and economical shop layout.

The maximum variation between any two features can be controlled by the tolerance on the dimension between the features, as shown in Figure 14.59*c*.

Tolerancing an Angular Surface
Where an angular surface is defined by a linear and angular dimension, the surface must lie between the planes represented by the tolerance zone, as shown in Figure 14.60.

The tolerance zone will be wider as the distance from the vertex of the angle increases. To avoid a tolerance zone of this shape, the basic angle method shown in Figure 14.61 is used. This method defines a tolerance zone with parallel boundaries at the basic angle. The accuracy of the produced angle must assure that no part of the resulting surface exceeds this boundary.

DIMENSIONING TO ENSURE A SINGLE INTERPRETATION OF A DESIGN

There is a growing demand for greater reliability of mass-produced goods. Maintaining desired reliability, among other factors, requires the elimination of ambiguities that often arise from the methods employed in dimensioning components that make up a unit or an assembly of units. What does the engineer try to accomplish when selecting dimensions for a part or component? Surely the intent would be to dimension the parts so that all concerned—engineer, manufacturer, and inspector—will interpret the design in the *same way*.

In our previous work we have had some experience in dimensioning simple geometric shapes and parts that are largely combinations of simple shapes. In addition, we determined limit dimensions of mating parts (i.e., shafts and bearings).

Now we will consider dimensioning with respect to precision and reliability, so as to ensure a single interpretation of a design.

EXAMPLE 1
Suppose we have a rectangular shape dimensioned as shown in Figure 14.62*a*. Each dimension has a tolerance of ±0.50 mm. How shall we interpret this drawing? What tolerance zone is defined by ±0.50 mm? Do we have any assurance that the manufacturer and the inspector will interpret the specifications in the same way? Several possible interpretations are shown in Figure 14.62*b*, 14.62*c*, and 14.62*d*. Remember that the manufactured part will not be perfectly straight and smooth. What interpretation would be meaningful? We will not identify correct, or incorrect, interpretation here. Instead, we assume that the interpretation shown in Figure 14.62*d* represents the *intent of the designer*. This interpretation imposes the

singular requirement that the profile of the part falls within the frames established by the limit dimensions (see Figure 14.63). How should we dimension the drawing to ensure this interpretation? We will answer this question after discussing Example 2 and the importance of *datums*.

EXAMPLE 2
Consider the cylindrical part shown in Figure 14.64. A casual inspection of the drawing would cause little concern. The part seems to be adequately defined by the dimensions. A more careful analysis, however, points out a serious problem of interpretation. Observe hole *h*, which

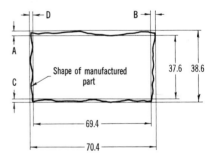

Figure 14.63 Interpretation of Figure 14.62*d*

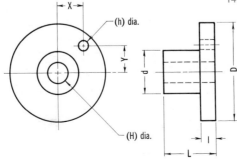

Figure 14.64 Example 2. Cylindrical part.

is shown to be located by distances X and Y from the centerlines. Centerlines of what? As the part passes through the shop and on to inspection, it does not have a pair of fixed centerlines going along with it. How, then, does the worker locate hole *h*? How does the inspector verify that the hole is properly located? Should the centerlines be established by the hole *H*, cylinder *d*, cylinder *D*, or by some method that does not involve the features of the part? The design engineer, who knows the functional requirements of the part, is the only person qualified to answer these questions; indeed, it is his or her sole responsibility to provide the dimensions that will establish the *single* interpretation necessary to manufacture and verify a functional part or specifically, *to establish and identify the datums properly*.

Datums

Datums are points, lines, planes, cylinders, etc., assumed to be exact for computation purposes, from which the location of features of a part may be established. This definition is accurate, but it is a limited version of the total datum concept. A comprehensive exploration of the datum idea must include considerations *pertinent to the manufacture and verification* of the parts described by the drawing.

Return to the cylindrical part (Figure 14.64) and *note the influence of function on the choice of a datum*. Observe two possible functional requirements illustrated in Figure 14.65. Can the same datum be used to locate hole *h* in each

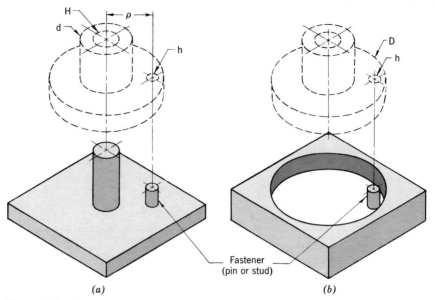

Figure 14.65 Choice of datum and its influence on function.

case? The functional requirement for application (Figure 14.65*a*) is as follows: (1) hole *H* must engage the projecting cylinder of the mating part; and (2) the fastener must pass through hole *h* to permit assembly. In this case the datum for the location of hole *h* must be established by the feature diameter *H*. Using either cylinder *d* or cylinder *D* as the datum feature to locate *h* would jeopardize the probability of assembly, because *the parts would not have a common datum*. For example, if the feature *d* was used to establish the datum for application (a), the distance ρ would be affected by the difference in position of the centerline of diameter *d* and the centerline of diameter *H*. This variation would add to the tolerance of location between diameters *H* and *h* (and the projecting cylinder and fastener on the mating part).

This additional variation may well prohibit assembly. By making a similar analysis, the student

should identify the proper datum feature [diameter *D* for application, (Figure 14.65*b*)]. It is possible to imagine an application wherein diameter *d* would properly establish the datum for locating hole *h*.

This example involves a very simple part. The magnitude of the problem will increase greatly as the parts become more complex.

Let us now list some concepts that will enable you, the student, to better understand and establish datums properly.

1. A *datum* is the origin of a measurement that locates or otherwise relates a feature to the datum feature or features. The primary function of a datum is to establish a point to serve as the origin of a measurement. In addition, the datum planes establish the *orientation* (direction) of the dimensions, as shown in Figure 14.66.

The desired sequence of datum features is specified in the feature control symbol, as shown in Figure

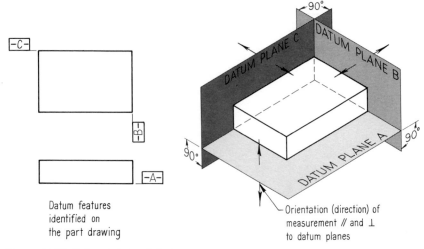

Datum features
identified on
the part drawing

Orientation (direction) of
measurement ∥ and ⊥
to datum planes

Figure 14.66 Datum interpretation.

14.67*a*. The datum features are identified as surfaces *D*, *E*, and *F*. As indicated in Figure 14.67*b*, these surfaces are design requirements needed for proper assembly and functioning of the part. Since surfaces *D*, *E*, and *F* are the primary, secondary, and tertiary datum features, respectively, they appear in that order in the feature control symbol. Depending on the type of tolerance (position or form) and the relationship required, it may be necessary to reference only one or two datum features. The primary datum feature relates the part to the datum reference frame by bringing a minimum of three points into contact with the first datum plane (see Figure 14.68*a*). The part is further related to the frame by bringing at least two points of the secondary datum feature into contact with the second datum plane (see Figure 14.68*b*). The relationship is completed by bringing at least one point of the tertiary datum feature into contact with the third datum plane (see Figure 14.68*c*). *As measurements are made from datum planes, positioning of the datum features in re-*

lation to these planes insures a common basis for measurements.

2. *The actual (physical) datum will rarely exist on the part*; it is usually a feature (or features) *on the processing tool and on the inspection equipment.* Machine tool tables, axes of spindles, surface plates, axes and surfaces of inspection equipment, etc., are examples of features that are physical datums.

3. The features of processing and verification equipment used as physical datums are considered to be "perfect." It is necessary, of course, to check the accuracy of a physical datum carefully to justify its use as

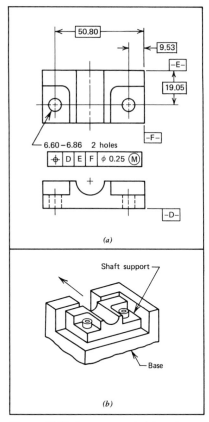

Figure 14.67 Sequence of datum features relates part to datum reference frame. (ANSI Y14.5-1973)

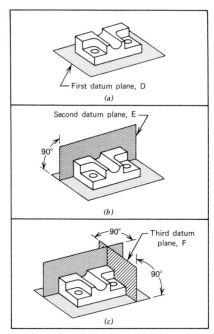

Figure 14.68 Example of part where datum features are flat surfaces. (ANSI Y14.5-1973)

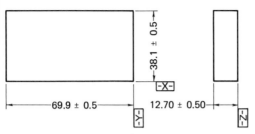

Figure 14.69 Rectangular shape of Figure 14.62a dimensioned with datums to ensure a single interpretation.

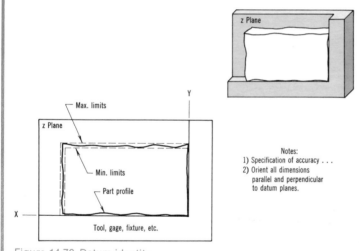

Notes:
1) Specification of accuracy . . .
2) Orient all dimensions parallel and perpendicular to datum planes.

Figure 14.70 Datum identity.

a datum (perfection). Although it is true that the use of these features will introduce an error, such equipment will usually be at least 10 times more precise than the part. The error introduced by this practice is therefore comparatively small.

4. The datum feature (on the part) should be carefully selected to (a) guarantee the proper function of the part, (b) allow the physical datum to be established on the processing and inspection equipment, and (c) insure the use of the same theoretical datum for manufacture, verification (inspection), and assembly.

The following examples, which illustrate the application of these concepts, will allow the student to observe the clarifying effect on interpretation of dimensions when datums are specified.

EXAMPLE 1
The dimensions of the rectangular shape shown in Figure 14.62a would be restricted to a single interpretation if datum planes were established, as shown in Figure 14.69. As an additional precaution, particularly when dealing with critical and expensive parts, it may be desirable to define the datum notation. Identification and definition of datums are frequently included in drafting room manuals (DRM), process standards, etc., which are used by the company. An example of the datum identity for our problem is shown in Figure 14.70. The part would not generally be included, but it is shown here to illustrate the definition.

EXAMPLE 2

Let us consider the "conventional" dimensions of a simple piece and the interpretation of the dimensions (see Figure 14.71). The piece is a plate with four holes. The design engineer's intent is to provide a part that may be fastened to an "identical" plate by four 12-mm (M12 × 1.75) diameter bolts. The edges of the plates need not match. At first glance, the drawing seems to be complete except for the size dimensions of the holes. These dimensions have been omitted intentionally. What is the correct interpretation of the dimensions?

1. Does the 20.00 mm ± 1.00 mm dimension apply to both holes A and 3 from the surface Y?

2. Does the 20.00 mm ± 1.00 mm dimension apply to both holes A and 1 from surface X?

3. Are holes 1, 2, and 3 located from hole A as a pattern within the tolerance ±0.25 mm?

4. Can holes 1 and 2 each vary from surface Y by the total amount of the tolerance ±1.00 mm + (±0.25 mm)?

5. What is the relation of surface X to surface Y?

6. From what point are measurements taken?

We now realize that interpretations of the drawing can vary, depending on the answers to these questions and others that could be raised. We can eliminate ambiguities by applying *the true position dimensioning method* with proper datum consideration.

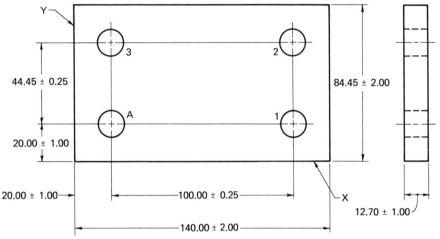

Figure 14.71 Example 2. Meaning of dimensions?

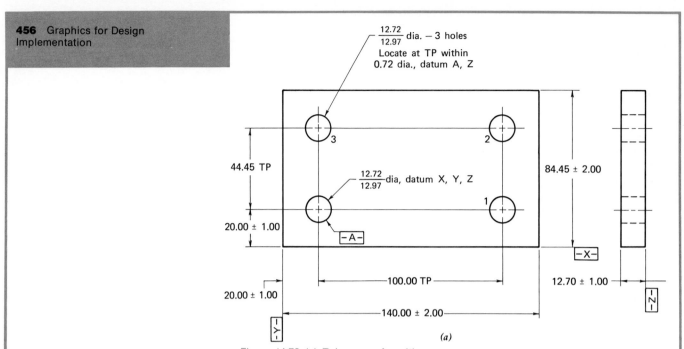

Figure 14.72 (a) Tolerance of position.

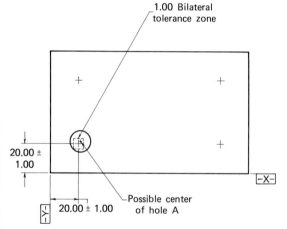

Figure 14.72 (b) Possible center of hole A.

Figure 14.72 (c) Positional tolerance zone, 0.72 mm diameter, for holes 1, 2, and 3.

Tolerance of Position

Study Figure 14.72a. The coordinate dimensions (20.00 mm ± 1.00 mm) locate hole A from the datum established by features X and Y. This is shown in Figure 14.72b. *Regardless of where the center of hole A falls within the 2.00-mm square tolerance zone, it is still the datum.* It is conceivable that the center could fall on a corner of the square tolerance zone. This allows us to increase the tolerance zone to a circle whose diameter is a diagonal of the square. This increase, about 57%, means that the rejection of parts is reduced by the number of parts in which a feature falls in the area between the square and the circle. *Hole A then becomes the datum feature for the remaining holes 1, 2, and 3.*

True position of hole 3, for example, is 44.45 mm from datum A on a line passing through datum A and perpendicular to datum X. True position for hole 2 is the intersection of a line 100.00 mm to the right of datum A and perpendicular to datum X and a line 44.45 mm above datum A and parallel to datum X. True position for hole 1 is 100.00 mm to the right of datum A on a line passing through datum A and parallel to datum X.

The diameter (0.72 mm) stated in the note in Figure 14.72a is the diameter of the positional tolerance zone for each of the holes 1, 2, and 3 (see Figure 14.72c). The use of datums, in this example, allows a large tolerance for the location of the pattern of holes relative to the edges of the plate. At the same time, it retains a desirable degree of precision for the location of the holes relative to one another. This agrees with the engineer's intent stated previously. In such cases, where the part is symmetrical, the worker must mark the part to identify the datum features. *This is necessary to allow the inspector to orient the plate so that the same datum features (X, Y, Z, and A) are used for verification that were used in production.*

Clearance Holes. The assembly of the plates is dependent on the size of the holes through each of the two plates. How shall we determine the size of the holes?

The datum and true position points of each plate can be aligned when placed one on the other, since the plates are dimensioned the same.

In their extreme positions, a pair of corresponding holes (one in the upper plate and the other in the lower plate) must have a minimum diameter of 12.72 mm in order to accommodate a 12-mm diameter bolt. This is clearly seen in Figure 14.73. The maximum deviation allowed from true position (TP) for *each plate* is the radius (0.36 mm) of the positional tolerance zone; therefore, the total deviation with respect to both plates is 0.72 mm. This value plus the bolt diameter establishes the minimum hole size in each plate or, 0.72 mm + 12.00 mm = 12.72 mm, *the minimum diameter of the clearance holes.*

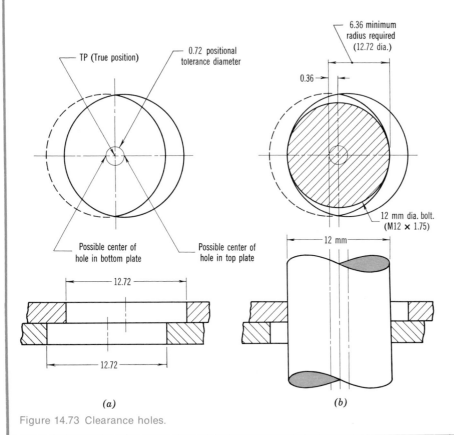

(a)   (b)

Figure 14.73 Clearance holes.

EXAMPLE 3

Cylindrical surfaces of parts bear no resemblance to datum planes. A cylindrical datum feature, however, is associated with two datum planes represented by centerlines drawn at right angles and intersecting at the axis of the cylindrical datum feature. An application of the three-plane concept to a cylindrical part is shown in Figure 14.74. In Figure 14.74*a* the datum feature *K* is a sealing surface to which the bolt clearance holes are normal. It has been selected as the primary datum for design purposes. The holes are related to cylindrical datum feature *M* and are dimensioned from the centerlines through the center of *M*. Since the part has thickness, the centerlines represent planes through the part center, as shown in Figure 14.74*b*. These planes represent the second and third datum planes, X and Y, respectively. The *clearance* holes are located in relation to these three datum planes.

In cases where a diameter is designated as a datum feature, the line formed by the intersection of two planes is the datum axis. In Figure 14.74 the sequence of planes X and Y is immaterial, since rotation of the pattern about the datum axis has no effect on the function of the part. *The feature control symbol references two datum features:* (1) *K*, the primary datum feature, a flat surface associated with the first datum plane; and (2) *M*, the secondary datum feature (cylindrical), which is associated with the second and third datum planes (or datum axis).

In the preceding Example 2 we used the expression "true position" and dimensions followed by the letters TP (see Figure 14.72). And in Example 3, we have mentioned the "feature control symbol." It is important that we know what these terms mean.

True Position

"True position" denotes the theoretically exact position of a feature established by basic dimensions. A basic (BSC) or true position (TP) dimension is shown symbolically by enclosing each such dimension in a rectangle, as shown in Figures 14.68*a* and 14.74*a*. True position dimensions are used to specify the *design intent* by defining the perfect configuration. Tolerances are then applied to the theoretically correct configuration, because we know that it is impossible to achieve perfection (see Figures 14.72 and 14.73).

Feature Control Symbol

A position or form tolerance is given by means of the feature control system, which consists of a rectangular frame that contains the geometric characteristic symbol, followed by the allowable tolerance. A vertical line separates the symbol from the tolerance. Where applicable, the tolerance value should be preceded by the diameter symbol and followed by the symbol Ⓜ for "maximum material condition" (MMC) or by the symbol Ⓢ for "regardless of feature size" (RFS), as shown in Figure 14.75.

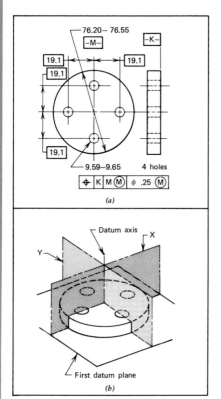

Figure 14.74 Example of part with cylindrical datum feature.

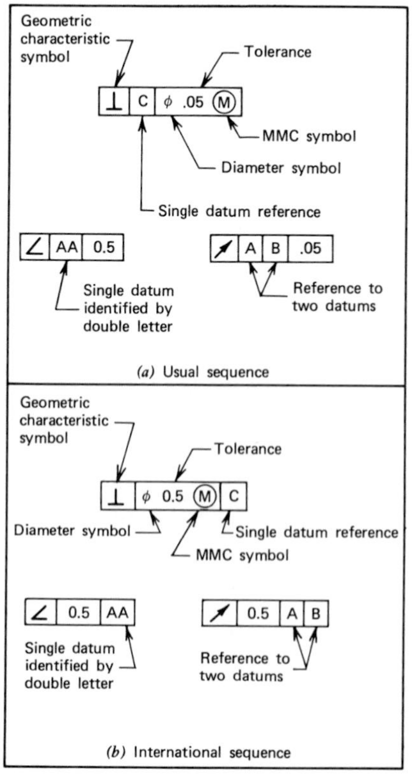

Figure 14.75 Feature control symbols incorporating datum references.

		Characteristic	Symbol	Notes
Individual features	Form tolerances	Straightness	—	1
		Flatness	▱	1
		Roundness (circularity)	○	
		Cylindricity	⌭	
Individual or related features		Profile of a line	⌒	2
		Profile of a surface	⌓	2
Related features		Angularity	∠	
		Perpendicularity (squareness)	⊥	
		Parallelism	//	3
	Location tolerances	Position	⊕	
		Concentricity	◎	3, 7
		Symmetry	≡	5
	Runout tolerances	Circular	↗	4
		Total	↗	4, 6

Note: 1) The symbol ~ formerly denoted flatness.

The symbol ⌒ or — formerly denoted flatness and straightness

2) Considered "related" features where datums are specified.

3) The symbol ∥ and ◉ formerly denoted parallelism and concentricity, respectively.

4) the symbol ↗ without the qualifier "CIRCULAR" formerly denoted total runout.

5) Where symmetry applies, it is preferred that the position symbol be used.

6) "TOTAL" must be specified under the feature control symbol.

7) Consider the use of position or runout.

Where existing drawings using the above former symbols are continued in use, each former symbol denotes that geometric characteristic which is applicable to the specific type of feature shown.

(a)

Figure 14.76 *(a)* Geometric characteristic symbols.

Tolerances of Form and Runout

Tolerances of form control the amount of variation permitted in the geometry of the feature. Tolerances of size and of position frequently control form to a certain degree; therefore, it is important to consider the influences of these tolerances before specifying tolerances of form. Consideration should be given to established shop practices, which may be sufficiently reliable to provide the specified accuracy and thereby obviate the call for certain form tolerances on design drawings. To specify tolerances of form, notes or symbols may be used; symbols are preferred. Geometric characteristic symbols are shown in Figure 14.76*a* and 14.76*b*.

Term	Abbreviation	Symbol
Maximum material condition	MMC	Ⓜ
Regardless of feature size	RFS	Ⓢ
Diameter	Dia	φ
Projected tolerance zone	Tol zone proj	Ⓟ
Reference	Ref	(1.250)
Basic	BSC	⎹3.875⎸

(b)

Figure 14.76 *(b)* Other symbols (ANSI Y14.5-1973)

Form tolerances control the conditions of *straightness, flatness, cylindricity, roundness, parallelism, perpendicularity, angularity, profile of a line, and profile of a surface.*

Examples of specifications and their interpretation of tolerances of form are shown in Figures 14.77, 14.78, and 14.79.

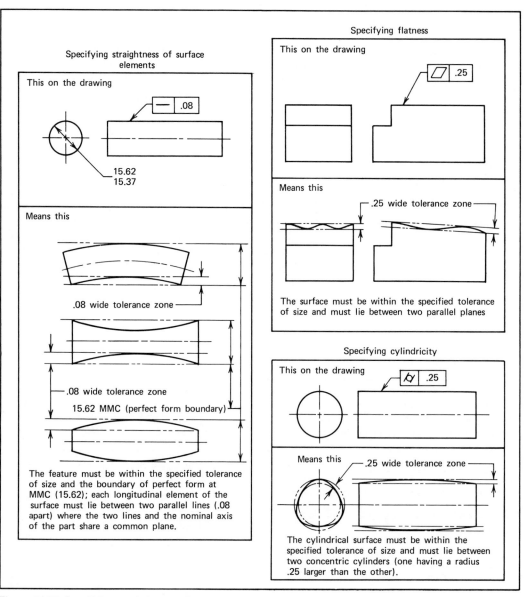

Figure 14.77 Specifying straightness, flatness, and cylindricity.

Specifying roundness for a cylinder or cone

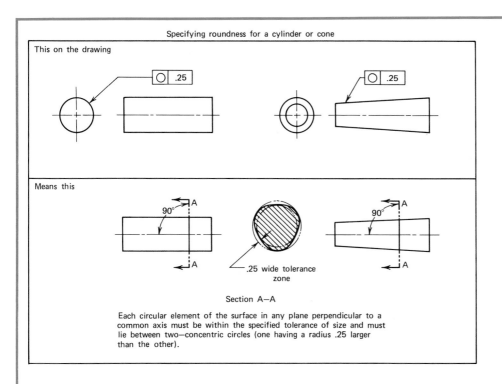

This on the drawing

⊙ .25

⊙ .25

Means this

90° ⟵A

.25 wide tolerance zone

90° ⟵A

Section A–A

Each circular element of the surface in any plane perpendicular to a common axis must be within the specified tolerance of size and must lie between two-concentric circles (one having a radius .25 larger than the other).

Specifying parallelism for a plane surface

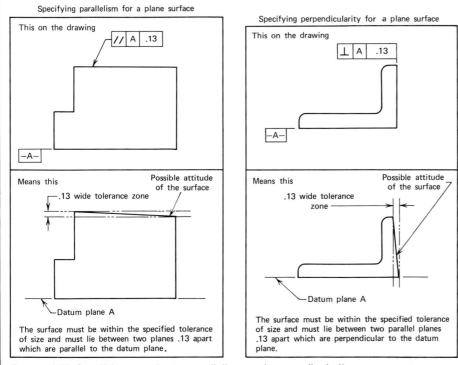

This on the drawing

// | A | .13

–A–

Means this

Possible attitude of the surface

.13 wide tolerance zone

Datum plane A

The surface must be within the specified tolerance of size and must lie between two planes .13 apart which are parallel to the datum plane.

Specifying perpendicularity for a plane surface

This on the drawing

⊥ | A | .13

–A–

Means this

Possible attitude of the surface

.13 wide tolerance zone

Datum plane A

The surface must be within the specified tolerance of size and must lie between two parallel planes .13 apart which are perpendicular to the datum plane.

Figure 14.78 Specifying roundness, parallelism, and perpendicularity.

Specifying angularity for a plane surface

This on the drawing

$$\angle \mid A \mid .38$$

30°

—A—

Means this

.38 wide tolerance zone

30°

Possible attitude of the surface

—Datum plane A

The surface must be within the specified tolerance of size and must lie between two parallel planes .38 apart which are inclined at 30° to the datum plane.

Profile of a line and size control

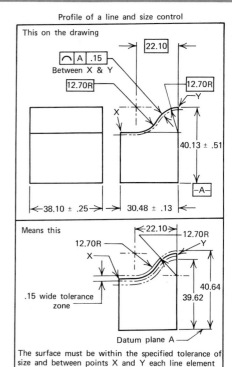

This on the drawing

⌒ | A | .15
Between X & Y

22.10

12.70R

12.70R

X

Y

40.13 ± .51

—A—

38.10 ± .25   30.48 ± .13

Means this

22.10   12.70R

12.70R   Y

X

.15 wide tolerance zone

40.64

39.62

Datum plane A

The surface must be within the specified tolerance of size and between points X and Y each line element of the surface at any cross section must lie between two profile boundaries .15 apart in relation to datum plane A.

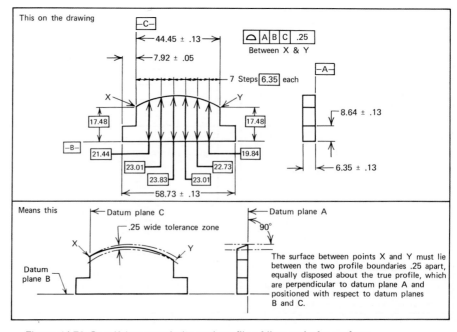

This on the drawing

—C—

44.45 ± .13

7.92 ± .05

⌒ | A | B | C | .25
Between X & Y

7 Steps 6.35 each

—A—

X   Y

17.48   17.48

—B—

21.44   19.84

23.01   22.73

23.83   23.01

58.73 ± .13

8.64 ± .13

6.35 ± .13

Means this

Datum plane C

.25 wide tolerance zone

X   Y

Datum plane B

Datum plane A

90°

The surface between points X and Y must lie between the two profile boundaries .25 apart, equally disposed about the true profile, which are perpendicular to datum plane A and positioned with respect to datum planes B and C.

Figure 14.79 Specifying angularity and profile of line and of a surface.

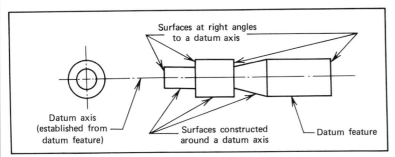

Figure 14.80 Features applicable to
runout tolerancing.

Runout Tolerance

Runout is a composite tolerance that is used to control the functional relationship of one or more features of a part to a datum axis. Features that are applicable to runout tolerancing are shown in Figure 14.80. Each considered feature must be within its runout tolerance when rotated about the datum axis.

Types of Runout Control

The two types of runout control are circular runout and total runout. Circular runout is normally a less complex requirement than total runout. It provides composite control of circular elements of a surface. The tolerance is applied independently at any circular measuring point as the part is fully rotated. An example is shown in Figure 14.81. Total amount provides composite control of all surface elements. The tolerance is applied simultaneously to all circular and profile measuring positions as the part is fully rotated (see Figure 14.82).

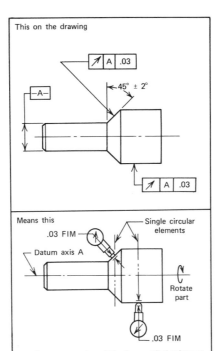

This on the drawing

Means this

The features must be within the specified tolerance of size. At any measuring position, each circular element of these surfaces must be within the specified runout tolerance (.03 full indicator movement) when the part is rotated 360° about the datum axis with the indicator fixed in a position normal to the true geometric surface. (This does not control the profile elements of these surfaces. Only the circular elements are controlled.) Whether the indicator is oriented normal to the actual surface or the true geometric (theoretically exact) surface will cause only a slight "cosine error" change in the magnitude of the FIM reading.

Figure 14.81 Specifying circular runout relative to a datum diameter.

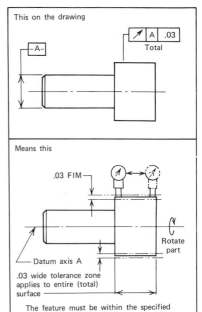

This on the drawing

Means this

.03 wide tolerance zone applies to entire (total) surface

The feature must be within the specified tolerance of size. The entire surface must lie within the specified runout tolerance zone (.03 full indicator movement) when the part is rotated 360° about the datum axis with the indicator placed at every location along the surface in a position normal to the true geometric surface, without reset of the indicator. (This controls the cumulative profile and circular elements of the entire surface.) Whether the indicator is oriented normal to the actual surface or the true geometric (theoretically exact) surface will cause only a slight "cosine error" change in the magnitude of the FIM reading.

Figure 14.82 Specifying total runout relative to a datum diameter.

Symmetry

Symmetry is a condition in which a feature (or features) is symmetrically oriented about the center plane of a datum feature (see Figure 14.83).

The Meaning of RFS as Related to Positional Tolerancing

When RFS is applied to the positional tolerance of circular features, the axis of each feature must be located within the specified positional tolerance, regardless of the size of the feature. For example, the six holes in Figure 14.84 may vary in size from 25.385 to 25.400 mm. The individual hole diameters could measure 25.387, 25.390, 25.392, 25.395, or 25.400 mm. To minimize spacing errors, each hole must be located within the specified positional tolerance *regardless of the size of that hole*. A hole at LMC (i.e., diameter 25.400 mm) is as accurately located as a hole at MMC (i.e., diameter 25.385 mm). The positional control is more restrictive than the MMC concept that is usually applied to positional tolerancing. Functional design requirements may necessitate RFS application to both the hole pattern and datum feature. That is, it may be necessary to require the axis of the datum feature, datum diameter *B*, to be the datum axis for the holes in the pattern, regardless of the datum feature size. The RFS application does not permit any shift between the datum axis and the pattern of feature as a group, where the datum feature departs from MMC.

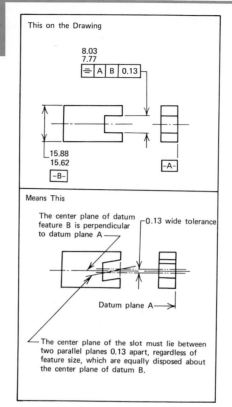

Figure 14.83 Specifying symmetry, datum and related feature RFS.

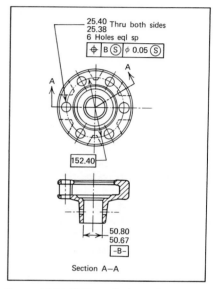

Figure 14.84 RFS applied to a feature and datum.

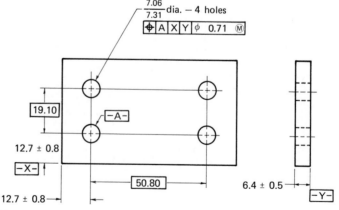

Figure 14.85 MMC applied to tolerance of position.

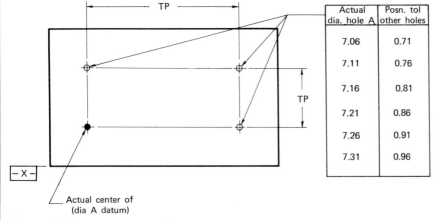

Actual dia. hole A	Posn. tol other holes
7.06	0.71
7.11	0.76
7.16	0.81
7.21	0.86
7.26	0.91
7.31	0.96

Figure 14.86 Interpretation of Ⓜ in Figure 14.85.

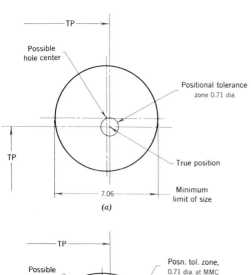

(a)

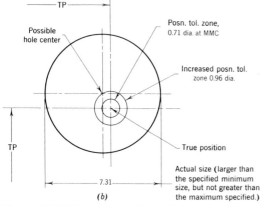

(b)

Figure 14.87 Exaggerated views of the upper right-hand hole in Figure 14.85.

Positional Tolerance (MMC)

EXAMPLE 1

Consider the application of MMC to tolerances of position. The feature control symbol in Figure 14.85 includes the positional tolerance diameter (0.71 mm) and the symbol Ⓜ for MMC. The interpretation of Ⓜ is given in the note in Figure 14.86. How were the values in the tabular note obtained? Before we answer this question, it would be best to review the analysis that has been made to determine clearance hole size (see Figure 14.73).

Now let us consider the upper right-hand hole in Figure 14.85 (other holes could have been selected). An exaggerated view of this hole is shown in Figure 14.87*a*. Note that the positional tolerance is 0.71-mm diameter when the hole is 7.06-mm diameter. *This is the MMC.*

From our previous study of positional tolerancing, we know that the actual center of the hole must lie within the positional tolerance zone, 0.71-mm diameter. Now suppose (Figure 14.87*b*) that the actual size of the hole is 7.31-mm diameter. We may increase the positional tolerance zone to 0.96-mm diameter. The *increase* is the difference between the actual diameter of the hole and the specified minimum; in this case it is 7.31 mm − 7.06 mm = 0.25 mm. Therefore, the positional tolerance zone has a diameter equal to 0.71 mm + 0.25 mm = 0.96 mm. If the actual size of the hole were any value (between the limits of size), there is an allowable increase in positional tolerance (see Figure 14.86).

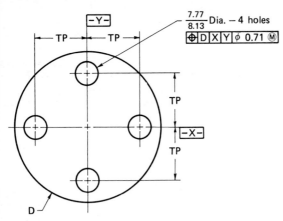

Figure 14.88 Circular plate.

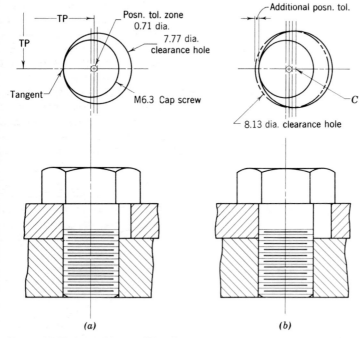

(a) *(b)*

Figure 14.89 Assembly condition for the two parts.

EXAMPLE 2

The circular plate in Figure 14.88 has been designed to assemble with a part in which four threaded holes have been properly located by dimensions similar to those of the plate. The *design intent* is that four 6.3-mm (M6.3 × 1) cap screws will pass through the clearance holes in the plate and engage the threaded holes in the mating part to fasten the two pieces. *An additional requirement is that complete interchangeability is possible.*

Examine the assembly condition for the two parts. When the centerline positions of a threaded hole (and hence the cap screw) and of a corresponding clearance hole are at extreme locations within the tolerance zone, the cap screw shank (body) will be tangent to the clearance hole of minimum diameter 7.77 mm. This "worst" condition is shown in Figure 14.89a.

Now suppose that the clearance hole diameter is larger than 7.77 mm (i.e., 8.13 mm). In this case, the center of the clearance hole could be at point C, as shown in Figure 14.89b, and still provide for functional assembly. Note that the positional tolerance zone is increased *without affecting the proper mating of the two parts.*

If the design engineer had not specified MMC, inspectors would reject the functional part shown in Figure 14.89b, because the clearance hole center did not fall within the tolerance zone specified.

Hole Axes in Relation to Positional Tolerance Zones

Where a hole is at MMC, its axis must fall within a cylindrical tolerance zone whose axis is located at true position and has a diameter equal to the positional tolerance. This is shown in Figure 14.90*a* and 14.90*b*. The cylindrical tolerance zone also defines the limits of variation in the attitude of the axis of the hole in relation to the datum surface. This is shown in Figure 14.90*c*. Note that only when the feature is at MMC that the specified positional tolerance is applicable.

Note also that a three-dimensional tolerance zone need not be cylindrical. In the example in Figure 14.83, which specifies symmetry, the three-dimensional tolerance zone is a right prism.

WORKING DRAWINGS

Working drawings and specifications provide the necessary information to implement the design. The specifications may be included on the drawings or prepared as a separate document. The use of a separate document is quite common in projects of considerable magnitude. The engineer in charge is responsible for the accuracy of the drawings and the specifications, although checking of the drawings and specifications is usually done by the engineer's staff.

Each part of the design must be accurately drawn and dimensioned to reflect the intent of the designer. Such drawings are known as *detail* working drawings. *Assembly* drawings show how the various parts should fit.

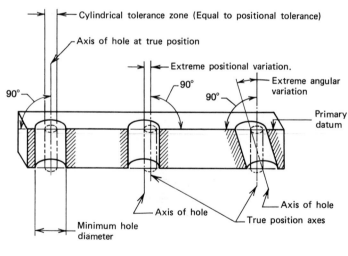

Axis of hole is coincident with true position axis

(a)

Axis of hole is located at extreme position to the left of true position axis (but within tolerance zone)

(b)

Axis of hole is inclined to extreme position within tolerance zone

(c)

Note that the length of the tolerance zone is equal to the length of the feature, unless otherwise specified on the drawing.

Figure 14.90 Hole axes in relation to positional tolerance zones.

Examples of detail drawings and of assembly drawings are shown in Figures 14.91a to 14.91f and 14.92 to 14.96. (Review Figure 14.76a and 14.76b so you can better interpret the drawings.)

Figure 14.91a is an assembly drawing of a self-aligning goniometer (instrument for measuring angles). "Measurements of joint motions during walking are useful for gait analysis and evaluation. Electrogoniometers are convenient for making such measurements because results are obtained immediately, with no requirement for data reduction."* Some of the other working drawings are shown in Figures 14.91b, 14.91c, and 14.91d. A schematic representation of the goniometer is shown in Figure 14.91e, and a photograph of the attachment frames and coupling linkages is shown in Figure 14.91f.

Figure 14.92a shows a detail of a **cell,** Figure 14.92b shows an **end flange,** and Figure 14.92c shows a **window flange** designed by the Lawrence Livermore Laboratory, Livermore, California. Figure 14.93a shows a **deployment tube assembly,** Figure 14.93b shows the **inner tube assembly,** Figure 14.93c shows the **deployment plate;** and Figure 14.93d shows the outer tube of a design project prepared by Sandia Laboratories, Livermore, California. Two additional examples of designs at these laboratories are shown in Figure 14.94a (**housing**) and 14.94b (**cylinder**).

Examples of metric drawings of designs by the General Electric Company are shown in Figures 14.95a through 14.95e and of a design by the Caterpillar Tractor Company in Figure 14.96.

This is just a small sample of the use of SI metric drawings. Many corporations (General Motors, Ford, Chrysler, International Harvester, Honeywell, IBM, Westinghouse, RCA, Caterpillar Tractor, 3M, etc.) use the SI metric system. Most companies will have converted from the English to the SI metric system by 1990.

The presentation of dimensioning in this chapter should help you to understand and appreciate the care that the engineer must exercise to express design intent with the confidence that the interpretations of the dimensions and specifications made by the manufacturer and inspector will be the same as those prescribed by the engineer.

As you progress in your further studies of engineering design, it would be helpful to review this chapter, especially the discussions and examples concerned with general tolerancing, dimensioning to ensure a single interpretation of a design, datum referencing, tolerances of location, and tolerances of form and runout. *In addition, become sufficiently familiar with feature and control symbols so that you can enhance your ability to "read" modern engineering design drawings.* "Read" the practical drawings in Figures 14.91 to 14.96 to test your understanding of the designer's intent.

Frequent practice in reading technical drawings and in making understandable technical freehand sketches (often prepared to clarify the meaning of complex portions of orthographic drawings) will enable you to develop competencies that will be invaluable in an engineering career.

---

* From L. W. Lamoreux, "A Self-Aligning Goniometer for Simplified Gait Analysis," Senior Development Engineer, Biomechanics Laboratory, University of California, Berkeley, 1978.

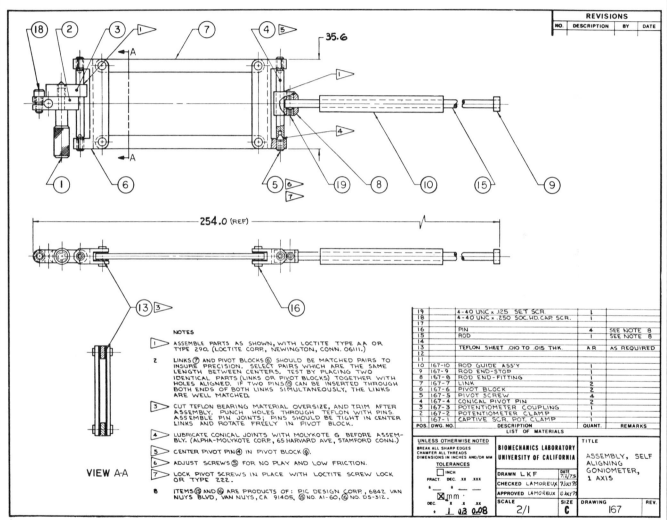

Figure 14.91 (a) Assembly drawing of
self-aligning goniometer.

(a)

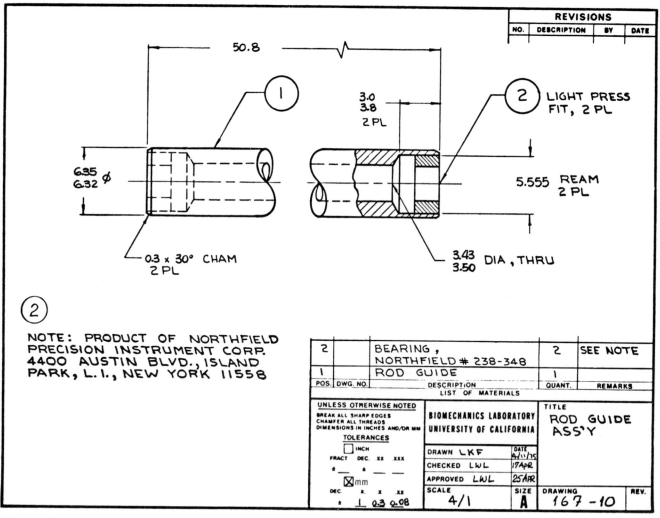

Figure 14.91 (b) Assembly drawing of the rod guide.

(b)

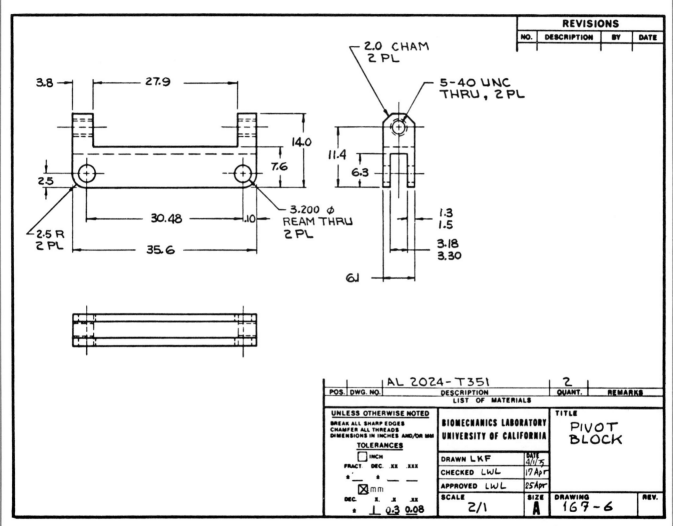

Figure 14.91 (c) Detail drawing of the pivot block.

(c)

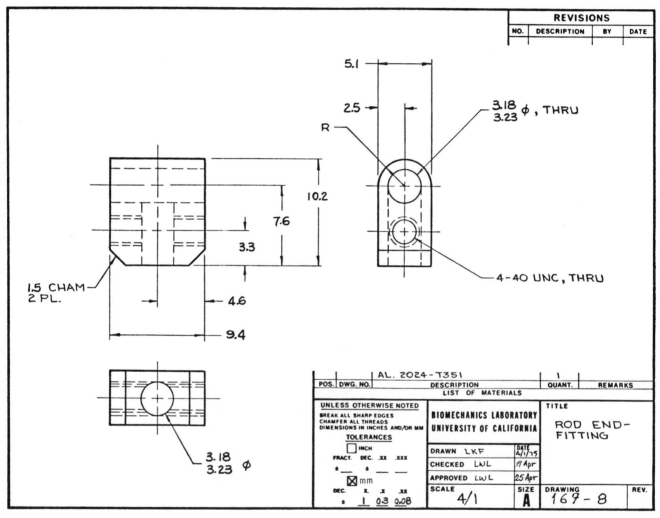

Figure 14.91 (*d*) Detail drawing of the
rod *end fitting*.

(*d*)

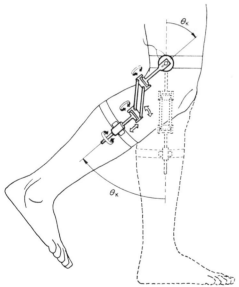

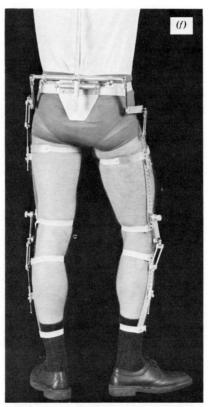

Figure 14.91 (e) Schematic representation of the self-aligning goniometer. A specially devised coupling linkage transmits the flexion angle θ_k, from the shank to a rotary transducer mounted on the thigh while simultaneously absorbing all other relative motions. An exaggerated misalignment between the transducer axis and the knee axis shows how the motions indicated by broad arrows permit the linkage to accommodate misalignment. This accommodation also insures that the device cannot restrict motion of the anatomic joint.

Figure 14.91 (f) Attachment frames and coupling linkages fitted to a research subject. (Courtesy the Biomechanics Laboratory, University of California, Berkeley)

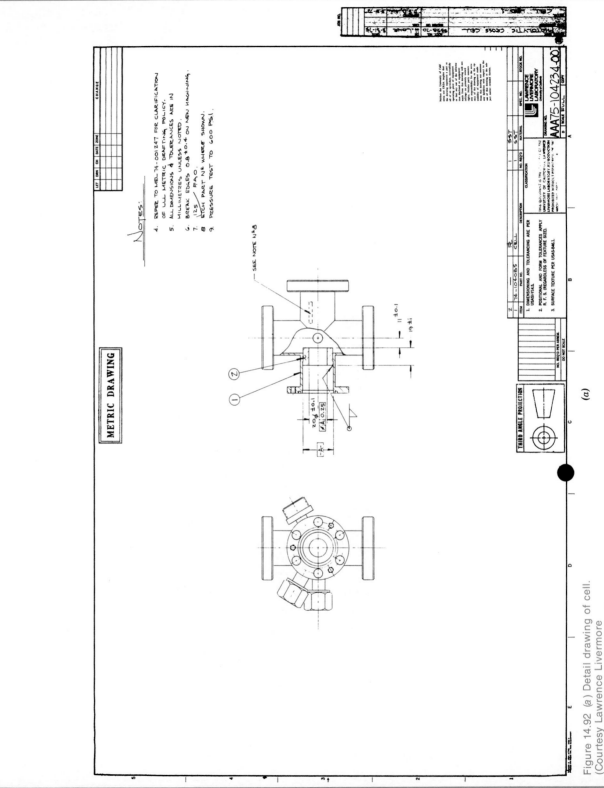

Figure 14.92 (a) Detail drawing of cell.
(Courtesy Lawrence Livermore
Laboratory, University of California,
and U.S.D.O.E.)

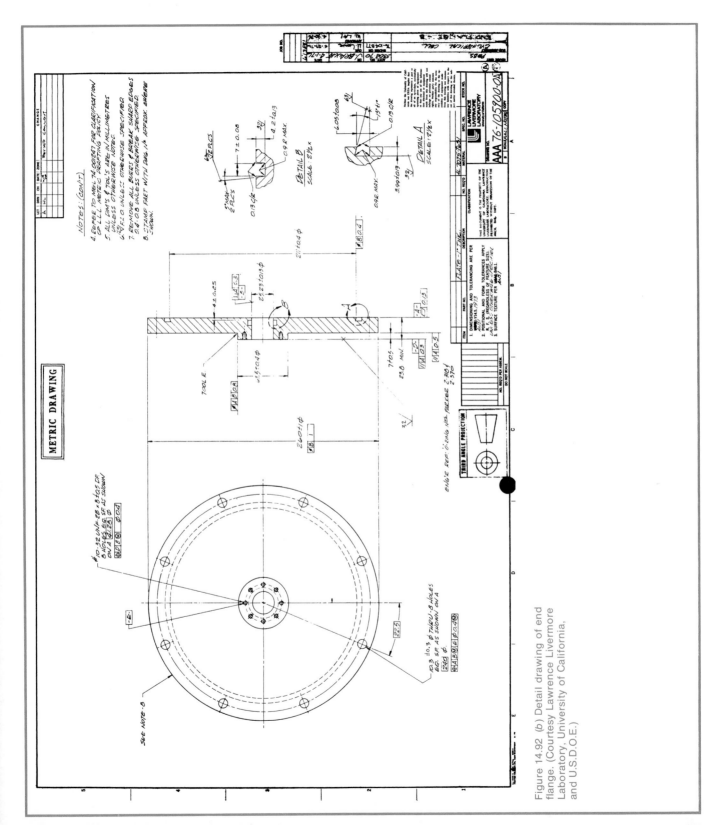

Figure 14.92 (b) Detail drawing of end flange. (Courtesy Lawrence Livermore Laboratory, University of California, and U.S.D.O.E.)

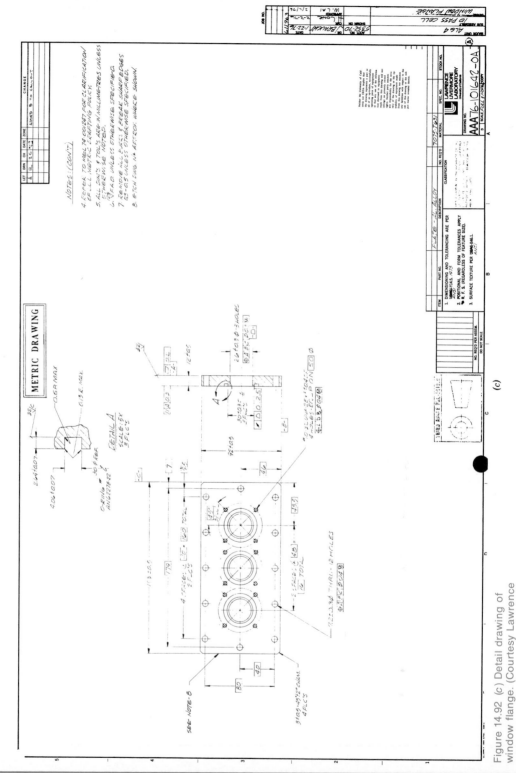

Figure 14.92 (c) Detail drawing of
window flange. (Courtesy Lawrence
Livermore Laboratory, University of
California, and U.S.D.O.E.)

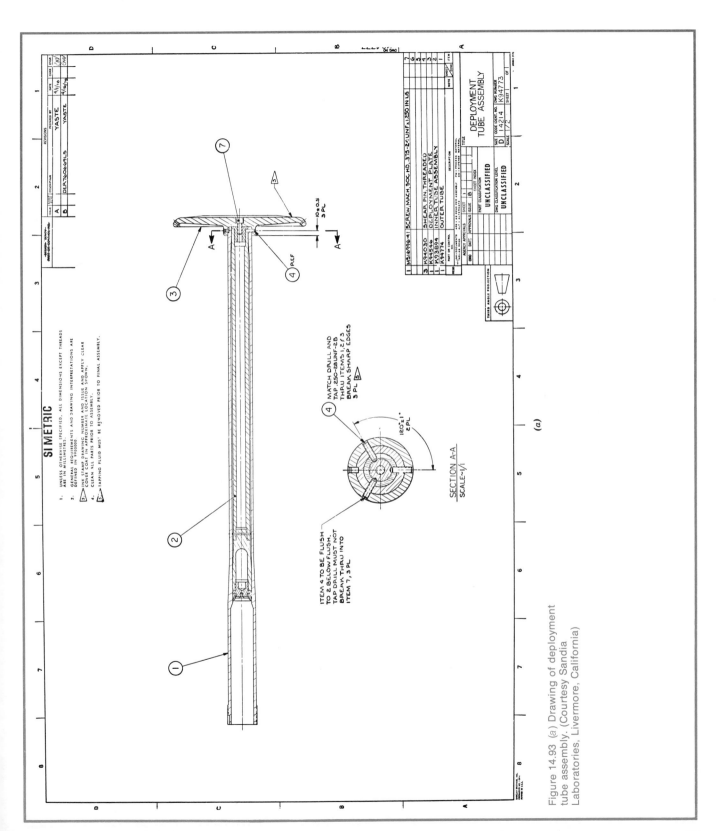

Figure 14.93 (a) Drawing of deployment tube assembly. (Courtesy Sandia Laboratories, Livermore, California)

477

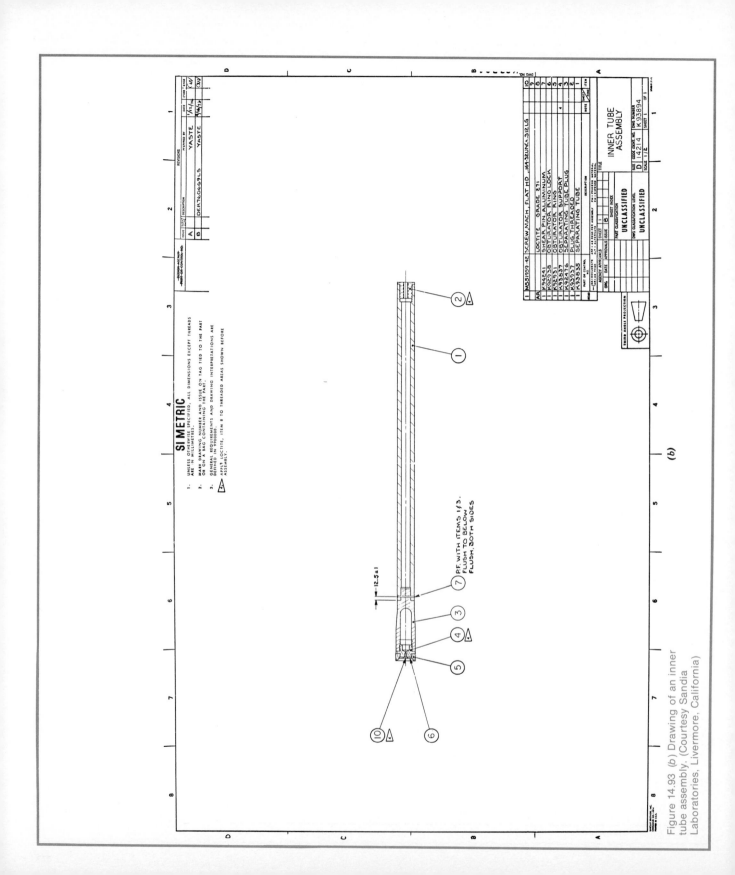

Figure 14.93 (b) Drawing of an inner tube assembly. (Courtesy Sandia Laboratories, Livermore, California)

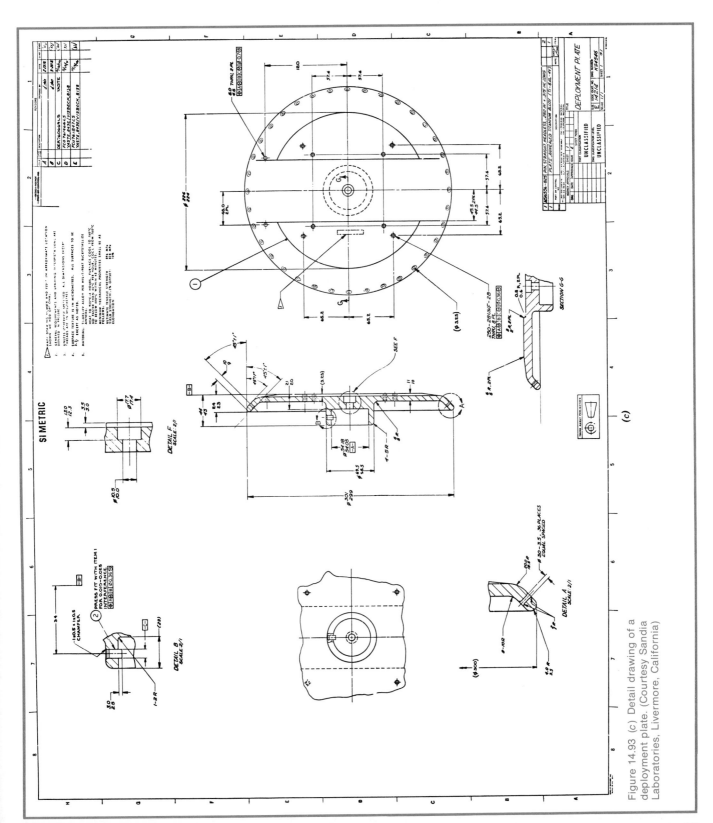

Figure 14.93 (c) Detail drawing of a deployment plate. (Courtesy Sandia Laboratories, Livermore, California)

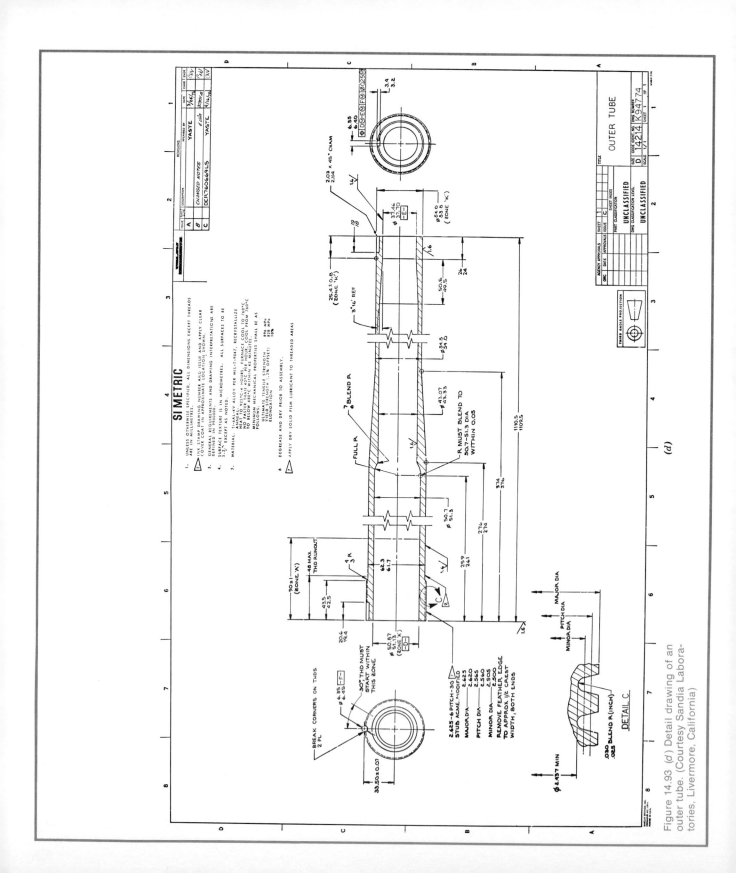

Figure 14.93 (d) Detail drawing of an outer tube. (Courtesy Sandia Laboratories, Livermore, California)

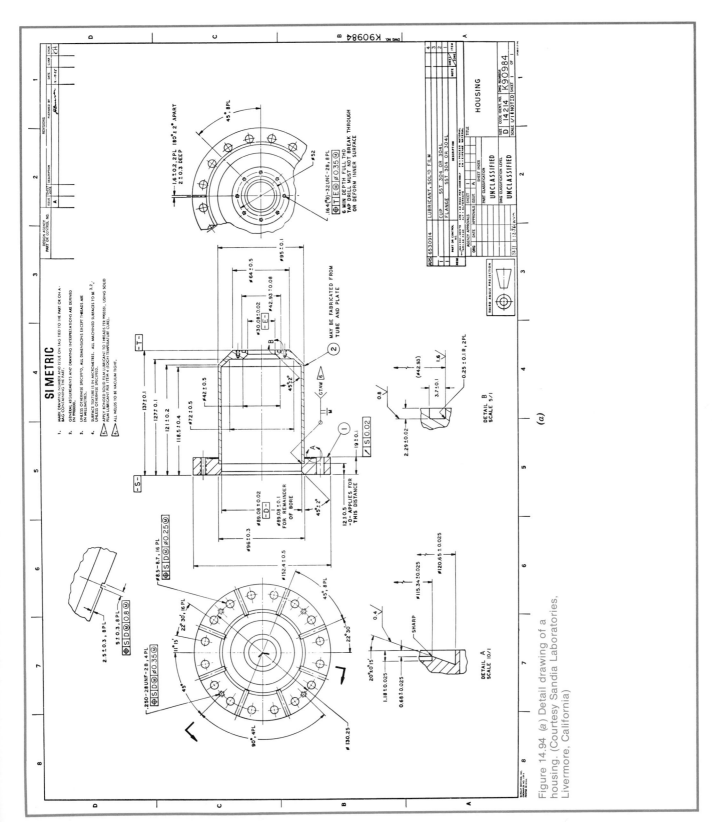

Figure 14.94 (a) Detail drawing of a
housing. (Courtesy Sandia Laboratories,
Livermore, California)

481

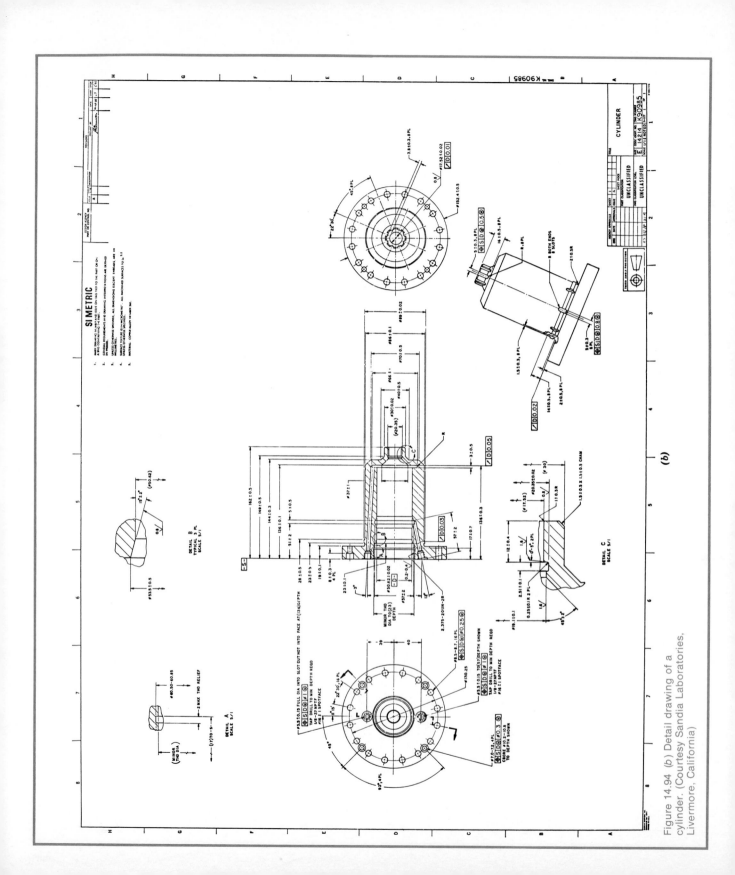

Figure 14.94 (b) Detail drawing of a cylinder. (Courtesy Sandia Laboratories, Livermore, California)

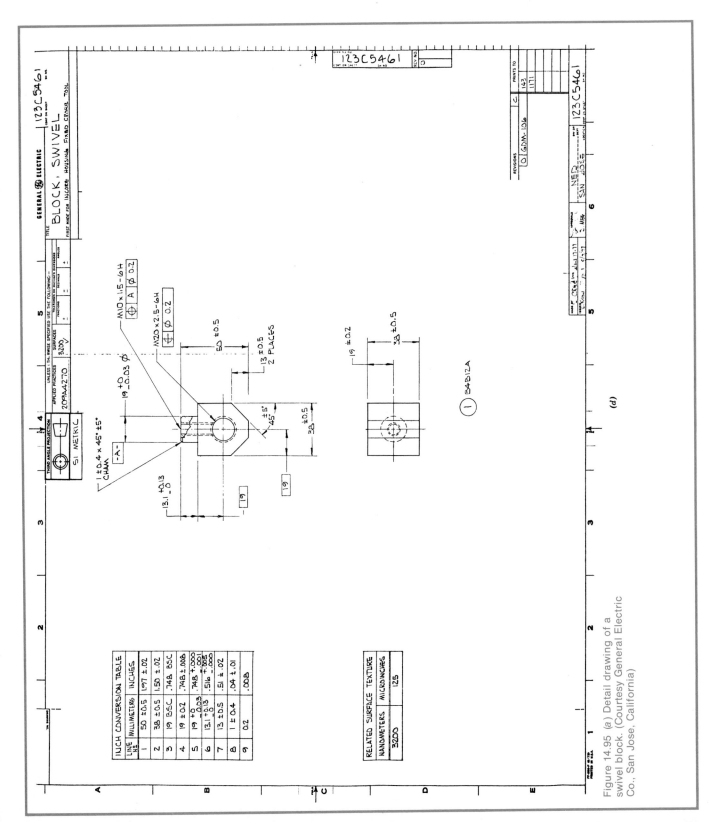

Figure 14.95 (a) Detail drawing of a swivel block. (Courtesy General Electric Co., San Jose, California)

483

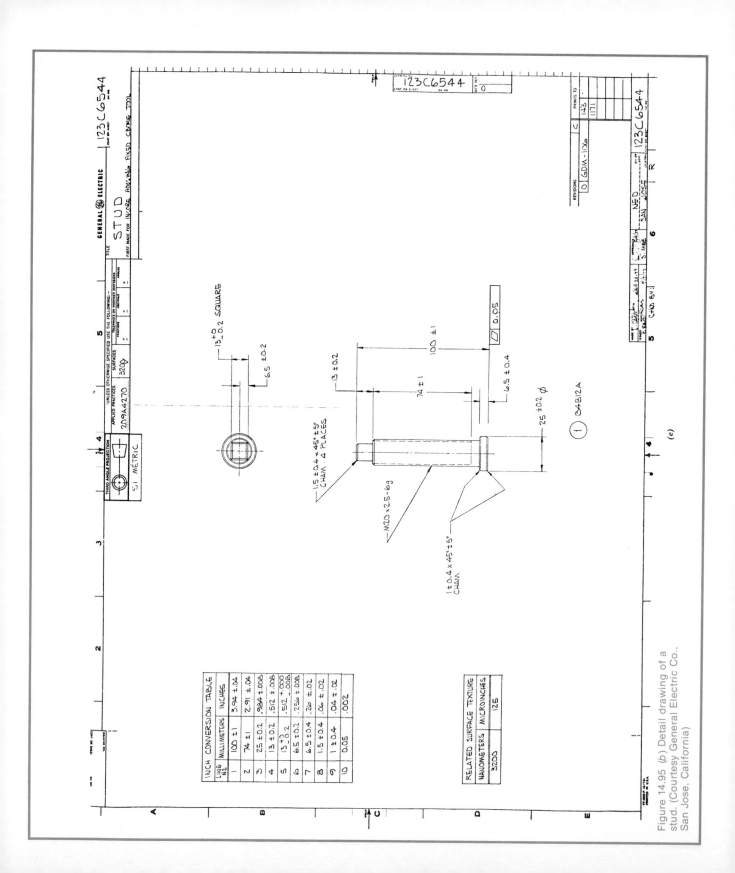

Figure 14.95 (b) Detail drawing of a stud. (Courtesy General Electric Co., San Jose, California)

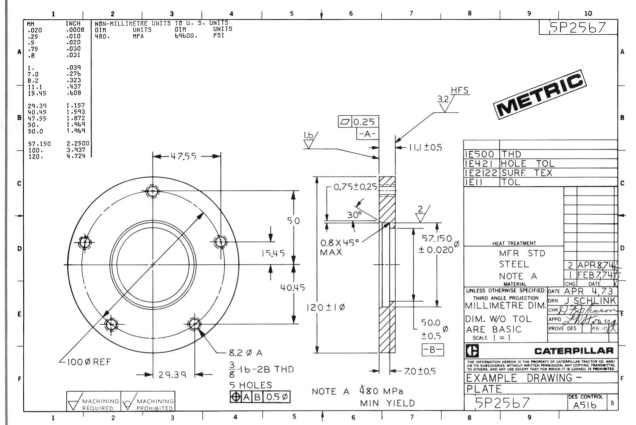

Figure 14.96 Detail drawing of a plate.
(Courtesy Caterpillar Tractor Co.,
East Peoria, Illinois)

Exercises

1. Determine the limits of size for a hole designated 25E7. Show calculations.

2. Determine the limits of size for a hole designated 60T8. Show calculations.

3. Determine the limits of size for a shaft designated 120g8. Show calculations.

4. Determine the limits of size for a shaft designated 50h7. Show calculations.

Determine the limit dimensions for shaft and bearing designs that satisfy the following requirements.

5. Basic size: 50.000 mm

Fit desired: minimum clearance = 0.050 mm; maximum clearance = 0.150 mm

Hole Basis, Clearance Fit

Prepare a sketch showing the limit dimensions.

6. Basic size: 120.000 mm

Fit desired: maximum interference = 0.195 mm; minimum interference = 0.135 mm.

Hole Basis, Interference Fit

Prepare a sketch showing the limit dimensions.

7. Basic size: 70.000 mm

Fit desired: minimum clearance = 0.060 mm; maximum clearance = 0.140 mm

Shaft Basis, Clearance Fit

Prepare a sketch showing the limit dimensions.

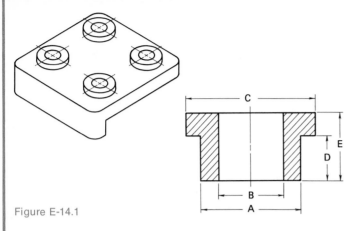

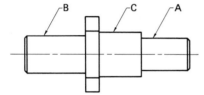

Figure E-14.1

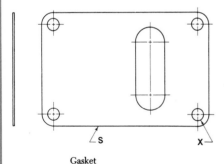

Figure E-14.2

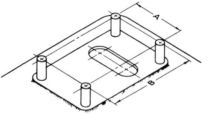

Gasket

Figure E-14.3

8. Basic Size: 130.000 mm

Fit desired: maximum interference = 0.110 mm; minimum interference = 0.045 mm

Shaft Basis, Interference Fit

Prepare a sketch showing the limit dimensions.

9. Determine the limit dimensions for the drill jig bushing shown in Figure E-14.1.

Design Requirements

(a) Diameter *A*—Fit H7/p6. Basic hole diameter = 40.000 mm.

(b) Diameter *B*—Fit H6/f6. Clearance for 6.2-mm drill.

(c) Diameter *C*—Nominal 45 mm (not critical).

(d) Diameter $D = 12.70\,{}^{+0.00}_{-0.25}$

(e) Diameter *E*—Nominal 20 mm (not critical).

(f) Diameter *B* must be concentric to *A* within 0.013 TIR.

10. Determine the limit dimensions for the following.

(a) Shaft A, the mating part of which has a nominal diameter of 14.00 mm and a H7/e7 fit;

(b) Shaft B, the mating part of which has a nominal diameter of 16.00 mm and a H7/j6 fit.

(c) Shaft C, the mating part of which has a nominal diameter of 19.00 mm and a H7/s6 fit.

See Figure E-14.2.

11. Dimension the gasket shown in Figure E-14.3. Use true position (TP) dimensions, MMC, for the stud holes. Use datums *X* and *S* for all dimensions.

The *design specifications* are:

(a) 0.80-mm rubber, silicone gasket stock (±0.12 tolerance on thickness).

(b) Gasket must fit over four 9.53-mm diameter studs on part shown. Stud spacing A = 75.00 ± 0.38; B = 125.00 ± 0.38.

Assume nominal overall dimensions of the gasket.

12. Study the object shown in Figure E-14.4.

(a) What surfaces would you use as datums, in addition to the one that is shown? Make a freehand sketch of the object and then add datum symbols $\boxed{-B-}$ and $\boxed{-C-}$ if needed.

(b) What does the symbol $\boxed{\oplus}\,\boxed{A}\,\boxed{\phi0.86\,\text{Ⓜ}}$ mean?

(c) What is the meaning of $\boxed{126.00}$?

(d) What does the symbol $\boxed{\perp}\,\boxed{A}\,\boxed{0.25}$ mean?

13. Study the slotted bar shown in Figure E-14.5. Draw a sketch that will show the meaning of the information shown in the figure.

14. Draw a sketch that will shown the meaning of the information shown in Figure E-14.6.

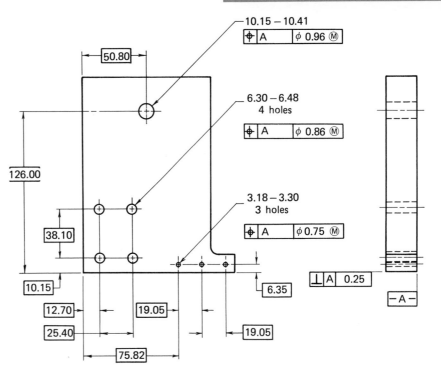

Figure E-14.4

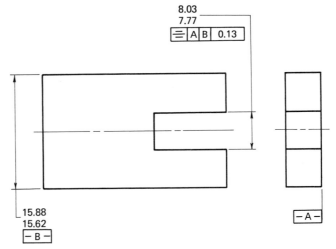

Figure E-14.5

15. Draw a sketch that will show the meaning of the symbol shown in Figure E-14.7a. Also prepare a sketch to show the meaning of the symbol shown in Figure E-14.7b.

16. Draw a sketch that will show the meaning of the information shown in Figure E-14.8.

17. Figure E-14.9 shows a drawing for one of two identical plates to be fastened with four 12.7-mm maximum diameter bolts. Note that the clearance holes at minimum diameter will pass a 12.7-mm diameter bolt, and that at maximum diameter they are equal to 13.7 mm. Also note that the positional tolerance is specified as zero at MMC. The positional tolerance allowed, however, will be in direct proportion to the actual clearance hole size.

Prepare a table that will show the clearance hole diameters and the corresponding positional tolerance diameters over the total range of values from 12.7 to 13.7 mm with intervals of 0.1 mm.

18. Refer to Figure E-14.10: What is the interpretation of the following?

(a) 8 holes ⌑Equispaced⌑

$\varnothing$ 11 $^{+\ 0.05}_{\ \ \ 0}$

⌖ ⌀ 0.01 Ⓜ A B Ⓜ

(b) ∥ 0.05 A

(c) ◻ 0.02

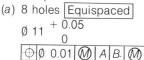
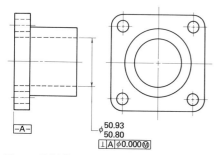

—A—

$\phi\ \frac{50.93}{50.80}$

⟂ A ϕ0.000Ⓜ

Figure E-14.8

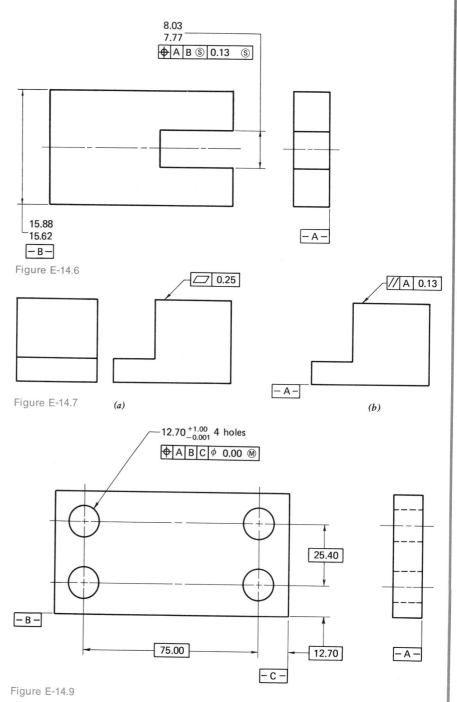

8.03
7.77

⌖ A B Ⓢ 0.13 Ⓢ

15.88
15.62

—B—

Figure E-14.6

◻ 0.25

∥ A 0.13

—A—

Figure E-14.7   (a)   (b)

12.70 $^{+1.00}_{-0.001}$ 4 holes

⌖ A B C ϕ 0.00 Ⓜ

—B—

25.40

75.00   12.70   —A—

—C—

Figure E-14.9

—A—

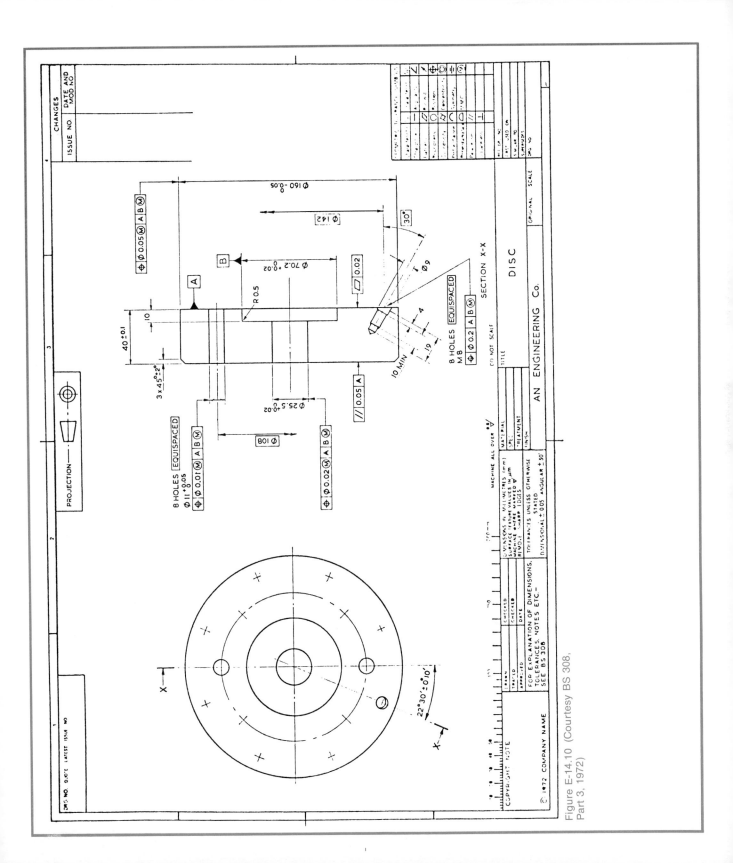

Figure E-14.10 (Courtesy BS 308,
Part 3, 1972)

19. Refer to Figure E-14.11. What is the interpretation of the following?

(a) ⊙ | ⌀ 0.03 Ⓜ | A Ⓜ

(b) ⌝ | 0.02 | A

(c) 4 holes ⌀ 6.5 / 6.4

⊕ | ⌀ 0.1 Ⓜ | A Ⓜ

20. Refer to Figure E-14.12. What is the interpretation of the following?

(a) 5 holes

(b) ⊕ | A | B | 0.50 ⌀

20

(c) ▱ | 0.25

-A-

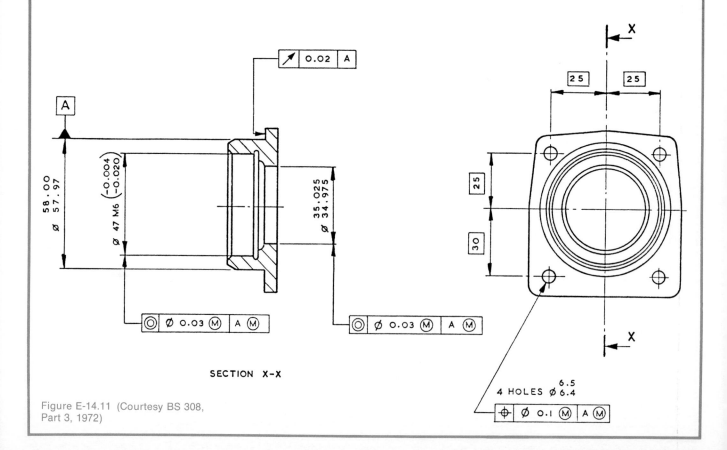

SECTION X-X

Figure E-14.11 (Courtesy BS 308, Part 3, 1972)

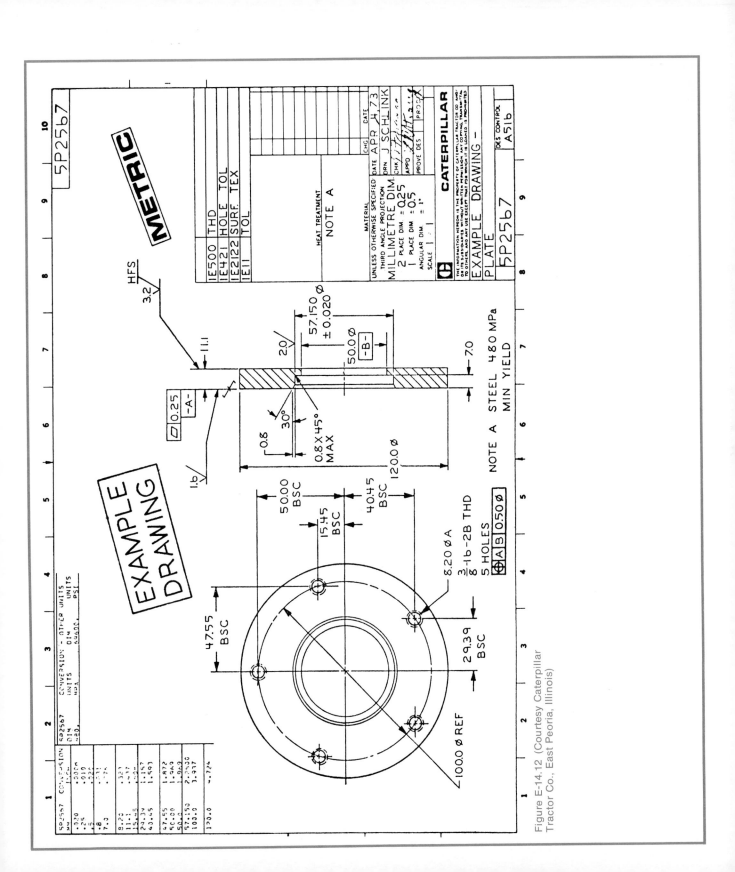

Figure E-14.12 (Courtesy Caterpillar
Tractor Co., East Peoria, Illinois)

21. Prepare detail and assembly working drawings of the shaft bearing pedestal in Figure E-14.13. Details may be drawn freehand.

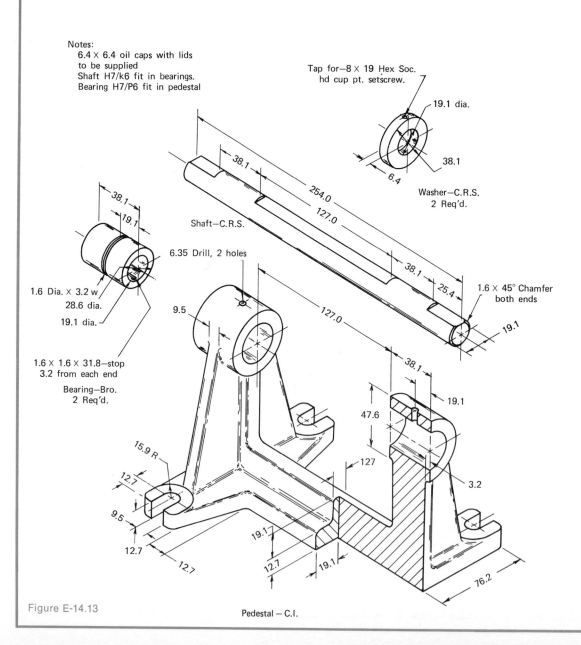

Notes:
  6.4 × 6.4 oil caps with lids
  to be supplied
  Shaft H7/k6 fit in bearings.
  Bearing H7/P6 fit in pedestal

Tap for—8 × 19 Hex Soc.
hd cup pt. setscrew.

19.1 dia.

38.1

6.4

Washer—C.R.S.
2 Req'd.

38.1

19.1

254.0

127.0

Shaft—C.R.S.

38.1

25.4

38.1

1.6 × 45° Chamfer
both ends

19.1

38.1

19.1

6.35 Drill, 2 holes

9.5

127.0

1.6 Dia. × 3.2 w
28.6 dia.
19.1 dia.

1.6 × 1.6 × 31.8—stop
3.2 from each end

Bearing—Bro.
2 Req'd.

47.6

127

3.2

15.9 R

12.7

9.5

12.7

12.7

19.1

12.7

19.1

76.2

Pedestal — C.I.

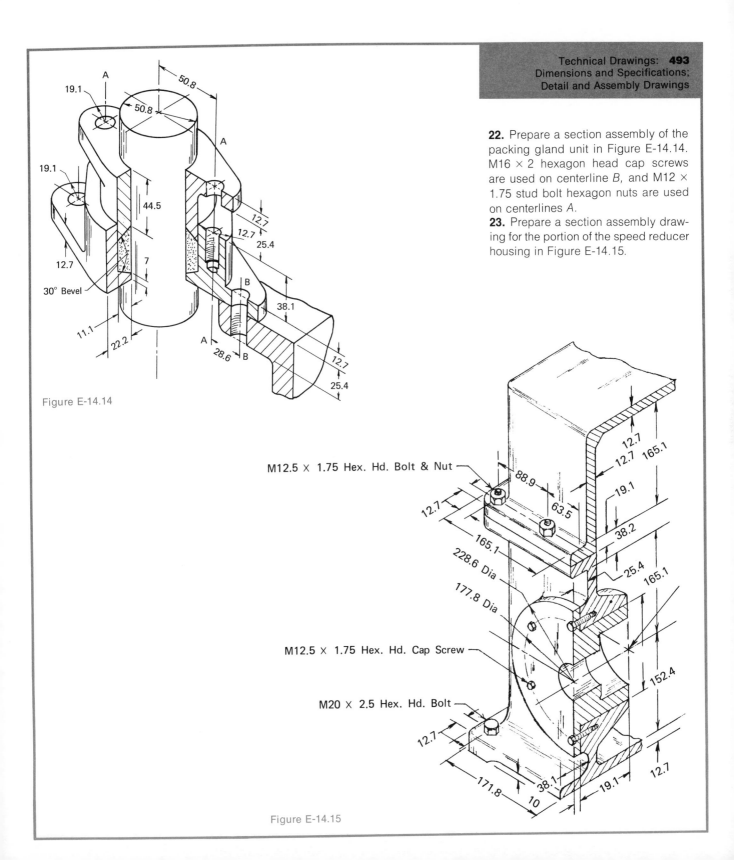

Figure E-14.14

22. Prepare a section assembly of the packing gland unit in Figure E-14.14. M16 × 2 hexagon head cap screws are used on centerline *B*, and M12 × 1.75 stud bolt hexagon nuts are used on centerlines *A*.

23. Prepare a section assembly drawing for the portion of the speed reducer housing in Figure E-14.15.

A

19.1

50.8

50.8

A

19.1

44.5

12.7

12.7

25.4

12.7

7

B

30° Bevel

38.1

11.1

22.2

A

28.6

B

12.7

25.4

M12.5 × 1.75 Hex. Hd. Bolt & Nut

88.9

63.5

12.7

12.7

165.1

19.1

38.2

12.7

165.1

25.4

165.1

228.6 Dia

177.8 Dia

M12.5 × 1.75 Hex. Hd. Cap Screw

152.4

M20 × 2.5 Hex. Hd. Bolt

12.7

171.8

38.1

10

19.1

12.7

Figure E-14.15

24. Prepare a section assembly of the
collet chuck holder in Figure E-14.16.
The parts are fastened with six equally
spaced socket head cap screws. The
internal threads are $4\frac{1}{4}$-16 UN-2B.

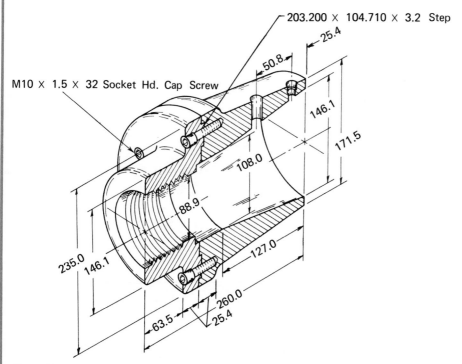

Figure E-14.16

25. Prepare detail and assembly drawings of the conveyor support in Figure E-14.17.

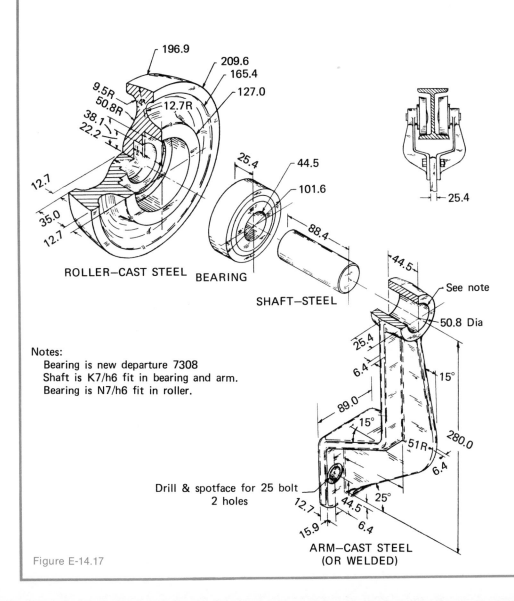

196.9

209.6
165.4
127.0

9.5R
50.8R
12.7R
38.1
22.2

25.4
44.5
101.6

12.7
35.0
12.7

88.4

ROLLER–CAST STEEL   BEARING

SHAFT–STEEL

25.4

44.5

See note

50.8 Dia

Notes:
 Bearing is new departure 7308
 Shaft is K7/h6 fit in bearing and arm.
 Bearing is N7/h6 fit in roller.

25.4
6.4
89.0
15°
15°

280.0

51R
6.4

Drill & spotface for 25 bolt
2 holes

12.7   44.5   25°

15.9   6.4

ARM–CAST STEEL
(OR WELDED)

Figure E-14.17

Chapter 15

Technical Reports; Design Documentation; Patents

INTRODUCTION

Industry's complaint is still valid: "Engineers can't write good reports".

The late Charles Kettering, former vice-president in charge of General Motors Research Laboratories, in a discussion of the importance of English to the engineer in industry, pointed out that:

Success in engineering and research depends as much upon the ability to present an idea convincingly as it does upon the ability to perform calculations or experiments. You may perform the most miraculous experiment in the laboratory, yet you have not contributed anything to the advancement of knowledge until you have transmitted your results to others. It is only by speech and writing that the discoveries made in the laboratories are made useful. Scientific men too often look upon writing reports or making talks as an irksome part of their job and do as little of it as possible. Engineering work is not finished until the results are clearly recorded and presented to others.

Edward Wells, a vice-president of the Boeing Company, expressed his views on the importance of technical report writing by stating:

I believe that one of the most difficult tasks faced by professional engineers either as individuals or in organized groups is that of developing and maintaining effective communication with others. An important element of communication is that provided by technical reports, for quite often the technical report is the only medium available for transmitting necessary technical information from the originator to those who may most effectively use it. An individual competent to make significant engineering and scientific contributions to the state of the art should also be competent to portray accurately the technical details of his findings. Without this capability, the individual contribution, important as it may seem to its originator, may very well be passed by in favor of ideas less valuable but more ably presented. An excellent report should not be counted upon to gain acceptance of a poor idea, but a poor report all too often results in the loss of an excellent idea—a loss affecting the originator, his associates, and society as a whole.

Margaret A. Maas, southeastern editor of *Design News*, made the following observations of the writing ability of engineers.

It has never ceased to amaze me how an engineer known for his technical competence can be so completely incompetent when someone puts a pencil in his hand. Tell him to write a technical paper or instruction manual on the work he is doing and, though he may be an expert in his field and hold one or more degrees, he is suddenly reduced to a stumbling, fumbling high-school student who finds it impossible to communicate. An instruction sheet on how to set the dials on a new test instrument becomes a verbose dissertation that makes a patent disclosure read like a comic book.

Why this metamorphosis? Why does the engineer find it so difficult to talk about the thing he knows so well? Set him down in front of you and he'll be able to give you instructions on how to use the same instrument and you'll have no difficulty understanding him. Why can't he put what he has to say on paper?

There will be many journalists and writers who will disagree with me, but I am convinced that the engineer can't write for one reason only—he has been told that he can't write. The engineer is expected to be illiterate. I have been told over and over by writers dealing with engineers, "Well, you know engineers can't write." Nonsense! The engineer can't write because he thinks he can't. He has been led to believe that good writing is some occult art that requires lengthy words and esoteric statements. Little does he realize that good writing is just the opposite. The more understandable and easier to read it is the better.

OVERCOMING THE "I CAN'T WRITE" SYNDROME

Maas developed a few techniques that can help you overcome this syndrome. Try them!

Point 1. "Tell it like it is." Literally pretend that you are talking to someone about the subject you are writing about and put those exact words down on paper. *Do not worry about getting the perfect word to describe something. Just get those words on paper exactly the way you would say them.* Go back later and clean up your grammar and punctuation and substitute better wording.

When it comes to cleanup time, if some slang does a good job in describing what you mean, leave it in. There is enough dull technical writing in this world, and a sensible use of slang, if necessary, can make your writing more interesting.

Point 2. Don't Begin. Having trouble finding a beginning? Well, do not look for one; jump right into the middle or into whatever part of the discussion comes to mind first. You often will be surprised when you read the material over that what seemed like starting in the middle ended up being not such a bad place to start after all. A lengthy introduction is often excess baggage that you could just as well do without. If you do need an introduction, tack it on after you have finished. It is a lot easier to work on a difficult introduction when you know you have the bulk of the writing behind you.

Point 3. Meander. Do not worry about writing in a logical order. Put down the information as it comes to mind. You can always play "paper dolls" afterward and cut out and paste the sections in a more coherent arrangement. The important thing is to get the information on paper. Shifting it around is easy.

No doubt that you will profit from these three points when you write a letter, a project proposal, a progress report, a final project report, or a technical paper. They have worked for us, and they can work for you.

WRITING THE PROJECT PROPOSAL

Introduction

It is important to identify the reader of your proposal. Is it to be read by an engineer, someone in business, an investor, or a customer (or by your instructor)? To be most effective, your writing must satisfy the interest threshold of the reader.

Proposals may vary in length from a few pages to hundreds of pages, depending on the magnitude of the project. In general, a proposal will include:

1. A clear statement of the problem.
2. Answers to questions such as:
(a) What is the purpose of, or need for, the proposed undertaking?
(b) Is it feasible?
(c) How will the problem be attacked?
(d) Do you (or your organization) have the equipment and personnel to do the job?
(e) How long will it take to do it?
(f) How much will it cost?
(g) What benefits to society will be derived to justify the investment?

Student's Proposals

For your purpose in dealing with relatively simple design projects it will be sufficient to write a short proposal that could be addressed to your instructor. Students' project proposals should be typed, placed in a suitable folder, with an appropriate title on the cover, and submitted at the time and date specified by the instructor. The following examples (from students) are typical:

EXAMPLE 1

Month ____ Day ____ Year ____

Professor _____

Department of _____

University of _____

Dear Professor _____:

I propose to design low-cost housing for an underdeveloped area. The world's housing situation is a cause for great concern, not only because too many people in underdeveloped areas are poorly housed, but also because the situation is getting worse. Despite the advances in science and technology, the rates of population increase and urbanization are accelerating faster than man's ability to resolve problems of overcrowding and of inferior housing.

Because of the time involved in this project my researching into all aspects of this problem would be impossible. Therefore, I have picked a city in Peru, Arequipa. According to the United Nations' publication, "World Housing Conditions and Estimated Housing Requirements," 40 percent of Arequipa's population lives in slums (barriadas). My intention is to design a low-cost house that may be used to ease the housing burdens of that city. As a consequence of my study of Arequipa's housing needs and its economic resources I have set the following specifications:

1. Cost less than $1,000
2. Built principally by the inhabitants with minimal technical supervision.
3. Built primarily from local materials.
4. Minimal amount of equipment and prefabrication.
5. Prototype must be easily expanded in size and in number of rooms.

In addition, I plan to consider the following:

1. Building methods to be used.
2. Climate conditions.
3. Geography of the site.
4. Insect, bacterial, and chemical conditions.

I am deeply interested in this design project and believe I can produce a worthwhile solution within the time alloted to this phase of our course. Your comments and suggestions will be appreciated.

Yours truly,
(*Signature of student*)

EXAMPLE 2

Month _____ Day _____ Year _____

Professor _____

Department of _____

University of _____

Dear Professor _____:

I propose to prepare a design for temporary furniture for use by students who plan to live in apartments during the school year. Those students who have limited funds would benefit from the rental of unfurnished apartments if they could purchase temporary furniture at low cost. I believe that such students could save approximately $50 per month.

I plan to prepare a design for a lounge-type chair; however, this is only one of several pieces of furniture that could be designed. The lounge-type chair will serve as an example. I have given considerable thought to the problem and believe the following requirements should be satisfied:

1. Cost not to exceed twenty-five dollars.
2. Assembly must be simple.
3. Should be made of disposable material which does not require hauling to a city dump.
4. Must be light in weight.
5. Must be stable and structurally sound.
6. Should be aesthetically pleasing.
7. Should be comfortable for the average-sized person.
8. Should not require use of bolts, screws, or other metal-type fasteners.

I considered the type of material that could be used to meet the requirements. Materials that I believe have merit are plastic, fiberglass, and cardboard. Preliminary probing favors the use of cardboard. Although fiberglass or some plastics have better structural properties, I felt that cardboard was better suited to meet the requirement of disposability, weight, cost, etc.

I propose to develop a design of a friction-fitted lounge-type chair using corrugated cardboard. I believe I can do it and meet the stated requirements.

Yours truly,
(Signature of student)

EXAMPLE 3

Month ____ Day ____ Year ____

Professor _____

Department of _____

Community College of _____

Dear Professor _____ :

I propose to design a pen for use by quadriplegics. They have a great deal of trouble in picking up, supporting, and writing with any commercial writing instrument because of their almost complete paralytic condition. A pen specifically designed to compensate for the quadriplegic's handicaps would be invaluable. For maximum usefulness, such a pen, I believe, would have to meet the following requirements:

1. *Portability*. The pen should be reasonably easy and convenient to carry, whether in a pocket or elsewhere.
2. *Availability and usability*. The pen should be reasonably quickly available, and should be easy and convenient to use in everyday practice.
3. *Ability to support itself in the writing position*. The quadriplegic can provide no grip to support the pen while writing, so the pen (or device which is to serve as a pen) must hold itself in the writing position with the fingers providing only a very small stabilizing force (if necessary).
4. *Comfort*. The pen should be as comfortable to use as any normal pen.
5. *Cost*. The pen should have a price in the range of any normal, good quality pen.
As an added convenience (though certainly not a necessity to the design):
6. *Automation*. The pen, if "ball point" in nature, could open and close automatically.

If the final design is to be marketable, i.e., stay within the price limitations while still functioning effectively and efficiently, it should probably be made primarily of plastic materials. The loads placed on the pen would be small and could easily be supported by suitable plastics.

I plan to develop several alternative solutions and discuss their merits with several knowledgeable persons in the field of bioengineering. I have already spoken with a few men in this area and have been encouraged to undertake this design task in which I have a deep interest.

Yours truly,
(*Signature of student*)

Requests for Proposals

In a number of cases a major contractor (e.g., an aircraft company) will request proposals from several subcontractors. The customer's (the major contractor) statement of the work, the specifications, etc., define in as much detail as possible what the subcontractor is expected to do. In most instances, the successful subcontractor (i.e., Company X) must clearly demonstrate that:

1. The proposed problem solution meets or exceeds the customer's requirements.
2. The design is better than those of competitors.
3. The design emphasizes the key features the customer desires in a product (reliable performance, low cost, negligible risk, potential for growth, compressed time schedules, etc.).
4. The management and technical teams are thoroughly capable, enthusiastic, and confident of success.
5. Company X's facilities are suitable in all respects and are available without interference with other programs.
6. Company X is the best place to get the job done.*

Research Project Proposal

Each year many colleges and schools of engineering submit proposals to agencies such as the National Science Foundation, National Aeronautics and Space Administration, Office of Research Services, Office of Naval Research, National Institute of Health, and the U.S. Public Health Service to obtain funds to support research activities. Usually the proposed pro-

---

* Items 1 to 6 are based on the article, "Argue Your Way to Success," *Machine Design*, May 13, 1965.

ject involves a principal investigator (a member of the faculty), graduate students, and technicians. The form of the proposal may vary, depending on the requirements of the sponsoring agency. In general, however, most of the proposals will include:

1. *Title Page.* This page shows the name of the agency to which the proposal is directed, the title of the proposed project, the proposed starting date, the amount of money requested, the duration of the investigation in months, the names of the principal investigator (or investigators), the department head, and the institutional administrative officer.

2. *Abstract.*

3. *Introduction.*

4. *Statement of the Problem.*

5. *Scope of the Problem.*

6. *Proposed Technical Program.*

7. *References.* A list of technical papers that are related to the proposed research area.

8. *Facilities Available.*

9. *Graduate Studies.* A statement such as the following is significant. "The proposed research program will be suitable for the participation of several students. Phases of the study could be utilized for theses work at the M.S. and Ph.D. levels."

10. *Supervision.* A statement such as "The proposed research will be conducted under the direct supervision of the principal investigator" is sufficient.

11. *Technical Background of the Principal Investigator.* This usually includes a listing of the investigator's (a) education, (b) academic activity, (c) nonacademic activity (i.e., consulting work), and (d) pertinent publications.

12. *Budget.* The budget shows the classification and number of employees, their time to be devoted to the project, their salaries; overhead charges, supplies and expense (materials, fringe benefits, etc.); equipment and facilities expense; and the total cost of the proposed project for the duration of the investigation.

FINAL PROJECT REPORT

The final project report should be typed. It should contain the following items, but not necessarily in this order.

1. *Letter of Transmittal.* The transmittal letter is usually short. It is addressed to the project manager, the chief engineer, the company president, or the sponsoring agency, as the case may be. The letter usually states that the report is being submitted to the reader for evaluation and action. The statement of the problem, brief comments on the content of the report, and conclusions could be included. The letter may be part of the report. Most often, however, it is separate, and accompanies the report.

A student's report would probably be directed to an instructor. The letter of transmittal may state, for example, that the report of the student's project (e.g., "Low-Cost Housing for an Underdeveloped Area") is herewith submitted for consideration and evaluation by the instructor. A few short paragraphs could be devoted to the statement of the problem, its relevance, and a brief description of the report contents.

2. *Title Page.* The title of the project should be descriptive, explicit, and brief. In some cases a subtitle may be added to clarify the meaning. The title is usually placed about $2\frac{1}{2}$ in. (60 mm) below the top of the page. The name of the author (or authors) is placed beneath the title. Additional ma-

terial that may be included is shown in Figure 15.1 (for a research report) and in Figure 15.2 (for a student's project).

3. *Table of Contents*. The table of contents lists the sections of the report and their corresponding page numbers.

4. *Abstract*. A short abstract (approximately 150 words) should appear on the first page. (In some reports the abstract precedes the table of contents.) The purpose of the abstract is to present a clear indication of the object and scope of the investigation and to highlight the significant findings. When the abstract is well prepared, it will stimulate the reader to read the full text of the report or, if it is read by a busy executive of a company, it will motivate that person to assign the reading of the full report to qualified staff members, who subsequently will make recommendations concerning the action that should be taken by the company.

5. *Introduction*. The introduction generally includes (a) background information that is essential to the establishment of the need for a new or improved device, structure, or system; (b) identification of the problem; (c) the purpose of the project; and (d) the nature of the investigation to solve the problem.

6. *Methodology*. What approach was used in solving the problem? A description of the method of attack on the problem should be given.

7. *Major Portion*. The major portion of the report would contain details of the approach and method used in solving the problem and other information necessary to

ENERGY AND POWER IN HUMAN WALKING

M. Y. ZARRUGH

BIOMECHANICS LABORATORY
UNIVERSITY OF CALIFORNIA
BERKELEY ■ SAN FRANCISCO

THIS STUDY WAS SUPPORTED BY
VETERANS ADMINISTRATION CONTRACT V101(134)P-336

JUNE 1976   TECHNICAL REPORT 62

Figure 15.1 Title page for a research report. (Original size 8½" x 11")

REPORT ON DESIGN PROJECT: COIN CHANGER FOR THE MECHANICAL ENGINEERING STUDENT LOUNGE

SUBMITTED
to
STUDENT SECTION
ASME
JIM POLK * CHAIRMAN

MAY 29, 1973

By
John Hughes
Mike Guidon
Bob Hamry

Figure 15.2 Title page for a student's project report. (Original size 8½" x 11")

support the results obtained. Charts, graphs, schematics, and photographs should be used to present significant information clearly (test data, performance data, etc.). Such pictorial representations are much easier to comprehend and interpret than word descriptions or tables. Remember that "one picture is worth a thousand words." In some cases, essential short tables may be included.

8. *Results and Recommendations*. Although the abstract highlights the significant findings of the investigation, it is good practice to elaborate the results. It may be desirable to point out what further investigations could be made to enhance the growth potential of the end product. For example, Boeing Company engineers have pointed out that it is possible to modify the design of the 747 to increase the passenger capacity to 500 persons. There are several examples of design modifications that have increased passenger loads considerably (i.e., variations in the designs of the Boeing 727s and of the McDonnell-Douglas DC-8s).

9. *Acknowledgments*. Give the names and official titles of persons who were helpful. State their contributions to your effort to solve the problem. List the names of companies that may have supplied technical information, materials, or equipment that was used in the project.

10. *Bibliography*. References to cited literature should be made as follows.

(*a*) If you cite three or four references use footnotes.

(*b*) If you cite more than four references use a bibliography at the end of the report. Each reference should be numbered consecutively. The numbers should appear in parentheses. References are usually arranged in the following manner.

(*1*) *Book*. Author (include initials), title of book, publisher, city, number of edition, year of publication, and page numbers.

(*2*) *Magazine or Periodical Article*. Author (include initials), title of article, full name of magazine or periodical, volume, number, year, and page numbers. (Check the requirements of specific journals.)

11. *Appendix*. The appendix generally includes tables, charts, sketches, drawings for prototypes, final drawings, bill of material, and cost data. (Some reports include a list of figures; others place the list after the table of contents.) The appendix is useful in documenting the design and in identifying patentable features. However, the inclusion of significant charts, diagrams, and short tables in the major portion of the report is not precluded. The appendix, in a sense, stores the input information on which the relevant results depend.

12. *Completed Report*. The completed report should be placed in an appropriate folder or cover. For student design projects a folder approximately $9 \times 11\frac{1}{2}$ in. (228×292 mm) is adequate. The cover could contain the information shown on the title page.

Design Documentation

Once a design project has been completed it is difficult to recall the original design parameters and the thinking that led to the choices that were made during the various aspects of the project. You may have wondered why a designer solved the problems inherent in the project in a particular way. Often, engineers look back at their designs and may wonder why themselves.

During a summer program conducted for engineering faculty, an aerospace company included dis-

cussions with their design engineers. It was frustrating to find that the recall by the designers, when asked why they chose certain components in the design of a control mechanism, was so meager. A typical answer was: "Well, what difference does it make, it works, doesn't it?" Hardly a good answer. If the design had been documented the engineer could have provided an answer based on thinking and analysis at the time of involvement with the project.

Design History

A design history written at the time that decisions and the reasons for making them are fresh in mind will preserve the information for reference when needed.

J. L. Mariotti, Design Engineer for the Automatic Electric Co., Northlake, Illinois, said that "The formula for a Design History is something like a recipe for hash. It is not assembled, it accumulates as the design progresses."

How can we document our design? We can write a report of the design history that includes the following elements.

1. *An Introduction*. Assume, for example, that the reader is a new engineer, perhaps a recent engineering graduate who, a year and a half from now, may be called on to modify your design as a consequence of consumer feedback. Include:

(*a*) A statement of the problem.

(*b*) A definition of the real problem.

(*c*) Where the product is used.

(*d*) What it does.

2. *Design Parameters.*

(a) Operational factors.

(b) Human factors.

(c) Size, shape, function.

(d) Environmental consideration.

3. *Design Considerations.*

(a) How the parameters were satisfied.

(b) How the constraints were met.

(c) How the limitations were resolved.

(d) What the basic approaches were, and why a specific approach was used.

(e) Choices of components and why.

(f) Any unusual aspects of the design?

(g) References to drawings (give identification numbers), tests, design notebook, etc.

4. *Significant Modifications.*

(a) Problems revealed by the prototype tests.

(b) How the problems were resolved.

5. *Recommendations.*

(a) Improvements—possible cost reduction (e.g., in more efficient manufacturing methods), simpler assembly techniques, etc.

(b) Technical advances—new materials, new processes, etc.

(c) Commentary—*the engineer who modifies the original design*

should write a supplement to the original design documentation.

6. *Patentability.*

(a) Patentable features—What were they, if any? What action was taken?

(b) What features were considered to be original and novel?

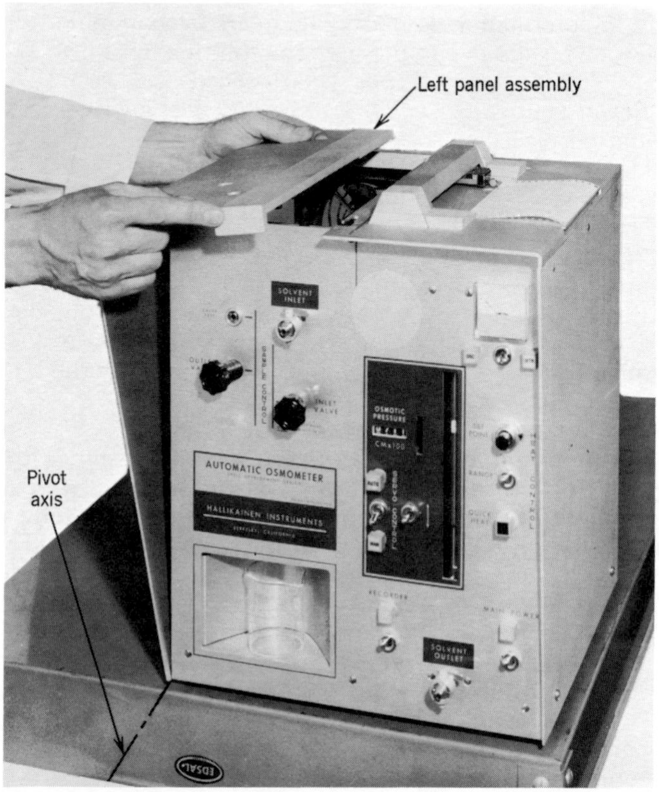

Design Documentation Example

An example of design documentation deals with the product design of an automatic osmometer, an analytical instrument* that is used for molecular weight determination of large, chain molecules. Problems considered in the design of the *left panel assembly*, which is shown in Figure 15.3, include the following.

1. *Design Objectives.*

(a) Access to the inside of the cabinet without tools.

(b) Left panel assembly as part of unified cabinet design.

2. *Prior Decisions Governing the Design of the Left Panel Assembly.*

(a) *Material and Finish.* Choice was limited by these criteria:

(1) Resistant to solvents.

(2) Shearable and formable with relatively light shop tools.

(3) Attractive appearance.

Decision:

(1) 5052-H32 aluminum alloy sheet.

(2) 0.090 in. thick.

(3) Straight grain and clear anodized finish

(4) Fasteners (screws and/or rivets) to be few and as inconspicuous as possible.

(b) *Shape and Form.* The fundamental "square" look of the cabinet with bent-metal top corners and tapered front profile had been established, thus governing the main outline and the fore-and-aft bend

Figure 15.3 Left panel assembly (Courtesy Hallikainen Instrument Co.).

* Invented by the Shell Development Co. The Hallikainen Instrument Co., Richmond, California, undertook the product design.

of the left panel. (The proportions of the "hood" type design had been decided from a full size *cardboard mock-up*. The exact dimensions chosen became a matter of subjective aesthetics, with the limitation that the front overhang could not extend so far that it blocked the view of the controls when the operator was standing slightly to the side of the instrument.)

(*c*) *Hook*. The lower left end of the left panel would require a self-contained hook to catch onto the end plate. The *details* of the hook, however, were relatively unaffected by the major decisions already made, so more will be said below about the evolution of this hook.

(*d*) *Miscellaneous*. Several other decisions affecting the left panel will be mentioned but not discussed. The latch details had already been worked out, and so were the size and positions of holes in the top surface for the glass thermometer and for the funnel through which sample fluid is admitted to the osmometry cell.

3. *Hook Design—Lower Left Edge of Left Panel Assembly*. Theoretically, the left panel had to hook on somehow to the base so as to be restrained against upward and outward motion after being installed. Note the sketches (Figure 15.4), which illustrate the concept prior to working out the details. In approaching the hook design, there was a certain amount of freedom remaining in the design details of the end plate. Thus the final design of the end plate—or whatever anchorage there was to be on the base—was intimately tied to the design of the hook.

At first a number of freehand sketches were made, one idea leading to another, as seen in Figures 15.5 to 15.9. Then several true-size, accurately drawn sections

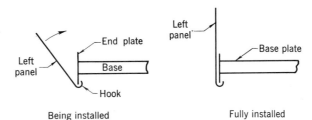

Figure 15.4 Hook design concept.

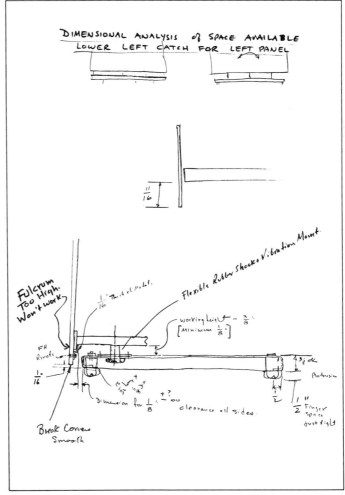

Figure 15.5 Sketches for dimensional analysis
(Courtesy Hallikainen Instrument Co.).

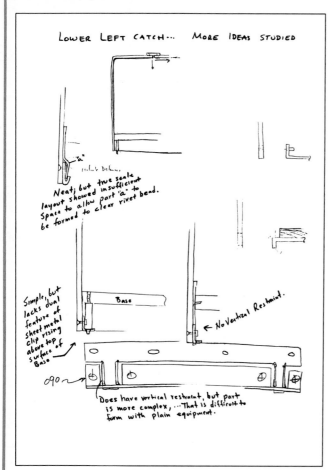

Figure 15.6 Sketches for lower left catch
(Courtesy Hallikainen Instrument Co.).

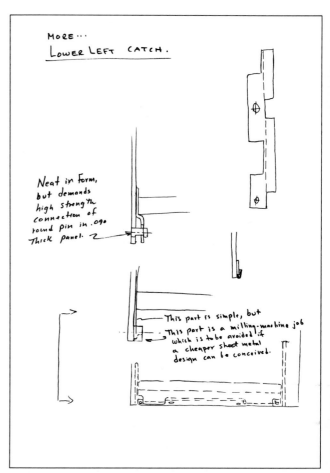

Figure 15.7 More sketches for lower left catch
(Courtesy Hallikainen Instrument Co.).

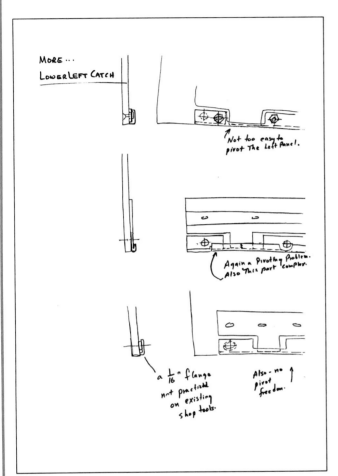

Figure 15.8 Still more sketches for lower left catch
(Courtesy Hallikainen Instrument Co.).

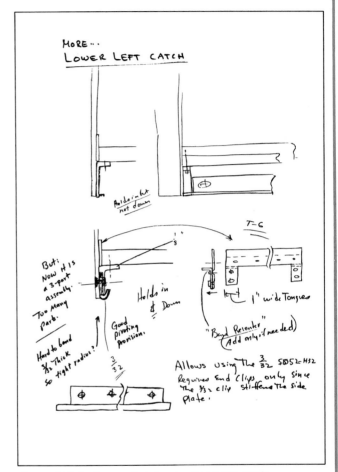

Figure 15.9 And more sketches for lower left catch
(Courtesy Hallikainen Instrument Co.).

were made on the osmometer design layout, a portion of which is reproduced in Figure 15.10*a*. A pictorial is shown in Figure 15.10*b*. Some of the earlier concepts would have been cheaper to produce but, on reflection, they were seen to have at least one clear flaw: the pivot axis was too high. The panel would have to *pivot* from near its *bottom* to avoid interference with the front and rear panels, so only those candidate designs having a pivot axis at the same nominal elevation as the bottom edge of the front and rear panels were likely to be workable.

Many minor features, and drawbacks, were evaluated as the flow of ideas progressed. One was a "human factors" concern. The instrument would be lifted with the hands under the side panels, and so the lower edge of the left panel assembly would have to be sturdy enough to take the load and smooth enough not to be uncomfortable. Another concern was overall cabinet symmetry. An *integral* hook was considered unacceptable, because it would destroy the external symmetry of the cabinet.

Eventually, the number of acceptable alternates, among the various concepts proposed, was narrowed down to the point where decisions could be made. Since it was decided the hook was not to be integral with the panel, it would be separate and fastened with rivets, the least conspicuous fastener. The rivet alloy was chosen for the type having the closest color shade to alloy 5052 (the panel) when clear anodized. It was thought best to anodize the entire panel assembly *after* riveting. This meant that, because ferrous metals are incompatible with aluminum anodizing solutions, the clip material had to be aluminum. The exact type of aluminum chosen for the hook was 0.063 in. 5052-H32, since it was a material and size already used elsewhere in the design.

One aspect of the end plate—the part attached to the $\frac{1}{2}$ in. thick aluminum base—took shape early. The upper edge of the end plate would extend $\frac{5}{16}$ in. above the base and thus provide a recess for receiving the bottom of the spring cover. The bottom portion of the end plate was designed to match the hook. It would be jogged $\frac{3}{32}$ in. so as to offer a horizontal spring-load when the left panel was installed, and it would have cutouts to clear the rivet heads used in assembling the hook.

4. *Production and Functional Experience*

(*a*) *General*. The catch-and-hold between the base and the left panel functioned just as intended. In placing the left panel on the osmometer cabinet, it catches neatly and securely at the bottom. Then it is swung up and to the right, engaging the latch. The latch bar is pushed in and the left panel is solidly in place, capable of supporting the osmometer firmly when the instrument is hand-transported. Some 20 to 30 osmometers have been manufactured, with many other design changes necessitated by experience. No fundamental change was required in the hook-and-end plate design, however, there were minor changes, some of which are mentioned in the following paragraphs.

(*b*) *Attachment of Hook to Panel*. Experience with the first osmometer resulted in changing the material thickness of the hook from 0.063 to 0.032 in. The thinner material, recommended by the Shop, was easier to bend into the **U** shape. The thinner gauge was sturdy enough, but only with the addition of two more fastenings. This addition of two rivets also meant two more cutouts on the end plate.

Soon after, the shop concluded that it lacked the particular skills and/or tooling to produce quality appearing flush riveted assemblies on a routine basis. So a stainless steel oval head screw (and nut) was substituted for the rivet.

(*c*) *Finish*. After the first three instruments were manufactured, a cost saving change was made affecting all the aluminum panels. An air-driven grit blasting was substituted for the more costly hand graining. Test panels of this blast finish were, of course, analyzed in advance. They demonstrated a pleasing and uniform appearance which, at the time, was all that seemed necessary. So the change was authorized for a run of six osmometers. In time, however, it was discovered that the grit-blasted surfaces of the osmometer cabinets became impossibly smudged and dingy. Just from ordinary room dust and handling, the pores of the metal filled with dirt and could be cleaned only with a brush and vigorous scrubbing. The grit blasting was a mistake, and on the succeeding production run the grained finish was reinstated. Unidirectional graining is easy to wipe clean, a "plus" feature of this type of surface not previously recognized by the engineer. This experience came at the cost of having to refinish some of the six cabinets, plus the intangible cost of the loss in product confidence by those customers (fortunately few) who received the grit-blasted cabinets.

(*d*) *End Plate Material*. The end

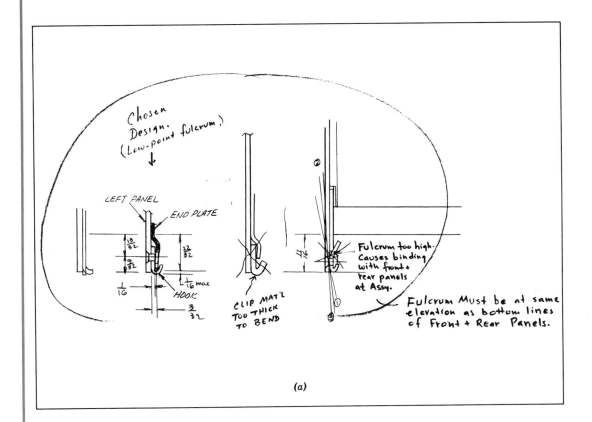

(a)

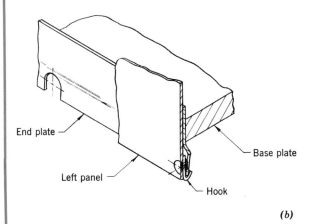

(b)

Figure 15.10 Selected design—low-point fulcrum
(Courtesy Hallikainen Instrument Co.).

plate on the base was originally specified cold-rolled steel, nickel plated, in the belief that this was the least expensive acceptable combination of material and finish. However, since the shop was experienced in cutting and forming stainless steel, they requested a change to the latter to save the logistics and delay costs entailed in

Figure 15.11 Osmometer, showing funnel to admit a sample solution. (Courtesy Hallikainen Instrument Co.)

sending end plates out to a vendor for nickel plating, even though other parts were routinely sent out for nickel plating (but not always concurrent with the day end plates were ready for plating). Because factors other than material alone were involved, and because the part represented a very small material cost, it was impractical to attempt a logical cost comparison between the two material alternatives. A change was made to stainless steel. This was a case of simply accepting the "feel" or judgment of the shop supervisor on which was the cheapest, overall.

(e) *Final Product*. Overall features of the automatic osmometer are:

(1) Speed, in comparison to manual methods.

(2) Precision, due to Shell Development's ingenious, patented, servo.

(3) Simplicity of operation, in part a contribution of Hallikainen's easy-access cabinet, and arrangement of all controls on the front panel.

(4) Direct readout of osmotic pressure on a numerical counter.

(5) Record of approach to equilibrium on a strip chart recorder, completely redesigned and enlarged by Hallikainen.

(6) Easy replacement of solvent after solute permeation, an original Shell feature.

(7) Attractive appearance, because of anodized and grained aluminum with color-anodized front panel artwork.

(8) Compact.

(9) Operates up to 135° C for analysis of polymers insoluble at room temperatures.

5. *Operational Principles*. A brief explanation of the operational principles may be of interest. The operator puts a sample solution of 5 to 10 ml into the funnel and then

isolates it in the cell (Figure 15.11) by means of the inlet and outlet valves. After 5 to 10 min, the osmotic pressure is read on a mechanical counter to one-hundredth of a centimeter, over a 10-cm range. A built-in recorder, the pen of which is directly driven by the balancing servomechanism, enables the operator to observe the balancing process and to ascertain that equilibrium has been established; solute permeation can be detected by a decrease of osmotic pressure with time. The osmometer cell consists of two cavities separated by a semipermeable membrane. On one side is the reference solvent; on the other is the sample solution. The cavity containing the sample solution is bounded by a thin, metal diaphragm that responds to changes of volume. Displacement of this diaphragm because of solvent flow through the membrane is sensed as an electrical capacity change in an oscillator circuit, causing the servomechanism to adjust the solvent head (static pressure) for zero osmotic flow. The speed with which this instrument makes a determination results not only in improved productivity but also in increased accuracy. The reason for this is because errors caused by small solute molecules permeating the membrane increase with time. These characteristics enable osmometry to become a practical routine method for analyzing polymers (such as polyethylene) in the molecular weight range 5000 to 500,000.

PATENTS

Introduction

You may believe that your design of a device, machine, or system is novel, unique, useful, and can be marketed profitably. Moreover, you may believe that your design or some features of it may be patentable. Hopefully, you have maintained a record of your continuous and persistent efforts in the development of your design from the initial idea stage to its completion. The record is invaluable in establishing your claims if other persons try to exploit your work or contend that they developed the invention before you did. Infringement of your patent rights, assuming that you have been granted a patent, can be tested in the courts (infringement suits arise in about one out of 100 patents).

To be recognized as a valid document, your written record must be in a *bound* notebook whose pages are numbered consecutively; it should be dated, signed, and witnessed by two persons who are not related to you in any way. Your entries should be witnessed every two or three weeks. It is preferable to use ink in making your entries. *Do not erase mistakes;* merely cross out the errors or discarded alternatives and continue to make corrections and necessary changes. Further development and refinement of your ideas should lead to implementation by model or prototype construction and testing. Your records of the *"reduction of your ideas to practical application"* should be witnessed by persons who are trustworthy and who can demonstrate that they clearly understood your invention and how it functioned. (A notary public should attest to the signatures of the witnesses.)

Procedure to Obtain a Patent

How does one proceed to obtain a patent? Considerable information can be obtained from publications by the U.S. Department of Commerce Patent and Trademark Office. Purchases can be made from the Superintendent of Documents,

U.S. Printing Office, Washington, D.C. 20402; field offices of the U.S. Department of Commerce (most large cities have a field office; consult your telephone directory); or local U.S. government bookstores. A list of pertinent publications includes the following titles.

> *Questions and Answers about Patents* (free)
> *General Information Concerning Patents* (nominal price)
> *Patents and Inventions: An Information Aid for Inventors* (nominal price)
> *Guide for Patent Draftsmen* (nominal price)
> *The U.S. Patent Office's "Disclosure Document Program"* (free)
> *Directory of Registered Patent Attorneys and Agents* (nominal price)
> *Patent Laws: A Compilation of Patent Laws in Force* (nominal price)

Questions and Answers about Patents

Answers to questions frequently asked about patents are most helpful to designers, inventors, and others who are concerned with patent grants. The following questions and answers have been selected from the first two publications in the preceding list.

What is a patent?

A patent is a grant issued by the United States Government giving an inventor the right to exclude all others from making, using, or selling his invention within the United States, its territories and possessions.

For how long a term of years is a patent granted?

Seventeen years from the date on which it is issued; except for patents on ornamental designs, which are granted for terms of $3\frac{1}{2}$, 7, or 14 years.

May the term of a patent be extended?

Only by special act of Congress, and this occurs very rarely and only in most exceptional circumstances.

Does the patentee continue to have any control over use of the invention after his patent expires?

No. Anyone has the free right to use an invention covered in an expired patent, so long as he does not use features covered in other unexpired patents in doing so.

What can be patented?

The patent law specifies the general field of subject matter that can be patented, and the conditions under which a patent may be obtained.

In the language of the statute, any person who "invents or discovers any new and useful process, machine, manufacture, or composition of matter, or any new and useful improvements thereof, may obtain a patent," subject to the conditions and requirements of the law. By the word "process" is meant a process or method, and new processes, primarily industrial or technical processes, may be patented. The term "machine" used in the statute needs no explanation. The term "manufacture" refers to articles which are made, and includes all manufactured articles. The term "composition of matter" relates to chemical compositions and may include mixtures of ingredients as well as new chemical compounds. These classes of subject matter taken together include practically everything which is made by man and the processes for making them.

On what subject matter may a patent be granted?

A patent may be granted to the inventor or discoverer of any new and useful process, machine, manufacture, or composition of matter, or any new and useful improvement thereof, or on any distinct and new variety of plant, other than a tuber-propagated plant, which is asexually reproduced, or on any new, original, and ornamental design for an article of manufacture.

On what subject matter may a patent not be granted?

A patent may not be granted on a useless device, on printed matter, on a method of doing business, on an improvement in a device which would be obvious to a person skilled in the art, or on a machine which will not operate, particularly on an alleged perpetual motion machine.

Is it advisable to conduct a search of patents and other records before applying for a patent?

Yes; if it is found that the device is shown in some prior patent it is useless to make application. By making a search beforehand the expense involved in filing a needless application is often saved.

Where can a search be conducted?

In the Search Room of the Patent and Trademark Office at Crystal Plaza, 2021 Jefferson Davis Hwy., Arlington, Virginia. Classified and numerically arranged sets of United States and foreign patents are kept there for public use.

Can I make a search or obtain technical information from patents at locations other than the Patent and Trademark Office Search Room?

Yes. The following libraries have sets of patent copies numerically arranged in bound volumes, and these patents may be used for search or other information purposes.

Locations of libraries which have printed copies of U.S. patents arranged in numerical order.

Albany, N. Y.	University of State of New York
Atlanta, Ga.	Georgia Tech Library*
Boston, Mass.	Public Library
Buffalo, N. Y.	Buffalo and Erie County Public Library
Chicago, Ill.	Public Library
Cincinnati, Ohio	Public Library
Cleveland, Ohio	Public Library
Columbus, Ohio	Ohio State University Library
Detoit, Mich.	Public Library
Kansas City, Mo.	Linda Hall Library*
Los Angeles, Calif.	Public Library
Madison, Wis.	State Historical Society of Wisconsin
Milwaukee, Wis.	Public Library
Newark, N. J.	Public Library
New York, N. Y.	Public Library
Philadelphia, Pa.	Franklin Institute
Pittsburgh, Pa.	Carnegie Library
Providence, R. I.	Public Library
St. Louis, Mo.	Public Library
Stillwater, Okla.	Oklahoma A. & M. College Library
Sunnyvale, Calif.	Public Library**
Toledo, Ohio	Public Library

* Collection incomplete.
** Arranged by subject matter, collection dates from January 2, 1962.
(*Note*. Check the library on your campus.)

How can technical information be found in a library collection of patents arranged in bound volumes in numerical order?

You must first find out from the *Manual of Classification* in the library the Patent and Trademark Office classes and subclasses which cover the field of your invention or interest. You can then, by referring to microfilm reels or volumes of the Index of Patents in the library identify the patents in these subclasses and, thence, look at them in the bound volumes. Further information on this subject may be found in the leaflet Obtaining Information from Patents a copy of which may be requested from the Patent and Trademark Office.

Will the Patent and Trademark Office make searches for individuals to help them decide whether to file patent applications?

No. But it will assist inventors who come to the Patent and Trademark Office by helping them to find the proper patent classes in which to make their searches. For a reasonable fee it will furnish lists of patents in any class and subclass, and copies of these lists may be purchased for $1.00 each.

How does one apply for a patent?

By making the proper application to the Commissioner of Patents, Washington, D.C., 20231.

What is the best way to prepare an application?

As the preparation and prosecution of an application are highly complex proceedings they should preferably be conducted by an attorney trained in this specialized practice. The Patent and Trademark Office therefore advises inventors to employ a patent at-torney or agent who is registered in the Patent and Trademark Office. No attorney or agent not registered in the Patent and Trademark Office may prosecute applications. The pamphlet entitled *Roster of Attorneys and Agents Registered to Practice before the U.S. Patent and Trademark Office* contains an alphabetical list of persons on the Patent and Trademark Office Register.

Of what does a patent application consist?

An application fee, a petition, a specification and claims describing and de-fining the invention, an oath or declaration, and a drawing if the invention can be illustrated. (*Authors' note.* An example of one of the forms that may be used follows. Other forms for joint inventors, and for an administrator of a deceased inventor are shown in the pamphlet, *General Information Concerning Patents.*)

To the Commissioner of Patents:

    Your petitioner, _____, a citizen of the United States and a resident of _____, State of _____, whose postoffice address is _____, prays that letters patent may be granted to him for the improvement in _____, set forth in the following specification; and he hereby appoints _____, of _____, (Registration No. _____), his attorney (or agent) to prosecute this application and to transact all business in the Patent Office connected therewith. (If no power of attorney is to be included in the application, omit the appointment of the attorney.)

    [The specification, which includes the description of the invention and the claims, is written here.]

_____, the above-named petitioner, being sworn (or affirmed), deposes and says that he is a citizen of the United States and resident of _____, State of _____, that he verily believes himself to be the original, first, and sole inventor of the improvement in _____ described and claimed in the foregoing specification; that he does not know and does not believe that the same was ever known or used before his invention thereof, or patented or described in any printed publication in any country before his invention thereof, or more than one year prior to this application, or in public use or on sale in the United States more than one year prior to this application; that said invention has not been patented in any country foreign to the United States on an application filed by him or his legal representatives or assigns more than twelve months prior to this application; and that no application for patent on said invention has been filed by him or his representatives or assigns in any country foreign to the United States, except as follows: _____.

_____
           (Inventor's full signature)

State of _____
County of _____
ss:
    Sworn to and subscribed before me this _____ day of _____,
19__.
    [SEAL]
         _____
          (Signature of notary or officer)

_____
           (Official character)

What are the Patent and Trademark Office fees in connection with filing of an application for patent and issuance of the patent?

A filing fee of $65 plus certain additional charges for claims depending on their number and the manner of their presentation are required when the application is filed. A final or issue fee of $100 plus certain printing charges are also required if the patent is to be granted. This final fee is not required until your application is allowed by the Patent and Trademark Office.

Are models required as part of the application?

Only in the most exceptional cases. The Patent and Trademark Office has the power to require that a model be furnished, but rarely exercises it.

Is there any danger that the Patent and Trademark Office will give others information contained in my application while it is pending?

No. All patent applications are maintained in the strictest secrecy until the patent is issued. After the patent is issued, however, the Patent and Trademark Office file containing the application and all correspondence leading up to issuance of the patent is made available in the Patent and Trademark Office Search Room for inspection by anyone, and copies of these files may be purchased from the Patent and Trademark Office.

May I write to the Patent and Trademark Office directly about my application after it is filed?

The Patent and Trademark Office will answer an applicant's inquiries as to the status of the application, and inform him whether his application has been rejected, allowed, or is awaiting action by the Patent and Trademark Office. However, if you have a patent attorney or agent the Patent and Trademark Office cannot correspond with both you and the attorney concerning the merits of your application. All comments concerning your invention should be forwarded through your patent attorney or agent.

What happens when two inventors apply separately for a patent for the same invention?

An "interference" is declared and testimony may be submitted to the Patent and Trademark Office to determine which inventor is entitled to the patent. Your attorney or agent can give you further information about this if it becomes necessary.

I have been making and selling my invention for the past 13 months and have not filed any patent application. Is it too late for me to apply for patent?

Yes. A valid patent may not be obtained if the invention was in public use or on sale in this country for more than one year prior to the filing of your patent application. Your own use and sale of the invention for more than a year before your application is filed will bar your right to a patent just as effectively as though this use and sale had been done by someone else.

I published an article describing my invention in a magazine 13 months ago. Is it too late to apply for patent?

Yes. The fact that you are the author of the article will not save your patent application. The law provides that the inventor is not entitled to a patent if the invention has been described in a printed publication anywhere in the world more than a year before his patent application is filed.

If two or more persons work together to make an invention, to whom will the patent be granted?

If each had a share in the ideas forming the invention, they are joint inventors and a patent will be issued to them jointly on the basis of a proper patent application filed by them jointly. If on the other hand one of these persons has provided all of the ideas of the invention, and the other has only followed instructions in making it, the person who contributed the ideas is the sole inventor and the patent application and patent should be in his name alone.

May a patent be granted if an inventor dies before filing his application?

Yes; the application may be filed by the inventor's executor or administrator.

While in England this summer, I found an article on sale which was very ingenious and has not been introduced into the United States or patented or described. May I obtain a United States patent on this invention?

No. A United States patent may be obtained only by the true inventor, not by someone who learns of an invention of another.

May the inventor sell or otherwise transfer his right to his patent or patent application to someone else?

Yes. He may sell all or any part of his interest in the patent application or patent to anyone by a properly worded assignment. The application must be filed in the Patent and Trademark Office as the invention of the true inventor, however, and not as the invention of the person who has purchased the invention from him.

If I obtain a patent on my invention will that protect me against the claims of others who assert that I am infringing their patents when I make, use, or sell my own invention?

No. There may be a patent of a more basic nature on which your invention is an improvement. If your invention is a detailed refinement or feature of such a basically protected invention, you may not use it without the consent of the patentee, just as no one will have the right to use your patented improvement without your consent. You should seek competent legal advice before starting to make or sell or use your invention commercially, even though it is protected by a patent granted to you.

Disclosure Document Program

A new service is provided for inventors by the U.S. Patent and Trademark Office for the acceptance and preservation, for a two-year period, of disclosure documents as evidence of the dates of conception of inventions. Details are available in a two-page announcement, *Disclosure Document Program*, which can be obtained from the U.S. Department of Commerce Patent and Trademark Office, Washington, D.C. 20231, or from the Field Office of the U.S. Department of Commerce.

The Sunnyvale, California, Patent Library provides the only subject-classified patent collection (*Complete Copies of U.S. Patents Dating from January 1962 to the Present. . . . First Patent Number for 1962—#3,015,103*) in the United States outside of Washington, D.C. Each week's newly issued patents are rearranged by the Sunnyvale Library staff into main classes.

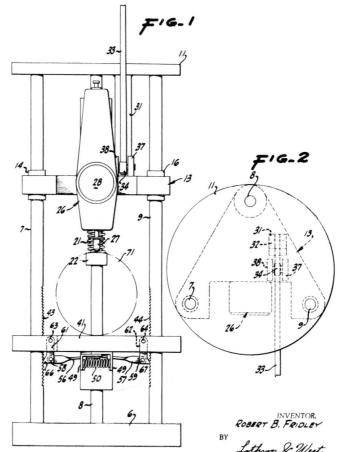

Figure 15.12 Copy of patent for firmness tester for fruit.

Patent Example

An example of a patent on a *firmness tester for fruit* is shown in Figure 15.12. The application was filed on June 9, 1966. The patent was granted on October 7, 1969. It is not unusual for patent attorneys to communicate with the patent office for three or more years before a patent is finally granted.

Included in the example in Figure 15.12 (Patent Number 3,470,-737) is (1) an abstract of the disclosure, (2) detailed description and specifications of the invention with accompanying drawings (drawn in accordance with the requirements specified in the pamphlet, *Guide for Patent Draftsmen*), and (3) the four claims that were allowed by the Patent Office.

It is interesting to note that the first patent was issued in 1790 and, since then, over 4 million patents have been granted. Information on patents can be obtained from the official *Gazette* of the U.S. Patent Office. The *Gazette* is issued each Tuesday, simultaneously with the weekly issue of the patents. It contains an abstract and a selected figure of the drawings. Copies of the *Gazette* may be found in the public libraries of the larger cities and usually in university libraries.

Commentary

Some inventors are capable of processing all the paper work and maintaining effective communication with the patent examiners; however, in most cases it is advisable to employ the service of a capable patent attorney who has the education and experience to expedite the granting of a patent, assuming that your invention is novel and useful.

Virtually all large corporations maintain a staff of patent attorneys to process the claims of the company.

Oct. 7, 1969   R. B. FRIDLEY   3,470,737

FIRMNESS TESTER FOR FRUIT

Filed June 9, 1966   3 Sheets-Sheet 2

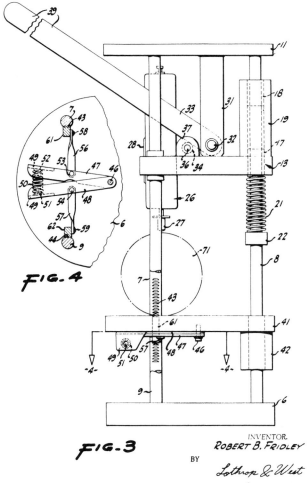

FIG. 4

FIG. 3

INVENTOR.
ROBERT B. FRIDLEY
BY
Lothrop & West
ATTORNEYS

Oct. 7, 1969   R. B. FRIDLEY   3,470,737

FIRMNESS TESTER FOR FRUIT

Filed June 9, 1966   3 Sheets-Sheet 3

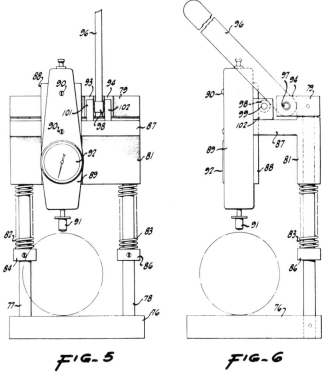

FIG. 5

FIG. 6

INVENTOR.
ROBERT B. FRIDLEY
BY
Lothrop & West
ATTORNEYS

1

3,470,737
FIRMNESS TESTER FOR FRUIT
Robert B. Fridley, Davis, Calif., assignor to The Regents
of the University of California, Berkeley, Calif.
Filed June 9, 1966, Ser. No. 556,354
Int. Cl. G01n *3/48*
U.S. Cl. 73—81   4 Claims

ABSTRACT OF THE DISCLOSURE

A firmness tester for fruit has a fruit support near the bottom of an upright frame carrying a force measuring gauge above the fruit support and for movement downwardly toward the support. A plunger extending downwardly from the gauge is pressed against a fruit on the support when a high ratio lever on the frame is manually actuated to move the gauge downwardly toward the plunger.

My invention relates to means useful in testing a fruit specimen primarily to determine whether the lot, of which it is a representative example, is of sufficient maturity for picking and for other related purposes.

There are available various devices for testing the firmness of fruit in order to get an idea of the ripeness of the fruit or its suitability for harvesting, and many of these are utilized in the field. They are usually not sufficiently accurate for work such as may be done by laboratory or technical investigators, by farm advisers and the like, who require results of considerable consistency and precision. While some testers give suitably precise indications, they are too elaborate and too expensive to permit of use in the field.

Many of the fruit testers available at the present time depend for their accuracy not only on the particular fruit being tested, but also on the skill of the operator. Some operators are able to operate carefully and smoothly and consistently, whereas others are wont to use excessive force or speed in the operation of the tester and so get variable results. Another variable is that the firmness of a fruit is affected by a time factor. That is to say, the firmness characteristics of the fruit may change under the pressure of a testing plunger so that it is of considerable importance to have all operators operate the plunger at substantially the same rate at all times.

A further difficulty is that the tester and fruit may not be related to each other with sufficient accuracy. The pressure exerted is not always normal to the surface of the fruit, or the fruit may move during testing so that the results are not consistent.

It is therefore an object of my invention to provide a firmness tester for fruit which is of considerable accuracy and can be utilized both in the laboratory and in the field.

Another object of the invention is to provide a firmness tester for fruit which is so designed as to operate substantially at the same speed at all times and with all operators.

Another object of the invention is to provide a firmness tester which will engage the fruit so that the fruit is held stationary during testing with little effort.

Another object of the invention is to afford smooth operation of the tester so that variations in firmness do not materially affect the operation.

A still further object of the invention is in general to provide an improved firmness tester for fruit.

Other objects together with the foregoing are attained in the embodiments of the invention described in the accompanying description and illustrated in the accompanying drawings, in which:

FIGURE 1 is a front elevation of one form of firmness tester for fruit pursuant to the invention;

2

FIGURE 2 is a plan of the structure shown in FIGURE 1;

FIGURE 3 is a side elevation of the firmness tester shown in FIGURE 1;

FIGURE 4 is a cross section, the plane of which is indicated by the line 4—4 of FIGURE 3;

FIGURE 5 is a front elevation of a modified form of fruit tester pursuant to the invention; and

FIGURE 6 is a side elevation of the form of device illustrated in FIGURE 5.

In one preferred form which has met with practical success, the firmness tester is provided with a generally rectangular base 6 designed to rest on a table or other comparable support. Upstanding from the base are three columns 7, 8 and 9 arranged parallel to each other and at their upper ends connected by a top plate 11 so that an upright frame is provided. Designed to slide on the various columns is a platform 13. This is a plate generally of triangular form having a pair of bushings 14 and 16 engaging the columns 7 and 9. The platform 13, in addition to the bushings 14 and 16, is provided with bushings 17 and 18 disposed near the ends of a sleeve 19 part of and upstanding from the platform.

In order to urge the carriage uwardly away from the base 6, a helical spring 21 is provided. At one end this rests on a collar 22 adjustably fastened on the column 8 and at the other end the spring bears against the nether side of the platform 13. This mechanism accurately locates the platform for sliding movement along the various columns and the spring exerts an adjustable upward force on the platform.

Mounted on the platform 13 adjacent the forward end thereof is a force measuring gauge 26. This may be a proprietary item available on the open market. It includes a casing from which a plunger 27 projects. A dial 28 on the casing indicates the motion of plunger 27 against the expelling force of a spring (not shown). The gauge 26 is fastened to the platform 13 so that the plunger 27 is disposed substantially between and parallel to the columns 7 and 9 at an appropriate distance away from the base 6.

In order simultaneously to move the platform 13 and the body of the gauge 26, I provide a special means. This includes a fulcrum block 31 depending from the top plate 11 and bifurcated near its lower end to receive a pivot pin 32 having its axis crosswise of the machine. Fulcrumed on the cross pin 32 is a relatively long, manually operable lever 33. The lever bears upon a roller 34 journalled on a cross pin 36 secured in bosses 37 and 38 upstanding from the carriage 13. The handle is extended to provide a grip portion 39 at a considerable distance from the fulcrum 32. In fact, it has been found convenient to make the distance between the fulcrum 32 and the handle portion about sixteen times as great as the distance between the fulcrum 32 and the roller 34.

When the user grasps the handle 33 and manually depresses the outermost end of the lever 33, he has a large mechanical advantage in moving the carriage 13 against the urgency of the spring 21 and in lowering the gauge 26 bodily. The amount of movement which can be accomplished easily in a given unit of time produces a similar but much smaller movement of the carriage 13 in that same time. Because of the large amount of leverage, the control of the carriage is quite precise and cannot vary a great deal in time since the ability of a user to move the lever downwardly from its uppermost position is not great. Variations in resistance are not material since the operator can easily overcome all of them.

Since fruit of various sizes is available for testing in this device, I provide means for submitting the fruit almost directly to the plunger 27. Designed to slide on the columns 7, 8 and 9 very much as the carriage 13 slides

3

is a table **41** of plate-like configuration having an extended boss **42** closely engaging around the lower portion of the column **8**. The table is movable toward and away from the base **6**.

Means are provided for holding the table **41** in any selected location. For that reason, the columns **7** and **9** in their portions within the normal table travel are provided with a series of indentations **43** and **44** to define ratchet teeth. Mounted to swing on a vertical pivot pin **46** extending beneath and set in the table are levers **47** and **48** having depending finger pads **49** urged apart by a coil spring **50** spanning the space between the levers and held against dislodgement by protuberances **51** and **52** formed in the finger pads **49**. Connected to the levers **47** and **48** by pins **53** and **54** are twisted links **56** and **57**. These in turn are connected by pins **58** and **59** to pawls **61** and **62** designed to swing on pins **63** and **64** mounted in cut-out portions of the table **41**. The pawls **61** and **62** have sharpened points **66** and **67** designed to interengage between the ratchet teeth **43** and **44** in any one of various selected locations.

The user by pinching the finger pads **49** of the levers **47** and **48** against the urgency of the spring **50** rocks the pawls **61** and **62** toward each other, thus disengaging their points **66** and **67** from the adjacent columns **7** and **9**. The table can then be manually lifted or lowered to any selected location, whereupon the levers **47** and **48** are released and the spring **50** again causes the pawl points **66** and **67** to engage between the adjacent ratchet teeth. The table is thus locked at the selected height. While the table can be lifted at any time without bothering to disengage the levers **47** and **48**, it cannot be lowered without operating the levers **47** and **48**.

In the use of this device, the table is set at an appropriate height to accommodate a fruit **71** to be tested for firmness. The fruit as nearly as possible is placed with one of its approximate diameters immediately continuing the axis of the plunger **27** and with a normal portion of the fruit surface parallel to the plunger bottom surface. It is important to have the fruit close to the bottom of the plunger and to support the fruit firmly in position on the table **41**. When the fruit has been so positioned and is being held, the user grasps the outer end **39** of the lever **33** and moves it downwardly. This motion by almost any user is naturally accomplished only at a fairly consistent, reasonable rate. The downward motion of the plunger **27** is approximately only one-sixteenth as far and at one-sixteenth the rate. In actual practice, a wide variety of users by natural inclination all approximate a steady, fairly standardized rate of approach of the plunger to the fruit.

When the plunger encounters the surface of the fruit, the increasing resistance of the fruit, although felt by the user, does not greatly change the rate of approach since the extra force required is reduced so much by the leverage. Variations in encountered resistance, therefore, do not change the speed of operation of the lever in any significant amount. After the plunger **27** is in abutment with the fruit, the gauge **28** indicates the value of the pressure being exerted. The lever **33** may be continued in its downward direction until after the plunger **27** pierces the outside of the fruit and so gives an indication of the firmness of the fruit. From data acquired in this fashion, the investigator, grower or buyer can determine how near to maturity the fruit may be.

When the test is over, the user releases the lever. The spring **21** restores the platform **13** to an upper position approximating that shown in FIGURE 1. It is easy to remove the tested fruit **71**, discard it and replace it with another fruit to be tested. If these fruits are substantially the same size, no intermediate adjustment of the table **41** must be made. However, if the fruit varies substantially in size from one test to another, then the table **41** is correspondingly moved so that in each instance the upper exposed portion of the fruit is closely beneath the lower

4

end of the plunger **27** so that a movement of the lever **33** within the permissible range will afford an accurate indication of the firmness of the fruit.

In some instances a tester of even simpler form serves many purposes equally well. As shown in FIGURES 5 and 6, this form of firmness tester includes a base **76** from the rear of which a pair of columns **77** and **78** extend upwardly to a cross bridge **79** pinned in place. Slidable on the columns is a carriage **81** pushed upwardly by a pair of helical springs **82** and **83**. Each of the springs rests on its own adjustable collar **84** or **86** so that the spring tension can be set at the desired value. The carriage **81** has a forward extension **87** with a pad **88** depending therefrom and to which a gauge **89** is fastened by screws **90**. A number of holes (not shown) in the pad **88** provide for limited vertical position change of the gauge **89** relative to the pad **88**. The gauge may be the same as shown in the preceding figures and includes a depending plunger **91** and a dial indicator **92**.

In order that the carriage **81** can be actuated, the bridge **79** carries a pair of forwardly projecting ears **93** and **94** in which a lever **96** of considerable length is fulcrumed by means of a pivot pin **97**. The lever bears against a roller **98** on a cross pin **99** mounted in ears **101** and **102** upstanding from the carriage **81**.

In the operation of this structure, the procedure is as before except that there is no movable table **41** to support the fruit. The fruit is supported directly upon the base **76**. This is readily accomplished when the fruit being tested is of a nearly uniform size. Following the positioning and holding of the fruit, the long lever **96** is operated by the user and this produces a nearly constant rate of actuation of the carriage **81**. The carriage is forced downwardly until the plunger **91** encounters and pierces the tested fruit. The indications of the dial **92** afford the numerical results.

What is claimed is:

1. A firmness tester for fruit comprising a base on which a fruit can be rested, a frame upstanding from said base, a platform, means for mounting said platform on said frame above said base for sliding movement toward and away from said base, means for urging said platform upwardly away from said base, a force measuring gauge having a movable plunger projecting therefrom, means for mounting said gauge on said platform with said plunger extending downwardly toward said base from a position above said fruit, a hand lever, pivot means pivoting one end of said hand lever to said frame and roller means on said platform adjacent said pivot means and engageable by said lever to produce a relatively small movement of said platform downwardly relative to said base for a relatively large movement of the other end of said hand lever relative to said base.

2. A firmness tester for fruit as in claim **1** in which said lever provides a motion ratio of the order of sixteen to one between the motion of the lever end and the motion of said platform.

3. A firmness tester as in claim **1** in which said plunger depends below and is solely supported by said gauge.

4. A firmness tester for fruit comprising a base on which a fruit can be rested, a frame upstanding from said base, a platform, means for mounting said platform on said frame above said base for sliding movement toward and away from said base, means for urging said platform upwardly away from said base, a force measuring gauge having a movable plunger projecting therefrom, means for mounting said gauge on said platform with said plunger extending downwardly toward said base from a position above said fruit, a hand lever, and means interrelating said hand lever with said frame and said platform to produce a relatively small movement of said platform downwardly relative to said base for a relatively large movement of said lever relative to said base, said frame including at least one column and said platform including at least one sleeve encompassing said column,

Exercises

1. How can you overcome the "I Can't Write" syndrome?

2. Write a letter to a company of your choice that will get you an interview for a possible summer job in some phase of engineering. (At the end of your freshman year, perhaps a job as, for example, a design detailer, a shop aide, or a construction "flunkie" would be appropriate. More advanced work at the end of your sophomore and junior years, possibly with the same company, could enhance your work experiences).

3. Read a newspaper article written by the science editor. Can you improve the article? Try the "scissors" technique; that is, Xerox the article, cut it up into the paragraphs as written, and rearrange the paragraphs into a sequence that you believe improves upon the original format. Why do you believe that the new arrangement is more effective?

4. Select a recent technical article in a journal such as *Machine Design, Design News, Product Engineering,* or *Civil Engineering* (available in your en-gineering library). Xerox the article and rearrange the paragraphs by "scissoring and pasting" to produce a more effective format. Add subheadings and some cursive writing if you believe this will enhance the effectiveness of the article.

5. Write a project proposal, addressed to a sponsoring agency, to support your request for funding the design of a prototype windmill to generate electricity for *home* use.

6. Write a proposal, addressed to a sponsoring agency, to support your "pet" project, assuming it is of value to our society.

7. Get a copy of a current proposal (your instructor or department chairman can arrange the procurement). Read the proposal carefully. Do you understand it? If not, question the proposers. Assuming that you now have a good grasp of the proposal, what changes, if any, would you recommend to make the proposal more effective?

8. Product liability must be seriously considered by the designer, the manufacturer, the distributor, etc. Lawsuits concerning faulty and hazardous prod-ucts abound. To what extent does "design documentation" protect the designers? Visit some of your local manufacturing or construction companies; discuss design documentation with the engineers who are responsible for this activity; get a sample design documentation if possible, and read it carefully. Discuss questions you might have concerning product liability with the company's attorneys. What recommendation as to changes in the design could you offer that might enhance the design and thereby reduce possible hazards?

9. Contact a local patent attorney's office and obtain current information on the costs of obtaining a patent. Then write a report.

10. If you believe that you have designed a novel and useful product and have built a working model that demonstrates the product's novelty and usefulness, you may wish to pursue the potential of your product if it is economically feasible. If that is the case, outline the steps that would be taken to obtain a patent.

APPENDICES

Contents

Appendix A
Line Conventions; Lettering; and Geometric Constructions

LINE CONVENTIONS

Line Symbols (See Figure A-1)

Thickness of lines will vary according to the size and type of drawing. Where lines are close together, for example, the lines are drawn thinner. For most pencil work, *two* widths of lines are adequate: *medium thick* for visible, hidden, and cutting plane lines; and *thin* for center, extension, dimension, and section lines. In order to produce good, legible print, all pencil lines should be clean, dense black, and uniform.

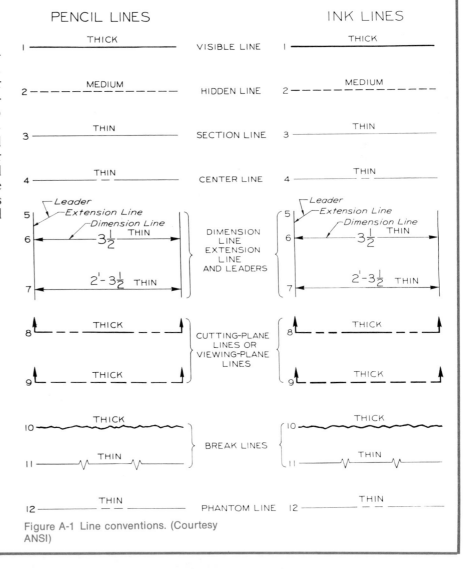

Figure A-1 Line conventions. (Courtesy ANSI)

Hidden Lines (See Figure A-2)

Hidden lines should begin and end with a dash in contact with the visible or hidden line from which they start or end, except as shown in the figure. Hidden lines should not be used unless they add to the clearness of the interpretation of the design drawing.

Section Lining (See Figure A-3)

Thin, solid lines should be used for section lining. Where more than one part is shown, the directions of the lines should be changed as illustrated in the figure for parts A, B, and C.

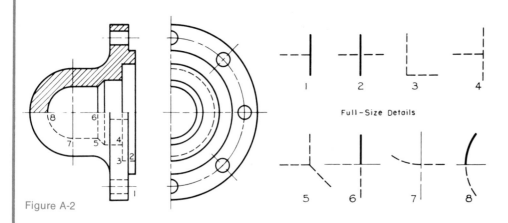

Figure A-2

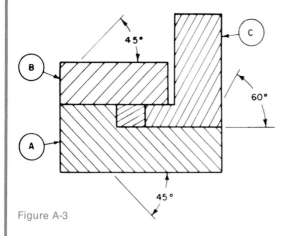

Figure A-3

Lettering (See Figures A-4 and A-5a)

The single-stroke Gothic style shown in the figures is used on most engineering design drawings. This style is easy to use and provides a means for rapid execution.

Above all, the most important requirement is legibility. Both the vertical and inclined styles are acceptable. The trend, however, is toward the use of the vertical style. In many industries notes are typed on the drawing.

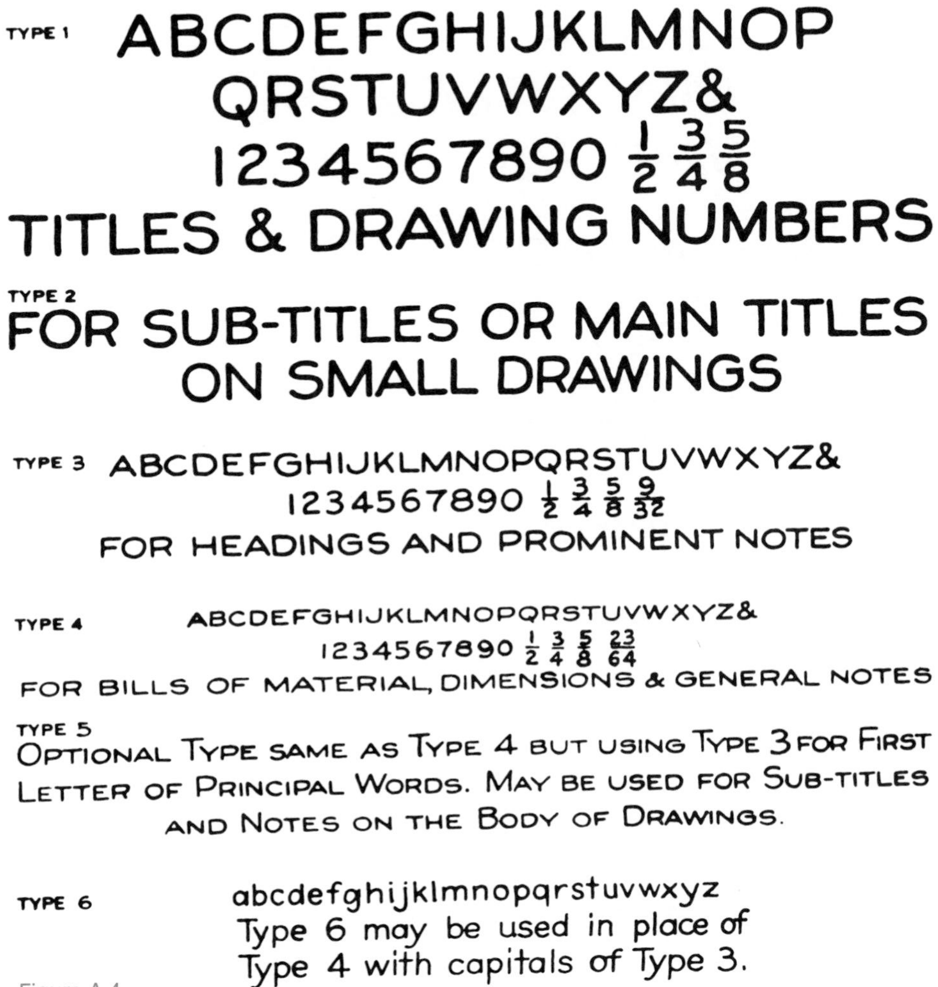

TYPE 1

ABCDEFGHIJKLMNOP QRSTUVWXYZ& 1234567890 $\frac{1}{2}\frac{3}{4}\frac{5}{8}$ TITLES & DRAWING NUMBERS

TYPE 2

FOR SUB-TITLES OR MAIN TITLES ON SMALL DRAWINGS

TYPE 3

ABCDEFGHIJKLMNOPQRSTUVWXYZ&
1234567890 $\frac{1}{2}\frac{3}{4}\frac{5}{8}\frac{9}{32}$
FOR HEADINGS AND PROMINENT NOTES

TYPE 4

ABCDEFGHIJKLMNOPQRSTUVWXYZ&
1234567890 $\frac{1}{2}\frac{3}{4}\frac{5}{8}\frac{23}{64}$
FOR BILLS OF MATERIAL, DIMENSIONS & GENERAL NOTES

TYPE 5

Optional Type same as Type 4 but using Type 3 for First Letter of Principal Words. May be used for Sub-titles and Notes on the Body of Drawings.

TYPE 6

abcdefghijklmnopqrstuvwxyz
Type 6 may be used in place of
Type 4 with capitals of Type 3.

Figure A-4

TYPE 1

ABCDEFGHIJKLMNOP 
QRSTUVWXYZ&
1234567890 $\frac{1}{2}$ $\frac{3}{4}$ $\frac{5}{8}$ $\frac{7}{16}$
TO BE USED FOR MAIN TITLES
& DRAWING NUMBERS

TYPE 2

ABCDEFGHIJKLMNOPQR
STUVWXYZ&
1234567890 $\frac{13}{64}$ $\frac{5}{8}$ $\frac{1}{2}$
TO BE USED FOR SUB-TITLES

TYPE 3

ABCDEFGHIJKLMNOPQRSTUVWXYZ&
1234567890 $\frac{1}{2}$ $\frac{3}{4}$ $\frac{5}{8}$ $\frac{7}{16}$
FOR HEADINGS AND PROMINENT NOTES

TYPE 4

ABCDEFGHIJKLMNOPQRSTUVWXYZ&
1234567890 $\frac{1}{2}$ $\frac{1}{4}$ $\frac{3}{8}$ $\frac{5}{16}$ $\frac{7}{32}$ $\frac{1}{8}$
FOR BILLS OF MATERIAL, DIMENSIONS & GENERAL NOTES

TYPE 5
OPTIONAL TYPE SAME AS TYPE 4 BUT USING TYPE 3 FOR FIRST
LETTER OF PRINCIPAL WORDS. MAY BE USED FOR SUB-TITLES &
NOTES ON THE BODY OF DRAWINGS.

TYPE 6
abcdefghijklmnopqrstuvwxyz
Type 6 may be used in place of
Type 4 with capitals of Type 3

Figure A-5a

Microfont Lettering

The most important requirement for lettering as used on drawings is legibility. The second is ease and rapidity of execution. In order to meet these requirements for drawings that are microfilmed, the National Microfilm Association's Drafting Standards Committee designed microfilm lettering. Approved specimens of microfont lettering are shown in Figure A-5*b*.

GEOMETRIC CONSTRUCTIONS

Many engineering designs make use of well-known geometric shapes and geometric constructions. In the design of components, for example, geometric shapes such as triangles, squares, trapezoids, pentagons, hexagons, circles, ellipses, parabolas, hyperbolas, and spirals are quite common. In addition, a number of geometric constructions are found quite useful in engineering design, in the construction of scales for nomograms, and in graphical solutions of problems arising in engineering and science. Most students are already familiar with many of the simple geometric shapes and elementary geometric constructions.

A number of these are presented *for review*, and several others are included that will be found helpful in effecting graphical solutions.

A. The Straight Line and Its Division

1. To Divide a Line Segment into a Specified Number of Equal Parts (Figure A-6)
Suppose line segment *AB* is given and that it is required to divide *AB* into seven equal parts.

Solution

Through point *A* draw line *m* and then lay off seven equal distances, starting at point *A*. Join points *E* and *B* and then draw parallels to *EB* through the points on *m*. The points of intersection of these parallels with line segment *AB* divide it into the required seven equal parts. Why is this true?

Suggestion

In laying off distances on a line segment, use needle-point dividers, alternating the rotation of the dividers as shown in Figure A-7.

ABCDEFGHIJKLMNO

PQRSTUVWXYZ

1234567890

ABCDEFGHIJKLMNO
PQRSTUVWXYZ.,:;
1234567890=÷+−±
@&☆?#×"%'()[]°!
¢$/_ ∠∞Δ≈~∅⊥<>μα
√σδΣΥΠΒθωΩ∴∥

ABCDEFGHIJKLMNO
PQRSTUVWXYZ.,:;
1234567890=÷+−±
@&☆?#×"%'()[]°!
¢$/_ ∠∞Δ≈~∅⊥<>μα
√σδΣΥΠΒθωΩ∴∥

1/8"

3 MM

THIS IS AN EXAMPLE OF 5/32" MICROFONT.
IT WAS ESPECIALLY CREATED BY THE NATIONAL
MICROFILM ASSOCIATION'S STANDARDS COMMITTEES.
"MICROFONT" IS RECOMMENDED AS THE LETTERING
STYLE TO BE USED IN THE PREPARATION OF
ALL ENGINEERING DRAWINGS.

Figure A-5*b*

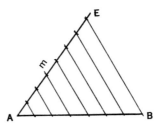

Figure A-6

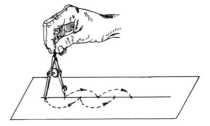

Figure A-7

2. To Divide a Line Segment into a Given Ratio (Figure A-8)

Let AB represent the given line segment and let the given ratio be 4:5:7. Draw line m through point A and at any convenient angle with AB. On m lay off distances $AC = 4$ units, $CD = 5$ units, $DE = 7$ units. Draw BE and then draw lines through C and D parallel to BE, cutting AB in points C' and D', which determine the required segments of AB.

3. To Construct a Line Segment that is the Mean Proportional (Geometric Mean) to Two Given Line Segments (Figure A-9)

Suppose the given segments are m and n. On line AB lay off consecutive segments equal to m and n. Construct a semicircle on the total length $(m + n)$ as a diameter. At the common point of the segments, construct a perpendicular to the diameter. Line g is the required mean proportional. The student should prove this by showing that $g^2 = m \times n$.

4. To Divide a Straight Line in Extreme and Mean Proportion (Figure A-10)

The given line is AB. At point B lay off line BC at 90° to AB and equal to $AB/2$. With C as center and radius CB, draw an arc cutting line AC at point D. With A as center and radius AD, draw an arc cutting line AB at point E, which divides line AB in extreme and mean proportion (i.e., the square on segment AE is equal to the rectangle having sides AB and EB).

B. The Construction of Triangles and Regular Polygons

1. To Construct a Triangle with Known Lengths of the Sides (Figure A-11)

Suppose m, n, and s are the given lengths. With the end points of m as centers and radii equal to n and s, respectively, draw intersecting arcs, locating point A. Join A with the end points of m to complete the construction of the triangle.

Is it possible to construct a triangle with 7, $3\frac{1}{2}$, and 3 cm as the lengths?

2. To Construct a Right Triangle when the Lengths of the Hypotenuse and One Side Are Known (Figure A-12)

Let line m and n represent the given lengths. Construct a semicircle with diameter AB equal to length m. With A as center (could use B) and radius equal to length n, draw an arc cutting the semicircle at point C. Triangle ABC is the required right triangle. Why is this true?

3. To Inscribe an Equilateral Triangle within a Circle Having a Given Diameter, AB (Figure A-13)

With center O and radius equal to

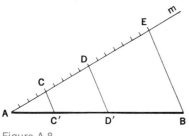

Figure A-8

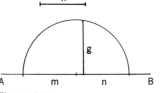

Figure A-9

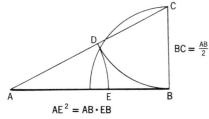

$AE^2 = AB \cdot EB$

Figure A-10

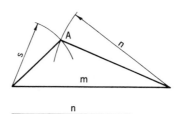

Figure A-11

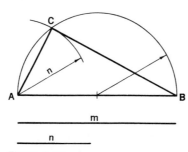

Figure A-12

A ———————|——————— B
          M

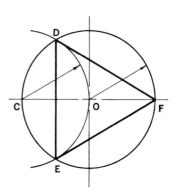

Figure A-13

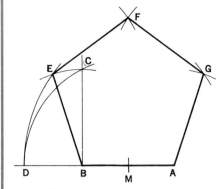

Figure A-14

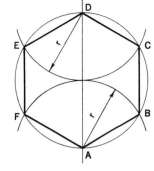

Figure A-16

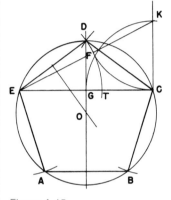

Figure A-15

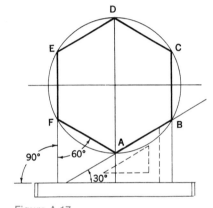

Figure A-17

AM (*M* is midpoint of *AB*), draw the circle. With *C* as center and the same length of radius, draw an arc cutting the circle in points *D* and *E*. Join points *D*, *E*, and *F*. The required triangle is *DEF*.

4. To Construct a Regular Pentagon when the Length AB of the Sides Is Known (Figure A-14)

First construct *BC* = *AB*, and perpendicular to *AB*. With *M* (midpoint of *AB*) as center and *MC* as radius, draw an arc cutting *AB* extended at point *D*. Now with *A* as center and *AD* as radius, draw an arc; and, with *B* as center and radius *BA*, draw an intersecting arc to locate point *E*. Line *BE* is a side of the regular pentagon. The construction for locating point *F* and *G* is fairly obvious. *ABEFGA* is the

required pentagon. (The solution of this problem is based on the fact that the larger segment of a diagonal of the pentagon, when divided in extreme and mean proportion, is the length of a side of the pentagon. Note that point *B* divides *AD* in extreme and mean proportion and that *AB* is the larger segment of *AD*.)

5. To Construct a Regular Pentagon when the Length EC of a Diagonal Is known (Figure A-15)

First divide *EC* in extreme and mean proportion ($ET^2 = EC \times TC$). With *E* and *C* as centers and radius *ET*, draw arcs that intersect at point *D*. Draw the perpendicular bisector of *ED* and locate point *O* on the vertical line passing through *D*. Point *O* is the center of the cir-

cle (radius *OE*) that circumscribes the pentagon.

With *E* and *C* as centers and a radius equal to *CD* (or *ED*), draw arcs cutting the circle at points *A* and *B*. Draw the necessary lines to form pentagon *ABCDEA*.

6. To Inscribe a Hexagon within a Given Circle (Figure A-16)

With *A* and *D* as centers and a radius equal to the radius of the circle, draw arcs that intersect the given circle in points *B*, *F*, *C*, and *E*. The required hexagon is *ABCDEF*. An alternative construction is shown in Figure A-17.

7. To Construct a Regular Polygon Having n Sides (Figure A-18)

The polygon in this example is a nonagon (nine equal sides), and *AB* is the given length of each side. With *B* as center and *AB* as radius, describe a semicircle and, by trial, divide it into nine equal parts. Starting from point *T*, locate the second division mark, *C*. Locate point *O*, the center of the circumscribing circle. (This is easily done by finding the intersection of the perpendicular bisectors of *AB* and *BC*.) Draw the circle with center *O* and radius *OA* and complete the nonagon.

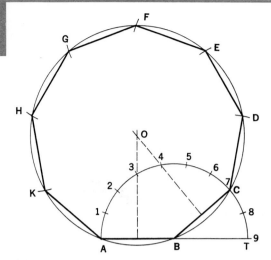

Figure A-18

C. Circles and Their Tangents

1. To Draw a Circle Through Three Given Points (Figures A-19 and A-20)

Let us assume *A*, *B*, and *C* as the given points. Draw the perpendicular bisectors of lines *AB* and *BC*. The intersections of the bisectors is point *O*, the center of the circle which passes through the three given points.

Now suppose that the center of the circle is inaccessible (Figure A-20). Again the given points are *A*, *B*, and *C*. Draw arcs with *A* and *C* as centers and radius equal to *AC*. Now draw lines *ABD* and *CBE*. On arc *AE* lay off, from point *E*, relatively short equal distances 1_U, 2_U, 3_U, etc., above *E* and, similarly, equal distances 1_L, 2_L, 3_L, etc., below *E*.

In like manner lay off distances 1_U, 2_U, 3_U, etc., above *D* and 1_L, 2_L, 3_L, etc., below *D*. Now draw lines A-3_U and C-3_L. Their intersection is point 3. Similarly locate point 3′ and, in like manner, additional points. The required circle

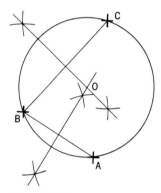

Figure A-19

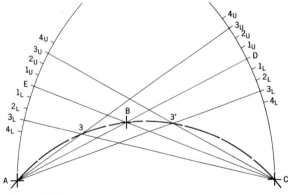

Figure A-20

will pass through points A, 3, B, 3′, and such additional points as desired. (The proof is left to the reader.)

2. To Inscribe a Circle in a Given Square, Using the Method of Intersecting Rays (Figure A-21)

If a plane parallel to the base of a right circular cone cuts the cone, the intersection is a circle (Figure A-21a). Points on the circle may be located by finding the intersection of rays C-1, C-2, etc., drawn through point C (Figure A-21b), with the corresponding rays D-1, D-2, etc., drawn through point D. Rectangle AEFB is half of the given square.

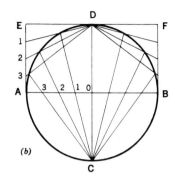
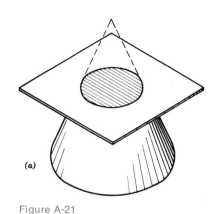

Figure A-21

3. To Draw Lines Tangent to a Circle, and Passing Through a Given Point, A (Figure A-22)

Locate point M, the midpoint of line OA. With M as center and radius MO, draw an arc cutting the given circle at the points of tangency, T and T′. The tangent lines are AT and AT′.

4. To Draw a Line Tangent to the Arc of a Circle, and Passing Through a Given Point, P (Figure A-23)

First draw through point P secant line PAB. Extend this line to point C such that PA = PC. Now find the mean proportional between PA and PB. This is shown as PD. With P as center and radius PD, draw an arc cutting the given arc at point T. Line PT is the required tangent. (The construction shown in Figure A-23 is based on the fact that $PT^2 = PA \times PB$.)

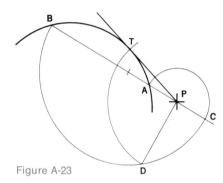

Figure A-22

Figure A-23

5. To Draw Tangents to Two Given Circles (Figure A-24)

On line O-O′, which joins the centers of the circles, lay off from O′, distance $O'B = (R - r)$, and $O'C = (R + r)$. With O′ as center and radii O′B and O′C, draw arcs·

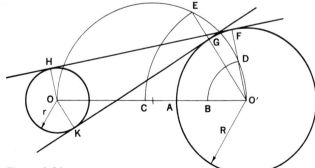

Figure A-24

that intersect the semicircle on O-O' at points D and E, respectively. Draw $O'D$ to intersect circle O' at F and, similarly, draw $O'E$ to intersect circle O' at G. Through center O draw OH parallel to $O'F$ and OK parallel to $O'G$. Lines FH and GK are tangent to both circles.

6. To Draw an Arc of Given Radius, r, Tangent to Two Given Lines m and n (Figure A-25)

The center O of the required arc is at the intersection of lines m' and

n', which are parallel to lines m and n, respectively, at distance r. The construction is clearly shown in the figure.

7. To Draw Arcs of Radius r, Tangent to a Given Line, m, and a Given Circle, O (Figure A-26)

With O as center and radius $(R + r)$, draw an arc cutting line m' (line m' is parallel to m at distance r) at points P and Q, which are the centers of the required arcs.

8. To Draw Arcs of Radius r, Tangent to Two Given Circles (Figure A-27)

With center O and radius $(R + r)$, draw an arc and, with center O' and radius $(R' + r)$, draw an arc.

The intersections of the two arcs are P and Q, which are the centers of the required arcs.

9. To Rectify a Circular Arc (Figure A-28)

Arc Om in Figure A-28a is first divided into a number of short equal segments, in this case 6. These distances, such as 0-1, 1-2, etc., are laid off on tangent OT, as 0-1', 1'-2', etc. Length 0-6' is (very nearly) the length of the arc from 0 to 6.

An alternative solution is shown in Figure A-28b. With K as center and KB as radius, an arc is drawn to intersect the tangent, t, at point C. AC is a close approximation to the length of the given arc m.

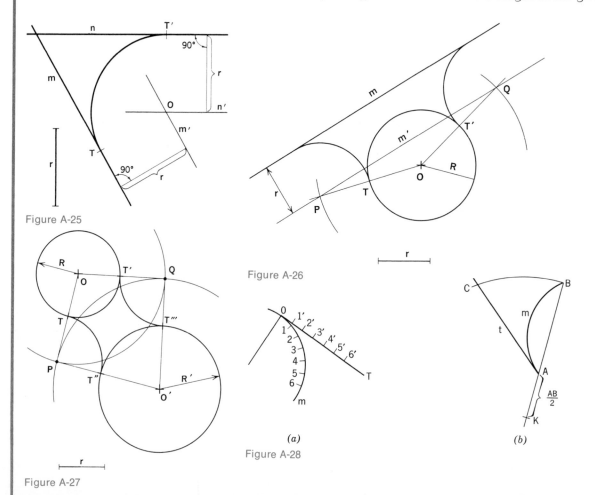

Figure A-25

Figure A-26

Figure A-27

(a)

(b)

Figure A-28

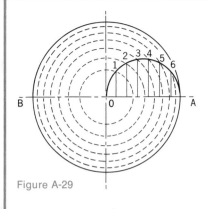

Figure A-29

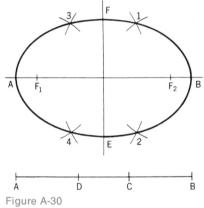

Figure A-30

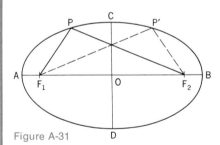

Figure A-31

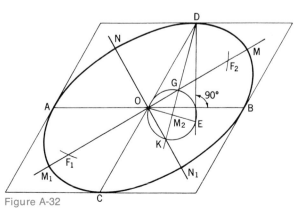

Figure A-32

10. To Divide a Circle into Seven Equal Parts by Concentric Circles (Figure A-29)

First, draw a semicircle on OA as diameter. Then divide OA into seven equal parts and construct verticals to intersect the semicircle in points 1, 2, etc. Finally, draw the required concentric circles with radii 0-1, 0-2, 0-3, etc.

D. Conic Sections

The Ellipse

The *ellipse* may be defined as the locus of all coplanar* points, the sum of whose distances from two fixed points (foci) is a constant.

1. To Construct an Ellipse when the Foci F_1 and F_2 and the Constant Distant AB are Given (Figure A-30)

With F_1 as center and radius AC (any portion of AB), an arc is drawn. Now with F_2 as center and radius CB, an arc is drawn intersecting the first arc in points 1 and 2, which are two points on the ellipse.

This construction is repeated for the location of additional points. For example, with F_1 as center and radius AD, an arc is drawn, and with F_2 as center and radius DB, an intersecting arc is drawn, thus locating two additional points 3 and 4. The smooth curve passing

---

* In the same plane.

---

through these points and others (not shown) is the ellipse. The major and minor axes are AB and EF, respectively.

2. To draw an Ellipse by the Pin-and-String Method when the Major Axis, AB, and the Minor Axis, CD, are Given (Figure A-31)

With center C and radius equal to OA draw an arc intersecting the major axis at points F_1 and F_2, which are the foci. Now fix the ends of a string at points F_1 and F_2 such that the length of the string is equal to AB. For any point on the ellipse, such as point P or P', the sum of distances PF_1 and PF_2 (or of $P'F_1$ and $P'F_2$) remains equal to the constant length of the string. Therefore, the ellipse is easily drawn by maintaining taut segments of the string as a pencil (or other marking device) is used to draw the curve.

3. To Draw an Ellipse when Two Conjugate Axes are Given (Figure A-32)

Let us assume lines AB and CD as the given conjugate axes (each axis is parallel to the tangents to the ellipse at the end points of the other axis). Lines drawn through points C and D, parallel to axis AB, are tangent to the ellipse at these points; and, similarly, lines drawn through points A and B, parallel to axis CD, are tangent to the ellipse at points A and B. The parallelogram formed by the four tangents circumscribes the required ellipse. Once the major and minor axes are located we can establish the positions of the foci and then describe the ellipse. The following construc-

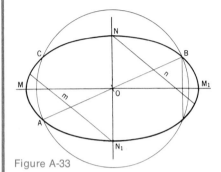

Figure A-33

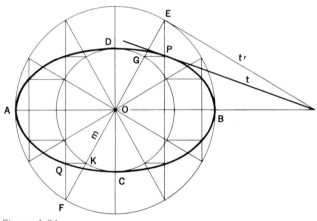

Figure A-34

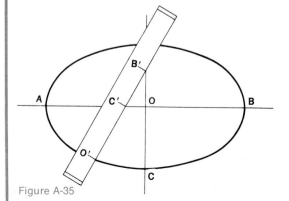

Figure A-35

tion is used to locate the major and minor axes of the ellipse. Through D draw DE perpendicular to AB and equal to OB. Draw OE and describe a circle with radius M_2O (or M_2E). Now draw line DM_2 to intersect the circle at points G and K. The minor axis, NN_1, of the ellipse contains line KO, and the major axis, MM_1, contains line GO. Length ON (or ON_1), the semiminor axis, is equal to length DG, and OM (or OM_1), the semimajor axis, is equal to length DK. The foci are easily located. With N as center and radius OM, describe an arc to intersect the major axis at the foci F and F_1. The ellipse may now be constructed.

4. To Determine the Major and Minor Axes, and the Foci of a Given Ellipse (Figure A-33)

First draw two parallels such as m and n. Now draw AB, which bisects lines m and n. Locate O, the midpoint of AB. With O as center and radius OA, draw a circle to intersect the ellipse at points C and B. Through O draw a line NN_1 parallel to line CA and draw line MM_1 perpendicular to CA. Lines NN_1 and MM_1 are the minor and major axes, respectively.

5. To Draw an Ellipse by the Concentric Circle Method, Given the Lengths of the Major and Minor Axes (Figure A-34)

Lines AB and CD are the major and minor axes, respectively. Through point O, the center of the ellipse, draw radial lines such as m to intersect the concentric circles having radii OB and OC in points E, F, G, and K. Through points E and F draw vertical lines to intersect the horizontals drawn through G and K, in points P and Q, which are two points on the ellipse. Repeat this construction for additional points and then draw a smooth curve through these points to form the ellipse.

The tangent, t, at point P passes through point R, which is the intersection of the tangent t' at point E of the major circle and the major axis extended.

6. To Draw an Ellipse by the Trammel Method when the Major and Minor Axes are Given (Figure A-35)

First a strip is marked with distance $O'B'$ equal to OB and $O'C'$ equal to OC. Now the strip is moved so that point B' travels along the minor axis while C' moves along the major axis. For any such position, O' will locate a point on the ellipse.

7. To Draw an Ellipse by the Use of Circular Arcs when the Axes are Given

This will result in a close approximation to a true ellipse (Figure A-36). Join points A and C. Lay off distance CD equal to CE (where $CE = (OA - OC)$). Now draw the perpendicular bisector of AD and locate points G and K. With G as center and radius GA, describe arc TAT_1. With K as center and radius KT, describe arc TCT_2. Center G' and radius $G'B$ are used to draw arc T_2BT_3; and center K' and radius $K'T_3$ are used to draw arc T_3T_1.

8. To Inscribe, an Ellipse in a Given Rectangle (Figure A-37)

Divide OA into a number of equal parts (four are shown) and then divide AE into the same number of equal parts. Now draw rays D-1, D-2, etc., to intersect the corresponding rays C-1, C-2, etc., in points that lie on the ellipse. The construction shown may be repeated for the other quarters of the rectangle in order to obtain additional points on the ellipse. The pictorial shows a right circular cone intersected by an inclined plane that cuts all the elements of the cone. The intersection is an ellipse.

The Parabola

A parabola is a plane curve any point of which is the same distance from a point called the focus as it is from a straight line known as the directrix.

1. To Locate Points on a Parabola when the Focus, F, and the Directrix, d, are Given (Figure A-38)

Points such as 1 and 2 are determined by locating the intersection of line s (any line parallel to the directrix) and an arc having center F and a radius equal to the distance between the parallel lines d and s. Now it is quite apparent that points 1 and 2 are the same distance from F, the focus, as they are from d, the directrix. The tangent, t, at point P bisects the angle KPF.

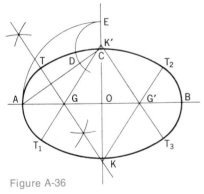

Figure A-36

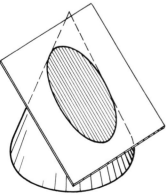

Figure A-37

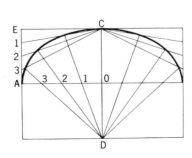

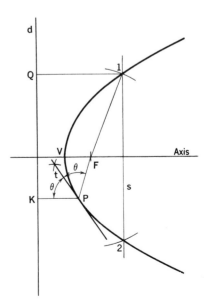

Figure A-38

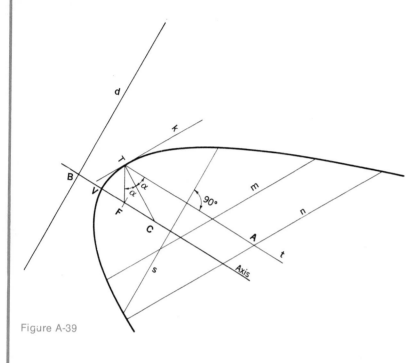

Figure A-39

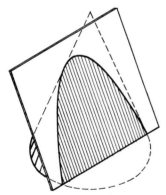

Figure A-40

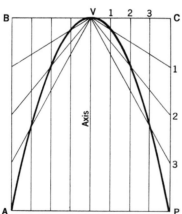

2. To Determine the Axis, Focus, and Directrix of a Given Parabola (Figure A-39)

The axis is located in the following manner. Draw two parallel chords such as m and n. The line t joining the midpoints of these chords is parallel to the axis. Now introduce a line such as s perpendicular to line t. The required axis is the perpendicular bisector of line s. The focus, F, is located by making angle FTC equal to angle CTA. Point C is the intersection of the axis, with the perpendicular to tangent line k at point T. The directrix, d, is perpendicular to the axis and at a distance from V equal to VF, that is, $VB = VF$.

3. To Construct a Parabola, Given the Axis, Vertex V, and a Point P Through which the Parabola Passes (Figure A-40)

First draw rectangle $PABC$. Now divide CP and CV into the same number of equal parts. Introduce lines parallel to the axis and passing through points 1, 2, and 3 on side VC. Draw rays V-1, V-2, and V-3. Finally, locate the points in which the parallels intersect the corresponding rays, (the parallel through point 1 intersects ray V-1, etc.). The curve through the points thus located is the parabola. The pictorial shows a right circular cone intersected by a plane parallel to an element of the cone. The intersection is a parabola.

The Hyperbola

The *hyperbola* may be defined as the locus of all coplanar points the differences of whose distances from two fixed points (foci) is a constant.

1. To Construct a Hyperbola when the Foci, F_1 and F_2, and the Constant Distance, AB, are Given (Figure A-41)

With F_1 as center and a radius greater than F_1B, an arc is drawn. Now, with F_2 as center and a radius that is equal to the difference between the first radius and length AB, an arc is drawn to intersect the first arc in points P and Q, which are two points on the hyperbola. It is clearly seen that $F_1P - F_2P = AB$ and that $F_1Q - F_2Q = AB$. Additional points may be found in a similar manner. The smooth curve passing through the points is the hyperbola. Note that the curve has two branches that are symmetrical with respect to the axes. The asymptotes pass through the center O and are tangent to the curve at infinity. They are located by joining point O the center of the hyperbola with points K and K'. These points are found by locating the intersections of the verticals through points A and B with the circle of radius OF_1. The tangent, t, at point P bisects the angle F_2PF_1.

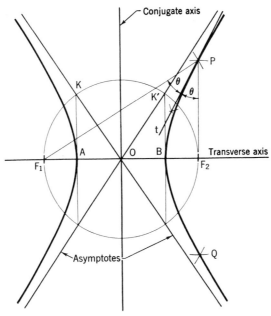

Figure A-41

2. To Construct a Rectangular Hyperbola (Asymptotes are at Right Angles), given the Asymptotes m and n, and one Point P on the Curve (Figure A-42)

Draw lines k and t through point P, respectively parallel to n and m. Select any point Q on line k and then draw line OQ. Locate point R, the intersection of OQ and t. Draw a horizontal line through point R and a vertical line through point Q. The intersection of these lines is point S, a point on the hyperbola. In a similar manner additional points are located.

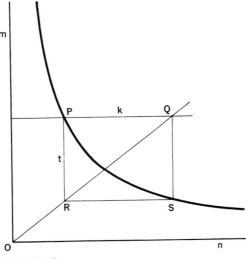

Figure A-42

3. To Construct a Hyperbola, Given the Transverse Axis AB, and a Point P on the Curve (Figure A-43)

First construct the rectangle *PCDE*. Now divide side *EP* into a number of equal parts (four are shown) and the right half of side *CP* into the same number of equal parts. Find the intersection of rays *A*-1, *A*-2, etc., with the corresponding rays *B*-1, *B*-2, etc. Repeat the procedure for the left half of the rectangle. The smooth curve that passes through the points thus located is one branch of the hyperbola. The other branch may be determined in a similar manner. The pictorial

shows a right circular cone intersected by a plane parallel to the axis of the cone. The intersection is a hyperbola (one branch shown).

Pascal and His Theorem

In 1640, at the age of 16, Pascal discovered the relationship that "the opposite sides of a hexagon, which is inscribed in a conic, intersect in points that lie on one line." For example, in Figure A-44, the sides of the inscribed hexagon are 1-2, 2-3, 3-4, 4-5, 5-6, and 6-1. Opposite sides 1-2 and 4-5 intersect at point *L*; opposite sides 2-3 and 5-6 intersect at point *M*; and opposite sides 3-4 and 6-1 intersect at point *N*. The line that passes through points *L*, *M*, and *N* is known as Pascal's line.

1. To Locate a Sixth Point on a Conic when Five Points are known (Figure A-45)

Let us suppose that points 1, 2, 3, 4, and 5 are known and that it is required to locate a sixth point, *K*. Basing our construction on Pascal's theorem, we can first establish point *L*, which is the intersection of the opposite sides 1-2 and 4-5 of the inscribed hexagon 1, 2, 3, 4, 5, *K*. Now we know that another pair of opposite sides is 2-3 and 5-*K*. Therefore, we may draw any line through point 5 to intersect side 2-3 in point *M*. We know that point *K* is somewhere on line 5-*M*. We also know that Pascal's line passes through points *L* and *M*. The third pair of opposite sides is 3-4 and *K*-1. If we join points 3 and 4, then line 3-4 must intersect Pascal's line in a point, *N*, through which side *K*-1 must pass. It is now seen that lines

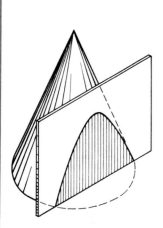

Figure A-43

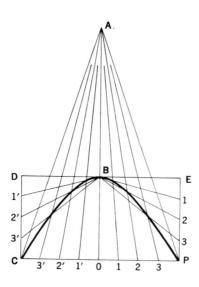

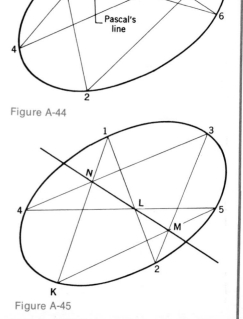

Figure A-44

Figure A-45

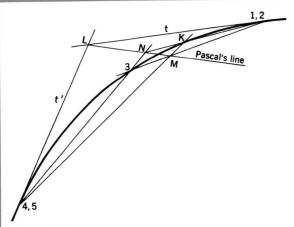

Figure A-46

1-N and 5-M must intersect at point K, which is a sixth point on the conic. Additional points may be located by drawing other lines through point 5 to intersect side 2-3 in a new M-point, and then repeating the described construction.

2. To Locate Points on a Conic that Passes Through a Given Point and is Tangent to Given Lines at Two Points (*Figure A-46*)

It is assumed that the conic passes through the given point 3 and tangent to lines t and t′ at points 2 and 4, respectively. We recall (Figure A-45) that when five points on a conic are known it is possible to locate a sixth point. How can we reduce the problem shown in Figure A-46 to the previous one? How shall we establish five known points on the conic, when apparently only three are given?

If point 5, for example (Figure A-45), were moved along the curve until it coincided with point 4, chord 4-5 would become a tangent to the curve at point 4. Therefore we can show point 5, coincident with point 4 in Figure A-46. Similarly, points 2 and 1 are coincident.

Now the intersection of opposite sides 1-2(t) and 4-5(t′) of the inscribed hexagon 1, 2, 3, 4, 5, K (K is a sixth point on the conic) is point L. The intersection of opposite sides, 2-3 and 5-K, is point M. If we take M as any point on side 2-3, we know that point K is somewhere on line 5-M. The line joining points L and M is a Pascal line. We know that the opposite sides, 3-4 and K-1, must meet on the Pascal line; therefore, the intersection of side 3-4 with the Pascal line is point N, through which side 1-K must pass. Therefore the intersection of lines 1-N and 5-M is point K, another point on the conic. Additional points may be located in a similar manner.

3. To Construct a Tangent to a Given Conic at a Point of the Conic (*Figure A-47*)

Let us construct the tangent to the conic at point P. Inscribe a hexagon 1, 2, 3, 4, 5, 6 such that side 5-6 will be the tangent. Two points L and M of the Pascal line are located by finding the intersection of sides 1-2 and 4-5, and of sides 3-4 and 6-1, respectively. Sides 2-3 and 5-6 must meet on the Pascal line, at point N, which is located by finding the intersection of side 2-3 with the Pascal line. The required tangent is line NP.

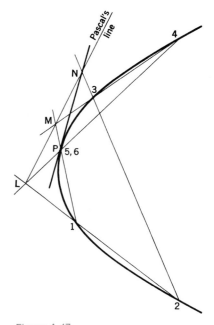

Figure A-47

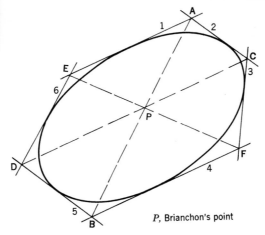

Figure A-48

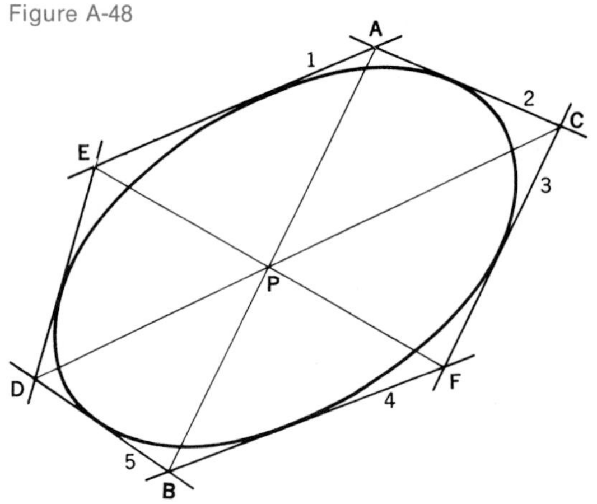

Brianchon's Theorem

This theorem, which is most useful in locating tangents to a conic, states that "the three lines joining the three pairs of opposite vertices of a hexagon circumscribed about a conic meet in a point."

In Figure A-48, line *AB* joins the pair of opposite vertices determined by tangents 1 and 2, and 4 and 5. Similarly, line *CD* joins the pair of opposite vertices determined by tangents 2 and 3 and 5 and 6. Lines *AB* and *CD* intersect in Brianchon's point, *P*. Line *EF*, which joins the remaining pair of opposite vertices, also passes through point *P*.

1. To Determine a Sixth Tangent to a Conic when Five Tangents (*Sides of the Circumscribed Hexagon*) are Given (*Figure A-49*)

Suppose that tangents 1, 2, 3, 4, and 5 are given. It is required to locate a sixth side of a circumscribed hexagon. Line *AB*, which joins one pair of opposite vertices, is easily determined. Now through point *C* (the intersection of tangents 2 and 3) a line is drawn intersecting *AB* at point *P* (Brianchon's point). Line *CP* intersects tangent 5 at point *D*. Line *FP* intersects tangent 1 at point *E*. Line *DE* is the required tangent or sixth side of the circumscribed hexagon.

2. To Determine the Point of Contact of a Tangent to a Conic (*Figure A-50*)

Suppose we wish to locate the point of tangency of tangent 1. If tangent 2 approaches tangent 1 as a limiting position, the point of intersection of tangents 1 and 2 approaches the contact point of tan-

Figure A-49

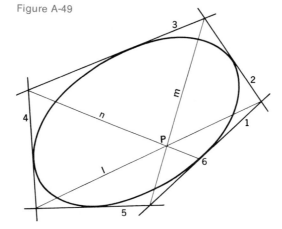

Figure A-50

gent 1 with the conic. Let us denote this point as point 6. Now we may make use of Brianchon's point to locate point 6. The intersection of lines l and m determines Brianchon's point, P. Line n passes through point P and the intersection of tangents 3 and 4. Point 6, the required point of contact of tangent 1 with the conic, is at the intersection of line n with the tangent line 1.

3. To Determine Additional Tangents to a Conic when Four Tangents and the Point of Contact on One of them are known (*Figure A-51*)

The four given tangents are 1, 2, 3,

and 4, and the point of contact on tangent 2 is point T. Line m is determined by joining the pair of opposite vertices that are given. Tangent 2 is actually two tangents that intersect at point T. Now through point T a line n is drawn to intersect line m at point P (a Brianchon's point), and tangent 4 at point A. Line l is drawn through the common point of tangents 2 and 3 and point P to intersect tangent 1 at point B. Line AB is an additional tangent to the conic. In a similar manner more tangents may be determined.

4. To Determine Additional Tangents to a Conic when Three Tangents and the Points of Contact on Two of them are known (*Figure A-52*)

The three known tangents are 1, 2, and 3, and the points of contact, points S and T. Line m joins points S and T. Now through the intersection of tangents 1 and 2 draw a line n to intersect line m at point P (a Brianchon's point) and to intersect tangent 3 at point A. Finally, draw line l through the common point of tangents 2 and 3, and point P to intersect tangent 1 at point B. Line AB is an additional tangent to the conic.

Additional Useful Geometric Constructions

1. To Draw a Line Through a Given Point, P, and the Inaccessible Copoint of Given Lines m and n (*Figure A-53*)

Draw a line such as s to intersect m and n at points Q and R, respectively. Form triangle PQR. Now select a point such as Q' on line m and draw through Q' lines respectively parallel to QP and QR. Through R' draw a line parallel to RP and form the triangle $P'Q'R'$. The line, PP', is the solution.

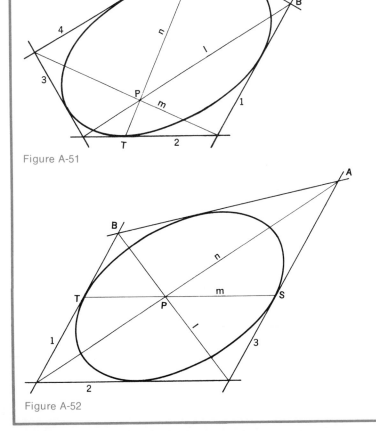

Figure A-51

Figure A-52

Figure A-53

2. To Draw a Line Perpendicular to a Given Line AB and through the Inaccessible Copoint of Given Lines m and n that pass through Points A and B, Respectively (Figure A-54)

Through points B and A draw lines respectively perpendicular to m and n and intersecting at point P. Now construct the required line through point P perpendicular to line AB.

3. To Divide a Given Quadrilateral ABCD into Two Equal Areas by a Line that passes through one of the Corners, A (Figure A-55)

Draw line BD and locate its midpoint, M. Draw a line through point M parallel to diagonal AC to intersect side BC at point E. Line AE divides the quadrilateral into equal areas AEB and $AECD$.

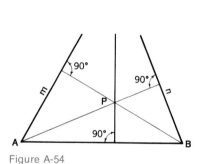

Figure A-54

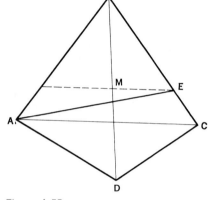

Figure A-55

Appendix B

English Unit and
Metric Unit Tables

ENGLISH UNIT TABLES

Table 1 Unified and American Screw Threads (Inch Units) Coarse, Fine, Extra-Fine, 8-, 12-, and 16-Thread Series

Sizes	Basic Major Diameter	Coarse UNC; NC Thds. per in.	Coarse Tap Drill	Fine UNF; NF Thds. per in.	Fine Tap Drill	Extra-Fine UNEF; NEF Thds. per in.	Extra-Fine Tap Drill	8-Thread 8 UN; 8 N Thds. per in.	8-Thread Tap Drill	12-Thread 12 UN; 12 N Thds. per in.	12-Thread Tap Drill	16-Thread 16 UN; 16 N Thds. per in.	16-Thread Tap Drill
0	0.0600			80	3/64								
1	0.0730	64	53	72	53								
2	0.0860	56	50	64	50								
3	0.0990	48	47	56	45								
4	0.1120	40	43	48	42								
5	0.1250	40	38	44	37								
6	0.1380	32	36	40	33								
8	0.1640	32	29	36	29								
10	0.1900	24	25	32	21								
12	0.2160	24	16	28	14	32	13						
1/4	0.2500	20	7	28	3	32	7/32						
5/16	0.3125	18	F	24	I	32	9/32						
3/8	0.3750	16	5/16	24	Q	32	11/32					UNC	5/16
7/16	0.4375	14	U	20	25/64	28	13/32					16	3/8
1/2	0.5000	13	27/64	20	29/64	28	15/32					16	7/16
9/16	0.5625	12	31/64	18	33/64	24	33/64			UNC	31/64	16	1/2
5/8	0.6250	11	17/32	18	37/64	24	37/64			12	35/64	16	9/16
11/16	0.6875					24	41/64			12	39/64	16	5/8
3/4	0.7500	10	21/32	16	11/16	20	45/64			12	43/64	UNF	11/16
13/16	0.8125					20	49/64			12	47/64	16	3/4
7/8	0.8750	9	49/64	14	13/16	20	53/64			12	51/64	16	13/16
15/16	0.9375					20	57/64			12	55/64	16	7/8
1	1.0000	8	7/8	12	59/64	20	61/64	UNC	7/8	UNF	59/64	16	15/16
1 1/16	1.0625					18	1	8	15/16	12	63/64	16	1
1 1/8	1.1250	7	63/64	12	1 3/64	18	1 3/64	8	1	UNF	1 3/64	16	1 1/16
1 3/16	1.1875					18	1 5/64	8	1 1/16	12	1 7/64	16	1 1/8
1 1/4	1.2500	7	1 7/64	12	1 11/64	18	1 3/16	8	1 1/8	UNF	1 11/64	16	1 3/16
1 5/16	1.3125					18	1 17/64	8	1 3/16	12	1 15/64	16	1 1/4

Series with graded pitches — Coarse UNC; NC, Fine UNF; NF, Extra-Fine UNEF; NEF

Series with constant pitches — 8-Thread 8 UN; 8 N, 12-Thread 12 UN; 12 N, 16-Thread 16 UN; 16 N

American Standard Unified and American (National) Screw Thread series — continued (Compiled from ANSI B1.1‑1960)

Nominal Size	Basic Major Dia.	Coarse (UNC) Thds/In.	Coarse Tap Drill	Fine (UNF) Thds/In.	Fine Tap Drill	Extra‑Fine (UNEF) Thds/In.	Extra‑Fine Tap Drill	8‑Thread Series Thds/In.	8‑Thread Tap Drill	12‑Thread Series Thds/In.	12‑Thread Tap Drill	16‑Thread Series Thds/In.	16‑Thread Tap Drill
1 3/8	1.3750	6	1 7/32	12	1 19/64	18	1 5/16	8	1 1/4	UNF	1 19/64	16	1 5/16
1 7/16	1.4375					18	1 3/8	8	1 5/16	12	1 23/64	16	1 3/8
1 1/2	1.5000	6	1 11/32	12	1 27/64	18	1 7/16	8	1 3/8	UNF	1 27/64	16	1 7/16
1 9/16	1.5625					18	1 1/2	8	1 7/16	12	1 31/64	16	1 1/2
1 5/8	1.6250					18	1 9/16	8	1 1/2	12	1 35/64	16	1 9/16
1 11/16	1.6875					18	1 5/8	8	1 9/16	12	1 39/64	16	1 5/8
1 3/4	1.7500	5	1 9/16					8	1 5/8	12	1 43/64	16	1 11/16
1 13/16	1.8125							8	1 11/16	12	1 47/64	16	1 3/4
1 7/8	1.8750							8	1 3/4	12	1 51/64	16	1 13/16
1 15/16	1.9375							8	1 13/16	12	1 55/64	16	1 7/8
2	2.0000	4 1/2	1 25/32					8	1 7/8	12	1 59/64	16	1 15/16
2 1/8	2.1250							8	2	12	2 3/64	16	2 1/16
2 1/4	2.2500	4 1/2	2 1/32					8	2 1/8	12	2 11/64	16	2 3/16
2 3/8	2.3750							8	2 1/4	12	2 19/64	16	2 5/16
2 1/2	2.5000	4	2 1/4					8	2 3/8	12	2 27/64	16	2 7/16
2 5/8	2.6250							8	2 1/2	12	2 35/64	16	2 9/16
2 3/4	2.7500	4	2 1/2					8	2 5/8	12	2 43/64	16	2 11/16
2 7/8	2.8750							8	2 3/4	12	2 51/64	16	2 13/16
3	3.0000	4	2 3/4					8	2 7/8	12	2 59/64	16	2 15/16
3 1/8	3.1250							8	3	12	3 3/64	16	3 1/16
3 1/4	3.2500	4	3					8	3 1/8	12	3 11/64	16	3 3/16
3 3/8	3.3750							8	3 1/4	12	3 19/64	16	3 5/16
3 1/2	3.5000	4	3 1/4					8	3 3/8	12	3 27/64	16	3 7/16
3 5/8	3.6250							8	3 1/2	12	3 35/64	16	3 9/16
3 3/4	3.7500	4	3 1/2					8	3 5/8	12	3 43/64	16	3 11/16
3 7/8	3.8750							8	3 3/4	12	3 51/64	16	3 13/16
4	4.0000	4	3 3/4					8	3 7/8	12	3 59/64	16	3 15/16
4 1/8	4.1250							8	4	12	4 3/64	16	4 1/16
4 1/4	4.2500							8	4 1/8	12	4 11/64	16	4 3/16
4 3/8	4.3750							8	4 1/4	12	4 19/64	16	4 5/16
4 1/2	4.5000							8	4 3/8	12	4 27/64	16	4 7/16
4 5/8	4.6250							8	4 1/2	12	4 35/64	16	4 9/16
4 3/4	4.7500							8	4 5/8	12	4 43/64	16	4 11/16
4 7/8	4.8750							8	4 3/4	12	4 51/64	16	4 13/16
5	5.0000							8	4 7/8	12	4 59/64	16	4 15/16
5 1/8	5.1250							8	5	12	5 3/64	16	5 1/16
5 1/4	5.2500							8	5 1/8	12	5 11/64	16	5 3/16

(Compiled from ANSI B1.1‑1960)

Table 2 Cap Screws

Hexagon head cap screws

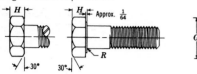

Nominal Size	Width across Flats, F (Max)	Width across Corners, G (Max)	Height, H (Nom)
1/4	7/16	**0.505**	5/32
5/16	1/2	**0.577**	13/64
3/8	9/16	**0.650**	15/64
7/16	5/8	**0.722**	9/32
1/2	3/4	**0.866**	5/16
9/16	13/16	**0.938**	23/64
5/8	15/16	**1.083**	25/64
3/4	1 1/8	**1.299**	15/32
7/8	1 5/16	**1.516**	35/64
1	1 1/2	**1.732**	39/64
1 1/8	1 11/16	**1.949**	11/16
1 1/4	1 7/8	**2.165**	25/32
1 3/8	2 1/16	**2.382**	27/32
1 1/2	2 1/4	**2.598**	15/16

(Courtesy ANSI)

Threads shall be coarse-, fine-, or 8-thread series, class 2A for plain (unplated) cap screws. For plated cap screws, the diameters may be increased by the amount of class 2A allowance. Thickness or quality of plating shall be measured or tested on the side of the head.

All dimensions given in inches.

BOLD TYPE INDICATES PRODUCTS UNIFIED DIMENSIONALLY WITH BRITISH AND CANADIAN STANDARDS.

Minimum thread length shall be twice the diameter plus 1/4 in. for lengths up to and including 6 in.; twice the diameter plus 1/2 in. for lengths over 6 in. The tolerance shall be plus 3/16 in. or 2 1/2 threads, whichever is greater. On products that are too short for minimum thread lengths the distance from the bearing surface of the head to the first complete thread shall not exceed the length of 2 1/2 threads, as measured with a ring thread gage, for sizes up to and including 1 in., and 3 1/2 threads for sizes larger than 1 in.

Dimensions of fillister head cap screws

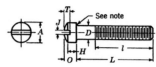

Nominal Size, D	Head Diameter, A (Max)	Height of Head, H (Max)	Total Height of Head, O (Max)	Width of Slot, J (Max)	Depth of Slot, T (Max)
1/4	0.375	0.172	0.216	0.075	0.097
5/16	0.437	0.203	0.253	0.084	0.115
3/8	0.562	0.250	0.314	0.094	0.143
7/16	0.625	0.297	0.368	0.094	0.168
1/2	0.750	0.328	0.412	0.106	0.188
9/16	0.812	0.375	0.466	0.118	0.214
5/8	0.875	0.422	0.521	0.133	0.240
3/4	1.000	0.500	0.612	0.149	0.283
7/8	1.125	0.594	0.720	0.167	0.334
1	1.312	0.656	0.802	0.188	0.372

(Courtesy ANSI)

All dimensions are given in inches.
The radius of the fillet at the base of the head:

For sizes 1/4 to 3/8 in. incl. is 0.016 min and 0.031 max.
7/16 to 9/16 in. incl. is 0.016 min and 0.047 max.
5/8 to 1 in. incl. is 0.031 min and 0.062 max.

Dimensions of round head cap screws

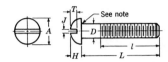

Nominal Size, D	Head Diameter, A (Max)	Height of Head, H (Max)	Width of Slot, J (Max)	Depth of Slot, T (Max)
¼	0.437	0.191	0.075	0.117
⁵⁄₁₆	0.562	0.246	0.084	0.151
³⁄₈	0.625	0.273	0.094	0.168
⁷⁄₁₆	0.750	0.328	0.094	0.202
½	0.812	0.355	0.106	0.219
⁹⁄₁₆	0.937	0.410	0.118	0.253
⅝	1.000	0.438	0.133	0.270
¾	1.250	0.547	0.149	0.337

(Courtesy ANSI)

All dimensions are given in inches.
Radius of the fillet at the base of the head:
   For sizes ¼ to ³⁄₈ in. incl. is 0.016 min and 0.031 max.
           ⁷⁄₁₆ to ⁹⁄₁₆ in. incl. is 0.016 min to 0.047 max.
           ⅝ to 1 in. incl. is 0.031 min and 0.062 max.

Dimensions of flat head cap screws

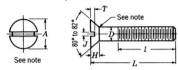

Nominal Size, D	Head Diameter, A (Max)	Height of Head, H (Average)	Width of Slot, J (Max)	Depth of Slot, T (Max)
¼	0.500	0.140	0.075	0.069
⁵⁄₁₆	0.625	0.176	0.084	0.086
³⁄₈	0.750	0.210	0.094	0.103
⁷⁄₁₆	0.8125	0.210	0.094	0.103
½	0.875	0.210	0.106	0.103
⁹⁄₁₆	1.000	0.245	0.118	0.120
⅝	1.125	0.281	0.133	0.137
¾	1.375	0.352	0.149	0.171
⅞	1.625	0.423	0.167	0.206
1	1.875	0.494	0.188	0.240

(Courtesy ANSI)

All dimensions are given in inches.
The maximum head diameters, A, are extended to the theoretical sharp corners.
The radius of the fillet at the base of the head shall not exceed twice the pitch of the screw thread.

Table 3 Machine Screws

Dimensions for hexagon-head machine screws, plain or slotted

Trimmed Head   Upset Head

| Nominal Size | Basic Diameter, D | Standard Trimmed or Upset Head | | Optional Upset Type Head for Special Requirements | | Height of Head, H (Max) | Width of Slot, J (Max) | Depth of Slot, T (Max) |
		Head Diameter, A (Max)	Across Corners, W (Min)	Head Diameter, A (Max)	Across Corners, W (Min)			
2	0.0860	0.125	0.134	. . . . .	. . . . .	0.050	. . . . .	. . . . .
3	0.0990	0.187	0.202	. . . . .	. . . . .	0.055	. . . . .	. . . . .
4	0.1120	0.187	0.202	0.219	0.238	0.060	0.039	0.036
5	0.1250	0.187	0.202	0.250	0.272	0.070	0.043	0.042
6	0.1380	0.250	0.272	. . . . .	. . . . .	0.080	0.048	0.046
8	0.1640	0.250	0.272	0.312	0.340	0.110	0.054	0.066
10	0.1900	0.312	0.340	. . . . .	. . . . .	0.120	0.060	0.072
12	0.2160	0.312	0.340	0.375	0.409	0.155	0.067	0.093
¼	0.2500	0.375	0.409	0.437	0.477	0.190	0.075	0.101
⁵⁄₁₆	0.3125	0.500	0.548	. . . . .	. . . . .	0.230	0.084	0.122
³⁄₈	0.3750	0.562	0.616	. . . . .	. . . . .	0.295	0.094	0.156

(Courtesy ANSI)

All dimensions are given in inches.
Hexagon-head machine screws are usually not slotted. The slot is optional.

Dimensions of round head machine screws

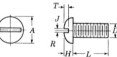

(2 in. and under)   (Over 2 in.)

Nominal Size	Diameter of Screw, D (Max)	Head Diameter, A (Max)	Height of Head, H (Max)	Width of Slot, J (Max)	Depth of Slot, T (Max)
0	0.060	0.113	0.053	0.023	0.039
1	0.073	0.138	0.061	0.026	0.044
2	0.086	0.162	0.069	0.031	0.048
3	0.099	0.187	0.078	0.035	0.053
4	0.112	0.211	0.086	0.039	0.058
5	0.125	0.236	0.095	0.043	0.063
6	0.138	0.260	0.103	0.048	0.068
8	0.164	0.309	0.120	0.054	0.077
10	0.190	0.359	0.137	0.060	0.087
12	0.216	0.408	0.153	0.067	0.096
¼	0.250	0.472	0.175	0.075	0.109
⁵⁄₁₆	0.3125	0.590	0.216	0.084	0.132
³⁄₈	0.375	0.708	0.256	0.094	0.155
⁷⁄₁₆	0.4375	0.750	0.328	0.094	0.196
½	0.500	0.813	0.355	0.106	0.211
⁹⁄₁₆	0.5625	0.938	0.410	0.118	0.242
⁵⁄₈	0.625	1.000	0.438	0.133	0.258
¾	0.750	1.250	0.547	0.149	0.320

(Courtesy ANSI)

All dimensions are given in inches.
Head dimensions for sizes ⁷⁄₁₆ in. and larger are in agreement with round head cap screw dimensions.
The diameter of the unthreaded portion of machine screws shall be not less than the minimum pitch diameter nor more than the maximum major diameter of the thread.
The radius of the fillet at the base of the head shall not exceed one-half the pitch of the screw thread.

Table 3   (*continued*)

Dimensions of oval head and flat head machine screws

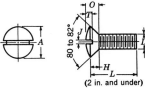 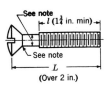

Nominal Size	Max Diameter of Screw, D	Head Diameter, A (Max Sharp)	Height of Head, H (Max)	Total Height of Head, O (Max)	Width of Slot, J (Max)	Depth of Slot, T	
						Oval (Max)	Flat (Max)
0	0.060	0.119	0.035	0.056	0.023	0.030	0.015
1	0.073	0.146	0.043	0.068	0.026	0.038	0.019
2	0.086	0.172	0.051	0.080	0.031	0.045	0.023
3	0.099	0.199	0.059	0.092	0.035	0.052	0.027
4	0.112	0.225	0.067	0.104	0.039	0.059	0.030
5	0.125	0.252	0.075	0.116	0.043	0.067	0.034
6	0.138	0.279	0.083	0.128	0.048	0.074	0.038
8	0.164	0.332	0.100	0.152	0.054	0.088	0.045
10	0.190	0.385	0.116	0.176	0.060	0.103	0.053
12	0.216	0.438	0.132	0.200	0.067	0.117	0.060
¼	0.250	0.507	0.153	0.232	0.075	0.136	0.070
5⁄16	0.3125	0.635	0.191	0.290	0.084	0.171	0.088
3⁄8	0.375	0.762	0.230	0.347	0.094	0.206	0.106
7⁄16	0.4375	0.812	0.223	0.345	0.094	0.210	0.103
½	0.500	0.875	0.223	0.354	0.106	0.216	0.103
9⁄16	0.5625	1.000	0.260	0.410	0.118	0.250	0.120
5⁄8	0.625	1.125	0.298	0.467	0.133	0.285	0.137
¾	0.750	1.375	0.372	0.578	0.149	0.353	0.171

(Courtesy ANSI)

Table 3 (*continued*)

Dimensions of fillister head machine screws

  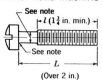

(2 in. and under)   (Over 2 in.)

Nominal Size	Diameter of Screw, D (Max)	Head Diameter, A (Max)	Height of Head, H (Max)	Total Height of Head, O (Max)	Width of Slot, J (Max)	Depth of Slot, T (Max)
0	0.060	0.096	0.045	0.059	0.023	0.025
1	0.073	0.118	0.053	0.071	0.026	0.031
2	0.086	0.140	0.062	0.083	0.031	0.037
3	0.099	0.161	0.070	0.095	0.035	0.043
4	0.112	0.183	0.079	0.107	0.039	0.048
5	0.125	0.205	0.088	0.120	0.043	0.054
6	0.138	0.226	0.096	0.132	0.048	0.060
8	0.164	0.270	0.113	0.156	0.054	0.071
10	0.190	0.313	0.130	0.180	0.060	0.083
12	0.216	0.357	0.148	0.205	0.067	0.094
¼	0.250	0.414	0.170	0.237	0.075	0.109
⁵⁄₁₆	0.3125	0.518	0.211	0.295	0.084	0.137
³⁄₈	0.375	0.622	0.253	0.355	0.094	0.164
⁷⁄₁₆	0.4375	0.625	0.265	0.368	0.094	0.170
½	0.500	0.750	0.297	0.412	0.106	0.190
⁹⁄₁₆	0.5625	0.812	0.336	0.466	0.118	0.214
⅝	0.625	0.875	0.375	0.521	0.133	0.240
¾	0.750	1.000	0.441	0.612	0.149	0.281

(Courtesy ANSI)

All dimensions are given in inches.
The diameter of the unthreaded portion of machine screws shall not be less than the minimum pitch diameter nor more than the maximum major diameter of the thread. The radius of the fillet at the base of the head shall not exceed one-half the pitch of the screw thread.

Table 4 Set Screws

Square head set screws

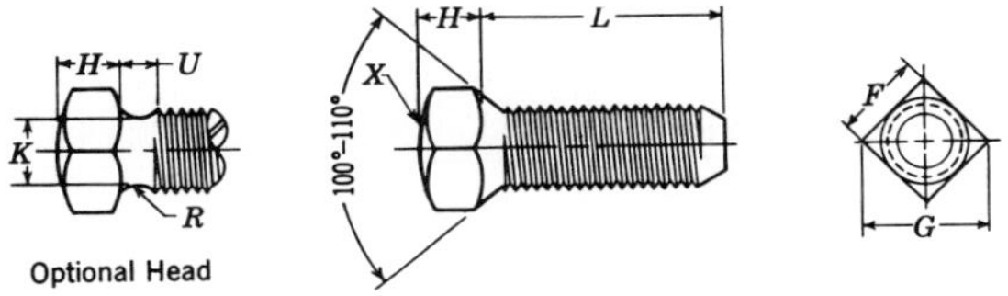

H
U
K
R
Optional Head
100°–110°
X
H
L
F
G

Nominal Size		Width across Flats, F (Max)	Width across Corners, G (Min)	Height of Head, H (Nom)	Diameter of Neck Relief, K (Max)	Radius of Head, X (Nom)	Rad of Neck Relief, R (Max)	Width of Neck Relief, U (Max)
10	0.190	0.1875	0.247	9/64	0.145	15/32	0.027	0.083
12	0.216	0.216	0.292	5/32	0.162	35/64	0.029	0.091
1/4	0.250	0.250	0.331	3/16	0.185	5/8	0.032	0.100
5/16	0.3125	0.3125	0.415	15/64	0.240	25/32	0.036	0.111
3/8	0.3750	0.375	0.497	9/32	0.294	15/16	0.041	0.125
7/16	0.4375	0.4375	0.581	21/64	0.345	1 3/32	0.046	0.143
1/2	0.500	0.500	0.665	3/8	0.400	1 1/4	0.050	0.154
9/16	0.5625	0.5625	0.748	27/64	0.454	1 13/32	0.054	0.167
5/8	0.6250	0.625	0.833	15/32	0.507	1 9/16	0.059	0.182
3/4	0.750	0.750	1.001	9/16	0.620	1 7/8	0.065	0.200
7/8	0.875	0.875	1.170	21/32	0.731	2 3/16	0.072	0.222
1	1.000	1.000	1.337	3/4	0.838	2 1/2	0.081	0.250
1 1/8	1.125	1.125	1.505	27/32	0.939	2 13/16	0.092	0.283
1 1/4	1.250	1.250	1.674	15/16	1.064	3 1/8	0.092	0.283
1 3/8	1.375	1.375	1.843	1 1/32	1.159	3 7/16	0.109	0.333
1 1/2	1.500	1.500	2.010	1 1/8	1.284	3 3/4	0.109	0.333

(Courtesy ANSI)

All dimensions given in inches.
Threads shall be coarse-, fine-, or 8-thread series, class 2A. Square head set screws 1/4 in. size and larger are normally stocked in coarse thread series only.
Square head set screws shall be made from alloy or carbon steel suitably hardened. Screws made from nonferrous material or corrosion-resisting steel shall be made from a material mutually agreed upon by manufacturer and user.

Table 4 *(continued)*

Square head set screw points

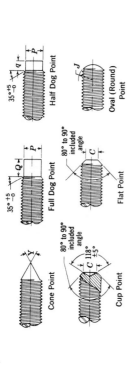

Cone Point Cup Point Full Dog Point Half Dog Point Flat Point Oval (Round) Point

Cup and Flat Point Diameter, C | *Oval (Round)* | *Full Dog, Half Dog, Pivot Point*

Nominal Size	Cup and Flat Point Diameter, C	Oval (Round) Point Radius, J	Full Dog, Half Dog, Pivot Point		
(Nom)	(Max)	(Nom)	Diameter, P (Max)	Full Dog Pivot, Q	Half Dog Pivot, q
10	0.102	0.141	0.127	0.090	0.045
12	0.115	0.156	0.144	0.110	0.055
1/4	0.132	0.188	0.156	0.125	0.063
5/16	0.172	0.234	0.203	0.156	0.078
3/8	0.212	0.281	0.250	0.188	0.094
7/16	0.252	0.328	0.297	0.219	0.109
1/2	0.291	0.375	0.344	0.250	0.125
9/16	0.332	0.422	0.391	0.281	0.140
5/8	0.371	0.469	0.469	0.313	0.156
3/4	0.450	0.563	0.563	0.375	0.188
7/8	0.530	0.656	0.656	0.438	0.219
1	0.609	0.750	0.750	0.500	0.250
1 1/8	0.689	0.844	0.844	0.562	0.281
1 1/4	0.767	0.938	0.938	0.625	0.312
1 3/8	0.848	1.031	1.031	0.688	0.344
1 1/2	0.926	1.125	1.125	0.750	0.375

(Courtesy ANSI)

All dimensions given in inches.
Pivot points are similar to full dog point except that the point is rounded by a radius equal to J.
Where usable length of thread is less than the nominal diameter, half-dog point shall be used.
When length equals nominal diameter or less, Y = 118 deg ± 2 deg; when length exceeds nominal diameter, Y = 90 deg ± 2 deg.

Table 4 *(continued)*

Dimensions of fluted and hexagonal socket headless set screws

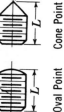
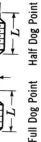

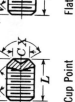

Half Dog Point Full Dog Point Cone Point Oval Point Flat Point Cup Point

Section through Hexagonal Socket Section through Fluted Socket

D	C	R	Y		P	Q	q	Number of Flutes	J	M	N	J (Hex)
	Cup and Flat Point Diameter (Mean)	Oval Point Radius	Cone Point Angle		Full Dog Point and Half Dog Point				Socket Diameter, Minor (Max)	Socket Diameter, Major (Max)	Socket Land Width (Max)	Socket Width across Flats (Max)
Nominal Diameter			118° ± 2° for these Lengths and Under	90° ± 2° for these Lengths and Over	Diameter (Max)	Full	Half					
5	1/16	3/32	1/8	3/16	0.083	0.06	0.03	4	0.053	0.071	0.022	0.0635
6	.069	7/64	1/8	3/16	0.092	0.07	0.03	4	0.056	0.079	0.023	0.0635
8	5/64	1/8	3/16	1/4	0.109	0.08	0.04	6	0.082	0.098	0.022	0.0791
10	3/32	9/64	3/16	1/4	0.127	0.09	0.04	6	0.098	0.115	0.025	0.0947
12	7/64	5/32	3/16	1/4	0.144	0.11	0.06	6	0.098	0.115	0.025	0.0947
1/4	1/8	3/16	1/4	5/16	5/32	1/8	1/16	6	0.128	0.149	0.032	0.1270
5/16	11/64	15/64	5/16	3/8	13/64	5/32	5/64	6	0.163	0.188	0.039	0.1582
3/8	13/64	9/32	3/8	7/16	1/4	3/16	3/32	6	0.190	0.221	0.050	0.1895
7/16	15/64	21/64	7/16	1/2	19/64	7/32	7/64	6	0.221	0.256	0.060	0.2207
1/2	9/32	3/8	1/2	9/16	11/32	1/4	1/8	6	0.254	0.298	0.068	0.2520
9/16	5/16	27/64	9/16	5/8	25/64	9/32	9/64	6	0.254	0.298	0.068	0.2520
5/8	23/64	15/32	5/8	3/4	15/32	5/16	5/32	6	0.319	0.380	0.092	0.3155
3/4	7/16	9/16	3/4	7/8	9/16	3/8	3/16	6	0.386	0.463	0.112	0.3780
7/8	33/64	21/32	7/8	1	21/32	7/16	7/32	6	0.509	0.604	0.138	0.5030
1	19/32	3/4	1	1 1/8	3/4	1/2	1/4	6	0.535	0.631	0.149	0.5655
1 1/8	43/64	27/32	1 1/8	1 1/4	27/32	9/16	9/32	6	0.604	0.709	0.168	0.5655
1 1/4	3/4	15/16	1 1/4	1 1/2	15/16	5/8	5/16	6	0.685	0.801	0.189	0.6290
1 3/8	53/64	1 1/32	1 3/8	1 5/8	1 1/32	11/16	11/32	6	0.744	0.869	0.207	0.6290
1 1/2	29/32	1 1/8	1 1/2	1 3/4	1 1/8	3/4	3/8	6	0.828	0.970	0.231	0.7540
1 3/4	1 1/16	1 5/16	1 3/4	2	1 5/16	7/8	7/16	6	1.007	1.275	0.298	1.0040
2	1 7/32	1 1/2	2	2 1/4	1 1/2	1	1/2	6	1.007	1.275	0.298	1.0040

(Courtesy ANSI)

All dimensions in inches.
Where usable length of thread is less than nominal diameter, half dog point shall be used.
Length (L). The length of the screw shall be measured overall on a line parallel to the axis. The difference between consecutive lengths shall be as follows:

(a) for screw lengths 1/4 to 5/8 in., difference = 1/16 in.
(b) for screw lengths 5/8 to 1 in., difference = 1/8 in.
(c) for screw lengths 1 to 4 in., difference = 1/4 in.
(d) for screw lengths 4 to 6 in., difference = 1/2 in.

Table 5 Square Bolts

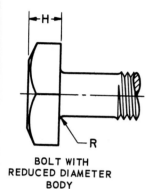

BOLT WITH
REDUCED DIAMETER
BODY

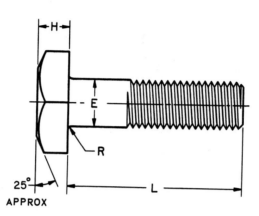

25°
APPROX

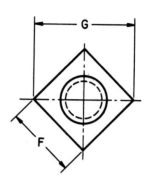

Dimensions of square bolts

Nominal Size[1] or Basic Product Dia		Body Dia[2] E	Width Across Flats F			Width Across Corners G		Height H			Radius of Fillet R
		Max	Basic	Max	Min	Max	Min	Basic	Max	Min	Max
1/4	0.2500	0.260	3/8	0.3750	0.362	0.530	0.498	11/64	0.188	0.156	0.031
5/16	0.3125	0.324	1/2	0.5000	0.484	0.707	0.665	13/64	0.220	0.186	0.031
3/8	0.3750	0.388	9/16	0.5625	0.544	0.795	0.747	1/4	0.268	0.232	0.031
7/16	0.4375	0.452	5/8	0.6250	0.603	0.884	0.828	19/64	0.316	0.278	0.031
1/2	0.5000	0.515	3/4	0.7500	0.725	1.061	0.995	21/64	0.348	0.308	0.031
5/8	0.6250	0.642	15/16	0.9375	0.906	1.326	1.244	27/64	0.444	0.400	0.062
3/4	0.7500	0.768	1 1/8	1.1250	1.088	1.591	1.494	1/2	0.524	0.476	0.062
7/8	0.8750	0.895	1 5/16	1.3125	1.269	1.856	1.742	19/32	0.620	0.568	0.062
1	1.0000	1.022	1 1/2	1.5000	1.450	2.121	1.991	21/32	0.684	0.628	0.093
1 1/8	1.1250	1.149	1 11/16	1.6875	1.631	2.386	2.239	3/4	0.780	0.720	0.093
1 1/4	1.2500	1.277	1 7/8	1.8750	1.812	2.652	2.489	27/32	0.876	0.812	0.093
1 3/8	1.3750	1.404	2 1/16	2.0625	1.994	2.917	2.738	29/32	0.940	0.872	0.093
1 1/2	1.5000	1.531	2 1/4	2.2500	2.175	3.182	2.986	1	1.036	0.964	0.093

(Courtesy ANSI B18.2.1-1972)

All dimensions given in inches.
Bold type indicates products unified dimensionally with British and Canadian standards.
Bolt need not be finished on any surface except threads.
Minimum thread length shall be twice the basic bolt diameter plus 0.25 in. for lengths up to and including 6 in., and twice the basic diameter plus 0.50 in. for lengths over 6 in. Bolts too short for the formula thread length shall be threaded as close to the head as practical.
Threads shall be in the Unified coarse thread series (UNC Series), Class 2A.

Table 6   Square Nuts

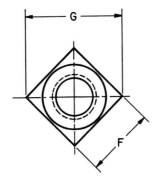

Dimensions of square nuts

Nominal Size[1] or Basic Major Dia of Thread		Width Across Flats F			Width Across Corners G		Thickness H		
		Basic	Max	Min	Max	Min	Basic	Max	Min
1/4	0.2500	7/16	0.4375	0.425	0.619	0.584	7/32	0.235	0.203
5/16	0.3125	9/16	0.5625	0.547	0.795	0.751	17/64	0.283	0.249
3/8	0.3750	5/8	0.6250	0.606	0.884	0.832	21/64	0.346	0.310
7/16	0.4375	3/4	0.7500	0.728	1.061	1.000	3/8	0.394	0.356
1/2	0.5000	13/16	0.8125	0.788	1.149	1.082	7/16	0.458	0.418
5/8	0.6250	1	1.0000	0.969	1.414	1.330	35/64	0.569	0.525
3/4	0.7500	1 1/8	1.1250	1.088	1.591	1.494	21/32	0.680	0.632
7/8	0.8750	1 5/16	1.3125	1.269	1.856	1.742	49/64	0.792	0.740
1	1.0000	1 1/2	1.5000	1.450	2.121	1.991	7/8	0.903	0.847
1 1/8	1.1250	1 11/16	1.6875	1.631	2.386	2.239	1	1.030	0.970
1 1/4	1.2500	1 7/8	1.8750	1.812	2.652	2.489	1 3/32	1.126	1.062
1 3/8	1.3750	2 1/16	2.0625	1.994	2.917	2.738	1 13/64	1.237	1.169
1 1/2	1.5000	2 1/4	2.2500	2.175	3.182	2.986	1 5/16	1.348	1.276

(Courtesy ANSI B18.2.2-1972)

All dimensions given in inches.
Threads shall be coarse-threaded series, Class 2B.

Table 7   Hexagon Bolts

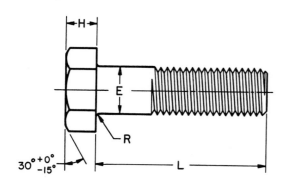

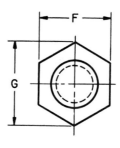

Dimensions of hex bolts

Nominal Size[1] or Basic Product Dia		Body Dia[2] E	Width Across Flats F			Width Across Corners G		Height H			Radius of Fillet R
		Max	Basic	Max	Min	Max	Min	Basic	Max	Min	Max
1/4	0.2500	0.260	7/16	0.4375	0.425	0.505	0.484	11/64	0.188	0.150	0.031
5/16	0.3125	0.324	1/2	0.5000	0.484	0.577	0.552	7/32	0.235	0.195	0.031
3/8	0.3750	0.388	9/16	0.5625	0.544	0.650	0.620	1/4	0.268	0.226	0.031
7/16	0.4375	0.452	5/8	0.6250	0.603	0.722	0.687	19/64	0.316	0.272	0.031
1/2	0.5000	0.515	3/4	0.7500	0.725	0.866	0.826	11/32	0.364	0.302	0.031
5/8	0.6250	0.642	15/16	0.9375	0.906	1.083	1.033	27/64	0.444	0.378	0.062
3/4	0.7500	0.768	1 1/8	1.1250	1.088	1.299	1.240	1/2	0.524	0.455	0.062
7/8	0.8750	0.895	1 5/16	1.3125	1.269	1.516	1.447	37/64	0.604	0.531	0.062
1	1.0000	1.022	1 1/2	1.5000	1.450	1.732	1.653	43/64	0.700	0.591	0.093
1 1/8	1.1250	1.149	1 11/16	1.6875	1.631	1.949	1.859	3/4	0.780	0.658	0.093
1 1/4	1.2500	1.277	1 7/8	1.8750	1.812	2.165	2.066	27/32	0.876	0.749	0.093
1 3/8	1.3750	1.404	2 1/16	2.0625	1.994	2.382	2.273	29/32	0.940	0.810	0.093
1 1/2	1.5000	1.531	2 1/4	2.2500	2.175	2.598	2.480	1	1.036	0.902	0.093
1 3/4	1.7500	1.785	2 5/8	2.6250	2.538	3.031	2.893	1 5/32	1.196	1.054	0.125
2	2.0000	2.039	3	3.0000	2.900	3.464	3.306	1 11/32	1.388	1.175	0.125
2 1/4	2.2500	2.305	3 3/8	3.3750	3.262	3.897	3.719	1 1/2	1.548	1.327	0.188
2 1/2	2.5000	2.559	3 3/4	3.7500	3.625	4.330	4.133	1 21/32	1.708	1.479	0.188
2 3/4	2.7500	2.827	4 1/8	4.1250	3.988	4.763	4.546	1 13/16	1.869	1.632	0.188
3	3.0000	3.081	4 1/2	4.5000	4.350	5.196	4.959	2	2.060	1.815	0.188
3 1/4	3.2500	3.335	4 7/8	4.8750	4.712	5.629	5.372	2 3/16	2.251	1.936	0.188
3 1/2	3.5000	3.589	5 1/4	5.2500	5.075	6.062	5.786	2 5/16	2.380	2.057	0.188
3 3/4	3.7500	3.858	5 5/8	5.6250	5.437	6.495	6.198	2 1/2	2.572	2.241	0.188
4	4.0000	4.111	6	6.0000	5.800	6.928	6.612	2 11/16	2.764	2.424	0.188

(Courtesy ANSI B18.2.1-1972)

All dimensions given in inches.
Bold type indicates product features unified dimensionally with British and Canadian standards.
Bolt need not be finished on any surface except threads.
Threads shall be in the Unified coarse thread series (UNC Series), Class 2A.

Table 8   Hex Flat Nuts and Hex Flat Jam Nuts

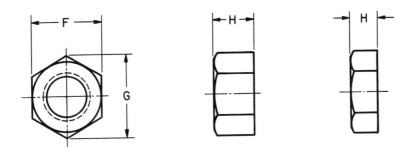

Dimensions of hex flat nuts and hex flat jam nuts

Nominal Size[1] or Basic Major Dia of Thread		Width Across Flats F			Width Across Corners G		Thickness Hex Flat Nuts H			Thickness Hex Flat Jam Nuts H		
		Basic	Max	Min	Max	Min	Basic	Max	Min	Basic	Max	Min
1 1/8	**1.1250**	1 11/16	1.6875	1.631	1.949	1.859	1	1.030	0.970	5/8	0.655	0.595
1 1/4	**1.2500**	1 7/8	1.8750	1.812	2.165	2.066	1 3/32	1.126	1.062	3/4	0.782	0.718
1 3/8	**1.3750**	2 1/16	2.0625	1.994	2.382	2.273	1 13/64	1.237	1.169	13/16	0.846	0.778
1 1/2	**1.5000**	2 1/4	2.2500	2.175	2.598	2.480	1 5/16	1.348	1.276	7/8	0.911	0.839

(Courtesy ANSI B18.2.2-1972)

All dimensions are in inches.
Bold type indicates products unified dimensionally with British and Canadian standards.
Threads shall be in the Unified coarse thread series, Class 2B.

Table 9 Regular Semifinished Hexagon Bolts

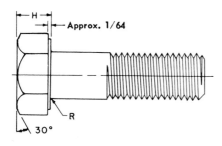

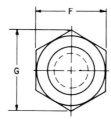

Dimensions of regular semifinished hexagon bolts

Nominal Size or Basic Major Diameter of Thread		Body Diam*	Width Across Flats F			Width Across Corners G		Height H			Radius of Fillet R	
		Max	Max (Basic)		Min	Max	Min	Nom	Max	Min	Min	Max
1/4	0.2500	0.260	7/16	0.4375	0.425	0.505	0.484	5/32	0.163	0.150	0.009	0.031
5/16	0.3125	0.324	1/2	0.5000	0.484	0.577	0.552	13/64	0.211	0.195	0.009	0.031
3/8	0.3750	0.388	9/16	0.5625	0.544	0.650	0.620	15/64	0.243	0.226	0.009	0.031
7/16	0.4375	0.452	5/8	0.6250	0.603	0.722	0.687	9/32	0.291	0.272	0.009	0.031
1/2	0.5000	0.515	3/4	0.7500	0.725	0.866	0.826	5/16	0.323	0.302	0.009	0.031
5/8	0.6250	0.642	15/16	0.9375	0.906	1.083	1.033	25/64	0.403	0.378	0.021	0.062
3/4	0.7500	0.768	1 1/8	1.1250	1.088	1.299	1.240	15/32	0.483	0.455	0.021	0.062
7/8	0.8750	0.895	1 5/16	1.3125	1.269	1.516	1.447	35/64	0.563	0.531	0.031	0.062
1	1.0000	1.022	1 1/2	1.5000	1.450	1.732	1.653	39/64	0.627	0.591	0.062	0.093
1 1/8	1.1250	1.149	1 11/16	1.6875	1.631	1.949	1.859	11/16	0.718	0.658	0.062	0.093
1 1/4	1.2500	1.277	1 7/8	1.8750	1.812	2.165	2.066	25/32	0.813	0.749	0.062	0.093
1 3/8	1.3750	1.404	2 1/16	2.0625	1.994	2.382	2.273	27/32	0.878	0.810	0.062	0.093
1 1/2	1.5000	1.531	2 1/4	2.2500	2.175	2.598	2.480	15/16	0.974	0.902	0.062	0.093

(Courtesy ANSI)

All dimensions given in inches. Thread shall be coarse-thread series, class 2A. Semifinished bolt is processed to produce a flat bearing surface under head only.

Table 10   Finished Hexagon Bolts

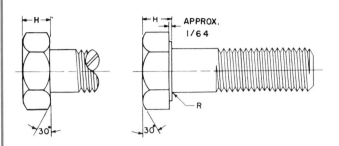
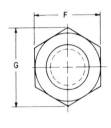

Dimensions of finished hexagon bolts

Nominal Size or Basic Major Diameter of Thread		Body Diameter Min (Maximum Equal to Nominal Size)	Width Across Flats F			Width Across Corners G		Height H			Radius of Fillet R	
			Max (Basic)		Min°	Max	Min	Nom	Max	Min	Max	Min
¼	**0.2500**	**0.2450**	⁷⁄₁₆	**0.4375**	**0.428**	**0.505**	**0.488**	⁵⁄₃₂	**0.163**	**0.150**	**0.023**	**0.009**
⁵⁄₁₆	**0.3125**	**0.3065**	½	**0.5000**	**0.489**	**0.577**	**0.557**	¹³⁄₆₄	**0.211**	**0.195**	**0.023**	**0.009**
⅜	**0.3750**	**0.3690**	⁹⁄₁₆	**0.5625**	**0.551**	**0.650**	**0.628**	¹⁵⁄₆₄	**0.243**	**0.226**	**0.023**	**0.009**
⁷⁄₁₆	**0.4375**	**0.4305**	⅝	**0.6250**	**0.612**	**0.722**	**0.698**	⁹⁄₃₂	**0.291**	**0.272**	**0.023**	**0.009**
½	**0.5000**	**0.4930**	¾	**0.7500**	**0.736**	**0.866**	**0.840**	⁵⁄₁₆	**0.323**	**0.302**	**0.023**	**0.009**
⁹⁄₁₆	**0.5625**	**0.5545**	¹³⁄₁₆	**0.8125**	**0.798**	**0.938**	**0.910**	²³⁄₆₄	**0.371**	**0.348**	**0.041**	**0.021**
⅝	**0.6250**	**0.6170**	¹⁵⁄₁₆	**0.9375**	**0.922**	**1.083**	**1.051**	²⁵⁄₆₄	**0.403**	**0.378**	**0.041**	**0.021**
¾	**0.7500**	**0.7410**	1⅛	**1.1250**	**1.100**	**1.299**	**1.254**	¹⁵⁄₃₂	**0.483**	**0.455**	**0.041**	**0.021**
⅞	**0.8750**	**0.8660**	1⁵⁄₁₆	**1.3125**	**1.285**	**1.516**	**1.465**	³⁵⁄₆₄	**0.563**	**0.531**	**0.062**	**0.041**
1	**1.0000**	**0.9900**	1½	**1.5000**	**1.469**	**1.732**	**1.675**	³⁹⁄₆₄	**0.627**	**0.591**	**0.093**	**0.062**
1⅛	**1.1250**	**1.1140**	1¹¹⁄₁₆	**1.6875**	**1.631**	**1.949**	**1.859**	¹¹⁄₁₆	**0.718**	**0.658**	**0.093**	**0.062**
1¼	**1.2500**	**1.2390**	1⅞	**1.8750**	**1.812**	**2.165**	**2.066**	²⁵⁄₃₂	**0.813**	**0.749**	**0.093**	**0.062**
1⅜	**1.3750**	**1.3630**	2¹⁄₁₆	**2.0625**	**1.994**	**2.382**	**2.273**	²⁷⁄₃₂	**0.878**	**0.810**	**0.093**	**0.062**
1½	**1.5000**	**1.4880**	2¼	**2.2500**	**2.175**	**2.598**	**2.480**	¹⁵⁄₁₆	**0.974**	**0.902**	**0.093**	**0.062**

(Courtesy ANSI)

All dimensions given in inches.
Bold type indicates products unified dimensionally with British and Canadian standards.
"Finished" in the title refers to the quality of manufacture and the closeness of tolerance and does not indicate that surfaces are completely machined.
Threads shall be coarse-, fine-, or 8-thread series, class 2A for plain (unplated) bolts. For plated bolts, the diameters may be increased by the amount of class 2A allowance.

Table 11   Hex Nuts and Hex Jam Nuts

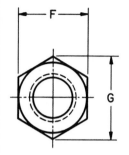
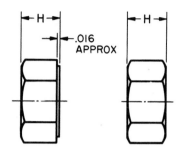
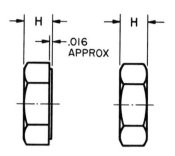

Dimensions of hex nuts and hex jam nuts

Nominal Size or Basic Major Dia of Thread		Width Across Flats F			Width Across Corners G		Thickness Hex Nuts H			Thickness Hex Jam Nuts H		
		Basic	Max	Min	Max	Min	Basic	Max	Min	Basic	Max	Min
1/4	0.2500	7/16	0.4375	0.428	0.505	0.488	7/32	0.226	0.212	5/32	0.163	0.150
5/16	0.3125	1/2	0.5000	0.489	0.577	0.557	17/64	0.273	0.258	3/16	0.195	0.180
3/8	0.3750	9/16	0.5625	0.551	0.650	0.628	21/64	0.337	0.320	7/32	0.227	0.210
7/16	0.4375	11/16	0.6875	0.675	0.794	0.768	3/8	0.385	0.365	1/4	0.260	0.240
1/2	0.5000	3/4	0.7500	0.736	0.866	0.840	7/16	0.448	0.427	5/16	0.323	0.302
9/16	0.5625	7/8	0.8750	0.861	1.010	0.982	31/64	0.496	0.473	5/16	0.324	0.301
5/8	0.6250	15/16	0.9375	0.922	1.083	1.051	35/64	0.559	0.535	3/8	0.387	0.363
3/4	0.7500	1 1/8	1.1250	1.088	1.299	1.240	41/64	0.665	0.617	27/64	0.446	0.398
7/8	0.8750	1 5/16	1.3125	1.269	1.516	1.447	3/4	0.776	0.724	31/64	0.510	0.458
1	1.0000	1 1/2	1.5000	1.450	1.732	1.653	55/64	0.887	0.831	35/64	0.575	0.519
1 1/8	1.1250	1 11/16	1.6875	1.631	1.949	1.859	31/32	0.999	0.939	39/64	0.639	0.579
1 1/4	1.2500	1 7/8	1.8750	1.812	2.165	2.066	1 1/16	1.094	1.030	23/32	0.751	0.687
1 3/8	1.3750	2 1/16	2.0625	1.994	2.382	2.273	1 11/64	1.206	1.138	25/32	0.815	0.747
1 1/2	1.5000	2 1/4	2.2500	2.175	2.598	2.480	1 9/32	1.317	1.245	27/32	0.880	0.808

(Courtesy ANSI B18.2.2-1972)

All dimensions are in inches.
Bold type indicates products unified dimensionally with British and Canadian standards.
Threads shall be in the Unified coarse, fine, or 8-thread series, Class 2B.

Table 12   Hex Slotted Nuts

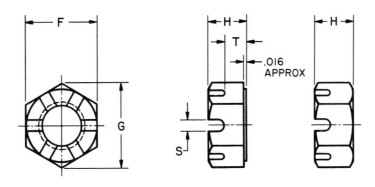

Dimensions of hex slotted nuts

Nominal Size or Basic Major Dia of Thread		Width Across Flats F			Width Across Corners G		Thickness H			Unslotted Thickness T		Width of Slot S	
		Basic	Max	Min	Max	Min	Basic	Max	Min	Max	Min	Max	Min
1/4	0.2500	7/16	0.4375	0.428	0.505	0.488	7/32	0.226	0.212	0.14	0.12	0.10	0.07
5/16	0.3125	1/2	0.5000	0.489	0.577	0.557	17/64	0.273	0.258	0.18	0.16	0.12	0.09
3/8	0.3750	9/16	0.5625	0.551	0.650	0.628	21/64	0.337	0.320	0.21	0.19	0.15	0.12
7/16	0.4375	11/16	0.6875	0.675	0.794	0.768	3/8	0.385	0.365	0.23	0.21	0.15	0.12
1/2	0.5000	3/4	0.7500	0.736	0.866	0.840	7/16	0.448	0.427	0.29	0.27	0.18	0.15
9/16	0.5625	7/8	0.8750	0.861	1.010	0.982	31/64	0.496	0.473	0.31	0.29	0.18	0.15
5/8	0.6250	15/16	0.9375	0.922	1.083	1.051	35/64	0.559	0.535	0.34	0.32	0.24	0.18
3/4	0.7500	1 1/8	1.1250	1.088	1.299	1.240	41/64	0.665	0.617	0.40	0.38	0.24	0.18
7/8	0.8750	1 5/16	1.3125	1.269	1.516	1.447	3/4	0.776	0.724	0.52	0.49	0.24	0.18
1	1.0000	1 1/2	1.5000	1.450	1.732	1.653	55/64	0.887	0.831	0.59	0.56	0.30	0.24
1 1/8	1.1250	1 11/16	1.6875	1.631	1.949	1.859	31/32	0.999	0.939	0.64	0.61	0.33	0.24
1 1/4	1.2500	1 7/8	1.8750	1.812	2.165	2.066	1 1/16	1.094	1.030	0.70	0.67	0.40	0.31
1 3/8	1.3750	2 1/16	2.0625	1.994	2.382	2.273	1 11/64	1.206	1.138	0.82	0.78	0.40	0.31
1 1/2	1.5000	2 1/4	2.2500	2.175	2.598	2.480	1 9/32	1.317	1.245	0.86	0.82	0.46	0.3 7

(Courtesy ANSI B18.2.2-1972)

All dimensions are in inches.
Bold type indicates product features unified dimensionally with British and Canadian standards.
Threads shall be in the Unified coarse, fine, or 8-thread series, Class 2B.

Table 13   Hex Thick Nuts

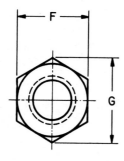
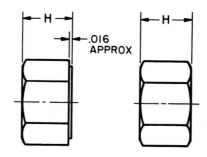

Dimensions of hex thick nuts

Nominal Size or Basic Major Dia of Thread		Width Across Flats F			Width Across Corners G		Thickness H		
		Basic	Max	Min	Max	Min	Basic	Max	Min
1/4	0.2500	7/16	0.4375	0.428	0.505	0.488	9/32	0.288	0.274
5/16	0.3125	1/2	0.5000	0.489	0.577	0.557	21/64	0.336	0.320
3/8	0.3750	9/16	0.5625	0.551	0.650	0.628	13/32	0.415	0.398
7/16	0.4375	11/16	0.6875	0.675	0.794	0.768	29/64	0.463	0.444
1/2	0.5000	3/4	0.7500	0.736	0.866	0.840	9/16	0.573	0.552
9/16	0.5625	7/8	0.8750	0.861	1.010	0.982	39/64	0.621	0.598
5/8	0.6250	15/16	0.9375	0.922	1.083	1.051	23/32	0.731	0.706
3/4	0.7500	1 1/8	1.1250	1.088	1.299	1.240	13/16	0.827	0.798
7/8	0.8750	1 5/16	1.3125	1.269	1.516	1.447	29/32	0.922	0.890
1	1.0000	1 1/2	1.5000	1.450	1.732	1.653	1	1.018	0.982
1 1/8	1.1250	1 11/16	1.6875	1.631	1.949	1.859	1 5/32	1.176	1.136
1 1/4	1.2500	1 7/8	1.8750	1.812	2.165	2.066	1 1/4	1.272	1.228
1 3/8	1.3750	2 1/16	2.0625	1.994	2.382	2.273	1 3/8	1.399	1.351
1 1/2	1.5000	2 1/4	2.2500	2.175	2.598	2.480	1 1/2	1.526	1.474

(Courtesy ANSI B18.2.2-1972)

All dimensions are in inches.
Threads shall be in the Unified coarse, fine, or 8-thread series (UNC, UNF, or 8UN), Class 2B.

Table 14   Hex Thick Slotted Nuts

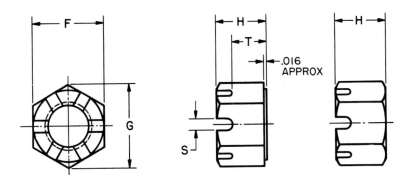

Dimensions of hex thick slotted nuts

Nominal Size or Basic Major Dia of Thread		Width Across Flats F			Width Across Corners G		Thickness H			Unslotted Thickness T		Width of Slot S	
		Basic	Max	Min	Max	Min	Basic	Max	Min	Max	Min	Max	Min
1/4	0.2500	7/16	**0.4375**	**0.428**	**0.505**	**0.488**	9/32	**0.288**	**0.274**	0.20	0.18	0.10	0.07
5/16	0.3125	1/2	**0.5000**	**0.489**	**0.577**	**0.557**	21/64	**0.336**	**0.320**	0.24	0.22	0.12	0.09
3/8	0.3750	9/16	**0.5625**	**0.551**	**0.650**	**0.628**	13/32	**0.415**	**0.398**	0.29	0.27	0.15	0.12
7/16	0.4375	11/16	**0.6875**	**0.675**	**0.794**	**0.768**	29/64	**0.463**	**0.444**	0.31	0.29	0.15	0.12
1/2	0.5000	3/4	**0.7500**	**0.736**	**0.866**	**0.840**	9/16	**0.573**	**0.552**	0.42	0.40	0.18	0.15
9/16	0.5625	7/8	**0.8750**	**0.861**	**1.010**	**0.982**	39/64	**0.621**	**0.598**	0.43	0.41	0.18	0.15
5/8	0.6250	15/16	**0.9375**	**0.922**	**1.083**	**1.051**	23/32	**0.731**	**0.706**	0.51	0.49	0.24	0.18
3/4	0.7500	1 1/8	**1.1250**	**1.088**	**1.299**	**1.240**	13/16	**0.827**	**0.798**	0.57	0.55	0.24	0.18
7/8	0.8750	1 5/16	**1.3125**	**1.269**	**1.516**	**1.447**	29/32	**0.922**	**0.890**	0.67	0.64	0.24	0.18
1	1.0000	1 1/2	**1.5000**	**1.450**	**1.732**	**1.653**	1	**1.018**	**0.982**	0.73	0.70	0.30	0.24
1 1/8	1.1250	1 11/16	**1.6875**	**1.631**	**1.949**	**1.859**	1 5/32	**1.176**	**1.136**	0.83	0.80	0.33	0.24
1 1/4	1.2500	1 7/8	**1.8750**	**1.812**	**2.165**	**2.066**	1 1/4	**1.272**	**1.228**	0.89	0.86	0.40	0.31
1 3/8	1.3750	2 1/16	**2.0625**	**1.994**	**2.382**	**2.273**	1 3/8	**1.399**	**1.351**	1.02	0.98	0.40	0.31
1 1/2	1.5000	2 1/4	**2.2500**	**2.175**	**2.598**	**2.480**	1 1/2	**1.526**	**1.474**	1.08	1.04	0.46	0.37

(Courtesy ANSI B18.2.2-1972)

All dimensions are in inches.
Bold type indicates product features unified dimensionally with British and Canadian standards.
Threads shall be in the Unified coarse, fine, or 8-thread series (UNC, UNF or 8UN series), Class 2B, Unification of the fine thread products is limited to sizes 1 in. and under.

Table 15   Hexagon Socket Head Shoulder Screws

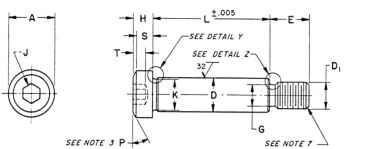

Dimensions of hexagon socket head shoulder screws

Nominal Size	D Shoulder Diameter		A Head Diameter		H Head Height		S Head Side Height	D_1 Nominal Thread Size	E Thread Length
	Max	Min	Max	Min	Max	Min	Min		
1/4	0.2480	0.2460	3/8	0.357	3/16	0.182	0.157	10—24	0.375
5/16	0.3105	0.3085	7/16	0.419	7/32	0.213	0.183	1/4—20	0.438
3/8	0.3730	0.3710	9/16	0.543	1/4	0.244	0.209	5/16—18	0.500
1/2	0.4980	0.4960	3/4	0.729	5/16	0.306	0.262	3/8—16	0.625
5/8	0.6230	0.6210	7/8	0.853	3/8	0.368	0.315	1/2—13	0.750
3/4	0.7480	0.7460	1	0.977	1/2	0.492	0.421	5/8—11	0.875
1	0.9980	0.9960	1 5/16	1.287	5/8	0.616	0.527	3/4—10	1.000
1 1/4	1.2480	1.2460	1 3/4	1.723	3/4	0.741	0.633	7/8—9	1.125
See Notes	1		2					13	6

(Courtesy ANSI B18.3-1972)

NOTES:
(1) *Shoulder.* The shoulder refers to the unthreaded portion of the screw and the maximum diameter shall conform to the nominal screw diameter less 0.002 in.
(2) *Head diameter.* The head may be plain or knurled at the option of the manufacturer.
(3) *Head chamfer.* The head shall be flat and chamfered. The flat shall be normal to the axis of the screw and the chamfer P shall be at an angle of 30 deg to 45 deg with the surface of the flat. The edge between the flat and the chamfer may be slightly rounded.
(4) *Length.* The length of the screw shall be measured, on a line parallel to the axis, from the plane of the bearing surface under the head to the plane of the shoulder at the threaded end.
(5) *Standard lengths.* The difference between consecutive lengths of standard screws shall be:

Nominal screw length	Standard length increment
$\frac{1}{4}$ to $\frac{3}{4}$	$\frac{1}{8}$
$\frac{3}{4}$ to 5	$\frac{1}{4}$
Over 5	$\frac{1}{2}$

(6) *Thread length tolerance.* The tolerance on thread length E shall be unilateral. For screw sizes 1/4 through 3/8 in., inclusive, the tolerance shall be minus 0.020 in.; and for sizes larger than 3/8 in., the tolerance shall be minus 0.030 in.
(7) *Screw point chamfer.* The point shall be flat and chambered. The flat shall be normal to the axis of the screw. The chamfer shall extend slightly below the root of the thread and the edge between the flat and chamfer may be slightly rounded. The included angle of the point should be approximately 90 deg.

Table 16   Wrench Openings for Square and Hex Bolts and Screws

Nominal Size of Wrench also Basic (Maximum) Width Across Flats of Bolt and Screw Heads		Allowance between Bolt or Screw Head and Jaws of Wrench	Wrench Openings			Square Bolt / Hex Bolt / Hex Cap Screw (Finished Hex Bolt) / Lag Screw	Heavy Hex Bolt / Heavy Hex Screw / Heavy Hex Structural Bolt
			Min	Tol	Max		
9/32	0.2812	0.002	0.283	0.005	0.288	No. 10	
5/16	0.3125	0.003	0.316	0.006	0.322		
11/32	0.3438	0.003	0.347	0.006	0.353		
3/8	0.3750	0.003	0.378	0.006	0.384	1/4*	
7/16	0.4375	0.003	0.440	0.006	0.446	1/4	
1/2	0.5000	0.004	0.504	0.006	0.510	5/16	
9/16	0.5625	0.004	0.566	0.007	0.573	3/8	
5/8	0.6250	0.004	0.629	0.007	0.636	7/16	
11/16	0.6875	0.004	0.692	0.007	0.699		
3/4	0.7500	0.005	0.755	0.008	0.763	1/2	
13/16	0.8125	0.005	0.818	0.008	0.826	9/16	
7/8	0.8750	0.005	0.880	0.008	0.888		1/2
15/16	0.9375	0.006	0.944	0.009	0.953	5/8	
1	1.0000	0.006	1.006	0.009	1.015		
1 1/16	1.0625	0.006	1.068	0.009	1.077		5/8
1 1/8	1.1250	0.007	1.132	0.010	1.142	3/4	
1 1/4	1.2500	0.007	1.257	0.010	1.267		3/4
1 5/16	1.3125	0.008	1.320	0.011	1.331	7/8	
1 3/8	1.3750	0.008	1.383	0.011	1.394		
1 7/16	1.4375	0.008	1.446	0.011	1.457		7/8
1 1/2	1.5000	0.008	1.508	0.012	1.520	1	
1 5/8	1.6250	0.009	1.634	0.012	1.646		1
1 11/16	1.6875	0.009	1.696	0.012	1.708	1 1/8	
1 13/16	1.8125	0.010	1.822	0.013	1.835		1 1/8
1 7/8	1.8750	0.010	1.885	0.013	1.898	1 1/4	
2	2.0000	0.011	2.011	0.014	2.025		1 1/4
2 1/16	2.0625	0.011	2.074	0.014	2.088	1 3/8	
2 3/16	2.1875	0.012	2.200	0.015	2.215		1 3/8
2 1/4	2.2500	0.012	2.262	0.015	2.277	1 1/2	
2 3/8	2.3750	0.013	2.388	0.016	2.404		1 1/2
2 7/16	2.4375	0.013	2.450	0.016	2.466	1 5/8	
2 9/16	2.5625	0.014	2.576	0.017	2.593		1 5/8
2 5/8	2.6250	0.014	2.639	0.017	2.656	1 3/4	
2 3/4	2.7500	0.014	2.766	0.017	2.783		1 3/4
2 13/16	2.8125	0.015	2.827	0.018	2.845	1 7/8	
2 15/16	2.9375	0.016	2.954	0.019	2.973		1 7/8
3	3.0000	0.016	3.016	0.019	3.035	2	
3 1/8	3.1250	0.017	3.142	0.020	3.162		2
3 3/8	3.3750	0.018	3.393	0.021	3.414	2 1/4	
3 1/2	3.5000	0.019	3.518	0.022	3.540		2 1/4
3 3/4	3.7500	0.020	3.770	0.023	3.793	2 1/2	
3 7/8	3.8750	0.020	3.895	0.023	3.918		2 1/2
4 1/8	4.1250	0.022	4.147	0.025	4.172	2 3/4	
4 1/4	4.2500	0.022	4.272	0.025	4.297		2 3/4
4 1/2	4.5000	0.024	4.524	0.026	4.550	3	
4 5/8	4.6250	0.024	4.649	0.027	4.676		3
4 7/8	4.8750	0.025	4.900	0.028	4.928	3 1/4	
5	5.0000	0.026	5.026	0.029	5.055		
5 1/4	5.2500	0.027	5.277	0.030	5.307	3 1/2	
5 3/8	5.3750	0.028	5.403	0.031	5.434		
5 5/8	5.6250	0.029	5.654	0.032	5.686	3 3/4	
5 3/4	5.7500	0.030	5.780	0.033	5.813		
6	6.0000	0.031	6.031	0.034	6.065	4	

(Courtesy ANSI)

All dimensions given in inches.
* Square bolt and lag screw only.
Wrenches shall be marked with the "Nominal Size of Wrench" which is equal to the basic (maximum) width across flats of the corresponding bolt or screw head.
Allowance (minimum clearance between maximum width across flats of bolt or screw head and jaws of wrench equals 0.005W + 0.001). Tolerance on wrench opening equals plus (0.005W + 0.004 from minimum) W equals nominal size of wrench.

Table 17 Formulas for Bolt and Screw Heads

Product	Width Across Flats		Height of Head	
	Basic[1]	Tolerance (Minus)	Basic[2]	Tolerance (Plus or Minus)
Square Bolt Lag Screw	$1\frac{1}{2}$ D	0.050D	$\frac{2}{3}$ D	0.016D + 0.012
Hex Bolt	Size Width $\frac{1}{4}$ $1\frac{1}{2}$D + $\frac{1}{16}$ $\frac{5}{16}$ to 4 $1\frac{1}{2}$D	0.050D	Size Height $\frac{1}{4}$ to $\frac{7}{16}$ $\frac{5}{8}$D + $\frac{1}{64}$ $\frac{1}{2}$ to $\frac{7}{8}$ $\frac{5}{8}$D + $\frac{1}{32}$ 1 to $1\frac{7}{8}$ $\frac{5}{8}$D + $\frac{1}{16}$ 2 to $3\frac{3}{4}$ $\frac{5}{8}$D + $\frac{1}{8}$ 4 $\frac{5}{8}$D + $\frac{3}{16}$	Plus tolerance only 0.016D + 0.012 Minus tolerance adjusted so that minimum head height is same as minimum head height of hex cap screw (finished hex bolt).
Hex Cap Screw (Finished Hex Bolt)	$\frac{1}{4}$ $1\frac{1}{2}$D + $\frac{1}{16}$ $\frac{5}{16}$ to 3 $1\frac{1}{2}$D	Size Tolerance $\frac{1}{4}$ to $\frac{5}{8}$ 0.015D + 0.006 $\frac{3}{4}$ to 1 0.025D + 0.006 $1\frac{1}{8}$ to 3 0.050D	$\frac{1}{4}$ to $\frac{7}{8}$ $\frac{5}{8}$D 1 to $1\frac{7}{8}$ $\frac{5}{8}$D - $\frac{1}{64}$ 2 to $2\frac{3}{4}$ $\frac{5}{8}$D - $\frac{1}{32}$ 3 $\frac{5}{8}$D	Size Tolerance $\frac{1}{4}$ to 1 0.015D + 0.003 $1\frac{1}{8}$ to 4 0.016D + 0.012
Heavy Hex Bolt	$1\frac{1}{2}$D + $\frac{1}{8}$	0.050D	Same as for * Hex Bolt	Same as for* Hex Bolt
Heavy Hex Screw Heavy Hex Structural Bolt	$1\frac{1}{2}$D + $\frac{1}{8}$	0.050D	Same as for** Hex Cap Screw- Finished Hex Bolt	Same as for** Hex Cap Screw (Finished Hex Bolt)

(Courtesy ANSI)

All dimensions given in inches.

* In 1960 head heights for heavy hex bolts were reduced. Prior to 1960 head heights were 3D/4 + $\frac{1}{16}$ in. Plus tolerance was 0.016D + 0.012 in. Minus tolerance was adjusted to that minimum head height would be the same as minimum head height of heavy hex screw.

** In 1960 head heights for heavy hex screws were reduced. Prior to 1960 head heights were 3D/4 + $\frac{1}{32}$ in. for sizes $\frac{1}{2}$ to $\frac{7}{8}$ in.; 3D/4 for sizes 1 to $1\frac{7}{8}$ in.; and 3D/4 − $\frac{1}{16}$ in. for sizes 2 to 3 in. Tolerances on head height for all sizes were plus and minus 0.016D + 0.012 in.

[1] Adjusted to sixteenths.

[2] $\frac{1}{4}$ to 1 in. sizes adjusted to sixty-fourths. $1\frac{1}{8}$ to $2\frac{1}{2}$ in. sizes adjusted upward to thirty-seconds. $2\frac{3}{4}$ to 4 in. sizes adjusted upward to sixteenths.

For all square bolt heads, maximum width across corners equals 1.4142 × F (max) and minimum width across corners equals 1.373 × F (min). For all hexagon bolt or screw heads, maximum width across corners equals 1.1547 × F (max) and minimum width across corners equals 1.14 × F (min).

D = Basic bolt or screw size.

F = Width across flats.

Table 18   Wrench Openings for Nuts

Nominal Size of Wrench also Basic (Maximum) Width Across Flats of Nuts		Allowance between Nut Flats and Jaws of Wrench	Wrench Openings			Square Nut	Hex Flat / Hex Flat Jam / Hex / Hex Jam / Hex Slotted / Hex Thick / Hex Thick Slotted / Hex Castle	Heavy Square / Heavy Hex Flat / Heavy Hex Flat Jam / Heavy Hex / Heavy Hex Jam / Heavy Hex Slotted
			Min	Tol	Max			
7/16	0.4375	0.003	0.440	0.006	0.446	1/4	1/4	
1/2	0.5000	0.004	0.504	0.006	0.510		5/16	1/4
9/16	0.5625	0.004	0.566	0.007	0.573	5/16	3/8	5/16
5/8	0.6250	0.004	0.629	0.007	0.636	3/8		
11/16	0.6875	0.004	0.692	0.007	0.699		7/16	3/8
3/4	0.7500	0.005	0.755	0.008	0.763	7/16	1/2	7/16
13/16	0.8125	0.005	0.818	0.008	0.826	1/2		
7/8	0.8750	0.005	0.880	0.008	0.888		9/16	1/2
15/16	0.9375	0.006	0.944	0.009	0.953		5/8	9/16
1	1.0000	0.006	1.006	0.009	1.015	5/8		
1 1/16	1.0625	0.006	1.068	0.009	1.077			5/8
1 1/8	1.1250	0.007	1.132	0.010	1.142	3/4	3/4	
1 1/4	1.2500	0.007	1.257	0.010	1.267			3/4
1 5/16	1.3125	0.008	1.320	0.011	1.331	7/8	7/8	
1 3/8	1.3750	0.008	1.383	0.011	1.394			
1 7/16	1.4375	0.008	1.446	0.011	1.457			7/8
1 1/2	1.5000	0.008	1.508	0.012	1.520	1	1	
1 5/8	1.6250	0.009	1.634	0.012	1.646			1
1 11/16	1.6875	0.009	1.696	0.012	1.708	1 1/8	1 1/8	
1 13/16	1.8125	0.010	1.822	0.013	1.835			1 1/8
1 7/8	1.8750	0.010	1.885	0.013	1.898	1 1/4	1 1/4	
2	2.0000	0.011	2.011	0.014	2.025			1 1/4
2 1/16	2.0625	0.011	2.074	0.014	2.088	1 3/8	1 3/8	
2 3/16	2.1875	0.012	2.200	0.015	2.215			1 3/8
2 1/4	2.2500	0.012	2.262	0.015	2.277	1 1/2	1 1/2	
2 3/8	2.3750	0.013	2.388	0.016	2.404			1 1/2
2 7/16	2.4375	0.013	2.450	0.016	2.466			
2 9/16	2.5625	0.014	2.576	0.017	2.593			1 5/8
2 5/8	2.6250	0.014	2.639	0.017	2.656			
2 3/4	2.7500	0.014	2.766	0.017	2.783			1 3/4
2 13/16	2.8125	0.015	2.827	0.018	2.845			
2 15/16	2.9375	0.016	2.954	0.019	2.973			1 7/8
3	3.0000	0.016	3.016	0.019	3.035			
3 1/8	3.1250	0.017	3.142	0.020	3.162			2
3 3/8	3.3750	0.018	3.393	0.021	3.414			
3 1/2	3.5000	0.019	3.518	0.022	3.540			2 1/4
3 3/4	3.7500	0.020	3.770	0.023	3.793			
3 7/8	3.8750	0.020	3.895	0.023	3.918			2 1/2
4 1/8	4.1250	0.022	4.147	0.025	4.172			
4 1/4	4.2500	0.022	4.272	0.025	4.297			2 3/4
4 1/2	4.5000	0.024	4.524	0.026	4.550			
4 5/8	4.6250	0.024	4.649	0.027	4.676			3
4 7/8	4.8750	0.025	4.900	0.028	4.928			
5	5.0000	0.026	5.026	0.029	5.055			3 1/4
5 1/4	5.2500	0.027	5.277	0.030	5.307			3 1/2
5 3/8	5.3750	0.028	5.403	0.031	5.434			3 1/2
5 5/8	5.6250	0.029	5.654	0.032	5.686			3 3/4
5 3/4	5.7500	0.030	5.780	0.033	5.813			3 3/4
6	6.0000	0.031	6.031	0.034	6.065			4
6 1/8	6.1250	0.032	6.157	0.035	6.192			4

(Courtesy ANSI)

All dimensions given in inches.
Wrenches shall be marked with the "Nominal Size of Wrench" which is equal to the basic (maximum) width across flats of the corresponding nut.
Allowance (minimum clearance) between maximum width across flats of the nut and jaws of wrench equals $(0.0005W + 0.001)$. Tolerance on wrench opening equals plus $(0.0005W + 0.004$ from minimum). W equals nominal size of wrench.

Table 19 Formulas for Nuts

Type of Nut	Width Across Flats				Thickness of Nut			
	Basic¹		Tolerance (Minus)		Basic²		Tolerance (Plus or Minus)	
	Size	Width	Size	Tolerance	Size	Thickness	Size	Tolerance
Hex Hex Slotted	¼ ⁵⁄₁₆ to 1 ½	1 ½D + ¹⁄₁₆ 1 ½D	¼ to ⁵⁄₈ ¾ to 1 ½	0.015D + 0.006 0.050D	¼ to ⁵⁄₈ ¾ to 1 ⅛ 1 ¼ to 1 ½	⁷⁄₈D ⁷⁄₈D − ¹⁄₆₄ ⁷⁄₈D − ¹⁄₃₂	¼ to ⁵⁄₈ ¾ to 1 ½	0.015D + 0.003 0.016D + 0.012
Hex Jam	¼ ⁵⁄₁₆ to 1 ½	1 ½D + ¹⁄₁₆ 1 ½D	¼ to ⁵⁄₈ ¾ to 1 ½	0.015D + 0.006 0.050D	¼ to ⁵⁄₈ ¾ to 1 ⅛ 1 ¼ to 1 ½	(See Table) ½D + ³⁄₆₄ ½D + ³⁄₃₂	¼ to ⁵⁄₈ ¾ to 1 ½	0.015D + 0.003 0.016D + 0.012
Hex Thick Hex Thick Slotted Hex Castle	¼ ⁵⁄₁₆ to 1 ½	1 ½D + ¹⁄₁₆ 1 ½D	¼ to ⁵⁄₈ ¾ to 1 ½	0.015D + 0.006 0.050D	(See Table)		0.015D + 0.003	
Square Hex Flat	¼ to ⁵⁄₈ ¾ to 1 ½	1 ½D + ¹⁄₁₆ 1 ½D	0.050D		⁷⁄₈D		0.016D + 0.012	
Hex Flat Jam	1 ⅛ to 1 ½	1 ½	0.050D		1 ⅛ 1 ¼ to 1 ½	½D + ¹⁄₁₆ ½D + ⅛	0.016D + 0.012	
Heavy Square Heavy Hex Flat	1 ½D + ⅛		0.050D		D		Plus tolerance only 0.016D + 0.012 Minus tolerance adjusted so that minimum thickness is equal to minimum thickness of heavy hex nut.	
Heavy Hex Flat Jam	1 ½D + ⅛		0.050D		¼ to 1 ⅛ 1 ¼ to 2 ¼ 2 ½ to 4	½D + ¹⁄₁₆ ½D + ⅛ ½D + ¼	Plus tolerance only 0.016D + 0.012 Minus tolerance adjusted so that minimum thickness is equal to minimum thickness of heavy hex jam nut.	
Heavy Hex Heavy Hex Slotted	1 ½D + ⅛		0.050D		¼ to 1 ⅛ 1 ¼ to 2 2 ¼ to 3 3 ¼ to 4	D − ¹⁄₆₄ D − ¹⁄₃₂ D − ³⁄₆₄ D − ¹⁄₁₆	0.016D + 0.012	
Heavy Hex Jam	1 ½D + ⅛		0.050D		¼ to 1 ⅛ 1 ¼ to 2 2 ¼ 2 ½ to 3 3 ¼ to 4	½D + ³⁄₆₄ ½D + ³⁄₃₂ ½D + ⁵⁄₆₄ ½D + ¹³⁄₆₄ ½D + ³⁄₁₆	0.016D + 0.012	

(Courtesy ANSI)

All dimensions given in inches.
¹ Adjusted to sixteenths.
² ¼ to 1 in. sizes adjusted to sixty-fourths. 1⅛ to 2½ in. sizes adjusted upward to thirty-seconds. 2¾ to 4 in. sizes adjusted upward to sixteenths.
For all square nuts, maximum width across corners equals 1.4142 × F (max) and minimum width across corners equals 1.373 × F (min).
For all hex nuts, maximum width across corners equals 1.1547 × F (max) and minimum width across corners equals 1.14 × F (min).
D = Nominal nut size.
F = Width across flats.

Table 20   Plain Washers—Type B

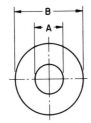 

Dimensions of type B plain washers

Nominal Washer Size**		Series	Inside Diameter A			Outside Diameter B			Thickness C		
			Basic	Tolerance		Basic	Tolerance		Basic	Max	Min
				Plus	Minus		Plus	Minus			
No. 0	0.060	Narrow	0.068	0.000	0.005	0.125	0.000	0.005	0.025	0.028	0.022
		Regular	0.068	0.000	0.005	0.188	0.000	0.005	0.025	0.028	0.022
		Wide	0.068	0.000	0.005	0.250	0.000	0.005	0.025	0.028	0.022
No. 1	0.073	Narrow	0.084	0.000	0.005	0.156	0.000	0.005	0.025	0.028	0.022
		Regular	0.084	0.000	0.005	0.219	0.000	0.005	0.025	0.028	0.022
		Wide	0.084	0.000	0.005	0.281	0.000	0.005	0.032	0.036	0.028
No. 2	0.086	Narrow	0.094	0.000	0.005	0.188	0.000	0.005	0.025	0.028	0.022
		Regular	0.094	0.000	0.005	0.250	0.000	0.005	0.032	0.036	0.028
		Wide	0.094	0.000	0.005	0.344	0.000	0.005	0.032	0.036	0.028
No. 3	0.099	Narrow	0.109	0.000	0.005	0.219	0.000	0.005	0.025	0.028	0.022
		Regular	0.109	0.000	0.005	0.312	0.000	0.005	0.032	0.036	0.028
		Wide	0.109	0.008	0.005	0.406	0.008	0.005	0.040	0.045	0.036
No. 4	0.112	Narrow	0.125	0.000	0.005	0.250	0.000	0.005	0.032	0.036	0.028
		Regular	0.125	0.008	0.005	0.375	0.008	0.005	0.040	0.045	0.036
		Wide	0.125	0.008	0.005	0.438	0.008	0.005	0.040	0.045	0.036
No. 5	0.125	Narrow	0.141	0.000	0.005	0.281	0.000	0.005	0.032	0.036	0.028
		Regular	0.141	0.008	0.005	0.406	0.008	0.005	0.040	0.045	0.036
		Wide	0.141	0.008	0.005	0.500	0.008	0.005	0.040	0.045	0.036
No. 6	0.138	Narrow	0.156	0.000	0.005	0.312	0.000	0.005	0.032	0.036	0.028
		Regular	0.156	0.008	0.005	0.438	0.008	0.005	0.040	0.045	0.036
		Wide	0.156	0.008	0.005	0.562	0.008	0.005	0.040	0.045	0.036
No. 8	0.164	Narrow	0.188	0.008	0.005	0.375	0.008	0.005	0.040	0.045	0.036
		Regular	0.188	0.008	0.005	0.500	0.008	0.005	0.040	0.045	0.036
		Wide	0.188	0.008	0.005	0.625	0.015	0.005	0.063	0.071	0.056
No. 10	0.190	Narrow	0.203	0.008	0.005	0.406	0.008	0.005	0.040	0.045	0.036
		Regular	0.203	0.008	0.005	0.562	0.008	0.005	0.040	0.045	0.036
		Wide	0.203	0.008	0.005	0.734*	0.015	0.007	0.063	0.071	0.056
No. 12	0.216	Narrow	0.234	0.008	0.005	0.438	0.008	0.005	0.040	0.045	0.036
		Regular	0.234	0.008	0.005	0.625	0.015	0.005	0.063	0.071	0.056
		Wide	0.234	0.008	0.005	0.875	0.015	0.007	0.063	0.071	0.056
¼	0.250	Narrow	0.281	0.015	0.005	0.500	0.015	0.005	0.063	0.071	0.056
		Regular	0.281	0.015	0.005	0.734*	0.015	0.007	0.063	0.071	0.056
		Wide	0.281	0.015	0.005	1.000	0.015	0.007	0.063	0.071	0.056

(Courtesy ANSI)

* The 0.734 in. outside diameter avoids washers which could be used in coin operated devices.
** Nominal washer sizes are intended for use with comparable nominal screw or bolt sizes.
Inside and outside diameters shall be concentric within at least the inside diameter tolerance.
Washers shall be flat within 0.005 in. for basic outside diameters up to and including 0.875 in. and within 0.010 in. for larger outside diameters.

Table 20   (*continued*)

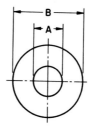 

Dimensions of type B plain washers (*continued*)

Nominal Washer Size**		Series	Inside Diameter A			Outside Diameter B			Thickness C		
			Basic	Tolerance		Basic	Tolerance		Basic	Max	Min
				Plus	Minus		Plus	Minus			
5/16	0.312	Narrow	0.344	0.015	0.005	0.625	0.015	0.005	0.063	0.071	0.056
		Regular	0.344	0.015	0.005	0.875	0.015	0.007	0.063	0.071	0.056
		Wide	0.344	0.015	0.005	1.125	0.015	0.007	0.063	0.071	0.056
3/8	0.375	Narrow	0.406	0.015	0.005	0.734*	0.015	0.007	0.063	0.071	0.056
		Regular	0.406	0.015	0.005	1.000	0.015	0.007	0.063	0.071	0.056
		Wide	0.406	0.015	0.005	1.250	0.030	0.007	0.100	0.112	0.090
7/16	0.438	Narrow	0.469	0.015	0.005	0.875	0.015	0.007	0.063	0.071	0.056
		Regular	0.469	0.015	0.005	1.125	0.015	0.007	0.063	0.071	0.056
		Wide	0.469	0.015	0.005	1.469*	0.030	0.007	0.100	0.112	0.090
1/2	0.500	Narrow	0.531	0.015	0.005	1.000	0.015	0.007	0.063	0.071	0.056
		Regular	0.531	0.015	0.005	1.250	0.030	0.007	0.100	0.112	0.090
		Wide	0.531	0.015	0.005	1.750	0.030	0.007	0.100	0.112	0.090
9/16	0.562	Narrow	0.594	0.015	0.005	1.125	0.015	0.007	0.063	0.071	0.056
		Regular	0.594	0.015	0.005	1.469*	0.030	0.007	0.100	0.112	0.090
		Wide	0.594	0.015	0.005	2.000	0.030	0.007	0.100	0.112	0.090
5/8	0.625	Narrow	0.656	0.030	0.007	1.250	0.030	0.007	0.100	0.112	0.090
		Regular	0.656	0.030	0.007	1.750	0.030	0.007	0.100	0.112	0.090
		Wide	0.656	0.030	0.007	2.250	0.030	0.007	0.160	0.174	0.146
3/4	0.750	Narrow	0.812	0.030	0.007	1.375	0.030	0.007	0.100	0.112	0.090
		Regular	0.812	0.030	0.007	2.000	0.030	0.007	0.100	0.112	0.090
		Wide	0.812	0.030	0.007	2.500	0.030	0.007	0.160	0.174	0.146
7/8	0.875	Narrow	0.938	0.030	0.007	1.469*	0.030	0.007	0.100	0.112	0.090
		Regular	0.938	0.030	0.007	2.250	0.030	0.007	0.160	0.174	0.146
		Wide	0.938	0.030	0.007	2.750	0.030	0.007	0.160	0.174	0.146
1	1.000	Narrow	1.062	0.030	0.007	1.750	0.030	0.007	0.100	0.112	0.090
		Regular	1.062	0.030	0.007	2.500	0.030	0.007	0.160	0.174	0.146
		Wide	1.062	0.030	0.007	3.000	0.030	0.007	0.160	0.174	0.146
1 1/8	1.125	Narrow	1.188	0.030	0.007	2.000	0.030	0.007	0.100	0.112	0.090
		Regular	1.188	0.030	0.007	2.750	0.030	0.007	0.160	0.174	0.146
		Wide	1.188	0.030	0.007	3.250	0.030	0.007	0.160	0.174	0.146
1 1/4	1.250	Narrow	1.312	0.030	0.007	2.250	0.030	0.007	0.160	0.174	0.146
		Regular	1.312	0.030	0.007	3.000	0.030	0.007	0.160	0.174	0.146
		Wide	1.312	0.045	0.010	3.500	0.045	0.010	0.250	0.266	0.234

(Courtesy ANSI)

* The 0.734 in. and 1.469 in. outside diameters avoid washers which could be used in coin operated devices.

** Nominal washer sizes are intended for use with comparable nominal screw or bolt sizes. Inside and outside diameters shall be concentric within at least the inside diameter tolerance.

Washers shall be flat within 0.005 in. for basic outside diameters up to and including 0.875 in., and within 0.010 in. for larger outside diameters.

Table 20   (*continued*)

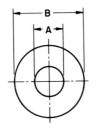

Dimensions of type B plain washers (*continued*)

Nominal Washer Size*		Series	Inside Diameter A			Outside Diameter B			Thickness C		
			Basic	Tolerance		Basic	Tolerance		Basic	Max	Min
				Plus	Minus		Plus	Minus			
1⅜	1.375	Narrow	1.438	0.030	0.007	2.500	0.030	0.007	0.160	0.174	0.146
		Regular	1.438	0.030	0.007	3.250	0.030	0.007	0.160	0.174	0.146
		Wide	1.438	0.045	0.010	3.750	0.045	0.010	0.250	0.266	0.234
1½	1.500	Narrow	1.562	0.030	0.007	2.750	0.030	0.007	0.160	0.174	0.146
		Regular	1.562	0.045	0.010	3.500	0.045	0.010	0.250	0.266	0.234
		Wide	1.562	0.045	0.010	4.000	0.045	0.010	0.250	0.266	0.234
1⅝	1.625	Narrow	1.750	0.030	0.007	3.000	0.030	0.007	0.160	0.174	0.146
		Regular	1.750	0.045	0.010	3.750	0.045	0.010	0.250	0.266	0.234
		Wide	1.750	0.045	0.010	4.250	0.045	0.010	0.250	0.266	0.234
1¾	1.750	Narrow	1.875	0.030	0.007	3.250	0.030	0.007	0.160	0.174	0.146
		Regular	1.875	0.045	0.010	4.000	0.045	0.010	0.250	0.266	0.234
		Wide	1.875	0.045	0.010	4.500	0.045	0.010	0.250	0.266	0.234
1⅞	1.875	Narrow	2.000	0.045	0.010	3.500	0.045	0.010	0.250	0.266	0.234
		Regular	2.000	0.045	0.010	4.250	0.045	0.010	0.250	0.266	0.234
		Wide	2.000	0.045	0.010	4.750	0.045	0.010	0.250	0.266	0.234
2	2.000	Narrow	2.125	0.045	0.010	3.750	0.045	0.010	0.250	0.266	0.234
		Regular	2.125	0.045	0.010	4.500	0.045	0.010	0.250	0.266	0.234
		Wide	2.125	0.045	0.010	5.000	0.045	0.010	0.250	0.266	0.234
2¼	2.250	Narrow	2.375	0.045	0.010	4.000	0.045	0.010	0.250	0.266	0.234
		Regular	2.375	0.045	0.010	5.000	0.045	0.010	0.250	0.266	0.234
		Wide	2.375	0.065	0.010	5.500	0.065	0.010	0.375	0.393	0.357
2½	2.500	Narrow	2.625	0.045	0.010	4.500	0.045	0.010	0.250	0.266	0.234
		Regular	2.625	0.065	0.010	5.500	0.065	0.010	0.375	0.393	0.357
		Wide	2.625	0.065	0.010	6.000	0.065	0.010	0.375	0.393	0.357
2¾	2.750	Narrow	2.875	0.045	0.010	5.000	0.045	0.010	0.250	0.266	0.234
		Regular	2.875	0.065	0.010	6.000	0.065	0.010	0.375	0.393	0.357
		Wide	2.875	0.065	0.010	6.500	0.065	0.010	0.375	0.393	0.357
3	3.000	Narrow	3.125	0.065	0.010	5.500	0.065	0.010	0.375	0.393	0.357
		Regular	3.125	0.065	0.010	6.500	0.065	0.010	0.375	0.393	0.357
		Wide	3.125	0.065	0.010	7.000	0.065	0.010	0.375	0.393	0.357

(Courtesy ANSI)

* Nominal washer sizes are intended for use with comparable nominal screw or bolt sizes. Inside and outside diameters shall be concentric within at least the inside diameter tolerance. Washers of sizes shown above shall be flat within 0.010 in.

Table 21   Medium Lock Washers

Dimensions of regular* helical spring lock washers

Nominal Washer Size		Inside Diameter A		Outside Diameter B	Washer Section	
					Width W	Thickness $\dfrac{T+t}{2}$
		Min	Max	Max**	Min	Min
No. 2	0.086	0.088	0.094	0.172	0.035	0.020
No. 3	0.099	0.101	0.107	0.195	0.040	0.025
No. 4	0.112	0.115	0.121	0.209	0.040	0.025
No. 5	0.125	0.128	0.134	0.236	0.047	0.031
No. 6	0.138	0.141	0.148	0.250	0.047	0.031
No. 8	0.164	0.168	0.175	0.293	0.055	0.040
No. 10	0.190	0.194	0.202	0.334	0.062	0.047
No. 12	0.216	0.221	0.229	0.377	0.070	0.056
¼	0.250	0.255	0.263	0.489	0.109	0.062
⁵⁄₁₆	0.312	0.318	0.328	0.586	0.125	0.078
⅜	0.375	0.382	0.393	0.683	0.141	0.094
⁷⁄₁₆	0.438	0.446	0.459	0.779	0.156	0.109
½	0.500	0.509	0.523	0.873	0.171	0.125
⁹⁄₁₆	0.562	0.572	0.587	0.971	0.188	0.141
⅝	0.625	0.636	0.653	1.079	0.203	0.156
¹¹⁄₁₆	0.688	0.700	0.718	1.176	0.219	0.172
¾	0.750	0.763	0.783	1.271	0.234	0.188
¹³⁄₁₆	0.812	0.826	0.847	1.367	0.250	0.203
⅞	0.875	0.890	0.912	1.464	0.266	0.219
¹⁵⁄₁₆	0.938	0.954	0.978	1.560	0.281	0.234
1	1.000	1.017	1.042	1.661	0.297	0.250
1¹⁄₁₆	1.062	1.080	1.107	1.756	0.312	0.266
1⅛	1.125	1.144	1.172	1.853	0.328	0.281
1³⁄₁₆	1.188	1.208	1.237	1.950	0.344	0.297
1¼	1.250	1.271	1.302	2.045	0.359	0.312
1⁵⁄₁₆	1.312	1.334	1.366	2.141	0.375	0.328
1⅜	1.375	1.398	1.432	2.239	0.391	0.344
1⁷⁄₁₆	1.438	1.462	1.497	2.334	0.406	0.359
1½	1.500	1.525	1.561	2.430	0.422	0.375

(Courtesy ANSI B27.1-1965)

* Formerly designated Medium Helical Spring Lock Washers.
** The maximum outside diameters specified allow for the commercial tolerances on cold drawn wire.

Table 22   Heavy Lock Washers

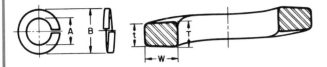

Dimensions of heavy helical spring lock washers

Nominal Washer Size		Inside Diameter A		Outside Diameter B	Washer Section	
					Width W	Thickness $\frac{T + t}{2}$
		Min	Max	Max*	Min	Min
No. 2	0.086	0.088	0.094	0.182	0.040	0.025
No. 3	0.099	0.101	0.107	0.209	0.047	0.031
No. 4	0.112	0.115	0.121	0.223	0.047	0.031
No. 5	0.125	0.128	0.134	0.252	0.055	0.040
No. 6	0.138	0.141	0.148	0.266	0.055	0.040
No. 8	0.164	0.168	0.175	0.307	0.062	0.047
No. 10	0.190	0.194	0.202	0.350	0.070	0.056
No. 12	0.216	0.221	0.229	0.391	0.077	0.063
¼	0.250	0.255	0.263	0.491	0.110	0.077
⁵⁄₁₆	0.312	0.318	0.328	0.596	0.130	0.097
⅜	0.375	0.382	0.393	0.691	0.145	0.115
⁷⁄₁₆	0.438	0.446	0.459	0.787	0.160	0.133
½	0.500	0.509	0.523	0.883	0.176	0.151
⁹⁄₁₆	0.562	0.572	0.587	0.981	0.193	0.170
⅝	0.625	0.636	0.653	1.093	0.210	0.189
¹¹⁄₁₆	0.688	0.700	0.718	1.192	0.227	0.207
¾	0.750	0.763	0.783	1.291	0.244	0.226
¹³⁄₁₆	0.812	0.826	0.847	1.391	0.262	0.246
⅞	0.875	0.890	0.912	1.494	0.281	0.266
¹⁵⁄₁₆	0.938	0.954	0.978	1.594	0.298	0.284
1	1.000	1.017	1.042	1.705	0.319	0.306
1¹⁄₁₆	1.062	1.080	1.107	1.808	0.338	0.326
1⅛	1.125	1.144	1.172	1.909	0.356	0.345
1³⁄₁₆	1.188	1.208	1.237	2.008	0.373	0.364
1¼	1.250	1.271	1.302	2.113	0.393	0.384
1⁵⁄₁₆	1.312	1.334	1.366	2.211	0.410	0.403
1⅜	1.375	1.398	1.432	2.311	0.427	0.422
1⁷⁄₁₆	1.438	1.462	1.497	2.406	0.442	0.440
1½	1.500	1.525	1.561	2.502	0.458	0.458
1⅝	1.625	1.650	1.686	2.693	0.491	0.458
1¾	1.750	1.775	1.811	2.818	0.491	0.458
1⅞	1.875	1.900	1.936	2.943	0.491	0.458
2	2.000	2.025	2.061	3.068	0.491	0.458
2¼	2.250	2.275	2.311	3.388	0.526	0.496
2½	2.500	2.525	2.561	3.638	0.526	0.496
2¾	2.750	2.775	2.811	3.888	0.526	0.496
3	3.000	3.025	3.061	4.138	0.526	0.496

(Courtesy ANSI B27.1-1965)

* The maximum outside diameters specified allow for the commercial tolerances on cold drawn wire.

Table 23   Internal Tooth Lock Washers

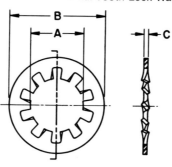

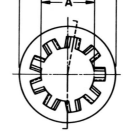

TYPE A   TYPE B

Dimensions of internal tooth lock washers

Nominal Washer Size		A Inside Diameter		B Outside Diameter		C Thickness	
		Min	Max	Max	Min	Max	Min
No. 2	0.086	0.089	0.095	0.200	0.175	0.015	0.010
No. 3	0.099	0.102	0.109	0.232	0.215	0.019	0.012
No. 4	0.112	0.115	0.123	0.270	0.255	0.019	0.015
No. 5	0.125	0.129	0.136	0.280	0.245	0.021	0.017
No. 6	0.138	0.141	0.150	0.295	0.275	0.021	0.017
No. 8	0.164	0.168	0.176	0.340	0.325	0.023	0.018
No. 10	0.190	0.195	0.204	0.381	0.365	0.025	0.020
No. 12	0.216	0.221	0.231	0.410	0.394	0.025	0.020
¼	0.250	0.256	0.267	0.478	0.460	0.028	0.023
⁵⁄₁₆	0.312	0.320	0.332	0.610	0.594	0.034	0.028
⅜	0.375	0.384	0.398	0.692	0.670	0.040	0.032
⁷⁄₁₆	0.438	0.448	0.464	0.789	0.740	0.040	0.032
½	0.500	0.512	0.530	0.900	0.867	0.045	0.037
⁹⁄₁₆	0.562	0.576	0.596	0.985	0.957	0.045	0.037
⅝	0.625	0.640	0.663	1.071	1.045	0.050	0.042
¹¹⁄₁₆	0.688	0.704	0.728	1.166	1.130	0.050	0.042
¾	0.750	0.769	0.795	1.245	1.220	0.055	0.047
¹³⁄₁₆	0.812	0.832	0.861	1.315	1.290	0.055	0.047
⅞	0.875	0.894	0.927	1.410	1.364	0.060	0.052
1	1.000	1.019	1.060	1.637	1.590	0.067	0.059
1⅛	1.125	1.144	1.192	1.830	1.799	0.067	0.059
1¼	1.250	1.275	1.325	1.975	1.921	0.067	0.059

Dimensions of heavy internal tooth lock washers

Nominal Washer Size		A Inside Diameter		B Outside Diameter		C Thickness	
		Min	Max	Max	Min	Max	Min
¼	0.250	0.256	0.267	0.536	0.500	0.045	0.035
⁵⁄₁₆	0.312	0.320	0.332	0.607	0.590	0.050	0.040
⅜	0.375	0.384	0.398	0.748	0.700	0.050	0.042
⁷⁄₁₆	0.438	0.448	0.464	0.858	0.800	0.067	0.050
½	0.500	0.512	0.530	0.924	0.880	0.067	0.055
⁹⁄₁₆	0.562	0.576	0.596	1.034	0.990	0.067	0.055
⅝	0.625	0.640	0.663	1.135	1.100	0.067	0.059
¾	0.750	0.768	0.795	1.265	1.240	0.084	0.070
⅞	0.875	0.894	0.927	1.447	1.400	0.084	0.075

(Courtesy ANSI B27.1-1965)

Table 24   External Tooth Lock Washers

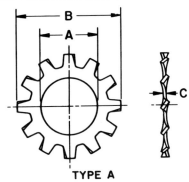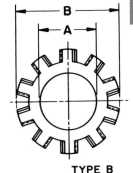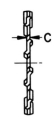

TYPE A   **TYPE B**

Dimensions of external tooth lock washers

Nominal Washer Size		A Inside Diameter		B Outside Diameter		C Thickness	
		Min	Max	Max	Min	Max	Min
No. 4	0.112	0.115	0.123	0.260	0.245	0.019	0.015
No. 6	0.138	0.141	0.150	0.320	0.305	0.022	0.016
No. 8	0.164	0.168	0.176	0.381	0.365	0.023	0.018
No. 10	0.190	0.195	0.204	0.410	0.395	0.025	0.020
No. 12	0.216	0.221	0.231	0.475	0.460	0.028	0.023
¼	0.250	0.256	0.267	0.510	0.494	0.028	0.023
⁵⁄₁₆	0.312	0.320	0.332	0.610	0.588	0.034	0.028
³⁄₈	0.375	0.384	0.398	0.694	0.670	0.040	0.032
⁷⁄₁₆	0.438	0.448	0.464	0.760	0.740	0.040	0.032
½	0.500	0.513	0.530	0.900	0.880	0.045	0.037
⁹⁄₁₆	0.562	0.576	0.596	0.985	0.960	0.045	0.037
⁵⁄₈	0.625	0.641	0.663	1.070	1.045	0.050	0.042
¹¹⁄₁₆	0.688	0.704	0.728	1.155	1.130	0.050	0.042
³⁄₄	0.750	0.768	0.795	1.260	1.220	0.055	0.047
¹³⁄₁₆	0.812	0.833	0.861	1.315	1.290	0.055	0.047
⁷⁄₈	0.875	0.897	0.927	1.410	1.380	0.060	0.052
1	1.000	1.025	1.060	1.620	1.590	0.067	0.059

(Courtesy ANSI B27.1-1965)

Table 25   Straight Shank Twist Drills

Drill Size	Decimal Equivalent Diameter	Drill Size	Decimal Equivalent Diameter	Drill Size	Decimal Equivalent Diameter	Drill Size	Decimal Equivalent Diameter
80	0.0135	7/64	0.1094	G	0.261	47/64	0.7344
79	0.0145	35	0.110	17/64	0.2656	3/4	0.7500
1/64	0.0156	34	0.111	H	0.266	49/64	0.7656
78	0.016	33	0.113	I	0.272	25/32	0.7812
77	0.018	32	0.116	J	0.277	51/64	0.7969
76	0.020	31	0.120	K	0.281	13/16	0.8125
75	0.021	1/8	0.1250	9/32	0.2812	53/64	0.8281
74	0.0225	30	0.1285	L	0.290	27/32	0.8437
73	0.024	29	0.136	M	0.295	55/64	0.8594
72	0.025	28	0.1405	19/64	0.2969	7/8	0.8750
71	0.026	9/64	0.1406	N	0.302	57/64	0.8906
70	0.028	27	0.144	5/16	0.3125	29/32	0.9062
69	0.0292	26	0.147	O	0.316	59/64	0.9219
68	0.031	25	0.1495	P	0.323	15/16	0.9375
1/32	0.0312	24	0.152	21/64	0.3281	61/64	0.9531
67	0.032	23	0.154	Q	0.332	31/32	0.9687
66	0.033	5/32	0.1562	R	0.339	63/64	0.9844
65	0.035	22	0.157	11/32	0.3437	1	1.0000
64	0.036	21	0.159	S	0.348	1 1/64	1.0156
63	0.037	20	0.161	T	0.358	1 1/32	1.0312
62	0.038	19	0.166	23/64	0.3594	1 3/64	1.0469
61	0.039	18	0.1695	U	0.368	1 1/16	1.0625
60	0.040	11/64	0.1719	3/8	0.375	1 5/64	1.0781
59	0.041	17	0.173	V	0.377	1 3/32	1.0937
58	0.042	16	0.177	W	0.386	1 7/64	1.1094
57	0.043	15	0.180	25/64	0.3906	1 1/8	1.1250
56	0.0465	14	0.182	X	0.397	1 9/64	1.1406
3/64	0.0468	13	0.185	Y	0.404	1 5/32	1.1562
55	0.052	3/16	0.1875	13/32	0.4062	1 11/64	1.1719
54	0.055	12	0.189	Z	0.413	1 3/16	1.1875
53	0.0595	11	0.191	27/64	0.4219	1 13/64	1.2031
1/16	0.0625	10	0.1935	7/16	0.4375	1 7/32	1.2187
52	0.0635	9	0.196	29/64	0.4531	1 15/64	1.2344
51	0.067	8	0.199	15/32	0.4687	1 1/4	1.2500
50	0.070	7	0.201	31/64	0.4844	1 9/32	1.2812
49	0.073	13/64	0.2031	1/2	0.5000	1 5/16	1.3125
48	0.076	6	0.204	33/64	0.5156	1 11/32	1.3437
5/64	0.0781	5	0.2055	17/32	0.5312	1 3/8	1.3750
47	0.0785	4	0.209	35/64	0.5469	1 13/32	1.4062
46	0.081	3	0.213	9/16	0.5625	1 7/16	1.4375
45	0.082	7/32	0.2187	37/64	0.5781	1 15/32	1.4687
44	0.086	2	0.221	19/32	0.5937	1 1/2	1.5000
43	0.089	1	0.228	39/64	0.6094	1 9/16	1.5625
42	0.0935	A	0.234	5/8	0.6250	1 5/8	1.6250
3/32	0.0937	15/64	0.2344	41/64	0.6406	1 11/16	1.6875
41	0.096	B	0.238	21/32	0.6562	1 3/4	1.7500
40	0.098	C	0.242	43/64	0.6719	1 13/16	1.8125
39	0.0995	D	0.246	11/16	0.6875	1 7/8	1.8750
38	0.1015	E & 1/4	0.250	45/64	0.7031	1 15/16	1.9375
37	0.104	F	0.257	23/32	0.7187	2	2.0000
36	0.1065						

(Courtesy ANSI)

Table 26   Dimensions of Sunk Keys*

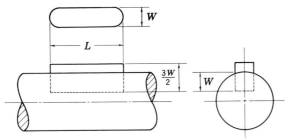

Key No.	L	W	Key No.	L	W
1	½	¹⁄₁₆	22	1⅜	¼
2	½	³⁄₃₂	23	1⅜	⁵⁄₁₆
3	½	⅛	F	1⅜	⅜
4	⅝	³⁄₃₂	24	1½	¼
5	⅝	⅛	25	1½	⁵⁄₁₆
6	⅝	⁵⁄₃₂	G	1½	⅜
7	¾	⅛	51	1¾	¼
8	¾	⁵⁄₃₂	52	1¾	⁵⁄₁₆
9	¾	³⁄₁₆	53	1¾	⅜
10	⅞	⁵⁄₃₂	26	2	³⁄₁₆
11	⅞	³⁄₁₆	27	2	¼
12	⅞	⁷⁄₃₂	28	2	⁵⁄₁₆
A	⅞	¼	29	2	⅜
13	1	³⁄₁₆	54	2¼	¼
14	1	⁷⁄₃₂	55	2¼	⁵⁄₁₆
15	1	¼	56	2¼	⅜
B	1	⁵⁄₁₆	57	2¼	⁷⁄₁₆
16	1⅛	³⁄₁₆	58	2½	⁵⁄₁₆
17	1⅛	⁷⁄₃₂	59	2½	⅜
18	1⅛	¼	60	2½	⁷⁄₁₆
C	1⅛	⁵⁄₁₆	61	2½	½
19	1¼	³⁄₁₆	30	3	⅜
20	1¼	⁷⁄₃₂	31	3	⁷⁄₁₆
21	1¼	¼	32	3	½
D	1¼	⁵⁄₁₆	33	3	⁹⁄₁₆
E	1¼	⅜	34	3	⅝

(Courtesy ANSI)

* Manufactured by Pratt and Whitney, Hartford, Conn.

Table 27 Woodruff Keys and Keyslots

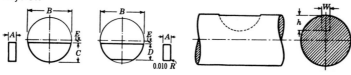

Key * Number	Nominal Size, A × B	Maximum Width of Key, A	Maximum Diameter of Key, B	Maximum Height of Key		Distance below Center, E	Keyslot	
				C	D		Maximum Width, W	Maximum Depth, h
204	$\frac{1}{16} \times \frac{1}{2}$	0.0635	0.500	0.203	0.194	$\frac{3}{64}$	0.0630	0.1718
304	$\frac{3}{32} \times \frac{1}{2}$	0.0948	0.500	0.203	0.194	$\frac{3}{64}$	0.0943	0.1561
305	$\frac{3}{32} \times \frac{5}{8}$	0.0948	0.625	0.250	0.240	$\frac{1}{16}$	0.0943	0.2031
404	$\frac{1}{8} \times \frac{1}{2}$	0.1260	0.500	0.203	0.194	$\frac{3}{64}$	0.1255	0.1405
405	$\frac{1}{8} \times \frac{5}{8}$	0.1260	0.625	0.250	0.240	$\frac{1}{16}$	0.1255	0.1875
406	$\frac{1}{8} \times \frac{3}{4}$	0.1260	0.750	0.313	0.303	$\frac{1}{16}$	0.1255	0.2505
505	$\frac{5}{32} \times \frac{5}{8}$	0.1573	0.625	0.250	0.240	$\frac{1}{16}$	0.1568	0.1719
506	$\frac{5}{32} \times \frac{3}{4}$	0.1573	0.750	0.313	0.303	$\frac{1}{16}$	0.1568	0.2349
507	$\frac{5}{32} \times \frac{7}{8}$	0.1573	0.875	0.375	0.365	$\frac{1}{16}$	0.1568	0.2969
606	$\frac{3}{16} \times \frac{3}{4}$	0.1885	0.750	0.313	0.303	$\frac{1}{16}$	0.1880	0.2193
607	$\frac{3}{16} \times \frac{7}{8}$	0.1885	0.875	0.375	0.365	$\frac{1}{16}$	0.1880	0.2813
608	$\frac{3}{16} \times 1$	0.1885	1.000	0.438	0.428	$\frac{1}{16}$	0.1880	0.3443
609	$\frac{3}{16} \times 1\frac{1}{8}$	0.1885	1.125	0.484	0.475	$\frac{5}{64}$	0.1880	0.3903
807	$\frac{1}{4} \times \frac{7}{8}$	0.2510	0.875	0.375	0.365	$\frac{1}{16}$	0.2505	0.2500
808	$\frac{1}{4} \times 1$	0.2510	1.000	0.438	0.428	$\frac{1}{16}$	0.2505	0.3130
809	$\frac{1}{4} \times 1\frac{1}{8}$	0.2510	1.125	0.484	0.475	$\frac{5}{64}$	0.2505	0.3590
810	$\frac{1}{4} \times 1\frac{1}{4}$	0.2510	1.250	0.547	0.537	$\frac{5}{64}$	0.2505	0.4220
811	$\frac{1}{4} \times 1\frac{3}{8}$	0.2510	1.375	0.594	0.584	$\frac{3}{32}$	0.2505	0.4690
812	$\frac{1}{4} \times 1\frac{1}{2}$	0.2510	1.500	0.641	0.631	$\frac{7}{64}$	0.2505	0.5160
1008	$\frac{5}{16} \times 1$	0.3135	1.000	0.438	0.428	$\frac{1}{16}$	0.3130	0.2818
1009	$\frac{5}{16} \times 1\frac{1}{8}$	0.3135	1.125	0.484	0.475	$\frac{5}{64}$	0.3130	0.3278
1010	$\frac{5}{16} \times 1\frac{1}{4}$	0.3135	1.250	0.547	0.537	$\frac{5}{64}$	0.3130	0.3908
1011	$\frac{5}{16} \times 1\frac{3}{8}$	0.3135	1.375	0.594	0.584	$\frac{3}{32}$	0.3130	0.4378
1012	$\frac{5}{16} \times 1\frac{1}{2}$	0.3135	1.500	0.641	0.631	$\frac{7}{64}$	0.3130	0.4848
1210	$\frac{3}{8} \times 1\frac{1}{4}$	0.3760	1.250	0.547	0.537	$\frac{5}{64}$	0.3755	0.3595
1211	$\frac{3}{8} \times 1\frac{3}{8}$	0.3760	1.375	0.594	0.584	$\frac{3}{32}$	0.3755	0.4065
1212	$\frac{3}{8} \times 1\frac{1}{2}$	0.3760	1.500	0.641	0.631	$\frac{7}{64}$	0.3755	0.4535

(Courtesy ANSI)

All dimensions given in inches.
* *Note:* Key numbers indicate the nominal key dimensions. The last two digits give the nominal diameter (B) in eighths of an inch and the digits preceding the last two give the nominal width (A) in thirty-seconds of an inch. Thus, 204 indicates a key $^2/_{32} \times {}^4/_8$ or $^1/_{16} \times {}^1/_2$ inches.

Table 28   Dimensions of Square and Flat Plain Parallel Stock Keys

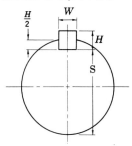
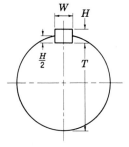

(Dimensions in Inches)

Shaft Diameter Range	Square Key, W x H	Flat Key, W x H	Shaft Diameter Range	Square Key, W x H	Flat Key, W x H
½ – ⁹⁄₁₆	⅛ x ⅛	⅛ x ³⁄₃₂	2⁵⁄₁₆–2¾	⅝ x ⅝	⅝ x ⁷⁄₁₆
⅝ – ⅞	³⁄₁₆ x ³⁄₁₆	³⁄₁₆ x ⅛	2⅞–3¼	¾ x ¾	¾ x ½
1⁵⁄₁₆–1¼	¼ x ¼	¼ x ³⁄₁₆	3⅜–3¾	⅞ x ⅞	⅞ x ⅝
1⁵⁄₁₆–1⅜	⁵⁄₁₆ x ⁵⁄₁₆	⁵⁄₁₆ x ¼	3⅞–4½	1 x 1	1 x ¾
1⁷⁄₁₆–1¾	⅜ x ⅜	⅜ x ¼	4¾–5½	1¼ x 1¼	1¼ x ⅞
1¹³⁄₁₆–2¼	½ x ½	½ x ⅜	5¾–6	1½ x 1½	1½ x 1

(Courtesy ANSI)

Stock keys are applicable to the general run of work.

Table 29   Square and Flat Plain Taper Stock Keys

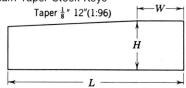

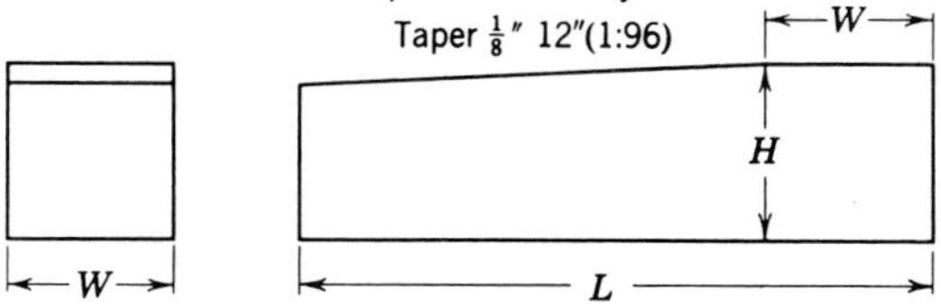

Taper ⅛″ 12″(1:96)

(Dimensions in Inches)

Shaft Diameter Range	Square Type Width Maximum, W	Square Type Height Minimum, H	Flat Type Width Maximum, W	Flat Type Height Minimum, H
½ – ⁹⁄₁₆	⅛	⅛	⅛	³⁄₃₂
⅝ – ⅞	³⁄₁₆	³⁄₁₆	³⁄₁₆	⅛
1⁵⁄₁₆–1¼	¼	¼	¼	³⁄₁₆
1⁵⁄₁₆–1⅜	⁵⁄₁₆	⁵⁄₁₆	⁵⁄₁₆	¼
1⁷⁄₁₆–1¾	⅜	⅜	⅜	¼
1¹³⁄₁₆–2¼	½	½	½	⅜
2⁵⁄₁₆–2¾	⅝	⅝	⅝	⁷⁄₁₆
2⅞–3¼	¾	¾	¾	½
3⅜–3¾	⅞	⅞	⅞	⅝
3⅞–4½	1	1	1	¾
4¾–5½	1¼	1¼	1¼	⅞
5¾–6	1½	1½	1½	1

(Courtesy ANSI)

The minimum stock length of keys is 4W, and the maximum stock length is 16W.

Table 29 (*continued*)

Dimensions of Square and Flat Gib-Head Taper Stock Keys

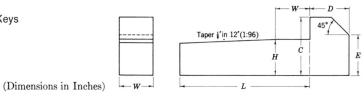

(Dimensions in Inches)

| Shaft Diameter Range | Square Type of Key | | | | | Flat Type of Key | | | | |
| | Key | | Gib-head | | | Key | | Gib-head | | |
	Maximum Width, W	Height, H	Height, C	Length, D	Height to Edge Chamfer, E	Maximum Width, W	Height, H	Height, C	Length, D	Height to Edge Chamfer, E
1/2 – 9/16	1/8	1/8	1/4	7/32	5/32	1/8	3/32	3/16	1/8	1/8
5/8 – 7/8	3/16	3/16	5/16	9/32	7/32	3/16	1/8	1/4	3/16	5/32
15/16–1 1/4	1/4	1/4	7/16	11/32	11/32	1/4	3/16	5/16	1/4	3/16
1 5/16 –1 3/8	5/16	5/16	9/16	13/32	13/32	5/16	1/4	3/8	5/16	1/4
1 7/16 –1 3/4	3/8	3/8	1 1/16	15/32	15/32	3/8	1/4	7/16	3/8	5/16
1 13/16–2 1/4	1/2	1/2	7/8	19/32	5/8	1/2	3/8	5/8	1/2	7/16
2 5/16 –2 3/4	5/8	5/8	1 1/16	23/32	3/4	5/8	7/16	3/4	5/8	1/2
2 7/8 –3 1/4	3/4	3/4	1 1/4	7/8	7/8	3/4	1/2	7/8	3/4	5/8
3 3/8 –3 3/4	7/8	7/8	1 1/2	1	1	7/8	5/8	1 1/16	7/8	3/4
3 7/8 –4 1/2	1	1	1 3/4	1 3/16	1 3/16	1	3/4	1 1/4	1	1 3/16
4 3/4 –5 1/2	1 1/4	1 1/4	2	1 7/16	1 7/16	1 1/4	7/8	1 1/2	1 1/4	1
5 3/4 –6	1 1/2	1 1/2	2 1/2	1 3/4	1 3/4	1 1/2	1	1 3/4	1 1/2	1 1/4

(Courtesy ANSI)

Stock keys are applicable to the general run of work. They are not intended to cover the finer applications where a close fit may be required. The minimum stock length of keys is 4W and the maximum stock length is 16W.

Table 30   Taper Pins

Dimensions of taper pins

Number	7/0	6/0	5/0	4/0	3/0	2/0	0	1	2	3	4	5	6	7	8	9	10
Size (Large End)	0.0625	0.0780	0.0940	0.1090	0.125						2500	0.2890	0.3410	0.4090	0.4920	0.5910	0.7060
Length, L																	
0.375	X	X															
0.500	X	X	X	X	X												
0.625	X	X	X	X	X												
0.750		X	X	X	X	X	X	X									
0.875					X	X	X	X	X	X							
1.000			X	X	X	X	X	X	X	X	X	X					
1.250						X	X	X	X	X	X	X	X				
1.500							X	X	X	X	X	X	X				
1.750								X	X	X	X	X	X				
2.000								X	X	X	X	X	X	X	X		
2.250									X	X	X	X	X	X	X		
2.500									X	X	X	X	X	X	X		
2.750										X	X	X	X	X	X	X	
3.000										X	X	X	X	X	X	X	
3.250													X	X	X	X	X
3.500													X	X	X	X	X
3.750													X	X	X	X	X
4.000														X	X	X	X
4.250															X	X	X
4.500															X	X	X
4.750															X	X	X
5.000																X	X
5.250																X	X
5.500																X	X
5.750																X	X
6.000																X	X

(Courtesy ANSI)

All dimensions are given in inches. Standard reamers are available for pins given above the line.
Pins Nos. 11 (size 0.8600), 12 (size 1.032), 13 (size 1.241), and 14 (1.523) are special sizes—hence their lengths are special.
To find small diameter of pin, multiply the length by 0.02083 and subtract the result from the large diameter.

TYPES	COMMERCIAL TYPE	PRECISION TYPE	
Sizes	7/0 to 14	7/0 to 10	0.0005 up to 1 in. long
Tolerance on Diameter	(+0.0013, −0.0007)	(+0.0013, −0.0007)	0.001 1¹/₁₆ to 2 in. long
Taper	¹/₄ In. per Ft	¹/₄ In. per Ft	0.002 2¹/₁₆ and longer
Length Tolerance	(±0.030)	(±0.030)	
Concavity Tolerance	None		

Table 31   Dimensions of Cotter Pins

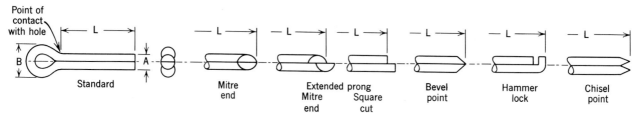

Diameter Nominal	Diameter A		Outside Eye Diameter B Min	Hole Sizes Recommended
	Max	Min		
0.031	0.032	0.028	1/16	3/64
0.047	0.048	0.044	3/32	1/16
0.062	0.060	0.056	1/8	3/64
0.078	0.076	0.072	5/32	3/32
0.094	0.090	0.086	3/16	7/64
0.109	0.104	0.100	7/32	1/8
0.125	0.120	0.116	1/4	9/64
0.141	0.134	0.130	9/32	3/32
0.156	0.150	0.146	3/16	11/64
0.188	0.176	0.172	3/8	13/64
0.219	0.207	0.202	7/16	15/64
0.250	0.225	0.220	1/2	17/64
0.312	0.280	0.275	5/8	5/16
0.375	0.335	0.329	3/4	3/8
0.438	0.406	0.400	7/8	7/16
0.500	0.473	0.467	1	1/2
0.625	0.598	0.590	1 3/4	5/8
0.750	0.723	0.715	1 1/2	3/4

(Courtesy ANSI)

All dimensions are given in inches.
A certain amount of leeway is permitted in the design of the head, however the outside diameters given should be adhered to.
Prongs are to be parallel, ends shall not be open.
Points may be blunt, bevel, extended prong, mitre, etc., and purchaser may specify type required.
Lengths shall be measured as shown on the above illustration. (L-Dimension)
Cotter pins shall be free from burrs or any defects that will affect their serviceability.

Table 32 Retaining Rings

See Fig. 2

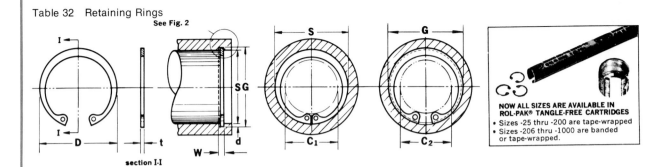

section I-I

NOW ALL SIZES ARE AVAILABLE IN ROL-PAK® TANGLE-FREE CARTRIDGES
• Sizes -25 thru -200 are tape-wrapped
• Sizes -206 thru -1000 are banded or tape-wrapped.

HOUSING DIA.			MIL-R-21248 MS 16625 INTERNAL SERIES N5000	TRUARC RING DIMENSIONS					GROOVE DIMENSIONS					APPLICATION DATA			
				Thickness t applies only to un-plated rings. For plated and stainless steel (Type H) rings, add .002″ to the listed maximum thickness. Maximum ring thickness will be at least .0002″ less than the listed minimum groove width (W).				Approx. weight per 1000 pieces	T.I.R. (total indicator reading) is the maximum allowable deviation of concentricity between groove and housing.				Nominal groove depth	CLEARANCE DIAMETER		ALLOW. THRUST LOAD (lbs.) Sharp corner abutment	
															When sprung into housing	When sprung into groove	RINGS	GROOVES
Dec. equiv. inch	Approx. fract. equiv. inch	Approx. mm		FREE DIA.		THICKNESS			DIAMETER		WIDTH					Safety factor = 4	Safety factor = 2
S	S	S	size—no.	D	tol.	t	tol.	lbs.	G	tol.	W	tol.	d	C_1	C_2	P_r	P_g
.250	1/4	6.4	N5000-25	.280		.015		.08	.268	±.001/.0015 T.I.R.	.018	+.002 -.000	.009	.115	.133	420	190
.312	5/16	7.9	N5000-31	.346		.015		.11	.330		.018		.009	.173	.191	530	240
.375	3/8	9.5	N5000-37	.415		.025		.25	.397	±.002 .002 T.I.R.	.029		.011	.204	.226	1050	350
.438	7/16	11.1	N5000-43	.482		.025		.37	.461		.029		.012	.23	.254	1220	440
.453	29/64	11.5	N5000-45	.498		.025		.43	.477		.029		.012	.25	.274	1280	460
.500	1/2	12.7	N5000-50	.548	+.010 -.005	.035		.70	.530		.039		.015	.26	.29	1980	510
.512	—	13.0	N5000-51	.560		.035		.77	.542	±.002 .004 T.I.R.	.039		.015	.27	.30	2030	520
.562	9/16	14.3	N5000-56	.620		.035		.86	.596		.039		.017	.275	.305	2220	710
.625	5/8	15.9	N5000-62	.694		.035		1.0	.665		.039		.020	.34	.38	2470	1050
.688	11/16	17.5	N5000-68	.763		.035		1.2	.732		.039	+.003 -.000	.022	.40	.44	2700	1280
.750	3/4	19.0	N5000-75	.831		.035		1.3	.796		.039		.023	.45	.49	3000	1460
.777	—	19.7	N5000-77	.859		.042		1.7	.825		.046		.024	.475	.52	4550	1580
.812	13/16	20.6	N5000-81	.901		.042		1.9	.862		.046		.025	.49	.54	4800	1710
.866	—	22.0	N5000-86	.961		.042		2.0	.920		.046		.027	.54	.59	5100	1980
.875	7/8	22.2	N5000-87	.971	+.015 -.010	.042	±.002	2.1	.931	±.003 .004 T.I.R.	.046		.028	.545	.60	5150	2080
.901	—	22.9	N5000-90	1.000		.042		2.2	.959		.046		.029	.565	.62	5350	2200
.938	15/16	23.8	N5000-93	1.041		.042		2.4	1.000		.046		.031	.61	.67	5600	2450
1.000	1	25.4	N5000-100	1.111		.042		2.7	1.066		.046		.033	.665	.73	5950	2800
1.023	—	26.0	N5000-102	1.136		.042		2.8	1.091		.046		.034	.69	.755	6050	3000
1.062	1 1/16	27.0	N5000-106	1.180		.050		3.7	1.130		.056		.034	.685	.75	7450	3050
1.125	1 1/8	28.6	N5000-112	1.249		.050		4.0	1.197		.056		.036	.745	.815	7900	3400
1.181	—	30.0	N5000-118	1.319		.050		4.3	1.255		.056		.037	.79	.86	8400	3700
1.188	1 3/16	30.2	N5000-118	1.319		.050		4.3	1.262		.056		.037	.80	.87	8400	3700
1.250	1 1/4	31.7	N5000-125	1.388		.050		4.8	1.330		.056		.040	.875	.955	8800	4250
1.259	—	32.0	N5000-125	1.388	+.025 -.020	.050		4.8	1.339		.056		.040	.885	.965	8800	4250
1.312	1 5/16	33.3	N5000-131	1.456		.050		5.0	1.396		.056		.042	.93	1.01	9300	4700
1.375	1 3/8	34.9	N5000-137	1.526		.050		5.1	1.461		.056		.043	.99	1.07	9700	5050
1.378	—	35.0	N5000-137	1.526		.050		5.1	1.464	±.004 .005 T.I.R.	.056	+.004 -.000	.043	.99	1.07	9700	5050
1.438	1 7/16	36.5	N5000-143	1.596		.050		5.8	1.528		.056		.045	1.06	1.15	10200	5500
1.456	—	37.0	N5000-145	1.616		.050		6.4	1.548		.056		.046	1.08	1.17	10300	5700
1.500	1 1/2	38.1	N5000-150	1.660		.050		6.5	1.594		.056		.047	1.12	1.21	10550	6000
1.562	1 9/16	39.7	N5000-156	1.734		.062		8.9	1.658		.068		.048	1.14	1.23	13700	6350
1.575	—	40.0	N5000-156	1.734	+.035 -.025	.062	±.003	8.9	1.671	±.005 .005 T.I.R.	.068		.048	1.15	1.24	13700	6350
1.625	1 5/8	41.3	N5000-162	1.804		.062		10.0	1.725		.068		.050	1.15	1.25	14200	6900
1.653	—	42.0	N5000-165	1.835		.062		10.4	1.755		.068		.051	1.17	1.27	14500	7200
1.688	1 11/16	42.9	N5000-168	1.874		.062		10.8	1.792		.068		.052	1.21	1.31	14800	7450

Table 32 *(continued)*

FIG. 1: MAXIMUM ALLOWABLE CORNER RADIUS (R_max.) AND CHAMFER (Ch_max.)

FIG. 2: ENLARGED DETAIL OF GROOVE PROFILE AND EDGE MARGIN (Z)

MAXIMUM BOTTOM RADII	
Ring Size	R
·25 thru ·100	.005
·102 thru ·1000	.010

FIG. 3: SUPPLEMENTARY RING DIMENSIONS

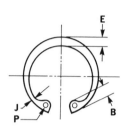

FIG. 4: MINIMUM GAP WIDTH (Ring installed in groove)

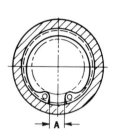

	SUPPLEMENTARY APPLICATION DATA				SUPPLEMENTARY RING DIMENSIONS								
INTERNAL SERIES **N5000**	Maximum allowable corner radii and chamfers of retained parts (Fig. 1)		Allow. assembly load with R max. or Ch max.	Edge margin (Fig. 2)	LUG		LARGE SECTION		SMALL SECTION		HOLE DIAMETER		MIN. GAP WIDTH (Fig. 4) Ring installed in groove
size—no.	$R_{max.}$	$Ch_{max.}$	P'_r(lbs.)	Z	B	tol.	E	tol.	J	tol.	P	tol.	A
N5000-25	.011	.0085	190	.027	.065		.025	±.002	.015	±.002	.031		.047
N5000-31	.016	.013	190	.027	.066		.033		.018		.031		.055
N5000-37	.023	.018	530	.033	.082		.040		.028		.041		.063
N5000-43	.027	.021	530	.036	.098	±.003	.049	±.003	.029	±.003	.041		.063
N5000-45	.027	.021	530	.036	.098		.050		.030		.047		.071
N5000-50	.027	.021	1100	.045	.114		.053		.035		.047		.090
N5000-51	.027	.021	1100	.045	.114		.053		.035		.047		.092
N5000-56	.027	.021	1100	.051	.132		.053		.035		.047		.095
N5000-62	.027	.021	1100	.060	.132		.060	±.004	.035	±.004	.062		.104
N5000-68	.027	.021	1100	.066	.132		.063		.036		.062	+.010 −.002	.118
N5000-75	.032	.025	1100	.069	.142		.070		.040		.062		.143
N5000-77	.035	.028	1650	.072	.146		.074		.044		.062		.145
N5000-81	.035	.028	1650	.075	.155		.077		.044		.062		.153
N5000-86	.035	.028	1650	.081	.155		.081		.045		.062		.172
N5000-87	.035	.028	1650	.084	.155		.084	±.005	.045	±.005	.062		.179
N5000-90	.038	.030	1650	.087	.155		.087		.047		.062		.188
N5000-93	.038	.030	1650	.093	.155		.091		.050		.062		.200
N5000-100	.042	.034	1650	.099	.155		.104		.052		.062		.212
N5000-102	.042	.034	1650	.102	.155		.106		.054		.062		.220
N5000-106	.044	.035	2400	.102	.180		.110		.055		.078		.213
(S=1.181) N5000-112	.047	.036	2400	.108	.180		.116		.057		.078		.232
(S=1.188) N5000-118	.047	.036	2400	.111	.180	±.005	.120		.058		.078		.226
(S=1.250) N5000-118	.047	.036	2400	.111	.180		.120		.058		.078		.245
(S=1.259) N5000-125	.048	.038	2400	.120	.180		.124		.062		.078		.265
N5000-125	.048	.038	2400	.120	.180		.124	±.006	.062	±.006	.078		.290
N5000-131	.048	.038	2400	.126	.180		.130		.062		.078		.284
(S=1.375) N5000-137	.048	.038	2400	.129	.180		.130		.063		.078		.297
(S=1.378) N5000-137	.048	.038	2400	.129	.180		.130		.063		.078	+.015 −.002	.305
N5000-143	.048	.038	2400	.135	.180		.133		.065		.078		.313
N5000-145	.048	.038	2400	.138	.180		.133		.065		.078		.320
N5000-150	.048	.038	2400	.141	.180		.133		.066		.078		.340
(S=1.562) N5000-156	.064	.050	3900	.144	.202		.157		.078		.078		.338
(S=1.575) N5000-156	.064	.050	3900	.144	.202		.157		.078		.078		.374
N5000-162	.064	.050	3900	.150	.227		.164	±.007	.082	±.007	.078		.339
N5000-165	.064	.050	3900	.153	.227		.167		.083		.078		.348
N5000-168	.064	.050	3900	.156	.227		.170		.085		.078		.357

Table 32 (continued)

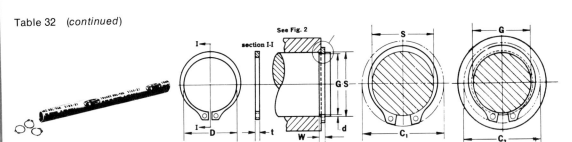

section I-I See Fig. 2

SHAFT DIAMETER			MIL-R-21248 MS 16624 EXTERNAL SERIES 5100	TRUARC RING DIMENSIONS					GROOVE DIMENSIONS					APPLICATION DATA			
Dec. equiv. inch	Approx fract. equiv. inch	Approx mm		FREE DIA.		THICKNESS		Approx. weight per 1000 pieces	DIAMETER		WIDTH		Nominal groove depth	CLEARANCE DIAMETER		ALLOW. THRUST LOAD (lbs.) Sharp Corner Abutment	
														When sprung over shaft	When sprung into groove	RINGS	GROOVES
S	S	S	size — no.	D	tol.	t	tol.	lbs.	G	tol.	W	tol.	d	C_1	C_2	P_r	P_g
.125	1/8	3.2	▲ 5100-12	.112		.010	±.001	.018	.117		.012	+.002 −.000	.004	.222	.214	110	35
.156	5/32	4.0	▲ 5100-15	.142		.010		.037	.146		.012		.005	.270	.260	130	55
.188	3/16	4.8	▲ 5100-18	.168	+.002 −.004	.015		.059	.175	±.0015 .0015 T.I.R.	.018		.006	.298	.286	240	80
.197	— —	5.0	▲ 5100-19	.179		.015		.063	.185		.018		.006	.319	.307	250	85
.219	7/32	5.6	▲ 5100-21	.196		.015		.074	.205		.018		.007	.338	.324	280	110
.236	15/64	6.0	▲ 5100-23	.215		.015		.086	.222		.018		.007	.355	.341	310	120
.250	1/4	6.4	• 5100-25	.225		.025		.21	.230		.029		.010	.45	.43	590	175
.276	— —	7.0	5100-27	.250		.025		.23	.255		.029		.010	.48	.46	650	195
.281	9/32	7.1	• 5100-28	.256		.025		.24	.261		.029		.010	.49	.47	660	200
.312	5/16	7.9	• 5100-31	.281		.025		.27	.290		.029		.011	.54	.52	740	240
.344	11/32	8.7	5100-34	.309		.025		.31	.321	±.002 .002 T.I.R.	.029		.011	.57	.55	800	265
.354	— —	9.0	5100-35	.320		.025		.35	.330		.029		.012	.59	.57	820	300
.375	3/8	9.5	• 5100-37	.338	+.002 −.005	.025		.39	.352		.029		.012	.61	.59	870	320
.394	— —	10.0	5100-39	.354		.025		.42	.369		.029		.012	.62	.60	940	335
.406	13/32	10.3	5100-40	.366		.025		.43	.382		.029		.012	.63	.61	950	350
.438	7/16	11.1	• 5100-43	.395		.025		.50	.412		.029		.013	.66	.64	1020	400
.469	15/32	11.9	5100-46	.428		.025		.54	.443		.029		.013	.68	.66	1100	450
.500	1/2	12.7	• 5100-50	.461		.035		.91	.468	±.002 .004 T.I.R.	.039	+.003 −.000	.016	.77	.74	1650	550
.551	— —	14.0	5100-55	.509		.035		.90	.519		.039		.016	.81	.78	1800	600
.562	9/16	14.3	• 5100-56	.521		.035		1.1	.530		.039		.016	.82	.79	1850	650
.594	19/32	15.1	5100-59	.550		.035	±.002	1.2	.559		.039		.017	.86	.83	1950	750
.625	5/8	15.9	• 5100-62	.579		.035		1.3	.588		.039		.018	.90	.87	2060	800
.669	— —	17.0	5100-66	.621		.035		1.4	.629		.039		.020	.93	.89	2200	950
.672	43/64	17.1	5100-66	.621		.035		1.4	.631		.039		.020	.93	.89	2200	950
.688	11/16	17.5	• 5100-68	.635	+.005 −.010	.042		1.8	.646	±.003 .004 T.I.R.	.046		.021	1.01	.97	3400	1000
.750	3/4	19.0	• 5100-75	.693		.042		2.1	.704		.046		.023	1.09	1.05	3700	1200
.781	25/32	19.8	5100-78	.722		.042		2.2	.733		.046		.024	1.12	1.08	3900	1300
.812	13/16	20.6	5100-81	.751		.042		2.5	.762		.046		.025	1.15	1.10	4000	1450
.875	7/8	22.2	5100-87	.810		.042		2.8	.821		.046		.027	1.21	1.16	4300	1650
.938	15/16	23.8	5100-93	.867		.042		3.1	.882		.046		.028	1.34	1.29	4650	1850
.984	63/64	25.0	5100-98	.910		.042		3.5	.926		.046		.029	1.39	1.34	4850	2000
1.000	1	25.4	5100-100	.925		.042		3.6	.940		.046		.030	1.41	1.35	4950	2100
1.023	— —	26.0	5100-102	.946		.042		3.9	.961		.046		.031	1.43	1.37	5050	2250
1.062	1 1/16	27.0	5100-106	.982	+.010 −.015	.050		4.8	.998	±.004 .005 T.I.R.	.056	+.004 −.000	.032	1.50	1.44	6200	2400
1.125	1 1/8	28.6	5100-112	1.041		.050		5.1	1.059		.056		.033	1.55	1.49	6600	2600

Thickness t applies only to un-plated rings. For plated and stainless steel (Type H) rings, add .002" to the listed maximum thickness. Maximum ring thickness will be at least .0002" less than the listed minimum groove width (W).

T.I.R. (total indicator reading) is the maximum allowable deviation of concentricity between groove and shaft.

Table 32 *(continued)*

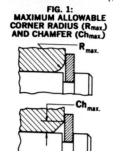

FIG. 1:
MAXIMUM ALLOWABLE CORNER RADIUS (R_{max.}) AND CHAMFER (Ch_{max.})

FIG. 2:
ENLARGED DETAIL OF GROOVE PROFILE AND EDGE MARGIN (Z)

FIG. 3:
SUPPLEMENTARY RING DIMENSIONS

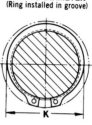

FIG. 4:
MAXIMUM GAGING DIAMETER
(Ring installed in groove)

FIG. 5:
LUG DESIGN
Sizes -12 thru -23

	SUPPLEMENTARY APPLICATION DATA					SUPPLEMENTARY RING DIMENSIONS								
EXTERNAL SERIES **5100**	Maximum allowable corner radii and chamfers of retained parts (Fig. 1)		Allow. assembly load with R max. or Ch max.	Edge margin (Fig. 2)	Calculated RPM limits (Std. ring mat'l.) Apply req'd. safety factor	(Fig. 3)								MAX. GAGING DIA. Ring installed in groove (Fig. 4)
						LUG		LARGE SECTION		SMALL SECTION		HOLE DIAMETER		
size — no.	R_{max.}	Ch_{max.}	P'_r (lbs.)	Z		B	tol.	E	tol.	J	tol.	P	tol.	K
▲ 5100-12	.010	.006	45	.012	80000	.046	±.002	.018	±.0015	.011	±.0015	.026		.148
▲ 5100-15	.015	.009	45	.015	80000	.054		.026		.016		.026		.189
▲ 5100-18	.014	.0085	105	.018	80000	.050		.025		.016		.025		.218
▲ 5100-19	.0145	.009	105	.018	80000	.056		.026	±.002	.016	±.002	.026		.229
▲ 5100-21	.015	.009	105	.021	80000	.056		.028		.017		.026		.252
▲ 5100-23	.0165	.010	105	.021	80000	.056		.030		.019		.026		.272
• 5100-25	.018	.011	470	.030	80000	.080		.035		.025		.041		.290
5100-27	.0175	.0105	470	.031	76000	.081		.035		.024		.041		.315
• 5100-28	.020	.012	470	.030	74000	.080		.038		.0255		.041		.326
• 5100-31	.020	.012	470	.033	70000	.087		.040		.026		.041		.357
5100-34	.021	.0125	470	.033	64000	.087		.042	±.003	.0265	±.003	.041		.390
5100-35	.023	.014	470	.036	62000	.087		.046		.029		.041		.405
• 5100-37	.026	.0155	470	.036	60000	.088		.050		.0305		.041		.433
5100-39	.027	.016	470	.037	56500	.087		.052		.031		.041		.452
5100-40	.0285	.017	470	.036	55000	.087	±.003	.054		.033		.041	+.010 −.002	.468
• 5100-43	.029	.0175	470	.039	50000	.088		.055		.033		.041		.501
5100-46	.031	.018	470	.039	42000	.088		.060		.035		.041		.540
• 5100-50	.034	.020	910	.048	40000	.108		.065		.040		.047		.574
5100-55	.027	.0165	910	.048	36000	.108		.053		.036		.047		.611
• 5100-56	.038	.023	910	.048	35000	.108		.072	±.004	.041	±.004	.047		.644
5100-59	.0395	.0235	910	.052	32000	.109		.076		.043		.047		.680
• 5100-62	.0415	.025	910	.055	30000	.110		.080		.045		.047		.715
(S=.669) 5100-66	.040	.024	910	.060	29000	.110		.082		.043		.047		.756
(S=.672) 5100-66	.040	.024	910	.060	29000	.110		.082		.043		.047		.758
• 5100-68	.042	.025	1340	.063	28000	.136		.084		.048		.052		.779
• 5100-75	.046	.0275	1340	.069	26500	.136	±.004	.092		.051		.052		.850
5100-78	.047	.028	1340	.072	25500	.136		.094		.052		.052		.883
5100-81	.047	.028	1340	.075	24500	.136		.096		.054		.052		.914
5100-87	.051	.0305	1340	.081	23000	.137		.104	±.005	.057	±.005	.052		.987
5100-93	.055	.033	1340	.084	21500	.166		.110		.063		.078		1.054
5100-98	.056	.0335	1340	.087	20500	.167		.114		.0645		.078	+.015 −.002	1.106
5100-100	.057	.034	1340	.090	20000	.167		.116		.065		.078		1.122
5100-102	.058	.035	1340	.093	19500	.168		.118		.066		.078		1.147
5100-106	.060	.036	1950	.096	19000	.181		.122	±.006	.069	±.006	.078		1.192
5100-112	.063	.038	1950	.099	18800	.182		.128		.071		.078		1.261

Table 33 Wire and Sheet-Metal Gages

No. of gage	American copper or B. & S. wire gage	British imperial wire gage	U. S. St'd. gage for plate	No. of gage	American copper or B. & S. wire gage	British imperial wire gage	U. S. St'd. gage for plate
0000000	. . .	0.5000	0.5000	20	0.0320	0.0360	0.0375
000000	0.5800	0.4640	0.4688	21	0.0285	0.0320	0.0344
00000	0.5165	0.4320	0.4375	22	0.0253	0.0280	0.0313
0000	0.4600	0.4000	0.4063	23	0.0226	0.0240	0.0281
000	0.4096	0.3720	0.3750	24	0.0201	0.0220	0.0250
00	0.3648	0.3480	0.3438	25	0.0179	0.0200	0.0219
0	0.3249	0.3240	0.3125	26	0.0159	0.0180	0.0188
1	0.2893	0.3000	0.2813	27	0.0142	0.0164	0.0172
2	0.2576	0.2760	0.2656	28	0.0126	0.0148	0.0156
3	0.2294	0.2520	0.2500	29	0.0113	0.0136	0.0141
4	0.2043	0.2320	0.2344	30	0.0100	0.0124	0.0125
5	0.1819	0.2120	0.2188	31	0.0089	0.0116	0.0109
6	0.1620	0.1920	0.2031	32	0.0080	0.0108	0.0102
7	0.1443	0.1760	0.1875	33	0.0071	0.0100	0.0094
8	0.1285	0.1600	0.1719	34	0.0063	0.0092	0.0086
9	0.1144	0.1440	0.1563	35	0.0056	0.0084	0.0078
10	0.1019	0.1280	0.1406	36	0.0050	0.0076	0.0070
11	0.0907	0.1160	0.1250	37	0.0045	0.0068	0.0066
12	0.0808	0.1040	0.1094	38	0.0040	0.0060	0.0063
13	0.0720	0.0920	0.0938	39	0.0035	0.0052	. . .
14	0.0641	0.0800	0.0781	40	0.0031	0.0048	. . .
15	0.0571	0.0720	0.0703	41	0.0028	0.0044	. . .
16	0.0508	0.0640	0.0625	42	0.0025	0.0040	. . .
17	0.0453	0.0560	0.0563	43	0.0022	0.0036	. . .
18	0.0403	0.0480	0.0500	44	0.0020	0.0032	. . .
19	0.0359	0.0400	0.0438	45	0.00176	0.0028	. . .

(Courtesy ANSI)

Table 34   Metric Hex Cap Screws and Hex Bolts

Table 1   Dimensions of hex cap screws

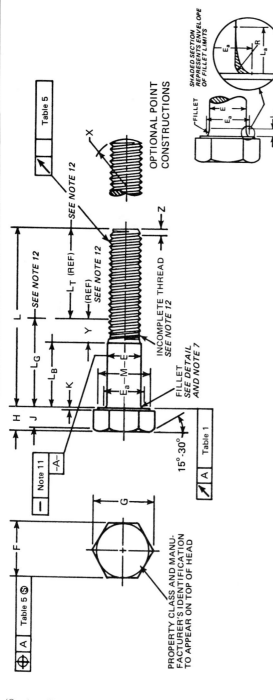

PROPERTY CLASS AND MANU-
FACTURER'S IDENTIFICATION
TO APPEAR ON TOP OF HEAD

OPTIONAL POINT
CONSTRUCTIONS

SHADED SECTION
REPRESENTS ENVELOPE
OF FILLET LIMITS

ENLARGED DETAIL OF FILLET

Nom Screw Size & Thread Pitch	E Body Diameter		F Width Across Flats		G Width Across Corners		H Head Height		J Wrenching Height	K Washer Face Thickness		M Washer Face Dia	Runout of Bearing Surface FIR	E_a Fillet Transition Dia	L_a Fillet Transition Length	R Radius of Fillet	L_T (Ref) Thread Length, basic			Y (Ref) Transition Thread Length
	Max	Min	Max	Min	Max	Min	Max	Min	Min	Max	Min	Min	Max	Max	Max	Min	Screw Lengths ≤125	Screw Lengths >125 and ≤200	Screw Lengths >200	Max
M5x0.8	5.00	4.82	8.00	7.78	9.24	8.87	3.65	3.35	2.4	0.5	0.2	7.0	0.22	5.7	1.2	0.2	16	22	35	4.0
M6.3x1	6.30	6.08	10.00	9.76	11.55	11.13	4.47	4.13	3.0	0.5	0.2	8.9	0.25	7.3	1.8	0.3	18.6	24.6	37.6	5.0
M8x1.25	8.00	7.78	13.00	12.73	15.01	14.51	5.50	5.10	3.7	0.6	0.3	11.6	0.28	9.2	2.0	0.4	22	28	41	6.2
M10x1.5	10.00	9.78	15.00	14.70	17.32	16.76	6.63	6.17	4.5	0.6	0.3	13.6	0.31	11.2	2.0	0.4	26	32	45	7.5
M12x1.75	12.00	11.73	18.00	17.67	20.78	20.14	7.76	7.24	5.2	0.6	0.3	16.6	0.35	13.2	3.0	0.4	30	36	49	8.8
M14x2	14.00	13.73	21.00	20.64	24.25	23.53	9.09	8.51	6.2	0.6	0.3	19.4	0.39	15.2	3.0	0.4	34	40	53	10.0
M16x2	16.00	15.73	24.00	23.61	27.71	26.92	10.32	9.68	7.0	0.8	0.4	22.4	0.43	17.7	3.0	0.6	38	44	57	10.0
M20x2.5	20.00	19.67	30.00	29.35	34.64	33.46	12.88	12.12	8.8	0.8	0.4	27.6	0.53	22.4	4.0	0.8	46	52	65	12.5
M24x3	24.00	23.67	36.00	35.25	41.57	40.19	15.44	14.56	10.5	0.8	0.4	32.9	0.63	26.4	4.0	0.8	54	60	73	15.0
M30x3.5	30.00	29.61	46.00	44.50	53.12	50.73	19.48	17.92	13.1	0.8	0.4	42.5	0.78	33.4	6.0	1.0	66	72	85	17.5
M36x4	36.00	35.61	55.00	53.20	63.51	60.65	23.38	21.62	15.8	0.8	0.4	50.8	0.93	39.4	6.0	1.0	78	84	97	20.0
M42x4.5	42.00	41.61	65.00	62.90	75.06	71.71	26.97	25.03	18.2	1.0	0.5	58.5	1.09	45.6	6.3	1.2	90	96	109	22.5
M48x5	48.00	47.61	75.00	72.60	86.60	82.76	31.07	28.93	21.0	1.0	0.5	67.5	1.25	52.6	8.0	1.5	102	108	121	25.0
M56x5.5	56.00	55.54	85.00	82.20	98.15	93.71	36.20	33.80	24.5	1.0	0.5	76.5	1.47	62.0	10.5	2.0	—	124	137	27.5
M64x6	64.00	63.54	95.00	91.80	109.70	104.65	41.32	38.68	28.0	1.0	0.5	85.5	1.69	70.0	10.5	2.0	—	140	153	30.0
M72x6	72.00	71.54	105.00	101.40	121.24	115.60	46.45	43.55	31.5	1.2	0.6	94.5	1.91	78.0	10.5	2.0	—	156	169	30.0
M80x6	80.00	79.54	115.00	111.00	132.79	126.54	51.58	48.42	35.0	1.2	0.6	103.5	2.13	86.0	10.5	2.0	—	172	185	30.0
M90x6	90.00	89.46	130.00	125.50	150.11	143.07	57.74	54.26	39.2	1.2	0.6	117.0	2.41	96.0	10.5	2.0	—	192	205	30.0
M100x6	100.00	99.46	145.00	140.00	167.43	159.60	63.90	60.10	43.4	1.2	0.6	130.5	2.69	107.0	12.2	2.5	—	212	225	30.0

(Courtesy IFI-506-1976)

Table 34 (*continued*)

Table 2 Maximum grip gaging lengths and minimum body lengths for hex cap screws

Nom Screw Size / L Nom	M5x0.8		M6.3x1		M8x1.25		M10x1.5		M12x1.75		M14x2		M16x2		M20x2.5		M24x3		M30x3.5		M36x4	
	LG Max	LB Min	LG Max	LB Min	LG Max	LB Min	LG Max	LB Min	LG Max	LB Min	LG Max	LB Min	LG Max	LB Min	LG Max	LB Min	LG Max	LB Min	LG Max	LB Min	LG Max	LB Min
8																						
10																						
12																						
14																						
16																						
20																						
25	9.0	5.0																				
30	9.0	5.0	11.4	6.4																		
35	19.0	15.0	16.4	11.4	13.0	6.8																
40	19.0	15.0	16.4	11.4	13.0	6.8	14.0	6.5														
45	29.0	25.0	26.4	21.4	23.0	16.8	19.0	11.5	15.0	6.2												
50	29.0	25.0	26.4	21.4	23.0	16.8	19.0	11.5	15.0	6.2	16.0	6.0										
(55)			36.4	31.4	33.0	26.8	29.0	21.5	25.0	16.2	21.0	11.0	17.0	7.0								
60			36.4	31.4	33.0	26.8	29.0	21.5	25.0	16.2	21.0	11.0	17.0	7.0								
(65)					43.0	36.8	39.0	31.5	35.0	26.2	31.0	21.0	27.0	17.0	19.0	6.5						
70					43.0	36.8	39.0	31.5	35.0	26.2	31.0	21.0	27.0	17.0	19.0	6.5						
(75)					53.0	46.8	49.0	41.5	45.0	36.2	41.0	31.0	37.0	27.0	29.0	16.5						
80					53.0	46.8	49.0	41.5	45.0	36.2	41.0	31.0	37.0	27.0	29.0	16.5	26.0	11.0				
(85)							59.0	51.5	55.0	46.2	51.0	41.0	47.0	37.0	39.0	26.5	31.0	16.0				
90							59.0	51.5	55.0	46.2	51.0	41.0	47.0	37.0	39.0	26.5	36.0	21.0	34.0	16.5		
100							74.0	66.5	70.0	61.2	66.0	56.0	62.0	52.0	54.0	41.5	46.0	31.0	44.0	26.5	32.0	12.0
110									80.0	71.2	76.0	66.0	72.0	62.0	64.0	51.5	56.0	41.0	54.0	36.5	42.0	22.0
120									90.0	81.2	86.0	76.0	82.0	72.0	74.0	61.5	66.0	51.0	58.0	40.5	46.0	26.0
130											90.0	80.0	86.0	76.0	78.0	65.5	70.0	55.0	68.0	50.5	56.0	36.0
140											100.0	90.0	96.0	86.0	88.0	75.5	80.0	65.0	78.0	60.5	66.0	46.0
150													106.0	96.0	98.0	85.5	90.0	75.0	88.0	70.5	76.0	56.0
160													116.0	106.0	108.0	95.5	100.0	85.0	98.0	80.5	86.0	66.0
(170)															118.0	105.5	110.0	95.0	108.0	90.5	96.0	76.0
180															128.0	115.5	120.0	105.0	118.0	100.5	106.0	86.0
(190)															138.0	125.5	130.0	115.0	128.0	110.5	116.0	96.0
200															148.0	135.5	140.0	125.0	135.0	117.5	123.0	103.0
220																	147.0	132.0	155.0	137.5	143.0	123.0
240																	167.0	152.0	175.0	157.5	163.0	143.0
260																			195.0	177.5	183.0	163.0
280																			215.0	197.5	203.0	183.0
300																						

(Courtesy IFI-506-1976)

NOTES:
1. All dimensions are in millimetres.
2. L is nominal length of screw; L_G is grip gaging length; L_B is body length.
3. Diameter-length combinations between the dashed lines are recommended. Lengths in parentheses are not recommended.
4. Screws with lengths above the solid line are threaded full length.

Table 35   Metric Hex Cap Screws and Hex Bolts

Table 6   Hex bolts

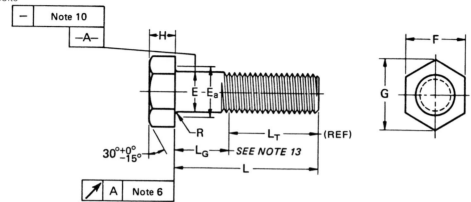

Nominal Bolt Size & Thread Pitch	E		F		G		H		E_a	R	L_T (Ref)		
	Body Diameter		Width Across Flats		Width Across Corners		Head Height		Fillet Transition Dia	Radius of Fillet	Thread Length (Basic)		
											Bolt Lengths ≤125	Bolt Lengths >125 and ≤200	Bolt Lengths >200
	Max	Min	Max	Min	Max	Min	Max	Min	Max	Min			
M5x0.8	5.48	4.52	8.00	7.75	9.24	8.84	3.88	3.35	5.8	0.2	16	22	35
M6.3x1	6.78	5.72	10.00	9.69	11.55	11.05	4.70	4.13	7.3	0.3	18.6	24.6	37.6
M8x1.25	8.58	7.42	13.00	12.60	15.01	14.36	5.73	5.10	9.2	0.4	22	28	41
M10x1.5	10.58	9.42	15.00	14.50	17.32	16.53	6.86	6.17	11.2	0.4	26	32	45
M12x1.75	12.70	11.30	18.00	17.40	20.78	19.84	7.99	7.24	13.2	0.4	30	36	49
M14x2	14.70	13.30	21.00	20.30	24.25	23.14	9.32	8.51	15.2	0.6	—	40	53
M16x2	16.70	15.30	24.00	23.20	27.71	26.45	10.56	9.68	17.8	0.6	—	44	57
M20x2.5	20.84	19.16	30.00	29.00	34.64	33.06	13.12	12.12	22.4	0.8	—	52	65
M24x3	24.84	23.16	36.00	34.80	41.57	39.67	15.68	14.56	26.4	0.8	—	60	73
M30x3.5	30.84	29.16	46.00	44.50	53.12	50.73	19.48	17.92	33.6	1.2	66	72	85
M36x4	37.00	35.00	55.00	53.20	63.51	60.65	23.38	21.72	39.6	1.2	78	84	97
M42x4.5	43.00	41.00	65.00	62.90	75.06	71.71	26.97	25.03	45.6	1.2	90	96	109
M48x5	49.00	47.00	75.00	72.60	86.60	82.76	31.07	28.93	52.6	1.5	102	108	121
M56x5.5	57.20	54.80	85.00	82.20	98.15	93.71	36.20	33.80	62.0	2.0	—	124	137
M64x6	65.52	62.80	95.00	91.80	109.70	104.65	41.32	38.68	70.0	2.0	—	140	153
M72x6	73.84	70.80	105.00	101.40	121.24	115.60	46.45	43.55	78.0	2.0	—	156	169
M80x6	82.16	78.80	115.00	111.00	132.79	126.54	51.58	48.42	86.0	2.0	—	172	185
M90x6	92.48	88.60	130.00	125.50	150.11	143.07	57.74	54.26	96.0	2.0	—	192	205
M100x6	102.80	98.60	145.00	140.00	167.43	159.60	63.90	60.10	107.0	2.5	—	212	225

(Courtesy IFI-506-1976)

NOTES:
1. **Dimensions.** All dimensions are in millimetres unless otherwise stated.
2. **Surface Condition.** Bolts need not be finished on any surface except threads.
3. **Top of Head.** Top of head shall be full form and chamfered or rounded with the diameter of chamfer circle or start of rounding being equal to the maximum width across flats, within a tolerance of minus 15 per cent.

Table 36   Metric Hex Nuts

Table 1   Dimensions of hex nuts

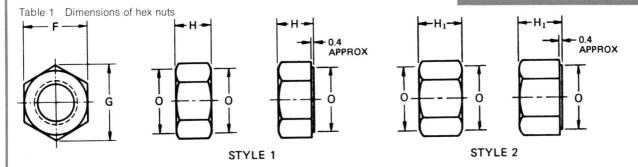

STYLE 1   STYLE 2

Nominal Nut Size and Thread Pitch	F Width Across Flats		G Width Across Corners		O Bearing Face Dia	H Nut Thickness Style 1		H₁ Nut Thickness Style 2		Runout of Bearing Surface FIR
	Max	Min	Max	Min	Min	Max	Min	Max	Min	Max
M1.6x0.35	3.20	3.02	3.70	3.44	2.5	—	—	1.3	1.1	—
M2x0.4	4.00	3.82	4.62	4.35	3.1	—	—	1.6	1.3	—
M2.5x0.45	5.00	4.82	5.77	5.49	4.1	—	—	2.0	1.7	—
M3x0.5	5.50	5.32	6.35	6.06	4.6	—	—	2.4	2.1	—
M3.5x0.6	7.00	6.78	8.08	7.73	6.0	—	—	2.8	2.5	—
M4x0.7	7.00	6.78	8.08	7.73	6.0	—	—	3.2	2.9	—
M5x0.8	8.00	7.78	9.24	8.87	7.0	4.5	4.2	5.3	5.0	0.30
M6.3x1	10.00	9.76	11.55	11.13	8.9	5.6	5.3	6.5	6.2	0.33
M8x1.25	13.00	12.73	15.01	14.51	11.6	6.6	6.2	7.8	7.4	0.36
M10x1.5	15.00	14.70	17.32	16.76	13.6	9.0	8.5	10.7	10.2	0.39
M12x1.75	18.00	17.67	20.78	20.14	16.6	10.7	10.2	12.8	12.3	0.42
M14x2	21.00	20.64	24.25	23.53	19.4	12.5	11.9	14.9	14.3	0.45
M16x2	24.00	23.61	27.71	26.92	22.4	14.5	13.9	17.4	16.8	0.48
M20x2.5	30.00	29.00	34.64	33.06	27.6	18.4	17.4	21.2	20.2	0.56
M24x3	36.00	34.80	41.57	39.67	32.9	22.0	20.9	25.4	24.3	0.64
M30x3.5	46.00	44.50	53.12	50.73	42.5	26.7	25.4	31.0	29.7	0.76
M36x4	55.00	53.20	63.51	60.65	50.8	32.0	30.5	37.6	36.1	0.89

(Courtesy IFI-507-1976)

NOTE: All dimensions are in millimetres.

Table 36 (*continued*)

Table 2 Dimensions of hex slotted nuts

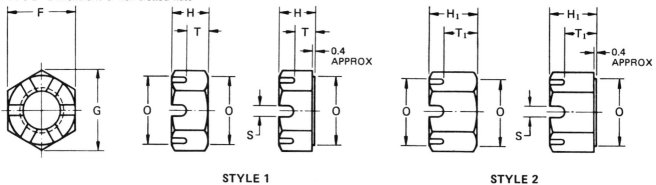

STYLE 1   STYLE 2

Nominal Nut Size and Thread Pitch	F Width Across Flats		G Width Across Corners		O Bearing Face Dia	H Nut Thickness Style 1		H₁ Nut Thickness Style 2		T Unslotted Thickness Style 1		T₁ Unslotted Thickness Style 2		S Width of Slot		Runout of Bearing Surface FIR
	Max	Min	Max	Min	Min	Max	Min	Max	Min	Max	Min	Max	Min	Max	Min	Max
M5x0.8	8.00	7.78	9.24	8.87	7.0	4.5	4.2	5.3	5.0	3.2	2.7	3.7	3.2	2.2	1.4	0.30
M6.3x1	10.00	9.76	11.55	11.13	8.9	5.6	5.3	6.5	6.2	3.9	3.4	4.5	4.0	2.8	2.0	0.33
M8x1.25	13.00	12.73	15.01	14.51	11.6	6.6	6.2	7.8	7.4	4.5	4.0	5.3	4.8	3.3	2.5	0.36
M10x1.5	15.00	14.70	17.32	16.76	13.6	9.0	8.5	10.7	10.2	6.0	5.5	7.1	6.6	3.6	2.8	0.39
M12x1.75	18.00	17.67	20.78	20.14	16.6	10.7	10.2	12.8	12.3	7.1	6.6	8.5	8.0	4.3	3.5	0.42
M14x2	21.00	20.64	24.25	23.53	19.4	12.5	11.9	14.9	14.3	8.2	7.7	9.8	9.3	4.3	3.5	0.45
M16x2	24.00	23.61	27.71	26.92	22.4	14.5	13.9	17.4	16.8	9.5	9.0	11.4	10.9	6.0	4.5	0.48
M20x2.5	30.00	29.00	34.64	33.06	27.6	18.4	17.4	21.2	20.2	12.1	11.3	13.9	13.1	6.0	4.5	0.56
M24x3	36.00	34.80	41.57	39.67	32.9	22.0	20.9	25.4	24.3	14.4	13.6	16.6	15.8	7.0	5.5	0.64
M30x3.5	46.00	44.50	53.12	50.73	42.5	26.7	25.4	31.0	29.7	17.3	16.5	20.1	19.3	9.3	7.0	0.76
M36x4	55.00	53.20	63.51	60.65	50.8	32.0	30.5	37.6	36.1	20.8	19.8	24.5	23.5	9.3	7.0	0.89

(Courtesy IFI-507-1976)

Table 36   (*continued*)

Table 3   Dimensions of large size hex nuts

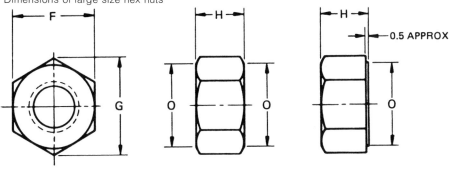

Nom Nut Size and Thread Pitch	F Width Across Flats		G Width Across Corners		H Nut Thickness		O Bearing Face Dia	Runout of Bearing Face FIR
	Max	Min	Max	Min	Max	Min	Min	Max
M42x4.5	65.00	62.90	75.06	71.71	34.5	32.6	58.5	1.02
M48x5	75.00	72.60	86.60	82.76	39.5	37.4	67.5	1.18
M56x5.5	85.00	82.20	98.15	93.71	46.2	43.8	76.5	1.33
M64x6	95.00	91.80	109.70	104.65	52.6	50.0	85.5	1.49
M72x6	105.00	101.40	121.24	115.60	59.3	56.4	94.5	1.65
M80x6	115.00	111.00	132.79	126.54	66.0	62.8	103.5	1.81
M90x6	130.00	125.50	150.11	143.07	75.3	71.8	117.0	2.04
M100x6	145.00	140.00	167.43	159.60	82.6	78.8	130.5	2.28

(Courtesy IFI-507-1976)

Table 37   12 Spline Flange Screws

Table 1   Dimensions of 12 spline flange screws

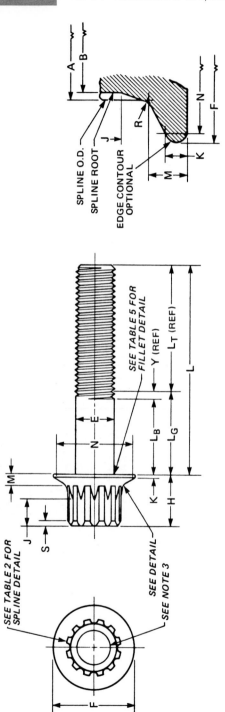

Nom Screw Size & Thread Pitch	Spline Size (See Table 2)	E Body Dia		F Flange Dia	N Bearing Circle Dia	K Flange Edge Thickness	M Flange Height	J Wrenching Height	H Head Height	S Chamfer Height	R Spline Junction Radius	L_T (Ref) Thread Length, Basic			Y (Ref) Transition Thread Length
		Max	Min	Max	Min	Min	Min	Min	Max	Max	Min	For Screw Lengths ≤125mm	For Screw Lengths >125mm and ≤200mm	For Screw Lengths >200mm	Max
M5x0.8	5	5.00	4.82	9.4	8.4	1.0	1.7	1.8	5.0	0.6	0.4	16.0	22.0	35.0	4.0
M6.3x1	6.3	6.30	6.08	11.8	10.7	1.2	2.2	2.3	6.3	0.8	0.5	18.6	24.6	37.6	5.0
M8x1.25	8	8.00	7.78	15.0	13.7	1.5	2.7	3.0	8.0	1.0	0.6	22.0	28.0	41.0	6.2
M10x1.5	10	10.00	9.78	18.6	17.1	2.0	3.4	3.8	10.0	1.2	0.7	26.0	32.0	45.0	7.5
M12x1.75	12	12.00	11.73	22.8	21.1	2.3	4.1	4.5	12.0	1.5	0.8	30.0	36.0	49.0	8.8
M14x2	14	14.00	13.73	26.4	24.5	2.7	4.8	5.4	14.0	1.8	0.9	34.0	40.0	53.0	10.0
M16x2	16	16.00	15.73	30.3	28.1	3.2	5.7	5.8	16.0	2.1	1.0	38.0	44.0	57.0	10.0
M20x2.5	20	20.00	19.67	37.4	34.9	4.1	7.2	7.2	20.0	2.5	1.2	46.0	52.0	65.0	12.5

(Courtesy IFI-511-1976)

Table 38   Metric Slotted and Recessed Head Machine Screws

Table 1

Nominal Screw Size	L_T Full Form Thread Length		Y Unthreaded Length Under Head				
	For Nominal Screw Lengths >Than	Min(1)	For Nominal Screw Lengths ≤Than(2)	Max(3)	For Nominal Screw Lengths		Max(4)
					>Than	≤Than	
M 2	30	25.0	6	0.40	6	30	0.8
M 2.5	30	25.0	8	0.45	8	30	0.9
M 3	30	25.0	9	0.50	9	30	1.0
M 3.5	50	38.0	10	0.60	10	50	1.2
M 4	50	38.0	12	0.70	12	50	1.4
M 5	50	38.0	15	0.80	15	50	1.6
M 6.3	50	38.0	19	1.00	19	50	2.0
M 8	50	38.0	24	1.25	24	50	2.5
M10	50	38.0	30	1.50	30	50	3.0
M12	50	38.0	36	1.75	36	50	3.5

(Courtesy IFI-513-1976)

NOTES:
1. These lengths shall apply unless otherwise specified by purchaser.
2. Tabulated values are equal to 3 times basic screw diameter, rounded to nearest mm.
3. Tabulated values are equal to 1 thread pitch.
4. Tabulated values are equal to 2 thread pitches.

Table 38 (*continued*)

Table 2 Metric slotted flat countersunk head machine screws

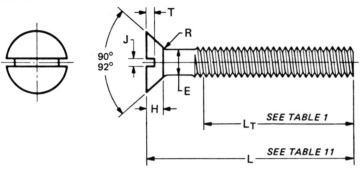

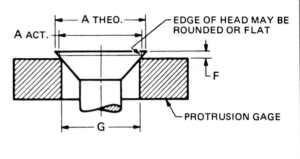

Nom Screw Size and Thread Pitch	E		A			H	R		J		T		F		G
	Body Dia		Head Diameter			Head Height	Fillet Radius		Slot Width		Slot Depth		Protrusion of Head Above Gage Dia		Gage Dia
			Theoretical Sharp		Actual										
	Max	Min	Max	Min	Min	Max Ref	Max	Min	Max	Min	Max	Min	Max	Min	
M2x0.4	2.00	1.65	4.40	3.90	3.60	1.20	0.8	0.2	0.7	0.5	0.6	0.4	0.79	0.52	2.82
M2.5x0.45	2.50	2.12	5.50	4.90	4.60	1.50	1.0	0.3	0.8	0.6	0.7	0.5	0.88	0.56	3.74
M3x0.5	3.00	2.58	6.60	5.80	5.50	1.80	1.2	0.3	1.0	0.8	0.9	0.6	0.98	0.55	4.65
M3.5x0.6	3.50	3.00	7.70	6.80	6.44	2.10	1.4	0.4	1.2	1.0	1.0	0.7	1.07	0.59	5.57
M4x0.7	4.00	3.43	8.65	7.80	7.44	2.32	1.6	0.4	1.4	1.2	1.1	0.8	1.09	0.63	6.48
M5x0.8	5.00	4.36	10.70	9.80	9.44	2.85	2.0	0.5	1.5	1.2	1.4	1.0	1.20	0.71	8.31
M6.3x1	6.30	5.51	13.50	12.30	11.87	3.60	2.5	0.6	1.9	1.6	1.8	1.3	1.41	0.77	10.69
M8x1.25	8.00	7.04	16.80	15.60	15.17	4.40	3.2	0.8	2.3	2.0	2.1	1.6	1.50	0.86	13.80
M10x1.5	10.00	8.86	20.70	19.50	18.98	5.35	4.0	1.0	2.8	2.5	2.6	2.0	2.58	1.91	15.54
M12x1.75	12.00	10.68	24.70	23.50	22.88	6.35	4.8	1.2	2.8	2.5	3.1	2.5	2.79	2.11	19.12

(Courtesy IFI-513-1976)

Table 38   (*continued*)

Table 3   Cross recess dimensions of flat countersunk head machine screws

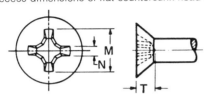

This type of recess has a large center opening, tapered wings, and blunt bottom, with all edges relieved or rounded.

TYPE 1

This type of recess has a large center opening, wide straight wings, and blunt bottom, with all edges relieved or rounded.

TYPE 1A

Nom Screw Size	Type 1										Type 1A									
	M		T		N	Driver Size	Recess Penetration Gaging Depth		M		T		N	Driver Size	Recess Penetration Gaging Depth	
	Recess Dia		Recess Depth		Recess Width				Recess Dia		Recess Depth		Recess Width			
	Max	Min	Max	Min	Min		Max	Min	Max	Min	Max	Min	Min		Max	Min
M 2	1.96	1.63	1.30	0.89	0.38	0	1.12	0.71	2.39	2.06	1.75	1.35	0.46	0	1.57	1.17
M 2.5	2.97	2.64	1.98	1.57	0.46	1	1.80	1.40	2.97	2.64	1.98	1.57	0.74	1	1.73	1.32
M 3	3.25	2.92	2.26	1.85	0.46	1	2.08	1.68	3.25	2.92	2.26	1.85	0.76	1	2.01	1.60
M 3.5	4.17	3.84	2.44	1.85	0.71	2	2.16	1.57	4.17	3.84	2.46	2.01	1.04	2	2.06	1.60
M 4	4.42	4.09	2.69	2.11	0.74	2	2.41	1.83	4.42	4.09	2.72	2.26	1.04	2	2.31	1.85
M 5	4.62	4.29	2.90	2.31	0.76	2	2.62	2.03	4.62	4.29	2.90	2.44	1.04	2	2.51	2.05
M 6.3	6.35	6.02	3.45	2.87	0.81	3	3.02	2.44	6.30	5.97	3.48	3.02	1.42	3	2.92	2.46
M 8	8.69	8.36	4.90	4.34	1.42	4	4.39	3.84	8.69	8.36	4.98	4.52	2.16	4	4.32	3.86
M10	9.60	9.27	5.84	5.28	1.57	4	5.33	4.78	9.60	9.27	5.92	5.46	2.18	4	5.23	4.77
M12	10.39	10.06	6.63	6.07	1.73	4	6.12	5.56	10.39	10.06	6.73	6.27	2.18	4	6.05	5.59

(Courtesy IFI-513-1976)

NOTE: Head dimensions not shown are the same as those of slotted heads given in Table 2.

Table 38   (*continued*)

Table 4   Slotted oval countersunk head machine screws

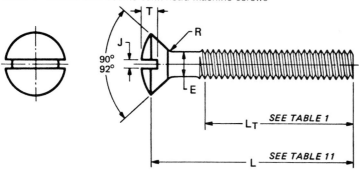

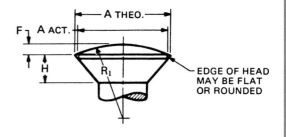

Nom Screw Size and Thread Pitch	E		A			H	F	R₁	R		J		T	
	Body Dia		Head Diameter			Head Side Height	Raised Head Height	Head Radius	Fillet Radius		Slot Width		Slot Depth	
			Theoretical Sharp		Actual									
	Max	Min	Max	Min	Min	Max Ref	Max	Approx	Max	Min	Max	Min	Max	Min
M2x0.4	2.00	1.65	4.40	3.90	3.60	1.20	0.50	3.8	0.8	0.2	0.7	0.5	1.0	0.8
M2.5x0.45	2.50	2.12	5.50	4.90	4.60	1.50	0.60	5.0	1.0	0.3	0.8	0.6	1.2	1.0
M3x0.5	3.00	2.58	6.60	5.80	5.50	1.80	0.75	5.7	1.2	0.3	1.0	0.8	1.5	1.2
M3.5x0.6	3.50	3.00	7.70	6.80	6.44	2.10	0.90	6.5	1.4	0.4	1.2	1.0	1.7	1.4
M4x0.7	4.00	3.43	8.65	7.80	7.44	2.32	1.00	7.8	1.6	0.4	1.4	1.2	1.9	1.6
M5x0.8	5.00	4.36	10.70	9.80	9.44	2.85	1.25	9.9	2.0	0.5	1.5	1.2	2.3	2.0
M6.3x1	6.30	5.51	13.50	12.30	11.87	3.60	1.60	12.2	2.5	0.6	1.9	1.6	3.0	2.6
M8x1.25	8.00	7.04	16.80	15.60	15.17	4.40	2.00	15.8	3.2	0.8	2.3	2.0	3.7	3.2
M10x1.5	10.00	8.86	20.70	19.50	18.98	5.35	2.50	19.8	4.0	1.0	2.8	2.5	4.5	4.0
M12x1.75	12.00	10.68	24.70	23.50	22.88	6.35	3.00	23.8	4.8	1.2	2.8	2.5	5.3	4.8

(Courtesy IFI-513-1976)

Table 39 Metric High Strength Structural Bolts, Nuts and Washers

Table 1 Dimensions of high strength structural bolts

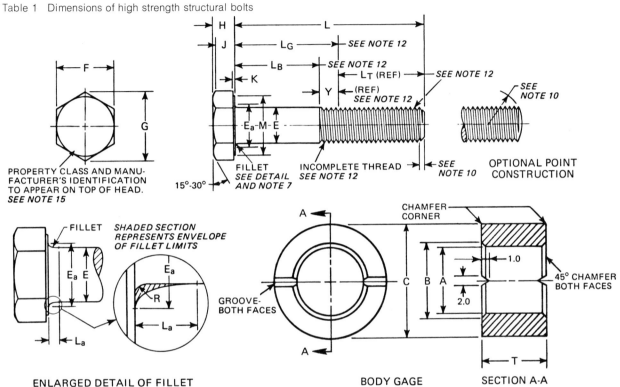

PROPERTY CLASS AND MANU-
FACTURER'S IDENTIFICATION
TO APPEAR ON TOP OF HEAD.
SEE NOTE 15

ENLARGED DETAIL OF FILLET

BODY GAGE

SECTION A-A

Nom Bolt Size and Thread Pitch	E Body Dia	F Width Across Flats		G Width Across Corners		H Head Height		J Wrenching Height	M Bearing Face Dia	K Washer Face Thickness	
	Min	Max	Min	Max	Min	Max	Min	Min	Min	Max	Min
M16x2	15.30	27.00	26.16	31.18	29.82	10.32	9.68	7.0	24.8	0.8	0.4
M20x2.5	19.16	34.00	33.00	39.26	37.62	12.88	12.12	8.8	30.5	0.8	0.4
M24x3	23.16	40.00	39.00	46.19	44.46	15.44	14.56	10.5	36.5	0.8	0.4
M30x3.5	29.16	50.00	49.00	57.74	55.86	19.48	17.92	13.1	46.5	0.8	0.4
M36x4	35.00	60.00	58.80	69.28	67.03	23.38	21.62	15.8	55.8	0.8	0.4
See Note	8			3		4		3	6		

Nom Bolt Size and Thread Pitch	Runout of Bearing Surface FIR	E_a Fillet Transition Dia	L_a Fillet Transition Length	R Fillet Radius	L_T (Ref) Basic Thread Length		Y (Ref) Transition Thread Length	A Body Gage	B	C	T
	Max	Max	Max	Min	Bolt Lengths ≤100	Bolt Lengths >100	Max	Hole Dia +0.00 −0.05	Counter-sink Dia +0.05 −0.00	Dia +0.5 −0.5	Thick-ness +0.5 −0.5
M16x2	0.45	18.2	3.0	0.6	33	38	6.0	18	18.2	27.0	16.0
M20x2.5	0.56	22.4	4.0	0.8	40	45	7.5	22	22.4	34.0	20.0
M24x3	0.67	26.4	4.0	1.0	45	50	9.0	26	26.4	40.0	24.0
M30x3.5	0.85	33.4	6.0	1.2	53	58	10.5	33	33.4	50.0	30.0
M36x4	1.01	39.4	6.0	1.5	60	65	12.0	39	39.4	60.0	36.0

(Courtesy IFI-526-1976)

Table 39 (*continued*)

Table 3 Metric hex nuts for high strength structural bolts

PROPERTY CLASS
AND MANUFACTURER'S
IDENTIFICATION TO
APPEAR ON ONE SURFACE.
SEE NOTE 9

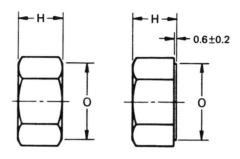

Nom Nut Size	F Width Across Flats		G Width Across Corners		O Bearing Face Dia	H Thickness		Runout of Bearing Surface FIR
	Max	Min	Max	Min	Min	Max	Min	Max
M16x2	27.00	26.16	31.18	29.82	24.8	17.4	16.8	0.45
M20x2.5	34.00	33.00	39.26	37.62	30.5	21.2	20.2	0.56
M24x3	40.00	39.00	46.19	44.46	36.5	25.4	24.3	0.67
M30x3.5	50.00	49.00	57.74	55.86	46.5	31.0	29.7	0.85
M36x4	60.00	58.80	69.28	67.03	55.8	37.6	36.1	1.01

(Courtesy IFI-526-1976)

NOTE: All dimensions are in millimetres.

Table 40   Clearance Holes for Metric Threaded Fasteners

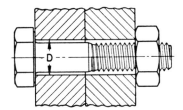

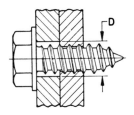

Nom Fastener Size	D — Clearance Hole Diameter, Basic		
	Close Clearance	Normal Clearance (Preferred)	Loose Clearance
1.6	1.75	1.9	2.1
2	2.2	2.4	2.6
2.5	2.7	2.9	3.1
3	3.2	3.4	3.6
3.5	3.7	4.0	4.2
4	4.2	4.5	4.8
5	5.3	5.6	6.0
6.3	6.7	7.1	7.5
8	8.5	9.0	9.5
10	10.5	11.0	12.0
12	12.5	13.0	14.0
14	14.5	15.0	16.0
16	16.5	17.5	18.5
20	21.0	22.0	23.0
24	25.0	26.0	27.0
30	31.0	33.0	35.0
36	37.0	39.0	41.0
42	43.0	45.0	48.0
48	50.0	52.0	56.0
56	58.0	62.0	66.0
64	66.0	70.0	74.0
72	74.0	78.0	82.0
80	82.0	86.0	91.0
90	93.0	96.0	101.0
100	104.0	107.0	112.0

Nom Screw Size	D — Clearance Hole Diameter, Basic		
	Close Clearance	Normal Clearance (Preferred)	Loose Clearance
2.2	2.4	2.6	2.8
2.9	3.1	3.3	3.6
3.5	3.8	4.0	4.4
4.2	4.5	4.8	5.2
4.8	5.2	5.4	5.8
5.5	5.8	6.2	6.5
6.3	6.7	7.1	7.5
8.0	8.5	9.0	9.5
9.5	10.0	10.5	11.5

1.0 SCOPE

1.1
This Recommended Practice presents clearance holes for metric threaded fasteners.

2.0 CLEARANCE HOLES

2.1
Basic clearance hole diameters for bolts, screws and studs are given in Table 1, and for tapping screws in Table 2.

2.1.1 Normal Clearance
Normal clearance hole sizes are preferred for general purpose applications, and should be specified unless special design considerations dictate the need for either a close or loose clearance hole.

2.1.2 Close Clearance
Close clearance hole sizes should be specified only where conditions such as critical alignment of assembled parts, wall thickness or other limitations necessitate use of a minimal hole.

2.1.3 Loose Clearance
Loose clearance hole sizes should be specified only for applications where maximum adjustment capability between components being assembled is necessary.

Table 41   Metric Screw and Washer Assemblies—Sems

Table 1   Dimensions of washers on internal tooth lock washer sems

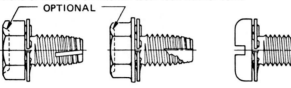

HEX HEAD
TYPE D TAPPING
SCREW

HEX WASHER HEAD
TYPE T TAPPING
SCREW

PAN HEAD
MACHINE
SCREW

Nom Screw Size (1)	Pan, Hex and Hex Washer Head Screws			
	Washer Outside Dia		Washer Thickness	
	Max	Min	Max	Min
2.9	7.00	6.50	0.45	0.30
3	7.35	6.85	0.45	0.30
3.5	7.50	7.00	0.55	0.40
4	8.75	8.25	0.55	0.40
4.2	8.75	8.25	0.60	0.45
4.8	9.70	9.20	0.65	0.50
5	10.50	10.00	0.65	0.50
5.5	10.50	10.00	0.65	0.50
6.3	12.15	11.65	0.70	0.55
8	15.50	14.75	0.85	0.70
9.5	17.70	16.95	1.00	0.80
10	17.70	16.95	1.00	0.80

(Courtesy IFI-531-1976)

Table 41   (*continued*)

Table 2   Dimensions of washers on external tooth lock washer sems

— OPTIONAL —

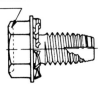

HEX HEAD
TYPE D TAPPING
SCREW

HEX WASHER HEAD
TYPE T TAPPING
SCREW

PAN HEAD
MACHINE
SCREW

Nom Screw Size (1)	Pan Head and Hex Head Screws				Hex Washer Head Screws			
	Washer Outside Dia		Washer Thickness		Washer Outside Dia		Washer Thickness	
	Max	Min	Max	Min	Max	Min	Max	Min
2.9	5.85	5.45	0.50	0.35	5.85	5.45	0.50	0.35
3	5.85	5.45	0.50	0.35	5.85	5.45	0.50	0.35
3.5	7.35	6.85	0.50	0.35	7.35	6.85	0.50	0.35
4	8.25	7.75	0.60	0.45	8.25	7.75	0.60	0.45
4.2	8.25	7.75	0.60	0.45	8.25	7.75	0.60	0.45
4.8	9.70	9.20	0.60	0.45	10.50	10.00	0.65	0.50
5	10.50	10.00	0.65	0.50	10.50	10.00	0.65	0.50
5.5	10.50	10.00	0.65	0.50	10.50	10.00	0.65	0.50
6.3	12.10	11.60	0.70	0.55	14.80	14.30	0.70	0.55
8	16.00	15.25	1.00	0.80	17.00	16.25	0.85	0.70
9.5	19.30	18.55	1.00	0.80	19.30	18.55	1.00	0.80
10	19.30	18.55	1.00	0.80	19.30	18.55	1.00	0.80

(Courtesy IFI-531-1976)

DESCRIPTION OF FITS
FOR TABLES 42 TO 46

Running and Sliding Fits

Running and sliding fits, for which limits of clearance are given in Table 42, are intended to provide a similar running performance, with suitable lubrication allowance, throughout the range of sizes. The clearance for the first two classes, used chiefly as slide fits, increase more slowly with diameter than the other classes, so that accurate location is maintained even at the expense of free relative motion.

These fits may be described briefly as follows.

RC 1 *Close sliding fits* are intended for the accurate location of parts that must assemble without perceptible play.

RC 2 *Sliding fits* are intended for accurate location, but with greater maximum clearance than class RC 1. Parts made to fit move and turn easily, but are not intended to run freely, and in the larger sizes may seize with small temperature changes.

RC 3 *Precision running fits* are about the closest fits that can be expected to run freely, and are intended for precision work at slow speeds and light journal pressures, but are not suitable where appreciable temperature differences are likely to be encountered.

RC 4 *Close running fits* are intended chiefly for running fits on accurate machinery with moderate surface speeds and journal pressures, where

accurate location and minimum play is desired.

RC 5 ⎰ *Medium running fits* are in-
RC 6 ⎱ tended for higher running speeds, or heavy journal pressures, or both.

RC 7 *Free running fits* are intended for use where accuracy is not essential, or where large temperature variations are likely to be encountered, or under both these conditions.

RC 8 ⎰ *Loose running fits* are in-
RC 9 ⎱ tended for use where materials, such as cold-rolled shafting and tubing, made to commercial tolerances are involved.

Locational Fits

Locational fits are intended to determine only the location of mating parts; they may provide rigid or accurate location, as with interfer-

Table 42 Running and Sliding Fits (Inch Units)
Limits are in thousandths of an inch. Limits for hole and shaft are applied algebraically to the basic size to obtain the limits of size for the parts. Data in bold face are in accordance with ABC agreements. Symbols H5, g5, etc., are Hole and Shaft designations.

Nominal Size Range Inches Over — To	Class RC 1 Limits of Clearance	Standard Limits Hole H5	Shaft g4	Class RC 2 Limits of Clearance	Standard Limits Hole H6	Shaft g5	Class RC 3 Limits of Clearance	Standard Limits Hole H7	Shaft f6	Class RC 4 Limits of Clearance	Standard Limits Hole H8	Shaft f7
0 — 0.12	0.1 0.45	+ 0.2 0	− 0.1 − 0.25	0.1 0.55	+ 0.25 0	− 0.1 − 0.3	0.3 0.95	+ 0.4 0	− 0.3 − 0.55	0.3 1.3	+ 0.6 0	− 0.3 − 0.7
0.12 — 0.24	0.15 0.5	+ 0.2 0	− 0.15 − 0.3	0.15 0.65	+ 0.3 0	− 0.15 − 0.35	0.4 1.12	+ 0.5 0	− 0.4 − 0.7	0.4 1.6	+ 0.7 0	− 0.4 − 0.9
0.24 — 0.40	0.2 0.6	0.25 0	− 0.2 − 0.35	0.2 0.85	+ 0.4 0	− 0.2 − 0.45	0.5 1.5	+ 0.6 0	− 0.5 − 0.9	0.5 2.0	+ 0.9 0	− 0.5 − 1.1
0.40 — 0.71	0.25 0.75	+ 0.3 0	− 0.25 − 0.45	0.25 0.95	+ 0.4 0	− 0.25 − 0.55	0.6 1.7	+ 0.7 0	− 0.6 − 1.0	0.6 2.3	+ 1.0 0	− 0.6 − 1.3
0.71 — 1.19	0.3 0.95	+ 0.4 0	− 0.3 − 0.55	0.3 1.2	+ 0.5 0	− 0.3 − 0.7	0.8 2.1	+ 0.8 0	− 0.8 − 1.3	0.8 2.8	+ 1.2 0	− 0.8 − 1.6
1.19 — 1.97	0.4 1.1	+ 0.4 0	− 0.4 − 0.7	0.4 1.4	+ 0.6 0	− 0.4 − 0.8	1.0 2.6	+ 1.0 0	− 1.0 − 1.6	1.0 3.6	+ 1.6 0	− 1.0 − 2.0
1.97 — 3.15	0.4 1.2	+ 0.5 0	− 0.4 − 0.7	0.4 1.6	+ 0.7 0	− 0.4 − 0.9	1.2 3.1	+ 1.2 0	− 1.2 − 1.9	1.2 4.2	+ 1.8 0	− 1.2 − 2.4
3.15 — 4.73	0.5 1.5	+ 0.6 0	− 0.5 − 0.9	0.5 2.0	+ 0.9 0	− 0.5 − 1.1	1.4 3.7	+ 1.4 0	− 1.4 − 2.3	1.4 5.0	+ 2.2 0	− 1.4 − 2.8
4.73 — 7.09	0.6 1.8	+ 0.7 0	− 0.6 − 1.1	0.6 2.3	+ 1.0 0	− 0.6 − 1.3	1.6 4.2	+ 1.6 0	− 1.6 − 2.6	1.6 5.7	+ 2.5 0	− 1.6 − 3.2
7.09 — 9.85	0.6 2.0	+ 0.8 0	− 0.6 − 1.2	0.6 2.6	+ 1.2 0	− 0.6 − 1.4	2.0 5.0	+ 1.8 0	− 2.0 − 3.2	2.0 6.6	+ 2.8 0	− 2.0 − 3.8

ence fits, or provide some freedom of location, as with clearance fits. Accordingly, they are divided into three groups: clearance fits, transition fits, and interference fits.

These fits are more fully described as follows.

LC *Locational clearance fits* are intended for parts that are normally stationary, but that can be freely assembled or disassembled. They run from snug fits for parts requiring accuracy of location, through the medium clearance fits for parts such as spigots, to the looser fastener fits where freedom of assembly is of prime importance.

LT *Transition fits* are a compromise between clearance and interference fits, for application where accuracy of location is important, but either a small amount of clearance or interference is permissible.

LN *Locational interference fits* are used where accuracy of location is of prime importance, and for parts requiring rigidity and alignment with special requirements for bore pressure. Such fits are not intended for parts designed to transmit frictional loads from one part to another by virtue of the tightness of fit, as these conditions are covered by force fits.

Force Fits

Force or shrink fits constitute a special type of interference fit, normally characterized by maintenance of constant bore pressures through the range of sizes. The interference therefore varies almost directly with diameter, and the difference between its minimum and maximum value is small, to maintain the resulting pressures within reasonable limits.

These fits may be described briefly as follows.

FN 1 *Light drive fits* are those requiring light assembly pressures, and produce more or less permanent assemblies. They are suitable for thin sections or long fits, or in cast-iron external members.

FN 2 *Medium drive fits* are suitable for ordinary steel parts, or for shrink fits on light sections. They are about the tightest fits that can be used with high-grade cast-iron external members.

FN 3 *Heavy drive fits* are suitable for heavier steel parts or for shrink fits in medium sections.

FN 4
FN 5 { *Force fits* are suitable for parts that can be highly stressed or, for shrink fits where the heavy pressing forces required are impractical.

Nominal Size Range Inches		Class RC 5			Class RC 6			Class RC 7			Class RC 8			Class RC 9		
		Limits of Clearance	Standard Limits		Limits of Clearance	Standard Limits		Limits of Clearance	Standard Limits		Limits of Clearance	Standard Limits		Limits of Clearance	Standard Limits	
Over	To		Hole H8	Shaft e7		Hole H9	Shaft e8		Hole H9	Shaft d8		Hole H10	Shaft c9		Hole H11	Shaft
0 — 0.12		0.6 1.6	+ 0.6 − 0	− 0.6 − 1.0	0.6 2.2	+ 1.0 − 0	− 0.6 − 1.2	1.0 2.6	+ 1.0 0	− 1.0 − 1.6	2.5 5.1	+ 1.6 0	− 2.5 − 3.5	4.0 8.1	+ 2.5 0	− 4.0 − 5.6
0.12— 0.24		0.8 2.0	+ 0.7 − 0	− 0.8 − 1.3	0.8 2.7	+ 1.2 − 0	− 0.8 − 1.5	1.2 3.1	+ 1.2 0	− 1.2 − 1.9	2.8 5.8	+ 1.8 0	− 2.8 − 4.0	4.5 9.0	+ 3.0 0	− 4.5 − 6.0
0.24— 0.40		1.0 2.5	+ 0.9 − 0	− 1.0 − 1.6	1.0 3.3	+ 1.4 − 0	− 1.0 − 1.9	1.6 3.9	+ 1.4 0	− 1.6 − 2.5	3.0 6.6	+ 2.2 0	− 3.0 − 4.4	5.0 10.7	+ 3.5 0	− 5.0 − 7.2
0.40— 0.71		1.2 2.9	+ 1.0 − 0	− 1.2 − 1.9	1.2 3.8	+ 1.6 − 0	− 1.2 − 2.2	2.0 4.6	+ 1.6 0	− 2.0 − 3.0	3.5 7.9	+ 2.8 0	− 3.5 − 5.1	6.0 12.8	+ 4.0 − 0	− 6.0 − 8.8
0.71— 1.19		1.6 3.6	+ 1.2 − 0	− 1.6 − 2.4	1.6 4.8	+ 2.0 − 0	− 1.6 − 2.8	2.5 5.7	+ 2.0 0	− 2.5 − 3.7	4.5 10.0	+ 3.5 0	− 4.5 − 6.5	7.0 15.5	+ 5.0 0	− 7.0 − 10.5
1.19— 1.97		2.0 4.6	+ 1.6 − 0	− 2.0 − 3.0	2.0 6.1	+ 2.5 − 0	− 2.0 − 3.6	3.0 7.1	+ 2.5 0	− 3.0 − 4.6	5.0 11.5	+ 4.0 0	− 5.0 − 7.5	8.0 18.0	+ 6.0 0	− 8.0 − 12.0
1.97— 3.15		2.5 5.5	+ 1.8 − 0	− 2.5 − 3.7	2.5 7.3	+ 3.0 − 0	− 2.5 − 4.3	4.0 8.8	+ 3.0 0	− 4.0 − 5.8	6.0 13.5	+ 4.5 0	− 6.0 − 9.0	9.0 20.5	+ 7.0 0	− 9.0 − 13.5
3.15— 4.73		3.0 6.6	+ 2.2 − 0	− 3.0 − 4.4	3.0 8.7	+ 3.5 − 0	− 3.0 − 5.2	5.0 10.7	+ 3.5 0	− 5.0 − 7.2	7.0 15.5	+ 5.0 0	− 7.0 − 10.5	10.0 24.0	+ 9.0 0	− 10.0 − 15.0
4.73— 7.09		3.5 7.6	+ 2.5 − 0	− 3.5 − 5.1	3.5 10.0	+ 4.0 − 0	− 3.5 − 6.0	6.0 12.5	+ 4.0 0	− 6.0 − 8.5	8.0 18.0	+ 6.0 0	− 8.0 − 12.0	12.0 28.0	+ 10.0 0	− 12.0 − 18.0
7.09— 9.85		4.0 8.6	+ 2.8 − 0	− 4.0 − 5.8	4.0 11.3	+ 4.5 − 0	− 4.0 − 6.8	7.0 14.3	+ 4.5 0	− 7.0 − 9.8	10.0 21.5	+ 7.0 0	− 10.0 − 14.5	15.0 34.0	+ 12.0 0	− 15.0 − 22.0

[Courtesy ANSI B4.1-1967 (R1974)]

Table 43 Locational Clearance Fits (Inch Units)

Limits are in thousandths of an inch. Limits for hole and shaft are applied algebraically to the basic size to obtain the limits of size for the parts. Data in bold face are in accordance with ABC agreements. Symbols H6, h5, etc., are Hole and Shaft designations.

Nominal Size Range Inches		Class LC 1			Class LC 2			Class LC 3			Class LC 4			Class LC 5		
		Limits of Clearance	Standard Limits		Limits of Clearance	Standard Limits		Limits of Clearance	Standard Limits		Limits of Clearance	Standard Limits		Limits of Clearance	Standard Limits	
Over	To		Hole H6	Shaft h5		Hole H7	Shaft h6		Hole H8	Shaft h7		Hole H10	Shaft h9		Hole H7	Shaft g6
0 —	0.12	0 / 0.45	+0.25 / −0	+0 / −0.2	0 / 0.65	+0.4 / −0	+0 / −0.25	0 / 1	+0.6 / −0	+0 / −0.4	0 / 2.6	+1.6 / −0	+0 / −1.0	0.1 / 0.75	+0.4 / −0	−0.1 / −0.35
0.12—	0.24	0 / 0.5	+0.3 / −0	+0 / −0.2	0 / 0.8	+0.5 / −0	+0 / −0.3	0 / 1.2	+0.7 / −0	+0 / −0.5	0 / 3.0	+1.8 / −0	+0 / −1.2	0.15 / 0.95	+0.5 / −0	−0.15 / −0.45
0.24—	0.40	0 / 0.65	+0.4 / −0	+0 / −0.25	0 / 1.0	+0.6 / −0	+0 / −0.4	0 / 1.5	+0.9 / −0	+0 / −0.6	0 / 3.6	+2.2 / −0	+0 / −1.4	0.2 / 1.2	+0.6 / −0	−0.2 / −0.6
0.40—	0.71	0 / 0.7	+0.4 / −0	+0 / −0.3	0 / 1.1	+0.7 / −0	+0 / −0.4	0 / 1.7	+1.0 / −0	+0 / −0.7	0 / 4.4	+2.8 / −0	+0 / −1.6	0.25 / 1.35	+0.7 / −0	−0.25 / −0.65
0.71—	1.19	0 / 0.9	+0.5 / −0	+0 / −0.4	0 / 1.3	+0.8 / −0	+0 / −0.5	0 / 2	+1.2 / −0	+0 / −0.8	0 / 5.5	+3.5 / −0	+0 / −2.0	0.3 / 1.6	+0.8 / −0	−0.3 / −0.8
1.19—	1.97	0 / 1.0	+0.6 / −0	+0 / −0.4	0 / 1.6	+1.0 / −0	+0 / −0.6	0 / 2.6	+1.6 / −0	+0 / −1	0 / 6.5	+4.0 / −0	+0 / −2.5	0.4 / 2.0	+1.0 / −0	−0.4 / −1.0
1.97—	3.15	0 / 1.2	+0.7 / −0	+0 / −0.5	0 / 1.9	+1.2 / −0	+0 / −0.7	0 / 3	+1.8 / −0	+0 / −1.2	0 / 7.5	+4.5 / −0	+0 / −3	0.4 / 2.3	+1.2 / −0	−0.4 / −1.1
3.15—	4.73	0 / 1.5	+0.9 / −0	+0 / −0.6	0 / 2.3	+1.4 / −0	+0 / −0.9	0 / 3.6	+2.2 / −0	+0 / −1.4	0 / 8.5	+5.0 / −0	+0 / −3.5	0.5 / 2.8	+1.4 / −0	−0.5 / −1.4
4.73—	7.09	0 / 1.7	+1.0 / −0	+0 / −0.7	0 / 2.6	+1.6 / −0	+0 / −1.0	0 / 4.1	+2.5 / −0	+0 / −1.6	0 / 10	+6.0 / −0	+0 / −4	0.6 / 3.2	+1.6 / −0	−0.6 / −1.6
7.09—	9.85	0 / 2.0	+1.2 / −0	+0 / −0.8	0 / 3.0	+1.8 / −0	+0 / −1.2	0 / 4.6	+2.8 / −0	+0 / −1.8	0 / 11.5	+7.0 / −0	+0 / −4.5	0.6 / 3.6	+1.8 / −0	−0.6 / −1.8

Nominal Size Range Inches (Over – To)	Class LC 6 Limits of Clearance	Class LC 6 Hole H9	Class LC 6 Shaft f8	Class LC 7 Limits of Clearance	Class LC 7 Hole H10	Class LC 7 Shaft e9	Class LC 8 Limits of Clearance	Class LC 8 Hole H10	Class LC 8 Shaft d9	Class LC 9 Limits of Clearance	Class LC 9 Hole H11	Class LC 9 Shaft c10	Class LC 10 Limits of Clearance	Class LC 10 Hole H12	Class LC 10 Shaft	Class LC 11 Limits of Clearance	Class LC 11 Hole H13	Class LC 11 Shaft
0 – 0.12	0.3 / 1.9	+1.0 / 0	−0.3 / −0.9	0.6 / 3.2	+1.6 / 0	−0.6 / −1.6	1.0 / 3.6	+1.6 / −0	−1.0 / −2.0	2.5 / 6.6	+2.5 / −0	−2.5 / −4.1	4 / 12	+4 / −0	−4 / −8	5 / 17	+6 / −0	−5 / −11
0.12 – 0.24	0.4 / 2.3	+1.2 / 0	−0.4 / −1.1	0.8 / 3.8	+1.8 / 0	−0.8 / −2.0	1.2 / 4.2	+1.8 / −0	−1.2 / −2.4	2.8 / 7.6	+3.0 / −0	−2.8 / −4.6	4.5 / 14.5	+5 / −0	−4.5 / −9.5	6 / 20	+7 / −0	−6 / −13
0.24 – 0.40	0.5 / 2.8	+1.4 / 0	−0.5 / −1.4	1.0 / 4.6	+2.2 / 0	−1.0 / −2.4	1.6 / 5.2	+2.2 / −0	−1.6 / −3.0	3.0 / 8.7	+3.5 / −0	−3.0 / −5.2	5 / 17	+6 / −0	−5 / −11	7 / 25	+9 / −0	−7 / −16
0.40 – 0.71	0.6 / 3.2	+1.6 / 0	−0.6 / −1.6	1.2 / 5.6	+2.8 / 0	−1.2 / −2.8	2.0 / 6.4	+2.8 / −0	−2.0 / −3.6	3.5 / 10.3	+4.0 / −0	−3.5 / −6.3	6 / 20	+7 / −0	−6 / −13	8 / 28	+10 / −0	−8 / −18
0.71 – 1.19	0.8 / 4.0	+2.0 / 0	−0.8 / −2.0	1.6 / 7.1	+3.5 / 0	−1.6 / −3.6	2.5 / 8.0	+3.5 / −0	−2.5 / −4.5	4.5 / 13.0	+5.0 / −0	−4.5 / −8.0	7 / 23	+8 / −0	−7 / −15	10 / 34	+12 / −0	−10 / −22
1.19 – 1.97	1.0 / 5.1	+2.5 / 0	−1.0 / −2.6	2.0 / 8.5	+4.0 / 0	−2.0 / −4.5	3.0 / 9.5	+4.0 / −0	−3.0 / −5.5	5 / 15	+6 / −0	−5 / −9	8 / 28	+10 / −0	−8 / −18	12 / 44	+16 / −0	−12 / −28
1.97 – 3.15	1.2 / 6.0	+3.0 / 0	−1.2 / −3.0	2.5 / 10.0	+4.5 / 0	−2.5 / −5.5	4.0 / 11.5	+4.5 / −0	−4.0 / −7.0	6 / 17.5	+7 / −0	−6 / −10.5	10 / 34	+12 / −0	−10 / −22	14 / 50	+18 / −0	−14 / −32
3.15 – 4.73	1.4 / 7.1	+3.5 / 0	−1.4 / −3.6	3.0 / 11.5	+5.0 / 0	−3.0 / −6.5	5.0 / 13.5	+5.0 / −0	−5.0 / −8.5	7 / 21	+9 / −0	−7 / −12	11 / 39	+14 / −0	−11 / −25	16 / 60	+22 / −0	−16 / −38
4.73 – 7.09	1.6 / 8.1	+4.0 / 0	−1.6 / −4.1	3.5 / 13.5	+6.0 / 0	−3.5 / −7.5	6 / 16	+6 / −0	−6 / −10	8 / 24	+10 / −0	−8 / −14	12 / 44	+16 / −0	−12 / −28	18 / 68	+25 / −0	−18 / −43
7.09 – 9.85	2.0 / 9.3	+4.5 / 0	−2.0 / −4.8	4.0 / 15.5	+7.0 / 0	−4.0 / −8.5	7 / 18.5	+7 / −0	−7 / −11.5	10 / 29	+12 / −0	−10 / −17	16 / 52	+18 / −0	−16 / −34	22 / 78	+28 / −0	−22 / −50

[Courtesy ANSI B4.1-1967 (R1974)]

Table 44   Locational Transition Fits (Inch Units)

Limits are in thousandths of an inch. Limits for hole and shaft are applied algebraically to the basic size to obtain the limits of size for the mating parts. Data in bold face are in accordance with ABC agreements. "Fit" represents the maximum interference (minus values) and the maximum clearance (plus values). Symbols H7, js6, etc., are Hole and Shaft designations.

Nominal Size Range Inches Over	To	Class LT 1 Fit	Hole H7	Shaft js6	Class LT 2 Fit	Hole H8	Shaft js7	Class LT 3 Fit	Hole H7	Shaft k6	Class LT 4 Fit	Hole H8	Shaft k7	Class LT 5 Fit	Hole H7	Shaft n6	Class LT 6 Fit	Hole H7	Shaft n7
0 —	0.12	−0.10 +0.50	+0.4 −0	+0.10 −0.10	−0.2 +0.8	+0.6 −0	+0.2 −0.2							−0.5 +0.15	+0.4 −0	+0.5 +0.25	−0.65 +0.15	+0.4 −0	+0.65 +0.25
0.12 —	0.24	−0.15 +0.65	+0.5 −0	+0.15 −0.15	−0.25 +0.95	+0.7 −0	+0.25 −0.25							−0.6 +0.2	+0.5 −0	+0.6 +0.3	−0.8 +0.2	+0.5 −0	+0.8 +0.3
0.24 —	0.40	−0.2 +0.8	+0.6 −0	+0.2 −0.2	−0.3 +1.2	+0.9 −0	+0.3 −0.3	−0.5 +0.5	+0.6 −0	+0.5 +0.1	−0.7 +0.8	+0.9 −0	+0.7 +0.1	−0.8 +0.2	+0.6 −0	+0.8 +0.4	−1.0 +0.2	+0.6 −0	+1.0 +0.4
0.40 —	0.71	−0.2 +0.9	+0.7 −0	+0.2 −0.2	−0.35 +1.35	+1.0 −0	+0.35 −0.35	−0.5 +0.6	+0.7 −0	+0.5 +0.1	−0.8 +0.9	+1.0 −0	+0.8 +0.1	−0.9 +0.2	+0.7 −0	+0.9 +0.5	−1.2 +0.2	+0.7 −0	+1.2 +0.5
0.71 —	1.19	−0.25 +1.05	+0.8 −0	+0.25 −0.25	−0.4 +1.6	+1.2 −0	+0.4 −0.4	−0.6 +0.7	+0.8 −0	+0.6 +0.1	−0.9 +1.1	+1.2 −0	+0.9 +0.1	−1.1 +0.2	+0.8 −0	+1.1 +0.6	−1.4 +0.2	+0.8 −0	+1.4 +0.6
1.19 —	1.97	−0.3 +1.3	+1.0 −0	+0.3 −0.3	−0.5 +2.1	+1.6 −0	+0.5 −0.5	−0.7 +0.9	+1.0 −0	+0.7 +0.1	−1.1 +1.5	+1.6 −0	+1.1 +0.1	−1.3 +0.3	+1.0 −0	+1.3 +0.7	−1.7 +0.3	+1.0 −0	+1.7 +0.7
1.97 —	3.15	−0.3 +1.5	+1.2 −0	+0.3 −0.3	−0.6 +2.4	+1.8 −0	+0.6 −0.6	−0.8 +1.1	+1.2 −0	+0.8 +0.1	−1.3 +1.7	+1.8 −0	+1.3 +0.1	−1.5 +0.4	+1.2 −0	+1.5 +0.8	−2.0 +0.4	+1.2 −0	+2.0 +0.8
3.15 —	4.73	−0.4 +1.8	+1.4 −0	+0.4 −0.4	−0.7 +2.9	+2.2 −0	+0.7 −0.7	−1.0 +1.3	+1.4 −0	+1.0 +0.1	−1.5 +2.1	+2.2 −0	+1.5 +0.1	−1.9 +0.4	+1.4 −0	+1.9 +1.0	−2.4 +0.4	+1.4 −0	+2.4 +1.0
4.73 —	7.09	−0.5 +2.1	+1.6 −0	+0.5 −0.5	−0.8 +3.3	+2.5 −0	+0.8 −0.8	−1.1 +1.5	+1.6 −0	+1.1 +0.1	−1.7 +2.4	+2.5 −0	+1.7 +0.1	−2.2 +0.4	+1.6 −0	+2.2 +1.2	−2.8 +0.4	+1.6 −0	+2.8 +1.2
7.09 —	9.85	−0.6 +2.4	+1.8 −0	+0.6 −0.6	−0.9 +3.7	+2.8 −0	+0.9 −0.9	−1.4 +1.6	+1.8 −0	+1.4 +0.2	−2.0 +2.6	+2.8 −0	+2.0 +0.2	−2.6 +0.4	+1.8 −0	+2.6 +1.4	−3.2 +0.4	+1.8 −0	+3.2 +1.4
9.85 —	12.41	−0.6 +2.6	+2.0 −0	+0.6 −0.6	−1.0 +4.0	+3.0 −0	+1.0 −1.0	−1.4 +1.8	+2.0 −0	+1.4 +0.2	−2.2 +2.8	+3.0 −0	+2.2 +0.2	−2.6 +0.6	+2.0 −0	+2.6 +1.4	−3.4 +0.6	+2.0 −0	+3.4 +1.4
12.41 —	15.75	−0.7 +2.9	+2.2 −0	+0.7 −0.7	−1.0 +4.5	+3.5 −0	+1.0 −1.0	−1.6 +2.0	+2.2 −0	+1.6 +0.2	−2.4 +3.3	+3.5 −0	+2.4 +0.2	−3.0 +0.6	+2.2 −0	+3.0 +1.6	−3.8 +0.6	+2.2 −0	+3.8 +1.6
15.75 —	19.69	−0.8 +3.3	+2.5 −0	+0.8 −0.8	−1.2 +5.2	+4.0 −0	+1.2 −1.2	−1.8 +2.3	+2.5 −0	+1.8 +0.2	−2.7 +3.8	+4.0 −0	+2.7 +0.2	−3.4 +0.7	+2.5 −0	+3.4 +1.8	−4.3 +0.7	+2.5 −0	+4.3 +1.8

[Courtesy ANSI B4.1-1967 (R1974)]

Table 45 Locational Interference Fits (Inch Units)

Limits are in thousandths of an inch. Limits for hole and shaft are applied algebraically to the basic size to obtain the limits of size for the parts. Data in bold face are in accordance with ABC agreements. Symbols H7, p6, etc., are Hole and Shaft designations.

Nominal Size Range Inches Over — To	Class LN 1 Limits of Interference	Class LN 1 Standard Limits Hole H6	Class LN 1 Standard Limits Shaft n5	Class LN 2 Limits of Interference	Class LN 2 Standard Limits Hole H7	Class LN 2 Standard Limits Shaft p6	Class LN 3 Limits of Interference	Class LN 3 Standard Limits Hole H7	Class LN 3 Standard Limits Shaft r6
0 — 0.12	**0** / **0.45**	+ 0.25 / − 0	+0.45 / +0.25	**0** / **0.65**	+ 0.4 / − 0	+ 0.65 / + 0.4	**0.1** / **0.75**	+ 0.4 / − 0	+ 0.75 / + 0.5
0.12 — 0.24	**0** / **0.5**	+ 0.3 / − 0	+0.5 / +0.3	**0** / **0.8**	+ 0.5 / − 0	+ 0.8 / + 0.5	**0.1** / **0.9**	+ 0.5 / 0	+ 0.9 / + 0.6
0.24 — 0.40	**0** / **0.65**	+ 0.4 / − 0	+0.65 / +0.4	**0** / **1.0**	+ 0.6 / − 0	+ 1.0 / + 0.6	**0.2** / **1.2**	+ 0.6 / − 0	+ 1.2 / + 0.8
0.40 — 0.71	**0** / **0.8**	+ 0.4 / − 0	+0.8 / +0.4	**0** / **1.1**	+ 0.7 / − 0	+ 1.1 / + 0.7	**0.3** / **1.4**	+ 0.7 / − 0	+ 1.4 / + 1.0
0.71 — 1.19	**0** / **1.0**	+ 0.5 / − 0	+1.0 / +0.5	**0** / **1.3**	+ 0.8 / − 0	+ 1.3 / + 0.8	**0.4** / **1.7**	+ 0.8 / − 0	+ 1.7 / + 1.2
1.19 — 1.97	**0** / **1.1**	+ 0.6 / − 0	+1.1 / +0.6	**0** / **1.6**	+ 1.0 / − 0	+ 1.6 / + 1.0	**0.4** / **2.0**	+ 1.0 / − 0	+ 2.0 / + 1.4
1.97 — 3.15	**0.1** / **1.3**	+ 0.7 / − 0	+1.3 / +0.8	**0.2** / **2.1**	+ 1.2 / − 0	+ 2.1 / + 1.4	**0.4** / **2.3**	+ 1.2 / − 0	+ 2.3 / + 1.6
3.15 — 4.73	**0.1** / **1.6**	+ 0.9 / − 0	+1.6 / +1.0	**0.2** / **2.5**	+ 1.4 / − 0	+ 2.5 / + 1.6	**0.6** / **2.9**	+ 1.4 / − 0	+ 2.9 / + 2.0
4.73 — 7.09	**0.2** / **1.9**	+ 1.0 / − 0	+1.9 / +1.2	**0.2** / **2.8**	+ 1.6 / − 0	+ 2.8 / + 1.8	**0.9** / **3.5**	+ 1.6 / − 0	+ 3.5 / + 2.5
7.09 — 9.85	**0.2** / **2.2**	+ 1.2 / − 0	+2.2 / +1.4	**0.2** / **3.2**	+ 1.8 / − 0	+ 3.2 / + 2.0	**1.2** / **4.2**	+ 1.8 / − 0	+ 4.2 / + 3.0
9.85 — 12.41	**0.2** / **2.3**	+ 1.2 / − 0	+2.3 / +1.4	**0.2** / **3.4**	+ 2.0 / − 0	+ 3.4 / + 2.2	**1.5** / **4.7**	+ 2.0 / − 0	+ 4.7 / + 3.5
12.41 — 15.75	**0.2** / **2.6**	+ 1.4 / − 0	+2.6 / +1.6	**0.3** / **3.9**	+ 2.2 / − 0	+ 3.9 / + 2.5	**2.3** / **5.9**	+ 2.2 / − 0	+ 5.9 / + 4.5
15.75 — 19.69	**0.2** / **2.8**	+ 1.6 / − 0	+2.8 / +1.8	**0.3** / **4.4**	+ 2.5 / − 0	+ 4.4 / + 2.8	**2.5** / **6.6**	+ 2.5 / − 0	+ 6.6 / + 5.0
19.69 — 30.09		+ 2.0 / − 0		0.5 / 5.5	+ 3 / − 0	+ 5.5 / + 3.5	4 / 9	+ 3 / − 0	+ 9 / + 7
30.09 — 41.49		+ 2.5 / − 0		0.5 / 7.0	+ 4 / − 0	+ 7.0 / + 4.5	5 / 11.5	+ 4 / − 0	+11.5 / + 9
41.49 — 56.19		+ 3.0 / − 0		1 / 9	+ 5 / − 0	+ 9 / + 6	7 / 15	+ 5 / − 0	+15 / +12
56.19 — 76.39		+ 4.0 / − 0		1 / 11	+ 6 / − 0	+11 / + 7	10 / 20	+ 6 / − 0	+20 / +16
76.39 — 100.9		+ 5.0 / − 0		1 / 14	+ 8 / − 0	+14 / + 9	12 / 25	+ 8 / − 0	+25 / +20
100.9 — 131.9		+ 6.0 / − 0		2 / 18	+10 / − 0	+18 / +12	15 / 31	+10 / − 0	+31 / +25
131.9 — 171.9		+ 8.0 / − 0		4 / 24	+12 / − 0	+24 / +16	18 / 38	+12 / − 0	+38 / +30
171.9 — 200		+10.0 / − 0		4 / 30	+16 / − 0	+30 / +20	24 / 50	+16 / − 0	+50 / +40

[Courtesy ANSI B4.1-1967 (R1974)]

Table 46 Force and Shrink Fits (Inch Units)

Limits are in thousandths of an inch. Limits for hole and shaft are applied algebraically to the basic size to obtain the limits of size for the parts. Data in bold face are in accordance with ABC agreements. Symbols H7, s6, etc., are Hole and Shaft designations.

Nominal Size Range Inches (Over – To)	FN 1 Limits of Interference	FN 1 Hole H6	FN 1 Shaft	FN 2 Limits of Interference	FN 2 Hole H7	FN 2 Shaft s6	FN 3 Limits of Interference	FN 3 Hole H7	FN 3 Shaft t6	FN 4 Limits of Interference	FN 4 Hole H7	FN 4 Shaft u6	FN 5 Limits of Interference	FN 5 Hole H8	FN 5 Shaft x7
0 – 0.12	0.05	+0.25	+ 0.5	0.2	+0.4	+ 0.85				0.3	+0.4	+ 0.95	0.3	+0.6	+ 1.3
	0.5	− 0	+ 0.3	0.85	− 0	+ 0.6				0.95	− 0	+ 0.7	1.3	− 0	+ 0.9
0.12 – 0.24	0.1	+0.3	+ 0.6	0.2	+0.5	+ 1.0				0.4	+0.5	+ 1.2	0.5	+ 0.7	+ 1.7
	0.6	− 0	+ 0.4	1.0	− 0	+ 0.7				1.2	− 0	+ 0.9	1.7	− 0	+ 1.2
0.24 – 0.40	0.1	+0.4	+ 0.75	0.4	+0.6	+ 1.4				0.6	+0.6	+ 1.6	0.5	+ 0.9	+ 2.0
	0.75	− 0	+ 0.5	1.4	− 0	+ 1.0				1.6	− 0	+ 1.2	2.0	− 0	+ 1.4
0.40 – 0.56	0.1	+0.4	+ 0.8	0.5	+0.7	+ 1.6				0.7	+ 0.7	+ 1.8	0.6	+ 1.0	+ 2.3
	0.8	− 0	+ 0.5	1.6	− 0	+ 1.2				1.8	− 0	+ 1.4	2.3	− 0	+ 1.6
0.56 – 0.71	0.2	+0.4	+ 0.9	0.5	+0.7	+ 1.6				0.7	+ 0.7	+ 1.8	0.8	+ 1.0	+ 2.5
	0.9	− 0	+ 0.6	1.6	− 0	+ 1.2				1.8	− 0	+ 1.4	2.5	− 0	+ 1.8
0.71 – 0.95	0.2	+0.5	+ 1.1	0.6	+0.8	+ 1.9				0.8	+0.8	+ 2.1	1.0	+ 1.2	+ 3.0
	1.1	− 0	+ 0.7	1.9	− 0	+ 1.4				2.1	− 0	+ 1.6	3.0	− 0	+ 2.2
0.95 – 1.19	0.3	+0.5	+ 1.2	0.6	+0.8	+ 1.9	0.8	+0.8	+ 2.1	1.0	+0·8	+ 2.3	1.3	+ 1.2	+ 3.3
	1.2	− 0	+ 0.8	1.9	− 0	+ 1.4	2.1	− 0	+ 1.6	2.3	− 0	+ 1.8	3.3	− 0	+ 2.5
1.19 – 1.58	0.3	+0.6	+ 1.3	0.8	+1.0	+ 2.4	1.0	+1.0	+ 2.6	1.5	+1.0	+ 3.1	1.4	+ 1.6	+ 4.0
	1.3	− 0	+ 0.9	2.4	− 0	+ 1.8	2.6	− 0	+ 2.0	3.1	− 0	+ 2.5	4.0	− 0	+ 3.0
1.58 – 1.97	0.4	+0.6	+ 1.4	0.8	+1.0	+ 2.4	1.2	+1.0	+ 2.8	1.8	+1.0	+ 3.4	2.4	+ 1.6	+ 5.0
	1.4	− 0	+ 1.0	2.4	− 0	+ 1.8	2.8	− 0	+ 2.2	3.4	− 0	+ 2.8	5.0	− 0	+ 4.0
1.97 – 2.56	0.6	+0.7	+ 1.8	0.8	+1.2	+ 2.7	1.3	+1.2	+ 3.2	2.3	+1.2	+ 4.2	3.2	+ 1.8	+ 6.2
	1.8	− 0	+ 1.3	2.7	− 0	+ 2.0	3.2	− 0	+ 2.5	4.2	− 0	+ 3.5	6.2	− 0	+ 5.0
2.56 – 3.15	0.7	+0.7	+ 1.9	1.0	+1.2	+ 2.9	1.8	+1.2	+ 3.7	2.8	+1.2	+ 4.7	4.2	+ 1.8	+ 7.2
	1.9	− 0	+ 1.4	2.9	− 0	+ 2.2	3.7	− 0	+ 3.0	4.7	− 0	+ 4.0	7.2	− 0	+ 6.0
3.15 – 3.94	0.9	+0.9	+ 2.4	1.4	+1.4	+ 3.7	2.1	+1.4	+ 4.4	3.6	+1.4	+ 5.9	4.8	+ 2.2	+ 8.4
	2.4	− 0	+ 1.8	3.7	− 0	+ 2.8	4.4	− 0	+ 3.5	5.9	− 0	+ 5.0	8.4	− 0	+ 7.0
3.94 – 4.73	1.1	+0.9	+ 2.6	1.6	+1.4	+ 3.9	2.6	+1.4	+ 4.9	4.6	+1.4	+ 6.9	5.8	+ 2.2	+ 9.4
	2.6	− 0	+ 2.0	3.9	− 0	+ 3.0	4.9	− 0	+ 4.0	6.9	− 0	+ 6.0	9.4	− 0	+ 8.0
4.73 – 5.52	1.2	+1.0	+ 2.9	1.9	+1.6	+ 4.5	3.4	+1.6	+ 6.0	5.4	+1.6	+ 8.0	7.5	+ 2.5	+11.6
	2.9	− 0	+ 2.2	4.5	− 0	+ 3.5	6.0	− 0	+ 5.0	8.0	− 0	+ 7.0	11.6	− 0	+10.0
5.52 – 6.30	1.5	+1.0	+ 3.2	2.4	+1.6	+ 5.0	3.4	+1.6	+ 6.0	5.4	+1.6	+ 8.0	9.5	+ 2.5	+13.6
	3.2	− 0	+ 2.5	5.0	− 0	+ 4.0	6.0	− 0	+ 5.0	8.0	− 0	+ 7.0	13.6	− 0	+12.0
6.30 – 7.09	1.8	+1.0	+ 3.5	2.9	+1.6	+ 5.5	4.4	+1.6	+ 7.0	6.4	+1.6	+ 9.0	9.5	+ 2.5	+13.6
	3.5	− 0	+ 2.8	5.5	− 0	+ 4.5	7.0	− 0	+ 6.0	9.0	− 0	+ 8.0	13.6	− 0	+12.0
7.09 – 7.88	1.8	+1.2	+ 3.8	3.2	+1.8	+ 6.2	5.2	+1.8	+ 8.2	7.2	+1.8	+10.2	11.2	+ 2.8	+15.8
	3.8	− 0	+ 3.0	6.2	− 0	+ 5.0	8.2	− 0	+ 7.0	10.2	− 0	+ 9.0	15.8	− 0	+14.0
7.88 – 8.86	2.3	+1.2	+ 4.3	3.2	+1.8	+ 6.2	5.2	+1.8	+ 8.2	8.2	+1.8	+11.2	13.2	+ 2.8	+17.8
	4.3	− 0	+ 3.5	6.2	− 0	+ 5.0	8.2	− 0	+ 7.0	11.2	− 0	+10.0	17.8	− 0	+16.0
8.86 – 9.85	2.3	+1.2	+ 4.3	4.2	+1.8	+ 7.2	6.2	+1.8	+ 9.2	10.2	+1.8	+13.2	13.2	+ 2.8	+17.8
	4.3	− 0	+ 3.5	7.2	− 0	+ 6.0	9.2	− 0	+ 8.0	13.2	− 0	+12.0	17.8	− 0	+16.0
9.85 – 11.03	2.8	+1.2	+ 4.9	4.0	+2.0	+ 7.2	7.0	+2.0	+10.2	10.0	+2.0	+13.2	15.0	+ 3.0	+20.0
	4.9	− 0	+ 4.0	7.2	− 0	+ 6.0	10.2	− 0	+ 9.0	13.2	− 0	+12.0	20.0	− 0	+18.0
11.03 – 12.41	2.8	+1.2	+ 4.9	5.0	+2.0	+ 8.2	7.0	+2.0	+10.2	12.0	+2.0	+15.2	17.0	+ 3.0	+22.0
	4.9	− 0	+ 4.0	8.2	− 0	+ 7.0	10.2	− 0	+ 9.0	15.2	− 0	+14.0	22.0	− 0	+20.0
12.41 – 13.98	3.1	+1.4	+ 5.5	5.8	+2.2	+ 9.4	7.8	+2.2	+11.4	13.8	+2.2	+17.4	18.5	+ 3.5	+24.2
	5.5	− 0	+ 4.5	9.4	− 0	+ 8.0	11.4	− 0	+10.0	17.4	− 0	+16.0	24.2	+ 0	+22.0
13.98 – 15.75	3.6	+1.4	+ 6.1	5.8	+2.2	+ 9.4	9.8	+2.2	+13.4	15.8	+2.2	+19.4	21.5	+ 3.5	+27.2
	6.1	− 0	+ 5.0	9.4	− 0	+ 8.0	13.4	− 0	+12.0	19.4	− 0	+18.0	27.2	− 0	+25.0
15.75 – 17.72	4.4	+1.6	+ 7.0	6.5	+2.5	+10.6	9.5	+2.5	+13.6	17.5	+2.5	+21.6	24.0	+ 4.0	+30.5
	7.0	− 0	+ 6.0	10.6	− 0	+ 9.0	13.6	− 0	+12.0	21.6	− 0	+20.0	30.5	− 0	+28.0
17.72 – 19.69	4.4	+1.6	+ 7.0	7.5	+2.5	+11.6	11.5	+2.5	+15.6	19.5	+2.5	+23.6	26.0	+ 4.0	+32.5
	7.0	− 0	+ 6.0	11.6	− 0	+10.0	15.6	− 0	+14.0	23.6	− 0	+22.0	32.5	− 0	+30.0

[Courtesy ANSI B4.1-1967 (R1974)]

Calculations for Limit Dimensions (Inch Units)

EXAMPLE 1 BASIC HOLE SYSTEM

The basic size of the hole is 1.500 in. and the class of fit is RC5. The portion of Table 42, p. 605, that applies to this case follows.

Nominal size range, inches	Limits of clearance	Standard limits	
		Hole	Shaft
1.19–1.97	2.0	+1.6	−2.0
	4.6	0	−3.0

(Limits are in thousandths of an inch)

The limit dimensions *for the hole* are:

$$1.500 + 0.000 = 1.500 \text{ as the minimum value, and}$$

$$1.5000 + 0.0016 = 1.5016 \text{ as the maximum value.}$$

The limit dimensions *for the shaft* are:

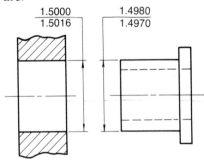

Figure 1

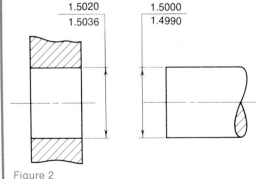

Figure 2

$$1.500 - 0.002 = 1.498 \text{ as the maximum value, and}$$

$$1.500 - 0.003 = 1.497 \text{ as the minimum value.}$$

The *allowance* is $1.500 - 1.498 = 0.002$, which is the difference between the smallest hole diameter and the largest shaft diameter, or the *tightest fit*. The *loosest fit* is the difference between the largest hole diameter and the smallest shaft diameter, or $1.5016 - 1.4970 = 0.0046$. Note that this value, 0.004, is equal to the allowance (0.002) plus both tolerances (each of which is 0.001). The graphic representation is shown in Figure 1.

Note that the tables just referred to were designed for the basic hole system, which is the preferred practice because it allows standard hole-forming tools to be used. However, there are cases in which it is necessary to maintain constant limit dimensions for the shaft on which several components may be fitted.

The bushing is the internal member which is regarded as the "shaft" in the table

EXAMPLE 2 BASIC SHAFT SYSTEM

The basic shaft size is 1.500 and the class of fit is RC5. To use the tabular values in Table 42, p. 605, directly, the basic hole size must first be determined. This is accomplished by *adding the allowance* to the basic shaft size. The allowance is shown as 2.0 (meaning 0.002) in the "limits of clearance column" (upper number). Therefore, the *basic hole size is* $1.500 + 0.002 = 1.502$. Now we can use the table as shown in Example 1.

The limit dimensions *for the hole* are:

$$1.5020 + 0.0016 = 1.5036, \text{ and}$$

$$1.5020 + 0.0000 = 1.5020.$$

The limit dimensions *for the shaft* are:

$$1.5020 - 0.0020 = 1.5000, \text{ and}$$

$$1.5020 - 0.0030 = 1.4990.$$

The allowance is $1.502 - 1.500 = 0.002$, which is the difference between the smallest hole diameter and the largest shaft diameter. The loosest fit is $1.5036 - 1.4990 = 0.0046$, which is the difference between the largest hole diameter and the smallest shaft diameter. The graphic representation is shown in Figure 2.

METRIC UNIT TABLES

Table 1 Lower Deviations for Holes for General Engineering Applications (Metric Units)

Tabulated values are (+) unless otherwise stated

μm

Basic size mm		Lower deviation for all tolerance grades											
Above	Up to and incl.	A* +	B* +	C +	CD +	D +	E +	EF +	F +	FG +	G +	H +	Js† ±
0	3	270	140	60	34	20	14	10	6	4	2	0	
3	6	270	140	70	46	30	20	14	10	6	4	0	
6	10	280	150	80	56	40	25	18	13	8	5	0	
10	18	290	150	95	—	50	32	—	16	—	6	0	
18	30	300	160	110	—	65	40	—	20	—	7	0	
30	40	310	170	120	—	80	50	—	25	—	9	0	
40	50	320	180	130	—	80	50	—	25	—	9	0	
50	65	340	190	140	—	100	60	—	30	—	10	0	
65	80	360	200	150	—	100	60	—	30	—	10	0	
80	100	380	220	170	—	120	72	—	36	—	12	0	
100	120	410	240	180	—	120	72	—	36	—	12	0	
120	140	460	260	200	—	145	85	—	43	—	14	0	
140	160	520	280	210	—	145	85	—	43	—	14	0	Deviation = ± (tolerance/2)
160	180	580	310	230	—	145	85	—	43	—	14	0	
180	200	660	340	240	—	170	100	—	50	—	15	0	
200	225	740	380	260	—	170	100	—	50	—	15	0	
225	250	820	420	280	—	170	100	—	50	—	15	0	
250	280	920	480	300	—	190	110	—	56	—	17	0	
280	315	1050	540	330	—	190	110	—	56	—	17	0	
315	355	1200	600	360	—	210	125	—	62	—	18	0	
355	400	1350	680	400	—	210	125	—	62	—	18	0	
400	450	1500	760	440	—	230	135	—	68	—	20	0	
450	500	1650	840	480	—	230	135	—	68	—	20	0	
500	630	—	—	—	—	260	145	—	76	—	22	0	
630	800	—	—	—	—	290	160	—	80	—	24	0	
800	1000	—	—	—	—	320	170	—	86	—	26	0	
1000	1250	—	—	—	—	350	195	—	98	—	28	0	
1250	1600	—	—	—	—	390	220	—	110	—	30	0	
1600	2000	—	—	—	—	430	240	—	120	—	32	0	
2000	2500	—	—	—	—	480	260	—	130	—	34	0	
2500	3150	—	—	—	—	520	290	—	145	—	38	0	

(Courtesy AS 1654–1974. Similar tables are also available in the *Interim General Motors Metric Engineering Standards* adapted from ISO Recommendation R 286–1672.)

* The deviations A and B for all grades of tolerance are not applicable to diameters smaller than 1 mm.

† For J_s, if the tolerance (obtained from Table 3) is greater than 50 μm and is an odd value, the two symmetrical deviations (± tolerance/2) shall be rounded to the nearest smaller whole number. No rounding applies to tolerances of 50 μm and smaller.

NOTE: Upper deviation is obtained by *adding* the appropriate tolerance value (Table 3) to the value of lower deviation.

Table 1 (*continued*)

Tabulated values are (−) unless otherwise stated   μm

Basic size mm		Tolerance class												
		K							M					
Above	Up to and incl.	6	7	8	9	10	11	12	6	7	8	9	10	11	12
0	3	6	10	14	25	40	60	100	8	12	16	27	42	62	102
3	6	6	9	13	—	—	—	—	9	12	16	34	52	79	124
6	10	7	10	16	—	—	—	—	12	15	21	42	64	96	156
10	18	9	12	19	—	—	—	—	15	18	25	50	77	117	187
18	30	11	15	23	—	—	—	—	17	21	29	60	92	138	218
30	50	13	18	27	—	—	—	—	20	25	34	71	109	169	259
50	80	15	21	32	—	—	—	—	24	30	41	85	131	201	311
80	120	18	25	38	—	—	—	—	28	35	48	100	153	233	363
120	180	21	28	43	—	—	—	—	33	40	55	115	175	265	415
180	250	24	33	50	—	—	—	—	37	46	63	132	202	307	477
250	315	27	36	56	—	—	—	—	41	52	72	150	230	340	540
315	400	29	40	61	—	—	—	—	46	57	78	161	251	381	591
400	500	32	45	68	—	—	—	—	50	63	86	178	273	423	653
500	630	44	70	110	175	280	440	700	70	96	136	201	306	466	726
630	800	50	80	125	200	320	500	800	80	110	155	230	350	530	830
800	1000	56	90	140	230	360	560	900	90	124	174	264	394	594	934
1000	1250	66	105	165	260	420	660	1050	106	145	205	300	460	700	1090
1250	1600	78	125	195	310	500	780	1250	126	173	243	358	548	828	1298
1600	2000	92	150	230	370	600	920	1500	150	208	288	428	658	978	1558
2000	2500	110	175	280	440	700	1100	1750	178	243	348	508	768	1168	1818
2500	3150	135	210	330	540	860	1350	2100	211	286	406	616	936	1426	2176

(Courtesy AS 1654–1974)

NOTE: Upper deviation is obtained by *adding* the appropriate tolerance value (Table 3) to the value of lower deviation.

Table 1 (*continued*)

Tabulated values are (−) unless otherwise stated   μm

Basic size mm		Tolerance class												
		N							P					
Above	Up to and incl.	6	7	8	9*	10*	11*	12*	6	7	8	9	10	11	12
		−	−	−	−	−	−	−	−	−	−	−	−	−	−
0	3	10	14	18	29	44	64	104	12	16	20	31	46	66	106
3	6	13	16	20	30	48	75	120	17	20	30	42	60	87	132
6	10	16	19	25	36	58	90	150	21	24	37	51	73	105	165
10	18	20	23	30	43	70	110	180	26	29	45	61	88	128	198
18	30	24	28	36	52	84	130	210	31	35	55	74	106	152	232
30	50	28	33	42	62	100	160	250	37	42	65	88	126	186	276
50	80	33	39	50	74	120	190	300	45	51	78	106	152	222	332
80	120	38	45	58	87	140	220	350	52	59	91	124	177	257	387
120	180	45	52	67	100	160	250	400	61	68	106	143	203	293	443
180	250	51	60	77	115	185	290	460	70	79	122	165	235	340	510
250	315	57	66	86	130	210	320	520	79	88	137	186	266	376	576
315	400	62	73	94	140	230	360	570	87	98	151	202	292	422	632
400	500	67	80	103	155	250	400	630	95	108	165	223	318	468	698
500	630	88	114	154	219	324	484	744	122	148	188	253	358	518	778
630	800	100	130	175	250	370	550	850	138	168	213	288	408	588	888
800	1000	112	146	196	286	416	616	956	156	190	240	330	460	660	1000
1000	1250	132	171	231	326	486	726	1116	186	225	285	380	540	780	1170
1250	1600	156	203	273	388	578	858	1328	218	265	335	450	640	920	1390
1600	2000	184	242	322	462	692	1012	1592	262	320	400	540	770	1090	1670
2000	2500	220	285	390	550	810	1210	1860	305	370	475	635	895	1295	1945
2500	3150	270	345	465	675	995	1485	2235	375	450	570	780	1100	1590	2340

(Courtesy AS 1654–1974)

* Tolerance classes N9, N10, N11 and N12 are not applicable to diameters smaller than 1 mm.

NOTE: Upper deviation is obtained by *adding* the appropriate tolerance value (Table 3) to the value of lower deviation.

Table 1 (*continued*)

Tabulated values are (−) unless otherwise stated   µm

Basic size mm		Tolerance class													
Above	Up to and incl.	R 6	7	8	9	10	11	12	S 6	7	8	9	10	11	12
		−	−	−	−	−	−	−	−	−	−	−	−	−	−
0	3	16	20	24	35	50	70	110	20	24	28	39	54	74	114
3	6	20	23	33	45	63	90	135	24	27	37	49	67	94	139
6	10	25	28	41	55	77	109	169	29	32	45	59	81	113	173
10	18	31	34	50	66	93	133	203	36	39	55	71	98	138	208
18	30	37	41	61	80	112	158	238	44	48	68	87	119	165	245
30	50	45	50	73	96	134	194	284	54	59	82	105	143	203	293
50	65	54	60	87	115	161	231	341	66	72	99	127	173	243	353
65	80	56	62	89	117	163	233	343	72	78	105	133	179	249	359
80	100	66	73	105	138	191	271	401	86	93	125	158	211	291	421
100	120	69	76	108	141	194	274	404	94	101	133	166	219	299	429
120	140	81	88	126	163	223	313	463	110	117	155	192	252	342	492
140	160	83	90	128	165	225	315	465	118	125	163	200	260	350	500
160	180	86	93	131	168	228	318	468	126	133	171	208	268	358	508
180	200	97	106	149	192	262	367	537	142	151	194	237	307	412	582
200	225	100	109	152	195	265	370	540	150	159	202	245	315	420	590
225	250	104	113	156	199	269	374	544	160	169	212	255	325	430	600
250	280	117	126	175	224	304	414	614	181	190	239	288	368	478	678
280	315	121	130	179	228	308	418	618	193	202	251	300	380	490	690
315	355	133	144	197	248	338	468	678	215	226	279	330	420	550	760
355	400	139	150	203	254	344	474	684	233	244	297	348	438	568	778
400	450	153	166	223	281	376	526	756	259	272	329	387	482	632	862
450	500	159	172	229	287	382	532	762	279	292	349	407	502	652	882
500	560	194	220	260	325	430	590	850	324	350	390	455	560	720	980
560	630	199	225	265	330	435	595	855	354	380	420	485	590	750	1010
630	710	225	255	300	375	495	675	975	390	420	465	540	660	840	1140
710	800	235	265	310	385	505	685	985	430	460	505	580	700	880	1180
800	900	266	300	350	440	570	770	1110	486	520	570	660	790	990	1330
900	1000	276	310	360	450	580	780	1120	526	560	610	700	830	1030	1370
1000	1120	316	355	415	510	670	910	1300	586	625	685	780	940	1180	1570
1120	1250	326	365	425	520	680	920	1310	646	685	745	840	1000	1240	1630
1250	1400	378	425	495	610	800	1080	1550	718	765	835	950	1140	1420	1890
1400	1600	408	455	525	640	830	1110	1580	798	845	915	1030	1220	1500	1970
1600	1800	462	520	600	740	970	1290	1870	912	970	1050	1190	1420	1740	2320
1800	2000	492	550	630	770	1000	1320	1900	1012	1070	1150	1290	1520	1840	2420
2000	2240	550	615	720	880	1140	1540	2190	1110	1175	1280	1440	1700	2100	2750
2240	2500	570	635	740	900	1160	1560	2210	1210	1275	1380	1540	1800	2200	2850
2500	2800	685	760	880	1090	1410	1900	2650	1385	1460	1580	1790	2110	2600	3350
2800	3150	715	790	910	1120	1440	1930	2680	1535	1610	1730	1940	2260	2750	3500

(Courtesy AS 1654–1974)

NOTE: Upper deviation is obtained by *adding* the appropriate tolerance value (Table 3) to the value of lower deviation.

Table 1 (*continued*)

Tabulated values are (−) unless otherwise stated   μm

Basic size mm		Tolerance class													
		T							U						
Above	Up to and incl.	6	7	8	9	10	11	12	6	7	8	9	10	11	12
		−	−	−	−	−	−	−	−	−	−	−	−	−	−
0	3	—	—	—	—	—	—	—	24	28	32	43	58	78	118
3	6	—	—	—	—	—	—	—	28	31	41	53	71	98	143
6	10	—	—	—	—	—	—	—	34	37	50	64	86	118	178
10	18	—	—	—	—	—	—	—	41	44	60	76	103	143	213
18	24	—	—	—	—	—	—	—	50	54	74	93	125	171	251
24	30	50	54	74	93	125	171	251	57	61	81	100	132	178	258
30	40	59	64	87	110	148	208	298	71	76	99	122	160	220	310
40	50	65	70	93	116	154	214	304	81	86	109	132	170	230	320
50	65	79	85	112	140	·186	256	366	100	106	133	161	207	277	387
65	80	88	94	121	149	195	265	375	115	121	148	176	222	292	402
80	100	106	113	145	178	231	311	441	139	146	178	211	264	344	474
100	120	119	126	158	191	244	324	454	159	166	198	231	284	364	494
120	140	140	147	185	222	282	372	522	188	195	233	270	330	420	570
140	160	152	159	197	234	294	384	534	208	215	253	290	350	440	590
160	180	164	171	209	246	306	396	546	228	235	273	310	370	460	610
180	200	186	195	238	281	351	456	626	256	265	308	351	421	526	696
200	225	200	209	252	295	365	470	640	278	287	330	373	443	548	718
225	250	216	225	268	311	381	486	656	304	313	356	399	469	574	744
250	280	241	250	299	348	428	538	738	338	347	396	445	525	635	835
280	315	263	272	321	370	450	560	760	373	382	431	480	560	670	870
315	355	293	304	357	408	498	628	838	415	426	479	530	620	750	960
355	400	319	330	383	434	524	654	864	460	471	524	575	665	795	1005
400	450	357	370	427	485	580	730	960	517	530	587	645	740	890	1120
450	500	387	400	457	515	610	760	990	567	580	637	695	790	940	1170
500	560	444	470	510	575	680	840	1100	644	670	710	775	880	1040	1300
560	630	494	520	560	625	730	890	1150	704	730	770	835	940	1100	1360
630	710	550	580	625	700	820	1000	1300	790	820	865	940	1060	1240	1540
710	800	610	640	685	760	880	1060	1360	890	920	965	1040	1160	1340	1640
800	900	676	710	760	850	980	1180	1520	996	1030	1080	1170	1300	1500	1840
900	1000	736	770	820	910	1040	1240	1580	1106	1140	1190	1280	1410	1610	1950
1000	1120	846	885	945	1040	1200	1440	1830	1216	1255	1315	1410	1570	1810	2200
1120	1250	906	945	1005	1100	1260	1500	1890	1366	1405	1465	1560	1720	1960	2350
1250	1400	1038	1085	1155	1270	1460	1740	2210	1528	1575	1645	1760	1950	2230	2700
1400	1600	1128	1175	1245	1360	1550	1830	2300	1678	1725	1795	1910	2100	2380	2850
1600	1800	1292	1350	1430	1570	1800	2120	2700	1942	2000	2080	2220	2450	2770	3350
1800	2000	1442	1500	1580	1720	1950	2270	2850	2092	2150	2230	2370	2600	2920	3500
2000	2240	1610	1675	1780	1940	2200	2600	3250	2410	2475	2580	2740	3000	3400	4050
2240	2500	1760	1825	1930	2090	2350	2750	3400	2610	2675	2780	2940	3200	3600	4250
2500	2800	2035	2110	2230	2440	2760	3250	4000	3035	3110	3230	3440	3760	4250	5000
2800	3150	2235	2310	2430	2640	2960	3450	4200	3335	3410	3530	3740	4060	4550	5300

(Courtesy AS 1654–1974)

NOTE: Upper deviation is obtained by *adding* the appropriate tolerance value (Table 3) to the value of lower deviation.

Table 1 (*continued*)

Tabulated values are (−) unless otherwise stated   μm

Basic size mm		Tolerance class													
		V							X						
Above	Up to and incl.	6 −	7 −	8 −	9 −	10 −	11 −	12 −	6 −	7 −	8 −	9 −	10 −	11 −	12 −
0	3	—	—	—	—	—	—	—	26	30	34	45	60	80	120
3	6	—	—	—	—	—	—	—	33	36	46	58	76	103	148
6	10	—	—	—	—	—	—	—	40	43	56	70	92	124	184
10	14	—	—	—	—	—	—	—	48	51	67	83	110	150	220
14	18	47	50	66	82	109	149	219	53	56	72	88	115	155	225
18	24	56	60	80	99	131	177	257	63	67	87	106	138	184	264
24	30	64	68	88	107	139	185	265	73	77	97	116	148	194	274
30	40	79	84	107	130	168	228	318	91	96	119	142	180	240	330
40	50	92	97	120	143	181	241	331	108	113	136	159	197	257	347
50	65	115	121	148	176	222	292	402	135	141	168	196	242	312	422
65	80	133	139	166	194	240	310	420	159	165	192	220	266	336	446
80	100	161	168	200	233	286	366	496	193	200	232	265	318	398	528
100	120	187	194	226	259	312	392	522	225	232	264	297	350	430	560
120	140	220	227	265	302	362	452	602	266	273	311	348	408	498	648
140	160	246	253	291	328	388	478	628	298	305	343	380	440	530	680
160	180	270	277	315	352	412	502	652	328	335	373	410	470	560	710
180	200	304	313	356	399	469	574	744	370	379	422	465	535	640	810
200	225	330	339	382	425	495	600	770	405	414	457	500	570	675	845
225	250	360	369	412	455	525	630	800	445	454	497	540	610	715	885
250	280	408	417	466	515	595	705	905	498	507	556	605	685	795	995
280	315	448	457	506	555	635	745	945	548	557	606	655	735	845	1045
315	355	500	511	564	615	705	835	1045	615	626	679	730	820	950	1160
355	400	555	566	619	670	760	890	1100	685	696	749	800	890	1020	1230
400	450	622	635	692	750	845	995	1225	767	780	837	895	990	1140	1370
450	500	687	700	757	815	910	1060	1290	847	860	917	975	1070	1220	1450

(Courtesy AS 1654−1974)

NOTE: Upper deviation is obtained by *adding* the appropriate tolerance value (Table 3) to the value of lower deviation.

Table 1 (*continued*)

Tabulated values are (−) unless otherwise stated

μm

Basic size mm		Tolerance class													
		Y							Z						
Above	Up to and incl.	6 −	7 −	8 −	9 −	10 −	11 −	12 −	6 −	7 −	8 −	9 −	10 −	11 −	12 −
0	3	—	—	—	—	—	—	—	32	36	40	51	66	86	126
3	6	—	—	—	—	—	—	—	40	43	53	65	83	110	155
6	10	—	—	—	—	—	—	—	48	51	64	78	100	132	192
10	14	—	—	—	—	—	—	—	58	61	77	93	120	160	230
14	18	—	—	—	—	—	—	—	68	71	87	103	130	170	240
18	24	72	76	96	115	147	193	273	82	86	106	125	157	203	283
24	30	84	88	108	127	159	205	285	97	101	121	140	172	218	298
30	40	105	110	133	156	194	254	344	123	128	151	174	212	272	362
40	50	125	130	153	176	214	274	364	147	152	175	198	236	296	386
50	65	157	163	190	218	264	334	444	185	191	218	246	292	362	472
65	80	187	193	220	248	294	364	474	223	229	256	284	330	400	510
80	100	229	236	268	301	354	434	564	273	280	312	345	398	478	608
100	120	269	276	308	341	394	474	604	325	332	364	397	450	530	660
120	140	318	325	363	400	460	550	700	383	390	428	465	525	615	765
140	160	358	365	403	440	500	590	740	433	440	478	515	575	665	815
160	180	398	405	443	480	540	630	780	483	490	528	565	625	715	865
180	200	445	454	497	540	610	715	885	540	549	592	635	705	810	980
200	225	490	499	542	585	655	760	930	595	604	647	690	760	865	1035
225	250	540	549	592	635	705	810	980	660	669	712	755	825	930	1100
250	280	603	612	661	710	790	900	1100	733	742	791	840	920	1030	1230
280	315	673	682	731	780	860	970	1170	813	822	871	920	1000	1110	1310
315	355	755	766	819	870	960	1090	1300	925	936	989	1040	1130	1260	1470
355	400	845	856	909	960	1050	1180	1390	1025	1036	1089	1140	1230	1360	1570
400	450	947	960	1017	1075	1170	1320	1550	1127	1140	1197	1255	1350	1500	1730
450	500	1027	1040	1097	1155	1250	1400	1630	1277	1290	1347	1405	1500	1650	1880

(Courtesy AS 1654–1974)

NOTE: Upper deviation is obtained by *adding* the appropriate tolerance value (Table 3) to the value of lower deviation.

Table 1 (*continued*)

Tabulated values are (−) unless otherwise stated μm

| Basic size mm | | Tolerance class | | | | | | | | | | | | |
| | | ZA | | | | | | ZB | | | | | | |
Above	Up to and incl.	6 −	7 −	8 −	9 −	10 −	11 −	12 −	6 −	7 −	8 −	9 −	10 −	11 −	12 −
0	3	38	42	46	57	72	92	132	46	50	54	65	80	100	140
3	6	47	50	60	72	90	117	162	55	58	68	80	98	125	170
6	10	58	61	74	88	110	142	202	73	76	89	103	125	157	217
10	14	72	75	91	107	134	174	244	98	101	117	133	160	200	270
14	18	85	88	104	120	147	187	257	116	119	135	151	178	218	288
18	24	107	111	131	150	182	228	308	145	149	169	188	220	266	346
24	30	127	131	151	170	202	248	328	169	173	193	212	244	290	370
30	40	159	164	187	210	248	308	398	211	216	239	262	300	360	450
40	50	191	196	219	242	280	340	430	253	258	281	304	342	402	492
50	65	239	245	272	300	346	416	526	313	319	346	374	420	490	600
65	80	287	293	320	348	394	464	574	373	379	406	434	480	550	660
80	100	350	357	389	422	475	555	685	460	467	499	532	585	665	795
100	120	415	422	454	487	540	620	750	540	547	579	612	665	745	875
120	140	488	495	533	570	630	720	870	638	645	683	720	780	870	1020
140	160	553	560	598	635	695	785	935	718	725	763	800	860	950	1100
160	180	618	625	663	700	760	850	1000	798	805	843	880	940	1030	1180
180	200	690	699	742	785	855	960	1130	900	909	952	995	1065	1170	1340
200	225	760	769	812	855	925	1030	1200	980	989	1032	1075	1145	1250	1420
225	250	840	849	892	935	1005	1110	1280	1070	1079	1122	1165	1235	1340	1510
250	280	943	952	1001	1050	1130	1240	1440	1223	1232	1281	1330	1410	1520	1720
280	315	1023	1032	1081	1130	1210	1320	1520	1323	1332	1381	1430	1510	1620	1820
315	355	1175	1186	1239	1290	1380	1510	1720	1525	1536	1589	1640	1730	1860	2070
355	400	1325	1336	1389	1440	1530	1660	1870	1675	1686	1739	1790	1880	2010	2220
400	450	1477	1490	1547	1605	1700	1850	2080	1877	1890	1947	2005	2100	2250	2480
450	500	1627	1640	1697	1755	1850	2000	2230	2127	2140	2197	2255	2350	2500	2730

(Courtesy AS 1654–1974)

NOTE: Upper deviation is obtained by *adding* the appropriate tolerance value (Table 3) to the value of lower deviation.

Table 1 (*continued*)

Tabulated values are (−) unless otherwise stated   μm

Basic size mm		Tolerance class						
		ZC						
Above	Up to and incl.	6 −	7 −	8 −	9 −	10 −	11 −	12 −
0	3	66	70	74	85	100	120	160
3	6	85	88	98	110	128	155	200
6	10	103	106	119	133	155	187	247
10	14	138	141	157	173	200	240	310
14	18	158	161	177	193	220	260	330
18	24	197	201	221	240	272	318	398
24	30	227	231	251	270	302	348	428
30	40	285	290	313	336	374	434	524
40	50	336	341	364	387	425	485	575
50	65	418	424	451	479	525	595	705
65	80	493	499	526	554	600	670	780
80	100	600	607	639	672	725	805	935
100	120	705	712	744	777	830	910	1040
120	140	818	825	863	900	960	1050	1200
140	160	918	925	963	1000	1060	1150	1300
160	180	1018	1025	1063	1100	1160	1250	1400
180	200	1170	1179	1222	1265	1335	1440	1610
200	225	1270	1279	1322	1365	1435	1540	1710
225	250	1370	1379	1422	1465	1535	1640	1810
250	280	1573	1582	1631	1680	1760	1870	2070
280	315	1723	1732	1781	1830	1910	2020	2220
315	355	1925	1936	1989	2040	2130	2260	2470
355	400	2125	2136	2189	2240	2330	2460	2670
400	450	2427	2440	2497	2555	2650	2800	3030
450	500	2627	2640	2697	2755	2850	3000	3230

(Courtesy AS 1654–1974)

NOTE: Upper deviation is obtained by *adding* the appropriate tolerance value (Table 3) to the value of lower deviation.

Table 2 Upper Deviations for Shafts for General Engineering Applications

Tabulated values are (−) unless otherwise stated μm

Basic size mm		Upper deviation for all tolerance grades											
Above	Up to and incl.	a*	b*	c	cd	d	e	ef	f	fg	g	h	js†
		−	−	−	−	−	−	−	−	−	−	−	±
0	3	270	140	60	34	20	14	10	6	4	2	0	
3	6	270	140	70	46	30	20	14	10	6	4	0	
6	10	280	150	80	56	40	25	18	13	8	5	0	
10	18	290	150	95	—	50	32	—	16	—	6	0	
18	30	300	160	110	—	65	40	—	20	—	7	0	
30	40	310	170	120	—	80	50	—	25	—	9	0	
40	50	320	180	130	—	80	50	—	25	—	9	0	
50	65	340	190	140	—	100	60	—	30	—	10	0	
65	80	360	200	150	—	100	60	—	30	—	10	0	
80	100	380	220	170	—	120	72	—	36	—	12	0	
100	120	410	240	180	—	120	72	—	36	—	12	0	
120	140	460	260	200	—	145	85	—	43	—	14	0	
140	160	520	280	210	—	145	85	—	43	—	14	0	
160	180	580	310	230	—	145	85	—	43	—	14	0	
180	200	660	340	240	—	170	100	—	50	—	15	0	
200	225	740	380	260	—	170	100	—	50	—	15	0	
225	250	820	420	280	—	170	100	—	50	—	15	0	
250	280	920	480	300	—	190	110	—	56	—	17	0	
280	315	1050	540	330	—	190	110	—	56	—	17	0	
315	355	1200	600	360	—	210	125	—	62	—	18	0	
355	400	1350	680	400	—	210	125	—	62	—	18	0	
400	450	1500	760	440	—	230	135	—	68	—	20	0	
450	500	1650	840	480	—	230	135	—	68	—	20	0	
500	630	—	—	—	—	260	145	—	76	—	22	0	
630	800	—	—	—	—	290	160	—	80	—	24	0	
800	1000	—	—	—	—	320	170	—	86	—	26	0	
1000	1250	—	—	—	—	350	195	—	98	—	28	0	
1250	1600	—	—	—	—	390	220	—	110	—	30	0	
1600	2000	—	—	—	—	430	240	—	120	—	32	0	
2000	2500	—	—	—	—	480	260	—	130	—	34	0	
2500	3150	—	—	—	—	520	290	—	145	—	38	0	

Deviation = ±(tolerance/2)

(Courtesy AS 1654−1974)

* The deviations a and b for all grades of tolerance are not applicable to diameters smaller than 1 mm.

† For j_s, if the tolerance (obtained from Table 5) is greater than 50 μm and is an odd value, the two symmetrical deviations (± tolerance/2) shall be rounded to the nearest smaller whole number. No rounding applies to tolerances of 50 μm and smaller.

NOTE: Lower deviation is obtained by *subtracting* the appropriate tolerance value (Table 3) from the value of upper deviation.

Table 2 (*continued*)

Tabulated values are (+) unless otherwise stated

μm

Basic size mm		Tolerance class													
		k							m						
Above	Up to and incl.	6 +	7 +	8 +	9 +	10 +	11 +	12 +	6 +	7 +	8 +	9 +	10 +	11 +	12 +
0	3	6	10	14	25	40	60	100	8	12	16	27	42	62	102
3	6	9	13	18	30	48	75	120	12	16	22	34	52	79	124
6	10	10	16	22	36	58	90	150	15	21	28	42	64	96	156
10	18	12	19	27	43	70	110	180	18	25	34	50	77	117	187
18	30	15	23	33	52	84	130	210	21	29	41	60	92	138	218
30	50	18	27	39	62	100	160	250	25	34	48	71	109	169	259
50	80	21	32	46	74	120	190	300	30	41	57	85	131	201	311
80	120	25	38	54	87	140	220	350	35	48	67	100	153	233	363
120	180	28	43	63	100	160	250	400	40	55	78	115	175	265	415
180	250	33	50	72	115	185	290	460	46	63	89	132	202	307	477
250	315	36	56	81	130	210	320	520	52	72	101	150	230	340	540
315	400	40	61	89	140	230	360	570	57	78	110	161	251	381	591
400	500	45	68	97	155	250	400	630	63	86	120	178	273	423	653
500	630	45	70	110	175	280	440	700	70	96	136	201	306	466	726
630	800	50	80	125	200	320	500	800	80	110	155	230	350	530	830
800	1000	56	90	140	230	360	560	900	90	124	174	264	394	594	934
1000	1250	66	105	165	260	420	660	1050	106	145	205	300	460	700	1090
1250	1600	78	125	195	310	500	780	1250	126	173	243	358	548	828	1298
1600	2000	92	150	230	370	600	920	1500	150	208	288	428	658	978	1558
2000	2500	110	175	280	440	700	1100	1750	178	243	348	508	768	1168	1818
2500	3150	135	210	330	540	860	1350	2100	211	286	406	616	936	1426	2176

(Courtesy AS 1654–1974)

NOTE: Lower deviation is obtained by *subtracting* the appropriate tolerance value (Table 3) from the value of upper deviation.

Table 2   (*continued*)

Tabulated values are (+) unless otherwise stated μm

Basic size mm		Tolerance class													
		n							p						
Above	Up to and incl.	6 +	7 +	8 +	9 +	10 +	11 +	12 +	6 +	7 +	8 +	9 +	10 +	11 +	12 +
0	3	10	14	18	29	44	64	104	12	16	20	31	46	66	106
3	6	16	20	26	38	56	83	128	20	24	30	42	60	87	132
6	10	19	25	32	46	68	100	160	24	30	37	51	73	105	165
10	18	23	30	39	55	82	122	192	29	36	45	61	88	128	198
18	30	28	36	48	67	99	145	225	35	43	55	74	106	152	232
30	50	33	42	56	79	117	177	267	42	51	65	88	126	186	276
50	80	39	50	66	94	140	210	320	51	62	78	106	152	222	332
80	120	45	58	77	110	163	243	373	59	72	91	124	177	257	387
120	180	52	67	90	127	187	277	427	68	83	106	143	203	293	443
180	250	60	77	103	146	216	321	491	79	96	122	165	235	340	510
250	315	66	86	115	164	244	354	554	88	108	137	186	266	376	576
315	400	73	94	126	177	267	397	607	98	119	151	202	292	422	632
400	500	80	103	137	195	290	440	670	108	131	165	223	318	468	698
500	630	88	114	154	219	324	484	744	122	148	188	253	358	518	778
630	800	100	130	175	250	370	550	850	138	168	213	288	408	588	888
800	1000	112	146	196	286	416	616	956	156	190	240	330	460	660	1000
1000	1250	132	171	231	326	486	726	1116	186	225	285	380	540	780	1170
1250	1600	156	203	273	388	578	858	1328	218	265	335	450	640	920	1390
1600	2000	184	242	322	462	692	1012	1592	262	320	400	540	770	1090	1670
2000	2500	220	285	390	550	810	1210	1860	305	370	475	635	895	1295	1945
2500	3150	270	345	465	675	995	1485	2235	375	450	570	780	1100	1590	2340

(Courtesy AS 1654–1974)

NOTE: Lower deviation is obtained by *subtracting* the appropriate tolerance value (Table 3) from the value of upper deviation.

Table 2 *(continued)*

Tabulated values are (+) unless otherwise stated

μm

Basic size mm		Tolerance class												
		r							s						
Above	Up to and incl.	6 +	7 +	8 +	9 +	10 +	11 +	12 +	6 +	7 +	8 +	9 +	10 +	11 +	12 +
0	3	16	20	24	35	50	70	110	20	24	28	39	54	74	114
3	6	23	27	33	45	63	90	135	27	31	37	49	67	94	139
6	10	28	34	41	55	77	109	169	32	38	45	59	81	113	173
10	18	34	41	50	66	93	133	203	39	46	55	71	98	138	208
18	30	41	49	61	80	112	158	238	48	56	68	87	119	165	245
30	50	50	59	73	96	134	194	284	59	68	82	105	143	203	293
50	65	60	71	87	115	161	231	341	72	83	99	127	173	243	353
65	80	62	73	89	117	163	233	343	78	89	105	133	179	249	359
80	100	73	86	105	138	191	271	401	93	106	125	158	211	291	421
100	120	76	89	108	141	194	274	404	101	114	133	166	219	299	429
120	140	88	103	126	163	223	313	463	117	132	155	192	252	342	492
140	160	90	105	128	165	225	315	465	125	140	163	200	260	350	500
160	180	93	108	131	168	228	318	468	133	148	171	208	268	358	508
180	200	106	123	149	192	262	367	537	151	168	194	237	307	412	582
200	225	109	126	152	195	265	370	540	159	176	202	245	315	420	590
225	250	113	130	156	199	269	374	544	169	186	212	255	325	430	600
250	280	126	146	175	224	304	414	614	190	210	239	288	368	478	678
280	315	130	150	179	228	308	418	618	202	222	251	300	380	490	690
315	355	144	165	197	248	338	468	678	226	247	279	330	420	550	760
355	400	150	171	203	254	344	474	684	244	265	297	348	438	568	778
400	450	166	189	223	281	376	526	756	272	295	329	387	482	632	862
450	500	172	195	229	287	382	532	762	292	315	349	407	502	652	882
500	560	194	220	260	325	430	590	850	324	350	390	455	560	720	980
560	630	199	225	265	330	435	595	855	354	380	420	485	590	750	1010
630	710	225	255	300	375	495	675	975	390	420	465	540	660	840	1140
710	800	235	265	310	385	505	685	985	430	460	505	580	700	880	1180
800	900	266	300	350	440	570	770	1110	486	520	570	660	790	990	1330
900	1000	276	310	360	450	580	780	1120	526	560	610	700	830	1030	1370
1000	1120	316	355	415	510	670	910	1300	586	625	685	780	940	1180	1570
1120	1250	326	365	425	520	680	920	1310	646	685	745	840	1000	1240	1630
1250	1400	378	425	495	610	800	1080	1550	718	765	835	950	1140	1420	1890
1400	1600	408	455	525	640	830	1110	1580	798	845	915	1030	1220	1500	1970
1600	1800	462	520	600	740	970	1290	1870	912	970	1050	1190	1420	1740	2320
1800	2000	492	550	630	770	1000	1320	1900	1012	1070	1150	1290	1520	1840	2420
2000	2240	550	615	720	880	1140	1540	2190	1110	1175	1280	1440	1700	2100	2750
2240	2500	570	635	740	900	1160	1560	2210	1210	1275	1380	1540	1800	2200	2850
2500	2800	685	760	880	1090	1410	1900	2650	1385	1460	1580	1790	2110	2600	3350
2800	3150	715	790	910	1120	1440	1930	2680	1535	1610	1730	1940	2260	2750	3500

(Courtesy AS 1654–1974)

NOTE: Upper deviation is obtained by *adding* the appropriate tolerance value (Table 3) to the value of lower deviation.

Table 2 (*continued*)

Tabulated values are (+) unless otherwise stated   μm

Basic size mm Above	Up to and incl.	t 6 +	t 7 +	t 8 +	t 9 +	t 10 +	t 11 +	t 12 +	u 6 +	u 7 +	u 8 +	u 9 +	u 10 +	u 11 +	u 12 +
0	3	—	—	—	—	—	—	—	24	28	32	43	58	78	118
3	6	—	—	—	—	—	—	—	31	35	41	53	71	98	143
6	10	—	—	—	—	—	—	—	37	43	50	64	86	118	178
10	18	—	—	—	—	—	—	—	44	51	60	76	103	143	213
18	24	—	—	—	—	—	—	—	54	62	74	93	125	171	251
24	30	54	62	74	93	125	171	251	61	69	81	100	132	178	258
30	40	64	73	87	110	148	208	298	76	85	99	122	160	220	310
40	50	70	79	93	116	154	214	304	86	95	109	132	170	230	320
50	65	85	96	112	140	186	256	366	106	117	133	161	222	277	387
65	80	94	105	121	149	195	265	375	121	132	148	176	222	292	402
80	100	113	126	145	178	231	311	441	146	159	178	211	264	344	474
100	120	126	139	158	191	244	324	454	166	179	198	231	284	364	494
120	140	147	162	185	222	282	372	522	195	210	233	270	330	420	570
140	160	159	174	197	234	294	384	534	215	230	253	290	350	440	590
160	180	171	186	209	246	306	396	546	235	250	273	310	370	460	610
180	200	195	212	238	281	351	456	626	265	282	308	351	421	526	696
200	225	209	226	252	295	365	470	640	287	304	330	373	443	548	718
225	250	225	242	268	311	381	486	656	313	330	356	399	469	574	744
250	280	250	270	299	348	428	538	738	347	367	396	445	525	635	835
280	315	272	292	321	370	450	560	760	382	402	431	480	560	670	870
315	355	304	325	357	408	498	628	838	426	447	479	530	620	750	960
355	400	330	351	383	434	524	654	864	471	492	524	575	665	795	1005
400	450	370	393	427	485	580	730	960	530	553	587	645	740	890	1120
450	500	400	423	457	515	610	760	990	580	603	637	695	790	940	1170
500	560	444	470	510	575	680	840	1100	644	670	710	775	880	1040	1300
560	630	494	520	560	625	730	890	1150	704	730	770	835	940	1100	1360
630	710	550	580	625	700	820	1000	1300	790	820	865	940	1060	1240	1540
710	800	610	640	685	760	880	1060	1360	890	920	965	1040	1160	1340	1640
800	900	676	710	760	850	980	1180	1520	996	1030	1080	1170	1300	1500	1640
900	1000	736	770	820	910	1040	1240	1580	1106	1140	1190	1280	1410	1610	1950
1000	1120	846	885	945	1040	1200	1440	1830	1216	1255	1315	1410	1570	1810	2200
1120	1250	906	945	1005	1100	1260	1500	1890	1366	1405	1465	1560	1720	1960	2350
1250	1400	1038	1085	1155	1270	1460	1740	2210	1528	1575	1645	1760	1950	2230	2700
1400	1600	1128	1175	1245	1360	1550	1830	2300	1678	1725	1795	1910	2100	2380	2850
1600	1800	1292	1350	1430	1570	1800	2120	2700	1942	2000	2080	2220	2450	2770	3350
1800	2000	1442	1500	1580	1720	1950	2270	2850	2092	2150	2230	2370	2600	2920	3500
2000	2240	1610	1675	1780	1940	2200	2600	3250	2410	2475	2580	2740	3000	3400	4050
2240	2500	1760	1825	1930	2090	2350	2750	3400	2610	2675	2780	2940	3200	3600	4250
2500	2800	2035	2110	2230	2440	2760	3250	4000	3035	3110	3230	3440	3760	4250	5000
2800	3150	2235	2310	2430	2640	2960	3450	4200	3335	3410	3530	3740	4060	4550	5300

(Courtesy AS 1654–1974)

NOTE: Lower deviation is obtained by *subtracting* the appropriate tolerance value (Table 3) from the value of upper deviation.

Table 2 *(continued)*

Tabulated values are (+) unless otherwise stated

μm

Basic size mm		Tolerance class												
		v							x					
Above	Up to and incl.	6 +	7 +	8 +	9 +	10 +	11 +	12 +	6 +	7 +	8 +	9 +	10 +	11 +	12 +
0	3	—	—	—	—	—	—	—	26	30	34	45	60	80	120
3	6	—	—	—	—	—	—	—	36	40	46	58	76	103	148
6	10	—	—	—	—	—	—	—	43	49	56	70	92	124	184
10	14	—	—	—	—	—	—	—	51	58	67	83	110	150	220
14	18	50	57	66	82	109	149	219	56	63	72	88	115	155	225
18	24	60	68	80	99	131	177	257	67	75	87	106	138	184	264
24	30	68	76	88	107	139	185	265	77	85	97	116	148	194	274
30	40	84	93	107	130	168	228	318	96	105	119	142	180	240	330
40	50	97	106	120	143	181	241	331	113	122	136	159	197	257	347
50	65	121	132	148	176	222	292	402	141	152	168	196	242	312	422
65	80	139	150	166	194	240	310	420	165	176	192	220	266	336	446
80	100	168	181	200	233	286	366	496	200	213	232	265	318	398	528
100	120	194	207	226	259	312	392	522	232	245	264	297	350	430	560
120	140	227	242	265	302	362	452	602	273	288	311	348	408	498	648
140	160	253	268	291	328	388	478	628	305	320	343	380	440	530	680
160	180	277	292	315	352	412	502	652	335	350	373	410	470	560	710
180	200	313	330	356	399	469	574	744	379	396	422	465	535	640	810
200	225	339	356	382	425	495	600	770	414	431	457	500	570	675	845
225	250	369	386	412	455	525	630	800	454	471	497	540	610	715	885
250	280	417	437	466	515	595	705	905	507	527	556	605	685	795	995
280	315	457	477	506	555	635	745	945	557	577	606	655	735	845	1045
315	355	511	532	564	615	705	835	1045	626	647	679	730	820	950	1160
355	400	566	587	619	670	760	890	1100	696	717	749	800	890	1020	1230
400	450	635	658	692	750	845	995	1225	780	803	837	895	990	1140	1370
450	500	700	723	757	815	910	1060	1290	860	883	917	975	1070	1220	1450

(Courtesy AS 1654–1974)

NOTE: Lower deviation is obtained by *subtracting* the appropriate tolerance value (Table 3) from the value of upper deviation.

Table 2   (*continued*)

Tabulated values are (+) unless otherwise stated   μm

Basic size mm		Tolerance class													
		y							z						
Above	Up to and incl.	6	7	8	9	10	11	12	6	7	8	9	10	11	12
		+	+	+	+	+	+	+	+	+	+	+	+	+	+
0	3	—	—	—	—	—	—	—	32	36	40	51	66	86	126
3	6	—	—	—	—	—	—	—	43	47	53	65	83	110	155
6	10	—	—	—	—	—	—	—	51	57	64	78	100	132	192
10	14	—	—	—	—	—	—	—	61	68	77	93	120	160	230
14	18	—	—	—	—	—	—	—	71	78	87	103	130	170	240
18	24	76	84	96	115	147	193	273	86	94	106	125	157	203	283
24	30	88	96	108	127	159	205	285	101	109	121	140	172	218	298
30	40	110	119	133	156	194	254	344	128	137	151	174	212	272	362
40	50	130	139	153	176	214	274	364	152	161	175	198	236	296	386
50	65	163	174	190	218	264	334	444	191	202	218	246	292	362	472
65	80	193	204	220	248	294	364	474	229	240	256	284	330	400	510
80	100	236	249	268	301	354	434	564	280	293	312	345	398	478	608
100	120	276	289	308	341	394	474	604	332	345	364	397	450	530	660
120	140	325	340	363	400	460	550	700	390	405	428	465	525	615	765
140	160	365	380	403	440	500	590	740	440	455	478	515	575	665	815
160	180	405	420	443	480	540	630	780	490	505	528	565	625	715	865
180	200	454	471	497	540	610	715	885	549	566	592	635	705	810	980
200	225	499	516	542	585	655	760	930	604	621	647	690	760	865	1035
225	250	549	566	592	635	705	810	980	669	686	712	755	825	930	1100
250	280	612	632	661	710	790	900	1100	742	762	791	840	920	1030	1230
280	315	682	702	731	780	860	970	1170	822	842	871	920	1000	1110	1310
315	355	766	787	819	870	960	1090	1300	936	957	989	1040	1130	1260	1470
355	400	856	877	909	960	1050	1180	1390	1036	1057	1089	1140	1230	1360	1570
400	450	960	983	1017	1075	1170	1320	1550	1140	1163	1197	1255	1350	1500	1730
450	500	1040	1063	1097	1155	1250	1400	1630	1290	1313	1347	1405	1500	1650	1880

(Courtesy AS 1654–1974)

NOTE: Lower deviation is obtained by *subtracting* the appropriate tolerance value (Table 3) from the value of upper deviation.

Table 2 (*continued*)

Tabulated values are (+) unless otherwise stated

μm

Basic size mm		Tolerance class													
		za							zb						
Above	Up to and incl.	6	7	8	9	10	11	12	6	7	8	9	10	11	12
		+	+	+	+	+	+	+	+	+	+	+	+	+	+
0	3	38	42	46	57	72	92	132	46	50	54	65	80	100	140
3	6	50	54	60	72	90	117	162	58	62	68	80	98	125	170
6	10	61	67	74	88	110	142	202	76	82	89	103	125	157	217
10	14	75	82	91	107	134	174	244	101	108	117	133	160	200	270
14	18	88	95	104	120	147	187	257	119	126	135	151	178	218	288
18	24	111	119	131	150	182	228	308	149	157	169	188	220	266	346
24	30	131	139	151	170	202	248	328	173	181	193	212	244	290	370
30	40	164	173	187	210	248	308	398	216	225	239	262	300	360	450
40	50	196	205	219	242	280	340	430	258	267	281	304	342	402	492
50	65	245	256	272	300	346	416	526	319	330	346	374	420	490	600
65	80	293	304	320	348	394	464	574	379	390	406	434	480	550	660
80	100	357	370	389	422	475	555	685	467	480	499	532	585	665	795
100	120	422	435	454	487	540	620	750	547	560	579	612	665	745	875
120	140	495	510	533	570	630	720	870	645	660	683	720	780	870	1020
140	160	560	575	598	635	695	785	935	725	740	763	800	860	950	1100
160	180	625	640	663	700	760	850	1000	805	820	843	880	940	1030	1180
180	200	699	716	742	785	855	960	1130	909	926	952	995	1065	1170	1340
200	225	769	786	812	855	925	1030	1200	989	1006	1032	1075	1145	1250	1420
225	250	849	866	892	935	1005	1110	1280	1079	1096	1122	1165	1235	1340	1510
250	280	952	972	1001	1050	1130	1240	1440	1232	1252	1281	1330	1410	1520	1720
280	315	1032	1052	1081	1130	1210	1320	1520	1332	1352	1381	1430	1510	1620	1820
315	355	1186	1207	1239	1290	1380	1510	1720	1536	1557	1589	1640	1730	1860	2070
355	400	1336	1357	1389	1440	1530	1660	1870	1686	1707	1739	1790	1880	2010	2220
400	450	1490	1513	1547	1605	1700	1850	2080	1890	1913	1947	2005	2100	2250	2480
450	500	1640	1663	1697	1755	1850	2000	2230	2140	2163	2197	2255	2350	2500	2730

(Courtesy AS 1654–1974)

NOTE: Lower deviation is obtained by *subtracting* the appropriate tolerance value (Table 3) from the value of upper deviation.

Table 2   (*continued*)

Tabulated values are (+) unless otherwise stated μm

Basic size mm		Tolerance class						
		zc						
Above	Up to and incl.	6	7	8	9	10	11	12
		+	+	+	+	+	+	+
0	3	66	70	74	85	100	120	160
3	6	88	92	98	110	128	155	200
6	10	106	112	119	133	155	187	247
10	14	141	148	157	173	200	240	310
14	18	161	168	177	193	220	260	330
18	24	201	209	221	240	272	318	398
24	30	231	239	251	270	302	348	428
30	40	290	299	313	336	374	434	524
40	50	341	350	364	387	425	485	575
50	65	424	435	451	479	525	595	705
65	80	499	510	526	554	600	670	780
80	100	607	620	639	672	725	805	935
100	120	712	725	744	777	830	910	1040
120	140	825	840	863	900	960	1050	1200
140	160	925	940	963	1000	1060	1150	1300
160	180	1025	1040	1063	1100	1160	1250	1400
180	200	1179	1196	1222	1265	1335	1440	1610
200	225	1279	1296	1322	1365	1435	1540	1710
225	250	1379	1396	1422	1465	1535	1640	1810
250	280	1582	1602	1631	1680	1760	1870	2070
280	315	1732	1752	1781	1830	1910	2020	2220
315	355	1936	1957	1989	2040	2130	2260	2470
355	400	2136	2157	2189	2240	2330	2460	2670
400	450	2440	2463	2497	2555	2650	2800	3030
450	500	2640	2663	2697	2755	2850	3000	3230

(Courtesy AS 1654–1974)

NOTE: Lower deviation is obtained by *subtracting* the appropriate tolerance value (Table 3) from the value of upper deviation.

Table 3   Numerical Values of Standard Tolerances Grades 6 to 12

μm

Basic size mm		Tolerance grade						
Above	Up to and incl.	6*	7	8	9	10	11	12
0	3	6	10	14	25	40	60	100
3	6	8	12	18	30	48	75	120
6	10	9	15	22	36	58	90	150
10	18	11	18	27	43	70	110	180
18	30	13	21	33	52	84	130	210
30	50	16	25	39	62	100	160	250
50	80	19	30	46	74	120	190	300
80	120	22	35	54	87	140	220	350
120	180	25	40	63	100	160	250	400
180	250	29	46	72	115	185	290	460
250	315	32	52	81	130	210	320	520
315	400	36	57	89	140	230	360	570
400	500	40	63	97	155	250	400	630
500	630	44	70	110	175	280	440	700
630	800	50	80	125	200	320	500	800
800	1000	56	90	140	230	360	560	900
1000	1250	66	105	165	260	420	660	1050
1250	1600	78	125	195	310	500	780	1250
1600	2000	92	150	230	370	600	920	1500
2000	2500	110	175	280	440	700	1100	1750
2500	3150	135	210	330	540	860	1350	2100

(Courtesy AS 1654–1974)

* Not recommended for fits in sizes above 500 mm.

NOTE: The tolerance grades in ISO/R 286 are designated ITO1, ITO, IT1, IT2 . . . IT16.

Table 3 (*continued*)

μm

Basic size mm		Tolerance grade										
Above	Up to and incl.	01	0	1	2	3	4	5	13	14*	15*	16*
0	3	0,3	0,5	0,8	1,2	2,0	3	4	140	250	400	600
3	6	0,4	0,6	1,0	1,5	2,5	4	5	180	300	480	750
6	10	0,4	0,6	1,0	1,5	2,5	4	6	220	360	580	900
10	18	0,5	0,8	1,2	2,0	3,0	5	8	270	430	700	1100
18	30	0,6	1,0	1,5	2,5	4,0	6	9	330	520	840	1300
30	50	0,6	1,0	1,5	2,5	4	7	11	390	620	1000	1600
50	80	0,8	1,2	2,0	3,0	5	8	13	460	740	1200	1900
80	120	1,0	1,5	2,5	4,0	6	10	15	540	870	1400	2200
120	180	1,2	2,0	3,5	5,0	8	12	18	630	1000	1600	2500
180	250	2,0	3,0	4,5	7,0	10	14	20	720	1150	1850	2900
250	315	2,5	4	6	8	12	16	23	810	1300	2100	3200
315	400	3,0	5	7	9	13	18	25	890	1400	2300	3600
400	500	4,0	6	8	10	15	20	27	970	1550	2500	4000
500	630	—	—	—	—	—	—	—	1100	1750	2800	4400
630	800	—	—	—	—	—	—	—	1250	2000	3200	5000
800	1000	—	—	—	—	—	—	—	1400	2300	3600	5600
1000	1250	—	—	—	—	—	—	—	1650	2600	4200	6600
1250	1600	—	—	—	—	—	—	—	1950	3100	5000	7800
1600	2000	—	—	—	—	—	—	—	2300	3700	6000	9200
2000	2500	—	—	—	—	—	—	—	2800	4400	7000	11000
2500	3150	—	—	—	—	—	—	—	3300	5400	8600	13500

(Courtesy AS 1654–1974)

* Not applicable to sizes below 1 mm.

NOTE: The tolerance grades in ISO/R 286 are designated ITO1, ITO, IT1, IT2 . . . IT16.

Table 4 Selected Fits—Hole Basis

Diagram to scale for 25 mm diameter

Holes (hatched) Shafts (solid)

		Clearance fits										Transition fits						Interference fits			
		H11 / c11		H9 / d10		H9 / e9		H8 / f7		H7 / g6		H7 / h6		H7 / k6		H7 / n6		H7 / p6		H7 / s6	
Basic size mm									**Upper and lower deviations for tolerance class**												
Above	Up to and incl.	H11 +	c11 −	H9 +	d10 −	H9 +	e9 −	H8 +	f7 −	H7 +	g6 −	H7 +	h6 −	H7 +	k6	H7 +	n6	H7 +	p6	H7 +	s6
0	3	60/0	60/120	25/0	20/60	25/0	14/39	14/0	6/16	10/0	2/8	10/0	6/0	10/0	6/0	10/0	10/4	10/0	12/6	10/0	20/14
3	6	75/0	70/145	30/0	30/78	30/0	20/50	18/0	10/22	12/0	4/12	12/0	8/0	12/0	9/1	12/0	16/8	12/0	20/12	12/0	27/19
6	10	90/0	80/170	36/0	40/98	36/0	25/61	22/0	13/28	15/0	5/14	15/0	9/0	15/0	10/1	15/0	19/10	15/0	24/15	15/0	32/23
10	18	110/0	95/205	43/0	50/120	43/0	32/75	27/0	16/34	18/0	6/17	18/0	11/0	18/0	12/1	18/0	23/12	18/0	29/18	18/0	39/28
18	30	130/0	110/240	52/0	65/149	52/0	40/92	33/0	20/41	21/0	7/20	21/0	13/0	21/0	15/2	21/0	28/15	21/0	35/22	21/0	48/35
30	40	160/0	120/280	62/0	80/180	62/0	50/112	39/0	25/50	25/0	9/25	25/0	16/0	25/0	18/2	25/0	33/17	25/0	42/26	25/0	59/43
40	50	160/0	130/290	62/0	80/180	62/0	50/112	39/0	25/50	25/0	9/25	25/0	16/0	25/0	18/2	25/0	33/17	25/0	42/26	25/0	59/43
50	65	190/0	140/330	74/0	100/220	74/0	60/134	46/0	30/60	30/0	10/29	30/0	19/0	30/0	21/2	30/0	39/20	30/0	51/32	30/0	72/53
65	80	190/0	150/340	74/0	100/220	74/0	60/134	46/0	30/60	30/0	10/29	30/0	19/0	30/0	21/2	30/0	39/20	30/0	51/32	30/0	78/59
80	100	220/0	170/390	87/0	120/260	87/0	72/159	54/0	36/71	35/0	12/34	35/0	22/0	35/0	25/3	35/0	45/23	35/0	59/37	35/0	93/71
100	120	220/0	180/400	87/0	120/260	87/0	72/159	54/0	36/71	35/0	12/34	35/0	22/0	35/0	25/3	35/0	45/23	35/0	59/37	35/0	101/79
120	140	250/0	200/450	100/0	145/305	100/0	85/185	63/0	43/83	40/0	14/39	40/0	25/0	40/0	28/3	40/0	52/27	40/0	68/43	40/0	117/92
140	160	250/0	210/460	100/0	145/305	100/0	85/185	63/0	43/83	40/0	14/39	40/0	25/0	40/0	28/3	40/0	52/27	40/0	68/43	40/0	125/100
160	180	250/0	230/480	100/0	145/305	100/0	85/185	63/0	43/83	40/0	14/39	40/0	25/0	40/0	28/3	40/0	52/27	40/0	68/43	40/0	133/108
180	200	290/0	240/530	115/0	170/355	115/0	100/215	72/0	50/96	46/0	15/44	46/0	29/0	46/0	33/4	46/0	60/31	46/0	79/50	46/0	151/122
200	225	290/0	260/550	115/0	170/355	115/0	100/215	72/0	50/96	46/0	15/44	46/0	29/0	46/0	33/4	46/0	60/31	46/0	79/50	46/0	159/130
225	250	290/0	280/570	115/0	170/355	115/0	100/215	72/0	50/96	46/0	15/44	46/0	29/0	46/0	33/4	46/0	60/31	46/0	79/50	46/0	169/140
250	280	320/0	300/620	130/0	190/400	130/0	110/240	81/0	56/108	52/0	17/49	52/0	32/0	52/0	36/4	52/0	66/34	52/0	88/56	52/0	190/158
280	315	320/0	330/650	130/0	190/400	130/0	110/240	81/0	56/108	52/0	17/49	52/0	32/0	52/0	36/4	52/0	66/34	52/0	88/56	52/0	202/170
315	355	360/0	360/720	140/0	210/440	140/0	125/265	89/0	62/119	57/0	18/54	57/0	36/0	57/0	40/4	57/0	73/37	57/0	98/62	57/0	226/190
355	400	360/0	400/760	140/0	210/440	140/0	125/265	89/0	62/119	57/0	18/54	57/0	36/0	57/0	40/4	57/0	73/37	57/0	98/62	57/0	244/208
400	450	400/0	440/840	155/0	230/480	155/0	135/290	97/0	68/131	63/0	20/60	63/0	40/0	63/0	45/5	63/0	80/40	63/0	108/68	63/0	272/232
450	500	400/0	480/880	155/0	230/480	155/0	135/290	97/0	68/131	63/0	20/60	63/0	40/0	63/0	45/5	63/0	80/40	63/0	108/68	63/0	292/252

(Courtesy AS 1654-1974)

Table 5 Selected Fits—Shaft Basis

Diagram to scale for 25 mm diameter

Holes (hatched)   Shafts (solid)

Fit types and tolerance classes (Shaft Basis):

Clearance fits						Transition fits		Interference fits	
C11/h11	D10/h9	E9/h9	F8/h7	G7/h6	H7/h6	K7/h6	N7/h6	P7/h6	S7/h6

Upper and lower deviations for tolerance class (values in µm; top number = upper deviation, bottom number = lower deviation)

Basic size (mm) Above	Up to and incl.	C11 +	h11 −	D10 +	h9 −	E9 +	h9 −	F8 +	h7 −	G7 +	h6 −	H7 +	h6 −	K7 +/−	h6 −	N7 −	h6 −	P7 −	h6 −	S7 −	h6 −
0	3	120/60	0/60	60/20	0/25	39/14	0/25	20/6	0/10	12/2	0/6	10/0	0/6	0/−10	0/6	4/14	0/6	6/16	0/6	14/24	0/6
3	6	145/70	0/75	78/30	0/30	50/20	0/30	28/10	0/12	16/4	0/8	12/0	0/8	+3/−9	0/8	4/16	0/8	8/20	0/8	15/27	0/8
6	10	170/80	0/90	98/40	0/36	61/25	0/36	35/13	0/15	20/5	0/9	15/0	0/9	+5/−10	0/9	4/19	0/9	9/24	0/9	17/32	0/9
10	18	205/95	0/110	120/50	0/43	75/32	0/43	43/16	0/18	24/6	0/11	18/0	0/11	+6/−12	0/11	5/23	0/11	11/29	0/11	21/39	0/11
18	30	240/110	0/130	149/65	0/52	92/40	0/52	53/20	0/21	28/7	0/13	21/0	0/13	+6/−15	0/13	7/28	0/13	14/35	0/13	27/48	0/13
30	40	280/120	0/160	180/80	0/62	112/50	0/62	64/25	0/25	34/9	0/16	25/0	0/16	+7/−18	0/16	8/33	0/16	17/42	0/16	34/59	0/16
40	50	290/130	0/160	180/80	0/62	112/50	0/62	64/25	0/25	34/9	0/16	25/0	0/16	+7/−18	0/16	8/33	0/16	17/42	0/16	34/59	0/16
50	65	330/140	0/190	220/100	0/74	134/60	0/74	76/30	0/30	40/10	0/19	30/0	0/19	+9/−21	0/19	9/39	0/19	21/51	0/19	42/72	0/19
65	80	340/150	0/190	220/100	0/74	134/60	0/74	76/30	0/30	40/10	0/19	30/0	0/19	+9/−21	0/19	9/39	0/19	21/51	0/19	48/78	0/19
80	100	390/170	0/220	260/120	0/87	159/72	0/87	90/36	0/35	47/12	0/22	35/0	0/22	+10/−25	0/22	10/45	0/22	24/59	0/22	58/93	0/22
100	120	400/180	0/220	260/120	0/87	159/72	0/87	90/36	0/35	47/12	0/22	35/0	0/22	+10/−25	0/22	10/45	0/22	24/59	0/22	66/101	0/22
120	140	450/200	0/250	305/145	0/100	185/85	0/100	106/43	0/40	54/14	0/25	40/0	0/25	+12/−28	0/25	12/52	0/25	28/68	0/25	77/117	0/25
140	160	460/210	0/250	305/145	0/100	185/85	0/100	106/43	0/40	54/14	0/25	40/0	0/25	+12/−28	0/25	12/52	0/25	28/68	0/25	85/125	0/25
160	180	480/230	0/250	305/145	0/100	185/85	0/100	106/43	0/40	54/14	0/25	40/0	0/25	+12/−28	0/25	12/52	0/25	28/68	0/25	93/133	0/25
180	200	530/240	0/290	355/170	0/115	215/100	0/115	122/50	0/46	61/15	0/29	46/0	0/29	+13/−33	0/29	14/60	0/29	33/79	0/29	105/151	0/29
200	225	550/260	0/290	355/170	0/115	215/100	0/115	122/50	0/46	61/15	0/29	46/0	0/29	+13/−33	0/29	14/60	0/29	33/79	0/29	113/159	0/29
225	250	570/280	0/290	355/170	0/115	215/100	0/115	122/50	0/46	61/15	0/29	46/0	0/29	+13/−33	0/29	14/60	0/29	33/79	0/29	123/169	0/29
250	280	620/300	0/320	400/190	0/130	240/110	0/130	137/56	0/52	62/17	0/32	52/0	0/32	+16/−36	0/32	14/66	0/32	36/88	0/32	138/190	0/32
280	315	650/330	0/320	400/190	0/130	240/110	0/130	137/56	0/52	62/17	0/32	52/0	0/32	+16/−36	0/32	14/66	0/32	36/88	0/32	150/202	0/32
315	355	720/360	0/360	440/210	0/140	265/125	0/140	151/62	0/57	75/18	0/36	57/0	0/36	+17/−40	0/36	16/73	0/36	41/98	0/36	169/226	0/36
355	400	760/400	0/360	440/210	0/140	265/125	0/140	151/62	0/57	75/18	0/36	57/0	0/36	+17/−40	0/36	16/73	0/36	41/98	0/36	187/244	0/36
400	450	840/440	0/400	480/230	0/155	290/135	0/155	165/68	0/63	83/20	0/40	63/0	0/40	+18/−45	0/40	17/80	0/40	45/108	0/40	209/272	0/40
450	500	880/480	0/400	480/230	0/155	290/135	0/155	165/68	0/63	83/20	0/40	63/0	0/40	+18/−45	0/40	17/80	0/40	45/108	0/40	229/292	0/40

(Courtesy AS 1654-1974)

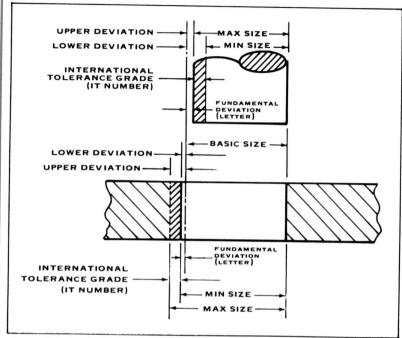

Figure B-1 Illustration of definitions (courtesy ANSI B4.2-1978).

Preferred Metric Limits and Fits—ANSI B4.2—1978

This new standard describes the ISO system of limits and fits for mating parts for general engineering usage in the United States. The selection of standard tolerance zones and preferred metric fits were based on international and national standards, including the AS 1654 (Australia), which was referred to in the preceding Tables 1–5. Some students may find the ANSI B4.2-1978 somewhat easier to use. Material from this standard has been selected to acquaint the student with its application to dimensioning of mating parts for various fits.

Definitions

The most important terms that relate to limits and fits are shown in Figure B-1. Terms are defined as follows.

(a) *Basic Size.* The size to which limits or deviations are assigned. The basic size is the same for both members of a fit. It is designated by the number 40 in 40H7.
(b) *Deviation.* The algebraic difference between a size and the corresponding basic size.
(c) *Upper Deviation.* The algebraical difference between the maximum limit of size and the corresponding basic size.
(d) *Lower Deviation.* The algebraic difference between the minimum limit of size and the corresponding basic size.
(e) *Fundamental Deviation.* That one of the two deviations closest to the basic size. It is designated by the letter H in 40H7.

(f) *Tolerance.* The difference between the maximum and minimum size limits on a part.
(g) *Tolerance Zone.* A zone representing the tolerance and its position in relation to the basic size.
(h) *International Tolerance Grade (IT).* A group of tolerances which vary depending on the basic size, but which provide the same relative level of accuracy within a given grade. It is designated by the number 7 in 40H7 (IT7).
(i) *Hole Basis.* The system of fits where the minimum hole size is basic. The fundamental deviation for a hole basis system is "H."
(j) *Shaft Basis.* The system of fits where the maximum shaft size is basic. The fundamental deviation for a shaft basis system is "h."
(k) *Clearance Fit.* The relationship between assembled parts when clearance occurs under all tolerance conditions.
(l) *Transition Fit.* The relationship between assembled parts when either a clearance or interference fit can result depending on the tolerance conditions of the mating parts.
(m) *Interference Fit.* The relationship between assembled parts when interference occurs under all tolerance conditions.

DESCRIPTION OF TOLERANCE DESIGNATION

An "International Tolerance grade establishes the magnitude of the tolerance zone or the amount of part size variation allowed for internal and external dimensions alike (see Figure B-1). *Tolerances are expressed in "grade numbers,"* which are consistent with International Tolerance grades identified by the prefix IT, i.e., "IT6," "IT11," etc. *A smaller grade number provides a smaller tolerance zone.*

A *fundamental deviation* establishes the position of the tolerance zone with respect to the basic size (see Figure B-1). Fundamental deviations are expressed by "tolerance position letters." Capital letters are used for internal dimensions, and lowercase or small letters are used for external dimensions.

Symbols

By combining the IT grade number and the tolerance position letter, the tolerance symbol is established which identifies the actual maximum and minimum limits of the part. The toleranced sizes are thus defined by the basic size of the part followed by a symbol composed of a letter and a number.

Examples

A fit is indicated by the basic size common to both components, followed by a symbol corresponding to each component, the internal part symbol preceding the external part symbol.

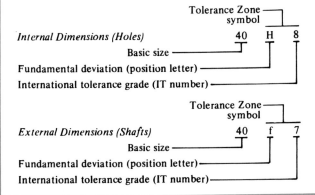

Example

Some methods of designating tolerances on drawing gages, etc. are shown in the following three examples.

a. 40H8   b. 40H8 $\left(\dfrac{40.039}{40.000}\right)$

c. $\dfrac{40.039}{40.000}$ (40H8)

(*Note.* Values in parentheses indicate reference only.)

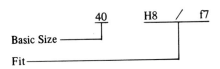

DESCRIPTION OF PREFERRED FITS

The standard B4.2-1978 describes ten *preferred* fits for both systems—hole basis and shaft basis. Figure B-2 defines the fits.

Preferred basic sizes of mating parts should be chosen, where possible, from the first choice sizes shown in the table.

Excerpted from the B4.2 standard are typical tabular values for:

Table A—Preferred *Hole Basis* Clearance Fits (page 639)

Table B—Preferred *Hole Basis* Transition and Interference Fits (page 641)

Table C—Preferred *Shaft Basis* Clearance Fits (page 643)

Table D—Preferred *Shaft Basis* Transition and Interference Fits (page 645)

Table E—Tolerance Zones for *Internal* (Hole) Dimensions (page 647)

Table F—Tolerance Zones for *External* (Shaft) Dimensions (page 648)

ISO SYMBOL		DESCRIPTION
Hole Basis	**Shaft Basis**	
H11/c11	C11/h11	*Loose running* fit for wide commercial tolerances or allowances on external members.
H9/d9	D9/h9	*Free running* fit not for use where accuracy is essential, but good for large temperature variations, high running speeds, or heavy journal pressures.
H8/f7	F8/h7	*Close running* fit for running on accurate machines and for accurate location at moderate speeds and journal pressures.
H7/g6	G7/h6	*Sliding* fit not intended to run freely, but to move and turn freely and locate accurately.
H7/h6	H7/h6	*Locational clearance* fit provides snug fit for locating stationary parts; but can be freely assembled and disassembled.
H7/k6	K7/h6	*Locational transition* fit for accurate location, a compromise between clearance and interference.
H7/n6	N7/h6	*Locational transition* fit for more accurate location where greater interference is permissible.
H7/p6[1]	P7/h6	*Locational interference* fit for parts requiring rigidity and alignment with prime accuracy of location but without special bore pressure requirements.
H7/s6	S7/h6	*Medium drive* fit for ordinary steel parts or shrink fits on light sections, the tightest fit usable with cast iron.
H7/u6	U7/h6	*Force* fit suitable for parts which can be highly stressed or for shrink fits where the heavy pressing forces required are impractical.

Clearance Fits — Transition Fits — Interference Fits (left margin)

More Clearance — More Interference (right margin)

[1] Transition fit for basic sizes in range from 0 through 3 mm.

Figure B-2 Description of preferred fits.

Preferred Sizes

Basic Size, mm		Basic Size, mm		Basic Size, mm	
First Choice	Second Choice	First Choice	Second Choice	First Choice	Second Choice
1		10		100	
	1.1		11		110
1.2		12		120	
	1.4		14		140
1.6		16		160	
	1.8		18		180
2		20		200	
	2.2		22		220
2.5		25		250	
	2.8		28		280
3		30		300	
	3.5		35		350
4		40		400	
	4.5		45		450
5		50		500	
	5.5		55		550
6		60		600	
	7		70		700
8		80		800	
	9		90		900
				1000	

Examples of the use of Tables A and B

1. Provide the limit dimensions for the shaft and bearing for a Clearance Fit, 40H7/g6. In Table A, locate the basic size 40; then proceed, horizontally, to the designation H7 and read 40.025 for the maximum dimension and 40.000 for the minimum value. Similarly, we can obtain the dimensions for the shaft, g6. These are 34.991 (max) and 39.975 (min). The *allowance* is the difference between the dimensions of the largest shaft and the smallest hole. In this case, it is $(40.000 - 39.991) = 0.009$ (see Figure B-3).

2. Provide the limit dimensions for a 40H7/k6 fit. Use Table B and follow the procedure used in 1, above (see Figure B-4).

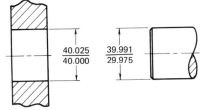

Figure B-3 *Clearance Fit*—40H7/g6
Allowance = 0.009 (see Table A, p. 640)

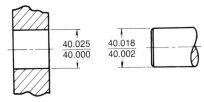

Figure B-4 *Transition Fit*—40H7/k6
Allowance = −0.018 (see Table B, p. 642)

3. Provide the limit dimensions for a 50H7/u6 fit. Again use Table B and follow the procedure used in 1 above (see Figure B-5).

4. Provide the limit dimensions for a 20G7/h6 fit—*Shaft Basis*. Use Table C and follow the procedure used previously. The dimensions for the hole are $\frac{20.028}{20.007}$ and for the shaft $\frac{20.000}{19.987}$ (see Figure B-6).

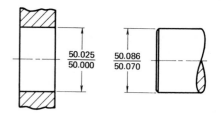

Note: Hole Basis; All dimensions in mm.
Figure B-5 *Interference Fit—*50H7/u6
Note: Hole Basis: All dimensions in mm.
Allowance = −0.086 (see Table B, p. 642)

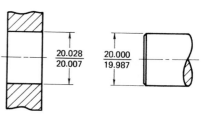

Figure B-6 *Clearance Fit—*20G7/h6 *Shaft Basis*

Dimensions in mm.

Table A   Preferred Hole Basis Clearance Fits

BASIC SIZE		LOCATIONAL TRANSN. Hole H7	Shaft k6	Fit	LOCATIONAL TRANSN. Hole H7	Shaft n6	Fit	LOCATIONAL INTERF Hole H7	Shaft p6	Fit	MEDIUM DRIVE Hole H7	Shaft s6	Fit	FORCE Hole H7	Shaft u6	Fit
1	MAX	1.010	1.006	0.010	1.010	1.010	0.006	1.010	1.012	0.004	1.010	1.020	-0.004	1.010	1.024	-0.008
	MIN	1.000	1.000	-0.006	1.000	1.004	-0.010	1.000	1.006	-0.012	1.000	1.014	-0.020	1.000	1.018	-0.024
1.2	MAX	1.210	1.206	0.010	1.210	1.210	0.006	1.210	1.212	0.004	1.210	1.220	-0.004	1.210	1.224	-0.008
	MIN	1.200	1.200	-0.006	1.200	1.204	-0.010	1.200	1.206	-0.012	1.200	1.214	-0.020	1.200	1.218	-0.024
1.6	MAX	1.610	1.606	0.010	1.610	1.610	0.006	1.610	1.612	0.004	1.610	1.620	-0.004	1.610	1.624	-0.008
	MIN	1.600	1.600	-0.006	1.600	1.604	-0.010	1.600	1.606	-0.012	1.600	1.614	-0.020	1.600	1.618	-0.024
2	MAX	2.010	2.006	0.010	2.010	2.010	0.006	2.010	2.012	0.004	2.010	2.020	-0.004	2.010	2.024	-0.008
	MIN	2.000	2.000	-0.006	2.000	2.004	-0.010	2.000	2.006	-0.012	2.000	2.014	-0.020	2.000	2.018	-0.024
2.5	MAX	2.510	2.506	0.010	2.510	2.510	0.006	2.510	2.512	0.004	2.510	2.520	-0.004	2.510	2.524	-0.008
	MIN	2.500	2.500	-0.006	2.500	2.504	-0.010	2.500	2.506	-0.012	2.500	2.514	-0.020	2.500	2.518	-0.024
3	MAX	3.010	3.006	0.010	3.010	3.010	0.006	3.010	3.012	0.004	3.010	3.020	-0.004	3.010	3.024	-0.008
	MIN	3.000	3.000	-0.006	3.000	3.004	-0.010	3.000	3.006	-0.012	3.000	3.014	-0.020	3.000	3.018	-0.024
4	MAX	4.012	4.009	0.011	4.012	4.016	0.004	4.012	4.020	0.000	4.012	4.027	-0.007	4.012	4.031	-0.011
	MIN	4.000	4.001	-0.009	4.000	4.008	-0.016	4.000	4.012	-0.020	4.000	4.019	-0.027	4.000	4.023	-0.031
5	MAX	5.012	5.009	0.011	5.012	5.016	0.004	5.012	5.020	0.000	5.012	5.027	-0.007	5.012	5.031	-0.011
	MIN	5.000	5.001	-0.009	5.000	5.008	-0.016	5.000	5.012	-0.020	5.000	5.019	-0.027	5.000	5.023	-0.031
6	MAX	6.012	6.009	0.011	6.012	6.016	0.004	6.012	6.020	0.000	6.012	6.027	-0.007	6.012	6.031	-0.011
	MIN	6.000	6.001	-0.009	6.000	6.008	-0.016	6.000	6.012	-0.020	6.000	6.019	-0.027	6.000	6.023	-0.031
8	MAX	8.015	8.010	0.014	8.015	8.019	0.005	8.015	8.024	0.000	8.015	8.032	-0.008	8.015	8.037	-0.013
	MIN	8.000	8.001	-0.010	8.000	8.010	-0.019	8.000	8.015	-0.024	8.000	8.023	-0.032	8.000	8.028	-0.037
10	MAX	10.015	10.010	0.014	10.015	10.019	0.005	10.015	10.024	0.000	10.015	10.032	-0.008	10.015	10.037	-0.013
	MIN	10.000	10.001	-0.010	10.000	10.010	-0.019	10.000	10.015	-0.024	10.000	10.023	-0.032	10.000	10.028	-0.037
12	MAX	12.018	12.012	0.017	12.018	12.023	0.006	12.018	12.029	0.000	12.018	12.039	-0.010	12.018	12.044	-0.015
	MIN	12.000	12.001	-0.012	12.000	12.012	-0.023	12.000	12.018	-0.029	12.000	12.028	-0.039	12.000	12.033	-0.044
16	MAX	16.018	16.012	0.017	16.018	16.023	0.006	16.018	16.029	0.000	16.018	16.039	-0.010	16.018	16.044	-0.015
	MIN	16.000	16.001	-0.012	16.000	16.012	-0.023	16.000	16.018	-0.029	16.000	16.028	-0.039	16.000	16.033	-0.044
20	MAX	20.021	20.015	0.019	20.021	20.028	0.006	20.021	20.035	-0.001	20.021	20.048	-0.014	20.021	20.054	-0.020
	MIN	20.000	20.002	-0.015	20.000	20.015	-0.028	20.000	20.022	-0.035	20.000	20.035	-0.048	20.000	20.041	-0.054
25	MAX	25.021	25.015	0.019	25.021	25.028	0.006	25.021	25.035	-0.001	25.021	25.048	-0.014	25.021	25.061	-0.027
	MIN	25.000	25.002	-0.015	25.000	25.015	-0.028	25.000	25.022	-0.035	25.000	25.035	-0.048	25.000	25.048	-0.061
30	MAX	30.021	30.015	0.019	30.021	30.028	0.006	30.021	30.035	-0.001	30.021	30.048	-0.014	30.021	30.061	-0.027
	MIN	30.000	30.002	-0.015	30.000	30.015	-0.028	30.000	30.022	-0.035	30.000	30.035	-0.048	30.000	30.048	-0.061

AMERICAN NATIONAL STANDARD
PREFERRED METRIC LIMITS AND FITS

Dimensions in mm.

Table A   Preferred Hole Basis Clearance Fits (Continued)

BASIC SIZE		LOOSE RUNNING			FREE RUNNING			CLOSE RUNNING			SLIDING			LOCATIONAL CLEARANCE		
		Hole H11	Shaft c11	Fit	Hole H9	Shaft d9	Fit	Hole H8	Shaft f7	Fit	Hole H7	Shaft g6	Fit	Hole H7	Shaft h6	Fit
40	MAX	40.160	39.880	0.440	40.062	39.920	0.204	40.039	39.975	0.089	40.025	39.991	0.050	40.025	40.000	0.041
	MIN	40.000	39.720	0.120	40.000	39.858	0.080	40.000	39.950	0.025	40.000	39.975	0.009	40.000	39.984	0.000
50	MAX	50.160	49.870	0.450	50.062	49.920	0.204	50.039	49.975	0.089	50.025	49.991	0.050	50.025	50.000	0.041
	MIN	50.000	49.710	0.130	50.000	49.858	0.080	50.000	49.950	0.025	50.000	49.975	0.009	50.000	49.984	0.000
60	MAX	60.190	59.860	0.520	60.074	59.900	0.248	60.046	59.970	0.106	60.030	59.990	0.059	60.030	60.000	0.049
	MIN	60.000	59.670	0.140	60.000	59.826	0.100	60.000	59.940	0.030	60.000	59.971	0.010	60.000	59.981	0.000
80	MAX	80.190	79.850	0.530	80.074	79.900	0.248	80.046	79.970	0.106	80.030	79.990	0.059	80.030	80.000	0.049
	MIN	80.000	79.660	0.150	80.000	79.826	0.100	80.000	79.940	0.030	80.000	79.971	0.010	80.000	79.981	0.000
100	MAX	100.220	99.830	0.610	100.087	99.880	0.294	100.054	99.964	0.125	100.035	99.988	0.069	100.035	100.000	0.057
	MIN	100.000	99.610	0.170	100.000	99.793	0.120	100.000	99.929	0.036	100.000	99.966	0.012	100.000	99.978	0.000
120	MAX	120.220	119.820	0.620	120.087	119.880	0.294	120.054	119.964	0.125	120.035	119.988	0.069	120.035	120.000	0.057
	MIN	120.000	119.600	0.180	120.000	119.793	0.120	120.000	119.929	0.036	120.000	119.966	0.012	120.000	119.978	0.000
160	MAX	160.250	159.790	0.710	160.100	159.855	0.345	160.063	159.957	0.146	160.040	159.986	0.079	160.040	160.000	0.065
	MIN	160.000	159.540	0.210	160.000	159.755	0.145	160.000	159.917	0.043	160.000	159.961	0.014	160.000	159.975	0.000
200	MAX	200.290	199.760	0.820	200.115	199.830	0.400	200.072	199.950	0.168	200.046	199.985	0.090	200.046	200.000	0.075
	MIN	200.000	199.470	0.240	200.000	199.715	0.170	200.000	199.904	0.050	200.000	199.956	0.015	200.000	199.971	0.000
250	MAX	250.290	249.720	0.860	250.115	249.830	0.400	250.072	249.950	0.168	250.046	249.985	0.090	250.046	250.000	0.075
	MIN	250.000	249.430	0.280	250.000	249.715	0.170	250.000	249.904	0.050	250.000	249.956	0.015	250.000	249.971	0.000
300	MAX	300.320	299.670	0.970	300.130	299.810	0.450	300.081	299.944	0.189	300.052	299.983	0.101	300.052	300.000	0.084
	MIN	300.000	299.350	0.330	300.000	299.680	0.190	300.000	299.892	0.056	300.000	299.951	0.017	300.000	299.968	0.000
400	MAX	400.360	399.600	1.120	400.140	399.790	0.490	400.089	399.938	0.208	400.057	399.982	0.111	400.057	400.000	0.093
	MIN	400.000	399.240	0.400	400.000	399.650	0.210	400.000	399.881	0.062	400.000	399.946	0.018	400.000	399.964	0.000
500	MAX	500.400	499.520	1.280	500.155	499.770	0.540	500.097	499.932	0.228	500.063	499.980	0.123	500.063	500.000	0.103
	MIN	500.000	499.120	0.480	500.000	499.615	0.230	500.000	499.869	0.068	500.000	499.940	0.020	500.000	499.960	0.000

Dimensions in mm.

Table B   Preferred Hole Basis Transition and Interference Fits

BASIC SIZE		LOOSE RUNNING Hole H11	LOOSE RUNNING Shaft c11	LOOSE RUNNING Fit	FREE RUNNING Hole H9	FREE RUNNING Shaft d9	FREE RUNNING Fit	CLOSE RUNNING Hole H8	CLOSE RUNNING Shaft f7	CLOSE RUNNING Fit	SLIDING Hole H7	SLIDING Shaft g6	SLIDING Fit	LOCATIONAL CLEARANCE Hole H7	LOCATIONAL CLEARANCE Shaft h6	LOCATIONAL CLEARANCE Fit
1	MAX	1.060	0.940	0.180	1.025	0.980	0.070	1.014	0.994	0.030	1.010	0.998	0.018	1.010	1.000	0.016
	MIN	1.000	0.880	0.060	1.000	0.955	0.020	1.000	0.984	0.006	1.000	0.992	0.002	1.000	0.994	0.000
1.2	MAX	1.260	1.140	0.180	1.225	1.180	0.070	1.214	1.194	0.030	1.210	1.198	0.018	1.210	1.200	0.016
	MIN	1.200	1.080	0.060	1.200	1.155	0.020	1.200	1.184	0.006	1.200	1.192	0.002	1.200	1.194	0.000
1.6	MAX	1.660	1.540	0.180	1.625	1.580	0.070	1.614	1.594	0.030	1.610	1.598	0.018	1.610	1.600	0.016
	MIN	1.600	1.480	0.060	1.600	1.555	0.020	1.600	1.584	0.006	1.600	1.592	0.002	1.600	1.594	0.000
2	MAX	2.060	1.940	0.180	2.025	1.980	0.070	2.014	1.994	0.030	2.010	1.998	0.018	2.010	2.000	0.016
	MIN	2.000	1.880	0.060	2.000	1.955	0.020	2.000	1.984	0.006	2.000	1.992	0.002	2.000	1.994	0.000
2.5	MAX	2.560	2.440	0.180	2.525	2.480	0.070	2.514	2.494	0.030	2.510	2.498	0.018	2.510	2.500	0.016
	MIN	2.500	2.380	0.060	2.500	2.455	0.020	2.500	2.484	0.006	2.500	2.492	0.002	2.500	2.494	0.000
3	MAX	3.060	2.940	0.180	3.025	2.980	0.070	3.014	2.994	0.030	3.010	2.998	0.018	3.010	3.000	0.016
	MIN	3.000	2.880	0.060	3.000	2.955	0.020	3.000	2.984	0.006	3.000	2.992	0.002	3.000	2.994	0.000
4	MAX	4.075	3.930	0.220	4.030	3.970	0.090	4.018	3.990	0.040	4.012	3.996	0.024	4.012	4.000	0.020
	MIN	4.000	3.855	0.070	4.000	3.940	0.030	4.000	3.978	0.010	4.000	3.988	0.004	4.000	3.992	0.000
5	MAX	5.075	4.930	0.220	5.030	4.970	0.090	5.018	4.990	0.040	5.012	4.996	0.024	5.012	5.000	0.020
	MIN	5.000	4.855	0.070	5.000	4.940	0.030	5.000	4.978	0.010	5.000	4.988	0.004	5.000	4.992	0.000
6	MAX	6.075	5.930	0.220	6.030	5.970	0.090	6.018	5.990	0.040	6.012	5.996	0.024	6.012	6.000	0.020
	MIN	6.000	5.855	0.070	6.000	5.940	0.030	6.000	5.978	0.010	6.000	5.988	0.004	6.000	5.992	0.000
8	MAX	8.090	7.920	0.260	8.036	7.960	0.112	8.022	7.987	0.050	8.015	7.995	0.029	8.015	8.000	0.024
	MIN	8.000	7.830	0.080	8.000	7.924	0.040	8.000	7.972	0.013	8.000	7.986	0.005	8.000	7.991	0.000
10	MAX	10.090	9.920	0.260	10.036	9.960	0.112	10.022	9.987	0.050	10.015	9.995	0.029	10.015	10.000	0.024
	MIN	10.000	9.830	0.080	10.000	9.924	0.040	10.000	9.972	0.013	10.000	9.986	0.005	10.000	9.991	0.000
12	MAX	12.110	11.905	0.315	12.043	11.950	0.136	12.027	11.984	0.061	12.018	11.994	0.035	12.018	12.000	0.029
	MIN	12.000	11.795	0.095	12.000	11.907	0.050	12.000	11.966	0.016	12.000	11.983	0.006	12.000	11.989	0.000
16	MAX	16.110	15.905	0.315	16.043	15.950	0.136	16.027	15.984	0.061	16.018	15.994	0.035	16.018	16.000	0.029
	MIN	16.000	15.795	0.095	16.000	15.907	0.050	16.000	15.966	0.016	16.000	15.983	0.006	16.000	15.989	0.000
20	MAX	20.130	19.890	0.370	20.052	19.935	0.169	20.033	19.980	0.074	20.021	19.993	0.041	20.021	20.000	0.034
	MIN	20.000	19.760	0.110	20.000	19.883	0.065	20.000	19.959	0.020	20.000	19.980	0.007	20.000	19.987	0.000
25	MAX	25.130	24.890	0.370	25.052	24.935	0.169	25.033	24.980	0.074	25.021	24.993	0.041	25.021	25.000	0.034
	MIN	25.000	24.760	0.110	25.000	24.883	0.065	25.000	24.959	0.020	25.000	24.980	0.007	25.000	24.987	0.000
30	MAX	30.130	29.890	0.370	30.052	29.935	0.169	30.033	29.980	0.074	30.021	29.993	0.041	30.021	30.000	0.034
	MIN	30.000	29.760	0.110	30.000	29.883	0.065	30.000	29.959	0.020	30.000	29.980	0.007	30.000	29.987	0.000

AMERICAN NATIONAL STANDARD
PREFERRED METRIC LIMITS AND FITS

Dimensions in mm.

Table B Preferred Hole Basis Transition and Interference Fits (Continued)

BASIC SIZE		LOCATIONAL TRANSN. Hole H7	Shaft k6	Fit	LOCATIONAL TRANSN. Hole H7	Shaft n6	Fit	LOCATIONAL INTERF. Hole H7	Shaft p6	Fit	MEDIUM DRIVE Hole H7	Shaft s6	Fit	FORCE Hole H7	Shaft u6	Fit
40	MAX	40.025	40.018	0.023	40.025	40.033	0.008	40.025	40.042	-0.001	40.025	40.059	-0.018	40.025	40.076	-0.035
	MIN	40.000	40.002	-0.018	40.000	40.017	-0.033	40.000	40.026	-0.042	40.000	40.043	-0.059	40.000	40.060	-0.076
50	MAX	50.025	50.018	0.023	50.025	50.033	0.008	50.025	50.042	-0.001	50.025	50.059	-0.018	50.025	50.086	-0.045
	MIN	50.000	50.002	-0.018	50.000	50.017	-0.033	50.000	50.026	-0.042	50.000	50.043	-0.059	50.000	50.070	-0.086
60	MAX	60.030	60.021	0.028	60.030	60.039	0.010	60.030	60.051	-0.002	60.030	60.072	-0.023	60.030	60.106	-0.057
	MIN	60.000	60.002	-0.021	60.000	60.020	-0.039	60.000	60.032	-0.051	60.000	60.053	-0.072	60.000	60.087	-0.106
80	MAX	80.030	80.021	0.028	80.030	80.039	0.010	80.030	80.051	-0.002	80.030	80.078	-0.029	80.030	80.121	-0.072
	MIN	80.000	80.002	-0.021	80.000	80.020	-0.039	80.000	80.032	-0.051	80.000	80.059	-0.078	80.000	80.102	-0.121
100	MAX	100.035	100.025	0.032	100.035	100.045	0.012	100.035	100.059	-0.002	100.035	100.093	-0.036	100.035	100.146	-0.089
	MIN	100.000	100.003	-0.025	100.000	100.023	-0.045	100.000	100.037	-0.059	100.000	100.071	-0.093	100.000	100.124	-0.146
120	MAX	120.035	120.025	0.032	120.035	120.045	0.012	120.035	120.059	-0.002	120.035	120.101	-0.044	120.035	120.166	-0.109
	MIN	120.000	120.003	-0.025	120.000	120.023	-0.045	120.000	120.037	-0.059	120.000	120.079	-0.101	120.000	120.144	-0.166
160	MAX	160.040	160.028	0.037	160.040	160.052	0.013	160.040	160.068	-0.003	160.040	160.125	-0.060	160.040	160.215	-0.150
	MIN	160.000	160.003	-0.028	160.000	160.027	-0.052	160.000	160.043	-0.068	160.000	160.100	-0.125	160.000	160.190	-0.215
200	MAX	200.046	200.033	0.042	200.046	200.060	0.015	200.046	200.079	-0.004	200.046	200.151	-0.076	200.046	200.265	-0.190
	MIN	200.000	200.004	-0.033	200.000	200.031	-0.060	200.000	200.050	-0.079	200.000	200.122	-0.151	200.000	200.236	-0.265
250	MAX	250.046	250.033	0.042	250.046	250.060	0.015	250.046	250.079	-0.004	250.046	250.169	-0.094	250.046	250.313	-0.238
	MIN	250.000	250.004	-0.033	250.000	250.031	-0.060	250.000	250.050	-0.079	250.000	250.140	-0.169	250.000	250.284	-0.313
300	MAX	300.052	300.036	0.048	300.052	300.066	0.018	300.052	300.088	-0.004	300.052	300.202	-0.118	300.052	300.382	-0.298
	MIN	300.000	300.004	-0.036	300.000	300.034	-0.066	300.000	300.056	-0.088	300.000	300.170	-0.202	300.000	300.350	-0.382
400	MAX	400.057	400.040	0.053	400.057	400.073	0.020	400.057	400.098	-0.005	400.057	400.244	-0.151	400.057	400.471	-0.378
	MIN	400.000	400.004	-0.040	400.000	400.037	-0.073	400.000	400.062	-0.098	400.000	400.208	-0.244	400.000	400.435	-0.471
500	MAX	500.063	500.045	0.058	500.063	500.080	0.023	500.063	500.108	-0.005	500.063	500.292	-0.189	500.063	500.580	-0.477
	MIN	500.000	500.005	-0.045	500.000	500.040	-0.080	500.000	500.068	-0.108	500.000	500.252	-0.292	500.000	500.540	-0.580

Table C   Preferred Shaft Basis Clearance Fits

Dimensions in mm.

BASIC SIZE		LOOSE RUNNING Hole C11	Shaft h11	Fit	FREE RUNNING Hole D9	Shaft h9	Fit	CLOSE RUNNING Hole F8	Shaft h7	Fit	SLIDING Hole G7	Shaft h6	Fit	LOCATIONAL CLEARANCE Hole H7	Shaft h6	Fit
1	MAX	1.120	1.000	0.180	1.045	1.000	0.070	1.020	1.000	0.030	1.012	1.000	0.018	1.010	1.000	0.016
	MIN	1.060	0.940	0.060	1.020	0.975	0.020	1.006	0.990	0.006	1.002	0.994	0.002	1.000	0.994	0.000
1.2	MAX	1.320	1.200	0.180	1.245	1.200	0.070	1.220	1.200	0.030	1.212	1.200	0.018	1.210	1.200	0.016
	MIN	1.260	1.140	0.060	1.220	1.175	0.020	1.206	1.190	0.006	1.202	1.194	0.002	1.200	1.194	0.000
1.6	MAX	1.720	1.600	0.180	1.645	1.600	0.070	1.620	1.600	0.030	1.612	1.600	0.018	1.610	1.600	0.016
	MIN	1.660	1.540	0.060	1.620	1.575	0.020	1.606	1.590	0.006	1.602	1.594	0.002	1.600	1.594	0.000
2	MAX	2.120	2.000	0.180	2.045	2.000	0.070	2.020	2.000	0.030	2.012	2.000	0.018	2.010	2.000	0.016
	MIN	2.060	1.940	0.060	2.020	1.975	0.020	2.006	1.990	0.006	2.002	1.994	0.002	2.000	1.994	0.000
2.5	MAX	2.620	2.500	0.180	2.545	2.500	0.070	2.520	2.500	0.030	2.512	2.500	0.018	2.510	2.500	0.016
	MIN	2.560	2.440	0.060	2.520	2.475	0.020	2.506	2.490	0.006	2.502	2.494	0.002	2.500	2.494	0.000
3	MAX	3.120	3.000	0.180	3.045	3.000	0.070	3.020	3.000	0.030	3.012	3.000	0.018	3.010	3.000	0.016
	MIN	3.060	2.940	0.060	3.020	2.975	0.020	3.006	2.990	0.006	3.002	2.994	0.002	3.000	2.994	0.000
4	MAX	4.145	4.000	0.220	4.060	4.000	0.090	4.028	4.000	0.040	4.016	4.000	0.024	4.012	4.000	0.020
	MIN	4.070	3.925	0.070	4.030	3.970	0.030	4.010	3.988	0.010	4.004	3.992	0.004	4.000	3.992	0.000
5	MAX	5.145	5.000	0.220	5.060	5.000	0.090	5.028	5.000	0.040	5.016	5.000	0.024	5.012	5.000	0.020
	MIN	5.070	4.925	0.070	5.030	4.970	0.030	5.010	4.988	0.010	5.004	4.992	0.004	5.000	4.992	0.000
6	MAX	6.145	6.000	0.220	6.060	6.000	0.090	6.028	6.000	0.040	6.016	6.000	0.024	6.012	6.000	0.020
	MIN	6.070	5.925	0.070	6.030	5.970	0.030	6.010	5.988	0.010	6.004	5.992	0.004	6.000	5.992	0.000
8	MAX	8.170	8.000	0.260	8.076	8.000	0.112	8.035	8.000	0.050	8.020	8.000	0.029	8.015	8.000	0.024
	MIN	8.080	7.910	0.080	8.040	7.964	0.040	8.013	7.985	0.013	8.005	7.991	0.005	8.000	7.991	0.000
10	MAX	10.170	10.000	0.260	10.076	10.000	0.112	10.035	10.000	0.050	10.020	10.000	0.029	10.015	10.000	0.024
	MIN	10.080	9.910	0.080	10.040	9.964	0.040	10.013	9.985	0.013	10.005	9.991	0.005	10.000	9.991	0.000
12	MAX	12.205	12.000	0.315	12.093	12.000	0.136	12.043	12.000	0.061	12.024	12.000	0.035	12.018	12.000	0.029
	MIN	12.095	11.890	0.095	12.050	11.957	0.050	12.016	11.982	0.016	12.006	11.989	0.006	12.000	11.989	0.000
16	MAX	16.205	16.000	0.315	16.093	16.000	0.136	16.043	16.000	0.061	16.024	16.000	0.035	16.018	16.000	0.029
	MIN	16.095	15.890	0.095	16.050	15.957	0.050	16.016	15.982	0.016	16.006	15.989	0.006	16.000	15.989	0.000
20	MAX	20.240	20.000	0.370	20.117	20.000	0.169	20.053	20.000	0.074	20.028	20.000	0.041	20.021	20.000	0.034
	MIN	20.110	19.870	0.110	20.065	19.948	0.065	20.020	19.979	0.020	20.007	19.987	0.007	20.000	19.987	0.000
25	MAX	25.240	25.000	0.370	25.117	25.000	0.169	25.053	25.000	0.074	25.028	25.000	0.041	25.021	25.000	0.034
	MIN	25.110	24.870	0.110	25.065	24.948	0.065	25.020	24.979	0.020	25.007	24.987	0.007	25.000	24.987	0.000
30	MAX	30.240	30.000	0.370	30.117	30.000	0.169	30.053	30.000	0.074	30.028	30.000	0.041	30.021	30.000	0.034
	MIN	30.110	29.870	0.110	30.065	29.948	0.065	30.020	29.979	0.020	30.007	29.987	0.007	30.000	29.987	0.000

AMERICAN NATIONAL STANDARD
PREFERRED METRIC LIMITS AND FITS

Dimensions in mm.

Table C   Preferred Shaft Basis Clearance Fits (Continued)

BASIC SIZE		LOOSE RUNNING Hole C11	Shaft h11	Fit	FREE RUNNING Hole D9	Shaft h9	Fit	CLOSE RUNNING Hole F8	Shaft h7	Fit	SLIDING Hole G7	Shaft h6	Fit	LOCATIONAL CLEARANCE Hole H7	Shaft h6	Fit
40	MAX	40.280	40.000	0.440	40.142	40.000	0.204	40.064	40.000	0.089	40.034	40.000	0.050	40.025	40.000	0.041
	MIN	40.120	39.840	0.120	40.080	39.938	0.080	40.025	39.975	0.025	40.009	39.984	0.009	40.000	39.984	0.000
50	MAX	50.290	50.000	0.450	50.142	50.000	0.204	50.064	50.000	0.089	50.034	50.000	0.050	50.025	50.000	0.041
	MIN	50.130	49.840	0.130	50.080	49.938	0.080	50.025	49.975	0.025	50.009	49.984	0.009	50.000	49.984	0.000
60	MAX	60.330	60.000	0.520	60.174	60.000	0.248	60.076	60.000	0.106	60.040	60.000	0.059	60.030	60.000	0.049
	MIN	60.140	59.810	0.140	60.100	59.926	0.100	60.030	59.970	0.030	60.010	59.981	0.010	60.000	59.981	0.000
80	MAX	80.340	80.000	0.530	80.174	80.000	0.248	80.076	80.000	0.106	80.040	80.000	0.059	80.030	80.000	0.049
	MIN	80.150	79.810	0.150	80.100	79.926	0.100	80.030	79.970	0.030	80.010	79.981	0.010	80.000	79.981	0.000
100	MAX	100.390	100.000	0.610	100.207	100.000	0.294	100.090	100.000	0.125	100.047	100.000	0.069	100.035	100.000	0.057
	MIN	100.170	99.780	0.170	100.120	99.913	0.120	100.036	99.965	0.036	100.012	99.978	0.012	100.000	99.978	0.000
120	MAX	120.400	120.000	0.620	120.207	120.000	0.294	120.090	120.000	0.125	120.047	120.000	0.069	120.035	120.000	0.057
	MIN	120.180	119.780	0.180	120.120	119.913	0.120	120.036	119.965	0.036	120.012	119.978	0.012	120.000	119.978	0.000
160	MAX	160.460	160.000	0.710	160.245	160.000	0.345	160.106	160.000	0.146	160.054	160.000	0.079	160.040	160.000	0.065
	MIN	160.210	159.750	0.210	160.145	159.900	0.145	160.043	159.960	0.043	160.014	159.975	0.014	160.000	159.975	0.000
200	MAX	200.530	200.000	0.820	200.285	200.000	0.400	200.122	200.000	0.168	200.061	200.000	0.090	200.046	200.000	0.075
	MIN	200.240	199.710	0.240	200.170	199.885	0.170	200.050	199.954	0.050	200.015	199.971	0.015	200.000	199.971	0.000
250	MAX	250.570	250.000	0.860	250.285	250.000	0.400	250.122	250.000	0.168	250.061	250.000	0.090	250.046	250.000	0.075
	MIN	250.280	249.710	0.280	250.170	249.885	0.170	250.050	249.954	0.050	250.015	249.971	0.015	250.000	249.971	0.000
300	MAX	300.650	300.000	0.970	300.320	300.000	0.450	300.137	300.000	0.189	300.069	300.000	0.101	300.052	300.000	0.084
	MIN	300.330	299.680	0.330	300.190	299.870	0.190	300.056	299.948	0.056	300.017	299.968	0.017	300.000	299.968	0.000
400	MAX	400.760	400.000	1.120	400.350	400.000	0.490	400.151	400.000	0.208	400.075	400.000	0.111	400.057	400.000	0.093
	MIN	400.400	399.640	0.400	400.210	399.860	0.210	400.062	399.943	0.062	400.018	399.964	0.018	400.000	399.964	0.000
500	MAX	500.880	500.000	1.280	500.385	500.000	0.540	500.165	500.000	0.228	500.083	500.000	0.123	500.063	500.000	0.103
	MIN	500.480	499.600	0.480	500.230	499.845	0.230	500.068	499.937	0.068	500.020	499.960	0.020	500.000	499.960	0.000

Dimensions in mm.

Table D Preferred Shaft Basis Transition and Interference Fits

BASIC SIZE		LOCATIONAL TRANSN.			LOCATIONAL TRANSN.			LOCATIONAL INTERF.			MEDIUM DRIVE			FORCE		
		Hole K7	Shaft h6	Fit	Hole N7	Shaft h6	Fit	Hole P7	Shaft h6	Fit	Hole S7	Shaft h6	Fit	Hole U7	Shaft h6	Fit
1	MAX	1.000	1.000	0.006	0.996	1.000	0.002	0.994	1.000	0.000	0.986	1.000	-0.008	0.982	1.000	-0.012
	MIN	0.990	0.994	-0.010	0.986	0.994	-0.014	0.984	0.994	-0.016	0.976	0.994	-0.024	0.972	0.994	-0.028
1.2	MAX	1.200	1.200	0.006	1.196	1.200	0.002	1.194	1.200	0.000	1.186	1.200	-0.008	1.182	1.200	-0.012
	MIN	1.190	1.194	-0.010	1.186	1.194	-0.014	1.184	1.194	-0.016	1.176	1.194	-0.024	1.172	1.194	-0.028
1.6	MAX	1.600	1.600	0.006	1.596	1.600	0.002	1.594	1.600	0.000	1.586	1.600	-0.008	1.582	1.600	-0.012
	MIN	1.590	1.594	-0.010	1.586	1.594	-0.014	1.584	1.594	-0.016	1.576	1.594	-0.024	1.572	1.594	-0.028
2	MAX	2.000	2.000	0.006	1.996	2.000	0.002	1.994	2.000	0.000	1.986	2.000	-0.008	1.982	2.000	-0.012
	MIN	1.990	1.994	-0.010	1.986	1.994	-0.014	1.984	1.994	-0.016	1.976	1.994	-0.024	1.972	1.994	-0.028
2.5	MAX	2.500	2.500	0.006	2.496	2.500	0.002	2.494	2.500	0.000	2.486	2.500	-0.008	2.482	2.500	-0.012
	MIN	2.490	2.494	-0.010	2.486	2.494	-0.014	2.484	2.494	-0.016	2.476	2.494	-0.024	2.472	2.494	-0.028
3	MAX	3.000	3.000	0.006	2.996	3.000	0.002	2.994	3.000	0.000	2.986	3.000	-0.008	2.982	3.000	-0.012
	MIN	2.990	2.994	-0.010	2.986	2.994	-0.014	2.984	2.994	-0.016	2.976	2.994	-0.024	2.972	2.994	-0.028
4	MAX	4.003	4.000	0.011	3.996	4.000	0.004	3.992	4.000	0.000	3.985	4.000	-0.007	3.981	4.000	-0.011
	MIN	3.991	3.992	-0.009	3.984	3.992	-0.016	3.980	3.992	-0.020	3.973	3.992	-0.027	3.969	3.992	-0.031
5	MAX	5.003	5.000	0.011	4.996	5.000	0.004	4.992	5.000	0.000	4.985	5.000	-0.007	4.981	5.000	-0.011
	MIN	4.991	4.992	-0.009	4.984	4.992	-0.016	4.980	4.992	-0.020	4.973	4.992	-0.027	4.969	4.992	-0.031
6	MAX	6.003	6.000	0.011	5.996	6.000	0.004	5.992	6.000	0.000	5.985	6.000	-0.007	5.981	6.000	-0.011
	MIN	5.991	5.992	-0.009	5.984	5.992	-0.016	5.980	5.992	-0.020	5.973	5.992	-0.027	5.969	5.992	-0.031
8	MAX	8.005	8.000	0.014	7.996	8.000	0.005	7.991	8.000	0.000	7.983	8.000	-0.008	7.978	8.000	-0.013
	MIN	7.990	7.991	-0.010	7.981	7.991	-0.019	7.976	7.991	-0.024	7.968	7.991	-0.032	7.963	7.991	-0.037
10	MAX	10.005	10.000	0.014	9.996	10.000	0.005	9.991	10.000	0.000	9.983	10.000	-0.008	9.978	10.000	-0.013
	MIN	9.990	9.991	-0.010	9.981	9.991	-0.019	9.976	9.991	-0.024	9.968	9.991	-0.032	9.963	9.991	-0.037
12	MAX	12.006	12.000	0.017	11.995	12.000	0.006	11.989	12.000	0.000	11.979	12.000	-0.010	11.974	12.000	-0.015
	MIN	11.988	11.989	-0.012	11.977	11.989	-0.023	11.971	11.989	-0.029	11.961	11.989	-0.039	11.956	11.989	-0.044
16	MAX	16.006	16.000	0.017	15.995	16.000	0.006	15.989	16.000	0.000	15.979	16.000	-0.010	15.974	16.000	-0.015
	MIN	15.988	15.989	-0.012	15.977	15.989	-0.023	15.971	15.989	-0.029	15.961	15.989	-0.039	15.956	15.989	-0.044
20	MAX	20.006	20.000	0.019	19.993	20.000	0.006	19.986	20.000	-0.001	19.973	20.000	-0.014	19.967	20.000	-0.020
	MIN	19.985	19.987	-0.015	19.972	19.987	-0.028	19.965	19.987	-0.035	19.952	19.987	-0.048	19.946	19.987	-0.054
25	MAX	25.006	25.000	0.019	24.993	25.000	0.006	24.986	25.000	-0.001	24.973	25.000	-0.014	24.960	25.000	-0.027
	MIN	24.985	24.987	-0.015	24.972	24.987	-0.028	24.965	24.987	-0.035	24.952	24.987	-0.048	24.939	24.987	-0.061
30	MAX	30.006	30.000	0.019	29.993	30.000	0.006	29.986	30.000	-0.001	29.973	30.000	-0.014	29.960	30.000	-0.027
	MIN	29.985	29.987	-0.015	29.972	29.987	-0.028	29.965	29.987	-0.035	29.952	29.987	-0.048	29.939	29.987	-0.061

AMERICAN NATIONAL STANDARD
PREFERRED METRIC LIMITS AND FITS

Table D Preferred Shaft Basis Transition and Interference Fits *(Continued)*

Dimensions in mm.

BASIC SIZE		LOCATIONAL TRANSN. Hole K7	Shaft h6	Fit	LOCATIONAL TRANSN. Hole N7	Shaft h6	Fit	LOCATIONAL INTERF. Hole P7	Shaft h6	Fit	MEDIUM DRIVE Hole S7	Shaft h6	Fit	FORCE Hole U7	Shaft h6	Fit
40	MAX	40.007	40.000	0.023	39.992	40.000	0.008	39.983	40.000	-0.001	39.966	40.000	-0.018	39.949	40.000	-0.035
	MIN	39.982	39.984	-0.018	39.967	39.984	-0.033	39.958	39.984	-0.042	39.941	39.984	-0.059	39.924	39.984	-0.076
50	MAX	50.007	50.000	0.023	49.992	50.000	0.008	49.983	50.000	-0.001	49.966	50.000	-0.018	49.939	50.000	-0.049
	MIN	49.982	49.984	-0.018	49.967	49.984	-0.033	49.958	49.984	-0.042	49.941	49.984	-0.059	49.914	49.984	-0.086
60	MAX	60.009	60.000	0.028	59.991	60.000	0.010	59.979	60.000	-0.002	59.958	60.000	-0.023	59.924	60.000	-0.067
	MIN	59.979	59.981	-0.021	59.961	59.981	-0.039	59.949	59.981	-0.051	59.928	59.981	-0.072	59.894	59.981	-0.106
80	MAX	80.009	80.000	0.028	79.991	80.000	0.010	79.979	80.000	-0.002	79.952	80.000	-0.029	79.909	80.000	-0.072
	MIN	79.979	79.981	-0.021	79.961	79.981	-0.039	79.949	79.981	-0.051	79.922	79.981	-0.078	79.879	79.981	-0.121
100	MAX	100.010	100.000	0.032	99.990	100.000	0.012	99.976	100.000	-0.002	99.942	100.000	-0.036	99.889	100.000	-0.089
	MIN	99.975	99.978	-0.025	99.955	99.978	-0.045	99.941	99.978	-0.059	99.907	99.978	-0.093	99.854	99.978	-0.146
120	MAX	120.010	120.000	0.032	119.990	120.000	0.012	119.976	120.000	-0.002	119.934	120.000	-0.044	119.869	120.000	-0.109
	MIN	119.975	119.978	-0.025	119.955	119.978	-0.045	119.941	119.978	-0.059	119.899	119.978	-0.101	119.834	119.978	-0.166
160	MAX	160.012	160.000	0.037	159.988	160.000	0.013	159.972	160.000	-0.003	159.915	160.000	-0.060	159.825	160.000	-0.150
	MIN	159.972	159.975	-0.028	159.948	159.975	-0.052	159.932	159.975	-0.068	159.875	159.975	-0.125	159.785	159.975	-0.215
200	MAX	200.013	200.000	0.042	199.986	200.000	0.015	199.967	200.000	-0.004	199.895	200.000	-0.076	199.781	200.000	-0.190
	MIN	199.967	199.971	-0.033	199.940	199.971	-0.060	199.921	199.971	-0.079	199.849	199.971	-0.151	199.735	199.971	-0.265
250	MAX	250.013	250.000	0.042	249.986	250.000	0.015	249.967	250.000	-0.004	249.877	250.000	-0.094	249.733	250.000	-0.238
	MIN	249.967	249.971	-0.033	249.940	249.971	-0.060	249.921	249.971	-0.079	249.831	249.971	-0.169	249.687	249.971	-0.313
300	MAX	300.016	300.000	0.048	299.986	300.000	0.018	299.964	300.000	-0.004	299.850	300.000	-0.118	299.670	300.000	-0.298
	MIN	299.964	299.968	-0.036	299.934	299.968	-0.066	299.912	299.968	-0.088	299.798	299.968	-0.202	299.618	299.968	-0.382
400	MAX	400.017	400.000	0.053	399.984	400.000	0.020	399.959	400.000	-0.005	399.813	400.000	-0.151	399.586	400.000	-0.378
	MIN	399.960	399.964	-0.040	399.927	399.964	-0.073	399.902	399.964	-0.098	399.756	399.964	-0.244	399.529	399.964	-0.471
500	MAX	500.018	500.000	0.058	499.983	500.000	0.023	499.955	500.000	-0.005	499.771	500.000	-0.189	499.483	500.000	-0.477
	MIN	499.955	499.960	-0.045	499.920	499.960	-0.080	499.892	499.960	-0.108	499.708	499.960	-0.292	499.420	499.960	-0.580

Table E Tolerance zones for Internal (Hole) Dimensions
(C13 through C8 and D12 through D7) (Dimensions in mm)

Dimensions in mm

BASIC SIZE		C13	C12	C11	C10	C9	C8	D12	D11	D10	D9	D8	D7
OVER 0 TO 3		+0.200 +0.060	+0.160 +0.060	+0.120 +0.060	+0.100 +0.060	+0.085 +0.060	+0.074 +0.060	+0.120 +0.020	+0.080 +0.020	+0.060 +0.020	+0.045 +0.020	+0.034 +0.020	+0.030 +0.020
OVER 3 TO 6		+0.250 +0.070	+0.190 +0.070	+0.145 +0.070	+0.118 +0.070	+0.100 +0.070	+0.088 +0.070	+0.150 +0.030	+0.105 +0.030	+0.078 +0.030	+0.060 +0.030	+0.048 +0.030	+0.042 +0.030
OVER 6 TO 10		+0.300 +0.080	+0.230 +0.080	+0.170 +0.080	+0.138 +0.080	+0.116 +0.080	+0.102 +0.080	+0.190 +0.040	+0.130 +0.040	+0.098 +0.040	+0.076 +0.040	+0.062 +0.040	+0.055 +0.040
OVER 10 TO 14		+0.365 +0.095	+0.275 +0.095	+0.205 +0.095	+0.165 +0.095	+0.138 +0.095	+0.122 +0.095	+0.230 +0.050	+0.160 +0.050	+0.120 +0.050	+0.093 +0.050	+0.077 +0.050	+0.068 +0.050
OVER 14 TO 18		+0.365 +0.095	+0.275 +0.095	+0.205 +0.095	+0.165 +0.095	+0.138 +0.095	+0.122 +0.095	+0.230 +0.050	+0.160 +0.050	+0.120 +0.050	+0.093 +0.050	+0.077 +0.050	+0.068 +0.050
OVER 18 TO 24		+0.440 +0.110	+0.320 +0.110	+0.240 +0.110	+0.194 +0.110	+0.162 +0.110	+0.143 +0.110	+0.275 +0.065	+0.195 +0.065	+0.149 +0.065	+0.117 +0.065	+0.098 +0.065	+0.086 +0.065
OVER 24 TO 30		+0.440 +0.110	+0.320 +0.110	+0.240 +0.110	+0.194 +0.110	+0.162 +0.110	+0.143 +0.110	+0.275 +0.065	+0.195 +0.065	+0.149 +0.065	+0.117 +0.065	+0.098 +0.065	+0.086 +0.065
OVER 30 TO 40		+0.510 +0.120	+0.370 +0.120	+0.280 +0.120	+0.220 +0.120	+0.182 +0.120	+0.159 +0.120	+0.330 +0.080	+0.240 +0.080	+0.180 +0.080	+0.142 +0.080	+0.119 +0.080	+0.105 +0.080
OVER 40 TO 50		+0.520 +0.130	+0.380 +0.130	+0.290 +0.130	+0.230 +0.130	+0.192 +0.130	+0.169 +0.130	+0.330 +0.080	+0.240 +0.080	+0.180 +0.080	+0.142 +0.080	+0.119 +0.080	+0.105 +0.080
OVER 50 TO 65		+0.600 +0.140	+0.440 +0.140	+0.330 +0.140	+0.260 +0.140	+0.214 +0.140	+0.186 +0.140	+0.400 +0.100	+0.290 +0.100	+0.220 +0.100	+0.174 +0.100	+0.146 +0.100	+0.130 +0.100
OVER 65 TO 80		+0.610 +0.150	+0.450 +0.150	+0.340 +0.150	+0.270 +0.150	+0.224 +0.150	+0.196 +0.150	+0.400 +0.100	+0.290 +0.100	+0.220 +0.100	+0.174 +0.100	+0.146 +0.100	+0.130 +0.100
OVER 80 TO 100		+0.710 +0.170	+0.520 +0.170	+0.390 +0.170	+0.310 +0.170	+0.257 +0.170	+0.224 +0.170	+0.470 +0.120	+0.340 +0.120	+0.260 +0.120	+0.207 +0.120	+0.174 +0.120	+0.155 +0.120
OVER 100 TO 120		+0.720 +0.180	+0.530 +0.180	+0.400 +0.180	+0.320 +0.180	+0.267 +0.180	+0.234 +0.180	+0.470 +0.120	+0.340 +0.120	+0.260 +0.120	+0.207 +0.120	+0.174 +0.120	+0.155 +0.120
OVER 120 TO 140		+0.830 +0.200	+0.600 +0.200	+0.450 +0.200	+0.360 +0.200	+0.300 +0.200	+0.263 +0.200	+0.545 +0.145	+0.395 +0.145	+0.305 +0.145	+0.245 +0.145	+0.208 +0.145	+0.185 +0.145
OVER 140 TO 160		+0.840 +0.210	+0.610 +0.210	+0.460 +0.210	+0.370 +0.210	+0.310 +0.210	+0.273 +0.210	+0.545 +0.145	+0.395 +0.145	+0.305 +0.145	+0.245 +0.145	+0.208 +0.145	+0.185 +0.145
OVER 160 TO 180		+0.860 +0.230	+0.630 +0.230	+0.480 +0.230	+0.390 +0.230	+0.330 +0.230	+0.293 +0.230	+0.545 +0.145	+0.395 +0.145	+0.305 +0.145	+0.245 +0.145	+0.208 +0.145	+0.185 +0.145
OVER 180 TO 200		+0.960 +0.240	+0.700 +0.240	+0.530 +0.240	+0.425 +0.240	+0.355 +0.240	+0.312 +0.240	+0.630 +0.170	+0.460 +0.170	+0.355 +0.170	+0.285 +0.170	+0.242 +0.170	+0.216 +0.170
OVER 200 TO 225		+0.980 +0.260	+0.720 +0.260	+0.550 +0.260	+0.445 +0.260	+0.375 +0.260	+0.332 +0.260	+0.630 +0.170	+0.460 +0.170	+0.355 +0.170	+0.285 +0.170	+0.242 +0.170	+0.216 +0.170
OVER 225 TO 250		+1.000 +0.280	+0.740 +0.280	+0.570 +0.280	+0.465 +0.280	+0.395 +0.280	+0.352 +0.280	+0.630 +0.170	+0.460 +0.170	+0.355 +0.170	+0.285 +0.170	+0.242 +0.170	+0.216 +0.170
OVER 250 TO 280		+1.110 +0.300	+0.820 +0.300	+0.620 +0.300	+0.510 +0.300	+0.430 +0.300	+0.381 +0.300	+0.710 +0.190	+0.510 +0.190	+0.400 +0.190	+0.320 +0.190	+0.271 +0.190	+0.242 +0.190
OVER 280 TO 315		+1.140 +0.330	+0.850 +0.330	+0.650 +0.330	+0.540 +0.330	+0.460 +0.330	+0.411 +0.330	+0.710 +0.190	+0.510 +0.190	+0.400 +0.190	+0.320 +0.190	+0.271 +0.190	+0.242 +0.190
OVER 315 TO 355		+1.250 +0.360	+0.930 +0.360	+0.720 +0.360	+0.590 +0.360	+0.500 +0.360	+0.449 +0.360	+0.780 +0.210	+0.570 +0.210	+0.440 +0.210	+0.350 +0.210	+0.299 +0.210	+0.267 +0.210
OVER 355 TO 400		+1.290 +0.400	+0.970 +0.400	+0.760 +0.400	+0.630 +0.400	+0.540 +0.400	+0.489 +0.400	+0.780 +0.210	+0.570 +0.210	+0.440 +0.210	+0.350 +0.210	+0.299 +0.210	+0.267 +0.210
OVER 400 TO 450		+1.410 +0.440	+1.070 +0.440	+0.840 +0.440	+0.650 +0.440	+0.595 +0.440	+0.537 +0.440	+0.860 +0.230	+0.630 +0.230	+0.480 +0.230	+0.385 +0.230	+0.327 +0.230	+0.293 +0.230
OVER 450 TO 500		+1.450 +0.480	+1.110 +0.480	+0.880 +0.480	+0.730 +0.480	+0.635 +0.480	+0.577 +0.480	+0.860 +0.230	+0.630 +0.230	+0.480 +0.230	+0.385 +0.230	+0.327 +0.230	+0.293 +0.230

Table F Tolerance Zones for External (Shaft) Dimensions
(c13 through c8 and c12 through d7) (Dimensions in mm)

Dimensions in mm

BASIC SIZE	c13	c12	c11	c10	c9	c8	d12	d11	d10	d9	d8	d7
Over 0 To 3	-0.060 / -0.200	-0.060 / -0.160	-0.060 / -0.120	-0.060 / -0.100	-0.060 / -0.085	-0.060 / -0.074	-0.020 / -0.120	-0.020 / -0.080	-0.020 / -0.060	-0.020 / -0.045	-0.020 / -0.034	-0.020 / -0.030
Over 3 To 6	-0.070 / -0.250	-0.070 / -0.190	-0.070 / -0.145	-0.070 / -0.118	-0.070 / -0.100	-0.070 / -0.088	-0.030 / -0.150	-0.030 / -0.105	-0.030 / -0.078	-0.030 / -0.060	-0.030 / -0.048	-0.030 / -0.042
Over 6 To 10	-0.080 / -0.300	-0.080 / -0.230	-0.080 / -0.170	-0.080 / -0.138	-0.080 / -0.116	-0.080 / -0.102	-0.040 / -0.190	-0.040 / -0.130	-0.040 / -0.098	-0.040 / -0.076	-0.040 / -0.062	-0.040 / -0.055
Over 10 To 14	-0.095 / -0.365	-0.095 / -0.275	-0.095 / -0.205	-0.095 / -0.165	-0.095 / -0.138	-0.095 / -0.122	-0.050 / -0.230	-0.050 / -0.160	-0.050 / -0.120	-0.050 / -0.093	-0.050 / -0.077	-0.050 / -0.068
Over 14 To 18	-0.095 / -0.365	-0.095 / -0.275	-0.095 / -0.205	-0.095 / -0.165	-0.095 / -0.138	-0.095 / -0.122	-0.050 / -0.230	-0.050 / -0.160	-0.050 / -0.120	-0.050 / -0.093	-0.050 / -0.077	-0.050 / -0.068
Over 18 To 24	-0.110 / -0.440	-0.110 / -0.320	-0.110 / -0.240	-0.110 / -0.194	-0.110 / -0.162	-0.110 / -0.143	-0.065 / -0.275	-0.065 / -0.195	-0.065 / -0.149	-0.065 / -0.117	-0.065 / -0.098	-0.065 / -0.086
Over 24 To 30	-0.110 / -0.440	-0.110 / -0.320	-0.110 / -0.240	-0.110 / -0.194	-0.110 / -0.162	-0.110 / -0.143	-0.065 / -0.275	-0.065 / -0.195	-0.065 / -0.149	-0.065 / -0.117	-0.065 / -0.098	-0.065 / -0.086
Over 30 To 40	-0.120 / -0.510	-0.120 / -0.370	-0.120 / -0.280	-0.120 / -0.220	-0.120 / -0.182	-0.120 / -0.159	-0.080 / -0.330	-0.080 / -0.240	-0.080 / -0.180	-0.080 / -0.142	-0.080 / -0.119	-0.080 / -0.105
Over 40 To 50	-0.130 / -0.520	-0.130 / -0.380	-0.130 / -0.290	-0.130 / -0.230	-0.130 / -0.192	-0.130 / -0.169	-0.080 / -0.330	-0.080 / -0.240	-0.080 / -0.180	-0.080 / -0.142	-0.080 / -0.119	-0.080 / -0.105
Over 50 To 65	-0.140 / -0.600	-0.140 / -0.440	-0.140 / -0.330	-0.140 / -0.260	-0.140 / -0.214	-0.140 / -0.186	-0.100 / -0.400	-0.100 / -0.290	-0.100 / -0.220	-0.100 / -0.174	-0.100 / -0.146	-0.100 / -0.130
Over 65 To 80	-0.150 / -0.610	-0.150 / -0.450	-0.150 / -0.340	-0.150 / -0.270	-0.150 / -0.224	-0.150 / -0.196	-0.100 / -0.400	-0.100 / -0.290	-0.100 / -0.220	-0.100 / -0.174	-0.100 / -0.146	-0.100 / -0.130
Over 80 To 100	-0.170 / -0.710	-0.170 / -0.520	-0.170 / -0.390	-0.170 / -0.310	-0.170 / -0.257	-0.170 / -0.224	-0.120 / -0.470	-0.120 / -0.340	-0.120 / -0.260	-0.120 / -0.207	-0.120 / -0.174	-0.120 / -0.155
Over 100 To 120	-0.180 / -0.720	-0.180 / -0.530	-0.180 / -0.400	-0.180 / -0.320	-0.180 / -0.267	-0.180 / -0.234	-0.120 / -0.470	-0.120 / -0.340	-0.120 / -0.260	-0.120 / -0.207	-0.120 / -0.174	-0.120 / -0.155
Over 120 To 140	-0.200 / -0.830	-0.200 / -0.600	-0.200 / -0.450	-0.200 / -0.360	-0.200 / -0.300	-0.200 / -0.263	-0.145 / -0.545	-0.145 / -0.395	-0.145 / -0.305	-0.145 / -0.245	-0.145 / -0.208	-0.145 / -0.185
Over 140 To 160	-0.210 / -0.840	-0.210 / -0.610	-0.210 / -0.460	-0.210 / -0.370	-0.210 / -0.310	-0.210 / -0.273	-0.145 / -0.545	-0.145 / -0.395	-0.145 / -0.305	-0.145 / -0.245	-0.145 / -0.208	-0.145 / -0.185
Over 160 To 180	-0.230 / -0.860	-0.230 / -0.630	-0.230 / -0.480	-0.230 / -0.390	-0.230 / -0.330	-0.230 / -0.293	-0.145 / -0.545	-0.145 / -0.395	-0.145 / -0.305	-0.145 / -0.245	-0.145 / -0.208	-0.145 / -0.185
Over 180 To 200	-0.240 / -0.960	-0.240 / -0.700	-0.240 / -0.530	-0.240 / -0.425	-0.240 / -0.355	-0.240 / -0.312	-0.170 / -0.630	-0.170 / -0.460	-0.170 / -0.355	-0.170 / -0.285	-0.170 / -0.242	-0.170 / -0.216
Over 200 To 225	-0.260 / -0.980	-0.260 / -0.720	-0.260 / -0.550	-0.260 / -0.445	-0.260 / -0.375	-0.260 / -0.332	-0.170 / -0.630	-0.170 / -0.460	-0.170 / -0.355	-0.170 / -0.285	-0.170 / -0.242	-0.170 / -0.216
Over 225 To 250	-0.280 / -1.000	-0.280 / -0.740	-0.280 / -0.570	-0.280 / -0.465	-0.280 / -0.395	-0.280 / -0.352	-0.170 / -0.630	-0.170 / -0.460	-0.170 / -0.355	-0.170 / -0.285	-0.170 / -0.242	-0.170 / -0.216
Over 250 To 280	-0.300 / -1.110	-0.300 / -0.820	-0.300 / -0.620	-0.300 / -0.510	-0.300 / -0.430	-0.300 / -0.381	-0.190 / -0.710	-0.190 / -0.510	-0.190 / -0.400	-0.190 / -0.320	-0.190 / -0.271	-0.190 / -0.242
Over 280 To 315	-0.330 / -1.140	-0.330 / -0.850	-0.330 / -0.650	-0.330 / -0.540	-0.330 / -0.460	-0.330 / -0.411	-0.190 / -0.710	-0.190 / -0.510	-0.190 / -0.400	-0.190 / -0.320	-0.190 / -0.271	-0.190 / -0.242
Over 315 To 355	-0.360 / -1.250	-0.360 / -0.930	-0.360 / -0.720	-0.360 / -0.590	-0.360 / -0.500	-0.360 / -0.449	-0.210 / -0.780	-0.210 / -0.570	-0.210 / -0.440	-0.210 / -0.350	-0.210 / -0.299	-0.210 / -0.267
Over 355 To 400	-0.400 / -1.290	-0.400 / -0.970	-0.400 / -0.760	-0.400 / -0.630	-0.400 / -0.540	-0.400 / -0.489	-0.210 / -0.780	-0.210 / -0.570	-0.210 / -0.440	-0.210 / -0.350	-0.210 / -0.299	-0.210 / -0.267
Over 400 To 450	-0.440 / -1.410	-0.440 / -1.070	-0.440 / -0.840	-0.440 / -0.690	-0.440 / -0.595	-0.440 / -0.537	-0.230 / -0.860	-0.230 / -0.630	-0.230 / -0.480	-0.230 / -0.385	-0.230 / -0.327	-0.230 / -0.293
Over 450 To 500	-0.480 / -1.450	-0.480 / -1.110	-0.480 / -0.880	-0.480 / -0.730	-0.480 / -0.635	-0.480 / -0.577	-0.230 / -0.860	-0.230 / -0.630	-0.230 / -0.480	-0.230 / -0.385	-0.230 / -0.327	-0.230 / -0.293

Appendix C
Abbreviations and Symbols

ABBREVIATIONS FOR USE ON DRAWINGS

Word	Abbr.	Word	Abbr.
Abampere, absolute ampere	ABAMP	Alternating Current Synchronous	ACS
Abrasive Resistant	ABRSV RES	Alternator	ALT
Absolute	ABS	Altimeter	ALTM
Accelerate	ACCEL	Altitude	ALT
Acetylene	ACET	Ambient	AMB
Acid Resisting	AR	American	AMER
Acoustic	ACST	American War Standard	AWS
Acre Foot	AC FT	American Wire Gage	AWG
Adapter	ADPT	Ammeter	AM.
Addendum	ADD.	Ampere	AMP
Aerodynamic	AERODYN	Ampere Turn	AT.
Aeronautic	AERO	Ampere-hour	AMP HR
Aeronautical Material Specifications	AMS	Ampere-hour Meter	AHM
Aeronautical Recommended Practice	ARP	Amphibian-Amphibious	AMPH
Afterburner	AB	Amplitude	AM.
Aileron	AIL.	Amplifier	AMPL
Air Blast Circuit Breaker	ABCB	Angstrom Unit	A
Air Break Switch	ABS	Anneal	ANL
Air Circuit Breaker	ACB	Anode	A
Air Force—Navy	AN.	Anodize	ANOD
Air Force—Navy Aeronautical	ANA	Antenna	ANT.
Air Force—Navy Civil	ANC	Anti-friction Bearing	AFB
Air Force—Navy Design	AND.	Anti-icing	AI
Air Horsepower	AHP	Antilogarithm	ANTILOG
Airborne	ABN	Apparatus	APP
Aircooled	ACLD	Apparent Watts	AW
Aircraft	ACFT	Arc Weld	ARC/W
Airplane	APL	Area	A
Airport	AP	Armature	ARM.
Airscoop	AS.	Armature Shunt	ASH.
Airspeed	A/S	Article	ART.
Air-to-air	A-A	As Required	AR
Air-to-ground	A-G	Assemble	ASSEM
Alarm	ALM	Assembly	ASSY
Alignment	ALIGN.	Astronomical Time Switch	ATS
Allowance	ALLOW.	Atmosphere	ATM
Alternating Current	AC	Atomic	AT

Word	Abbr.
Attenuator	ATTEN
Audio Frequency	AF
Automatic Direction Finder	ADF
Automatic Mixture Control	AMC
Automatic Phase Control	APC
Automatic Volume Control	AVC
Automotive	AUTOM
Auxiliary	AUX
Auxiliary Power Unit	APU
Auxiliary Switch (breaker) Normally Closed	ASC
Auxiliary Switch (breaker) Normally Open	ASO
Average	AVG
Aviation	AVI
Aviation Gas Turbine	AGT
Axial Flow	AX FL
Azimuth	AZ
Babbitt	BAB
Back Pressure	BP
Back to Back	B to B
Bacteriological	BACT
Balanced Voltage	BV
Ball Bearing	BB
Barometer	BAR
Barrel	BBL
Baume	BE
Bearing	BRG
Bell and Bell	B&B
Bell and Flange	B&F
Bell and Spigot	B&S
Bell Crank	BELCRK
Bench Mark	BM
Bending Moment	M
Between Centers	BC
Between Perpendiculars	BP
Bill of Material	B/M
Billion Electron Volts	BEV
Biochemical Oxygen Demand	BOD
Birmingham Wire Gage	BWG
Blower	BLO
Board Foot	FBM
Boiler Feed Pump	BFP
Boiler Feed Water	BFW
Boiler Horsepower	BHP
Bolt Circle	BC
Both Sides	BS
Bottoming	BOTMG
Boundary	BDY
Bracket	BRKT
Brake Horsepower	BHP
Brake Mean Effective Pressure	BMEP
Brazing	BRZG
Bridge	BRDG
Brinell Hardness	BH
Brinell Hardness Number	BHN
British Thermal Units	BTU
Broach	BRO

Word	Abbr.
Bronze	BRZ
Brown & Sharp	B&S
(Wire Gage, same as AWG)	
Bulkhead	BHD
Bureau of Standards	BU STD
Burnish	BNH
Bushing	BUSH.
Bushing Current Transformer	BCT
Buttock Line	BL
Cadmium Plate	CD PL
Calculate	CALC
Calibrate	CAL
Caliper	CLPR
Calked Joint	CAJ
Calking	CLKG
Calorie	CAL
Cap Screw	CAP. SCR
Capacitor	CAP.
Capitance	C
Carload	CL
Case Harden	CH
Casing	CSG
Cast Iron	CI
Cast Steel	CS
Casting	CSTG
Castle Nut	CAS NUT
Cathode-ray	CR
Cathode-ray Oscilloscope or Oscillograph	CRO
Cathode-ray Tube	CRT
Center	CTR
Center Line	₵ or CL
Center of Gravity	CG
Center of Pressure	CP
Center to Center	C to C
Centigrade	C
Centimeter	CM
Centimeter-Gram-Second System	CGS
Centimeters per Second	CMPS
Centipoises	CP
Centrifugal	CENT.
Centrifugal Force	CF
Ceramic	CER
Chamfer	CHAM
Chemically Pure	CP
Chromium Plate	CR PL
Chrome Vanadium	CR VAN
Cinematographic	CINE
Circle	CIR
Circuit	CKT
Circuit Breaker	CB
Circular Mil	CM
Circular Mils, Thousands	MCM
Circumference	CIRC
Clear	CLR
Clearance	CL
Clevis	CLV

Word	Abbr.
Clockwise	CW
Coaxial	COAX
Cockpit	CKPT
Coefficient	COEF
Coils per Slot	CPS
Cold Drawn Steel	CDS
Column	COL
Combustion	COMB
Commercial	COML
Communication	COMM
Commutator	COMM
Compressor	COMPR
Concentric	CONC
Condensate	CNDS
Condenser	COND
Conductor	COND
Conduit	CND
Cone Point	CP
Constant	CONST
Constant Current Transformer	CCT
Construction	CONST
Contact-making Voltmeter	CMVM
Contact Potential Difference	CPD
Control Switch	CS
Coolant	COOL.
Corrosion Resistant	CRE
Counter Clockwise	CCW
Counter Electromotive Force	CEMF
Counterbore	CBORE
Counter-radar Measures	CRM
Countersink	CSK
Cowling	COWL.
Cubic Feet per Minute	CFM
Cubic Foot	CU FT
Cubic Inch	CU IN.
Cubic Meter	CU M
Cubic Micron	CU MU
Cubic Yard	CU YD
Current Directional Relay	CDR
Current Transformer	CT
Current-limiting Resistor	CLR
Cycles per Minute	CPM
Cycles per Second	CPS
Cylinder	CYL
Damage Control	DC
Dash Pot	DP
Datum	DAT
Decibel	DB
Decontamination	DECONTN
Dedendum	DED
Deep Drawn	DD
Deflect	DEFL
Demand Meter	DM
Demodulator	DEM
Density	D
Design	DSGN

Word	Abbr.
Detail	DET
Diagonal	DIAG
Diagram	DIAG
Diameter	DIA
Diametrical Pitch	DP
Differential Time Relay	DIFF TR
Dimension	DIM.
Diode	DIO
Direct Current	DC
Direction Finder	DF
Displacement	DISPL
Distance	DIST
Double Extra Strong	XXSTR
Double Pole Both Connected	DPBC
Double Pole, Single Throw	DPST
Dowel	DWL
Drafting Room Manual	DRM
Drawbar Horsepower	DBHP
Drill	DR
Drop Forge	DF
Dynamic	DYN
Dynamotor	DYNM
Eccentric	ECC
Effective	EFF
Effective Horsepower	EHP
Efficiency	EFF
Ejector	EJECT.
Electric	ELEC
Electric Horsepower	EHP
Electromotive Force	EMF
Elevation	EL
Elevator	ELEV
Elongation	ELONG
Elongation in 2 Inches	EL2
Emergency	EMER
Empennage	EMP
Engineer	ENGR
Engineering	ENGRG
Engineering Change Order	ECO
Engineering Order	EO
Engineering Work Order	EWO
Equation	EQ
Equivalent	EQUIV
Estimate	EST
Exhaust	EXH
Exhaust Gas Temperature	EGT
Experiment	EXP
Explosive	XPL
Extension	EXT
Extra	EXT
Extra Heavy	X HVY
Extra Strong	X STR
Fabricate	FAB
Facsimile	FAX
Fahrenheit	F

Word	Abbr.
Fast Operating (Relay)	FO
Fast Release (Relay)	FR
Feet Board Measure	FBM
Feet per Minute	FPM
Feet per Second	FPS
Field Accelerator	FAC
Field Decelerator	FDE
Field Forcing (Decreasing)	FFD
Field Forcing (Increasing)	FFI
Figure	FIG.
Filament	FIL
Fillet	FIL
Fillister	FIL
Fillister Head	FILH
Filter	FLT
Finish	FIN.
Finish All Over	FAO
Fireproof	FPRF
Fitting	FTG
Flange	FLG
Flat Fillister Head	FFILH
Flat Head	FH
Flat Oval	FO
Fluorescent	FLUOR
Focus	FOC
Foot Candle	FC
Foot Pounds	FT LB
Force	F
Forged Steel	FST
Forging	FORG
Freeboard	FREEBD
Frequency	FREQ
Frequency, Extremely High	EHF
Frequency, High	HF
Frequency, Low	LF
Frequency, Medium	MF
Frequency Meter	FRM
Frequency Modulation	FM
Fuselage	FUS
Gage or Gauge	GA
Gallons per Minute	GPM
Gallons per Second	GPS
Galvanize	GALV
Gas	G
Gasket	GSKT
Gasoline	GASO
Glass Block	GLB
Glaze	GL
Government	GOVT
Gram	G
Gram-calorie	G-CAL
Graphic	GRAPH.
Gravity	G
Grommet	GROM
Ground-controlled Approach	GCA
Ground-position Indicator	GPI

Word	Abbr.
Ground-to-ground	G-G
Gyroscope	GYRO
Half Dog Point	½DP
Hanger	HGR
Hard Chromium	HD CR
Hardware	HDW
Head	HD
Headless	HDLS
Heat Resisting	HR
Heat Treat	HT TR
Heavy	HVY
Hexagon	HEX
Hexagonal Head	HXH
Hexagonal Socket	HXSOC
High Frequency	HF
High-speed	HS
High-speed Steel	HSS
High Tension	HT
High Voltage	HV
Highway	HWY
Horizontal	HOR
Horizontal Center Line	HCL
Horizontal Reference Line	HRL
Hundredweight	CWT
Hydraulic	HYD
Hydrostatic	HYDRO
Identify	IDENT
Illustrate	ILLUS
Impact	IMP
Inboard	INBD
Indicated Horsepower Hour	IHPH
Inductance or Induction	IND
Inductance-capacitance	LC
Inductance-capacitance Resistance	LCR
Inductance Coil	L
Inside Diameter	ID
Instrument	INST
Instrument Landing System	ILS
Interchangeable	INTCHG
Intercommunication	INTERCOM
Intercooler	INCLR
Iron Pipe	IP
Isometric	ISO
Job Order	JO
Joint	JT
Joule	J
Junction	JCT
Junction Box	JB
Kelvin	K
Keyseat	KST
Kilo	K
Kilocycle	KC
Kilocycles per Second	KCPS
Kilogram	KG

Word	Abbr.
Kilograms per Second	KGPS
Kilohm	K
Kiloliter	KL
Kilovolt-ampere	KVA
Kilowatt Hour	KWH
Kip (1000 lb)	K
Kips per Square Inch	KSI
Landing Gear	LG
Lateral	LAT
Length	LG
Length Over All	LOA
Linear	LIN
Liquid	LIQ
Liter	L
Logarithm	LOG.
Longeron	LONGN
Longitude	LONG.
Longitudinal Expansion Joint	LEJ
Lubricate	LUB
Machine Screw	MS
Machine Steel	MS
Magnet	MAG
Male & Female	M&F
Manual	MAN.
Manufacture	MFR
Manufactured	MFD
Manufacturing	MFG
Material	MATL
Material List	ML
Mechanical	MECH
Membrane	MEMB
Metal	MET.
Micro	μ or U
Microampere	μA or UA
Microangstrom	μA
Microfarad	μF or UF
Micrometer	MIC
Micro-micro	μ-μ or U-U
Micron	μ or U
Microphone	MIKE
Milliampere	MA
Million Gallons per Day	MGD
Milliwatt	MW
Minimum	MIN
Miscellaneous	MISC
Molecular Weight	MOL WT
Nacelle	NAC
National	NATL
National Aircraft Standards	NAS
National Electrical Code	NEC
No Good	NG
Nomenclature	NOM
Noon	M
Normalize	NORM
Not to Scale	NTS

Word	Abbr.
Oblique	OBL
Observe	OBS
Obsolete	OBS
Ohm	Ω
Ohmmeter	OHM
Oil Circuit Breaker	OCB
Oil Ring	OR.
Oil Seal	OSL
On Center	OC
Open-close-open	OCO
Operate	OPR
Optical	OPT
Orifice	ORF
Outboard	OUTBD
Outside Radius	OR
Out to Out	O to O
Oval Head	OVH
Oval Point	OVP
Overhead	OVHD
Overload	OVLD
Oxidized	OXD
Painted	PTD
Panel	PNL
Pantograph	PANT.
Pantry	PAN.
Parabola	PRB
Paraboloid	PRBD
Parallel	PAR.
Patent	PAT.
Penny (Nails, etc)	d
Pennyweight	dWT
Perpendicular	PERP
Perspective	PERS
Phase	PH
Phase Meter	PHM
Photograph	PHOTO
Piece	PC
Pipe Tap	PT
Pitch	P
Pitch Diameter	PD
Plain Washer	PW
Plastic	PLSTC
Plate	PL
Plotting	PLOT.
Pneumatic	PNEU
Point of Curve	PC
Point of Tangent	PT
Polar	POL
Polyphase	PYPH
Potential	POT.
Potentiometer	POT.
Pound	LB
Pounder	PDR
Pounds per Square Foot	PSF
Power	PWR
Power Circuit Breaker	PCB
Power Factor	PF

Word	Abbr.
Preheater	PHR
Printed Circuit	PCKT
Process	PROC
Procurement	PROC
Production	PROD
Profile	PF
Project	PROJ
Propeller	PROP
Pulley	PUL
Pulse-frequency	PF
Pulse-position Modulation	PPM
Pulses per Second	PPS
Punch	PCH
Purchase	PUR
Push-pull	P-P
Pyrometer	PYR
Quadrangle	QUAD
Quality	QUAL
Quantity	QTY
Quick-opening Device	QOD
Rabbet	RAB
Radar	RDR
Radar Counter Measure	RCM
Radial	RAD
Radius	R
Reactor	REAC
Ream	RM
Rankine	R
Recommend	RECM
Recovery	RECY
Reduce	RED.
Reference	RF
Reference Line	REF L
Reinforce	REINF
Relative Humidity	RH
Relief Valve	RV
Remote Control	RC
Remove	REM
Request	REQ
Required	REQD
Requisition	REQ
Residual	RESID
Resistance	RES
Revolutions per Minute	RPM
Right of Way	R/W
Rivet	RIV
Rocket	RKT
Rocket Launcher	RL
Rockwell Hardness	RH
Roentgen	R
Roller Bearing	RB
Root Diameter	RD
Root Mean Square	RMS
Root Sum Square	RSS
Rough	RGH
Round	RD

Word	Abbr.
Rubber	RUB.
Runout	RO
Safe Working Pressure	SWP
Sand Blast	SD BL
Saturate	SAT.
Saybolt Seconds Furol (Oil Viscosity)	SSF
Saybolt Seconds Universal (Oil Viscosity)	SSU
Schedule	SCH
Schematic	SCHEM
Scleroscope Hardness	SH
Screw	SCR
Sea Level	SL
Section	SECT
Selsyn	SELS
Semi-finished	SF
Set Screw	SS
Shaft	SFT
Sheathing	SHTHG
Shield	SHLD
Shop Order	SO
Shot Blast	SH BL
Shoulder	SHLD
Signal	SIG
Signal-to-noise Ratio	SNR
Silver Solder	SILS
Simplex	SX
Sink	SK
Sketch	SK
Sleeve	SLV
Sleeve Bearing	SB
Sliding Expansion Joint	SEJ
Slope	S
Socket	SOC
Solder	SLD
Sound	SND
Space	SP
Space Heater	SPH
Speaker	SPKR
Specific Gravity	SP GR
Specific Heat	SP HT
Specification	SPEC
Speedometer	SPEEDO
Spherical	SPHER
Spot Face	SF
Spot Face Other Side	SF-O
Spot Weld	SW
Spring	SPG
Square	SQ
Square Head	SQH
Stabilize	STAB
Stainless	STN
Stanchion	STAN
Starboard	STBD
Static Pressure	SP
Steam	ST
Steel	STL
Stock	STK

Word	Abbr.
Strength	STR
Structural	STR
Structural Carbon Steel Hard	SCSH
Substation	SUBSTA
Substitute	SUB
Substructure	SUBSTR
Supercharge	S-CHG
Superheater	SUPHTR
Supersede	SUPSD
Switch and Relay Types	
Single Pole Switch	SP SW
Single Pole Single Throw Switch	SPST SW
Single Pole Double Throw Switch	SPDT SW
Double Pole Switch	DP SW
Double Pole Single Throw Switch	DPST SW
Double Pole Double Throw Switch	DPDT SW
Triple Pole Switch	3P SW
Triple Pole Single Throw Switch	3PST SW
Triple Pole Double Throw Switch	3PDT SW
4 Pole Switch	4P SW
4 Pole Single Throw Switch	4PST SW
4 Pole Double Throw Switch	4PDT SW
etc	
Switchboard	SWBD
Switchgear	SWGR
Symmetrical	SYM
System	SYS
Tabulate	TAB.
Tachometer	TACH
Tail Landing Gear	TLG
Tank	TK
Taper	TPR
Technical	TECH
Telemeter	TLM
Temperature	TEMP
Tensile Strength	TS
Terminal	TERM.
Thermocouple	TC
Thermostat	THERMO
Thousand Pound	KIP
Tinned	TD
Tolerance	TOL
Tongue & Groove	T&G
Torque	TOR

Word	Abbr.
Total Indicator Reading	TIR
Transistor	TSTR
Transmission	XMSN
Transmitter	XMTR
Transportation	TRANS
Transverse	TRANSV
Treatment	TREAT.
Tubing	TUB.
Turbine	TURB
Turbine Drive	TD
Turbine Generator	TURBO GEN
Ultra-high Frequency	UHF
Unfinished	UNFIN
United States Standard	USS
Unless Otherwise Specified	UOS
Vacuum	VAC
Vacuum Tube	VT
Valve	V
Valve Box	VB
Velocity	V
Vertical	VERT
Vertical Center Line	VCL
Vertical Reference Line	VRL
Very-high Frequency	VHF
Viscosity	VISC
Volt	V
Voltammeter	VAM
Volume	VOL
Washer	WASH.
Water Cooled	WCLD
Watt	W
Watthour	WHR
Wavelength	WL
Withdrawn	W/D
Without	W/O
Without Equipment and Spare Parts	W/O E&SP
Woodruff	WDF
Working Point	WP
Working Pressure	WP
Wrought	WRT
Wrought Brass	W BRS
Wrought Iron	WI

ABBREVIATIONS FOR CHEMICAL SYMBOLS

Word	Abbr.	Word	Abbr.
Actinium	AC	Masurium	MA
Aluminum	AL	Mercury (hydrargyrum)	HG
Antimony (**stibium**)	SB	Molybdenum	MO
Argon	A	Neodymium	ND
Arsenic	AS	Neon	NE
Barium	BA	Nickel	NI
Beryllium (glucinum)	BE	Nitrogen	N
Bismuth	BI	Osmium	OS
Boron	B	Oxygen	O
Bromine	BR	Palladium	PD
Cadmium	CD	Phosphorus	P
Caesium	CS	Platinum	PT
Calcium	CA	Polonium	PO
Carbon	C	Potassium (kalium)	K
Cerium	CE	Praseodymium	PR
Chlorine	CL	Proactinium	PA
Chromium	CR	Radium	RA
Cobalt	CO	Radon (niton)	RN
Columbium (niobium)	CB	Rhenium	RE
Copper	CU	Rhodium	RH
Dysprosium	DY	Rubidium	RB
Erbium	ER	Ruthenium	RU
Europium	EU	Samarium	SM
Fluorine	F	Scandium	SC
Gadolinium	GD	Selenium	SE
Gallium	GA	Silicon	SI
Germanium	GE	Silver (**argentum**)	AG
Gold (aurum)	AU	Sodium (natrium)	NA
Hafnium	HF	Strontium	SR
Helium	HE	Sulfur	S
Holmium	HO	Tantalum	TA
Hydrogen	H	Tellurium	TE
Illinium	IL	Terbium	TB
Indium	IN	Thallium	TL
Iodine	I	Thorium	TH
Ionium	IO	Thulium	TM
Iridium	IR	Tin (**stannum**)	SN
Iron (ferrum)	FE	Titanium	TI
Krypton	KR	Tungsten (wolfranium)	W
Lanthanum	LA	Uranium	U
Lead (plumbum)	PB	Vanadium	V
Lithium	LI	Xenon	XE
Lutecium	LU	Ytterbium	YB
Magnesium	MG	Yttrium	YT
Manganese	MN	Zinc	ZN
		Zirconium	ZR

ABBREVIATIONS FOR ENGINEERING SOCIETIES

American Association of Engineers	AAE
American Boiler Manufacturers' Association & Affiliated Industries	ABMA
American Bureau of Shipping	ABS
Air Conditioning & Refrigerating Machinery Association	ACRMA
American Chemical Society	ACS
American Concrete Institute	ACI
American Electrochemical Society	AES
American Electroplaters Society	AES
American Engineering Council	AEC
American Foundrymen's Association	AFA
American Gas Association	AGA
American Gear Manufacturers' Association	AGMA
American Institute of Architects	AIA
American Institute of Chemical Engineers	AIChE
American Institute of Electrical Engineers	AIEE
American Institute of Mining & Metallurgical Engineers	AIMME
American Institute of Steel Construction	AISC
American Iron & Steel Institute	AISI
American Petroleum Institute	API
American Railway Engineering Association	AREA
American Railway Bridge & Building Association	ARBBA
American Society of Aeronautical Engineers	ASAE
American Society of Body Engineers	ASBE
American Society of Civil Engineers	ASCE
American Society of Engineers and Architects	ASEA
American Society of Heating & Ventilating Engineers	ASHVE
American Society of Lubricating Engineers	ASLE
American Society of Mechanical Engineers	ASME
American Society of Metals	ASM
American Society of Refrigerating Engineers	ASRE
American Society of Safety Engineers	ASSE
American Society of Sanitary Engineering	ASSE
American Society for Steel Treating	ASST
American Society for Testing Materials	ASTM
American Society of Tool Engineers	ASTE
American Standards Association	ASA
American Steel Foundrymen's Association	ASFA
American Transit Association	ATA
American Water Works Association	AWWA
American Welding Society	AWS
American Wood Preservers' Association	AWPA

Anti-friction Bearing Manufacturers' Association	AFBMA
Association of American Railroads	AAR
Association of American Steel Manufacturers	AASM
Association of Iron & Steel Engineers	AISE
Automobile Manufacturers' Association	AMA
Canadian Lumbermen's Association	CLA
Canadian Standards Association	CSA
Compressed Air Institute	CAI
Edison Electric Institute	EEI
Electrochemical Society	ES
Gas Appliances Manufacturers' Association	GAMA
Hydraulic Institute	HI
Illuminating Engineering Society	IES
Institute of Radio Engineers	IRE
Institute of Traffic Engineers	ITE
Insulated Power Cable Engineers' Association	IPCEA
Joint Electron Tube Engineering Council	JETEC
Manufacturers, Standardization Society of the Valve and Fittings Industry	MSS
National Advisory Committee for Aeronautics	NACA
National Aircraft Standards	NAS
National Bureau of Standards	NBS
National Association of Manufacturers	NAM
National Conservation Bureau	NCB
National Electrical Manufacturers' Association	NEMA
National Hardwood Lumber Association	NHLA
National Housing Agency	NHA
National Lumber Manufacturers' Association	NLMA
National Machine Tool Builders' Association	NMTBA
National Petroleum Association	NPA
National Safety Council	NSC
Oil Heat Institute of America	OHIA
Radio Manufacturers' Association	RMA
Refrigeration Equipment Manufacturers' Association	REMA
Society for the Advancement of Management	SAM
Society of Automotive Engineers	SAE
Society of Fire Engineers	SFE
Society of Industrial Engineers	SIE
Society of Military Engineers	SME
Society of Naval Architects and Marine Engineers	SNA&ME
Society of Tractor Engineers	STE
Standards Engineers' Society	SES
Underwriters' Laboratories, Inc	UL

GRAPHICAL SYMBOLS FOR PIPE FITTINGS AND VALVES

	Flanged	Screwed	Bell and Spigot	Welded	Soldered
1. Joint					
2. Elbow 90 deg					
3. Elbow—45 deg					
4. Elbow—Turned Up					
5. Elbow—Turned Down					
6. Elbow—Long Radius					
7. Side Outlet Elbow—Outlet Down					
8. Side Outlet Elbow—Outlet Up					
9. Base Elbow					
10. Double Branch Elbow					

(Courtesy ANSI)

	Flanged	Screwed	Bell and Spigot	Welded	Soldered
11. Single Sweep Tee					
12. Double Sweep Tee					
13. Reducing Elbow					
14. Tee					
15. Tee—Outlet Up					
16. Tee—Outlet Down					
17. Side Outlet Tee Outlet Up					
18. Side Outlet Tee Outlet Down					
19. **Cross**					
20. **Reducer, Concentric**					

(Courtesy ANSI)

	Flanged	Screwed	Bell and Spigot	Welded	Soldered
21. Reducer, Eccentric					
22. **Lateral**					
23. Gate Valve Elevation (See 169)					
24. Globe Valve Elevation (See 170)					
25. Angle Gate Valve Elevation (See 171)					
26. Angle Globe Valve Elevation (See 172)					
27. Check Valve					
28. Angle Check Valve					
29. Stop Cock					
30. Safety Valve					

(Courtesy ANSI)

GRAPHICAL SYMBOLS FOR ELECTRICAL DIAGRAMS

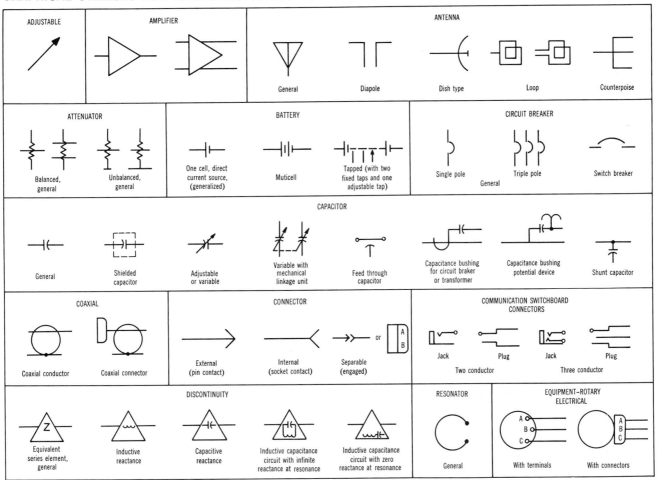

ADJUSTABLE

AMPLIFIER

ANTENNA

General

Diapole

Dish type

Loop

Counterpoise

ATTENUATOR

Balanced, general

Unbalanced, general

BATTERY

One cell, direct current source, (generalized)

Muticell

Tapped (with two fixed taps and one adjustable tap)

CIRCUIT BREAKER

Single pole

Triple pole

General

Switch breaker

CAPACITOR

General

Shielded capacitor

Adjustable or variable

Variable with mechanical linkage unit

Feed through capacitor

Capacitance bushing for circuit braker or transformer

Capacitance bushing potential device

Shunt capacitor

COAXIAL

Coaxial conductor

Coaxial connector

CONNECTOR

External (pin contact)

Internal (socket contact)

Separable (engaged)

or

A
B

COMMUNICATION SWITCHBOARD CONNECTORS

Jack

Plug

Jack

Plug

Two conductor

Three conductor

DISCONTINUITY

Equivalent series element, general

Inductive reactance

Capacitive reactance

Inductive capacitance circuit with infinite reactance at resonance

Inductive capacitance circuit with zero reactance at resonance

RESONATOR

General

EQUIPMENT–ROTARY ELECTRICAL

With terminals

With connectors

(Courtesy ANSI)

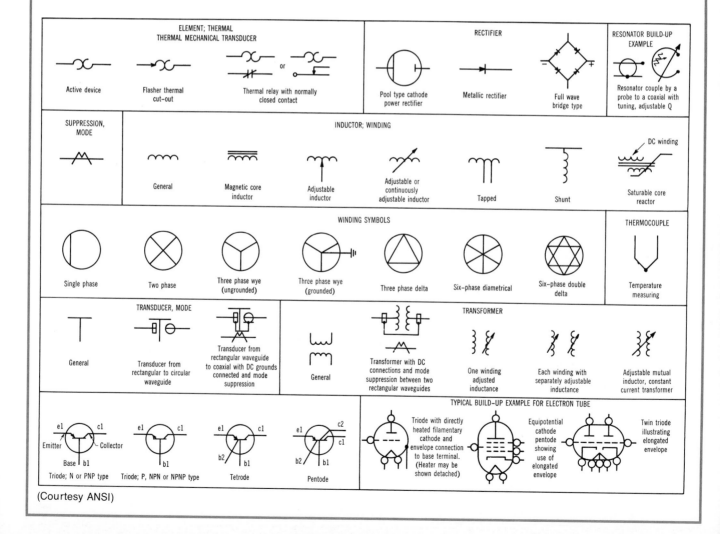

ELEMENT; THERMAL
THERMAL MECHANICAL TRANSDUCER

Active device

Flasher thermal
cut–out

Thermal relay with normally
closed contact

RECTIFIER

Pool type cathode
power rectifier

Metallic rectifier

Full wave
bridge type

RESONATOR BUILD-UP
EXAMPLE

Resonator couple by a
probe to a coaxial with
tuning, adjustable Q

SUPPRESSION,
MODE

INDUCTOR; WINDING

General

Magnetic core
inductor

Adjustable
inductor

Adjustable or
continuously
adjustable inductor

Tapped

Shunt

Saturable core
reactor

DC winding

WINDING SYMBOLS

Single phase

Two phase

Three phase wye
(ungrounded)

Three phase wye
(grounded)

Three phase delta

Six–phase diametrical

Six–phase double
delta

THERMOCOUPLE

Temperature
measuring

TRANSDUCER, MODE

General

Transducer from
rectangular to circular
waveguide

Transducer from
rectangular waveguide
to coaxial with DC grounds
connected and mode
suppression

TRANSFORMER

General

Transformer with DC
connections and mode
suppression between two
rectangular waveguides

One winding
adjusted
inductance

Each winding with
separately adjustable
inductance

Adjustable mutual
inductor, constant
current transformer

TYPICAL BUILD-UP EXAMPLE FOR ELECTRON TUBE

e1   c1
Emitter   Collector
Base   b1
Triode; N or PNP type

e1   c1
b1
Triode; P, NNP or NPNP type

e1   c1
b2   b1
Tetrode

e1   c2
c1
b2   b1
Pentode

Triode with directly
heated filamentary
cathode and
envelope connection
to base terminal.
(Heater may be
shown detached)

Equipotential
cathode
pentode
showing
use of
elongated
envelope

Twin triode
illustrating
elongated
envelope

(Courtesy ANSI)

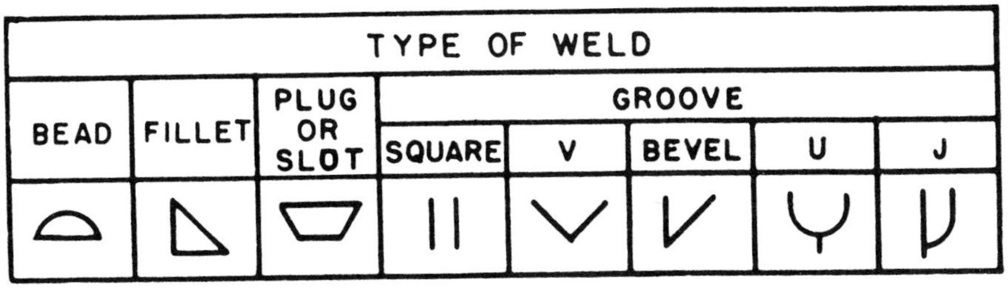

TYPE OF WELD							
BEAD	FILLET	PLUG OR SLOT	GROOVE				
			SQUARE	V	BEVEL	U	J
⌒	△	▱	‖	∨	∨	∪	∪

Figure C-1 Basic arc and gas weld symbols. (Courtesy American Welding Society)

GRAPHICAL SYMBOLS FOR WELDING

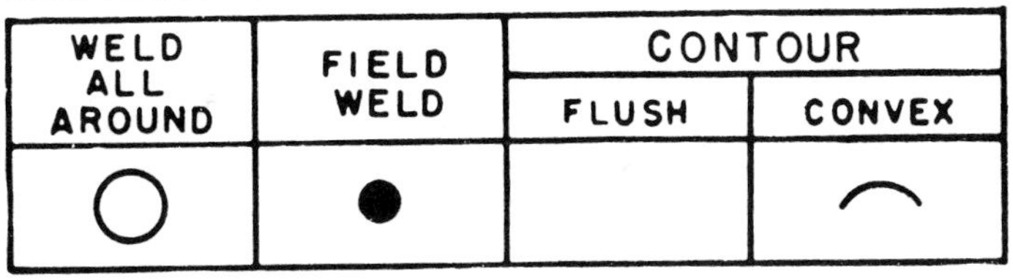

WELD ALL AROUND	FIELD WELD	CONTOUR	
		FLUSH	CONVEX
◯	●		⌒

Figure C-2 Supplementary symbols. (Courtesy American Welding Society)

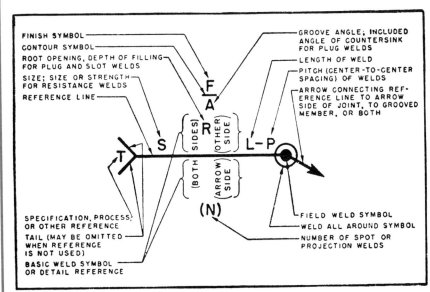

Figure C-3 Standard locations of elements of a welding symbol. (Courtesy American Welding Society)

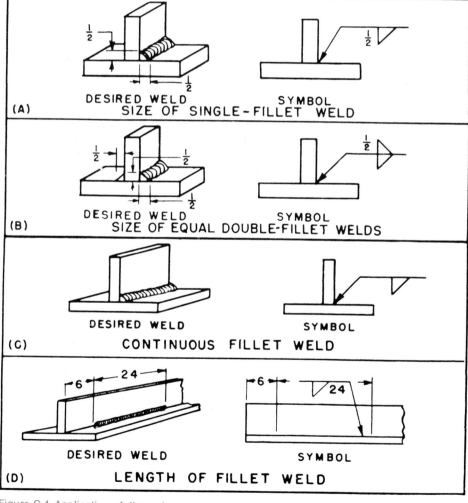

Figure C-4 Application of dimensions to
fillet welding symbols. (Courtesy
American Welding Society)

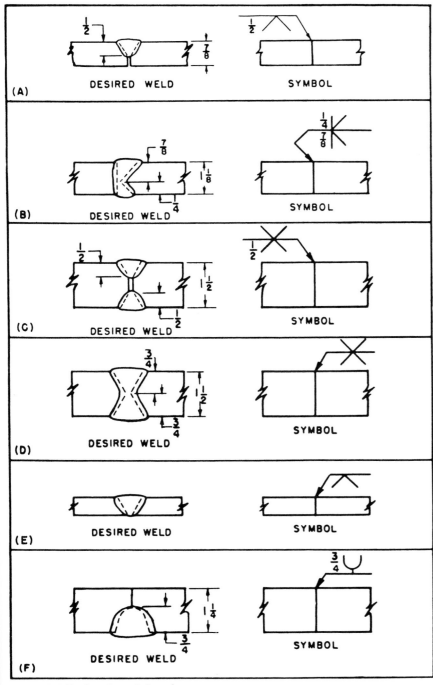

Figure C-5 Designation of size of groove welds with no specified root penetration. (Courtesy American Welding Society)

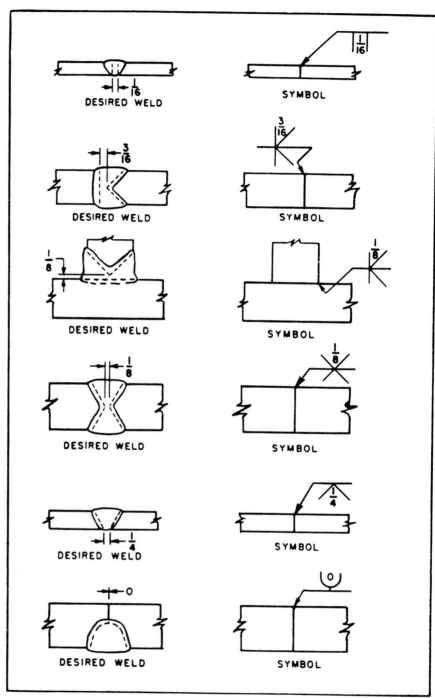

Figure C-6 Designation of root opening of groove welds. (Courtesy American Welding Society)

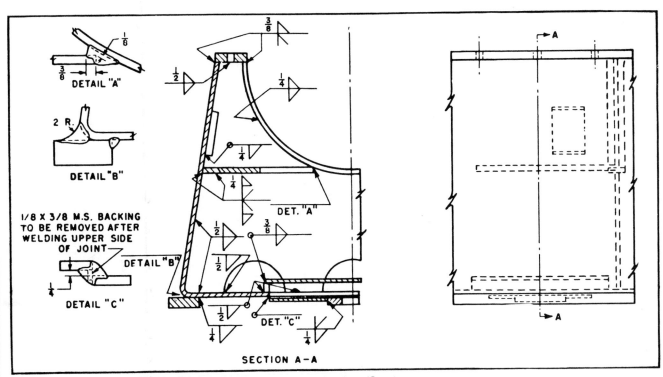

Figure C-7 Use of arc and gas welding symbols on machinery drawing. (Courtesy American Welding Society)

Figure C-8 Use of welding symbols on a structural drawing. (Courtesy American Welding Society)

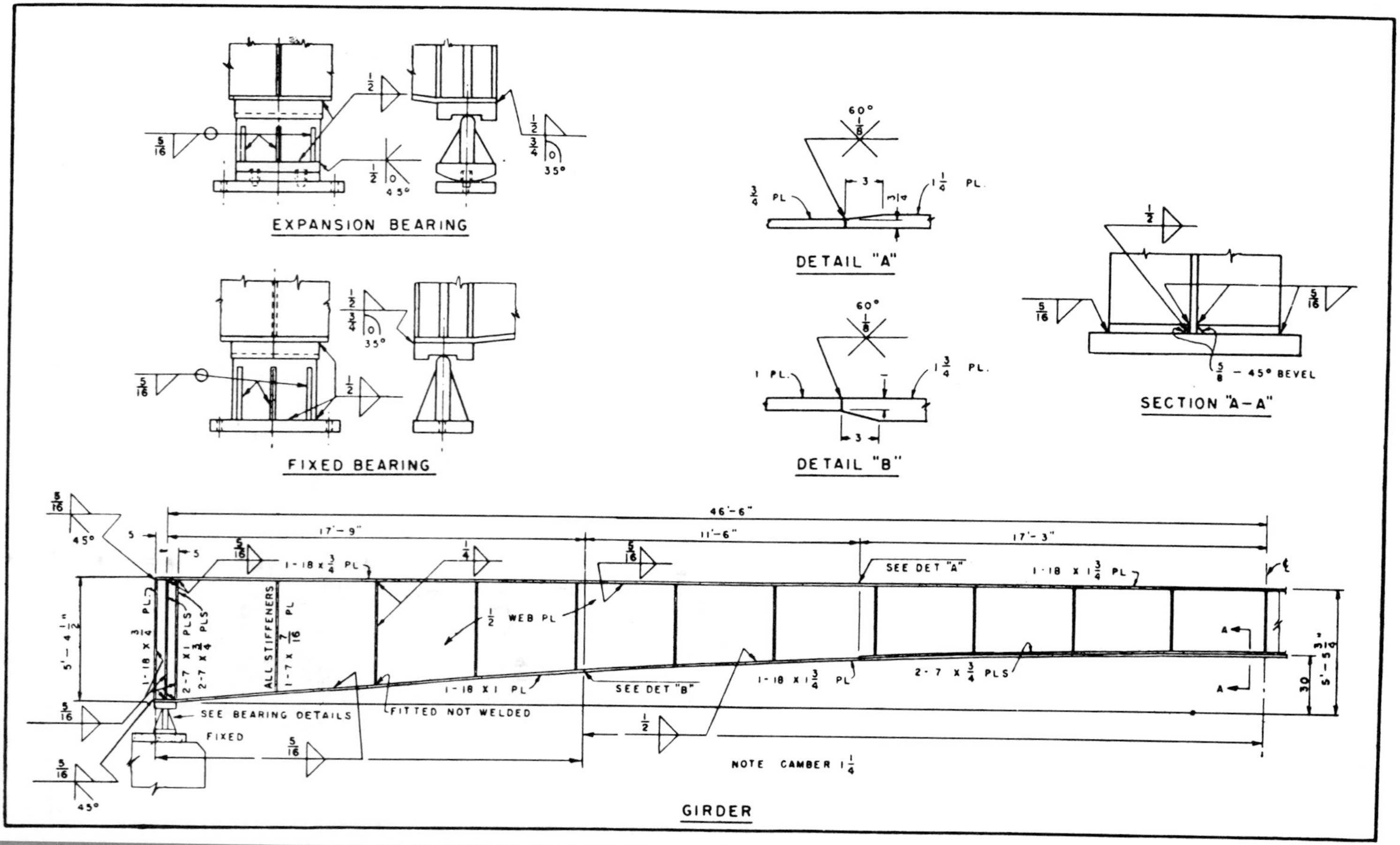

Appendix D

The ordinary vertical and horizontal reference planes divide all space into four compartments that are arbitrarily numbered from 1 to 4, as shown in Figure D-1.

If an object is placed in the first or third angles its projections on the two reference planes when superposed are on opposite sides of the ground line (the line of intersection of the H and F reference planes); therefore these views cannot overlap. If, however, the object is placed in either the second or fourth angles the superposed projections are on the same side of the ground line and may overlap.

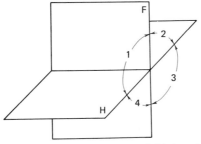

Figure D-1 Basis for first and third angle arrangement of views.

For this reason, *in engineering practice*, the object is usually placed in either the first or third angle. When the object is placed in the first angle, we say that we have first angle arrangement; when it is placed in the third angle, we have third angle arrangment. Note from Figure D-2 that the relative positions of observer, object, and picture plane for either top or front view are as follows.

First angle arrangement: observer, object, picture plane
Third angle arrangement: observer, picture plane, object

VISIBILITY OF LINES

As soon as the picture plane is other than one of the principal planes of reference, the words "first" and "third" angle have no significance. However, the question arises in every auxiliary (supplementary) view of a solid as to which lines are visible and which are not visible. Every view offers two possible ways of drawing the lines so as to present a plausible picture. It is therefore desirable to adopt some plan so that we may definitely know which of the two possible pictures is consistent with the views previously drawn.

If the drawing to start with was first angle arrangement, the sequence of auxiliary views is said to be drawn according to "first angle arrangement"; if the drawing to start with was third angle, the se-

quence of auxiliary views is said to be drawn according to "third angle arrangement." Although the auxiliary planes of projection have nothing to do with either the first or the third angles of space, the essentials that are retained in the two arrangements are the relative positions of observer, object, and picture plane.

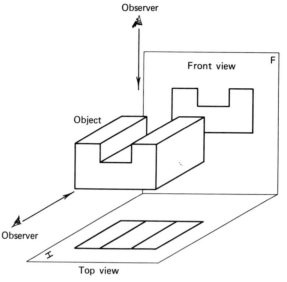

(a) First angle

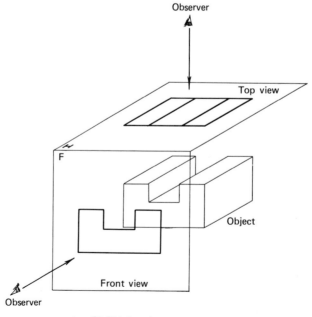

(b) Third angle

Figure D-2 First and third angle arrangement of views of the same object.

EXAMPLE

Let it be required to draw the ordinary top and front views of the object shown in Figure D-2, together with two auxiliary views in sequence according to both first and third angle arrangement.

The top view (Figure D-3) is the view on the *H* plane (appearing as line *H/F* in the front view) with the observer at *A* and looking at the *H* plane in the direction of the arrow. Note the relative position: observer object, picture plane *H*.

The front view is the view on the *F* plane (appearing as line *H/F* in the top view) with the observer at *B* and looking at plane *F* in the direction of the arrow. Note the relative position: observer, object, picture plane.

The first auxiliary view is the projection on plane 1 (appearing as line *F/1* in the front view) with the observer at *C* and looking at plane 1 in the direction of the arrow. Note the relative position: observer, object, picture plane 1. Also note that line 5-6 is visible and line 3-4 is not visible.

The second auxiliary view is the projection on plane 2 (appearing as line 1/2 in view 1) with the observer at *D* and looking at plane 2 in the direction of the arrow. Note the relative position: observer, object, picture plane 2. Also note that face 1-5-8 (which includes points 9, 13, 16, 12, and 4) is visible. After this face is drawn in, the visibility of the other lines is easily determined.

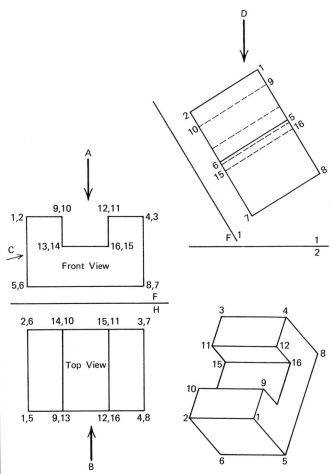

Figure D-3

Drawing the same object in third angle arrangement (Figure D-4), we observe that the top view is the view on the *H* plane (appearing as line *H/F* in the front view) with the observer at *A* and looking through plane *H* in the direction of the arrow. Note the relative position: observer, picture plane *H*, object.

The front view is the view on the *F* plane (appearing as line *F/H* in the top view) with the observer at *B* and looking at plane *F* in the direction of the arrow. Note the relative position: observer, picture plane *F*, object.

The first auxiliary view is the projection on plane 1 (appearing as line *F/1* in the front view) with the observer at *C* and looking at plane 1 in the direction of the arrow. Again, we have the relative positions of observer, picture plane 1, object. Also note that line 3-4 is visible and line 5-6 is invisible.

The second auxiliary view is the projection on plane 2 (appearing as line 1/2 in view 1) with the observer at *D* looking through the picture plane 2 at the object. Note that face 1-5-8 (which includes points 9, 13, 16, 12, and 4) is visible. There is no question concerning the visibility of line 7-8.

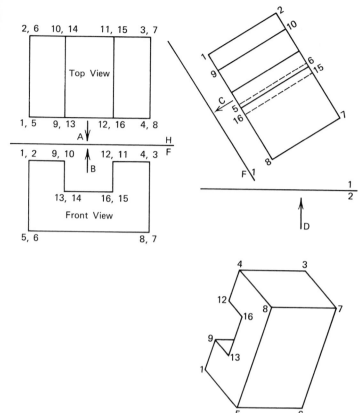

Figure D-4

Appendix E
Mathematical (Algebraic) Solutions of Space Problems

1. LENGTH OF A LINE SEGMENT (FIGURE E-1)

When the coordinates of the end points of the line segment are known or can be determined, the length of the segment can be computed.

In Figure E-1 it is clearly seen that the line segment AB can be regarded as the hypotenuse of a right triangle, the vertical side of which is equal to $(Z_B - Z_A)$ and the horizontal side equal to

$$\sqrt{(X_B - X_A)^2 + (Y_B - Y_A)^2}$$

From these values, d or

$$\overline{AB} = \sqrt{(X_B - X_A)^2 + (Y_B - Y_A)^2 + (Z_B - Z_A)^2}$$

If the coordinates of the points shown in the figure are $A(10, 6, 9)$ and $B(2, 16, 3)$, the true length of the line is

$$d = \sqrt{(2 - 10)^2 + (16 - 6)^2 + (3 - 9)^2}$$
$$= \sqrt{64 + 100 + 36}$$
$$= \sqrt{200} = 14.1 \text{ units}$$

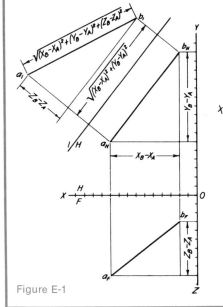

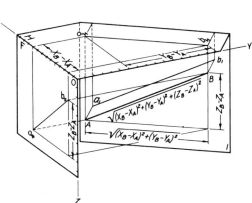

Figure E-1

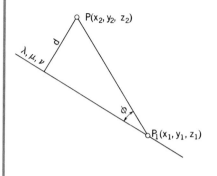

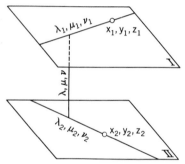

Figure E-3

2. PERPENDICULAR DISTANCE FROM A POINT TO A LINE (FIGURE E-2)

Let us assume that the given point is $P(x_2 \, y_2 \, z_2)$ and the line is given by the equations

$$\frac{x - x_1}{\lambda} = \frac{y - y_1}{\mu} = \frac{z - z_1}{v}$$

where the denominator values are the direction cosines of the line; that is, $\lambda = \cos \alpha;\ \mu = \cos \beta;\ v = \cos \gamma$ (α, β, and γ are the angles that the line makes with the X, Y, and Z axes, respectively).

From the figure it is evident that

$$
\begin{aligned}
d &= PP_1 \sin \phi = PP_1 \sqrt{1 - \cos^2 \phi} \\
&= PP_1 \sqrt{1 - \left[\lambda \cdot \frac{x_1 - x_2}{PP_1} + \mu \cdot \frac{y_1 - y_2}{PP_1} + v \cdot \frac{z_1 - z_2}{PP_1} \right]^2} \\
&= \sqrt{PP_1{}^2 - [\lambda(x_1 - x_2) + \mu(y_1 - y_2) + v(z_1 - z_2)]^2} \\
&= \sqrt{(x_1 - x_2)^2 + (y_1 - y_2)^2 + (z_1 - z_2)^2 - [\lambda(x_1 - x_2)} \\
&\qquad \overline{+ \mu(y_1 - y_2) + v(z_1 - z_2)]^2}
\end{aligned}
$$

3. DISTANCE FROM A POINT TO A PLANE

The given plane is $Ax + By + Cz + D = 0$ and the point is $P(x_1,\ y_1,\ z_1)$.

The plane through point P and parallel to the given plane is

$$Ax + By + Cz - Ax_1 - By_1 - Cz_1 = 0$$

The distance from the point to the given plane is the same as the distance between the parallel planes, or,

$$\text{Distance} = \frac{Ax_1 + By_1 + Cz_1 + D}{\sqrt{A^2 + B^2 + C^2}}$$

4. DISTANCE BETWEEN TWO SKEW LINES (FIGURE E-3)

The distance between two skew lines is the distance between the two parallel planes, each of which contains one of the lines and is perpendicular to the common perpendicular.

The equations of the two planes are

I. $(\lambda x + \mu y + v z = \lambda x_1 + \mu y_1 + v z_1)$
II. $(\lambda x + \mu y + v z = \lambda x_2 + \mu y_2 + v z_2)$

The distance between the two planes is the difference between their constant terms (the equations are in the normal form); therefore,

$$\text{Distance} = \lambda(x_1 - x_2) + \mu(y_1 - y_2) + v(z_1 - z_2)$$

In terms of the given elements, the preceding equation is

Distance =

$$\frac{(\mu_1 v_2 - \mu_2 v_1)(x_1 - x_2) + (v_1 \lambda_2 - v_2 \lambda_1)(y_1 - y_2) + (\lambda_1 \mu_2 - \lambda_2 \mu_1)(z_1 - z_2)}{[(\mu_1 v_2 - \mu_2 v_1)^2 + (v_1 \lambda_2 - v_2 \lambda_1)^2 + (\lambda_1 \mu_2 - \lambda_2 \mu_1)^2]^{1/2}}$$

which may be written in the form

$$\text{Distance} = \frac{\begin{vmatrix} x_1 - x_2 & y_1 - y_2 & z_1 - z_2 \\ \lambda_1 & \mu_1 & \nu_1 \\ \lambda_2 & \mu_2 & \nu_2 \end{vmatrix}}{[(\mu_1\nu_2 - \mu_2\nu_1)^2 + (\nu_1\lambda_2 - \nu_2\lambda_1)^2 + (\lambda_1\mu_2 - \lambda_2\mu_1)^2]^{1/2}}$$

which, in turn, may be expressed as

$$\text{Distance} = \frac{\begin{vmatrix} x_1 - x_2 & y_1 - y_2 & z_1 - z_2 \\ L_1 & M_1 & N_1 \\ L_2 & M_2 & N_2 \end{vmatrix}}{\left[\begin{vmatrix} M_1 N_1 \\ M_2 N_2 \end{vmatrix}^2 + \begin{vmatrix} N_1 L_1 \\ N_2 L_2 \end{vmatrix}^2 + \begin{vmatrix} L_1 M_1 \\ L_2 M_2 \end{vmatrix}^2 \right]^{1/2}}$$

(L_1, M_1, N_1, L_2, M_2, and N_2 are direction numbers that are proportional, respectively, to the direction cosines.)

5. LINE THROUGH A POINT AND PERPENDICULAR TO A GIVEN PLANE

It is required to determine the equations of the line that passes through point $P(X_1, Y_1, Z_1)$ and perpendicular to the plane

$$Ax + By + Cz + D = 0$$

A set of direction numbers of the required line is A, B, and C.

The equations of the line are

$$\frac{X - X_1}{A} = \frac{Y - Y_1}{B} = \frac{Z - Z_1}{C}$$

EXAMPLE

Suppose the point is $P(-8, 4, 1)$ and the plane is $(2x + y - 6z - 32 = 0)$. The equations of the line through P and perpendicular to the plane are

$$\frac{X + 8}{2} = \frac{Y - 4}{1} = \frac{Z - 1}{-6}$$

or

$$X - 2Y + 16 = 0$$
$$-6X - 2Z - 46 = 0$$

6. PLANE THROUGH A POINT AND PERPENDICULAR TO A GIVEN LINE

Let us suppose that the given point is $P(X_1, Y_1, Z_1)$ and that the direction of the given line is given by its direction cosines (λ, μ, ν) or its direction numbers (L, M, N). Now any plane that is perpendicular to the given line has an equation of the form

$$\lambda X + \mu Y + \nu Z = p \tag{1}$$

Since the plane must pass through point P, then

$$\lambda X_1 + \mu Y_1 + \nu Z_1 = p \tag{2}$$

From Equations 1 and 2 we obtain

$$\lambda X + \mu Y + \nu Z = \lambda X_1 + \mu Y_1 + \nu Z_1$$

which is the required plane.

The equation can be written as

$$\lambda(X - X_1) + \mu(Y - Y_1) + \nu(Z - Z_1) = 0$$

or

$$L(X - X_1) + M(Y - Y_1) + N(Z - Z_1) = 0$$

7. THE ANGLE BETWEEN TWO LINES (FIGURE E-4)

The angle between two nonintersecting lines is defined as the angle between their directions, or the angle between two lines parallel, respectively, to the two given lines and passing through any selected point.

Let the two given lines have direction cosines λ_1, μ_1, ν_1, and λ_2, μ_2, ν_2, respectively, and let ϕ represent the angle between them. Through the origin let us pass lines parallel to the two given lines. Let $P_1(x_1, y_1, z_1)$ be any point on one of these lines and $P_2(x_2, y_2, z_2)$ be any point on the other. Connect P_1 and P_2.

In triangle OP_1P_2, by virtue of the cosine law of trigonometry, we have

$$\overline{P_1P_2}^2 = \overline{OP_1}^2 + \overline{OP_2}^2 - 2OP_1 \cdot OP_2 \cos \phi$$

or

$$\cos \phi = \frac{\overline{OP_1}^2 + \overline{OP_2}^2 - \overline{P_1P_2}^2}{2 \cdot OP_1 \cdot OP_2} \tag{1}$$

Hence, by the distance formula,

$$\cos \phi = \frac{[x_1^2 + y_1^2 + z_1^2] + [x_2^2 + y_2^2 + z_2^2]}{2OP_1 \cdot OP_2}$$
$$- \frac{[(x_2 - x_1)^2 + (y_2 - y_1)^2 + (z_2 - z_1)^2]}{2OP_1 \cdot OP_2}$$

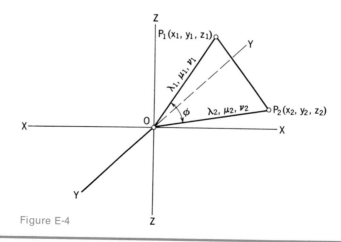

Figure E-4

$$= \frac{x_1^2 + y_1^2 + z_1^2 + x_2^2 + y_2^2 + z_2^2 - x_2^2}{2OP_1 \cdot OP_2}$$

$$+ \frac{2x_2 x_1 - x_1^2 - y_2^2 + 2y_1 y_2 - y_1^2 - z_2^2 + 2z_2 z_1 - z_1^2}{2OP_1 \cdot OP_2}$$

$$= \frac{2x_1 x_2 + 2y_1 y_2 + 2z_1 z_2}{2OP_1 \cdot OP_2} = \frac{x_1 x_2 + y_1 y_2 + z_1 z_2}{OP_1 \cdot OP_2}$$

$$= \frac{x_1}{OP_1} \cdot \frac{x_2}{OP_2} + \frac{y_1}{OP_1} \cdot \frac{y_2}{OP_2} + \frac{z_1}{OP_1} \cdot \frac{z_2}{OP_2}$$

But

$$\frac{x_1}{OP_1} = \frac{\text{Difference in the } x \text{ coordinates of points } O \text{ and } P_1}{\text{Distance } OP_1} = \lambda_1$$

And, similarly,

$$\frac{x_2}{OP_2} = \lambda_2, \frac{y_1}{OP_1} = \mu_1, \frac{y}{OP_2} = \mu_2, \text{ etc.}$$

Therefore,

$$\cos \phi = \lambda_1 \lambda_2 + \mu_1 \mu_2 + \nu_1 \nu_2$$

8. THE ANGLE BETWEEN TWO PLANES

The angle between two planes may be defined as the angle between normals to the two planes respectively.

Let the two planes be

$$A_1 x + B_1 y + C_1 z + D_1 = 0 \qquad (1)$$

and

$$A_2 x + B_2 y + C_2 z + D_2 = 0 \qquad (2)$$

Let ϕ represent the angle between them.

The direction cosines of the normal to the first plane are

$$\lambda_1 = \frac{A_1}{\sqrt{A_1^2 + B_1^2 + C_1^2}}$$

$$\mu_1 = \frac{B_1}{\sqrt{A_1^2 + B_1^2 + C_1^2}}$$

$$\nu_1 = \frac{C_1}{\sqrt{A_1^2 + B_1^2 + C_1^2}}$$

The direction cosines of the normal to the second plane are

$$\lambda_2 = \frac{A_2}{\sqrt{A_2^2 + B_2^2 + C_2^2}}$$

$$\mu_2 = \frac{B_2}{\sqrt{A_2^2 + B_2^2 + C_2^2}}$$

$$\nu_2 = \frac{C_2}{\sqrt{A_2^2 + B_2^2 + C_2^2}}$$

The formula for the angle between the planes then is

$$\cos \phi = \lambda_1 \lambda_2 + \mu_1 \mu_2 + \nu_1 \nu_2$$

$$= \frac{A_1 A_2 + B_1 B_2 + C_1 C_2}{\sqrt{A_1^2 + B_1^2 + C_1^2} \cdot \sqrt{A_2^2 + B_2^2 + C_2^2}}$$

9. THE ANGLE BETWEEN A LINE AND A PLANE

The angle between a line and a plane is *defined as the angle between the line and its projection on the plane.*

If α represents the angle between the line and its projection on the plane, and if β represents the angle between the line and a normal to the plane,

$$\alpha + \beta = 90°$$

If the given line has direction cosines λ_1, μ_1, ν_1 and if the given plane is

$$A_1 x + B_1 y + C_1 z + D_1 = 0$$

$$\cos \beta = \frac{A_1 \lambda_1 + B_1 \mu_1 + C_1 \nu_1}{\sqrt{A_1^2 + B_1^2 + C_1^2}}$$

Hence the formula for the required angle α is

$$\sin \alpha = \frac{A_1 \lambda_1 + B_1 \mu_1 + C_1 \nu_1}{\sqrt{A_1^2 + B_1^2 + C_1^2}}$$

10. THE ANGLES BETWEEN A GIVEN PLANE AND THE THREE REFERENCE PLANES

Let the given plane be

$$Ax + By + Cz + D = 0 \tag{1}$$

The equation of the horizontal plane ($z = 0$) may be written

$$0x + 0y + Kz = 0 \tag{2}$$

where K is any finite constant different from zero.

Applying the formula for the angle between the two planes,

$$\cos \phi = \frac{A_1 A_2 + B_1 B_2 + C_1 C_2}{\sqrt{A_1^2 + B_1^2 + C_1^2} \cdot \sqrt{A_2^2 + B_2^2 + C_2^2}}$$

to the two planes under consideration, we have

$$\cos H = \frac{A \cdot 0 + B \cdot 0 + C \cdot K}{\sqrt{A^2 + B^2 + C^2} \cdot \sqrt{O^2 + O^2 + K^2}}$$

$$= \frac{C}{\sqrt{A^2 + B^2 + C^2}}$$

Similarly,

$$\cos F = \frac{B}{\sqrt{A^2 + B^2 + C^2}}$$

$$\cos P = \frac{A}{\sqrt{A^2 + B^2 + C^2}}$$

The letters H, F, and P are used to designate the angles that the given plane makes with the H, F, and P planes, respectively.

11. THE ANGLES BETWEEN A GIVEN PLANE AND THE REFERENCE AXES

Let the given plane be

$$Ax + By + Cz + D = 0 \tag{1}$$

The formula for the angle between a plane and a line is

$$\sin \phi = \frac{A_1 \lambda_1 + B_1 \mu_1 + C_1 \nu_1}{\sqrt{A_1^2 + B_1^2 + C_1^2}} \tag{2}$$

The direction cosines of the x-axis are $\lambda_1 = 1$, $\mu_1 = 0$, $\nu_1 = 0$.

Letting α, β, and ζ represent the angles between the given plane and the x; y; and z-axes, respectively, we have

$$\sin \alpha = \frac{A \cdot 1 + B \cdot 0 + C \cdot 0}{\sqrt{A^2 + B^2 + C^2}}$$

$$= \frac{A}{\sqrt{A^2 + B^2 + C^2}}$$

$$\sin \beta = \frac{B}{\sqrt{A^2 + B^2 + C^2}}$$

$$\sin \zeta = \frac{C}{\sqrt{A^2 + B^2 + C^2}}$$

Comparing these formulas with those for the angles that the plane makes with the reference planes, we observe

$\sin \alpha = \cos P$
$\sin \beta = \cos F$
$\sin \zeta = \cos H$

or

$$\alpha + P = \beta + F = \zeta + H = 90°$$

In other words, if a plane makes a certain angle with a reference plane, it makes the complement of that angle with the reference axis that is the normal to the reference plane under consideration.

Appendix F

Graphical Solutions of Differential Equations

1. THE SLOPE-FIELD METHOD

Most engineering and science students are familiar with differential equations. The brief treatment presented here deals with the use of the "slope-field method" in solving ordinary differential equations of the first order.

EXAMPLE 1

Consider the differential equation

$$\frac{dy}{dx} = 2x + 2 \qquad (1)$$

It is recognized that Equation 1 may be integrated directly to give $y = x^2 + 2x + C$; however, it will be instructive to solve this equation by the slope-field method before dealing with equations that are not so easily solved by algebraic methods.

Suppose we let $x = 1$ in Equation (1); then

$$\frac{dy}{dx} = 4$$

On the line $x = 1$ a number of short lines having a slope of 4 are drawn (see Figure F-1). Additional slope lines are drawn for $x = 0$, $\frac{1}{2}$, $1\frac{1}{2}$, 2, etc. The totality of slope lines is the slope field.

Let us assume that a particular solution of the equation is the integral curve that passes through point C $(-1, 1)$. The curve is started at point C and sketched in, by eye, so that the slope of the curve at each value of x is the same as that of the slope lines drawn previously.

Other solutions may be sketched in similarly.

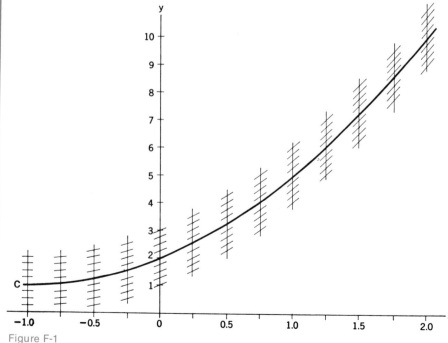

Figure F-1

EXAMPLE 2

Let us consider the differential equation

$$\frac{dy}{dx} = y - x^2 \qquad (2)$$

Again, let $x = 0, \frac{1}{2}, 1, 1\frac{1}{2}$, etc., and then let us draw corresponding slope lines, as shown in Figure F-2. For example, when $x = 0$,

$$\frac{dy}{dx} = y$$

The slopes of the short lines crossing $x = 0$ are equal in magnitude to the corresponding values of y. Observe that these slope lines pass through the point $(-1, 0)$. Once this is recognized, it becomes a very simple process to draw as many slope lines as one finds necessary. Again, when $x = 1$,

$$\frac{dy}{dx} = y - 1$$

The slope lines on $x = 1$ pass through the point $(0, 1)$.

Two solutions are shown in Figure F-2, one the integral curve passing through point $(0, 2.5)$, and the other the curve passing through point $(0, 1)$.

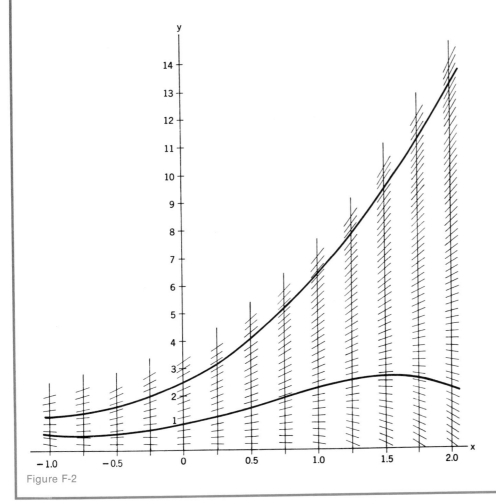

Figure F-2

EXAMPLE 3

Figure F-3 shows a small object that slides down the surface of a rough cylinder, starting from rest in the position defined by $\theta = 30°$. The free-body diagram shows the forces acting on the block for a general position θ. From the equations of motion, we obtain

$$W \sin \theta - fN = \frac{W}{g} a_t$$

and

$$W \cos \theta - N = \frac{W}{g} a_n = \frac{W}{g} \frac{v^2}{r}$$

Eliminating N between the two equations gives

$$a_t = g(\sin \theta - f \cos \theta) + \frac{fv^2}{r}$$

If the definitions $a_t = d^2s/dt^2 = rd^2\theta/dt^2$ and $v = ds/dt = r\, d\theta/dt$ are substituted, the equation becomes a second-order nonlinear equation. If, on the other hand, a_t is replaced by its equivalent, $a_t = v\, dv/ds = d(v^2)/2\, ds = d(v^2)/2r\, d\theta$, the equation becomes

$$\frac{d(v^2)}{d\theta} = 2gr(\sin \theta - f \cos \theta) + 2fv^2 \quad (3)$$

which is a linear, first-order equation in the variables v^2 and θ. The initial condition for Equation 3 is

$$v^2 = 0 \quad \text{when} \quad \theta = 30° \qquad (4)$$

The slope-field solution of the differential Equation 3, with its initial condition (Equation 4), is accomplished just as in the first two examples (see Figure F-4).

While application of the slope-field method herein illustrated is restricted to first-order equations—or to higher-order equations reducible to the first order (as in Example 3) —many practical problems fall within this group, so that the method has, in a number of instances, proved itself a useful tool in research.

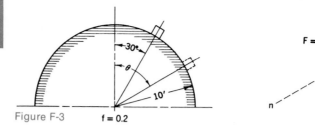

Figure F-3   f = 0.2

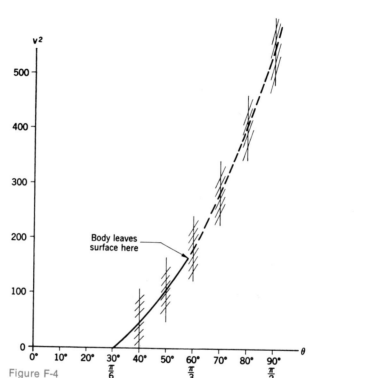

Body leaves surface here

Figure F-4

2. THE POLE-AND-RAY METHOD

In Chapter 11 we employed the pole-and-ray method for both integration and differentiation. Now let us see what the relationship is between graphical integration and differentiation. Assume (Figure F-5) that the given curve has been integrated by the pole-and-ray method. From the figure we obtain the relation

$$\frac{BC}{AC} = \frac{OQ}{OP} \tag{1}$$

from which

$$BC = \frac{AC \times OQ}{OP}$$

$$= \frac{\text{area of strip (shown shaded)}}{\text{pole distance}}$$

We also note from Equation 1 that

$$OQ = OP \times \frac{BC}{AC} = \begin{array}{l}\text{pole distance} \\ \text{times the aver-} \\ \text{age slope of the} \\ \text{integral curve} \\ \text{for interval } AC\end{array}$$

If the width AC approaches zero (A considered fixed), chord AB approaches the tangent to the integral curve at point A; and the ordinate of the given curve approaches the value of this slope, multiplied by the pole distance. *This shows that the given curve is the derived curve (curve of slopes) of the integral curve.*

When the pole distance is unity, OQ represents the value of the average slope for interval AC. Now we will use this relation to solve the following problems.

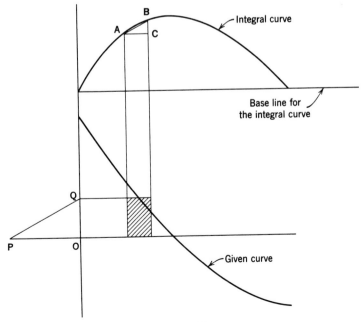

Figure F-5 Relationship between graphical integration and differentiation.

EXAMPLE 1

Let us consider the differential equation $dy/dx = 1 - X$. We wish to determine the integral curve that, when differentiated, will yield $dy/dx = 1 - X$. We will assume that X varies from 0 to 1 and that $\Delta X = 0.1$. First we plot the "curve of slopes" from the expression $dy/dx = 1 - X$. Then this curve is integrated. *The integral curve is the graphical solution of the differential equation.* Figure F-6 shows the solution for the condition that the curve passes through the point $(0, 0)$. In this elementary example, we can check the accuracy of the integral curve algebraically, since we can easily find the solution $Y = X - X^2/2 + C$. The advantages in using the pole-and-ray method, however, should be quite evident for such cases where an algebraic solution is not always feasible nor necessary.

EXAMPLE 2

Let us apply the pole-and-ray method to Equation 1, which was solved by the slope-field method.

The differential equation is

$$\frac{dy}{dx} = 2X + 2$$

We will assume that X varies from 0 to 2; that $\Delta X = 0.2$; and that the integral curve contains the point $(0, 2)$. As in the previous example, a plot of the "curve of slopes" is made. This curve is then integrated to establish the integral curve that is the graphical solution of the differential equation (see Figure F-7).

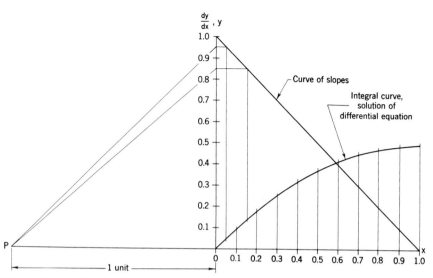

Figure F-6 Graphical solution of differential equation Example 1. Pole-and-ray method.

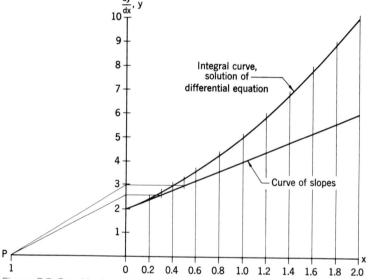

Figure F-7 Graphical solution of differential equation Example 2. Pole-and-ray method.

EXAMPLE 3

Let us consider the differential equation

$$\frac{dy}{dx} = y - x^2$$

We will assume that X varies from 0 to 2; that $\Delta X = 0.2$; and that the integral curve passes through point (0, 2). The graphical solution, a combination of "curves of slope" and "pole and ray," is shown in Figure F-8.

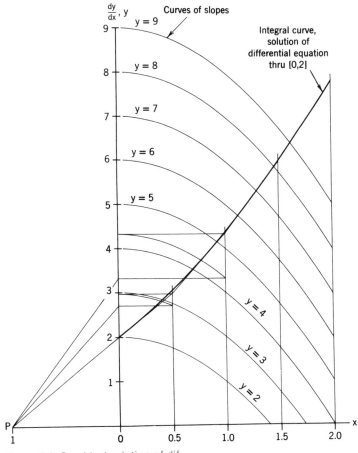

Figure F-8 Graphical solution of differential equation Example 3. Curves of slopes and pole-and-ray method.

Appendix G Useful Technical Terms and Human Factors

Allen Screws—Socket-type screws with a hexagonal socket in the head (see Figure G-1).

Figure G-1 Allen screws.

Alloy—Two or more metals mixed or combined to make a substance that is different from the pure metal; for example, copper + tin = bronze.

Alternating Current—An electric current that first increases to a maximum in one direction, decreases to zero, then increases in the opposite direction, and so on.

Ampere—Unit for measuring the rate of flow of an electric current.

Amplifier—Radio valve used to increase the amplitude of an alternating current, and so strengthen the sound.

Anneal—When glass or metal is shaped stresses are caused. Annealing reduces the stresses; for example, the glass (or metal) is heated, then cooled slowly to relieve the forces. The heat treatment may be to remove stresses; induce softness; refine the structure; or to alter toughness, electrical, magnetic, or other properties of the material.

Backlash—Looseness in a joint of a machine so that there is some free movement in one part before another part (which is joined to it) begins to move; for example, between gear teeth of mating gears.

Bascule—A bridge of which one end lifts up so that ships may pass; or both ends, thus opening in the middle.

Bearing—The part of a machine in which a revolving rod (or shaft) is held or that turns on a fixed rod (or shaft).

Bearing, Ball—One in which steel balls are used between the bearing surfaces to permit rolling action (see Figure G-2).

Figure G-2 Ball bearing.

Bearing, Roller—One in which cylindrical rolls are employed in place of

Figure G-3 Roller bearing.

steel balls. When rolls are quite small, the bearing is referred to as a *needle bearing* (see Figure G-3).

Bearing, Sleeve—(Journal, Sliding, Bushing), cylindrical shaped; sliding contact with shaft (see Figure G-4).

Figure G-4 (Courtesy Boston Gear Works)

Bell Crank—A solid lever having two fixed arms, and pivoted where they join (see Figure G-5).

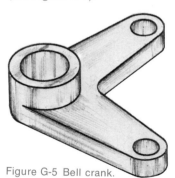

Figure G-5 Bell crank.

Block and Tackle—Usually an arrangement of ropes and pulleys to obtain a mechanical advantage (see Figure G-6).

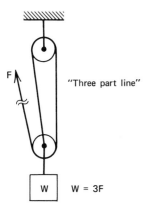

F

"Three part line"

W   W = 3F

Figure G-6

Blocks, Flanged—Housing supports for bearings. The mountings are flanged (see Figure G-7).

Flanged block
Figure G-7 (Courtesy Boston Gear Works)

Blocks, Gage—Hardened steel blocks accurately ground and finished to a specified dimension. Used in precision measurement.

Blocks, Lapping—Flat cast iron blocks with slots that crisscross the surface which, when impregnated with fine abrasives, is used to produce a very smooth (lapped) surface finish.

Blocks, Pillow—Housing supports for bearings. Cast iron pillow blocks are in common use. Pressed steel pillow blocks are also available (see Figure G-8).

Pillow block
Figure G-8 (Courtesy Boston Gear Works)

Bloom—Mass of melted metal or glass.
Blooming Mills—Machines for rolling out iron and steel into sheets.
Board Foot—Unit-measure of wood, $1 \times 12 \times 12$ in.
Boss—A circular raised portion above the surface of a part. Usually provides a bearing surface around a hole (see Figure G-9). Note the difference between a boss and a *pad* (see Figure G-10).

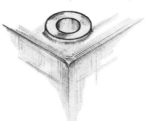

Figure G-9 Boss.

Figure G-10 Pad.

Brazing—A process for joining metals with alloys of copper and zinc. The alloys have melting temperatures quite below the melting temperatures of the metals.

Broach—A tool with a square or six-sided or special-shaped blade, smaller toward the point, pushed or pulled to make a hole a certain shape (see Figure G-11).

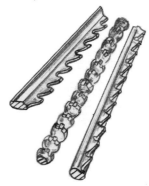

Figure G-11 Broaches.

Bulldozer—Machine like an army "tank" with a large plate in front used to push masses of earth, etc., in road-making and in clearing.

Bushing—A cylindrical lining used as a bearing for a shaft or similar parts. Also a removable liner, inserted into the guide holes of a drill jig, to position the drill (see Figure G-12).

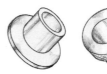

Figure G-12 Bushings.

Cam—A device used on a rotating shaft to transform rotary motion to lateral motion. There are several types of cams, such as disc (or plate), face, and cylindrical. The follower travels along the periphery of a disc cam; travels in a groove of a face cam; or in a groove cut in the outer surface of a cylindrical part to provide motion parallel to the axis of the cam (see Figure G-13).

Plate cam   Face cam

Figure G-13 Cams.

Capacitance—Measure of the power of a capacitor (electric condenser) to hold electricity. The measure (farads) equals the amount of electricity (coulombs) held divided by the voltage between the plates. Capacitance depends on the distance between the plates, the substance (air, paper, plastic) between the plates and the area of the plates.

Carburize—To cause the surface layer of steel to be combined with carbon, such as, by heating the steel in a box with carbon (charcoal) packed around it (= case hardening).

Casting—Any object made by pouring molten metal into a mold.

Cathode—A negative electrical plate. In an X-ray tube or radio valve, electrons pass from the cathode through the gas (or vacuum) to the anode (the positive plate).

Collet—A device to hold a tool or piece of stock during a machining operation (see Figure G-14).

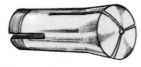

Figure G-14 Collets.

Condenser—(Electrical); also called a capacitor.

Conduction—Passing electricity along a substance (e.g., wire), passing heat along a substance (e.g., metal bar).

Conduit—A large pipe (e.g., underground pipe) to carry water or to contain electrical supply wires.

Current—Flow of water in one direction; flow of electricity along a wire. Direct current: electric flow in one direction. Alternating current: flow first in one and then in the other direction.

Counterbore—A tool to enlarge, to a specified depth, a previously drilled hole so that the bottom of the enlarged hole has a square shoulder (see Figure G-15).

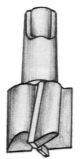

Figure G-15 Counterbore.

Countersink—A cone-shaped tool to provide a seat for a flat-head screw or rivet (see Figure G-16).

Figure G-16 Countersink.

Decibel—One tenth of a bel, a unit used to describe how many times one sound is more powerful than another.

Dielectric (substance)—A substance that prevents the flow of electricity (e.g., the air, paper).

Die Casting—A method for producing castings by forcing metal into a metallic mold or die under mechanical or pneumatic pressure.

Diode—A radio valve containing only two electrodes; cathode (negative) that is usually a hot wire, anode (positive) plate.

Elastic Limit—The largest load per unit area that will not produce a measurable permanent deformation after the load is released.

Electron—Negative electric charge that forms part of an atom.

Electronics—The study of instruments in which free electrons move (e.g., radio valves, cathode-ray tubes).

Electron-Volt—Unit for describing the energy of a moving electron.

Energy—The capacity of a body to do work.

Engineering—Engineering as a profession is concerned primarily with the design of circuits, machines, processes, and structures, or with combinations of these components into plants and systems, and the characteristic activity of full-fledged professional engineers is the prediction of performance and cost under specified conditions.

Fatigue—Change in structure of a material when subjected to repeated strain.

Fixture—A special tool used for holding work while performing operations on the work.

Flange—A rim that extends from the main part (e.g., the edge which stands out from a railway wheel so the wheel will remain on the track) (see Figure G-17).

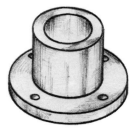

Figure G-17 Flange.

Frames—Composed of members similar to a truss, but some members may have bending forces (see Figure G-18).

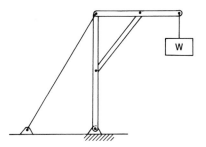

Figure G-18

Force—The action of one body on another; it changes, or tends to change, the state of rest or motion of the body acted on.

Flux—A substance used in soldering or brazing to promote the fusion of the metals.

Flux (magnetic)—The total amount of magnetic power passing through an area (e.g., through the coil of an electric generator).

Forging—Metal that is shaped or formed, while hot or cold, by a hammer, press, or drop hammer.

Gage—An instrument for determining the accuracy of specified manufactured parts.

Galvanizing—A process of coating metal parts by dipping the parts in a molten bath containing another metal (e.g., coating iron with zinc). Other methods can be used.

Gears, Spur—Used to transmit rotary motion between parallel shafts.

Gusset Plate—Metal plate fixed over two or more members meeting at an angle to strengthen the joint, (e.g., in fabricating a roof truss).

Helical, Worm—Used to transmit rotary motion between shafts perpendicular to each other but not in the same plane.

Helical, Bevel—Used to transmit rotary motion between shafts perpendicular to each other and in the same plane.

Hopper—A container with sides sloping in toward an opening in the bottom.

Hydraulic—Having to do with water in movement.

Hydraulic Machinery—*Cylinder* can convert fluid pressure into linear movement (see Figure G-19). *Motor* can convert fluid pressure into rotary motion. *Pump* can convert electrical or chemical (internal combustion) energy into pressurized hydraulic fluid. *Valve* controls directions of hydraulic fluid.

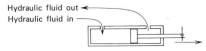

Figure G-19

Index Head—An attachment to milling machines that divides the circumference of a cylindrical part into a number of equal spaces.

Indicator, Dial—A measuring instrument with a graduated dial face on which the movement of an attached spindle that contacts the work is registered.

Inductance—When a changing current flows through a coil, pressure of electricity is caused in the coil.

Ion—An ion is an atom that lacks its full number of electrons and so is positive or has too many electrons and so is negative.

Isotopes—Atoms of different atomic weight, but chemically the same substance.

Jig—A special device used to guide the tool or to hold the material in the right position for cutting, shaping, or fitting together.

Joints, Pinned—Members are free to rotate if allowed to (see Figure G-20).

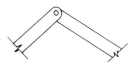

Figure G-20

Joints, Gusseted—A plate is utilized to complete the connection. Sometimes gusseted joints are considered to be pinned joints to simplify analysis (see Figure G-21).

Figure G-21

Journal—That part of a shaft that rotates in a bearing.

Knurling—Forming of fine ridges on the surface of a part to increase gripping power (see Figure G-22).

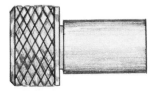

Figure G-22 Knurling.

Lathe—A machine tool with a horizontal spindle that supports and rotates a part while it is machined by a cutting tool that is supported on a cross-slide. The lathe is used primarily for turning cylindrical pieces.

Mach Number—A number that expresses the relation between speed of airflow and the speed of sound. A number less than one indicates speed less than the speed of sound.

Machinability—A term used to denote the relative ease with which parts can be machined.

Magnetron—A kind of radio valve used for producing high-frequency oscillations (swings of current backward and forward).

Microprocessor—A small computer that converts input information into instruction information. Excellent for control of processes and machines.

Microwaves—Very short radio waves (less than 20 cm).

Milling—An operation for removing metal from a piece by means of a revolving cutter that is a multiple-toothed cutter.

Milling, Straddle—An operation for milling opposite sides of one part at one time so that the surfaces will be parallel (see Figure G-23).

Figure G-23 Straddle milling.

Mold—A form into which molten metal is poured to produce a casting of a desired shape.

Moment—Tendency of a body to rotate about some axis because of the application of a force at some distance from the axis.

Nitriding—Hardening the surface of steel by passing ammonia gas over it.

Normalizing—Heating iron-base alloys above the critical temperature range and cooling in air at room temperature to reduce internal forces.

Ohm—The unit for measuring electrical resistance. A conductor has a resistance of 1 Ω when 1 A flows through it at a pressure of 1 V.

Pawl—A device used to prevent a toothed wheel from turning backward, or a device that stops, locks, or releases a mechanism (see Figure G-24).

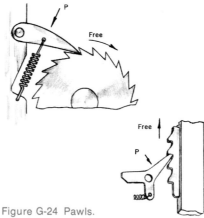

Figure G-24 Pawls.

Peen—A process to expand or stretch metal by hammering with a peen hammer or by a shot-peening method.

Pickle—A process for removing scale from castings or forgings by immersion in a water solution of sulphuric or nitric acids. Pickling conditions the surfaces for plating.

Planer—Machine tool for producing flat surfaces on metal parts.

Polish—To produce a smooth surface, usually by a polishing wheel of leather, felt, or wool to which a fine abrasive is glued.

Punch—A tool that pierces holes in metal. Also, a small hand tool used to set nails or to mark centers.

Quench—Cooling hot metal quickly by immersion in liquids or gases.

Rack—A toothed bar acting on (or acted on by) a gear wheel (see Figure G-25).

Figure G-25 Rack.

Radar—(Radio-Directing and Ranging), a way of finding the distance and direction of objects. Radio pulses are sent out and reflected back from the object. The return pulses are seen on a screen like that of television.

Reamer—A tool used to finish drilled holes. The fluted tool produces the finish by rotation through the drilled hole. Commonly used reamers are straight or tapered (see Figure G-26).

Figure G-26 Reamer.

Relief—A goove (e.g., a cut next to a shoulder) on a part to facilitate machining operations (e.g., grinding) (see Figure G-27).

Figure G-27

Resistor—Piece of wire (e.g., coil of high-resistance wire) used in an electric circuit to offer resistance to the flow of current.

Sandblasting—A process of cleaning castings by driving sand, under pressure, against the surfaces of the pieces. Also used to give a rough surface to glass.

Shaper—A machine tool that has a sliding ram; used to produce flat surfaces. The work is stationary. The cutting tool is carried on the ram, which has a reciprocating motion.

Shear—A tool for cutting metal. The cutting can be accomplished by two blades, one fixed and the other movable.

Sheave—Used in mechanical transmission with flexible rope and belts (see Figure G-28).

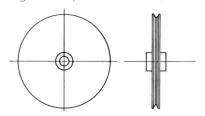

Figure G-28

Shim—A thin strip of metal (often laminated) placed between surfaces to permit adjustment for fit.

Spline—A key for a relatively long slot (see Figure G-29).

Figure G-29 Spline.

Spot-facing—A drilling operation, using a counterbore tool, to smooth the surface around a hole to provide a good bearing surface for a bolt or screw head.

Spot Weld—(Resistance Welding), the resistance of the metals to the current causes the temperature to rise quickly and the metal becomes plastic; the weld is completed by mechanical pressure from the electrodes (see Figure G-30).

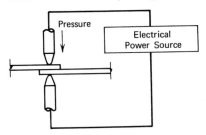

Figure G-30

Stochastic—Implies the presence of a random variable.

Strain—(Linear) The amount a member changes in length divided by the length of the member when the member is subjected to a force.

Stress—Force distributed over an area. Examples are pounds per square inch (inch-based units) and Pascals, which are Newtons per square meter (metric units).

Swaging—A method for forming a piece of cold metal by drawing, sqeezing, or hammering with a pair of dies shaped to the desired form. Swage blocks can be used to shape the part.

Tap—A tool for cutting internal threads. The tool has a thread cut on it

Figure G-31 Tap.

and is fluted to produce cutting edges and to remove chips (see Figure G-31).

Thermionics—The science dealing with the emission of electrons from hot bodies.

Toggle Mechanism—Usually a mechanical arrangement of members to produce a large force acting through a small distance from a relatively smaller force acting through a larger distance (see Figure G-32).

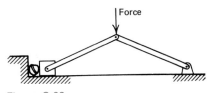

Figure G-32

Transducer—A power-transforming device for insertion between electrical, mechanical, or acoustic parts of systems of communication.

Transistor—An electronic device for rectification and/or amplification, consisting of a semiconducting material to which contact is made by three or more electrodes, which are usually metal points, or by soldered junctions.

Trusses—Composed of structural members arranged and pinned to form one or more triangles. All members are subjected to either tension or compression forces, but not to bending forces. (see Figure G-33).

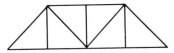

Figure G-33

Upsetting—A process for enlarging the diameter of a metal rod or the end of a tenon or stud. Rivet shanks are upset to provide material to form rivet heads in fastening parts.

Valves, Gate—Flow regulated by plate located perpendicular to direction of flow of fluid.

Valves, Check—Often a spring-loaded flapper that allows flow easily in one direction but closes when flow attempts to reverse. Examples include fluid, pumped into pressurized areas, a tire pump, and a well pump.

Welding—Method for joining two metal pieces by heating the joint to fusion temperature.

Note. For additional terms that relate to the manufacture of fasteners, see *Glossary of Terms for Mechanical Fasteners*, ASA B18.12-1962.

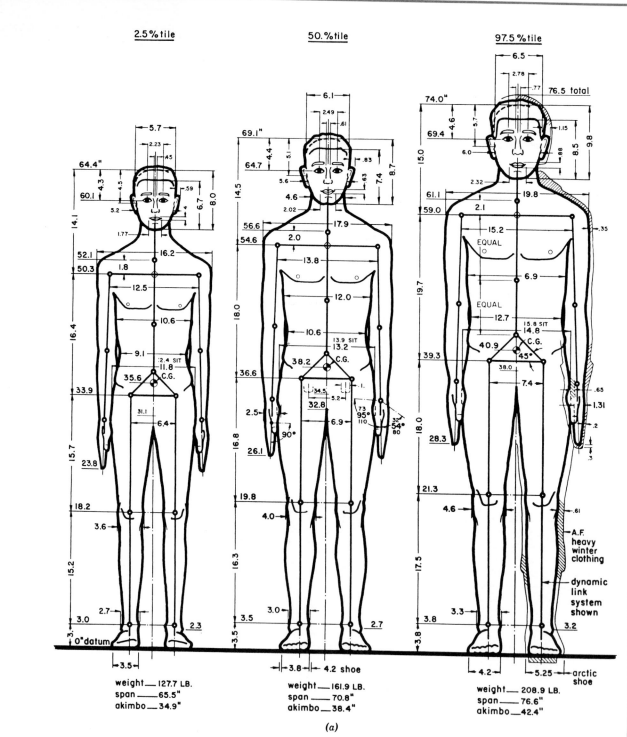

Figure G-34 Basic body dimensions. (Courtesy Henry Dreyfus and the Whitney Library of Design from *The Measure of Man,* an imprint of Watson-Guptil Publications, N.Y., 1967.) (*a*) Standing adult male.

Anthropometric Data—Deal with the measurement of the human body and its parts. The data shown in Figure G-34a, G-34b, G-34C, and G-34d accommodate 95% of the adult population in the United States. Figure G-34e shows the dimensions for children. The designer uses the data in determining, for example, ranges of motion of limbs and the spaces required to accommodate the needs of a worker.

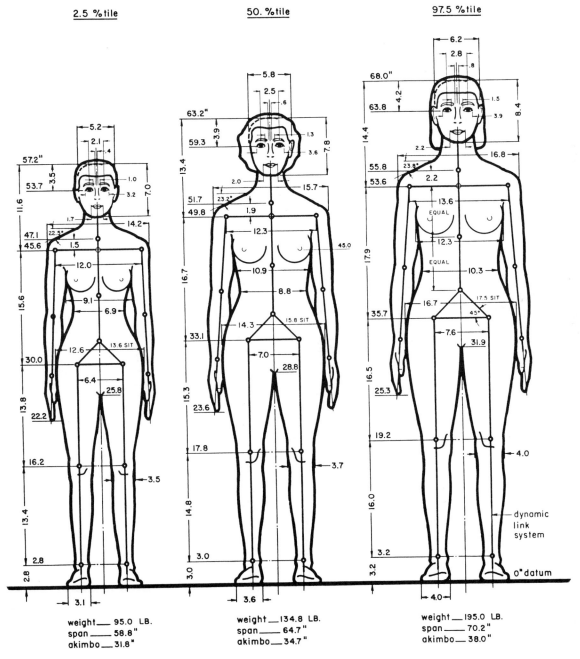

2.5 %tile

50. %tile

97.5 %tile

weight___ 95.0 LB.
span _____ 58.8"
akimbo__31.8"

weight___134.8 LB.
span_____ 64.7"
akimbo__34.7"

weight___195.0 LB.
span _____ 70.2"
akimbo__ 38.0"

Figure G-34 (b) Standing adult female.   (b)

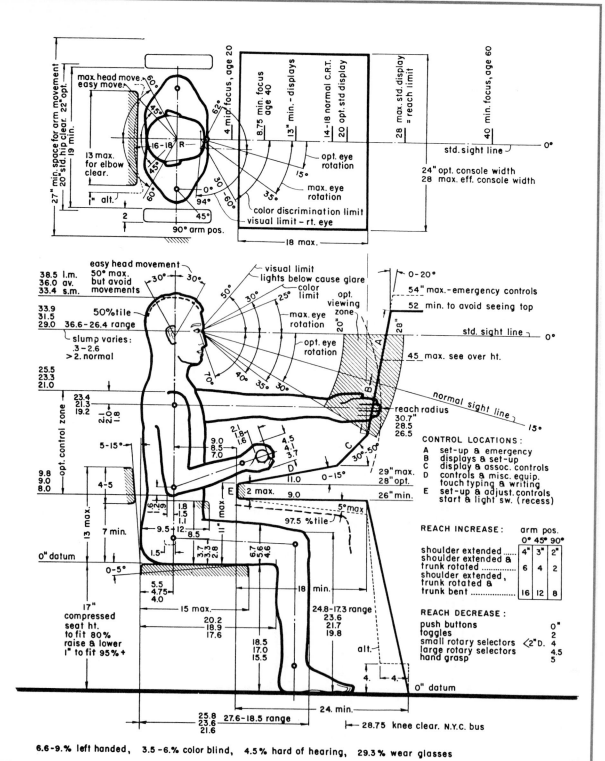

Figure G-34 (c) Adult male seated at console. (c)

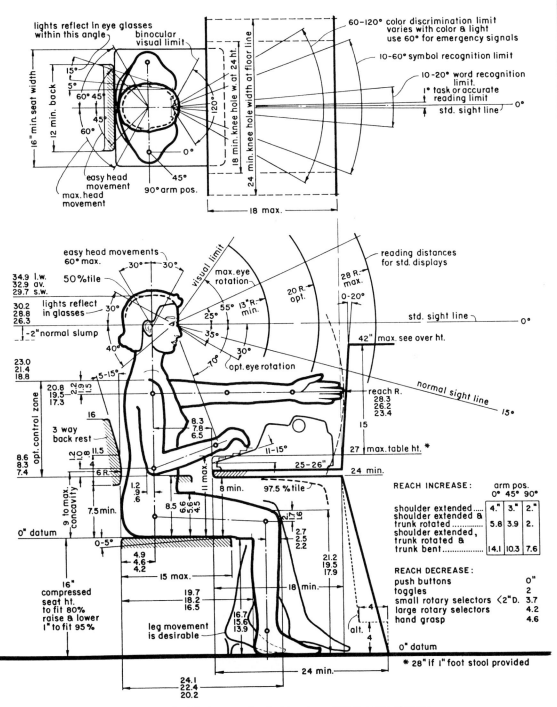

Figure G-34 (*d*) Adult female seated at console.

(*d*)

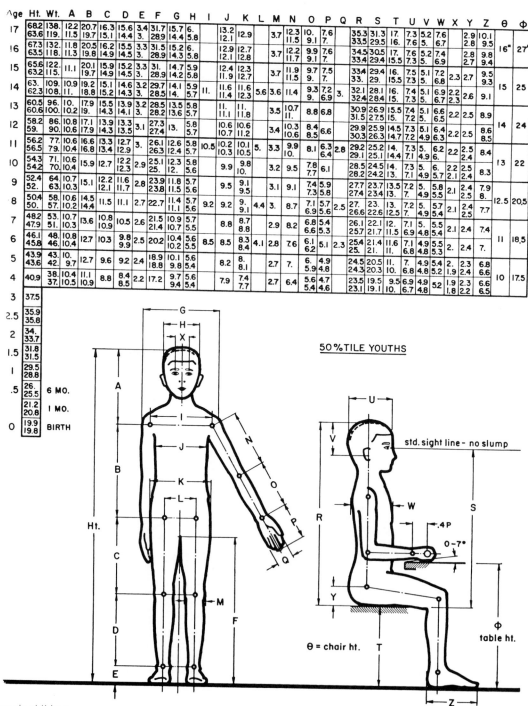

Figure G-34 (e) Male and female children. Top figure in box is data for boys, lower figure is for girls, and one figure applies to both.

(e)

Appendix H
Strength of Materials; Coefficients of Friction; Rolled Steel Structural Shapes; Beam Diagrams; and Solid Conductors

Table 1   Strength of Materials*

Since small variations in the composition of a material as well as any variation in the processing of the material may greatly affect its strength, it is difficult to give any general values that are useful. The values listed should be used with caution in any design work.

MATERIAL	ULTIMATE STRENGTH (Kips/Sq In.)			ELASTIC LIMIT (Kips/Sq In.) Tension and Compression	MODULUS OF ELASTICITY (Lb/Sq In.)
	Tension	Compression	Shear		
METALS					
Aluminum, commercial, 99% pure............	...	...	...	5	10,300,000
Cast....................	13	...	...	9	...
Rolled and annealed......	13.5	...	...	8.5	...
Hard drawn.............	30	...	...	20	...
Hard drawn wire.........	40	...	...	30	...
Aluminum casting alloys....	...	...	...	...	10,300,000
Sand casting............	19-46	...	14-33	9-36	...
Permanent mold casting..	24-48	...	14-36	9-42	...
Die cast................	29-45	...	...	13-25	...
Wrought aluminum alloys:					
Work hardened..........	13-57	...	9.5-33	5-45	...
Heat treated............	16-82	...	11-49	8-72	...
Brass:					
low, 80% Cu, 20% Zn....	44-74	...	...	14-59	15,000,000
red, 85% Cu, 15% Zn....	40-70	...	...	12-58	15,000,000
yellow, 65% Cu, 35% Zn..	47-75	...	...	15-60	15,000,000
Bronze, commercial:					
90% Cu, 10% Zn........	37-61	...	...	10-54	15,000,000
Copper 99.9% pure........	...	...	...	...	16,000,000
Annealed...............	32	...	...	10	...
Hard drawn.............	50	...	...	45	...
Iron, 99.97% pure, annealed.	40	...	...	20	29,700,000
Cold rolled.............	100	...	...	95	...
Hot rolled..............	48	...	...	30	...
Iron:					
cast....................	16-60	80-200	...	8-40	15,000,000
wrought................	45-55	25-35	35	25-35	28,000,000
Steel:					
Cold rolled.............	80-90	60	60-65	60	29,000,000
Mild alloys.............	50-100	30-60	40-80	30-60	29,000,000
High strength alloys......	100-290	...	...	...	29,000,000
Structural..............	65-80	...	...	30-40	29,000,000
NON-METALLIC MATERIALS					
Concrete (Portland cement).	...	2-5	...	...	2,500,000
Brick, common............	...	4	1	...	2,000,000
Masonry with Portland cement mortar.........	...	1	...	...	...
Masonry with lime mortar.	...	0.7	...	...	...
Brick, pressed.............	...	6	1	...	3,000,000
Masonry with Portland cement mortar.........	...	2	...	...	...
Masonry with lime mortar.	...	1.4	...	...	...
Brick, vitrified:					
Masonry with Portland cement mortar.........	...	2.8	...	...	...

* Courtesy Albert L. Hoag, Donald C. McNeese, Engineering and Technical Handbook, 1957. Reprinted by permission of Prentice Hall, Inc. Englewood Cliffs, N.J.

Table 1   (*continued*)

MATERIAL	ULTIMATE STRENGTH (Kips/Sq In.)			ELASTIC LIMIT (Kips/Sq In.) Tension and Compression	MODULUS OF ELASTICITY (Lb/Sq In.)
	Tension	Compression	Shear		
Terra cotta block..........	...	4	...	...	...
Masonry with Portland cement mortar........	...	3	...	...	...
BUILDING STONE					
Granite..................	...	15-26	1.8-2.8	...	7,500,000
Limestone...............	...	3.2-20	1.0-2.2	...	8,400,000
Marble..................	...	10-16	1.0-1.6	...	8,000,000
Sandstone...............	...	7-19	1.2-2.5	...	3,300,000
Slate...................	...	14-30	...	...	14,000,000
ROPE					
Rope, manila.............	8-9	...	...	...	...
Rope, wire, hemp core:					
Iron.....................	38	...	...	...	...
Cast steel...............	74	...	...	...	...
Ex-strong, cast steel......	86	...	...	...	...
Plow steel...............	94	...	...	...	...

Table 2 Strength Characteristics of Wood Air Dry, 12% Moisture Content, psi*

TYPE OF WOOD	Weight Lb/Cu Ft	BENDING			COMPRESSION			SHEAR
					Parallel to Grain		Perpendicular to Grain	Parallel to Grain
		Elast. Limit	Modulus of Rupture	Modulus of Elasticity	Ultimate	Elast. Limit	Elast. Limit	Ultimate
Alder...............	28	6,900	9,800	1,380,000	5,800	4,550	550	1,100
Ash, commercial white	41	8,900	14,600	1,680,000	7,300	5,600	1,500	1,900
Birch...............	44	10,100	16,700	2,000,000	8,300	6,200	1,250	2,000
Cedar:								
Alaska............	31	7,100	11,100	1,400,000	6,300	5,200	750	1,100
Eastern red........	33	3,800	8,800	880,000	6,000	...	1,150	...
Southern white.....	23	4,800	6,800	930,000	4,700	2,750	500	800
Western red.......	23	5,300	7,700	1,120,000	5,000	4,350	600	850
Cypress, southern.....	32	7,200	10,600	1,140,000	6,350	4,750	900	1,000
Douglas fir:								
Coast region........	34	8,100	11,700	1,920,000	7,400	6,450	900	1,150
Inland Empire......	31	7,400	11,300	1,610,000	6,700	5,500	950	1,200
Rocky Mtn. region...	30	6,300	9,600	1,400,000	6,050	4,650	800	1,050
Gum, red...........	34	8,100	11,900	1,490,000	5,800	4,700	850	1,600
Hemlock:								
Eastern...........	28	6,100	8,900	1,200,000	5,400	4,000	800	1,050
Western...........	29	6,800	10,100	1,490,000	6,200	5,350	700	1,150
Hickory.............	48	10,900	19,700	2,180,000	8,950	6,400	2,300	2,150
Larch, western.......	36	7,900	11,900	1,710,000	7,500	5,950	1,100	1,350
Locust, black........	48	12,800	19,400	2,050,000	10,200	6,800	2,250	2,450
Maple:								
Big leaf............	34	6,600	10,700	1,450,000	5,950	4,800	900	1,700
Black.............	40	8,300	13,300	1,620,000	6,700	4,600	1,250	1,800
Red..............	38	8,700	13,400	1,640,000	6,550	4,650	1,250	1,850
Silver.............	33	6,200	8,900	1,140,000	5,200	4,350	900	1,500
Sugar.............	44	9,500	15,800	1,830,000	7,800	5,400	1,800	2,300
Oak:								
Red..............	44	8,400	14,400	1,810,000	6,900	4,600	1,250	1,850
White.............	47	7,900	13,900	1,620,000	7,050	4,350	1,400	1,900
Pine:								
Northern white.....	25	6,000	8,800	1,280,000	4,850	3,700	550	850
Norway...........	34	9,400	12,500	1,800,000	7,350	5,350	850	1,250
Ponderosa.........	28	6,300	9,200	1,260,000	5,250	4,050	750	1,150
Southern yellow....	38	8,300	13,400	1,850,000	7,550	5,350	1,050	1,400
Sugar.............	25	5,700	8,000	1,200,000	4,750	4,150	600	1,050
Western white......	27	6,200	9,500	1,510,000	5,600	4,500	550	...
Poplar, yellow.......	28	6,100	9,200	1,500,000	5,300	3,550	600	1,100
Redwood...........	28	6,900	10,000	1,340,000	6,150	4,550	850	950
Spruce:								
Eastern...........	28	6,500	10,100	1,440,000	5,600	4,150	600	1,050
Engelmann........	23	6,000	8,500	1,160,000	4,600	3,600	650	1,000
Sitka.............	28	6,700	10,200	1,570,000	5,600	4,800	700	1,150
Walnut, black.......	38	10,500	14,600	1,680,000	7,600	5,600	1,250	1,350

* Courtesy Albert L. Hoag, Donald C. McNeese, Engineering and Technical Handbook, 1957. Reprinted by permission of Prentice Hall, Inc. Englewood Cliffs, N.J.

MATERIAL	STATIC		SLIDING	
	Dry	Lubricated	Dry	Lubricated
Aluminum on aluminum	1.05	...	1.4	...
Cast iron on asbestos-fabric brake material	...	...	0.35-0.40	...
Cast iron on brass	...	...	0.30	...
Cast iron on bronze	...	...	0.22	0.07-0.08
Cast iron on cast iron	1.10	...	0.15	0.06-0.10
Cast iron on copper	1.05	...	0.29	...
Cast iron on lead	...	...	0.43	...
Cast iron on leather	0.30-0.50	...	0.56	0.13-0.36
Cast iron on oak (parallel)	...	...	0.30-0.50	0.07-0.20
Cast iron on magnesium	...	...	0.25	...
Cast iron on steel, mild	...	0.18	0.23	0.133
Cast iron on tin	...	...	0.32	...
Cast iron on zinc	0.85	...	0.21	...
Earth on earth	0.25-1.0	...	...	...
Glass on glass	0.94	...	0.40	...
Hemp rope on wood	0.50-0.80	...	0.40-0.70	...
Nickel on nickel	1.10	...	0.53	0.12
Oak on leather (parallel)	0.50-0.60	...	0.30-0.50	...
Oak on oak (parallel)	0.62	...	0.48	0.16
Oak on oak (perpendicular)	0.54	...	0.32	0.07
Rubber tire on pavement	0.8-0.9	0.6-0.7*	0.75-0.85	0.5-0.7*
Steel on ice	0.03	...	0.01	...
Steel, hard, on babbit	0.42-0.70	0.08-0.25	0.33-0.35	0.05-0.16
Steel, hard on steel, hard	0.78	0.11-0.23	0.42	0.03-0.12
Steel, mild, on aluminum	0.61	...	0.47	...
Steel, mild on brass	0.51	...	0.44	...
Steel, mild on bronze	...	...	0.34	0.17
Steel, mild on copper	0.53	...	0.36	0.18
Steel, mild on steel, mild	0.74	...	0.57	0.09-0.19
Stone masonry on concrete	0.76	...	...	...
Stone masonry on ground	0.65	...	...	...
Wrought iron on bronze	0.19	0.07-0.08	0.18	...
Wrought iron on wrought iron	...	0.11	0.44	0.08-0.10

* Wet pavement.
† Courtesy Albert L. Hoag, Donald C. McNeese, Engineering and Technical Handbook, 1957. Reprinted by permission of Prentice Hall, Inc. Englewood Cliffs, N.J.

Table 4   Factors of Safety, Average Values*

MATERIAL	Constant Load	Varying Load	Impact Loading
Cast iron	6	10	20
Wrought iron	4	6	10
Structural steel	4	6	10
High strength steel	5	8	15
Timber	8	10	15
Brick and stone	15	25	40

* Courtesy Albert L. Hoag, Donald C. McNeese, Engineering and Technical Handbook, 1957. Reprinted by permission of Prentice Hall, Inc. Englewood Cliffs, N.J.

Table 5   Rolled Steel Structural Shapes*

AMERICAN STANDARD BEAMS
PROPERTIES FOR DESIGNING

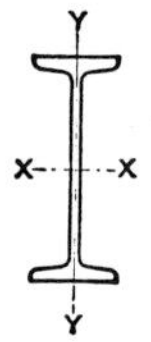

Nominal Size	Weight per Foot	Area	Depth	Flange Width	Flange Thickness	Web Thickness	Axis X-X I	Axis X-X S	Axis X-X r	Axis Y-Y I	Axis Y-Y S	Axis Y-Y r
In.	Lb.	In.²	In.	In.	In.	In.	In.⁴	In.³	In.	In.⁴	In.³	In.
24 x 7⅞	120.0	35.13	24.00	8.048	1.102	.798	3010.8	250.9	9.26	84.9	21.1	1.56
	105.9	30.98	24.00	7.875	1.102	.625	2811.5	234.3	9.53	78.9	20.0	1.60
24 x 7	100.0	29.25	24.00	7.247	.871	.747	2371.8	197.6	9.05	48.4	13.4	1.29
	90.0	26.30	24.00	7.124	.871	.624	2230.1	185.8	9.21	45.5	12.8	1.32
	79.9	23.33	24.00	7.000	.871	.500	2087.2	173.9	9.46	42.9	12.2	1.36
20 x 7	95.0	27.74	20.00	7.200	.916	.800	1599.7	160.0	7.59	50.5	14.0	1.35
	85.0	24.80	20.00	7.053	.916	.653	1501.7	150.2	7.78	47.0	13.3	1.38
20 x 6¼	75.0	21.90	20.00	6.391	.789	.641	1263.5	126.3	7.60	30.1	9.4	1.17
	65.4	19.08	20.00	6.250	.789	.500	1169.5	116.9	7.83	27.9	8.9	1.21
18 x 6	70.0	20.46	18.00	6.251	.691	.711	917.5	101.9	6.70	24.5	7.8	1.09
	54.7	15.94	18.00	6.000	.691	.460	795.5	88.4	7.07	21.2	7.1	1.15
15 x 5½	50.0	14.59	15.00	5.640	.622	.550	481.1	64.2	5.74	16.0	5.7	1.05
	42.9	12.49	15.00	5.500	.622	.410	441.8	58.9	5.95	14.6	5.3	1.08
12 x 5¼	50.0	14.57	12.00	5.477	.659	.687	301.6	50.3	4.55	16.0	5.8	1.05
	40.8	11.84	12.00	5.250	.659	.460	268.9	44.8	4.77	13.8	5.3	1.08
12 x 5	35.0	10.20	12.00	5.078	.544	.428	227.0	37.8	4.72	10.0	3.9	.99
	31.8	9.26	12.00	5.000	.544	.350	215.8	36.0	4.83	9.5	3.8	1.01
10 x 4⅝	35.0	10.22	10.00	4.944	.491	.594	145.8	29.2	3.78	8.5	3.4	.91
	25.4	7.38	10.00	4.660	.491	.310	122.1	24.4	4.07	6.9	3.0	.97
8 x 4	23.0	6.71	8.00	4.171	.425	.441	64.2	16.0	3.09	4.4	2.1	.81
	18.4	5.34	8.00	4.000	.425	.270	56.9	14.2	3.26	3.8	1.9	.84
7 x 3⅝	20.0	5.83	7.00	3.860	.392	.450	41.9	12.0	2.68	3.1	1.6	.74
	15.3	4.43	7.00	3.660	.392	.250	36.2	10.4	2.86	2.7	1.5	.78
6 x 3⅜	17.25	5.02	6.00	3.565	.359	.465	26.0	8.7	2.28	2.3	1.3	.68
	12.5	3.61	6.00	3.330	.359	.230	21.8	7.3	2.46	1.8	1.1	.72
5 x 3	14.75	4.29	5.00	3.284	.326	.494	15.0	6.0	1.87	1.7	1.0	.63
	10.0	2.87	5.00	3.000	.326	.210	12.1	4.8	2.05	1.2	.82	.65
4 x 2⅝	9.5	2.76	4.00	2.796	.293	.326	6.7	3.3	1.56	.91	.65	.58
	7.7	2.21	4.00	2.660	.293	.190	6.0	3.0	1.64	.77	.58	.59
3 x 2⅜	7.5	2.17	3.00	2.509	.260	.349	2.9	1.9	1.15	.59	.47	.52
	5.7	1.64	3.00	2.330	.260	.170	2.5	1.7	1.23	.46	.40	.53

Table 5 (*continued*)

AMERICAN STANDARD
CHANNELS

PROPERTIES FOR DESIGNING

Nominal Size	Weight per Foot	Area	Depth	Flange Width	Flange Average Thickness	Web Thickness	AXIS X-X I	AXIS X-X S	AXIS X-X r	AXIS Y-Y I	AXIS Y-Y S	AXIS Y-Y r	x
In.	Lb.	In.²	In.	In.	In.	In.	In.⁴	In.³	In.	In.⁴	In.³	In.	In.
*18 x 4	58.0	16.98	18.00	4.200	.625	.700	670.7	74.5	6.29	18.5	5.6	1.04	.88
	51.9	15.18	18.00	4.100	.625	.600	622.1	69.1	6.40	17.1	5.3	1.06	.87
	45.8	13.38	18.00	4.000	.625	.500	573.5	63.7	6.55	15.8	5.1	1.09	.89
	42.7	12.48	18.00	3.950	.625	.450	549.2	61.0	6.64	15.0	4.9	1.10	.90
15 x 3⅜	50.0	14.64	15.00	3.716	.650	.716	401.4	53.6	5.24	11.2	3.8	.87	.80
	40.0	11.70	15.00	3.520	.650	.520	346.3	46.2	5.44	9.3	3.4	.89	.78
	33.9	9.90	15.00	3.400	.650	.400	312.6	41.7	5.62	8.2	3.2	.91	.79
12 x 3	30.0	8.79	12.00	3.170	.501	.510	161.2	26.9	4.28	5.2	2.1	.77	.68
	25.0	7.32	12.00	3.047	.501	.387	143.5	23.9	4.43	4.5	1.9	.79	.68
	20.7	6.03	12.00	2.940	.501	.280	128.1	21.4	4.61	3.9	1.7	.81	.70
10 x 2⅝	30.0	8.80	10.00	3.033	.436	.673	103.0	20.6	3.42	4.0	1.7	.67	.65
	25.0	7.33	10.00	2.886	.436	.526	90.7	18.1	3.52	3.4	1.5	.68	.62
	20.0	5.86	10.00	2.739	.436	.379	78.5	15.7	3.66	2.8	1.3	.70	.61
	15.3	4.47	10.00	2.600	.436	.240	66.9	13.4	3.87	2.3	1.2	.72	.64
9 x 2½	20.0	5.86	9.00	2.648	.413	.448	60.6	13.5	3.22	2.4	1.2	.65	.59
	15.0	4.39	9.00	2.485	.413	.285	50.7	11.3	3.40	1.9	1.0	.67	.59
	13.4	3.89	9.00	2.430	.413	.230	47.3	10.5	3.49	1.8	.97	.67	.61
8 x 2¼	18.75	5.49	8.00	2.527	.390	.487	43.7	10.9	2.82	2.0	1.0	.60	.57
	13.75	4.02	8.00	2.343	.390	.303	35.8	9.0	2.99	1.5	.86	.62	.56
	11.5	3.36	8.00	2.260	.390	.220	32.3	8.1	3.10	1.3	.79	.63	.58
7 x 2⅛	14.75	4.32	7.00	2.299	.366	.419	27.1	7.7	2.51	1.4	.79	.57	.53
	12.25	3.58	7.00	2.194	.366	.314	24.1	6.9	2.59	1.2	.71	.58	.53
	9.8	2.85	7.00	2.090	.366	.210	21.1	6.0	2.72	.98	.63	.59	.55
6 x 2	13.0	3.81	6.00	2.157	.343	.437	17.3	5.8	2.13	1.1	.65	.53	.52
	10.5	3.07	6.00	2.034	.343	.314	15.1	5.0	2.22	.87	.57	.53	.50
	8.2	2.39	6.00	1.920	.343	.200	13.0	4.3	2.34	.70	.50	.54	.52
5 x 1¾	9.0	2.63	5.00	1.885	.320	.325	8.8	3.5	1.83	.64	.45	.49	.48
	6.7	1.95	5.00	1.750	.320	.190	7.4	3.0	1.95	.48	.38	.50	.49
4 x 1⅝	7.25	2.12	4.00	1.720	.296	.320	4.5	2.3	1.47	.44	.35	.46	.46
	5.4	1.56	4.00	1.580	.296	.180	3.8	1.9	1.56	.32	.29	.45	.46
3 x 1½	6.0	1.75	3.00	1.596	.273	.356	2.1	1.4	1.08	.31	.27	.42	.46
	5.0	1.46	3.00	1.498	.273	.258	1.8	1.2	1.12	.25	.24	.41	.44
	4.1	1.19	3.00	1.410	.273	.170	1.6	1.1	1.17	.20	.21	.41	.44

* Car and Shipbuilding Channel; not an American Standard.

Table 5 (*continued*)

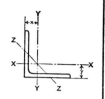

ANGLES
EQUAL LEGS

PROPERTIES FOR DESIGNING

Size	Thickness	Weight per Foot	Area	AXIS X-X AND AXIS Y-Y				AXIS Z-Z
				I	S	r	x or y	r
In.	In.	Lb.	In.²	In.⁴	In.³	In.	In.	In.
8 x 8	1⅛	56.9	16.73	98.0	17.5	2.42	2.41	1.56
	1	51.0	15.00	89.0	15.8	2.44	2.37	1.56
	⅞	45.0	13.23	79.6	14.0	2.45	2.32	1.57
	¾	38.9	11.44	69.7	12.2	2.47	2.28	1.57
	⅝	32.7	9.61	59.4	10.3	2.49	2.23	1.58
	⁹⁄₁₆	29.6	8.68	54.1	9.3	2.50	2.21	1.58
	½	26.4	7.75	48.6	8.4	2.50	2.19	1.59
6 x 6	1	37.4	11.00	35.5	8.6	1.80	1.86	1.17
	⅞	33.1	9.73	31.9	7.6	1.81	1.82	1.17
	¾	28.7	8.44	28.2	6.7	1.83	1.78	1.17
	⅝	24.2	7.11	24.2	5.7	1.84	1.73	1.18
	⁹⁄₁₆	21.9	6.43	22.1	5.1	1.85	1.71	1.18
	½	19.6	5.75	19.9	4.6	1.86	1.68	1.18
	⁷⁄₁₆	17.2	5.06	17.7	4.1	1.87	1.66	1.19
	⅜	14.9	4.36	15.4	3.5	1.88	1.64	1.19
	⁵⁄₁₆	12.5	3.66	13.0	3.0	1.89	1.61	1.19
5 x 5	⅞	27.2	7.98	17.8	5.2	1.49	1.57	.97
	¾	23.6	6.94	15.7	4.5	1.51	1.52	.97
	⅝	20.0	5.86	13.6	3.9	1.52	1.48	.98
	½	16.2	4.75	11.3	3.2	1.54	1.43	.98
	⁷⁄₁₆	14.3	4.18	10.0	2.8	1.55	1.41	.98
	⅜	12.3	3.61	8.7	2.4	1.56	1.39	.99
	⁵⁄₁₆	10.3	3.03	7.4	2.0	1.57	1.37	.99
4 x 4	¾	18.5	5.44	7.7	2.8	1.19	1.27	.78
	⅝	15.7	4.61	6.7	2.4	1.20	1.23	.78
	½	12.8	3.75	5.6	2.0	1.22	1.18	.78
	⁷⁄₁₆	11.3	3.31	5.0	1.8	1.23	1.16	.78
	⅜	9.8	2.86	4.4	1.5	1.23	1.14	.79
	⁵⁄₁₆	8.2	2.40	3.7	1.3	1.24	1.12	.79
	¼	6.6	1.94	3.0	1.1	1.25	1.09	.80

* Reproduced by permission of American Institute of Steel Construction.

Table 5   (*continued*)

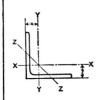

ANGLES
EQUAL LEGS

PROPERTIES FOR DESIGNING

Size	Thickness	Weight per Foot	Area	AXIS X–X AND AXIS Y–Y				AXIS Z–Z
				I	S	r	x or y	r
In.	In.	Lb.	In.²	In.⁴	In.³	In.	In.	In.
3½ x 3½	½	11.1	3.25	3.6	1.5	1.06	1.06	.68
	⁷⁄₁₆	9.8	2.87	3.3	1.3	1.07	1.04	.68
	³⁄₈	8.5	2.48	2.9	1.2	1.07	1.01	.69
	⁵⁄₁₆	7.2	2.09	2.5	.98	1.08	.99	.69
	¼	5.8	1.69	2.0	.79	1.09	.97	.69
3 x 3	½	9.4	2.75	2.2	1.1	.90	.93	.58
	⁷⁄₁₆	8.3	2.43	2.0	.95	.91	.91	.58
	³⁄₈	7.2	2.11	1.8	.83	.91	.89	.58
	⁵⁄₁₆	6.1	1.78	1.5	.71	.92	.87	.59
	¼	4.9	1.44	1.2	.58	.93	.84	.59
	³⁄₁₆	3.71	1.09	.96	.44	.94	.82	.59
2½ x 2½	½	7.7	2.25	1.2	.72	.74	.81	.49
	³⁄₈	5.9	1.73	.98	.57	.75	.76	.49
	⁵⁄₁₆	5.0	1.47	.85	.48	.76	.74	.49
	¼	4.1	1.19	.70	.39	.77	.72	.49
	³⁄₁₆	3.07	.90	.55	.30	.78	.69	.49
2 x 2	³⁄₈	4.7	1.36	.48	.35	.59	.64	.39
	⁵⁄₁₆	3.92	1.15	.42	.30	.60	.61	.39
	¼	3.19	.94	.35	.25	.61	.59	.39
	³⁄₁₆	2.44	.71	.27	.19	.62	.57	.39
	⅛	1.65	.48	.19	.13	.63	.55	.40
1¾ x 1¾	¼	2.77	.81	.23	.19	.53	.53	.34
	³⁄₁₆	2.12	.62	.18	.14	.54	.51	.34
	⅛	1.44	.42	.13	.10	.55	.48	.35
1½ x 1½	¼	2.34	.69	.14	.13	.45	.47	.29
	³⁄₁₆	1.80	.53	.11	.10	.46	.44	.29
	⅛	1.23	36	.08	.07	.47	.42	.30
1¼ x 1¼	¼	1.92	.56	.08	.09	.37	.40	.24
	³⁄₁₆	1.48	.43	.06	.07	.38	.38	.24
	⅛	1.01	.30	.04	.05	.38	.36	.25
1 x 1	¼	1.49	.44	.04	.06	.29	.34	.20
	³⁄₁₆	1.16	.34	.03	.04	.30	.32	.19
	⅛	.80	.23	.02	.03	.30	.30	.20

Table 6   Beam Diagrams

Symbols used: R = reaction, V = vertical shear, M = bending moment, δ = deflection.

1. Simple beam: concentrated load at center:

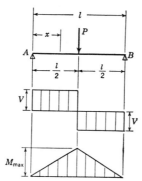

$$R_A = R_B = V = \frac{P}{2}$$

$$M_{max} \text{ (at load point)} = \frac{Pl}{4}$$

$$M_x \left(\text{when } x < \frac{l}{2}\right) = \frac{Px}{2}$$

$$\delta_{max} \text{ (at load point)} = \frac{Pl^3}{48EI}$$

$$\delta_x \left(\text{when } x < \frac{l}{2}\right) = \frac{Px}{48EI}(3l^2 - 4x^2)$$

2. Simple beam: two equal concentrated symmetrical loads:

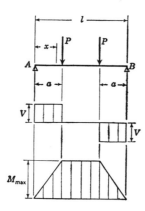

$$R_A = R_B = V = P$$

$$M_{max} \text{ (between loads)} = Pa$$

$$M_x \text{ (when } x < a) = Px$$

$$\delta_{max} \text{ (at center)} = \frac{Pa}{24EI}(3l^2 - 4a^2)$$

$$\delta_x \text{ (when } x < a) = \frac{Px}{6EI}(3la - 3a^2 - x^2)$$

$$\delta_x \text{ [when } x > a \text{ and } < (l - a)]$$
$$= \frac{Pa}{6EI}(3lx - 3x^2 - a^2)$$

3. Cantilever beam: concentrated load at any point:

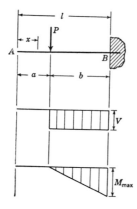

$$R_B = V_x \text{ (when } x > a) = P$$

$$M_{max} \text{ (at } B) = Pb$$

$$M_x \text{ (when } x > a) = P(x - a)$$

$$\delta_{max} \text{ (at } A) = \frac{Pb^2}{6EI}(3l - b)$$

$$\delta_1 \text{ (at point of load)} = \frac{Pb^3}{3EI}$$

$$\delta_x \text{ (when } x < a) = \frac{Pb^2}{6EI}(3l - 3x - b)$$

$$\delta_x \text{ (when } x > a) = \frac{P(l - x)^2}{6EI}(3b - l + x)$$

Table 6 *(continued)*

Beam Diagrams **705**

4. Simple beam: uniformly distributed load:

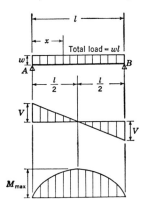

$$R_A = R_B = V = \frac{wl}{2}$$

$$V_x = w\left(\frac{l}{2} - x\right)$$

$$M_{\text{max}} \text{ (at center)} = \frac{wl^2}{8}$$

$$M_x = \frac{wx}{2}(l - x)$$

$$\delta_{\text{max}} \text{ (at center)} = \frac{5wl^4}{384EI}$$

$$\delta_x = \frac{wx}{24EI}(l^3 - 2lx^2 + x^3)$$

5. Beam fixed at both ends: concentrated load at center:

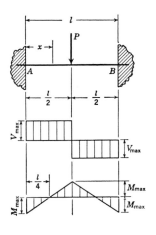

$$R_A = R_B = V_{\text{max}} = \frac{P}{2}$$

$$M_{\text{max}} \text{ (at } A, B, \text{ and ctr.)} = \frac{Pl}{8}$$

$$M_x\left(\text{when } x < \frac{l}{2}\right) = \frac{P}{8}(4x - l)$$

$$\delta_{\text{max}} \text{ (at center)} = \frac{Pl^3}{192EI}$$

$$\delta_x = \frac{Px^2}{48EI}(3l - 4x)$$

Table 7 Dimensions, Weights, and Resistance of Soft or Standard Annealed-Copper Wire and of Aluminum Wire of 61% Conductivity (Solid Conductors)

A.W.G. (B.&S.) NUMBER	Diameter, Mils	Cross-Section Area		Resistance, ohms per 1000 ft 20°C (68°F)		Weight, pounds per 1000 feet	
		Cir Mils	Sq Mils	Copper	Aluminum	Copper	Aluminum
	d	d²	0.7854d²				
0,000	460.0	211,600	166,200	0.04901	0.0804	640.5	195.0
000	409.6	167,800	131,800	0.06180	0.101	507.9	154.0
00	364.8	133,080	104,500	0.07793	0.128	402.8	122.0
0	324.9	105,530	82,890	0.09827	0.161	319.5	97.1
1	289.3	83,690	65,730	0.1239	0.203	253.3	76.9
2	257.6	66,370	52,130	0.1563	0.256	200.9	61.1
3	229.4	52,630	41,340	0.1970	0.323	159.3	48.4
4	204.3	41,740	32,780	0.2485	0.407	126.4	38.4
5	181.9	33,100	26,000	0.3133	0.514	100.2	30.5
6	162.0	26,250	20,620	0.3951	0.648	79.46	24.16
7	144.3	20,820	16,350	0.4982	0.817	63.02	19.16
8	128.5	16,510	12,970	0.6282	1.03	49.97	15.19
9	114.4	13,100	10,280	0.7921	1.30	39.63	12.05
10	101.9	10,380	8,153	0.9989	1.64	31.43	9.56
11	90.74	8,234	6,467	1.260	2.07	24.92	7.58
12	80.81	6,530	5,129	1.588	2.60	19.77	6.01
13	71.96	5,178	4,067	2.003	3.28	15.68	4.77
14	64.08	4,107	3,225	2.525	4.14	12.43	3.78
15	57.07	3,257	2,558	3.184	5.22	9.858	2.997
16	50.82	2,583	2,029	4.016	6.58	7.818	2.377
17	45.26	2,048	1,609	5.064	8.30	6.200	1.883
18	40.30	1,624	1,276	6.385	10.47	4.917	1.495
19	35.89	1,288	1,012	8.051	13.20	3.899	1.185
20	31.96	1,022	802.3	10.15	16.64	3.092	0.940
21	28.46	810.1	636.3	12.80	20.99	2.452	0.746
22	25.35	642.7	504.8	16.14	26.46	1.945	0.591
23	22.57	509.5	400.1	20.36	33.37	1.542	0.469
24	20.10	404.0	317.3	25.67	42.08	1.223	0.372
25	17.90	320.4	251.6	32.37	53.06	0.9699	0.295
26	15.94	254.0	199.5	40.81	67.0	0.7692	0.234
27	14.20	201.5	158.3	51.47	84.4	0.6100	0.185
28	12.64	159.8	125.5	64.90	106	0.4837	0.147
29	11.26	126.7	99.53	81.84	134	0.3836	0.117
30	10.03	100.5	78.93	103.2	169	0.3042	0.0924
31	8.93	79.71	62.60	130.1	213	0.2413	0.0733
32	7.95	63.20	49.64	164.1	269	0.1913	0.0581
33	7.08	50.13	39.37	206.9	339	0.1517	0.0461
34	6.30	39.74	31.21	260.9	428	0.1203	0.0365
35	5.61	31.52	24.76	329.0	540	0.09542	0.0290
36	5.00	25.00	19.64	414.8	681	0.07568	0.0230
37	4.453	19.83	15.57	523.1	858	0.06001	0.0182
38	3.965	15.72	12.35	659.6	1080	0.04759	0.0145
39	3.531	12.47	9.793	831.8	1360	0.03774	0.0115
40	3.145	9.888	7.766	1049	1720	0.02993	0.0091

The resistance of medium-hard drawn or hard-drawn wire may be obtained by multiplying the given values by the following factors:

Size (incl.)	Med. hard drawn	Hard drawn
4/0-1/0	1.024	1.029
1-18	1.035	1.040

Table 8   Typical Physical Properties of and Allowable Stresses for Some Common Materials (Metric Units)[a]

Material	Unit Mass $\times 10^3$ kg/m³	Ultimate Strength, MPa			Yield Strength,[g] MPa		Allow Stresses,[i] MPa		Elastic Moduli GPa		Coef. of Thermal Expans. $\times 10^{-6}$ per °C
		Tens.	Comp.[c]	Shear	Tens.[h]	Shear	Tens. or Comp.	Shear	Tens. or Comp.	Shear	
Aluminum alloy (extruded) 2014–T6	2.77	414	...	241	365	214			75	27.6	23.2
6061–T6		262	...	207	241	138			70	25.6	23.4
Cast iron Gray	7.64	210	825	...[e]					90	41	10.4
Malleable		370	...	330	250	165			170	83	12.1
Concrete[b] 0.70 water–cement ratio	2.41	...	20	...[e]			−9.31[j]	0.455	20	...	10.8
0.53 water–cement ratio		...	35	...			−15.5[j]	0.592	35	...	
Magnesium alloy, AM100A	1.80	275	...	145	150	...			45	17	25.2
Steel 0.2% Carbon (hot rolled)	7.83	450	...	330	250	165	±165	100	200	83	11.7
0.6% Carbon (hot rolled)		690	...	550	415	250					
0.6% Carbon (quenched)		825	...	690	515	310					
3½% Ni, 0.4% C		1380	...	1035	1035	620					
Wood Douglas Fir (coast)	0.50	...	51[d]	7.6[f]			±13.1[k]	0.825[f]	12.1	...	...
Southern Pine (long leaf)	0.58	...	58[d]	10[f]			±15.5[k]	0.930[f]	12.1	...	...

[Courtesy E. P. Popov, SI version, *Mechanics of Materials*, 2nd ed., copyright © 1978, pp. 570 (Table 1) and 572–578 (Tables 3–9). Reprinted by permission of Prentice-Hall, Inc., Englewood Cliffs, New Jersey.]

[a] Mechanical properties of metals depend not only on composition but also on heat treatment, previous cold working, etc. Data for wood are for clear 50-mm by 50-mm specimens at 12 percent moisture content. True values vary. Where SI values are not yet available, a soft conversion of values currently accepted in industry was used in constructing this table.

[b] Water-cement ratio by weight for concrete with a 75 to 100 mm slump. Values are for 28-day-old concrete.

[c] For short blocks only. For ductile materials the ultimate strength in compression is indefinite; may be assumed to be the same as that in tension.

[d] Compression parallel to grain on short blocks. Compression perpendicular to grain at proportional limit 6.56 MPa, 8.20 MPa, respectively Soft conversion of values from *Wood Handbook,* U.S. Dept. of Agriculture.

[e] Fails in diagonal tension.

[f] Parallel to grain.

[g] For most materials, 0.2 percent offset.

[h] For ductile materials compressive yield strength may be assumed the same.

[i] For static loads only. Much lower stresses required in machine design because of fatigue properties and dynamic loadings.

[j] No tensile stress is allowed in concrete.

[k] In bending only. Timber stresses are for select and dense grade.

Table 9 American Standard Steel Beams, S Shapes,
Properties for Designing (Metric Units)

Designation*	Area	Depth	Flange Width	Flange Thickness	Web Thickness	Axis X-X I	Axis X-X $\frac{I}{c}$	Axis X-X r	Axis Y-Y I	Axis Y-Y $\frac{I}{c}$	Axis Y-Y r
	mm²	d, mm	mm	mm		$\times 10^6$ mm⁴	$\times 10^3$ mm³	mm	$\times 10^6$ mm⁴	$\times 10^3$ mm³	mm
S 610 × 179	22 770	610	204	28.0	20.3	1260	4140	235	35.0	343	39.2
× 157.6	20 060	610	200	28.0	15.9	1180	3860	242	32.5	325	40.3
× 149	18 970	610	184	22.1	19.0	995	3260	229	19.9	216	32.4
× 134	17 100	610	181	22.1	15.8	937	3070	234	18.7	207	33.1
× 118.9	15 160	610	178	22.1	12.7	878	2880	241	17.6	198	34.1
S 510 × 141	18 000	508	183	23.3	20.3	670	2640	193	20.7	226	33.9
× 127	16 130	508	179	23.3	16.6	633	2490	198	19.2	215	34.5
× 112	14 260	508	162	20.1	16.3	533	2100	193	12.3	152	29.4
× 97.3	12 390	508	159	20.1	12.7	491	1930	199	11.4	144	30.3
S 460 × 104	13 290	457	159	17.6	18.1	385	1690	170	10.0	126	27.5
× 81.4	10 390	457	152	17.6	11.7	335	1460	179	8.66	114	28.9
S 380 × 74	9 480	381	143	15.8	14.0	202	1060	146	6.53	91.2	26.3
× 64	8 130	381	140	15.8	10.4	186	977	151	5.99	85.8	27.2
S 310 × 74	9 480	305	139	16.8	17.4	127	833	116	6.53	94.0	26.3
× 60.7	7 740	305	133	16.8	11.7	113	743	121	5.66	84.9	27.0
× 52	6 640	305	129	13.8	10.9	95.3	625	120	4.11	63.7	24.9
× 47.3	6 032	305	127	13.8	8.9	90.7	595	123	3.90	61.4	25.4
S 250 × 52	6 640	254	126	12.5	15.1	61.2	482	96.0	3.48	55.4	22.9
× 37.8	4 806	254	118	12.5	7.9	51.6	406	104	2.83	47.7	24.2
S 200 × 34	4 368	203	106	10.8	11.2	27.0	266	78.6	1.79	33.9	20.3
× 27.4	3 484	203	102	10.8	6.9	24.0	236	82.9	1.55	30.6	21.1
S 180 × 30	3 794	178	97	10.0	11.4	17.7	199	68.2	1.32	26.9	18.7
× 22.8	2 890	178	92	10.0	6.4	15.3	172	72.5	1.10	23.6	19.5
S 150 × 25.7	3 271	152	90	9.1	11.8	11.0	144	57.9	0.961	21.2	17.1
× 18.6	2 362	152	84	9.1	5.8	9.20	121	62.3	0.758	17.9	17.9
S 130 × 22.0	2 800	127	83	8.3	12.5	6.33	99.6	47.5	0.695	16.7	15.8
× 15	1 884	127	76	8.3	5.3	5.12	80.6	52.0	0.508	13.3	16.4
S 100 × 14.1	1 800	102	70	7.4	8.3	2.83	55.6	39.6	0.376	10.6	14.5
× 11.5	1 452	102	67	7.4	4.8	2.53	49.8	41.7	0.318	9.40	14.8
S 75 × 11.2	1 426	76	63	6.6	8.9	1.22	32.0	29.2	0.244	7.66	13.1
× 8.5	1 077	76	59⁻	6.6	4.3	1.05	27.5	31.2	0.189	6.40	13.3

[Courtesy E. P. Popov, SI version, *Mechanics of Materials*, 2nd Ed., copyright © 1978, pp. 570 (Table 1) and 572–578 (Tables 3–9). Reprinted by permission of Prentice-Hall, Inc., Englewood Cliffs, New Jersey.]

* American Standard I-shaped beams are referred to as S-shapes and are designated by the letter S followed by their depth in millimeters, with their mass in kilograms per lineal meter given last. For example, S 610 × 179 means that this S-shape is 610 mm deep and has a mass of 179 kg/m. These values are approximate conversions of those in the AISC tables into SI units.

Table 10   American Wide-Flange Steel Beams, W Shapes,
Properties for Designing (Abridged List) (Metric Units)

Designation	Area	Depth	Flange Width	Flange Thickness	Web Thickness	Axis X-X I	Axis X-X $\frac{I}{c}$	Axis X-X r	Axis Y-Y I	Axis Y-Y $\frac{I}{c}$	Axis Y-Y r
	mm^2	d, mm	mm	mm	mm	$\times 10^6$ mm^4	$\times 10^3$ mm^3	mm	$\times 10^6$ mm^4	$\times 10^3$ mm^3	mm
W 920 × 342	43 680	911	418	32.0	19.3	6240	13 700	378	391	1870	94.6
× 223	28 520	910	304	23.9	15.9	3760	8 260	363	112	739	62.8
W 840 × 298	38 000	838	400	29.2	18.2	4620	11 000	349	312	1560	90.6
× 193	24 710	841	292	21.7	14.7	2790	6·640	336	90.7	621	60.6
W 760 × 256	32 710	759	381	27.1	16.6	3290	8 680	317	249	1310	87.2
× 161	20 520	757	266	19.3	13.9	1860	4 910	301	60.8	456	54.4
W 690 × 216	27 550	683	355	24.8	15.2	2260	6 620	286	184	1040	81.8
× 140	17 870	684	254	19.0	12.4	1360	3 980	276	51.6	407	53.7
W 610 × 193	24 710	616	356	22.9	14.4	1670	5 430	260	172	965	83.3
× 149	19 030	610	305	19.7	11.9	1250	4 100	256	92.8	609	69.8
× 113	14 450	607	228	17.3	11.2	874	2 880	246	34.4	301	48.8
W 530 × 167	21 290	533	330	22.0	13.4	1090	4 090	226	132	799	78.7
× 122	15 610	530	228	20.2	12.7	733	2 770	217	39.8	350	50.5
× 92	11 810	533	209	15.6	10.2	554	2 080	217	23.9	229	45.0
× 82	10 450	528	209	13.3	9.5	475	1 800	213	20.1	193	43.9
W 460 × 143	18 190	461	298	21.1	13.0	699	3 030	196	93.7	628	71.7
× 95	12 190	454	221	17.4	10.2	437	1 930	189	31.6	285	50.9
× 82	10 450	460	191	16.0	9.9	371	1 610	188	18.7	196	42.3
× 74	9 480	457	191	14.5	9.1	334	1 460	188	16.7	176	42.0
× 67	8 520	454	190	12.7	8.5	294	1 300	186	14.5	153	41.2
× 52	6 640	450	152	10.9	7.6	214	949	179	6.45	84.7	31.2
W 410 × 143	18 190	415	293	22.2	13.6	566	2 730	176	93.2	637	71.6
× 131	16 710	410	292	20.2	12.8	508	2 470	174	84.1	576	70.9
× 80	11 030	403	215	16.4	10.3	311	1 550	168	27.2	253	49.6
× 74	9 480	413	180	16.0	9.7	273	1 330	170	15.4	172	40.4
× 54	6 840	403	178	10.9	7.6	186	924	165	10.2	114	38.5
× 39	4 950	398	140	8.8	6.4	125	628	159	3.99	57.2	28.4
W 360 × 476	60 710	427	424	53.2	48.0	1720	8 070	168	683	3220	106
× 202	25 580	375	374	27.0	16.8	662	3 530	160	236	1260	95.7
× 130	16 520	356	368	17.5	10.7	403	2 260	156	146	791	93.9
× 125	15 940	360	305	19.8	11.5	386	2 140	156	93.7	613	76.7
× 116	14 770	357	305	18.2	10.9	354	1 980	155	86.2	565	76.4

[Courtesy E. P. Popov, SI version, *Mechanics of Materials,* 2nd Ed., copyright © 1978, pp. 570 (Table 1) and 572–578 (Tables 3–9).
Reprinted by permission of Prentice-Hall, Inc., Englewood Cliffs, New Jersey.]

* American wide-flange I- or H-shaped steel beams are referred to as W shapes and are designated by the letter W
followed by their *nominal* depth in millimeters, with their mass in kilograms per lineal meter given last. For example, W 530 × 167 means that this W shape is 530 mm deep and has a mass of 167 kg/m. These values are approximate conversions of those in the AISC tables into SI units.

Table 10 *(continued)*

Designation	Area	Depth	Flange		Web Thick-ness	Axis X-X			Axis Y-Y		
			Width	Thick-ness		I	$\dfrac{I}{c}$	r	I	$\dfrac{I}{c}$	r
	mm²	d, mm	mm	mm	mm	$\times 10^6$ mm⁴	$\times 10^3$ mm³	mm	$\times 10^6$ mm⁴	$\times 10^3$ mm³	mm
W 360 × 110	14 060	360	256	19.9	11.4	332	1840	154	55.4	433	62.7
× 101	12 900	357	255	18.2	10.6	301	1690	153	50.4	395	62.5
× 91	11 550	353	254	16.3	9.6	267	1510	152	44.5	351	62.1
× 79	10 060	354	205	16.7	9.4	226	1270	150	23.9	234	48.8
× 64	8 130	347	203	13.4	7.8	179	1030	148	18.8	185	48.1
× 57	7 230	359	172	13.0	8.0	161	896	149	11.1	129	39.1
× 51	6 450	356	171	11.5	7.3	142	796	148	9.70	113	38.8
× 45	5 700	352	171	9.7	6.9	121	686	146	8.12	94.9	37.8
W 310 × 126	16 130	317	307	20.2	12.6	301	1900	137	97.8	636	77.9
× 97	12 320	308	305	15.4	9.9	222	1440	134	72.8	478	76.9
× 79	10 060	306	254	14.6	8.8	177	1160	133	40.0	315	63.0
× 60	7 610	303	203	13.1	7.5	129	851	130	18.4	181	49.1
× 54	6 840	311	167	13.7	7.7	117	752	131	10.6	127	39.4
× 46	5 890	307	166	11.8	6.7	99.5	648	130	8.99	108	39.1
× 40	5 129	304	165	10.2	6.0	84.9	559	129	7.62	92.3	38.5
W 250 × 167	21 230	289	265	31.7	19.2	299	2070	119	97.8	740	67.9
× 149	18 970	282	263	28.4	17.4	260	1840	117	86.2	656	67.4
× 132	16 900	276	261	25.3	15.6	226	1630	116	75.3	577	66.8
× 115	14 640	270	259	22.0	13.6	190	1410	114	63.7	492	65.9
× 89	11 420	260	256	17.3	10.5	143	1100	112	48.3	377	65.3
× 73	9 290	254	254	14.2	8.6	114	895	110	38.7	305	64.5
× 67	8 520	257	204	15.7	8.9	104	806	110	22.1	217	51.0
× 58	7 420	252	203	13.4	8.1	87.4	692	109	18.7	184	50.2
× 49	6 265	248	202	11.0	7.4	71.2	575	107	15.2	150	49.2
× 43	5 510	260	147	12.7	7.3	65.8	507	109	6.78	92.1	35.1
× 31	4 000	251	146	8.6	6.1	44.5	354	106	4.50	61.6	33.5
W 200 × 100	12 710	229	210	23.7	14.6	113	991	94.4	36.9	350	53.9
× 86	11 030	222	209	20.5	13.0	94.5	850	92.5	31.2	299	53.2
× 71	9 100	216	206	17.3	10.3	76.6	709	91.8	25.3	246	52.8
× 60	7 610	210	205	14.2	9.3	60.8	580	89.3	20.4	199	51.8
× 52	6 640	206	204	12.5	8.0	52.4	509	88.8	17.7	174	51.6
× 46	5 884	203	203	11.0	7.3	45.8	451	88.2	15.4	152	51.2
× 42	5 310	205	166	11.8	7.2	40.7	398	87.6	8.99	108	41.1
× 36	4 555	201	165	10.1	6.2	34.3	341	86.8	7.58	91.8	40.8
× 30	3 800	207	134	9.6	6.3	28.9	279	87.2	3.84	57.4	31.8
× 25	3 232	203	133	7.8	5.8	23.6	232	85.4	3.10	46.4	31.0

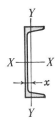

Table 11 American Standard Steel Channels, Properties for Designing (Metric Units)

Designation*	Area	Depth	Flange Width	Thick-ness	Web Thick-ness	Axis X-X I	Axis X-X $\frac{I}{c}$	Axis X-X r	Axis Y-Y I	Axis Y-Y $\frac{I}{c}$	Axis Y-Y r	Axis Y-Y x
	mm²	d, mm²	mm	mm	mm	$\times 10^6$ mm⁴	$\times 10^3$ mm³	mm	$\times 10^6$ mm⁴	$\times 10^3$ mm³	mm	mm
C 380 × 74	9480	381	94	16.5	18.2	168	883	133	4.58	97.0	22.0	20.3
× 60	7610	381	89	16.5	13.2	145	763	138	3.84	85.9	22.5	19.8
× 50.4	6426	381	86	16.5	10.2	131	688	143	3.38	78.4	23.0	20.0
C 310 × 45	5690	305	80	12.7	13.0	67.4	442	109	2.14	53.1	19.4	17.1
× 37	4742	305	77	12.7	9.8	59.9	393	112	1.86	48.1	19.8	17.1
× 30.8	3929	305	74	12.7	7.2	53.7	352	117	1.61	43.2	20.3	17.7
C 250 × 45	5690	254	76	11.1	17.1	42.9	338	86.8	1.64	42.6	17.0	16.5
× 37	4742	254	73	11.1	13.4	38.0	299	89.5	1.40	38.2	17.2	15.7
× 30	3794	254	69	11.1	9.6	32.8	259	93.0	1.17	33.6	17.6	15.4
× 22.8	2897	254	65	11.1	6.1	28.1	221	98.4	0.949	28.7	18.1	16.1
C 230 × 30	3794	229	67	10.5	11.4	25.3	222	81.7	1.01	30.0	16.3	14.8
× 22	2845	229	63	10.5	7.2	21.2	186	86.4	0.803	25.5	16.8	14.9
× 19.9	2542	229	61	10.5	5.9	19.9	174	88.6	0.733	23.7	17.0	15.3
C 200 × 27.9	3555	203	64	9.9	12.4	18.3	180	71.8	0.824	25.7	15.2	14.4
× 20.5	2606	203	59	9.9	7.7	15.0	148	75.9	0.637	21.4	15.6	14.0
× 17.1	2181	203	57	9.9	5.6	13.6	134	78.9	0.549	19.1	15.9	14.5
C 180 × 22.0	2794	178	58	9.3	10.6	11.3	127	63.7	0.574	19.7	14.3	13.5
× 18.2	2323	178	55	9.3	8.0	10.1	113	65.9	0.487	17.5	14.5	13.3
× 14.6	1852	178	53	9.3	5.3	8.87	99.7	69.2	0.403	15.2	14.8	13.7
C 150 × 19.3	2471	152	54	8.7	11.1	7.24	95.0	54.1	0.437	16.0	13.3	13.1
× 15.6	1994	152	51	8.7	8.0	6.33	83.0	56.3	0.360	13.9	13.4	12.7
× 12.2	1548	152	48	8.7	5.1	5.45	71.6	59.3	0.288	11.8	13.6	13.0
C 130 × 13.4	1703	127	47	8.1	8.3	3.70	58.3	46.6	0.263	11.0	12.4	12.1
× 10.0	1271	127	44	8.1	4.8	3.12	49.1	49.5	0.199	8.95	12.5	12.3
C 100 × 10.8	1374	102	43	7.5	8.2	1.91	37.6	37.3	0.180	8.23	11.4	11.7
× 8.0	1026	102	40	7.5	4.7	1.60	31.5	39.5	0.133	6.60	11.4	11.6
C 75 × 8.9	1135	76	40	6.9	9.0	0.862	22.6	27.5	0.127	6.26	10.6	11.6
× 7.4	948	76	37	6.9	6.6	0.770	20.2	28.5	0.103	5.40	10.4	11.1
× 6.1	781	76	35	6.9	4.3	0.691	18.1	29.8	0.082	4.58	10.2	11.1

[Courtesy E. P. Popov, SI version, *Mechanics of Materials,* 2nd Ed., copyright © 1978, pp. 570 (Table 1) and 572–578 (Tables 3–9). Reprinted by permission of Prentice-Hall, Inc., Englewood Cliffs, New Jersey.]

* American standard steel Channels are designated by the letter C followed by their depth in millimeters, with their mass is kilograms per lineal meter given last. For example, C 380 × 74 means that this channel is 380 mm deep and has a mass of 74 kg/m. These values are approximate conversions of those in the AISC tables into SI units.

Table 12 Steel Angles with Equal Legs, Properties for Designing (Abridged List) (Metric Units)

Size and Thickness	Mass per Meter	Area	Axes X-X and Y-Y				Axis Z-Z
			I	$\dfrac{I}{c}$	r	x or y	r
mm	kg	mm²	$\times 10^6$ mm⁴	$\times 10^3$ mm³	mm	mm	mm
L 203 × 203 × 28.6	84.7	10 770	40.8	287	61.5	61.2	39.6
× 25.4	75.9	9 680	37.0	259	61.9	60.2	39.6
× 22.2	67.0	8 520	33.1	230	62.4	58.9	39.9
× 19.0	57.9	7 360	29.0	200	62.8	57.9	40.1
× 15.9	48.7	6 200	24.7	169	63.1	56.6	40.1
× 14.3	44.0	5 600	22.5	153	63.4	56.1	40.4
× 12.7	39.3	5 000	20.2	137	63.6	55.6	40.4
L 152 × 152 × 25.4	55.7	7 100	14.8	141	45.6	47.2	29.7
× 22.2	49.3	6 277	13.3	125	46.0	46.2	29.7
× 19.0	42.7	5 445	11.7	110	46.4	45.2	29.7
× 15.9	36.0	4 587	10.1	92.9	46.9	43.9	30.0
× 14.3	32.6	4 148	9.20	84.4	47.1	43.4	30.0
× 12.7	29.2	3 710	8.28	75.5	47.3	42.7	30.0
× 11.1	25.6	3 265	7.37	66.8	47.5	42.2	30.2
× 9.5	22.2	2 813	6.41	57.9	47.7	41.7	30.6
× 7.9	18.5	2 355	5.41	48.6	47.9	41.1	30.5
L 127 × 127 × 22.2	40.5	5 148	7.41	85.0	37.9	39.9	24.7
× 19.0	35.1	4 477	6.53	73.9	38.2	38.6	24.8
× 15.9	29.8	3 781	5.66	63.3	38.7	37.6	24.8
× 12.7	24.1	3 065	4.70	51.9	39.2	36.3	25.0
× 11.1	21.3	2 697	4.16	45.6	39.3	35.8	25.0
× 9.5	18.3	2 329	3.64	39.7	39.5	35.3	25.1
× 7.9	15.3	1 955	3.09	33.5	39.7	34.8	25.2
L 102 × 102 × 19.0	27.5	3 510	3.19	46.0	30.2	32.3	19.8
× 15.9	23.4	2 974	2.77	39.4	30.5	31.2	19.8
× 12.7	19.0	2 419	2.31	32.3	30.9	30.0	19.9
× 11.1	16.8	2 135	2.07	28.7	31.1	29.5	19.9
× 9.5	14.6	1 845	1.81	25.0	31.4	29.0	20.0
× 7.9	12.2	1 548	1.54	21.1	31.6	28.4	20.1
× 6.4	9.8	1 252	1.27	17.1	31.8	27.7	20.2
L 89 × 89 × 12.7	16.5	2 097	1.52	24.4	26.9	26.9	17.3
× 11.1	14.6	1 852	1.36	21.7	27.1	26.4	17.4
× 9.5	12.6	1 600	1.19	18.9	27.3	25.7	17.4
× 7.9	10.7	1 348	1.02	16.0	27.5	25.1	17.5
× 6.4	8.6	1 090	.837	13.0	27.7	24.6	17.6
L 76 × 76 × 12.7	14.0	1 774	.924	17.6	22.8	23.7	14.8
× 11.1	12.4	1 568	.828	15.6	23.0	23.1	14.9
× 9.5	10.7	1 361	.733	13.7	23.2	22.6	14.9
× 7.9	9.1	1 148	.629	11.6	23.4	22.1	15.0
× 6.4	7.3	929	.516	9.42	23.6	21.4	15.0
× 4.8	5.5	703	.400	7.23	23.9	20.8	15.1
L 64 × 64 × 12.7	11.4	1 452	.512	11.9	18.8	20.5	12.4
× 9.5	8.7	1 116	.410	9.28	19.2	19.4	12.4
× 7.9	7.4	942	.353	7.90	19.4	18.8	12.4
× 6.4	6.1	768	.293	6.46	19.5	18.2	12.5
× 4.8	4.6	581	.228	4.96	19.8	17.6	12.6

[Courtesy E. P. Popov, SI version, *Mechanics of Materials,* 2nd Ed., copyright © 1978, pp. 570 (Table 1) and 572–578 (Tables 3–9). Reprinted by permission of Prentice-Hall, Inc., Englewood Cliffs, New Jersey.]

Table 13   Steel Angles with Unequal Legs, Properties for Designing
Abridged List) (Metric Units)

Size and Thickness	Mass per Meter	Area	Axis X-X				Axis Y-Y				Axis Z-Z	
			I	$\frac{I}{c}$	r	y	I	$\frac{I}{c}$	r	x	r	Tan α
			$\times 10^6$	$\times 10^3$			$\times 10^6$	$\times 10^3$				
mm	kg	mm²	mm⁴	mm³	mm	mm	mm⁴	mm³	mm	mm	mm	
L 203 × 152 × 25.4	65.5	8390	33.6	247	63.3	67.3	16.1	146	43.9	41.9	32.5	.543
× 19.0	50.1	6413	26.4	191	64.1	65.0	12.8	113	44.6	39.6	32.8	.551
× 12.7	34.1	4355	18.4	131	65.1	62.7	9.03	78.5	45.5	37.3	33.0	.558
L 203 × 102 × 25.4	55.4	7100	29.0	230	63.9	77.5	4.83	64.4	26.1	26.7	21.5	.247
× 19.0	42.5	5445	22.9	178	64.8	74.9	3.90	50.3	26.8	24.2	21.6	.258
× 12.7	29.0	3710	16.0	123	65.7	72.6	2.81	35.2	27.5	21.8	22.0	.267
L 152 × 102 × 19.0	35.0	4477	10.2	102	47.7	52.8	3.61	48.7	28.4	27.4	21.8	.428
× 12.7	24.0	3065	7.24	71.1	48.6	50.6	2.61	34.1	29.2	25.1	22.1	.440
L 127 × 76 × 12.7	19.0	2419	3.93	47.7	40.3	44.5	1.07	18.8	21.1	19.1	16.5	.357
× 9.5	14.5	1845	3.07	36.6	40.8	43.2	.849	14.6	21.5	17.9	16.6	.364
× 6.4	9.8	1252	2.13	25.1	41.2	42.2	.599	10.1	21.9	16.7	16.8	.371
L 102 × 89 × 12.7	17.6	2258	2.21	31.7	31.3	31.8	1.58	24.8	26.4	25.4	18.3	.750
× 9.5	13.5	1723	1.74	24.6	31.8	30.7	1.23	19.0	26.7	24.3	18.5	.755
× 6.4	9.2	1168	1.21	16.8	32.2	29.5	.870	13.2	27.3	23.1	18.6	.759
L 102 × 76 × 12.7	16.4	2097	2.10	31.0	31.7	33.8	1.01	18.6	21.9	21.9	16.2	.543
× 9.5	12.6	1600	1.65	23.9	32.1	32.5	.799	14.2	22.3	19.9	16.4	.551
× 6.4	8.6	1090	1.15	16.4	32.5	31.5	.566	9.84	22.8	18.7	16.5	.558
L 89 × 64 × 12.7	13.9	1774	1.35	23.1	27.6	30.5	.566	12.4	17.9	17.9	13.6	.486
× 11.1	12.3	1568	1.21	20.6	27.8	30.0	.512	11.1	18.1	17.3	13.6	.491
× 9.5	10.7	1361	1.07	17.9	28.0	29.5	.454	9.71	18.3	16.8	13.6	.496
× 7.9	9.0	1148	.912	15.2	28.2	29.0	.391	8.26	18.4	16.2	13.7	.501
× 6.4	7.3	929	.749	12.3	28.4	28.2	.323	6.75	18.7	15.6	13.8	.506
L 76 × 64 × 12.7	12.6	1613	.866	17.0	23.2	25.4	.541	12.2	18.3	19.1	13.2	.667
× 11.1	11.3	1426	.783	15.2	23.4	24.8	.491	10.9	18.6	18.5	13.2	.672
× 9.5	9.8	1239	.691	13.3	23.6	24.3	.433	9.50	18.7	17.9	13.3	.676
× 7.9	8.3	1045	.591	11.3	23.8	23.7	.374	8.10	18.9	17.3	13.3	.680
× 6.4	6.7	845	.487	9.18	24.0	23.1	.309	6.62	19.1	16.8	13.4	.684
× 4.8	5.1	643	.378	7.04	24.2	22.6	.240	5.08	19.3	16.2	13.5	.688
L 76 × 51 × 12.7	11.5	1452	.799	16.4	23.5	27.4	.280	7.77	13.9	14.8	10.9	.414
× 11.1	10.1	1290	.720	14.6	23.6	26.9	.253	6.94	14.0	14.2	10.9	.421
× 9.5	8.8	1116	.637	12.8	23.9	26.4	.226	6.09	14.2	13.7	10.9	.428
× 7.9	7.4	942	.549	10.9	24.2	25.9	.196	5.19	14.4	13.1	11.0	.435
× 6.4	6.1	768	.454	8.90	24.3	25.2	.163	4.26	14.6	12.5	11.0	.440
× 4.8	4.6	582	.350	6.80	24.5	24.6	.128	3.29	14.8	11.9	11.2	.446
L 64 × 51 × 9.5	7.9	1000	.380	8.95	19.5	21.1	.214	5.94	14.6	14.8	10.7	.614
× 7.9	6.7	845	.328	7.64	19.7	20.5	.186	5.07	14.8	14.2	10.7	.620
× 6.4	5.4	684	.272	6.26	20.0	20.0	.155	4.17	15.0	13.6	10.8	.626
× 4.8	4.2	522	.212	4.81	20.1	19.4	.121	3.21	15.2	13.1	10.8	.631

[Courtesy E. P. Popov, SI version, *Mechanics of Materials*, 2nd Ed., copyright © 1978, pp. 570 (Table 1) and 572–578 (Tables 3–9). Reprinted by permission of Prentice-Hall, Inc., Englewood Cliffs, New Jersey.]
[Courtesy E. P. Popov, SI version, *Mechanics of Materials*, 2nd Ed., copyright © 1978, pp. 570 (Table 1) and 572–578 (Tables 3–9). Reprinted by permission of Prentice-Hall, Inc., Englewood Cliffs, New Jersey.]

Table 14 Standard Steel Pipe (Metric Units)

Dimensions					Properties		
			Mass per Meter				
Outside Diam.	Inside Diam.	Thickness	Plain Ends	Thread & Cplg.	I	A	r
mm	mm	mm	kg	kg	$\times 10^6$ mm^4	mm^2	mm
10.3	6.8	1.73	.36	.37	.0004	46	3.0
13.7	9.2	2.24	.63	.64	.0012	81	3.9
17.1	12.5	2.31	.85	.85	.0029	108	5.2
21.3	15.8	2.77	1.26	1.26	.0071	161	6.6
26.7	20.9	2.87	1.68	1.68	.0154	215	8.5
33.4	26.6	3.38	2.50	2.50	.0362	319	10.7
42.2	35.1	3.56	3.38	3.39	.0812	432	13.7
48.3	40.9	3.68	4.05	4.06	.1290	515	15.8
60.3	52.5	3.91	5.43	5.48	.2772	694	20.0
73.0	62.7	5.16	8.62	8.66	.6368	1099	24.1
88.9	77.9	5.49	11.28	11.34	1.256	1437	29.6
101.6	90.1	5.74	13.56	13.69	1.993	1729	34.0
114.3	102.3	6.02	16.06	16.21	3.011	2048	38.3
141.3	128.2	6.55	21.76	22.04	6.310	2774	47.7
168.3	154.1	7.11	28.23	28.56	11.71	3601	57.0
219.1	202.7	8.18	42.49	42.87	30.17	5419	74.6
273.1	254.5	9.27	60.24	61.21	66.89	7684	93.3
323.9	304.8	9.53	73.75	75.46	116.3	9406	111.2

Table 15 Plastic Section Moduli Around the *X-X* Axis (Metric Units)

Shape	Plastic Modulus Z $\times 10^3$ mm^3	Shape	Plastic Modulus Z $\times 10^3$ mm^3
W 920 × 342	15 500	W 610 × 101	2880
W 840 × 329	13 700	W 530 × 101	2620
W 920 × 289	12 600	W 610 × 91	2490
W 920 × 271	11 800	W 610 × 82	2200
W 920 × 253	10 900	W 530 × 82	2060
W 920 × 238	10 200	W 460 × 82	1840
W 920 × 223	9 520	W 530 × 73	1770
W 840 × 210	8 420	W 530 × 65	1560
W 920 × 201	8 360	W 460 × 60	1280
W 840 × 193	7 650	W 410 × 60	1190
W 840 × 176	6 800	W 460 × 52	1090
W 760 × 173	6 190	S 310 × 74	1000
W 760 × 161	5 670	W 410 × 46	885
W 760 × 147	5 130	W 360 × 39	655
W 690 × 140	4 560	W 360 × 33	542
W 610 × 140	4 150	S 250 × 37.8	465
W 690 × 125	4 000	W 200 × 30	313
S 610 × 134	3 640	W 200 × 25	261
S 610 × 118.9	3 360	S 180 × 30	238

Appendix I Metric Practice and Conversion Tables

AN American National Standard ASTM/IEEE Standard Metric Practice[1]

1. SCOPE

1.1 This standard gives guidance for application of the modernized metric system in the United States. The International System of Units, developed and maintained by the General Conference on Weights and Measures (abbreviated CGPM from the official French name Conférence Générale des Poids et Mesures) is intended as a basis for worldwide standardization of measurement units. The name International System of Units and the international abbreviation SI[2] were adopted by the 11th CGPM in 1960. SI is a complete, coherent system that is being universally adopted.

1.2 Information is included on SI, a limited list of non-SI units recognized for use with SI units, and a list of conversion factors from non-SI to SI units, together with general guidance on proper style and usage.

[1] Courtesy ASTM/IEEE.
[2] From the French name, Le Système International d'Unités.

1.3 It is hoped that an understanding of the system and its characteristics, and careful use according to this standard, will help to avoid the degradation that has occurred in all older measurement systems.

2. SI UNITS AND SYMBOLS

2.1 Classes of Units. SI units are divided into three classes:

    base units
    supplementary units
    derived units

2.2 Base Units. SI is based on seven well-defined units which by convention are regarded as dimensionally independent:

Quantity	Unit	Symbol
Length	Meter	m
Mass	Kilogram	kg
Time	Second	s
Electric current	Ampere	A
Thermodynamic temperature[a]	Kelvin	K
Amount of substance	Mole	mol
Luminous intensity	Candela	cd

[a] For a discussion of Celsius temperature see 3.4.2.

2.3 Supplementary Units. The units listed below are called *supplementary units* and may be regarded either as base units or as derived units:

Quantity	Unit	Symbol
Plane angle	Radian	rad
Solid angle	Steradian	sr

2.4 Derived Units

2.4.1 Derived units are formed by combining base units, supplementary units, and other derived units according to the algebraic relations linking the corresponding quantities. The symbols for derived units are obtained by means of the mathematical signs for multiplication, division, and use of exponents. For example, the SI unit for velocity is the meter per second (m/s or $m \cdot s^{-1}$), and that for angular velocity is the radian per second (rad/s or $rad \cdot s^{-1}$).

2.4.2 Those derived SI units which have special names and symbols approved by the CGPM are listed below.

2.4.3 It is frequently advantageous to express derived units in terms of other derived units with special names; for example, the SI unit for electric dipole moment is usually expressed as C·m instead of A·s·m.

Quantity	Unit	Symbol	Formula
Frequency (of a periodic phenomenon)	Hertz	Hz	1/s
Force	Newton	N	kg · m/s²
Pressure, stress	Pascal	Pa	N/m²
Energy, work, quantity of heat	Joule	J	N · m
Power, radiant flux	Watt	W	J/s
Quantity of electricity, electric charge	Coulomb	C	A·s
Electric potential, potential difference, electromotive force	Volt	V	W/A
Capacitance	Farad	F	C/V
Electric resistance	Ohm	Ω	V/A
Conductance	Siemens	S	A/V
Magnetic flux	Weber	Wb	V · s
Magnetic flux density	Tesla	T	Wb/m²
Inductance	Henry	H	Wb/A
Luminous flux	Lumen	lm	cd · sr
Illuminance	Lux	lx	lm/m²
Activity (of radionuclides)	Becquerel[a]	Bq	1/s
Absorbed dose	Gray[a]	Gy	J/kg

[a] Adopted by the CGPM in 1975.

2.4.4 Some common derived units are listed in Table 1.

Table 1 Some Common Derived Units of SI

Quantity	Unit	Symbol
Acceleration	Meter per second squared	m/s²
Angular acceleration	Radian per second squared	rad/s²
Angular velocity	Radian per second	rad/s
Area	Square meter	m²
Concentration (of amount of substance)	Mole per cubic meter	mol/m³
Current density	Ampere per square meter	A/m²
Density, mass	Kilogram per cubic meter	kg/m³
Electric charge density	Coulomb per cubic meter	C/m³
Electric field strength	Volt per meter	V/m
Electric flux density	Coulomb per square meter	C/m²
Energy density	Joule per cubic meter	J/m³
Entropy	Joule per kelvin	J/K
Heat capacity	Joule per kelvin	J/K
Heat flux density } Irradiance	Watt per square meter	W/m²
Luminance	Candela per square meter	cd/m²
Magnetic field strength	Ampere per meter	A/m
Molar energy	Joule per mole	J/mol
Molar entropy	Joule per mole kelvin	J/(mol·K)
Molar heat capacity	Joule per mole kelvin	J/(mol·K)
Moment of force	Newton meter	N · m
Permeability	Henry per meter	H/m
Permittivity	Farad per meter	F/m
Radiance	Watt per square meter steradian	W/(m²·sr)
Radiant intensity	Watt per steradian	W/sr
Specific heat capacity	Joule per kilogram kelvin	J/(kg·K)
Specific energy	Joule per kilogram	J/kg
Specific entropy	Joule per kilogram kelvin	J/(kg·K)
Specific volume	Cubic meter per kilogram	m³/kg
Surface tension	Newton per meter	N/m
Thermal conductivity	Watt per meter kelvin	W/(m·K)
Velocity	Meter per second	m/s
Viscosity, dynamic	Pascal second	Pa·s
Viscosity, kinematic	Square meter per second	m²/s
Volume	Cubic meter	m³
Wavenumber	1 per meter	1/m

2.5 SI Prefixes (See 3.2 for Application)

2.5.1 The prefixes and symbols listed below are used to form names and symbols of the decimal multiples and submultiples of the SI units:

Multiplication factor	Prefix	Symbol
1 000 000 000 000 000 000 = 10^{18}	exa[a]	E
1 000 000 000 000 000 = 10^{15}	peta[a]	P
1 000 000 000 000 = 10^{12}	tera	T
1 000 000 000 = 10^{9}	giga	G
1 000 000 = 10^{6}	mega	M
1 000 = 10^{3}	kilo	k
100 = 10^{2}	hecto[b]	h
10 = 10^{1}	deka[b]	da
0.1 = 10^{-1}	deci[b]	d
0.01 = 10^{-2}	centi[b]	c
0.001 = 10^{-3}	milli	m
0.000 001 = 10^{-6}	micro	μ
0.000 000 001 = 10^{-9}	nano	n
0.000 000 000 001 = 10^{-12}	pico	p
0.000 000 000 000 001 = 10^{-15}	femto	f
0.000 000 000 000 000 001 = 10^{-18}	atto	a

[a] Adopted by the CGPM in 1975.
[b] To be avoided where possible. See 3.2.2.

2.5.2 These prefixes or their symbols are directly attached to names or symbols of units, forming multiples and submultiples of the units. In strict terms these must be called "multiples and submultiples of SI units," particularly in discussing the coherence of the system. (See 5.) In common parlance, the base units and derived units, along with their multiples and submultiples, are all called SI units.

3. APPLICATION OF THE METRIC SYSTEM

3.1 General. SI is the form of the metric system that is preferred for all applications. It is important that this modernized form of the metric system be thoroughly understood and properly applied. Obsolete metric units and practices are widespread, particularly in those countries that long ago adopted the metric system, and much usage is improper. This section gives guidance concerning the limited number of cases in which units outside SI are appropriately used, and makes recommendations concerning usage and style.

3.2 Application of SI Prefixes

3.2.1 General. In general the SI prefixes (2.5) should be used to indicate orders of magnitude, thus eliminating nonsignificant digits and leading zeros in decimal fractions, and providing a convenient alternative to the powers-of-ten notation preferred in computation. For example,

12 300 mm becomes 12.3 m
12.3×10^3 m becomes 12.3 km
0.00123 μA becomes 1.23 nA

3.2.2 Selection. When expressing a quantity by a numerical value and a unit, prefixes should preferably be chosen so that the numerical value lies between 0.1 and 1000. To minimize variety, it is recommended that prefixes representing powers of 1000 be used. However, three factors may justify deviation from the above:

3.2.2.1 In expressing area and volume, the prefixes hecto-, deka-, deci-, and centi- may be required, for example, square hectometer, cubic centimeter.

3.2.2.2 In tables of values of the same quantity, or in a discussion of such values within a given context, it is generally preferable to use the same unit multiple throughout.

3.2.2.3 For certain quantities in particular applications, one particular multiple is customarily used. For example, the millimeter is used for linear dimensions in mechanical engineering drawings even when the values lie far outside the range 0.1 to 1000 mm; the centimeter is often used for body measurements and clothing sizes.

3.2.3 Prefixes in Compound Units.[3] It is recommended that only one prefix be used in forming a multiple of a compound unit. Normally the prefix should be attached to a unit in the numerator. One exception to this occurs when the kilogram is one of the units.

[3] A compound unit is a derived unit expressed in terms of two or more units, that is, not expressed with a single special name.

EXAMPLES
V/m, *not* mV/mm, and MJ/kg, *not* kJ/g

3.2.4 Compound Prefixes. Compound prefixes, formed by the juxtaposition of two or more SI prefixes are not to be used. For example, use

1 nm, *not* 1 mμm
1 pF, *not* 1 $\mu\mu$F

If values are required outside the range covered by the prefixes, they should be expressed using powers of ten applied to the base unit.

3.2.5 Powers of Units. An exponent attached to a symbol containing a prefix indicates that the multiple or submultiple of the unit (the unit with its prefix) is raised to the power expressed by the exponent. For example:

1 cm^3 = $(10^{-2}$ m$)^3$ = 10^{-6} m^3
1 ns^{-1} = $(10^{-9}$ s$)^{-1}$ = 10^9 s^{-1}
1 mm^2/s = $(10^{-3}$ m$)^2$/s = 10^{-6} m^2/s

3.2.6 Unit of Mass (Kilogram). Among the base and derived units of SI, the unit of mass is the only one whose name, for historical reasons, contains a prefix. Names of decimal multiples and submultiples of the unit of mass are formed by attaching prefixes to the word *gram*.

3.2.7 Calculations. Errors in calculations can be minimized if the base and the coherent derived SI units are used and the resulting numerical values are expressed in powers-of-ten notation instead of using prefixes.

Table 2 Decimal and Metric Equivalents of Fractions of One Inch

Fraction	Decimal Equivalents	Metric Equivalents, mm	Fraction	Decimal Equivalents	Metric Equivalents, mm
1/64	0.015625	0.397	33/64	0.515625	13.096
1/32	0.03125	0.794	17/32	0.53125	13.493
3/64	0.046875	1.191	35/64	0.546875	13.890
1/16	0.0625	1.587	9/16	0.5625	14.287
5/64	0.078125	1.984	37/64	0.578125	14.684
3/32	0.09375	2.381	19/32	0.59375	15.081
7/64	0.109375	2.778	39/64	0.609375	15.478
1/8	0.1250	3.175	5/8	0.6250	15.875
9/64	0.140625	3.572	41/64	0.640625	16.272
5/32	0.15625	3.968	21/32	0.65625	16.668
11/64	0.171875	4.365	43/64	0.671875	17.065
3/16	0.1875	4.762	11/16	0.6875	17.462
13/64	0.203125	5.159	45/64	0.703125	17.859
7/32	0.21875	5.556	23/32	0.71875	18.256
15/64	0.234375	5.953	47/64	0.734375	18.653
1/4	0.2500	6.349	3/4	0.7500	19.050
17/64	0.265625	6.746	49/64	0.765625	19.447
9/32	0.28125	7.144	25/32	0.78125	19.843
19/64	0.296875	7.541	51/64	0.796875	20.240
5/16	0.3125	7.937	13/16	0.8125	20.637
21/64	0.328125	8.334	53/64	0.828125	21.034
11/32	0.34375	8.731	27/32	0.84375	21.431
23/64	0.359375	9.128	55/64	0.859375	21.828
3/8	0.3750	9.525	7/8	0.8750	22.225
25/64	0.390625	9.922	57/64	0.890625	22.622
13/32	0.40625	10.319	29/32	0.90625	23.018
27/64	0.421875	10.716	59/64	0.921875	23.415
7/16	0.4375	11.112	15/16	0.9375	23.812
29/64	0.453125	11.509	61/64	0.953125	24.209
15/32	0.46875	11.906	31/32	0.96875	24.606
31/64	0.484375	12.303	63/64	0.984375	25.003
1/2	0.5000	12.699	1	1.000	25.400

Table 3 Inch-Millimeter Equivalents

in	0	1	2	3	4	5	6	7	8	9
						mm				
0	0.0	25.4	50.8	76.2	101.6	127.0	152.4	177.8	203.2	228.6
10	254.0	279.4	304.8	330.2	355.6	381.0	406.4	431.8	457.2	482.6
20	508.0	533.4	558.8	584.2	609.6	635.0	660.4	685.8	711.2	736.6
30	762.0	787.4	812.8	838.2	863.6	889.0	914.4	939.8	965.2	990.6
40	1016.0	1041.4	1066.8	1092.2	1117.6	1143.0	1168.4	1193.8	1219.2	1244.6
50	1270.0	1295.4	1320.8	1346.2	1371.6	1397.0	1422.4	1447.8	1473.2	1498.6
60	1524.0	1549.4	1574.8	1600.2	1625.6	1651.0	1676.4	1701.8	1727.2	1752.6
70	1778.0	1803.4	1828.8	1854.2	1879.6	1905.0	1930.4	1955.8	1981.2	2006.6
80	2032.0	2057.4	2082.8	2108.2	2133.6	2159.0	2184.4	2209.8	2235.2	2260.6
90	2286.0	2311.4	2336.8	2362.2	2387.6	2413.0	2438.4	2463.8	2489.2	2514.6
100	2540.0	2565.4	2590.8	2616.2	2641.6	2667.0	2692.4	2717.8	2743.2	2768.6

NOTE: All values in this table are exact, based on the relation 1 in = 25.4 mm. By manipulation of the decimal point any decimal value or multiple of an inch may be converted to its exact equivalent in millimeters.

Table 4   Classified List of Units

To convert from	to	Multiply by
ACCELERATION		
ft/s²	meter per second² (m/s²)	3.048 000*E −01
free fall, standard (g)	meter per second² (m/s²)	9.806 650*E +00
gal	meter per second² (m/s²)	1.000 000*E −02
in/s²	meter per second² (m/s²)	2.540 000*E −02
ANGLE		
degree (angle)	radian (rad)	1.745 329 E −02
minute (angle)	radian (rad)	2.908 882 E −04
second (angle)	radian (rad)	4.848 137 E −06
AREA		
acre (US survey)	meter² (m²)	4.046 873 E +03
are	meter² (m²)	1.000 000*E +02
barn	meter² (m²)	1.000 000*E −28
circular mil	meter² (m²)	5.067 075 E −10
ft²	meter² (m²)	9.290 304*E −02
hectare	meter² (m²)	1.000 000*E +04
in²	meter² (m²)	6.451 600*E −04
mi² (international)	meter² (m²)	2.589 988 E +06
mi² (US survey)	meter² (m²)	2.589 998 E +06
section	meter² (m²)	[see footnote 12]
township	meter² (m²)	[see footnote 12]
yd²	meter² (m²)	8.361 274 E −01
BENDING MOMENT OR TORQUE		
dyne · em	newton meter (N · m)	1.000 000*E −07
kgf · m	newton meter (N · m)	9.806 650*E +00
ozf · in	newton meter (N · m)	7.061 552 E −03
lbf · in	newton meter (N · m)	1.129 848 E −01
lbf · ft	newton meter (N · m)	1.355 818 E +00
BENDING MOMENT OR TORQUE PER UNIT LENGTH		
lbf · ft/in	newton meter per meter (N · m/m)	5.337 866 E +01
lbf · in/in	newton meter per meter (N · m/m)	4.448 222 E +00
CAPACITY (See VOLUME)		
DENSITY (See MASS PER UNIT VOLUME)		
ELECTRICITY AND MAGNETISM[a]		
abampere	ampere (A)	1.000 000*E +01
abcoulomb	coulomb (C)	1.000 000*E +01
abfarad	farad (F)	1.000 000*E +09
abhenry	henry (H)	1.000 000*E −09
abmho	siemens (S)	1.000 000*E +09
abohm	ohm (Ω)	1.000 000*E −09
abvolt	volt (V)	1.000 000*E −08
ampere hour	coulomb (C)	3.600 000*E +03

[a] ESU means electrostatic cgs unit. EMU means electromagnetic cgs unit.

Table 4   *(continued)*

To convert from	to	Multiply by
EMU of capacitance	farad (F)	1.000 000*E + 09
EMU of current	ampere (A)	1.000 000*E + 01
EMU of electric potential	volt (V)	1.000 000*E − 08
EMU of inductance	henry (H)	1.000 000*E − 09
EMU of resistance	ohm (Ω)	1.000 000*E − 09
ESU of capacitance	farad (F)	1.112 650 E − 12
ESU of current	ampere (A)	3.335 6 E − 10
ESU of electric potential	volt (V)	2.997 9 E + 02
ESU of inductance	henry (H)	8.987 554 E + 11
ESU of resistance	ohm (Ω)	8.987 554 E + 11
faraday (based on carbon-12)	coulomb (C)	9.648 70 E + 04
faraday (chemical)	coulomb (C)	9.649 57 E + 04
faraday (physical)	coulomb (C)	9.652 19 E + 04
gamma	tesla (T)	1.000 000*E − 09
gauss	tesla (T)	1.000 000*E − 04
gilbert	ampere (A)	7.957 747 E − 01
maxwell	weber (Wb)	1.000 000*E − 08
mho	siemens (S)	1.000 000*E + 00
oersted	ampere per meter (A/m)	7.957 747 E + 01
ohm centimeter	ohm meter (Ω · m)	1.000 000*E − 02
ohm circular-mil per foot	ohm millimeter suqared per meter (Ω · mm²/m)	1.662 426 E − 03
statampere	ampere (A)	3.335 640 E − 10
statcoulomb	coulomb (C)	3.335 640 E − 10
statfarad	farad (F)	1.112 650 E − 12
stathenry	henry (H)	8.987 554 E + 11
statmho	siemens (S)	1.112 650 E − 12
statohm	ohm (Ω)	8.987 554 E + 11
statvolt	volt (V)	2.997 925 E + 02
unit pole	weber (Wb)	1.256 637 E − 07
ENERGY (Includes WORK)		
British thermal unit (International Table)	joule (J)	1.055 056 E + 03
British thermal unit (mean)	joule (J)	1.055 87 E + 03
British thermal unit (thermochemical)	joule (J)	1.054 350 E + 03
British thermal unit (39° F)	joule (J)	1.059 67 E + 03
British thermal unit (59° F)	joule (J)	1.054 80 E + 03
British thermal unit (60° F)	joule (J)	1.054 68 E + 03
calorie (International Table)	joule (J)	4.186 800*E + 00
calorie (mean)	joule (J)	4.190 02 E + 00
calorie (thermochemical)	joule (J)	4.184 000*E + 00
calorie (15° C)	joule (J)	4.185 80 E + 00
calorie (20° C)	joule (J)	4.181 90 E + 00
calorie (kilogram, International Table)	joule (J)	4.186 800*E + 03
calorie (kilogram, mean)	joule (J)	4.190 02 E + 03
calorie (kilogram, thermochemical)	joule (J)	4.184 000*E + 03
electronvolt	joule (J)	1.602 19 E − 19

Table 4 (*continued*)

To convert from	to	Multiply by
erg	joule (J)	1.000 000*E −07
ft · lbf	joule (J)	1.355 818 E +00
ft · poundal	joule (J)	4.214 011 E −02
kilocalorie (International Table)	joule (J)	4.186 800*E +03
kilocalorie (mean)	joule (J)	4.190 02 E +03
kilocalorie (thermochemical)	joule (J)	4.184 000*E +03
kW · h	joule (J)	3.600 000*E +06
therm	joule (J)	1.055 056 E +08
ton (nuclear equivalent of TNT)	joule (J)	4.184 E +09
W · h	joule (J)	3.600 000*E +03
W · s	joule (J)	1.000 000*E +00
ENERGY PER UNIT AREA TIME		
Btu (thermochemical)/ft² · s	watt per meter² (W/m²)	1.134 893 E +04
Btu (thermochemical)/ft² · min	watt per meter² (W/m²)	1.891 489 E +02
Btu (thermochemical)/ft² · h	watt per meter² (W/m²)	3.152 481 E +00
Btu (thermochemical)/in² · s	watt per meter² (W/m²)	1.634 246 E +06
cal (thermochemical)/cm² · min	watt per meter² (W/m²)	6.973 333 E +02
erg/cm² · s	watt per meter² (W/m²)	1.000 000*E −03
W/cm²	watt per meter² (W/m²)	1.000 000*E +04
W/in²	watt per meter² (W/m²)	1.550 003 E +03
FLOW (See MASS PER UNIT TIME or VOLUME PER UNIT TIME)		
FORCE		
dyne	newton (N)	1.000 000*E −05
kilogram-force	newton (N)	9.806 650*E +00
kilopond	newton (N)	9.806 650*E +00
kip (1000 lbf)	newton (N)	4.448 222 E +03
ounce-force	newton (N)	2.780 139 E −01
pound-force (lbf avoirdupois)	newton (N)	4.448 222 E +00
lbf/lb (thrust/weight [mass] ratio)	newton per kilogram (N/kg)	9.806 650 E +00
poundal	newton (N)	1.382 550 E −01
ton-force (2000 lbf)	newton (N)	8.896 444 E +03
FORCE PER UNIT AREA (See PRESSURE)		
FORCE PER UNIT LENGTH		
lbf/ft	newton per meter (N/m)	1.459 390 E +01
lbf/in	newton per meter (N/m)	1.751 268 E +02
HEAT		
Btu (International Table) · ft/h · ft² · ° F (*k*, thermal conductivity)	watt per meter kelvin (W/m · K)	1.730 735 E +00
Btu (thermochemical) · ft/h · ft² · ° F (*k*, thermal conductivity)	watt per meter kelvin (W/m · K)	1.729 577 E +00
Btu (International Table) · in/h · ft² · ° F (*k*, thermal conductivity)	watt per meter kelvin (W/m · K)	1.442 279 E −01

Table 4   (*continued*)

To convert from	to	Multiply by
Btu (thermochemical) · in/h · ft² · °F (*k*, thermal conductivity)	watt per meter kelvin (W/m · K)	1.441 314 E −01
Btu (International Table) · in/s · ft² · °F (*k*, thermal conductivity)	watt per meter kelvin (W/m · K)	5.192 204 E −02
Btu (thermochemical) · in/s · ft² · °F (*k*, thermal conductivity)	watt per meter kelvin (W/m · K)	5.188 732 E +02
Btu (International Table)/ft²	joule per meter² (J/m²)	1.135 653 E +04
Btu (thermochemical)/ft²	joule per meter² (J/m²)	1.134 893 E +04
Btu (International Table)/h · ft² · °F (*C*, thermal conductance)	watt per meter² kelvin (W/m² · K)	5.678 263 E +00
Btu (thermochemical)/h · ft² · °F (*C*, thermal conductance)	watt per meter² kelvin (W/m² · K)	5.674 466 E +00
Btu (International Table)/s · ft² · °F	watt per meter² kelvin (W/m² · K)	2.044 175 E +04
Btu (thermochemical)/s · ft² · °F	watt per meter² kelvin (W/m² · K)	2.042 808 E +04
Btu (International Table)/lb	joule per kilogram (J/kg)	2.326 000*E +03
Btu (thermochemical)/lb	joule per kilogram (J/kg)	2.324 444 E +03
Btu (International Table)/lb · °F (*c*, heat capacity)	joule per kilogram kelvin (J/kg · K)	4.186 800*E +03
Btu (thermochemical)/lb · °F (*c*, heat capacity)	joule per kilogram kelvin (J/kg · K)	4.184 000 E +03
cal (thermochemical)/cm · s · °C	watt per meter kelvin (W/m · K)	4.184 000*E +02
cal (thermochemical)/cm²	joule per meter² (J/m²)	4.184 000*E +04
cal (thermochemical)/cm² · min	watt per meter² (W/m²)	6.973 333 E +02
cal (thermochemical)/cm² · s	watt per meter² (W/m²)	4.184 000*E +04
cal (International Table)/g	joule per kilogram (J/kg)	4.186 800*E +03
cal (thermochemical)/g	joule per kilogram (J/kg)	4.184 000*E +03
cal (International Table)/g · °C	joule per kilogram kelvin (J/kg · K)	4.186 800*E +03
cal (thermochemical)/g · °C	joule per kilogram kelvin (J/kg · K)	4.184 000*E +03
cal (thermochemical)/min	watt (W)	6.973 333 E −02
cal (thermochemical)/s	watt (W)	4.184 000*E +00
clo	kelvin meter² per watt (K · m²/W)	2.003 712 E −01
°F · h · ft²/Btu (International Table) (*R*, thermal resistance)	kelvin meter² per watt (K · m²/W)	1.761 102 E −01
°F · h · ft²/Btu (thermochemical) (*R*, thermal resistance)	kelvin meter² per watt (K · m²/W)	1.762 280 E −01
ft²/h (thermal diffusivity)	meter² per second (m²/s)	2.580 640*E −05
LENGTH		
angstrom	meter (m)	1.000 000*E −10
astronomical unit	meter (m)	1.495 979 E +11
caliber (inch)	meter (m)	2.540 000*E −02
fathom	meter (m)	1.828 8 E +00
fermi (femtometer)	meter (m)	1.000 000*E −15
foot	meter (m)	3.048 000*E −01
foot (US survey)	meter (m)	3.048 006 E −01
inch	meter (m)	2.540 000*E −02
league	meter (m)	[see footnote 12]
light year	meter (m)	9.460 55 E +15

Table 4   (*continued*)

To convert from	to	Multiply by
microinch	meter (m)	2.540 000*E − 08
micron	meter (m)	1.000 000*E − 06
mil	meter (m)	2.540 000*E − 05
mile (international nautical)	meter (m)	1.852 000*E + 03
mile (UK nautical)	meter (m)	1.853 184*E + 03
mile (US nautical)	meter (m)	1.852 000*E + 03
mile (international)	meter (m)	1.609 344*E + 03
mile (statute)	meter (m)	1.609 3 E + 03
mile (US survey)	meter (m)	1.609 347 E + 03
parsec	meter (m)	3.085 678 E + 16
pica (printer's)	meter (m)	4.217 518 E − 03
point (printer's)	meter (m)	3.514 598*E − 04
rod	meter (m)	
yard	meter (m)	9.144 000*E − 01

LIGHT

footcandle	lux (lx)	1.076 391 E + 01
footlambert	candela per meter^2 (cd/m^2)	3.426 259 E + 00
lambert	candela per meter^2 (cd/m^2)	3.183 099 E + 03

MASS

carat (metric)	kilogram (kg)	2.000 000*E − 04
grain	kilogram (kg)	6.479 891*E − 05
gram	kilogram (kg)	1.000 000*E − 03
hundredweight (long)	kilogram (kg)	5.080 235 E + 01
hundredweight (short)	kilogram (kg)	4.535 924 E + 01
kgf · s^2/m (mass)	kilogram (kg)	9.806 650*E + 00
ounce (avoirdupois)	kilogram (kg)	2.834 952 E − 02
ounce (troy or apothecary)	kilogram (kg)	3.110 348 E − 02
pennyweight	kilogram (kg)	1.555 174 E − 03
pound (lb avoirdupois)	kilogram (kg)	4.535 924 E − 01
pound (troy or apothecary)	kilogram (kg)	3.732 417 E − 01
slug	kilogram (kg)	1.459 390 E + 01
ton (assay)	kilogram (kg)	2.916 667 E − 02
ton (long, 2240 lb)	kilogram (kg)	1.016 047 E + 03
ton (metric)	kilogram (kg)	1.000 000*E + 03
ton (short, 2000 lb)	kilogram (kg)	9.071 847 E + 02
tonne	kilogram (kg)	1.000 000*E + 03

MASS PER UNIT AREA

oz/ft^2	kilogram per meter^2 (kg/m^2)	3.051 517 E − 01
oz/yd^2	kilogram per meter^2 (kg/m^2)	3.390 575 E − 02
lb/ft^2	kilogram per meter^2 (kg/m^2)	4.882 428 E + 00

MASS PER UNIT CAPACITY (See MASS PER UNIT VOLUME)

MASS PER UNIT LENGTH

denier	kilogram per meter (kg/m)	1.111 111 E − 07
lb/ft	kilogram per meter (kg/m)	1.488 164 E + 00
lb/in	kilogram per meter (kg/m)	1.785 797 E + 01
tex	kilogram per meter (kg/m)	1.000 000*E − 06

Table 4   (*continued*)

To convert from	to	Multiply by
MASS PER UNIT TIME (Includes FLOW)		
perm (0° C)	kilogram per pascal second meter² (kg/Pa · s · m²)	
perm (23° C)	kilogram per pascal second meter² (kg/Pa · s · m²)	5.721 35 E − 11
perm · in (0° C)	kilogram per pascal second meter (kg/Pa · s · m)	5.745 25 E − 11
perm · in (23° C)	kilogram per pascal second meter (kg/Pa · s · m)	1.453 22 E − 12
		1.459 29 E − 12
lb/h	kilogram per second (kg/s)	1.259 979 E − 04
lb/min	kilogram per second (kg/s)	7.559 873 E − 03
lb/s	kilogram per second (kg/s)	4.535 924 E − 01
lb/hp · h		
(SFC, specific fuel consumption)	kilogram per joule (kg/J)	1.689 659 E − 07
ton (short)/h	kilogram per second (kg/s)	2.519 958 E − 01
MASS PER UNIT VOLUME (Includes DENSITY and MASS CAPACITY)		
grain (lb avoirdupois/7000)/gal (US liquid)	kilogram per meter³ (kg/m³)	1.711 806 E − 02
g/cm³	kilogram per meter³ (kg/m³)	1.000 000*E + 03
oz (avoirdupois)/gal (UK liquid)	kilogram per meter³ (kg/m³)	6.236 021 E + 00
oz (avoirdupois)/gal (US liquid)	kilogram per meter³ (kg/m³)	7.489 152 E + 00
oz (avoirdupois)/in³	kilogram per meter³ (kg/m³)	1.729 994 E + 03
lb/ft³	kilogram per meter³ (kg/m³)	1.601 846 E + 01
lb/in³	kilogram per meter³ (kg/m³)	2.767 990 E + 04
lb/gal (UK liquid)	kilogram per meter³ (kg/m³)	9.977 633 E + 01
lb/gal (US liquid)	kilogram per meter³ (kg/m³)	1.198 264 E + 02
lb/yd³	kilogram per meter³ (kg/m³)	5.932 764 E − 01
slug/ft³	kilogram per meter³ (kg/m³)	5.153 788 E + 02
ton (long)/yd³	kilogram per meter³ (kg/m³)	1.328 939 E + 03
POWER		
Btu (International Table)/h	watt (W)	2.930 711 E − 01
Btu (thermochemical)/h	watt (W)	2.928 751 E − 01
Btu (thermochemical)/min	watt (W)	1.757 250 E + 01
Btu (thermochemical)/s	watt (W)	1.054 350 E + 03
cal (thermochemical)/min	watt (W)	6.973 333 E − 02
cal (thermochemical)/s	watt (W)	4.184 000*E + 00
erg/s	watt (W)	1.000 000*E − 07
ft · lbf/h	watt (W)	3.766 161 E − 04
ft · lbf/min	watt (W)	2.259 697 E − 02
ft · lbf/s	watt (W)	1.355 818 E + 00
horsepower (550 ft · lbf/s)	watt (W)	7.456 999 E + 02
horsepower (boiler)	watt (W)	9.809 50 E + 03
horsepower (electric)	watt (W)	7.460 000*E + 02
horsepower (metric)	watt (W)	7.354 99 E + 02
horsepower (water)	watt (W)	7.460 43 E + 02
horsepower (UK)	watt (W)	7.457 0 E + 02

Table 4   (*continued*)

To convert from	to	Multiply by
kilocalorie (thermochemical)/min	watt (W)	6.973 333 E + 01
kilocalorie (thermochemical)/s	watt (W)	4.184 000*E + 03
ton (refrigeration)	watt (W)	3.516 800 E + 03
PRESSURE or STRESS (FORCE PER UNIT AREA)		
atmosphere (standard)	pascal (Pa)	1.013 250*E + 05
atmosphere (technical = 1 kgf/cm²)	pascal (Pa)	9.806 650*E + 04
bar	pascal (Pa)	1.000 000*E + 05
centimeter of mercury (0° C)	pascal (Pa)	1.333 22 E + 03
centimeter of water (4° C)	pascal (Pa)	9.806 38 E + 01
dyne/cm²	pascal (Pa)	1.000 000*E − 01
foot of water (39.2° F)	pascal (Pa)	2.988 98 E + 03
gram-force/cm²	pascal (Pa)	9.806 650*E + 01
inch of mercury (32° F)	pascal (Pa)	3.386 38 E + 03
inch of mercury (60° F)	pascal (Pa)	3.376 85 E + 03
inch of water (39.2° F)	pascal (Pa)	2.490 82 E + 02
inch of water (60° F)	pascal (Pa)	2.488 4 E + 02
kgf/cm²	pascal (Pa)	9.806 650*E + 04
kgf/m²	pascal (Pa)	9.806 650*E + 00
kgf/mm²	pascal (Pa)	9.806 650*E + 06
kip/in² (ksi)	pascal (Pa)	6.894 757 E + 06
millibar	pascal (Pa)	1.000 000*E + 02
millimeter of mercury (0° C)	pascal (Pa)	1.333 22 E + 02
poundal/ft²	pascal (Pa)	1.488 164 E + 00
lbf/ft²	pascal (Pa)	4.788 026 E + 01
lbf/in² (psi)	pascal (Pa)	6.894 757 E + 03
psi	pascal (Pa)	6.894 757 E + 03
torr (mm Hg, 0° C)	pascal (Pa)	1.333 22 E + 02
SPEED (See VELOCITY)		
STRESS (see PRESSURE)		
TEMPERATURE		
degree Celsius	kelvin (K)	$t_K = t_{°C} + 273.15$
degree Fahrenheit	degree Celsius	$t_{°C} = (t_{°F} - 32)/1.8$
degree Fahrenheit	kelvin (K)	$t_K = (t_{°F} + 459.67)/1.8$
degree Rankine	kelvin (K)	$t_K = t_{°R}/1.8$
kelvin	degree Celsius	$t_{°C} = t_K - 273.15$
TIME		
day (mean solar)	second (s)	8.640 000 E + 04
day (sidereal)	second (s)	8.616 409 E + 04
hour (mean solar)	second (s)	3.600 000 E + 03
hour (sidereal)	second (s)	3.590 170 E + 03
minute (mean solar)	second (s)	6.000 000 E + 01

Table 4 (*continued*)

To convert from	to	Multiply by
minute (sidereal)	second (s)	5.983 617 E+01
month (mean calendar)	second (s)	2.628 000 E+06
second (sidereal)	second (s)	9.972 696 E−01
year (calendar)	second (s)	3.153 600 E+07
year (sidereal)	second (s)	3.155 815 E+07
year (tropical)	second (s)	3.155 693 E+07

TORQUE (See BENDING MOMENT)

VELOCITY (Includes SPEED)

ft/h	meter per second (m/s)	8.466 667 E−05
ft/min	meter per second (m/s)	5.080 000*E−03
ft/s	meter per second (m/s)	3.048 000*E−01
in/s	meter per second (m/s)	2.540 000*E−02
km/h	meter per second (m/s)	2.777 778 E−01
knot (international)	meter per second (m/s)	5.144 444 E−01
mi/h (international)	meter per second (m/s)	4.470 400*E−01
mi/min (international)	meter per second (m/s)	2.682 240*E+01
mi/s (international)	meter per second (m/s)	1.609 344*E+03
mi/h (international)	kilometer per hour (km/h)[b]	1.609 344*E+00

VISCOSITY

centipoise	pascal second (Pa · s)	1.000 000*E−03
centistokes	meter^2 per second (m^2/s)	1.000 000*E−06
ft^2/s	meter^2 per second (m^2/s)	9.290 304*E−02
poise	pascal second (Pa · s)	1.000 000*E−01
poundal · s/ft^2	pascal second (Pa · s)	1.488 164 E+00
lb/ft · h	pascal second (Pa · s)	4.133 789 E−04
lb/ft · s	pascal second (Pa · s)	1.488 164 E+00
lbf · s/ft^2	pascal second (Pa · s)	4.788 026 E+01
rhe	1 per pascal second (1/Pa · s)	1.000 000*E+01
slug/ft · s	pascal second (Pa · s)	4.788 026 E+01
stokes	meter^2 per second (m^2/s)	1.000 000*E−04

VOLUME (Includes CAPACITY)

acre-foot (US survey)	meter^3 (m^3)	1.233 489 E+03
barrel (oil, 42 gal)	meter^3 (m^3)	1.589 873 E−01
board foot	meter^3 (m^3)	2.359 737 E−03
bushel (US)	meter^3 (m^3)	3.523 907 E−02
cup	meter^3 (m^3)	2.365 882 E−04
fluid ounce (US)	meter^3 (m^3)	2.957 353 E−05
ft^3	meter^3 (m^3)	2.831 685 E−02
gallon (Canadian liquid)	meter^3 (m^3)	4.546 090 E−03
gallon (UK liquid)	meter^3 (m^3)	4.546 092 E−03
gallon (US dry)	meter^3 (m^3)	4.404 884 E−03

[b] Although speedometers may read km/h, the correct SI unit is m/s.

Table 4 (*Continued*)

To convert from	to	Multiply by
gallon (US liquid)	meter³ (m³)	3.785 412 E −03
gill (UK)	meter³ (m³)	1.420 654 E −04
gill (US)	meter³ (m³)	1.182 941 E −04
in³	meter³ (m³)	1.638 706 E −05
liter	meter³ (m³)	1.000 000*E −03
ounce (UK fluid)	meter³ (m³)	2.841 307 E −05
ounce (US fluid)	meter³ (m³)	2.957 353 E −05
peck (US)	meter³ (m³)	8.809 768 E −03
pint (US dry)	meter³ (m³)	5.506 105 E −04
pint (US liquid)	meter³ (m³)	4.731 765 E −04
quart (US dry)	meter³ (m³)	1.101 221 E −03
quart (US liquid)	meter³ (m³)	9.463 529 E −04
stere	meter³ (m³)	1.000 000*E +00
tablespoon	meter³ (m³)	1.478 676 E −05
teaspoon	meter³ (m³)	4.928 922 E −06
ton (register)	meter³ (m³)	2.831 685 E +00
yd³	meter³ (m³)	7.645 549 E −01
VOLUME PER UNIT TIME (Includes FLOW)		
ft³/min	meter³ per second (m³/s)	4.719 474 E −04
ft³/s	meter³ per second (m³/s)	2.831 685 E −02
gal (US liquid)/hp · h		
(SFC, specific fuel consumption)	meter³ per joule (m³/J)	1.410 089 E −09
in³/min	meter³ per second (m³/s)	2.731 177 E −07
yd³/min	meter³ per second (m³/s)	1.274 258 E −02
gal (US liquid)/day	meter³ per second (m³/s)	4.381 264 E −08
gal (US liquid)/min	meter³ per second (m³/s)	6.309 020 E −05
WORK (See ENERGY)		

Index

DATE DUE
